TK 7816 .T47 2003

NEW ENGLAND INSTITUTE
OF TECHNOLOGY
LIBRARY

ELECTRONICS
FOR COMPUTER TECHNOLGY

— • DAVID TERRELL • —

THOMSON
DELMAR LEARNING

Australia   Canada   Mexico   Singapore   Spain   United Kingdom   United States

Electronics for Computer Technology
David Terrell

Business Unit Director:
Alar Elken

Executive Editor:
Sandy Clark

Senior Acquisitions Editor:
Gregory L. Clayton

Senior Development Editor:
Michelle Ruelos Cannistraci

Executive Marketing Manager:
Maura Theriault

Channel Manager:
Fair Huntoon

Marketing Coordinator:
Karen Smith

Executive Production Manager:
Mary Ellen Black

Production Manager:
Larry Main

Senior Project Editor:
Christopher Chien

Art/Design Coordinator:
David Arsenault

Technology Project Manager:
David Porush

Technology Project Specialist:
Kevin Smith

Editorial Assistant:
Jennifer Luck

COPYRIGHT © 2003 by Delmar Learning, a division of Thomson Learning, Inc. Thomson Learning™ is a trademark used herein under license.

Printed in the United States of America
1 2 3 4 5 XX 06 05 04 03 02

For more information contact
Delmar Learning
Executive Woods
5 Maxwell Drive, PO Box 8007,
Clifton Park, NY 12065-8007

Or find us on the World Wide Web at
http://www.delmar.com

ALL RIGHTS RESERVED. Certain portions of this work © 2000. No part of this work covered by the copyright hereon may be reproduced or used in any form or by any means—graphic, electronic, or mechanical, including photocopying, recording, taping, Web distribution, or information storage and retrieval systems—without the written permission of the publisher.

For permission to use material from the text or product, contact us by
Tel   (800) 730-2214
Fax   (800) 730-2215
www.thomsonrights.com

Library of Congress Cataloging-in-Publication Data:

ISBN: 0-7668-3872-2

NOTICE TO THE READER

Publisher does not warrant or guarantee any of the products described herein or perform any independent analysis in connection with any of the product information contained herein. Publisher does not assume, and expressly disclaims, any obligation to obtain and include information other than that provided to it by the manufacturer.

The reader is expressly warned to consider and adopt all safety precautions that might be indicated by the activities herein and to avoid all potential hazards. By following the instructions contained herein, the reader willingly assumes all risks in connection with such instructions.

The publisher makes no representation or warranties of any kind, including but not limited to, the warranties of fitness for particular purpose or merchantability, nor are any such representations implied with respect to the material set forth herein, and the publisher takes no responsibility with respect to such material. The publisher shall not be liable for any special, consequential, or exemplary damages resulting, in whole or part, from the readers' use of, or reliance upon, this material.

CONTENTS

Preface — vii

CHAPTER 1 Electronic Systems — 1
1.1 Representative Systems 1
1.2 System Notations 3
1.3 Physical System Hierarchy 9
1.4 System Connectivity 13
1.5 Elements of System-Level Troubleshooting 15
1.6 Circuit Simulation 20

CHAPTER 2 Basic Electronics and Units of Measure — 23
2.1 Electrical Quantities 23
2.2 Mechanical Quantities 38
2.3 Light and Other Waves 39
2.4 Magnetism and Electromagnetism 42

CHAPTER 3 Basic Components and Technical Notation — 49
3.1 Technical Notation 49
3.2 Wire and Cable 55
3.3 Electronic Components 63

CHAPTER 4 Circuits — 87
4.1 Basic Requirements for Current 87
4.2 Series Circuits 90
4.3 Parallel Circuits 101
4.4 Series-Parallel Circuits 108
4.5 Complex Circuits 118
4.6 Ground and Other Reference Points 119

CHAPTER 5 Circuit Troubleshooting — 129
5.1 Troubleshooting Series Circuits 130
5.2 Troubleshooting Parallel Circuits 138
5.3 Troubleshooting Series-Parallel Circuits 141
5.4 Troubleshooting Strategies 143

CHAPTER 6 Alternating Current — 153
6.1 Generation of Alternating Voltage 153
6.2 Sine Wave Characteristics 157

6.3 Working with Phase Angles 170
6.4 Circuit Analysis of AC Resistive Circuits 176
6.5 Alternating Voltage Applications 181

CHAPTER 7 Inductors, Capacitors, and Transformers — 191

7.1 Inductors 192
7.2 Transformers 204
7.3 Capacitors 215
7.4 *RC* and *RL* Circuits 231
7.5 *RLC* Circuits 241

CHAPTER 8 Semiconductor Technology — 261

8.1 Basic Atomic Theory 261
8.2 Semiconductor Theory 264
8.3 Semiconductor Junctions 275
8.4 Troubleshooting Semiconductors 283

CHAPTER 9 Diodes and Diode Circuits — 291

9.1 Diode Characteristics 291
9.2 Power Supply Applications 296
9.3 Miscellaneous Diode Applications 304
9.4 Special Diodes 311
9.5 Troubleshooting Diode Circuits 320

CHAPTER 10 Transistors and Transistor Circuits — 331

10.1 Bipolar Transistors 332
10.2 Junction Field-Effect Transistors 348
10.3 MOS Field-Effect Transistors 353
10.4 Transistor Applications 361
10.5 Troubleshooting Transistor Circuits 373

CHAPTER 11 Op Amps and Op Amp Circuits — 383

11.1 Op Amp Characteristics 383
11.2 Basic Amplifier Configurations 392
11.3 Op Amp Applications 403
11.4 Troubleshooting Op Amp Circuits 420

CHAPTER 12 Power Supply and Voltage-Regulator Circuits — 427

12.1 Voltage Regulation Fundamentals 428
12.2 Series Voltage Regulation 432
12.3 Shunt Voltage Regulation 437
12.4 Switching Voltage Regulation 438

12.5 Power Supply Protection Circuits 448
12.6 Troubleshooting Power Supply Circuits 452

CHAPTER 13 Thyristors and Optoelectronic Devices 465

13.1 Thyristor Characteristics 466
13.2 Thyristor Types and Applications 473
13.3 Troubleshooting Thyristors 485
13.4 Optoelectronic Devices 489

CHAPTER 14 Integrated Circuit Applications 507

14.1 Oscillator Circuits 508
14.2 Industrial Computer Applications 523
14.3 Troubleshooting Circuits Based on ICs 533

CHAPTER 15 Digital Electronics 541

15.1 Digital Concepts and Terminology 541
15.2 Combinational Logic 555
15.3 Sequential Logic 565
15.4 Interfacing Digital and Analog Systems 578
15.5 Troubleshooting Digital Systems 581

CHAPTER 16 Microprocessors and Computers 595

16.1 Basic Concepts and Terminology 595
16.2 Hardware 605
16.3 Software 611

CHAPTER 17 Telecommunications 621

17.1 Telecommunication: Information and Networks 622
17.2 Technical Characteristics 630
17.3 Wireless Telecommunication 645

APPENDIX A	Logic Families	655
APPENDIX B	Measurements with Electronic Test Equipment	657
APPENDIX C	Karnaugh Maps	661
APPENDIX D	Answers to Odd-Numbered Exercise and Review Problems and to All Circuit Explorations	665
	Glossary	709
	Index	731

LIST OF SYSTEM PERSPECTIVE FEATURES

Bipolar Transistors	344
Capacitors	220
Flip-Flops	574
Glue Logic	565

Integrated Circuits	523
Laser Diodes	493
Op Amps	403
Power Supplies	427
Series Circuits	97
Time and Frequency Domains	167
Transformers and Inductors	207
Transistors and Relays	362

PREFACE

INTENDED AUDIENCE

This book is designed for students with an ultimate career goal that focuses on computers, computer technology, and electronics technology. There are no technical prerequisites for success in the material presented other than basic math skills. The important concepts in basic electronics are thoroughly covered, including component identification and behavior, dc and ac circuit principles and analyses, solid-state devices, operational amplifiers, and both linear and switching voltage-regulator circuits. More advanced chapters are also included that serve as an excellent introduction to digital electronics, microprocessor technology, and telecommunications. The learner is provided with a rich blend of theory directly tied to practical system-level examples and troubleshooting strategies. This approach enables the student to immediately see the relevance of the various topics to long-range career goals in computer technology.

Computer and electronics technology students will enter a field that requires a diverse knowledge of both hardware (electrical) and software (programming) aspects of computers. Although there are some software topics in this book, the emphasis is on hardware. Computer hardware includes all of the physical parts of the computer and, in particular, the various electronic components within the computer. It also includes electronic devices connected to the computer such as printers, modems, video displays, sound systems, and so on. Additionally, computer technology encompasses much of modern industrial electronics such as robotics and other industrial control systems.

APPROACH

This book is unique in several important ways. First, it promotes system-level thinking in the learner. Second, skills in mathematics and circuit analysis are essential tools for success in electronics, and this book focuses on those specific areas that have direct impact on career success in computer technology and as an electronics technician. Third, troubleshooting practices, strategies, and techniques are presented throughout the text and include component-level, circuit-level, and system-level discussions.

System-Level Approach

The classic approach to basic electronics curricula presents each supporting topic as a stand-alone subject. Integration of these basic components into a practical system is expected to occur in more advanced courses or as a result of on-the-job experience. Historically, electronics curricula have placed heavy emphasis on mathematics and circuit analysis. The general expectation was that if you could calculate circuit values, you would therefore understand circuit operation. Similarly, you should be able to troubleshoot inoperable circuits to locate defective components. It has been the author's experience that this approach is no longer consistent with student goals and industry practices.

First, understanding the operation of complex electronics devices today requires system-level thinking. It is important to understand the general behavior of every electronic component, but understanding their interaction with other

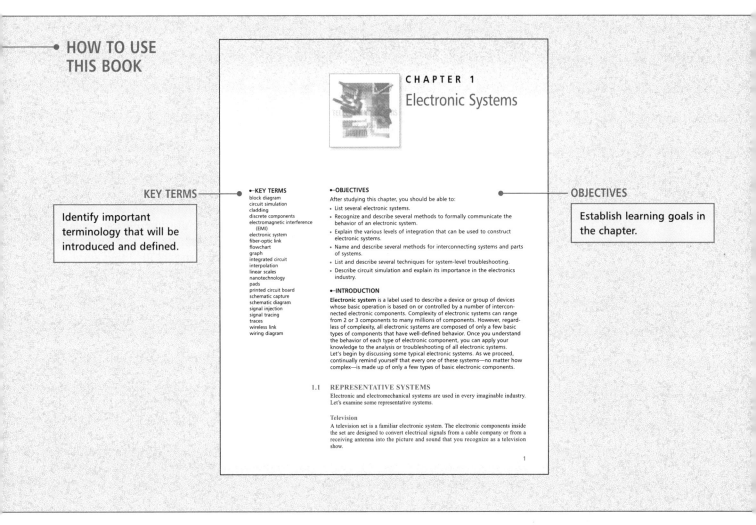

functional parts of the system is more relevant to a computer technology student than calculating the theoretical behavior of an isolated component.

Second, troubleshooting defective systems also demands system-level thinking. A computer system may consist of thousands or even tens of thousands of electronic components. Furthermore, the flow of electrical signals is not simple and linear as it used to be with less complex electronic devices such as transmitters, receivers, and amplifiers. Because of this complexity, the intricate signal paths, and the high-speed of operation, it is not practical to diagnose defects on a component level. At least, it is not practical initially. A system-level approach is required to reduce the number of possible defects to a reasonable value. In many cases that is all that is required, since the final "fix" involves replacing an entire circuit board or module. In those cases where component-level repairs are practical, system-level troubleshooting serves to localize the detailed troubleshooting effort.

This book cultivates system-level thinking in the student through several means. The first chapter focuses on system-level topics including representative systems, system notations, functional hierarchies, system connectivity, and system-level troubleshooting. As the various topics are presented, references are made to system applications. This helps put the topic being discussed into the context of an overall system. Finally, a System Perspective feature appears at many points throughout the book. This short narrative illustrates how the device or circuit being discussed is actually used in a practical, functional system such as a computer. Students find the material more interesting when the relevancy to their career goals is apparent.

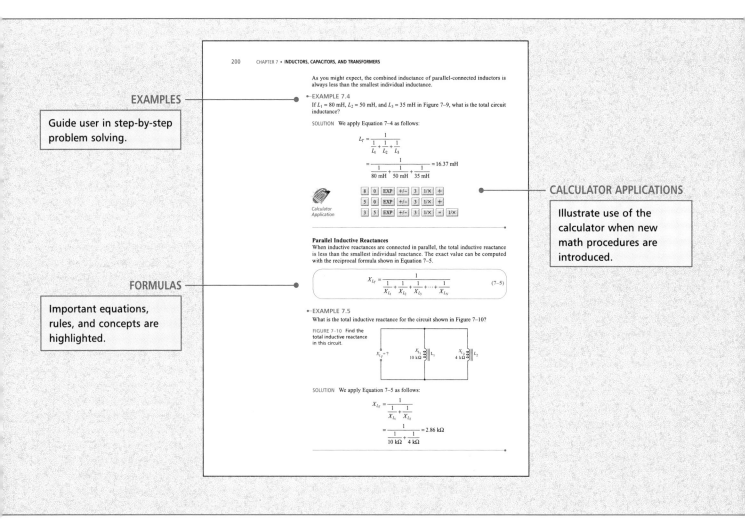

Mathematics and Circuit Analysis

One of the primary goals of the author is to communicate component and circuit behavior in an intuitive and logical way. Wherever practical, the behavior of components is discussed in terms of Ohm's and Kirchhoff's Laws and basic circuit principles. Numerical circuit analysis is always preceded by an intuitive discussion of component or circuit operation. Typical calculator sequences are provided wherever a new type of calculation is introduced.

Mathematics and numerical circuit analysis are included throughout the text wherever either of the following requirements is satisfied:

1. A graduate in the field of computer technology will likely be required to perform similar calculations on the job.
2. The calculations and numerical analyses are critical to the understanding of subsequent material or are generally required to interpret manufacturers' datasheets for electronic components.

Some specific examples of mathematics and circuit analyses that are presented in the book include the following: applications of Ohm's and Kirchhoff's Laws, power calculations, engineering notation, prefix conversions, resistive circuit analysis (series, parallel, and series-parallel), ac calculations, general reactance and impedance calculations (*RL*, *RC*, and *RLC* circuits), turns ratios, op amp voltage gain, cutoff frequencies for active filters, line and load regulation, number conversions (binary, decimal, and hexadecimal), and basic Boolean algebra.

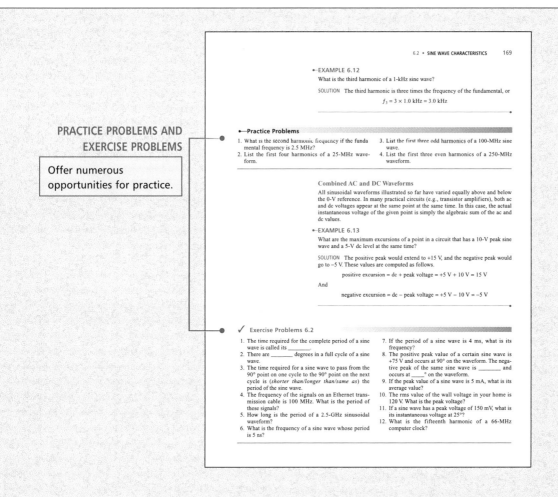

PRACTICE PROBLEMS AND EXERCISE PROBLEMS
Offer numerous opportunities for practice.

Other computations that are commonly found in standard electronics books, but which are not included in this book because they do not satisfy the above requirements, include the following: calculating the number of electrons in a given shell of an atom, converting horsepower to watts, determining the number of coulombs required to charge a capacitor to a certain voltage, detailed ac calculations for various transistor amplifier configurations, computing the effects of a wide range of nonideal op amp characteristics, and calculating the amount of jitter in a particular phase-locked loop.

While some or all of these topics are near and dear to the hearts of many electronics instructors (including the author), they simply don't satisfy the requirements for computer technology students in today's career environment.

Troubleshooting

Troubleshooting strategies are distributed throughout the book. Some are specifically oriented toward system-level troubleshooting, while others are focused on the testing of a specific component. In addition to the material integrated into the various parts of the book, Chapter 5 is entirely devoted to troubleshooting techniques.

The topics presented are important for the development of practical troubleshooting skills, but there is no substitute for actual experience. Therefore, the learner is encouraged to practice troubleshooting either in a laboratory setting or in a circuit simulation environment. Circuit Exploration exercises in each chapter provide opportunities for troubleshooting in a simulation or lab environment.

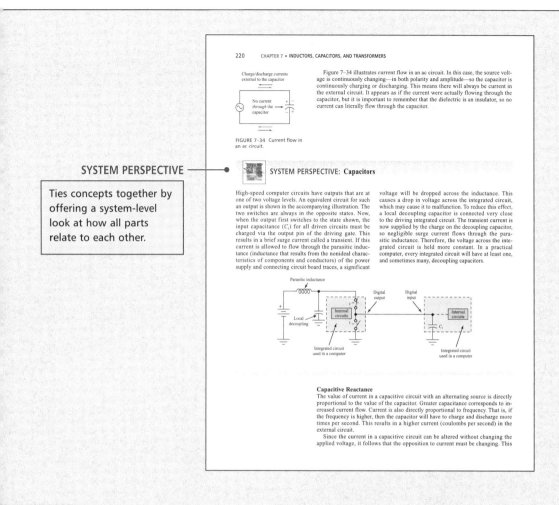

SYSTEM PERSPECTIVE — Ties concepts together by offering a system-level look at how all parts relate to each other.

ORGANIZATION

1. *Electronic Systems:* This introductory chapter lays the groundwork for a system-oriented approach to electronics. It includes representative systems and notation methods, such as block and schematic diagrams, and shows methods for interconnecting functional blocks. It also introduces the concept of a system hierarchy ranging from functional systems to electronic components to nanotechnology. System-level troubleshooting techniques are also included.

2. *Basic Electronics and Units of Measure:* Chapter 2 discusses atomic structure and defines basic electrical quantities such as charge, voltage, current, resistance, and power. Ohm's Law is also introduced. The system-level approach includes other system quantities such as mechanical (e.g., pressure and acceleration), light, magnetism, and electromagnetism.

3. *Basic Components and Technical Notation:* Powers of ten and engineering notation are presented in this topic. Interpretation, conversions, calculator operation, and the use of metric prefixes are included. The basic behavior of many electronic components (e.g., resistors, capacitors, inductors, transformers, switches, relays, transistors, and integrated circuits) is also introduced.

4. *Circuits:* The basic requirements for current are defined, and the four basic circuit topologies (series, parallel, series-parallel, and complex) are introduced. Numerical methods for analysis of series, parallel, and series-parallel circuits are provided. Additionally, the concept of ground and other reference points is presented.

CIRCUIT EXPLORATION

Explores computer simulation as a problem-solving tool. CD icon denotes MultiSIM circuit is available for troubleshooting.

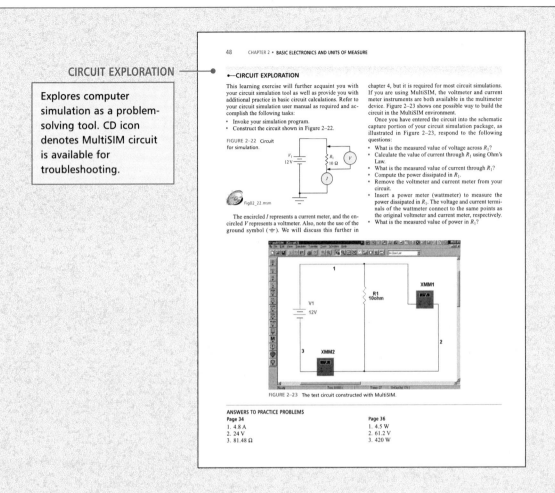

5. *Circuit Troubleshooting:* This chapter outlines specific troubleshooting procedures with application to series, parallel, and series-parallel circuits. It also emphasizes minimization of measurements and proper equipment selection.
6. *Alternating Current:* The generation and characteristics of alternating current waveforms are introduced with emphasis on sinusoidal waveforms. Sine wave calculations are developed, including basic phasor calculations using right-triangle relationships. Calculator operation is also reviewed.
7. *Inductors, Capacitors, and Transformers:* Chapter 7 presents the operational characteristics of inductors and capacitors. They are discussed both as individual components and in combination (e.g., *RC*, *RL*, and *RLC*). The basic principles of resonance are discussed as an integral part of the *RLC* circuit presentation. The construction and operation of transformers is discussed. This includes turns ratio calculations. Troubleshooting strategies are also outlined.
8. *Semiconductor Technology:* This chapter begins with basic semiconductor physics and progresses through current flow in a pn junction. Troubleshooting techniques for semiconductors are also included.
9. *Diodes and Diode Circuits:* The basic behavior of semiconductor diodes is discussed. The chapter also includes discussions on several diode applications including rectifier circuits, clipper circuits, isolation diodes, and others. The operational characteristics of several specialized diodes are highlighted, including zeners, varactors, Schottky diodes, current regulator diodes, tunnel diodes, and others. Troubleshooting techniques for diode circuits are presented.

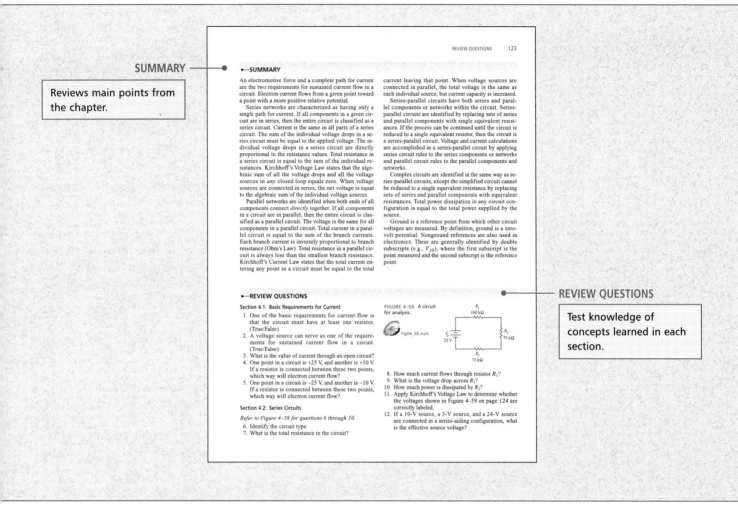

10. *Transistors and Transistor Circuits:* Chapter 10 presents the operational characteristics of bipolar and field-effect transistors (JFET and MOSFET). Biasing and amplifier configurations are discussed. Although some calculations are included, an intuitive understanding of transistor operation is emphasized. Transistor applications and troubleshooting methods are also discussed.

11. *Op Amps and Op Amp Circuits:* This chapter introduces op amp characteristics. Emphasis is on ideal behavior with reference to practical limitations on important parameters. A wide range of op amp applications is presented, including inverting and noninverting amplifiers, current-to-voltage converters, active rectifiers, integrators, comparators, and active filters. Troubleshooting techniques for op amp circuits are discussed.

12. *Power Supply and Voltage-Regulator Circuits:* The basic concepts of voltage regulation are introduced followed by operational discussions of several representative circuits. Series, shunt, and switching voltage-regulator circuits are included. A discussion of protection circuits for power supplies is presented along with troubleshooting techniques that are applicable to power supply circuits.

13. *Thyristors and Optoelectronic Devices:* Generalized thyristor behavior is introduced followed by more specific discussions of representative thyristors such as SCRs, triacs, diacs, and others. A number of optoelectronic devices are also discussed in this chapter, including standard LEDs, laser diodes, phototransistors, optoisolators, optointerrupters, and others. Troubleshooting strategies for thyristors and optoelectronic devices are presented.

14. *Integrated Circuit Applications:* This chapter discusses several representative applications for integrated circuits. Discussions include several oscillator circuits, phase-locked loops, and automotive and industrial applications. Troubleshooting techniques for IC-based circuits are highlighted.
15. *Digital Electronics:* Chapter 15 introduces digital electronics, including the binary numbering system, logic representation (e.g., truth tables, logic diagrams, Boolean algebra, and so on), combinational logic circuits, sequential logic circuits, and D/A and A/D conversions. Tools and methods for troubleshooting digital circuits are discussed.
16. *Microprocessors and Computers:* This chapter introduces the terminology and basic concepts associated with microprocessor-based systems. System architecture and an overview of computer software are included. The chapter discusses the functional block diagram of a typical notebook computer.
17. *Telecommunications:* The book concludes with a discussion of telecommunications concepts and methods. Emphasis in this chapter is on digital communications, but analog topics are also discussed. Classification of computer networks is presented. More technical topics such as modulation and multiplexing techniques are also introduced. The chapter concludes with a discussion of wireless technology including Bluetooth™ systems.

LEARNING FEATURES

Highly Readable

One of the most dominant learning features is the conversational writing style of the author. The material is written for easy interpretation by learners. Topic discussions are presented in a logical, intuitive, and commonsense way that comes from the author's experience both in industry and in education.

Instructional Aids

Learning **Objectives** are presented at the start of each chapter. They are written as behavioral objectives, so they clearly establish learning goals for the student. A list of **Key Terms** is also provided at the start of each chapter to further focus the learning goals. **Important rules, procedures, concepts, and formulas** are clearly highlighted in the body of the text. A typical **Calculator Application** Sequence is provided for each new type of mathematical problem. Numerous **Examples** are provided.

Each chapter includes **Exercise Problems** at the end of each major section. Additionally, there is a **Summary** and an extensive number of **Review Questions** at the end of each chapter. **Answers and solutions to the odd-numbered problems** are provided in Appendix D. Answers are also provided for all Circuit Exploration exercises.

Applications and System Perspective

There are many application examples sprinkled throughout the text. Some are an integral part of the chapter discussions, while others are highlighted in a **System Perspective** feature. These serve to promote and sustain learner motivation by emphasizing the relevance of the current material to their more distant career goals.

Circuit Exploration

Each chapter includes a **Circuit Exploration** feature. Here the learner is typically presented with an unfamiliar circuit whose operation is related to important points in the chapter. The learner is expected to construct the circuit (either in laboratory or in a circuit simulation environment) and to explore its operation. Often a list of tasks is provided to guide the learning activity, but the feature is generally intended to present a challenge to the learner. Troubleshooting practice should be an integral part of these exercises.

Computer Simulation

It is not essential for the learner to have access to a circuit simulation package in order to succeed with the material presented in this book. However, these packages can add another dimension to the learning process, and they should be utilized if at all possible. They are not a complete substitute for genuine hands-on experience with real devices, but they do allow a level of exploration and investigation into device and circuit operation that would be time-prohibitive in a practical laboratory environment. For those who would like to explore computer simulation tools, the accompanying **CD** includes the **Textbook Edition of MultiSIM** and **MultiSIM circuits** tied to the textbook. The icon shown to the left is placed beside those selected figures pre-created in MultiSIM.

References

An extensive glossary and a thorough index are welcome tools for student and instructor alike. Additionally, numerous industry references are provided in the form of component part numbers and URLs for manufacturers' web sites. Appendix A provides a brief summary and contrast of several logic family technologies, Appendix B offers a concise discussion of circuit measurement techniques, and Appendix C presents logic simplification with Karnaugh maps.

SUPPLEMENTS

The following materials are available for this text:

Instructor's Guide

An *Instructor's Guide* that includes detailed solutions to all problems, as well as suggestions and additional information related to certain problems, is available. The *Instructor's Guide* also provides Instructional Goals, System Perspective Ideas that can help illustrate important points in each chapter, Presentation Tools and Tips that offer supplementary teaching strategies and classroom exercises, and Student Worksheets. A worksheet that can be reproduced and used as a quiz, a homework assignment, or as a source of extra practice problems is provided for each chapter. Alternatively, the worksheets can be projected to serve as a powerful teaching tool during a class review session.

ISBN: 0-7668-3874-9

Lab Manual

The *Lab Manual* includes experiments and projects tied directly to the textbook.

ISBN: 0-7668-3873-0

e.resource

This is an educational resource that creates a truly electronic classroom. It is a CD-ROM containing tools and instructional resources that enrich your classroom and make your preparation time shorter. The elements of the *e.resource* link directly to the text and tie together to provide a unified instructional system.

Features contained in the *e.resource* include:

PowerPoint Presentation Slides
These slides provide the basis for a lecture outline that helps you present concepts and material. Key points and concepts can be graphically highlighted for student retention.

Exam View
This computerized testbank includes questions provided in multiple formats to assess student comprehension.

Image Library
Two hundred images selected from the textbook allow you to customize PowerPoint presentations or use them as transparency masters. The Image Library provides the ability to browse and search images with key words and allows quick and easy use.

Electronics Technology Website
Includes direct weblink to Delmar Learning's Electronics Technology website found at www.electronictech.com and to *Electronics for Computer Technology's* Online Companion for additional resources.
ISBN: 0-7668-3875-7

ABOUT THE AUTHOR

The author has over 35 years of electronics experience in both higher education and industrial environments. He has worked at multiple campuses for ITT Technical Institute in the capacities of Electronics Instructor, Evening Dean, and Director of Education. He also served on a number of national committees on electronics education. For several years he was the Senior Consulting Engineer and Manager of Laboratory Operations for TKC, Inc., an electronics engineering firm providing design guidance and engineering education to designers of high-speed digital equipment for major corporations around the world. He is currently employed by a leading manufacturer of industrial controls (PLCs) and industrial computers where he works as a staff EMC engineer to mitigate design problems in high-speed digital products. He also develops and conducts specialized training courses for technicians, engineers, and circuit-board designers. He has written numerous electronics books on various subjects, including basic electronics, op amps, digital technology, microprocessor technology, computer technology, and high-speed circuit design with emphasis on electromagnetic compatibility.

ACKNOWLEDGMENTS

The author and Delmar Learning would like to thank the following reviewers for their valuable feedback during the development of this project:

- Alan Brown, ECPI College of Technology, Hampton, VA
- Jim Duffey, Lewis & Clark Community College, Godfrey, IL
- Kenneth Lawell, Brown Institute, Mendota Heights, MN
- Tim Morgan, DeVry University, Kansas City, MO
- Tim Nichols, Computer Education Center at Cittone Institute, Philadelphia, PA
- Roger Peterson, Northland Community and Technical College, Thief River Falls, MN
- Lew Rakocy, DeVry University, Columbus, OH

We would also like to extend our thanks to the following contributors:

- Ernest Arney from ITT Technical Institute for his work on the Lab Manual
- John Reeder from Merced College for creating and testing the MultiSIM Circuit Files
- Don Arney from IVY Tech for creating the PowerPoint presentation slides
- Marcus Rasco from DeVry University at Irving, Texas for creating the computerized testbank

DEDICATION

I dedicate this book to my best friend, Linda. Her personal efforts on the original creation and coordination of the illustrations and art in this text were instrumental to the project. Additionally, her continuous encouragement and support were the inspiration to persevere on this massive project even though the light at the end of the tunnel was rarely visible. My sincere thanks and appreciation go to this impressive lady.

CHAPTER 1
Electronic Systems

•—KEY TERMS
block diagram
circuit simulation
cladding
discrete components
electromagnetic interference (EMI)
electronic system
fiber-optic link
flowchart
graph
integrated circuit
interpolation
linear scales
nanotechnology
pads
printed circuit board
schematic capture
schematic diagram
signal injection
signal tracing
traces
wireless link
wiring diagram

•—OBJECTIVES
After studying this chapter, you should be able to:
- List several electronic systems.
- Recognize and describe several methods to formally communicate the behavior of an electronic system.
- Explain the various levels of integration that can be used to construct electronic systems.
- Name and describe several methods for interconnecting systems and parts of systems.
- List and describe several techniques for system-level troubleshooting.
- Describe circuit simulation and explain its importance in the electronics industry.

•—INTRODUCTION
Electronic system is a label used to describe a device or group of devices whose basic operation is based on or controlled by a number of interconnected electronic components. Complexity of electronic systems can range from 2 or 3 components to many millions of components. However, regardless of complexity, all electronic systems are composed of only a few basic types of components that have well-defined behavior. Once you understand the behavior of each type of electronic component, you can apply your knowledge to the analysis or troubleshooting of all electronic systems.
Let's begin by discussing some typical electronic systems. As we proceed, continually remind yourself that every one of these systems—no matter how complex—is made up of only a few types of basic electronic components.

1.1 REPRESENTATIVE SYSTEMS
Electronic and electromechanical systems are used in every imaginable industry. Let's examine some representative systems.

Television
A television set is a familiar electronic system. The electronic components inside the set are designed to convert electrical signals from a cable company or from a receiving antenna into the picture and sound that you recognize as a television show.

Computers

Your personal computer is one specific example of a wide range of computer types. Computers are electronic systems that accept inputs, perform computations and other actions, and then produce outputs or change their operation based on the inputs. Typical inputs include keyboards, mouse-pointing devices, joysticks, the human voice, switches, sensors on industrial machines, telephone lines, radio signals, and data from the Internet. There can be many kinds of outputs such as printers, video displays, indicator lamps and signals to control machinery. Internal operations include computations, magnitude comparisons, and logical decisions.

Industrial Robotics

There are many types of industrial robots, but each is an electronic system that is controlling a mechanical device. Figure 1–1 shows a representative industrial robot. The electronics portion of the overall electromechanical system issues commands to control motors and valves that serve as muscles for the robot. Some robots, such as those used for surgical procedures, are capable of extreme precision and delicate movements. Others are powerful enough to manipulate many tons of steel during a manufacturing process. But, for robots on either extreme, it is electronic systems that control the movements of the robots to accomplish the required tasks.

Aircraft Electronics (Avionics)

A typical aircraft is brimming with electronic systems. Some are independent, stand-alone systems, while others communicate with each other to form larger systems. Examples of aircraft electronics systems include navigational, flight control,

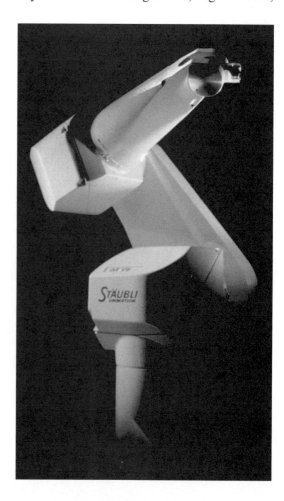

FIGURE 1–1 An industrial robot is an electromechanical system. (*Courtesy of Stäubli*)

stabilization, and communications systems; weapons control for military aircraft; electronic countermeasure systems to thwart enemy attack; environmental control systems; entertainment systems; and many more. It is important to remember, however, that all of these seemingly unrelated systems are built from the same basic blocks. The operation of many of these blocks/devices is presented in this book.

The Internet

The Internet might correctly be labeled as a system composed of other systems (sometimes called subsystems). As you know, computers all over the world can communicate with each other via the Internet. Each computer is a relatively complex system. When a particular computer system connects to the Internet via an Internet service provider, the local computer becomes an integral part of a worldwide system. The size and complexity of the overall Internet system is not only hard to imagine, but it is constantly changing as computers log on and off of the Net. In spite of this mind-boggling complexity, it is important for you to realize that any given part of the system consists of a combination of a few basic electronic components. You must understand the operation of these individual components to successfully analyze the performance of larger systems.

✓ **Exercise Problems 1.1**

1. Define an electronic system.
2. Most electronic systems are constructed from thousands of different types of electronic components. (True/False)
3. Name three electronic systems that were not specifically cited in the text.

1.2 SYSTEM NOTATIONS

Technical people such as technicians, engineers, and technologists must communicate detailed information about electronic systems. For example, a designer has to communicate system operation to the person who will be installing the system. Similarly, a team of designers and technologists needs to communicate the technical details of a new computer system as it is being developed, in order to make appropriate decisions. Communications of these sorts are greatly enhanced by using standard methods for representing systems. Depending on the type of system, the nature of the discussion or presentation, and the preference of the individuals, technical people use one or more methods to communicate the behavior of electronic systems: block diagrams, flowcharts, graphical data, wiring diagrams, and schematics. Each of these is described in the following sections.

Block Diagrams

A **block diagram** is a simplified representation of a system that indicates the relationships between functional sections or blocks in the system. Figure 1–2 on page 4 shows a simplified block diagram for a computer system. As you can see, each block on the diagram represents some well-defined function within the overall system. Lines and arrows are used between blocks to indicate the flow of information or control signals.

Many of the functional blocks in Figure 1–2 are connected together via the system data and address busses. A bus is a group of wires that all serve the same general purpose. For example, all wires in an address bus carry some portion of the address used to access memory. The lines that make up the data bus, by contrast, are used to transfer information (data) between functional blocks. In some cases, as illustrated in Figure 1–2, the block diagram is labeled to show the width (i.e., number of wires) of the bus. Notice that in the case of the mouse and keyboard, the

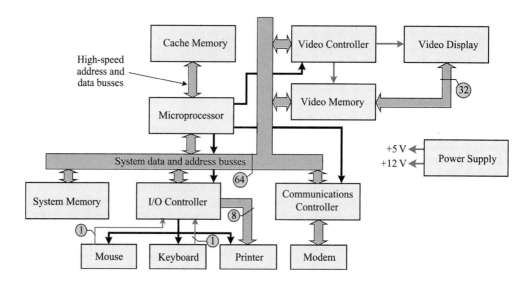

FIGURE 1–2 A simplified block diagram of a computer system.

width is indicated as ①. This type of connection—called serial communication—requires only a single path.

Control lines can also be illustrated on a block diagram. In the case of Figure 1–2, the control lines are bold, single arrows. They point from the controlling block to the block that is being controlled. Figure 1–2 also includes a block representing the system power supply. It is assumed that power is routed to all functional blocks of the system, so the connections are generally omitted to reduce clutter on the drawing.

You can understand the general nature of a system by studying its block diagram. Consider the following description of a simple process.

You know that when you press a key on your keyboard, the corresponding letter appears on the video display. The block diagram in Figure 1–2 provides insights into the activity that must occur behind the scenes to make this simple event happen. As you can see, there is no direct link between the keyboard and the video display. In short, the electrical impulses representing the letter to be displayed travel from the keyboard to the I/O controller (I/O is an abbreviation for input/output). From here it may move via the system data bus into system memory temporarily. It is then moved from system memory into video memory. Most of this activity is under the control of the microprocessor. Once the character is in video memory, the video controller generates the impulses necessary to cause the letter to be displayed on the video display.

You will learn the details of computer operation in a later course, but for now it is important to understand that a block diagram is an important tool for communicating the overall nature of a system and the general flow of information and control between major functional blocks.

Flowcharts

A **flowchart** is another way the behavior of a system can be communicated between technical people. Flowcharts are valuable tools used to:

- Program computers
- Describe the technical operation of an electronic system
- Troubleshoot complex equipment
- Explain system operation to users
- Optimize the design of a system

Figure 1–3 shows some of the common symbols that are used to draw flowcharts. These can be interconnected to describe the behavior of a system. The level of detail represented by a flowchart ranges from very general to extreme, depending on the purpose of the diagram.

Figure 1–4 shows a flowchart that might be used to help troubleshoot a computer system. It is presented here to illustrate the use of flowchart symbols and does not have sufficient detail for actual troubleshooting of a computer. Because flowcharts are so widely used for technical communications and documentation, it is important for you to know how to interpret them accurately.

As a general rule, block diagrams are more concerned with the electrical functionality of a system, whereas a flowchart is more commonly used to describe the behavior of the system from an operator's point of view. However, neither of these general uses is restrictive.

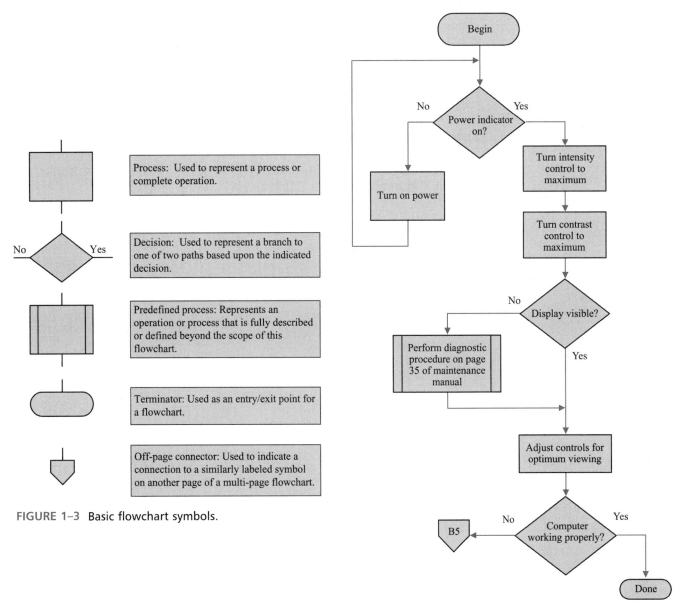

FIGURE 1–3 Basic flowchart symbols.

FIGURE 1–4 A simplified flowchart for troubleshooting a computer system.

Graphical Data

A **graph** is a tool used to communicate technical details about the behavior of an electronic system. Depending on the specific information to be described, there are usually many different types of graphs that can be employed.

Figure 1–5 shows several kinds of graphs that are widely used to describe electronic systems. For now, we will focus on how to read values from a given graph type.

Figure 1–5(a) is a line graph. Because both the vertical and horizontal scales have equally-spaced divisions, the scales are called **linear scales**. From any point on the horizontal axis, move straight up until you reach the plotted curve. From there, move directly to the left until you intersect the vertical scale. This point of intersection gives you the value on the vertical scale that corresponds to the initial value on the horizontal scale. Thus, for example, a wavelength of 600 nm for the device represented in Figure 1–5(a) corresponds to a relative emission of near 100%. When values fall between unnumbered divisions on a given scale, you must estimate the indicated value. This estimation process is called **interpolation**. In effect, you mentally add additional scale marks to assist you.

For example, halfway between 600 and 650 would be 625. Similarly, halfway between 600 and 625 would be 612.5, and so on. With a little practice, interpolation will become second nature to you.

The graph in Figure 1–5(b) is read in a similar manner except that the horizontal or x-axis is logarithmic. The nonuniform spacing easily identifies a logarithmic scale. Interpolation is more difficult with a logarithmic scale, since, for example, 2 is not midway between 1 and 3. Nonetheless, values are read from the graph in the same general way as from any other line graph. As a specific example, verify from the graph in Figure 1–5(b) that the relative output voltage is −25 dB at 100 MHz. Logarithmic scales are used when the range of the scale is quite large. For example, the horizontal scale in Figure 1–5(b) extends from 1 to 1000.

The graph shown in Figure 1–5(c) is a polar graph. This particular example shows the brightness of a light-emitting diode as viewed from different angles. A light-emitting diode is a device that emits light when electrical current passes through it. In the specific example shown, each of the concentric circles corresponds to a value of relative intensity beginning at 0 in the center and progressing to 120% on the outermost circle. Each radial extending from the center outward corresponds to the viewing angle. These range from 0 to 90° in the present example.

Interpretation of values on a polar graph is still accomplished in the same basic way as previously discussed. From any given point on one scale, follow the associated circle (for intensity) or the associated radial (for angular displacement) until

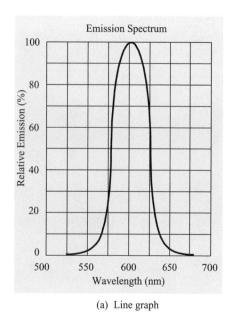

(a) Line graph

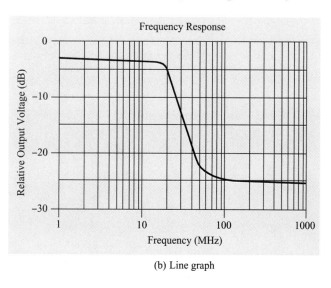

(b) Line graph

FIGURE 1–5 Many types of graphs are used to describe electronic systems.

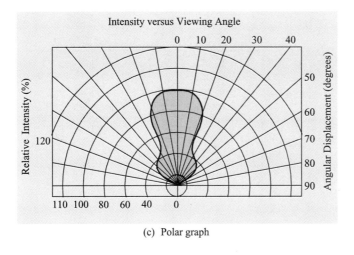
(c) Polar graph

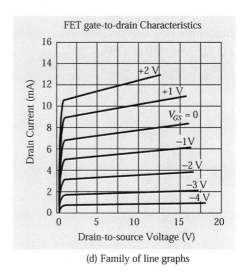
(d) Family of line graphs

FIGURE 1–5 (*continued*)

you intersect the plotted curve. Then follow the intersecting circle or radial to determine the corresponding value. For example, refer to Figure 1–5(c) and confirm that the relative intensity is 80% at 20° and approximately 40% at 35°.

Figure 1–5(d) shows another linear graph similar to Figure 1–5(a). However, the graph in Figure 1–5(d) shows a family of curves instead of just a single line. A single line graph shows the relationship between two variables such as wavelength and relative emission in Figure 1–5(a). A family of curves can show the effects of a third variable. In the example shown in Figure 1–5(d), each curve corresponds to a different value for the variable called V_{GS}. Once you have selected a particular V_{GS} curve, then you read the two corresponding values from the vertical and horizontal axes as usual.

•—EXAMPLE 1.1

Refer to Figure 1–5(d) and determine which of the following is *incorrect:*

a. $V_{GS} = 0$ V, Drain-to-source Voltage = 12.5 V, and Drain Current = 8 mA.
b. $V_{GS} = -3$ V, Drain-to-source Voltage = 10 V, and Drain Current = 2 mA.
c. $V_{GS} = +2$ V, Drain-to-source Voltage = 7.5 V, and Drain Current = 12 mA.
d. $V_{GS} = -2$ V, Drain-to-source Voltage = 5 V, and Drain Current = 6 mA.

SOLUTION You should be able to verify that the first three combinations are true, and the last combination is wrong.

Be sure you work as many problems as necessary to become proficient at reading graphs of all types, since they are so important to your electronics career.

Wiring Diagrams

A **wiring diagram** is a pictorial sketch that represents the components in a circuit. Additionally, each component is labeled for clarity. This type of diagram can often be understood by nontechnical people, since there is a very close relationship between the wiring diagram and the physical circuit or system. Figure 1–6 on page 8 shows a representative wiring diagram. All electrical interconnections between the various devices and terminals are shown with lines representing wires.

Schematic Diagrams

A **schematic diagram** is a graphical illustration based on symbols that shows how every electronic component is connected to other components to form a functional system. Schematic diagrams are one of the most important ways that the operation

FIGURE 1–6 A wiring diagram shows the physical connections in a system.

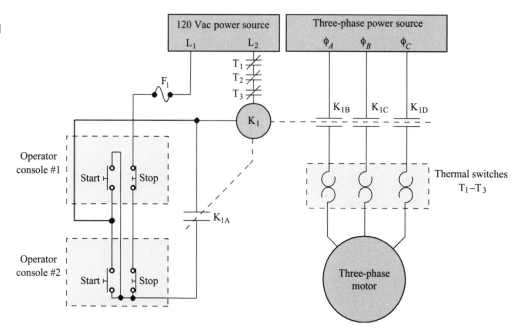

of an electronic system is communicated between technical people. Schematics are valuable troubleshooting tools, since they reveal the normal behavior of the circuit. Test instruments can measure the actual values in a circuit. Once you know the correct value and the actual value for a given circuit quantity, you can then make judgments to locate a defective component.

We will utilize schematic diagrams throughout this book. Figure 1–7 shows a few common symbols used to represent electronic components on a schematic diagram.

FIGURE 1–7 Symbols are used to represent electronic components on a schematic diagram.

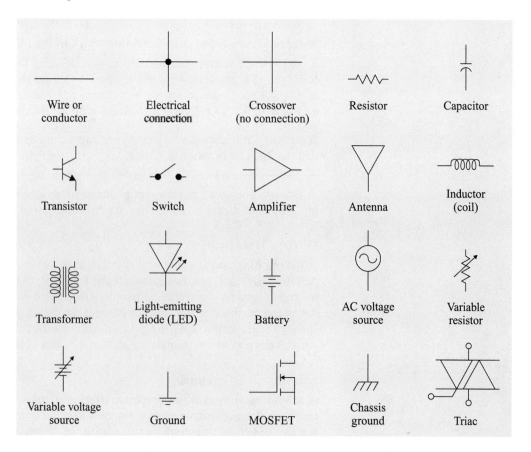

Figure 1–8 shows a representative schematic diagram using some of these symbols. Here a switch (S_1) is connected in a circuit with a voltage source (V_1), a light-emitting diode (LED_1), and two resistors (R_1 and R_2). If this circuit is constructed, the LED will illuminate when the switch is closed. Once you have learned the operation and characteristics of each basic component, then interpretation of a schematic diagram becomes relatively easy.

FIGURE 1–8 A schematic diagram is a valuable tool.

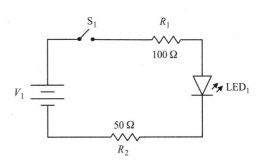

Schematic diagrams and wiring diagrams have a lot in common. In fact, in some cases they are indistinguishable. But, as a general rule, a wiring diagram maintains a very close relationship with the physical components, whereas a schematic focuses only on the electrical connections.

✓ Exercise Problems 1.2

1. A block diagram generally provides more technical detail than a schematic. (True/False)
2. Why are power supply connections not normally shown on a block diagram representation of an electronic system?
3. Refer to Figure 1–5(a). What is the relative emission when the wavelength is 575 nm?
4. Refer to Figure 1–5(b). What frequency produces a relative output voltage of −5 dB?
5. What is the name of a drawing that uses pictorial representations of the components and the electrical connections between them?

1.3 PHYSICAL SYSTEM HIERARCHY

We have said that the most complex electronic system is still constructed using the same basic components as a simple system. In this section, we will take a closer look at several levels of system building blocks.

Electronic Components

Several basic electronic components will be introduced in the next chapter. For now, it is important to understand that electronic components are fundamental building blocks for all electronic products. Each component has certain well-defined electrical characteristics. The physical appearance of electronic components, by contrast, is quite varied. Figure 1–9 on page 10 shows several types of resistors. Resistors and other electronic components are available in much larger and much smaller sizes than those pictured in Figure 1–9.

Electronic Circuits and Printed Circuit Boards

It is possible to build complete functional systems based entirely on individual electronic components (called **discrete components**). The various components are mounted on a thin (typically 0.062″) sheet of nonconductive material called a **printed circuit board** or PCB (also called a PWA for printed wiring assembly). Thin copper strips called **traces** (or runs) and **pads** (or lands) are bonded to the

FIGURE 1-9 (a) Resistors vary in value and physical appearance. Their value is indicated with a printed number or with color-coded bands. (b) The schematic symbol for a resistor.

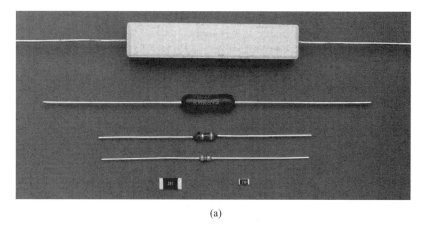

(a)   (b)

base material and serve to provide the electrical interconnection of the various components. For extremely simple circuits, the traces may be on only one side of the PCB. More complex systems utilize a laminated stack with alternating layers of traces and nonconducting board material. Typical boards range from six to twelve interconnect layers, but many boards have over twenty layers.

The level of system complexity based on discrete components is important to understand and appreciate, but you should also be aware that for most practical, modern systems, this type of circuit construction would be considered obsolete. This approach is mostly being replaced by other methods that allow far greater circuit densities (i.e., more functional circuitry in a smaller space). Figure 1–10, for example, shows a PCB with both discrete and integrated components.

FIGURE 1-10 An electronic system composed of discrete components.

Integrated Circuits

An **integrated circuit** or IC is essentially a thin slice of silicon (often called a chip) perhaps 0.05″ to 0.2″ on a side that has thousands (even millions) of basic electronic components. The components are functionally similar to their discrete counterparts, but they are many times smaller. The various components are interconnected within the IC by thin conducting traces of a material such as gold or aluminum. The silicon chip and the metallized interconnecting traces serve the same functional purpose that a printed circuit board serves for discrete components.

Manufacturing environments for integrated circuits utilize high-quality clean rooms that filter the air to remove any dust or contaminants that might otherwise interfere with the operation of the miniature components. Once the desired circuits have been constructed on the chip, it is placed into a larger package. Figure 1–11 shows several representative integrated circuit packages. The package is sealed to protect the sensitive components. The rugged external leads, which connect via fine gold wires to the internal components, provide a means for connecting the IC into an external system.

FIGURE 1–11 Two types of integrated circuits. (*Courtesy of Agilent Technologies*)

A single integrated circuit can easily hold complex circuits with functionality that previously required many individual printed circuit boards with discrete components. The range of available IC functions is essentially boundless (since new ones are continually being introduced), but representative functions include amplifiers, computer memory, microprocessors, and industrial control devices.

Integrated circuits can be classified by their integration complexity. A group of interconnected transistors (discussed in chapter 10) called a logic gate (discussed in chapter 15) is the basic building block for digital integrated circuits. Table 1–1 shows several levels of integration and the number of equivalent logic gates contained within the integrated circuit package. If you multiply the number of logic gates by ten, you will have an estimate of the total number of electronic components in the package.

TABLE 1–1 Levels of integration for integrated circuits.

Category	Abbreviation	Number of Logic Gates
Small-scale integration	SSI	1–10
Medium-scale integration	MSI	11–99
Large-scale integration	LSI	100–999
Very large-scale integration	VLSI	1000–999,999
Ultra large-scale integration	ULSI	1,000,000 or more

Functional Modules

Electronics designers use printed circuit boards to interconnect a number of integrated circuits and discrete components to form a functional module. The module may be a stand-alone system—such as an AM/FM radio—or it may be part of a larger system. Figure 1–12 shows several functional modules designed to be inserted into a computer. Modules such as the ones shown in Figure 1–12 provide functions such as interfaces between the computer and the Internet, video capture cards to allow video images to be merged into documents, communication links to industrial machines, and global positioning system (GPS) devices that serve to pinpoint a location anywhere on earth.

FIGURE 1–12 An array of functional modules. (*Courtesy of National Instruments*)

Systems on a Chip—Electronic

The same basic technology that allows integration of basic electronic components into integrated circuits can also be extended to put complete electronic systems on a single silicon chip. Just as integrated circuits combine many discrete components into a single IC, a system on a chip combines the functions of several standard integrated circuits into a single chip. It is now possible, for example, to buy complete computer systems on a single chip. The functions provided by the single package previously required many individual integrated circuits. An engineering calculator is another familiar example of a system on a chip.

Systems on a Chip—Electromechanical

Much research currently underway is aimed at the integration of electronic and mechanical systems into a single silicon chip. The machines on this atomic level are constructed one molecule or even one atom at a time. Several types of machines have been demonstrated including motors, valves, levers, and gears. This atomic level of system construction is termed **nanotechnology**. It is an exciting field with huge promises, but for now, most of the focus is on research. As these breakthroughs are transformed into practical products, we can expect to see incredible changes in our daily lives. Figure 1–13 on page 13 illustrates the size of these electromechanical systems by showing a dust mite crawling over the top of an integrated machine.

✓ Exercise Problems 1.3

1. Which has the greatest component density: discrete components or integrated circuits?
2. What do the abbreviations PCB and PWA represent?
3. What is the name used to describe building of atomic-level electromechanical systems?

FIGURE 1–13 A dust mite crawling over an integrated machine built using nanotechnology. (*Courtesy of Sandia National Laboratories, USA*)

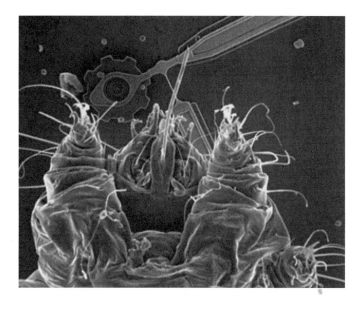

1.4 SYSTEM CONNECTIVITY

As we saw previously, electronic systems require connections between the various parts making up the system. This is true whether the parts are discrete electronic components, integrated circuits, or complete functional modules. This section presents several ways that the various parts of a complete functional system can be interconnected.

Wires

Perhaps conductive wire is the most well-known method for providing electrical interconnection between electronic parts. The wire is usually made of copper although other materials such as gold, silver, and aluminum are also used. For many applications, the conductive wire is covered with a nonconductive outer covering (called insulation) to prevent accidental connections from occurring if two or more wires contact each other. We will discuss several types of wire in more detail in chapter 3, but generally, the conductors used to provide interconnection of electronic parts are either solid or stranded as illustrated in Figure 1–14. Both types serve the same basic purpose, but each has advantages in certain applications.

FIGURE 1–14 (a) Solid-conductor wire and (b) stranded wire.

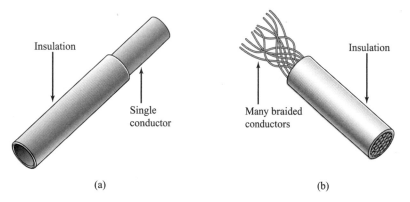

Printed Circuit Board Traces

The copper traces on a printed circuit board provide electrical connectivity between the various parts mounted on the PCB. The thickness of the traces is usually in the range of 0.0007" to 0.0028" with typical widths on the order of 0.006" to

0.2″ although none of these dimensions should be considered as limits. The PCB traces are visible in the photo shown in Figure 1–10.

Infrared Links

Many electronic systems are effectively connected to each other through infrared (IR) links. The basic scheme is illustrated in Figure 1–15.

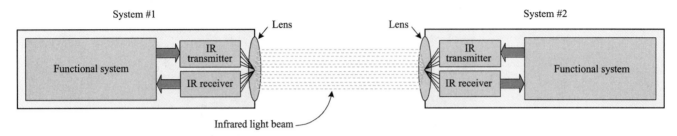

FIGURE 1–15 An infrared link can connect two operational systems.

Each system has an infrared transmitter and an infrared receiver. An infrared transmitter accepts information and commands from the rest of the functional system and converts them into corresponding intensity variations in an infrared light beam. By contrast, an IR receiver detects variations in an incoming infrared light beam and converts those changes into corresponding electrical signals that can be interpreted by the rest of the system. As illustrated in Figure 1–15, lenses are used to focus the beam.

Once an IR link has been established, the two systems are effectively coupled to each other. They can pass commands and information back and forth as needed to accomplish their intended task. Common examples that utilize IR links include cordless computer mice, television remote controls (generally one-direction only), and personal data assistants (PDAs) that are designed to be linked to a larger computer.

Fiber-Optic Links

A **fiber-optic link** consists of thin glass threads, which have a very dense, clear inner core, and a less dense outer glass covering called **cladding**. This is illustrated in Figure 1–16.

FIGURE 1–16 Optical fiber construction.

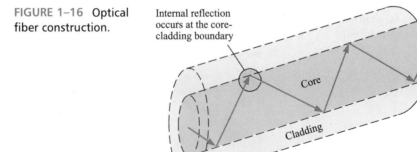

When light enters the core on one end of a long fiber, it travels to the other end with only slight losses. Most of the light is reflected at the junction of the core and cladding material. It may be helpful to think of the core-cladding interface as a circular mirror that reflects all light inside of the fiber. In most practical cases, several optical fibers are bound closely together with an overall outer protective covering.

Many applications use a laser as the source of light on one end of a fiber-optic bundle. The laser is turned on and off millions of times per second to encode the information that is to be transmitted over the fiber-optic link. The information then travels down the optical link as light energy. When it reaches the far end of the fiber-optic path, a light-sensitive device converts the pulsing laser light back into electrical signals that are replicates of the original information that was used to control the transmitting laser. In this way, two electronic systems can be effectively linked together for the purpose of exchanging commands and information.

Fiber-optic links have a number of advantages over wire and infrared links for many applications. Outstanding advantages include the ability to carry many more simultaneous information channels (e.g., phone conversations or computer data transfers) and immunity to interference. Most cable TV companies are now replacing their existing wire links with fiber-optic links. This provides a better quality signal, it is less affected by electrical noise called **electromagnetic interference (EMI)**, and it allows additional services to be offered. One particularly interesting service that is generally available when cable TV companies move to optical links is high-speed connection between your computer and the Internet. This allows much faster operation than communication via a phone line.

Radio Links

If you replace the IR transmitters and receivers in Figure 1–15 with radio transmitters and receivers (with antennas instead of lenses), you will have a block diagram of a radio frequency (RF) **wireless link**. This is a very popular type of link for computer equipment. Examples include remote access to the Internet via satellite communication, transmission and reception of faxes from an aircraft, and communication between industrial computers used to control machines without the need to run physical wires between the various devices. As with the other links previously presented, once a link has been established, the two systems are effectively connected and are able to exchange commands and information.

✓ Exercise Problems 1.4

1. What is the purpose of the nonconductive outer covering sometimes found on wires?
2. The thickness of printed circuit board traces is typically in the range of 0.25″–0.5″. (True/False)
3. One of the more outstanding advantages of optical-fiber links over conventional copper wire links is immunity to electrical interference. (True/False)

1.5 ELEMENTS OF SYSTEM-LEVEL TROUBLESHOOTING

The ability to diagnose and correct malfunctions in electronic systems is an important skill for a career in computer technology or electronics in general. This section will introduce you to some of the strategies that you must master in order to troubleshoot electronic systems efficiently.

Block-Diagram Thinking

When troubleshooting a complicated electronic system, it is easy to be overwhelmed by the complexity. Many systems are simply too complex for most people to be able to visualize all of the system details and how the various parts interact with each other. Additionally, the electrical impulses at a given point in a system may be changing millions or even billions of times per second. This makes it very difficult to classify a particular impulse as "normal" or "abnormal" without using special test equipment. One important way to avoid being overwhelmed by system complexity and still be able to logically troubleshoot the system is to view the system from a block diagram perspective.

Block diagrams show the relationships between functional blocks. They show where the inputs come from, and they show where the outputs go. However, they do not provide insights into how a particular function is actually implemented. Figure 1–17 shows a block diagram for a simple public address system. It is important to realize that the block diagram for a given electronic system remains the same regardless of whether the system is physically realized with discrete components, integrated circuits, or some higher level of system integration.

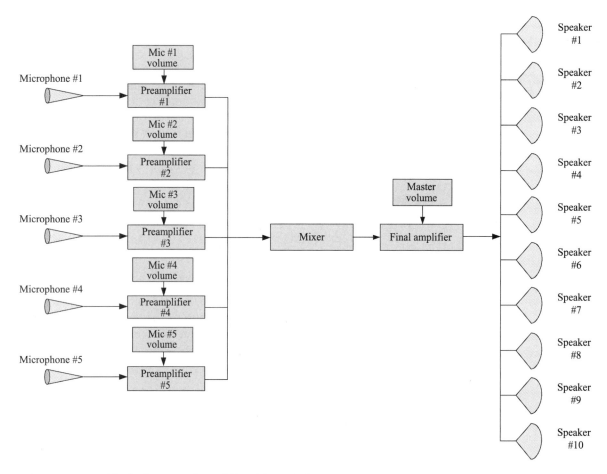

FIGURE 1–17 A block diagram of a public address (PA) system.

When troubleshooting, you should try to keep a clear picture of the system block diagram in front of you (either physically or mentally depending upon the complexity of the system and your familiarity with it). By observing (i.e., measuring with test equipment) the signals entering a given functional part of the circuit and contrasting them with the signals leaving that block, you can locate which functional block is defective. In complex systems such as computers, the speed and intricacies of the system obscure this simple approach. Nonetheless, maintaining a clear image of the system block diagram and the functional relationships it describes is an important tool when troubleshooting electronic systems regardless of complexity.

Monitoring Signals (Signal Tracing)

Signal tracing is a troubleshooting technique that requires you to monitor circuit quantities (typically voltages) at various sequential points in the system. At each monitored point, you must make a judgment to classify the actual measured value as good or bad with respect to the correct or expected value. If a given measurement is classified as bad, then you can conclude that the defect lies at some earlier stage in the system. On the other hand, if a given measurement is good, then the defect lies in some subsequent stage of the system. This troubleshooting strategy is especially well suited to electronic systems that are sequential in nature. In effect, you follow (i.e., trace) the signal through the system until you reach a point where the signal is no longer what it should be. Signal tracing can quickly localize defects in many kinds of systems.

Forcing Signals (Signal Injection)

Signal injection is a troubleshooting technique that requires you to inject a known good signal at some point in the system. If all subsequent portions of the system function normally, the defect lies ahead of the injection point. On the other hand, if the injected signal fails to make all subsequent portions of the system operate correctly, then the defect lies after the point of injection. As with signal tracing, signal injection is well suited to electronic systems that are essentially sequential in nature. This includes many types of systems or portions of systems such as radio, television, amplifiers, and distributed systems such as computers connected to a common network in an office environment.

Diagnostic Software

One of the most powerful troubleshooting tools for many electronic systems is diagnostic software. This is specially designed software that generally requires only limited system capability in order to execute successfully. The software can locate defects throughout the system by stimulating the circuits and then monitoring them for a correct response to the stimulus.

For example, diagnostic software is often used to locate defects in computer memories. The software itself requires very little memory for it to run properly. Once the software is running, it stores various numbers in all of the available memory locations. It then returns to those same locations and checks the value of the stored number. If the memory is working correctly, the stored value should be the same as the value that was originally placed into the memory location. The software interprets any discrepancies as memory defects.

Diagnostic software is probably the most common troubleshooting tool for many types of electronic systems. In many cases, the software is designed to run on the same system that is being diagnosed. However, in order to execute properly, the system must have some degree of functionality. If it does not, then the software cannot be used, and other troubleshooting strategies must be employed initially. In other cases, the diagnostic software is actually executed on a separate electronic system. An automobile car analyzer is a good example of this latter case. Here the analyzer has an internal computer that runs diagnostic software. The software monitors the operation of the computer in the automobile and evaluates its performance.

Power of Observation

Probably the single, most important troubleshooting tool that you can have in your "tool kit" is the power of observation. Many defects—even in complex electronic systems—can be successfully diagnosed if the symptoms of the malfunction are carefully observed. This is not "cheating." Rather, it is an important skill that you should utilize to its fullest extent.

For example, while troubleshooting a computer keyboard, you observe that the keystrokes are being sensed by the computer (i.e., characters are displayed on the video screen each time a key is pressed), but the wrong character is displayed. With a block diagram of the keyboard circuitry and the given observation, you can eliminate a major portion of the keyboard circuitry as having suspected defects. For example, you know it cannot be a severed cable between the keyboard and the computer; if the cable were cut, then no characters could be transferred to the computer.

Observation includes three other important troubleshooting strategies:

- User interview
- Front-panel milking
- Review of history records

By *interviewing* the user or operator of an electronic system, you can often gain insights that will lead you to the defective portion of the system. For example, the user may describe a peculiar sound, odor, or system behavior that occurred just before the defect became evident. This is important information.

Front-panel milking is the intentional operation of all front-panel controls on the defective system while observing the behavior of the system. Careful observation of the behavior can lead you to the defective part of the system.

For example, suppose you are troubleshooting a printer for a computer system. The symptom is that no paper is ejected from the printer under computer control. To help diagnose the problem, you could use front-panel milking. One of the controls that is typically found on a printer is a manual paper feed or form feed button. If you press this button and a sheet of paper is ejected from the printer, then you can eliminate a major portion of the printer as being defective. You now know, for instance, that the motor, which pulls the paper through the printer, has to be good.

If you have the responsibility for troubleshooting defects in electronic systems, then you are well advised to maintain a logbook that records symptoms and defects. Regardless of system complexity, most of the malfunctions tend to be repetitive. In other words, although there may be 1,000 *possible* defects in a given system, there may be only fifty of those that actually occur during the life of the product. Further, 80 percent of those fifty defects that do occur, result from only three different kinds of malfunctions. While the numbers cited here are for illustration only, it is well known that most electronic systems have recurring malfunctions. If you have a logbook that lists the *maintenance history of a system* that you must maintain, then the chances are excellent that most problems can be solved instantly by simply looking up what you found to be the defect on previous occasions. Figure 1–18 shows a page from one person's logbook.

Here's an example you may recognize. Tell any auto mechanic that your car makes a certain sound when accelerating, or vibrates a certain way, or pulls to one side when braking, or any other verbal description of a defect, and the odds are good that the mechanic will quickly announce a list of probable causes. Granted, the mechanic may not be referring to a physical logbook, but his or her experience is serving the same purpose. This is a powerful troubleshooting tool.

Substitution: Benefits and Risks

Once you have localized a defect in an electronic system, you can substitute a replacement component. You then reevaluate the performance of the system to determine whether you have correctly identified the defect. Substitution can be used at virtually any level of system integration. For example, if a home computer system doesn't work properly, you can substitute the monitor, keyboard, mouse, printer, computer, and other system components one at a time until the system works properly. This is not a recommended method in most cases, but clearly it would succeed in positively identifying the defective component. Similarly, if you have localized the defect to the computer proper, then you might substitute the various circuit boards with replacement circuit boards until the system works properly. Again, the defect would be located. Following this same strategy, if you have localized the defect to a particular portion of a printed circuit board, you can substitute the parts in that portion of the board, one at a time, until the system is operational.

Although component substitution does ultimately lead to positive identification of the defect, it must be used wisely in order to be effective and practical. First, if

FIGURE 1–18 A logbook can be a valuable troubleshooting tool.

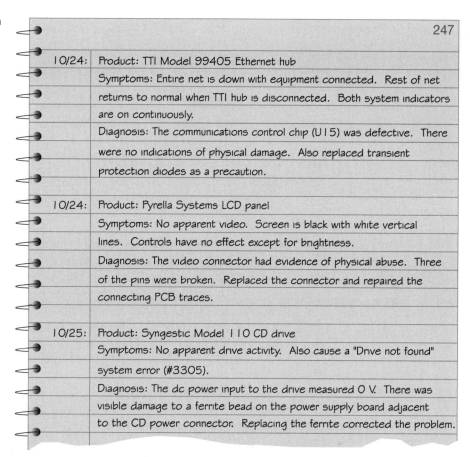

substitution were your only method of troubleshooting, then you would be required to have a huge inventory of replacement components. Clearly, this is not practical. Second, many electronic components are fragile and can be damaged when removing them from a printed circuit board. In fact, the printed circuit board itself is fairly easy to damage in the process of removing a component. Therefore, you will always want to use other troubleshooting strategies to localize the defect. Reserve substitution for "plug-in" components or components that you believe to be defective based on other diagnostic measurements.

Finally, in order for substitution to be an effective troubleshooting tool, the replacement components *must* be known to be good. If you have a stack of "old" parts lying around your work area that may or may not be any good, they are of limited value when used for substitution. In short, always be absolutely certain that the replacement part is good, or your conclusions may be wrong.

Consider this example. Many electronic components that are used for modern systems are fragile and easily damaged. The damage is not always visible to the eye. Many of these components cost pennies. Therefore, if you remove one of these parts because you suspect it is defective but find that it was not, then *throw it away*. Replace the part with a new one. Failure to follow this practice can result in your inadvertently introducing new defects into the system in addition to those that were already there. It is not worth the risk when the cost of the component is negligible.

Another related point may occur when you have two or more identical systems available. In some cases, you can remove a component (typically pluggable) from the good system and substitute it into the defective system. If the problem is corrected, you have identified the defect. Care must be used in this situation, however, since you are risking damage to the substituted component or module. If damage does occur, then you will have two defective systems. You may have great difficulty justifying this result to a customer.

✓ Exercise Problems 1.5

1. What is the major advantage of using block-diagram thinking when troubleshooting a complex system?
2. Contrast signal tracing and signal injection strategies.
3. Describe front-panel milking as a troubleshooting strategy.

1.6 CIRCUIT SIMULATION

Circuit simulation is a technique wherein computer software simulates the behavior of an electronic circuit or complete system. In this way, a new design can be optimized and debugged before actually constructing the circuit with physical parts. This represents a huge savings in both time and material cost. Circuit simulation is also valuable in the learning environment because you can quickly "build" and analyze the behavior of new circuits without the need for a fully-equipped laboratory. You can use circuit simulation as a way to practice troubleshooting electronic circuits. Throughout this book you will be given the opportunity to enhance your learning by solving a number of problems using circuit simulation. The problems can be solved on any general-purpose circuit simulator software package, but the examples are given using MultiSIM™ by Electronics Workbench, since this is such a popular package in both schools and industry.

Schematic Capture

Schematic capture is a process whereby new circuits are constructed in a virtual laboratory environment. You basically choose the needed components from a library of components and place them according to your preference. Once the components have been positioned, you then "wire" the various components with lines that represent electrical connections. Once you have constructed the schematic, you can save it in a file for subsequent use.

Virtual Instruments

Just as in a real laboratory, the virtual laboratory environment allows you to connect various types of test instruments to the circuit. You will study equipment such as voltmeters, ohmmeters, oscilloscopes, logic analyzers, and spectrum analyzers later in your studies. For now, realize that all of these instruments (and others) can be simulated in the circuit simulation environment.

Circuit Analysis

Circuit simulators are also capable of performing a number of mathematical analyses to better characterize the behavior of the circuit. One type of analysis, for example, randomly varies the values of each component within specified limits and computes the effect on circuit performance. This analysis helps predict the circuit's behavior when built with imperfect components that are subjected to environmental variations.

✓ Exercise Problems 1.6

1. Circuit simulation uses actual laboratory test equipment. (True/False)
2. The schematic representation of a circuit serves as the input to a circuit simulation software package. (True/False)
3. Circuit simulation software can mathematically analyze the behavior of circuit designs. (True/False)
4. In a circuit simulation environment, the meters, oscilloscopes, logic analyzers, and other test equipment are called _____ _____.

SUMMARY

There are endless examples of electronic and electromechanical systems. Examples include computers, robotic systems, radios, television, radar equipment, and environmental control systems. The behavior and electrical characteristics of a system can be communicated in many ways including block diagrams, flowcharts, graphs, wiring diagrams, and schematic diagrams.

Regardless of system complexity, all electronic systems are constructed from a relatively small number of electronic component types. It is essential for you to understand the behavior and characteristics of each basic component in order to be able to analyze and troubleshoot complex systems. The characteristics of many of the basic components will be presented in later chapters.

Circuits may be physically constructed from discrete components or from a number of increasingly dense techniques such as circuit boards, integrated circuits, functional modules, systems on a chip, and nanotechnology. Electronic systems can be connected to each other through many different methods including conductive wire, PCB traces, optical links (e.g., infrared and fiber optics with lasers), and wireless RF links.

Success in troubleshooting defective electronic systems requires an understanding of electronic components and circuits (i.e., you must be able to define normal operation). Additionally, you need a systematic troubleshooting procedure. Some of the techniques that are valuable troubleshooting tools include block-diagram thinking, signal tracing, signal injection, diagnostic software, and observation.

Circuit simulation uses computer software to create a virtual electronic laboratory environment. Virtual circuits can be constructed (schematic capture), measured (virtual test equipment), and analyzed (numerical analyses) to determine the behavior of similar circuits constructed from real parts.

REVIEW QUESTIONS

Section 1.1: Representative Systems

1. Which of the following could be classified as an electronic system? Circle all that apply.
   a. 2-way radio
   b. notebook computer
   c. electronic calculator
   d. yardstick
2. Which of the following are probably classified as electromechanical systems? Circle all that apply.
   a. child's wagon
   b. automobile
   c. airplane
   d. CD player

Section 1.2: System Notations

3. Which type of system notation provides a general overview of the system operation?
4. Which type of system description provides the greatest level of technical detail?
5. Which types of system documentation can be used as troubleshooting tools?

Section 1.3: Physical System Hierarchy

6. What is nanotechnology?
7. Discrete electronic components provide greater circuit density (more functions in less space) than integrated circuits. (True/False)

Section 1.4: System Connectivity

8. Name two ways that electronic systems can be electrically interconnected without any physical connections such as wires.
9. Explain how lasers can be used to provide system connectivity.
10. What type of system connectivity is used between cell phones?

Section 1.5: Elements of System-Level Troubleshooting

11. Why is block-diagram thinking an important troubleshooting tool?
12. Why is it important to interview the operator or user when troubleshooting an electronic system?
13. Signal injection is a troubleshooting method where the *normal* signal is monitored and followed through the various stages of a circuit until a stage is reached that no longer processes the signal correctly. (True/False)

Section 1.6: Circuit Simulation

14. What is the first major step in circuit simulation?
15. Circuit simulation uses computer software to monitor the outputs from actual laboratory instruments when they are connected to circuits constructed in the laboratory. (True/False)

►─ CIRCUIT EXPLORATION

You will want to use circuit simulation as part of your electronics education whether or not it is an integral part of your school's program. The learning activities listed in this section prepare you for use of the simulation tool in later chapters. It is not necessary to have any knowledge of electronics to accomplish the listed tasks.

- Determine what circuit simulator software is used or recommended by your school, and find out how to access it. If necessary, buy your own package. Student versions are quite reasonable—free in many cases.
- Learn how to start the program.
- Learn how to choose electronic components from the component libraries.
- Learn how to interconnect components.
- Learn how to save your schematic to a computer file.

CHAPTER 2
Basic Electronics and Units of Measure

•—KEY TERMS
acceleration
ampere
candlepower
charge
conductance
conductor
conventional current
coulomb
current flow
difference of potential
electromagnetic field
electromotive force
electron
electron current
electrostatic field
flux
flux density
free electron
frequency
insulator
intensity
ion
ionization
magnetic domains
magnetic field
magnetizing force
magnetomotive force
negative ion
ohm
optoelectronics
period
permeability
positive ion
power
pressure
reluctance
resistance
semiconductors
shell
speed
superconductivity
superconductors
tesla
valence electron
velocity
volt
voltage
wavelength
weber

•—OBJECTIVES
After studying this chapter, you should be able to:
- Define each of the following quantities, state its unit of measure, state its abbreviation, and write its symbol where applicable:

Acceleration	Conductance	Current
Magnetic field	Flux	Flux density
Frequency	Intensity	Magnetizing force
Permeability	Position	Power
Pressure	Reluctance	Resistance
Speed	Time	Voltage
Wavelength		

- Contrast conductors, semiconductors, superconductors, and insulators.
- Describe the nature of magnetic and electromagnetic fields.
- Describe the interaction of like or opposite magnetic poles.
- Apply the left-hand rule to a current-carrying conductor to determine the direction of the magnetic field.
- State Ohm's Law and solve related problems.
- State the "Ohm's Law" of magnetic circuits.

•—INTRODUCTION
As you study the material presented in this chapter, you will begin to lay the foundation of your electronics career. The chapter contains a lot of material, some elements of which may seem unrelated to each other and even to electronics. However, be assured that all of the material is relevant to your career, and it is absolutely essential that you master the vocabulary, symbols, and calculations presented in this chapter. They are the basis for all subsequent studies.

2.1 ELECTRICAL QUANTITIES

This section introduces you to a number of electrical quantities and will serve to familiarize you with the definitions, units of measure, abbreviations, and symbols associated with the quantities.

Atomic Structure
All matter is composed of increasingly smaller building blocks as you may have learned in previous science classes. Table 2–1 serves to summarize some of the key terms concerning the building blocks of matter:

TABLE 2–1 Basic building blocks of matter.

Name	Description
Mixture	A material composed of more than one element or compound whose dissimilar atoms or molecules are not chemically bound together.
Compound	A material that consists of two or more elements that are chemically bonded.
Molecule	The smallest part of a compound that still exhibits the properties of the compound.
Element	A material whose atoms are all identical. An element cannot be subdivided by chemical means.
Atom	The smallest part of an element that still has the properties of the element.
Electron	A negatively-charged atomic particle that orbits the nucleus of all atoms.
Proton	A positively-charged particle in the nucleus of all atoms.
Neutron	An atomic particle located in the nucleus of an atom and having no charge.

In the study of electronics, we are quite interested in the behavior of subatomic particles. Electrons are of special interest. You may recall from earlier science studies that an atom can be viewed as a miniature planetary system similar in concept to our own solar system. This is illustrated in Figure 2–1. Each atom has a dense nucleus, which contains two major subatomic particles: protons and neutrons. Neutrons and protons have similar masses. For our immediate purposes, the mass of a particle can be thought of as being equivalent to the weight of the particle.

FIGURE 2–1 Atoms have a planetary structure similar in some ways to our own solar system.

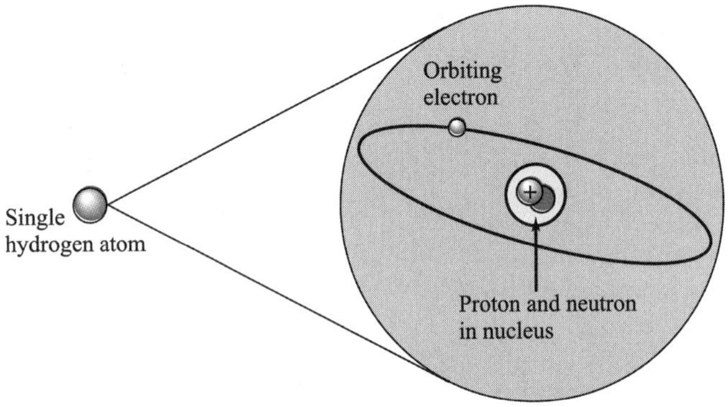

Electrons orbit the nucleus of an atom. An electron has a minute mass compared to the masses of protons and neutrons. A given type of atom (e.g., hydrogen or silicon) has a definite number of orbiting electrons and a definite number of protons and neutrons. Additionally, there are equal numbers of protons (+) and electrons (−), so the atom has no net charge under normal conditions.

The structure of an atom conforms to some very definite rules. For example, electrons can only travel in orbits that have certain dimensions. Each general allowable orbit level is called a **shell**. Only a certain number of electrons can orbit in a given shell. The more energy an electron has, the higher its orbit. In order to change orbits, an electron must either gain energy and go to a higher orbit, or give up energy and fall to a lower orbit. If an outer-orbit electron acquires sufficient energy, it can break away from its parent atom. Disassociated electrons are called **free electrons**. Free electrons play key roles in electronics.

When an electron in the outermost orbit (also called a **valence electron**) escapes orbit and becomes free, the parent atom is left with more protons than electrons. Thus, it has a positive electric charge. We call this positively-charged atom a

positive ion. Similarly, if a free electron falls into orbit around a previously neutral atom, the atom takes on a net negative charge. It is called a **negative ion**. An **ion**, then, is a charged atomic particle; an atom with unequal numbers of electrons and protons is one example of an ion. The act of forming ions is called **ionization**. This is another important factor in the operation of many devices (e.g., batteries, neon lamps, and transistors).

Classes of Materials

Materials can be classified according to their ability to pass electrical current. The number of electrons in the outer orbit largely determines this property.

Insulators

The maximum number of electrons that can be in the outer orbit (the valence band) of an atom is eight. If an atom has a full valence band, it is very stable and does not take part in chemical reactions. Additionally, the valence electrons are tightly bound in an atom with a full (or nearly full) valence band. This means there are few free electrons in such a material. This type of material is called an **insulator**. Insulators are used in electrical circuits as a way to prevent the flow of electricity. As you will soon see, the flow of electricity is actually the flow of free electrons and is called current flow. Some common insulators used in electronics are mica, plastic, rubber, Teflon, and air.

Conductors

Other materials such as copper, gold, aluminum, and most other metals have loosely-bound valence electrons. Copper, for example, has only a single outer-orbit electron, which is held loosely to the atom. It takes little energy to free the electron. In fact, even at room temperature, many of the valence electrons in copper are free. The abundance of free electrons in a material such as copper makes it easy to cause the electrons to move through the material in a controlled way. This movement of electrons is called **current flow**. A material that has many free electrons available and, therefore, readily allows current flow, is called a **conductor**.

Semiconductors

There are other materials that are neither good conductors nor good insulators. These are called **semiconductors** and have four valence electrons. Semiconductors such as silicon and germanium are essential for the fabrication of transistors, integrated circuits, and other electronic devices.

Superconductors

There is a special class of materials called **superconductors**. These materials exhibit a property or state called **superconductivity**. They offer no opposition to current flow and are essentially perfect conductors. To date, this ideal behavior has only been demonstrated at extremely cold temperatures (e.g., −196°C or less). Substantial research is ongoing to develop room-temperature superconductors.

Charge

An electrical **charge** is created when a material has more (or less) electrons than protons. Electrons are negatively-charged particles, and protons are positively-charged particles. If we cause any substance (for example, a piece of paper) to have an excess of electrons, then we say the substance has a negative charge. Similarly, if a substance has a deficiency of electrons (i.e., more protons than electrons), then we say the material is positively charged. This idea is illustrated in Figure 2–2 on page 26.

FIGURE 2–2 A material with excess electrons is negatively charged. A material with a deficiency of electrons is positively charged.

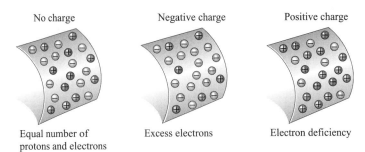

Attraction and Repulsion by Charges

You may recall an important and fundamental electrical law from your early science studies:

> **ELECTRICAL LAW** Like charges repel each other. Unlike charges attract each other.

Figure 2–3 illustrates this important concept. The balls shown in Figure 2–3 are very lightweight spheres called pith balls. They are shown suspended by threads so they are free to swing. Figure 2–3 shows the effects of the attractive and repulsive forces caused by unlike and like polarity charges, respectively. Figure 2–3 also shows that no deflection occurs when there is no charge on the balls. The region near a charged body where another charged body is affected (i.e., is attracted or repelled) is called an **electrostatic field**.

FIGURE 2–3 Unlike charges attract each other, while like charges repel.

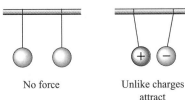

Unit of Measure for Charge

Charge is measured in coulombs (abbreviated C). One **coulomb** is equal to the charge caused by an accumulation (or deficiency) of 6,250,000,000,000,000,000 electrons. The symbol for charge is Q.

$$1\text{C} = 6{,}250{,}000{,}000{,}000{,}000{,}000 \text{ electrons} \quad (2\text{–}1)$$

•—EXAMPLE 2.1

If a certain material has an excess of 30,625,000,000,000,000,000 electrons, what is its charge?

SOLUTION

$$Q = \frac{\text{No. of electrons}}{6{,}250{,}000{,}000{,}000{,}000{,}000}$$

$$= \frac{30{,}625{,}000{,}000{,}000{,}000{,}000}{6{,}250{,}000{,}000{,}000{,}000{,}000}$$

$$= 4.9 \text{ C}$$

Calculator Application

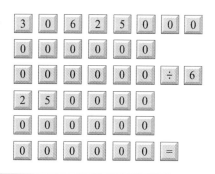

In chapter 3, we will discuss engineering notation, which makes calculations such as this much easier to write and to perform.

Voltage

When some distance separates two bodies with unequal charges, there is the potential for doing work. If, for example, a charged particle (such as an electron) were released within the electrostatic field associated with the two charged bodies, the electrostatic field would move the particle. In this particular example, the negatively-charged electron would be attracted by the more positive charge and repelled by the more negative charge. When there is a *difference* in the charge between any two points, we refer to it as a **difference of potential** (or alternately, a potential difference). Thus, a difference of potential has the ability to do work—it can move charged particles.

Unit of Measure for Voltage

Potential difference is expressed in terms of **voltage**. The unit of measure for voltage is the **volt**. The abbreviation for voltage is V; the symbol for volts is V.

•—EXAMPLE 2.2

Write an expression to describe a potential difference (label it V_1) of ten volts between two points.

SOLUTION

$$V_1 = 10 \text{ V}$$

Electromotive Force

The potential difference between two charged points may diminish as charges are transferred between the points. This is what happens in a lightning bolt. There is a charge difference between the earth and the clouds. At some point, there is a huge transfer of electrons between the two points, which is seen as a lightning bolt. This reduces the potential difference, so the lightning strike cannot be maintained. By contrast, other sources of potential difference do not decay when charges are transferred. The battery in your car is used to transfer charges through the various electrical devices in the car. It is designed to maintain a relatively constant potential difference of 12 V regardless of how many charges are transferred. A potential difference that is maintained as the charges are transferred is called an **electromotive force**.

Current Flow

Figure 2–4 shows a cross-sectional view of a length of copper wire. As discussed in chapter 1, copper wire is used extensively in electronic systems to interconnect the various components. The inset in Figure 2–4 shows a simplified representation of a copper atom. A neutral copper atom has a total of 29 electrons that are

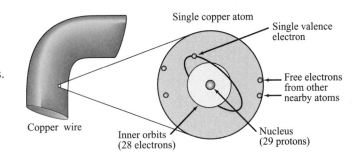

FIGURE 2–4 Cross-sectional view of a copper wire showing the valence electrons and the free electrons.

distributed into four major shells having 2, 8, 18, and 1 electrons each. At room temperature there is enough thermal energy present to ionize many of the copper atoms. Thus, it follows that at room temperature there is an abundance of free electrons within the body of the copper wire.

Visualizing Current
Try to visualize being inside of the copper wire. Imagine what you would see. First, you should notice that there is a lot of open space. The atoms and free electrons in the copper wire are tiny compared to the vast space between copper atoms. If, for example, you visualize the nucleus of the copper atom as being the size of a tennis ball, then the valence electron orbits would have approximately six-mile diameters. The electron itself would be a very tiny object. As you mentally travel through the copper wire, you will see many valence electrons skipping out of orbit and zooming off as free electrons. Sometimes one of the free electrons approaches a positively-charged copper atom (i.e., one whose valence electron has already left) and falls into its valence orbit for awhile. At any given time, however, there are many free electrons sailing *randomly* through the vast space between the copper atoms.

Figure 2–5 illustrates the results when we put a negative charge (many excess electrons) on one end of the copper wire and a positive charge (positive ions) on the other end. The charges are generated by the chemical action in the battery. The positive charge on one end of the wire attracts all of the electrons in the copper wire. Similarly, the negative charge on the opposite end of the wire repels every electron in the length of wire. Although the battery's charges also affect any ionized copper atoms, the copper atoms are massive and relatively immobile so they remain in position.

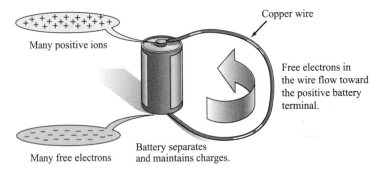

FIGURE 2–5 The positive charge attracts the free electrons while the negative charge repels them. This causes electron current flow.

The charges on the ends of the wire are not sufficiently high to pull the lower-level electrons out of orbit. But, whenever one of the valence electrons escapes the orbit of a copper atom, it is immediately affected by the charges on the ends of the wire. Since the electron has little mass and is very mobile (i.e., free), it races toward the positive terminal of the battery.

Electron Current and Conventional Current
As electrons in the copper wire leave the vicinity of the negative battery terminal as illustrated in Figure 2–5, they leave positive ions behind. These positive ions attract other electrons from the abundance of free electrons on the negative end of

the battery. Similarly, when a free electron reaches the positive end of the copper wire, it is attracted by the positive ions in the battery and leaves the wire. The continuous *directed* movement of electrons is called **electron current**. This process will continue as long as there is an excess of free electrons on one end of the wire and an abundance of positive ions on the other end. This is exactly what a battery does; it maintains a supply of free electrons at the negative terminal of the battery and an abundance of positive ions at the positive terminal. Throughout the remainder of this text, the term current shall refer to electron current. Therefore, current flows from negative to positive. Figure 2–6(a) illustrates electron current flow.

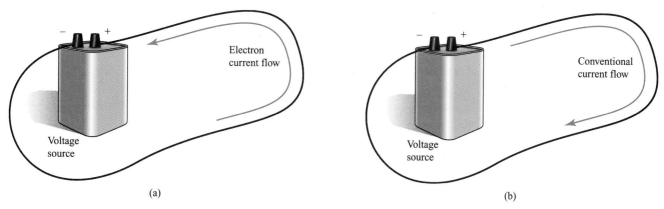

FIGURE 2–6 (a) Electron current flows from negative to positive since it is the movement of negative charges (electrons). (b) Conventional current flows from positive to negative since it represents the direction a positive charge would move.

For many applications, particularly when a circuit is modeled mathematically, it is not essential to be concerned about the *physical* nature of the moving charges. Rather, it is convenient to speak in terms of current without specifying the exact nature of the current.

Scientists, physicists, and engineers have adopted the general term *current* to refer to the movement of positive charges. Current identified in this way is called **conventional current**. It follows that conventional current would be considered as flowing from positive to negative since that is the direction a positive charge would move. Conventional current is represented in Figure 2–6(b).

Do not make the mistake of arguing with someone about which way current flows. As illustrated in Figure 2–6, it depends on the definition of current that is being used. Again, be reminded that electron current is assumed throughout this text, but you should be able to interpret technical literature that uses either electron current or conventional current.

Unit of Measure for Current

You will recall that charge is a measure (excess or lack) of the number of electrons in a particular body. Current is a movement of charges. It is a measure of the number of electrons per second that flow past a given point. More specifically, the unit of measure for current (abbreviated I) is the ampere (A). One **ampere** is the amount of current that flows when one coulomb (Q) flows past a point in one second (t). Equation 2–2 expresses this mathematically:

$$I = \frac{Q}{t} \tag{2-2}$$

where I is in amperes, Q is in coulombs, and t is in seconds. This is an important formula that relates current, time, and charge movement.

•—EXAMPLE 2.3

How much current is flowing in a wire if 3.5 coulombs pass a particular point every two seconds?

SOLUTION
We apply Equation 2–2 as follows:

$$I = \frac{Q}{t}$$

$$= \frac{3.5 \text{ C}}{2 \text{ s}} = 1.75 \text{ A}$$

Resistance

As current flows through a material, it encounters opposition. In the case of metallic conductors, for example, the free electrons have frequent collisions as they find their way between the atoms in the wire. These collisions and the attraction of the various atoms present an opposition to current flow. This opposition to current flow is called **resistance**. In order to produce a current with a given magnitude through a material, an electromotive force that has sufficient energy to overcome the resistance of the material must be applied across the material.

Unit of Measure for Resistance

Resistance (abbreviated R) is measured in ohms (Ω). An **ohm** is defined as the amount of resistance required to limit the current flow through a material to one ampere when one volt is applied across the material. Practical values of resistance used in electronics range from thousandths of an ohm to hundreds of millions of ohms. This range is not meant to be all inclusive; there are applications that require smaller and larger resistances.

Ohm's Law

Possibly the most important relationship you will ever study in electronics is the relationship between voltage, current and resistance in a circuit. Georg Simon Ohm, a German teacher, formally expressed this relationship in 1827. It is now called Ohm's Law and can be stated as follows:

> ➤ **OHM'S LAW** The current that flows in a circuit is directly proportional to the voltage across the circuit and inversely proportional to the resistance in the circuit.

This relationship is shown mathematically in Equation 2–3.

$$I = \frac{V}{R}$$
or
$$I = \frac{E}{R}$$

(2–3)

The selection of V (voltage) or E (electromotive force) to represent voltage is largely a personal choice. In most cases, the "correct" choice is defined by the person who signs your paycheck! For the purpose of this text, we shall generally use V as the symbol for voltage.

Current Is Directly Proportional to Voltage

Let us first examine the part of Ohm's Law that says current is directly proportional to voltage. Figure 2–7(a) shows an electric circuit composed of a voltage source and a resistance. It has a certain amount of current, which could be computed with Equation 2–3. Figure 2–7(b) shows the same basic circuit, but the applied voltage is higher. The higher voltage causes more current to flow in the circuit. This should seem reasonable to you since voltage is the force that causes current to flow and resistance is what limits current flow. If there is an increased force trying to cause current flow but the opposition to current flow (resistance) remains constant, then it should seem logical that the current flow will increase. Figure 2–7(c) illustrates the effect of reducing the voltage to a value below that in Figure 2–7(a).

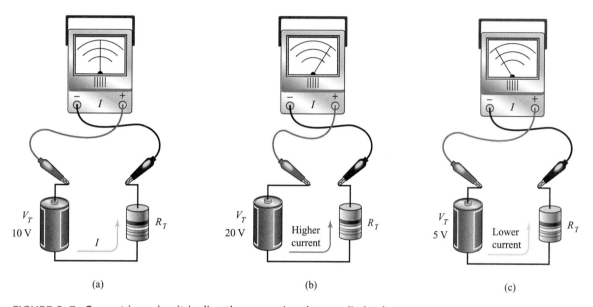

FIGURE 2–7 Current in a circuit is directly proportional to applied voltage.

The relationship between current and voltage is termed *linearly proportional*. This important relationship is shown graphically in Figure 2–8. The horizontal scale of the graph shown in Figure 2–8 represents the voltage in the circuit. The

FIGURE 2–8 A graph showing the linearly proportional relationship between applied voltage and the resulting current flow in a fixed-resistance circuit.

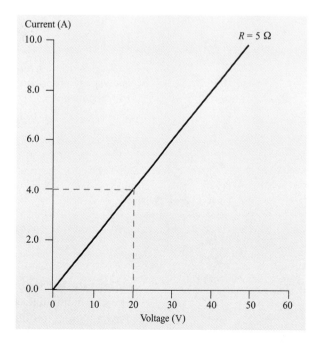

vertical scale shows the resulting current flow. From any given voltage value, you can move vertically (as indicated by the dotted lines in Figure 2–8) until you intersect the *resistance line,* and then move horizontally from that point until you intersect the current axis. The point of intersection on the current axis indicates the amount of current that flows with the given applied voltage and the value of resistance represented by the resistance line.

Current Is Inversely Proportional to Resistance

Figure 2–9 examines the second part of Ohm's Law, which states that current flow is inversely proportional to resistance. Figure 2–9(a) shows that a voltage source connected across a certain resistance will cause a given amount of current to flow. We can compute the value of current by using Equation 2–3. Figure 2–9(b) shows that current becomes less when the resistance in the circuit is increased. This should seem reasonable since resistance is a measure of the opposition to current flow. Finally, Figure 2–9(c) illustrates that current in the circuit will increase if the opposition (resistance) is reduced.

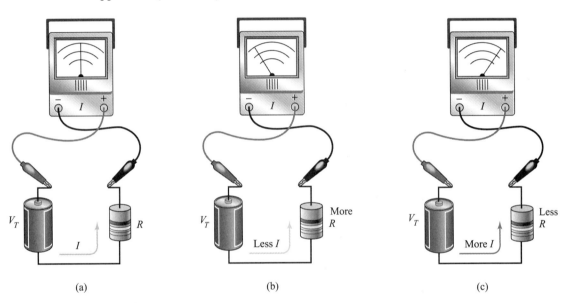

FIGURE 2–9 More resistance causes less current flow.

The inverse relationship between current and resistance is shown graphically in Figure 2–10.

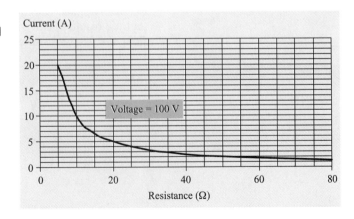

FIGURE 2–10 Current is inversely proportional to resistance.

Alternate Forms of Ohm's Law

We can manipulate the basic Ohm's Law equation (Equation 2–3), as we can with any equation, to obtain the alternate forms shown as Equations 2–4 and 2–5.

$$R = \frac{V}{I} \tag{2-4}$$

$$V = IR \tag{2-5}$$

These equations (Equation 2–3, Equation 2–4, and Equation 2–5) are so fundamental and used so frequently that you should commit all three forms to memory as soon as possible.

Compute *I* When *V* and *R* Are Known
If you know the values of voltage and resistance in a circuit, then you can calculate the value of current by applying Equation 2–3.

•─EXAMPLE 2.4
How much current flows through a 25-Ω resistor with 10 V across it?

SOLUTION We apply Ohm's Law (Equation 2–3) as follows:

$$I = \frac{V}{R}$$
$$= \frac{10 \text{ V}}{25 \text{ }\Omega} = 0.4 \text{ amperes}$$

Calculator Application

•─EXAMPLE 2.5
If an electric circuit has 120 V applied and has a total resistance of 20 Ω, how much current is flowing in the circuit?

SOLUTION We apply Ohm's Law (Equation 2–3) as follows:

$$I = \frac{V}{R}$$
$$= \frac{120 \text{ V}}{20 \text{ }\Omega} = 6.0 \text{ amperes}$$

Compute *R* When *V* and *I* Are Known
If you know the current and voltage in an electric circuit, then the resistance can be computed by applying Equation 2–4. Again, the units of measure for voltage, current, and resistance are volts, amperes, and ohms, respectively.

•─EXAMPLE 2.6
If a certain electronic component allows 0.25 A to flow when 35 V are across it, what is its resistance?

SOLUTION Ohm's Law (Equation 2–4) provides our answer as follows:

$$R = \frac{V}{I}$$
$$= \frac{35 \text{ V}}{0.25 \text{ A}} = 140 \text{ }\Omega$$

Calculator Application

•—EXAMPLE 2.7

How much resistance is required in an electric circuit to restrict the current flow to 8.5 A when 240 V are applied?

SOLUTION We apply Equation 2–4.

$$R = \frac{V}{I}$$

$$= \frac{240 \text{ V}}{8.5 \text{ A}} = 28.24 \text{ }\Omega$$

Compute V When I and R Are Known

If you know the current and resistance values in an electric circuit, then you can calculate the value of applied voltage with Ohm's Law (Equation 2–5).

•—EXAMPLE 2.8

If a circuit in a computer has a resistance of 200 Ω, how much voltage must be connected across it to cause 0.005 A of current to flow?

SOLUTION We apply Equation 2–5 as follows:

$$V = IR$$

$$= 0.005 \text{ A} \times 200 \text{ }\Omega = 1.0 \text{ V}$$

•—EXAMPLE 2.9

How much voltage is required to cause 2.5 A of current to flow through an electronic component that has a resistance of 5 Ω?

SOLUTION

$$V = IR$$

$$= 2.5 \text{ A} \times 5 \text{ }\Omega = 12.5 \text{ V}$$

Calculator Application

•—Practice Problems

Apply Ohm's Law to solve the following problems:

1. How much current flows through a robotic controller if it has 12 V across it and an internal resistance of 2.5 Ω?
2. If a circuit in a system printer has 240 Ω of resistance and 0.1 A flowing through it, how much voltage is across it?
3. What value of resistance is required in a circuit to limit the current to 2.7 A when 220 V are applied?

Power

Electrical **power** is a measure of the rate at which energy is used. Electrical power is often evidenced as heat. All electrical devices such as lightbulbs, computers, and microwave transmitters dissipate power.

Unit of Measurement for Electrical Power

Electrical power is measured in watts (W). It is the rate of using electrical energy.

Power Calculations Based on Ohm's Law

When working with electronic circuits, it is convenient to compute power in terms of other known circuit quantities such as voltage, current, and resistance. There are three basic formulas that we will use extensively for computing electrical power in a circuit. First, the product of voltage and current gives the power dissipated in a component or an entire circuit. This is expressed formally as Equation 2–6.

$$P = IV \qquad (2\text{–}6)$$

We can use this basic power equation and the Ohm's Law relationships previously cited to develop two more important power relationships.

$$P = IV$$
$$= \frac{V}{R} \times V \quad \text{(substitute } \frac{V}{R} \text{ for } I\text{)}$$
$$= \frac{V^2}{R}$$

and,

$$P = IV$$
$$= I \times IR \quad \text{(substitute } IR \text{ for } V\text{)}$$
$$= I^2 R$$

These two important power relationships are given formally as Equations 2–7 and 2–8.

$$P = \frac{V^2}{R} \qquad (2\text{–}7)$$

$$P = I^2 R \qquad (2\text{–}8)$$

Equations 2–6, 2–7, and 2–8 are the basic power formulas that we will use throughout the remainder of this text. We will refer to these collectively as the *power formulas*.

In many cases, these calculations are used to determine the amount of electrical power that is converted to heat. For example, when electron current flows through a conductor, the free electrons lose energy as they collide with atoms in the material. Much of the energy loss is converted to heat, so the material heats up. Figure 2–11 shows the application of these equations to an electrical circuit.

Your electric toaster and electric oven are two excellent examples of electrical energy being consumed as power. When current passes through the heating elements, it encounters opposition or resistance. Much of the electrical energy

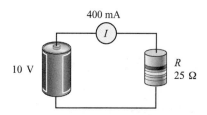

FIGURE 2–11 The amount of electrical power in a circuit is related to the voltage, current, and resistance in the circuit.

$P = IV$	$P = I^2 R$
$P = 400 \text{ mA} \times 10 \text{ V}$	$P = (400 \text{ mA})^2 \times 25 \text{ }\Omega$
$P = 4 \text{ W}$	$P = 4 \text{ W}$

$$P = \frac{V^2}{R}$$
$$P = \frac{(10 \text{ V})^2}{25 \text{ }\Omega}$$
$$P = 4 \text{ W}$$

supplied to the heating element is converted to heat; some is converted to light energy as the element glows red hot.

•—EXAMPLE 2.10

How much power is expended in an electrical circuit when 10 V are connected across a 25-Ω resistance?

SOLUTION We can apply Equation 2–7 directly as follows:

$$P = \frac{V^2}{R}$$

$$= \frac{10^2 \text{ V}}{25 \, \Omega} = 4 \text{ W}$$

•—EXAMPLE 2.11

How much current must flow through a 100,000-Ω resistance in order to produce 5 W of power?

SOLUTION We apply Equation 2–8 as follows:

$$P = I^2 R, \text{ or}$$

$$I = \sqrt{\frac{P}{R}}$$

$$= \sqrt{\frac{5 \text{ W}}{100,000 \, \Omega}} = 0.00707 \text{ A}$$

•—Practice Problems

1. How much power is dissipated when 0.3 A flows through a 50-Ω resistance?
2. How much voltage is required to produce 25 W in a 150-Ω resistance?
3. If a certain electrical device has 120 V applied and has 3.5 A of current flow, how much power is dissipated?

Conductance

Recall that resistance is a measure of the opposition to current flow. **Conductance**, by contrast, is the inverse of resistance. It expresses how readily current can flow through a material. The unit of measure for conductance (G) is siemens (S). The relationship between resistance and conductance can be expressed mathematically with Equation 2–9:

$$G = \frac{1}{R} \quad (2\text{–}9)$$

where G is in siemens (or an older unit of measure called mhos) and R is in ohms.

•—EXAMPLE 2.12

How much conductance does an electronic device have if its resistance is 50 Ω?

SOLUTION We apply Equation 2–9:

$$G = \frac{1}{R}$$

$$= \frac{1}{50 \, \Omega} = 0.02 \text{ S}$$

Electrical Quantities, Abbreviations, Units, and Symbols

In later chapters, we will further explore the various electrical quantities introduced in this section as well as several others. For now, it is important to know that there are many quantities in the field of electronics that must be measured or expressed. Each quantity has a unit of measurement (e.g., feet, inches, pounds), an abbreviation, and a symbol. To be successful, you will need to know the unit, abbreviation, and symbol for each quantity used in your field of interest.

Table 2–2 lists many of the basic electrical quantities along with their units, abbreviations, and symbols. Table 2–2 also provides a representative expression for each quantity that illustrates the use of the abbreviation and symbol. Some of the abbreviations shown in the example column have subscripts. These are used to distinguish multiple instances of the same quantity. Thus, for example, V_1, V_2, and V_O might represent three different voltages.

TABLE 2–2 Basic electrical quantities with their units, abbreviations, symbols, and examples of usage.

Quantity	Symbol	Unit	Abbreviation	Example
Capacitance	C	farad	F	$C_3 = 0.002$ F
Charge	Q	coulomb	C	$Q_1 = 2.6$ C
Conductance	G	siemen	S	$G_5 = 0.01$ S
Current	I	ampere	A	$I_2 = 3.9$ A
Frequency	f	hertz	Hz	$f = 60$ Hz
Impedance	Z	ohm	Ω	$Z_3 = 2700 \, \Omega$
Inductance	L	henry	H	$L_1 = 0.005$ H
Power	P	watt	W	$P_2 = 150$ W
Reactance	X	ohm	Ω	$X_L = 257 \, \Omega$
Resistance	R	ohm	Ω	$R_S = 3900 \, \Omega$
Time	t	second	s	$t_0 = 0.003$ s
Voltage	V	volt	V	$V_4 = 12$ V

✓ Exercise Problems 2.1

1. Express the following using the proper abbreviation and symbol:
   a. Voltage of 25.2 volts
   b. Frequency of 375 hertz
   c. Power of 5 watts
   d. Impedance of 3000 ohms
   e. Time of 0.09 seconds
   f. Current of 12.5 amperes
2. Materials that are good conductors have loosely-bound valence-band electrons. (True/False)
3. What class of materials behaves like an ideal conductor and offers no opposition to current flow?
4. If all of the valence-band electrons in a particular atom become free electrons, the atom itself becomes a _____-charged _____.
5. Like charges _____.
6. What is the name given to a potential difference that is maintained as charges are transferred?

2.2 MECHANICAL QUANTITIES

Practical electronic systems often measure and/or control mechanical quantities. Examples include such things as the speed of a rotating shaft, the pressure of a manufacturing process, or the physical position of a robotic arm. This section will briefly cite a few of the more common mechanical quantities, their units of measure, and their abbreviations.

Force

You know intuitively that force is what pushes or pulls on a physical object in order to move it. More formally, force is any action on an object that tends to change its position in a specific direction. For many practical purposes, force is measured in pounds.

Pressure

Monitoring and controlling pressure is a common application for electronic systems. You may be familiar with automotive computers that monitor oil pressure, manifold pressure, and, in some cases, tire pressure. **Pressure** is force distributed over an area. We can express this mathematically with Equation 2–10.

$$P = \frac{F}{A} \qquad (2\text{--}10)$$

Force (F) can be expressed in several units of measure, but the most common is pounds. Similarly, area (A) can be expressed as in^2, ft^2, $meter^2$, cm^2, and others. The units selected to express force and area, therefore, determine the units of measure for pressure. Probably the most familiar to many people is pounds per square inch (lb/in^2), or simply psi. The tire pressure in your car tires, for example, might be 28 psi.

Position

Measurement and control of physical position generally require the expression of one or more dimensions in terms of some linear unit of length. Examples include inches (in), millimeters (mm), or feet (ft). Figure 2–12 illustrates how a computer might communicate a target position to a robotic arm.

In the example shown in Figure 2–12, a reference point with coordinates of 0,0,0 has been established for the system. Any other point can be specified in terms of this reference point. In the case shown, a target point is specified as 3 units in the x-axis direction, 5 units in the y-axis direction, and −2 units in the z-axis direction.

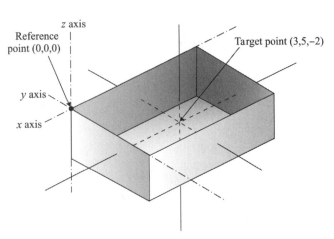

FIGURE 2–12 Position can be expressed as a set of three coordinates.

Representative systems utilize positioning in one dimension (e.g., height only), two dimensions (positioning on a flat surface), three dimensions (as illustrated in Figure 2–12), and more dimensions, which might include, for example, the angle of approach to the target point (e.g., a computer-controlled drill or a paint sprayer nozzle).

Speed and Velocity

Many practical electronic systems monitor and control the speed of machinery. Examples include automated tram systems at airports, variable-speed fans for environmental control systems, and shaft speeds for many types of industrial equipment.

Speed is a measure of how fast an object changes position without regard to specific direction. **Velocity** is a similar measure, but it implies movement in a specific direction. For the purposes of this text, we will use the two terms interchangeably unless specifically noted.

Speed is measured in distance per unit time for nonrotating objects. Examples include miles per hour (mph), feet per second (fps), and so on. In the case of rotating objects, speed is normally expressed as revolutions per unit time. Thus, a shaft might be rotating at a speed of 1000 revolutions per minute (1000 rpm).

Acceleration

Acceleration is a term used to describe the rate of change of speed (or velocity). Suppose that you were driving a car at 10 mph. You press on the accelerator and increase your speed to 50 mph. The amount of acceleration describes how quickly your speed changes from 10 mph to 50 mph. If the acceleration is low, then it may take some time to reach the higher speed. On the other hand, if there is a high value of acceleration, you may be thrown back in your seat as the car increases to the higher speed in a relatively short time.

The unit of measure for acceleration is speed per unit time. In the automobile example, we might measure acceleration in terms of miles per hour per second or mph/s. If, for example, the car was going 5 mph faster for every passing second, then we could say the acceleration was 5 mph per second.

Electronic systems often measure and control the acceleration of machines. Consider the electronic system used to control an elevator in a tall building. The elevator car must move fairly fast in order to travel the full distance in a reasonable time period. However, its acceleration and deceleration (slowing down or negative acceleration) are controlled to prevent discomfort to the passengers. Thus, it slowly increases its speed from a standstill to full speed. It moves at full speed until it is close to its destination. Finally, it slows down gradually before coming to a full stop. In order for this familiar action to occur, the electronic control system for the elevator must be able to measure and subsequently control acceleration.

✓ **Exercise Problems 2.2**

1. Pounds per square inch is a common unit of measure for pressure. (True/False)
2. Pressure can be expressed in terms of centimeters per second (cm/s). (True/False)
3. What does the abbreviation rpm represent?
4. Could a value of 100 feet per second per second (100 ft/s/s or $100 \frac{ft/s}{s}$) be a measure of acceleration for some object?

2.3 LIGHT AND OTHER WAVES

Light plays a critical role in the operation of many electronic and electromechanical systems. Some systems must only distinguish between the presence and absence of light. Other systems measure and/or control the brightness of light.

Still other electronic systems respond to and/or manipulate the color of light. Electronic devices that are designed to sense some characteristic of light, or alternatively to emit light having certain characteristics, are called optoelectronic components. The overall term used to describe all of these types of components is **optoelectronics**.

There are many other waves besides light waves that are critical to the operation of electronic systems. A few of these include sound waves, ultrasonic waves (used in burglar alarms and medical applications), radio waves (e.g., television broadcast, cell phones, and satellite communications), and many forms of radiation (e.g., X rays).

Intensity

As you may know, not all light is visible to the human eye. However, for the purposes of the immediate discussion, we will use visible light as an example. **Intensity** of light is simply a measure of how bright the light is. The intensity of light is normally measured in **candlepower** (candela).

Acoustical waves or sound waves are often measured in decibels (dB). Intensity of a sound wave is simply a measure of how loud it is. You will learn to perform calculations and express measurements in terms of decibels later in your education. For now, however, refer to Table 2–3 for a list of familiar sounds and their approximate intensities.

TABLE 2–3   Representative sound sources and their corresponding intensities in decibels.

Sound Source	Approximate Intensity (dB)
Threshold of hearing	0
Office environment	25–30
Major street traffic	65–70
Power woodworking tools	90–100
Threshold of pain	120

Wavelength and Frequency

The color of visible light and the tone of a sound wave are determined by a property called **frequency** or, inversely, wavelength. Light waves, radio waves, sound waves, and even waves in water exhibit many common properties. Figure 2–13 illustrates a basic wave and some of its important characteristics.

FIGURE 2–13
Characteristics of a wave.

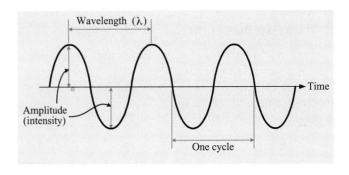

A wave consists of repetitive peaks (also called crests) and valleys (also called troughs). Waves move outward from their source at a velocity determined by the characteristics of the wave and the medium through which the waves are moving.

If you could somehow freeze time for a moment and closely examine a wave, you would find a pattern similar to that shown in Figure 2–13. The time axis also corresponds to physical distance from the source at any given time, since the wave moves physically away from the source.

Suppose, for example, we examined a sound wave that was somehow frozen in time for our examination. Each of the uppermost peaks on the waveform in Figure 2–13 would be identified by regions in space where the air molecules were compressed (i.e., high-pressure areas). Low-pressure areas, by contrast, would identify the lowermost peaks. The physical distance between corresponding points on two consecutive waves is called **wavelength**. It is a measure of the distance a wave travels during the time it takes to complete a full cycle of the wave. How much time is required for a complete cycle of the wave and how fast the wave travels determine this distance. The time required for a complete cycle of a wave is called the **period** of the wave. The reciprocal of a wave's period is its frequency (measured in hertz). This is expressed formally in Equation 2–11:

$$f = \frac{1}{t} \qquad (2\text{–}11)$$

where t is the period of the wave in seconds.

Wavelength can be expressed mathematically with Equation 2–12:

$$\lambda = \frac{v}{f} \qquad (2\text{–}12)$$

where v is the velocity of the wave, and f is the frequency. In the majority of cases, we are primarily interested in the approximate speeds of two types of waves: acoustical waves and electromagnetic waves. Acoustical waves include audio and ultasonic applications. Electromagnetic waves include radio waves, visible and invisible light waves, and other forms of radiation. Acoustical waves travel through air at a velocity of approximately 1,128 feet per second, whereas electromagnetic waves move through air at about 186,411 miles per second (300,000,000 meters per second). These velocities are different through other materials, but these are good estimates for many purposes. You may be familiar with the practice of counting slowly (one-second intervals) between the time you see a lightning flash and the time you hear the corresponding clap of thunder. Since light travels so much faster than sound, its travel time from the lightning event to your eye can be ignored. Therefore, as you count, you are determining the number of seconds required for the sound to travel to your ear from the place where the lightning occurred. Since sound travels at just over 1000 feet per second, it will travel roughly one mile every five seconds. So, if you divide your resulting count by five, you can estimate the approximate distance (miles) to the lightning event.

•—EXAMPLE 2.13

What is the wavelength of a 500-Hz sound wave, and how far will it travel in one second?

SOLUTION   Apply Equation 2–12.

$$\lambda = \frac{v}{f}$$

$$= \frac{1{,}128 \text{ ft/s}}{500 \text{ Hz}} = 2.26 \text{ ft}$$

Calculator Application

Remember, all sound waves travel approximately 1,128 feet per second, so in one second it will travel 1,128 feet.

✓ Exercise Problems 2.3

1. The intensity of light is a measure of its color. (True/False)
2. The music at an indoor rock concert would normally be close to 0 dB. (True/False)
3. What is the wavelength of a 100,000,000-Hz radio signal?
4. What is the frequency of a wave whose period is 0.001 seconds?
5. What is the period of an ultrasonic wave whose frequency is 50,000 Hz?

2.4 MAGNETISM AND ELECTROMAGNETISM

Magnetic and electromagnetic fields both play important roles in many kinds of electronic and electromechanical systems. Devices that rely on magnetic and electromagnetic fields include such things as motors, generators, speakers, microphones, computer memories, radio receivers, computer-controlled valves, and satellites. This section introduces some of the key terms associated with magnetic applications.

Magnetic Fields

We have all experienced some of the effects of magnetic fields. We know that a magnet will attract pieces of metal when they are brought near the magnet. We know that two magnets affect each other *even before they touch each other.* These effects are due to the presence of an invisible, but measurable region around a magnet called a **magnetic field**. A magnetic field, then, is the region around a magnet where another magnet would experience an interaction (attraction or repulsion). Although a magnetic field is not a physical object, it is helpful to visualize the field as being made up of many individual lines. Figure 2–14 shows a pictorial representation of the magnetic lines around a bar magnet.

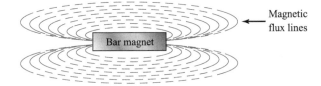

FIGURE 2–14 The magnetic field around a bar magnet.

Magnetic Flux

The lines around the magnet in Figure 2–14 are called lines of **flux** or magnetic flux lines. The lines are invisible, but the effects of the magnetic flux are easily demonstrated. Each flux line is elastic, but continuous. Additionally, flux lines never intersect or touch each other. The unit of magnetic flux is the **weber**. One weber corresponds to 100,000,000 flux lines. Flux in a magnetic circuit is analogous to current in an electrical circuit. The symbol for magnetic flux is ϕ.

Magnetic Poles

As shown in Figure 2–15, the flux lines are concentrated near the ends of the magnet. The regions where the *external* flux lines are most concentrated are called the poles of the magnet.

FIGURE 2–15 Flux lines are concentrated at the poles and extend externally from north pole to south pole.

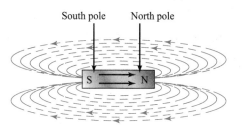

Flux lines extended externally from north pole to south pole.

Since the two poles of a magnet exhibit different characteristics, it is necessary to distinguish between them. If a magnet is suspended so that it is free to move, it will align itself with the earth's magnetic field. The end of the magnet that points toward the North pole of the earth is called the *north-seeking pole* or simply north pole. The other end of the magnet is called the magnet's south pole. North and south poles always appear in pairs. No one has ever shown that a single, isolated magnetic pole can exist. Further, like magnetic poles repel each other, while unlike poles attract each other.

Flux Density

Flux density is not concerned with total flux. Rather, it is a measure of how many flux lines appear in a certain area (i.e., how closely packed the lines are). The unit of measurement for flux density is the **tesla** (T). One tesla corresponds to a flux density of one weber per square meter.

Magnetic Domains

Some materials can be magnetized; others cannot. Materials that can be magnetized have regions called **magnetic domains**. Each magnetic domain acts like a miniature bar magnet. The length of a domain is only a fraction of a millimeter. Figure 2–16(a) shows that the domains in an unmagnetized material are pointed in random directions. Their fields do not aid each other.

Figure 2–16(b) shows the same material after it has been magnetized. Here, the domains have all rotated to point in the same direction. Their fields are now additive, and the overall material takes on the same magnetic polarity as the collective polarity of the individual domains.

It takes energy to cause a domain to rotate or change orientation. This energy loss is often evidenced as heat in electromagnetic devices such as transformers and motors.

Magnetomotive Force

Magnetomotive force (mmf) refers to the energy source that actually creates the flux; it is the equivalent of electromotive force (emf) in electrical circuits. The symbol for magnetomotive force is $\mathcal{F}$. Its unit of measure is ampere-turns (A · t).

Magnetizing Force

Magnetizing force (H), also called magnetic field intensity, is closely related to the magnetomotive force, but it includes a physical dimension. It is convenient to consider the closed loop formed by flux lines as a magnetic circuit. If the length of that circuit is increased, then we know intuitively that the field intensity will be reduced. The unit of measure for magnetizing force is ampere-turns per meter (A · t/m).

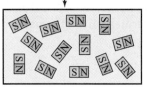

Domains have random orientations.

No net magnetic field in unmagnetized material.

(a)

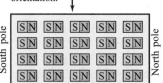

Domains have a common orientation.

Net magnetic field has same polarity as domains.

(b)

FIGURE 2–16 Magnetic materials have magnetic domains.

Reluctance

Reluctance is to magnetic circuits as resistance is to electrical circuits. That is, reluctance opposes the passage of magnetic flux. Thus, a material with a higher reluctance will have less flux for a given magnetomotive force. The symbol for reluctance is $\mathfrak{R}$, and the unit of measure is A · t per weber.

Ohm's Law for Magnetic Circuits

We know that magnetomotive force, reluctance, and flux in a magnetic circuit are comparable to voltage, resistance, and current, respectively, in an electrical circuit. This leads to the "Ohm's Law" equation for magnetic circuits expressed as Equation 2–13. It is read as magnetomotive force divided by reluctance.

$$\phi = \frac{\mathcal{F}}{\mathfrak{R}} \qquad (2\text{–}13)$$

Permeability

The ability of a material to pass magnetic flux is called its **permeability**. It is easier to pass a magnetic field through a material with a higher permeability. Permeability in a magnetic circuit is similar to conductivity in an electrical circuit. The standard symbol for permeability is μ.

Relative Permeability

Permeability as described in the preceding paragraph is called *absolute permeability*. It is more common to express the permeability of a material in terms of *relative permeability*. More specifically, the permeability of the material is compared to the permeability of a vacuum (or air for practical purposes). The relative permeability (μ_r) of a material, then, describes how easily the material passes magnetic flux *as compared to a vacuum*. We can calculate the relative permeability of a material with Equation 2–14.

$$\mu_r = \frac{\mu}{\mu_v} \qquad (2\text{–}14)$$

The units of measure cancel in this division problem. That is, relative permeability of a material is a simple ratio and has no units of measure. Values of typical materials range from slightly less than one to values as high as 100,000 for some ferromagnetic alloys.

Magnetic versus Electromagnetic Fields

Our discussion thus far has been limited to magnetic fields associated with permanent magnets. A permanent magnet retains its magnetism after the magnetizing field has been removed. Many electromechanical devices rely on the use of temporary magnets. These magnets exhibit the characteristics of a magnet as long as they are in a magnetizing field. However, if the magnetizing field is removed, then the domains in the material return to their original, random arrangement. We say that the material is no longer magnetized. Temporary magnets generally rely on the flow of electrons through a wire to create the magnetizing field.

Figure 2–17 shows that the direction of the magnetic field is determined by the direction of current flow. Additionally, the intensity of the magnetic field is determined by the magnitude of the current.

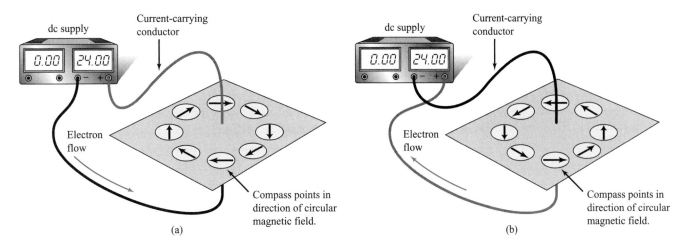

FIGURE 2–17 The direction of the magnetic field is determined by the direction of the electron current.

For the purposes of this chapter, we shall consider an **electromagnetic field** to be a magnetic field that is produced by the flow of electron current. The terms magnetic field and electromagnetic field are often used interchangeably when discussing electromagnetic devices.

Many technicians use a memory aid called the *left-hand rule for conductors*, which can be used to determine the direction of the magnetic field around a current-carrying conductor. Figure 2–18 illustrates the left-hand rule for conductors.

FIGURE 2–18 The left-hand rule for conductors.

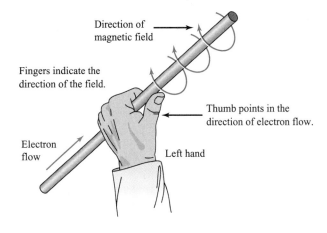

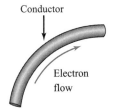

FIGURE 2–19 What is the direction of the magnetic field?

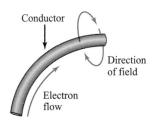

FIGURE 2–20 Magnetic field polarity is determined by the direction of current flow.

The left-hand rule for conductors works as follows:

1. Grasp the wire with the left hand so that the thumb is pointing in the direction of electron current.
2. As the fingers curl around the wire, they will point in the direction of the magnetic field.

•—EXAMPLE 2.14

Determine the direction of the magnetic field for the conductor shown in Figure 2–19.

SOLUTION Mentally grasp the wire with the left thumb pointing in the direction of electron current flow. Your fingers will indicate the direction of the magnetic field. In this case, the field goes over the top and down behind the wire as illustrated in Figure 2–20.

✓ Exercise Problems 2.4

1. What is the name given to describe the region near a magnet where another magnet is attracted or repelled?
2. What is the unit of measure for flux density?
3. Describe the difference between the arrangement of the domains in a magnetized and an unmagnetized material.
4. What is the quantity in a magnetic circuit that is analogous to resistance in an electrical circuit?
5. The motor in a CD drive uses a type of magnetic material that readily allows magnetic flux to pass through it. This material must have a (*low/high*) relative permeability.

•—SUMMARY

Matter is composed of atoms. Atoms are characterized as having a dense nucleus with positively-charged protons and neutral neutrons. Smaller, negatively-charged electrons orbit the nucleus in definite bands called shells. An atom with equal numbers of protons and electrons has no net charge. Atoms with more or less electrons than protons have a net negative or positive charge, respectively. Charged atoms are called ions. Creation of charged particles is called ionization. Like charges repel each other, while unlike charges attract.

Insulators are materials whose valence electrons are tightly bound. They are characterized as very poor conductors of electric current. Conductors, by contrast, have loosely-bound valence electrons and readily pass current. Materials whose ability to pass currents is midway between good conductors and good insulators are called semiconductors. Finally, materials that exhibit the characteristics of an ideal conductor are called superconductors. At present, superconductivity can only be achieved at extremely cold temperatures.

Voltage (measured in volts) is a force that causes current (measured in amperes) to flow in an electric circuit. For the purposes of this book, current (I) is considered to be the directed flow of electrons. As electrons move through a material, they encounter opposition called resistance (R). Resistance is measured in ohms. Ohm's Law describes the relationship between current, voltage, and resistance in a circuit.

Power (P) is a measure of the rate at which energy is used. Power is measured in watts (W). In many cases, electrical power is evidenced as heat in a component.

Figure 2–21 shows a memory aid that some students like to use to help remember the relationships between current, voltage, resistance, and power. However, these are so important that you should commit them to memory as soon as practical.

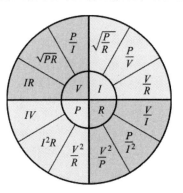

FIGURE 2–21 A memory aid for Ohm's Law and the power equations.

Electronics systems often measure and control mechanical quantities such as pressure, physical position, speed, and acceleration. Similarly, electronics systems also measure and control brightness of light, intensity of sound, and other characteristics of acoustical and electromagnetic waves. There is an inverse relationship between the period and frequency of a wave. Acoustical waves travel through air at about 1,128 ft/s, whereas electromagnetic waves travel about 300,000,000 m/s.

Many devices and systems use magnetic and electromagnetic fields. These fields consist of magnetic lines of force called flux. They exist around a magnet (or electromagnet) and interact with similar fields from other magnets. Regions with the highest flux density are called poles. There are north and south poles. Like poles repel each other, while unlike poles attract.

Magnetomotive force, reluctance, and flux in a magnetic circuit are analogous to voltage, resistance, and current, respectively, in an electrical circuit. This relationship is sometimes called the Ohm's Law for magnetic circuits.

REVIEW QUESTIONS

Section 2.1: Electrical Quantities

1. What is the smallest part of an element that still has the properties of the element?
2. Which atomic particle is characterized as having a negative charge?
3. Two isolated or free electrons would tend to (*repel/attract*) each other.
4. Of the three atomic particles (protons, neutrons, and electrons), which has the least mass?
5. If a valence electron breaks free from a neutral atom, the atom becomes a negative ion. (True/False)
6. What is the name given to the class of materials that are very poor conductors of current flow?
7. If an object with a charge of +50 C is brought near another object having a −50 C charge, the two objects will attract each other. (True/False)
8. Compute the amount of current that flows through a computer monitor if the monitor requires a voltage of 120 V and has an internal resistance of 300 Ω.
9. A certain component inside of a high-speed modem used to connect to the Internet requires a current of 0.15 A. If it has a difference of potential of 3 V across its terminals, what is its resistance?
10. The nameplate on a special-purpose computer states that it consumes 250 W of power when plugged into a 12-V source. How much current does the computer require?
11. Which of the following computer modems would have the higher effective internal resistance: (1) a device that operates on 12 V and draws 0.1 A, or (2) a device that operates on 5 V and draws 0.25 A?
12. The heat in a component is determined by the power it dissipates. Which of the following components would be warmer to the touch: (1) a 100-Ω component with 5 V across it, or (2) a 500-Ω component with 0.5 A of current through it?
13. The greater conductance a material has, the greater the opposition to current flow. (True/False)
14. Write the expression "frequency equals 250 hertz" using the proper symbol and abbreviation.
15. Write the expression "resistance equals 120 ohms" using the proper symbol and abbreviation.

Section 2.2: Mechanical Quantities

16. Pressure is a measure of force distributed over an area. (True/False)
17. Some computer-controlled positioning systems measure and control the physical position of mechanical objects in two or more dimensions. (True/False)
18. Velocity is the another term to describe acceleration. (True/False)
19. Rpm is often used as a unit of measure for the speed of a rotating shaft or wheel. (True/False)

Section 2.3: Light and Other Waves

20. Radio waves and light waves are two examples of acoustical waves. (True/False)
21. The volume or apparent loudness of an audio wave is often measured in decibels. (True/False)
22. Frequency or wavelength determines the color of a light wave. (True/False)
23. The time required for a complete cycle of a wave is called its _____.
24. Electromagnetic waves travel through air at approximately _____, while acoustical waves travel through air at about _____.
25. What is the frequency of a radio wave whose period is 0.0005 seconds?
26. What is the wavelength (expressed in meters) of a wireless communications link whose basic operating frequency is 2,400,000,000 Hz?

Section 2.4: Magnetism and Electromagnetism

27. Magnetic fields consist of invisible lines of force called magnetic _____.
28. If the north pole of one magnet is brought close to the north pole of a second magnet, the two magnets will (*attract/repel*) each other.
29. Tesla is the unit of measure for _____.
30. Flux density is a measure of the total flux created by a magnet. (True/False)
31. What can be said about a material whose magnetic domains are arranged in random order?
32. The mmf in a magnetic circuit is analogous to the emf in an electrical circuit. (True/False)
33. What property in a magnetic circuit is analogous to resistance in an electrical circuit?
34. What is the symbol used to represent permeability in a magnetic circuit?
35. What are the units of measure for relative permeability?
36. Visualize a wire extending out of the page in this book such that electron current is flowing directly toward you. What is the direction of the resulting magnetic field? (Clockwise or counterclockwise)

CIRCUIT EXPLORATION

This learning exercise will further acquaint you with your circuit simulation tool as well as provide you with additional practice in basic circuit calculations. Refer to your circuit simulation user manual as required and accomplish the following tasks:

- Invoke your simulation program.
- Construct the circuit shown in Figure 2–22.

FIGURE 2–22 Circuit for simulation.

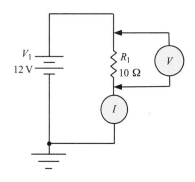

Fig02_22.msm

The encircled *I* represents a current meter, and the encircled *V* represents a voltmeter. Also, note the use of the ground symbol ($\perp$). We will discuss this further in chapter 4, but it is required for most circuit simulations. If you are using MultiSIM, the voltmeter and current meter instruments are both available in the multimeter device. Figure 2–23 shows one possible way to build the circuit in the MultiSIM environment.

Once you have entered the circuit into the schematic capture portion of your circuit simulation package, as illustrated in Figure 2–23, respond to the following questions:

- What is the measured value of voltage across R_1?
- Calculate the value of current through R_1 using Ohm's Law.
- What is the measured value of current through R_1?
- Compute the power dissipated in R_1.
- Remove the voltmeter and current meter from your circuit.
- Insert a power meter (wattmeter) to measure the power dissipated in R_1. The voltage and current terminals of the wattmeter connect to the same points as the original voltmeter and current meter, respectively.
- What is the measured value of power in R_1?

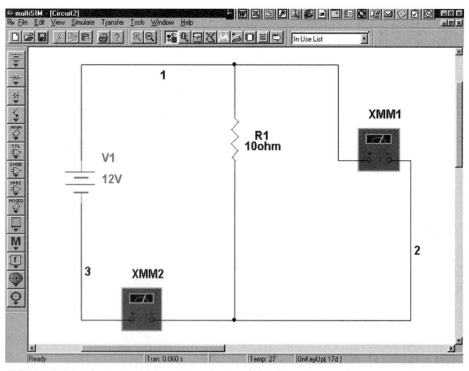

FIGURE 2–23 The test circuit constructed with MultiSIM.

ANSWERS TO PRACTICE PROBLEMS

Page 34
1. 4.8 A
2. 24 V
3. 81.48 Ω

Page 36
1. 4.5 W
2. 61.2 V
3. 420 W

CHAPTER 3

Basic Components and Technical Notation

•—KEY TERMS

ampacity
ampere-hour rating
armature
breakdown voltage
cable
coaxial cable
drain wire
engineering notation
magnet wire
momentary-contact switch
ohmmeter
pole
potentiometer
power of ten
power supply
precision resistor
programmable logic controller
resistor tolerance
response time
rheostat
ribbon cable
short circuit
spiraling
surface-mount technology
surge
temperature coefficient
throw
transient

•—OBJECTIVES

After studying this chapter, you should be able to:

- Utilize metric prefixes in the expression of multiple and submultiple units.
- Convert between various engineering units and prefixes.
- Identify various types of wire and cable.
- Calculate the circular mil area of a wire with a known diameter.
- Calculate the resistance of a given length of copper wire.
- Describe each of the following and include at least one unique characteristic:

Capacitor	Relay	Transformer
Diode	Resistor	Transistor
Inductor	Switch	Voltage source
Integrated circuit		

•—INTRODUCTION

Circuit quantities and component values in a typical electronics circuit have extreme ranges. For example, it is not uncommon to have values such as 1,200,000,000 Hz and 0.000000000047 F in the same circuit. Clearly, calculating such extreme values can be quite tedious and prone to error even with the aid of a calculator. Similarly, written and verbal communication of technical information is complicated when values such as these form an integral part of the discussion. For these reasons, technicians and engineers utilize metric prefixes and engineering notation to simplify both communication and computation. We will explore both of these techniques in this chapter.

This chapter also examines most of the basic electronic components that form the basis for all electronic circuits. We will explore each of these in greater technical depth as we progress, but for now, we will focus on the symbols and general characteristics of the various components.

3.1 TECHNICAL NOTATION

Engineering notation is simply a method of expressing very large and very small numbers in forms that are more easily interpreted and that simplify calculations. One police radar frequency, for example, is 10,525,000,000 hertz. Some of the voltages found in a radar receiver, however, might be in the range of 0.0000034 volt. The solution of problems using numbers with this wide range can be very cumbersome—even with a calculator. This section introduces a technical notation

49

that will *greatly* simplify your calculations. Since engineering notation is based on powers of ten, we will begin with a brief overview of this mathematical topic.

Powers of Ten

The terminology **power of ten** refers to the value that results when the number ten is raised to a power. Table 3–1 shows several powers of ten, their numerical values, and their equivalent forms.

TABLE 3–1 Some examples of powers of ten.

Power of Ten	Value	Equivalent
10^{-3}	.001	$\frac{1}{10 \times 10 \times 10}$
10^{-2}	.01	$\frac{1}{10 \times 10}$
10^{-1}	.1	$\frac{1}{10}$
10^{0}	1	1
10^{1}	10	10
10^{2}	100	10×10
10^{3}	1000	$10 \times 10 \times 10$

Any number can be expressed as a series of digits multiplied times a power of ten. The following procedure will allow you to express any number using powers of ten:

1. Move the decimal point to any desired position while counting the number of positional changes.
2. If the decimal point was moved to the left, the exponent will be positive. If the decimal point was moved to the right, the exponent will be negative.
3. The value of the exponent will be the same as the number of positional changes made by the decimal point.

This is a *very* important concept. Let's work some example problems to illustrate the conversion process.

EXAMPLE 3.1

Convert the following numbers to equivalent numbers expressed in power-of-ten form.

a. 123.5
b. 0.0042
c. 0.0226
d. 78,200,000
e. −2.77

SOLUTION There are an infinite number of correct answers to the preceding problems. Let's look at several alternatives.

a.

Original Number	Direction	Count	Result
123.5	left	2	1.235×10^{2}
123.5	right	2	12350×10^{-2}
123.5	left	5	$.001235 \times 10^{5}$
123.5	none	0	123.5×10^{0}
123.5	right	3	123500×10^{-3}

b.

Original Number	Direction	Count	Result
0.0042	right	4	42×10^{-4}
0.0042	right	6	4200×10^{-6}
0.0042	left	2	0.000042×10^{2}

c. Possible correct answers include: 22.6×10^{-3}, 0.000226×10^{2}, and 0.0226×10^{0}
d. Possible correct answers include: 78.2×10^{6}, 0.782×10^{8}, and $78,200,000,000 \times 10^{-3}$
e. Possible correct answers include: -0.277×10^{1}, -2770×10^{-3}, and $-2,770,000,000 \times 10^{-9}$. Note that the sign of the original number is simply assigned to the converted number.

Sometimes the original number will be expressed as a power of ten, and it is desired to express it as a different power of ten. The process here is identical to that just described except that your count starts with the current power of ten. The procedure is described below:

1. Subtract the current exponent of ten from the new exponent of ten.
2. If the result of step 1 is positive, then move the decimal point to the left.
3. If the result of step 1 is negative, then move the decimal point to the right.
4. In either case (step 2 or 3), move the decimal point the number of positions indicated by the magnitude of the result in step 1.

●—EXAMPLE 3.2

Make the indicated conversions:

a. 144×10^{2} converts to _____ $\times 10^{4}$
b. -0.09×10^{-1} converts to _____ $\times 10^{-5}$
c. $327,000 \times 10^{3}$ converts to _____ $\times 10^{6}$
d. 0.002×10^{4} converts to $200,000 \times 10^{——}$

SOLUTION

a. 144×10^{2} converts to 1.44×10^{4}
b. -0.09×10^{-1} converts to -900×10^{-5}
c. $327,000 \times 10^{3}$ converts to 327×10^{6}
d. 0.002×10^{4} converts to $200,000 \times 10^{-4}$

●—Practice Problems

Make the indicated conversions:

1. 0.005×10^{6} converts to _____ $\times 10^{0}$
2. 100.6×10^{-4} converts to _____ $\times 10^{-6}$
3. 100.6×10^{-4} converts to _____ $\times 10^{6}$
4. -247×10^{3} converts to $-2.47 \times 10^{——}$

Engineering Notation

Engineering notation is similar to standard powers-of-ten operations but with minor restrictions. For engineering notation, we can move the decimal point to any convenient position *as long as the resulting exponent is either zero or a multiple of 3* (e.g., −6, −3, 0, 3, 6, 9). The procedure can be stated as follows:

1. Write the number using powers of ten.
2. Move the decimal point left or right while increasing or decreasing, respectively, the exponent.
3. The final exponent must be zero or a number that is evenly divisible by three.

This is an important way of expressing numbers in electronics. Powers of ten that are evenly divisible by three have prefixes that can replace them to further simplify the expression of quantities. Any given number can be expressed several different ways and still qualify as a valid form of engineering notation. The problem to be solved and your personal preference determine the "best" choice for an exponent. For now, let us apply the practice of selecting a base number between 1 and 999.

•—EXAMPLE 3.3

Convert the following numbers to engineering notation. For purposes of this example, choose the form that produces a base number between 1 and 999.

a. 129×10^5
b. 0.0056×10^{-1}
c. -33.9×10^7

SOLUTION

a. We move the decimal one place to the left, which causes our exponent to count up to six. The result expressed in engineering notation form is 12.9×10^6.
b. The preferred base number in this case is 560. The result is 560×10^{-6}.
c. In this case we shall move the decimal point one place to the right, which will decrease our exponent to six. The result in engineering notation is -339×10^6.

Converting Between Engineering Units

Sometimes a value is expressed using engineering notation, but you want to express it in a different form. For example, you may have a value expressed as 25×10^{-9}, but you prefer to have it expressed using an exponent of −12 or perhaps −6. The conversion process is identical to that previously discussed for powers-of-ten manipulations. The following problems provide an opportunity for you to practice.

•—Practice Problems

1. Convert 25×10^{-6} to _____ $\times 10^{-3}$.
2. Convert 0.001×10^{-6} to _____ $\times 10^{-12}$.
3. Convert 105×10^3 to _____ $\times 10^0$.
4. Convert 0.045×10^9 to _____ $\times 10^6$.
5. Convert 0.2×10^6 to _____ $\times 10^3$.

Calculator Sequences

This section provides a brief summary with respect to calculator operations using powers of ten and/or engineering notation.

•—EXAMPLE 3.4

Perform the following calculations on an engineering calculator:

a. $125.8 \times 10^3 \times 0.6$
b. $62.9 \times 10^{-6} \div 29 \times 10^{-5}$
c. $0.7 \times 10^4 + 81.2 \times 10^3$
d. $219 \times 10^{-2} - 0.5$

SOLUTION Figure 3–1 illustrates a calculator sequence as performed on a standard engineering calculator. Figure 3–2 shows a similar sequence for engineering calculators such as those made by Hewlett Packard that use Reverse Polish Notation (RPN). In either case, the correct answers are as follows:

a. 75.48×10^3
b. 216.897×10^{-3}
c. 88.2×10^3
d. 1.69

a. [1] [2] [5] [.] [8] [EEX] [3] [×] [.] [6] [=]
b. [6] [2] [.] [9] [EEX] [+/−] [6] [÷] [2] [9] [EEX] [+/−] [5] [=]
c. [.] [7] [EEX] [4] [+] [8] [1] [.] [2] [EEX] [3] [=]
d. [2] [1] [9] [EEX] [+/−] [2] [−] [.] [5] [=]

FIGURE 3–1 Calculator sequence for powers-of-ten calculations using a standard calculator.

a. [1] [2] [5] [.] [8] [EEX] [3] [ENTER] [.] [6] [×]
b. [6] [2] [.] [9] [EEX] [+/−] [6] [ENTER] [2] [9] [EEX] [+/−] [5] [÷]
c. [.] [7] [EEX] [4] [ENTER] [8] [1] [.] [2] [EEX] [3] [+]
d. [2] [1] [9] [EEX] [+/−] [2] [ENTER] [.] [5] [−]

FIGURE 3–2 Calculator sequence for powers-of-ten calculations using an RPN calculator.

Prefixes for Units of Measure

Since it is extremely cumbersome to communicate (either verbally or in writing) the extreme values of numbers that are routinely used in electronics, it is common to use metric prefixes to further simplify the expression of very small and very large values. Metric prefixes are used in place of powers of ten. They have the same meaning, but they are generally easier to verbalize and to write.

Table 3–2 shows the powers of ten (engineering notation in most cases) that are most commonly used in electronics along with the equivalent prefixes, symbols, and magnitudes.

TABLE 3–2 The powers of ten used in electronics have equivalent prefixes and symbols.

Power of Ten	Prefix	Symbol	Magnitude
10^{-15}	femto	f	one-quadrillionth
10^{-12}	pico	p	one-trillionth
10^{-9}	nano	n	one-billionth
10^{-6}	micro	μ	one-millionth
10^{-3}	milli	m	one-thousandth
10^{-2}	centi	c	one-hundredth
10^{0}	Whole units need no prefix or symbol.		
10^{3}	kilo	k	thousand
10^{6}	mega	M	million
10^{9}	giga	G	billion
10^{12}	tera	T	trillion

When you first begin working with powers of ten and prefixes, they often seem to complicate, rather than simplify, the operation. However, if you practice, you will not only master the techniques, but you will soon realize their labor-saving effects. You *must* be proficient with prefixes and powers of ten, or you will be working in a world that speaks a foreign language.

We can use a prefix to express a quantity by applying the following procedure:

1. Write the quantity in engineering notation.
2. Replace the power of ten with the equivalent prefix.

Let's master this important process by example and practice.

•—EXAMPLE 3.5

Write the following quantities using the nearest standard prefix as listed in Table 3–2.

a. 120×10^3 volts
b. 47×10^{-12} farads
c. 200×10^{-5} henries
d. 2.9×10^{-2} meters
e. 1.02×10^{10} hertz

SOLUTION

a. 120 kilovolts
b. 47 picofarads
c. 2000 microhenries
d. 2.9 centimeters
e. 10.2 gigahertz

•—EXAMPLE 3.6

Write the following quantities using the nearest standard symbol as listed in Table 3–2.

a. 125×10^{-3} V
b. 5×10^3 Hz
c. 0.001×10^{-11} F
d. 10.2×10^{-9} s
e. 3300×10^5 Ω

SOLUTION

a. 125 mV
b. 5 kHz
c. 0.01 pF
d. 10.2 ns
e. 330 MΩ

•—Practice Problems

Write the following quantities using the nearest standard symbol listed in Table 3–2.

1. 56×10^3 Ω
2. 22×10^{-4} A
3. 500×10^4 W
4. 12×10^{-6} s
5. 1.75×10^8 Hz

We can also convert from one prefix to another by using the following procedure:

1. Replace the given prefix with the equivalent power of ten.
2. Relocate the decimal (left or right) while you adjust the power of ten.

3. When the power of ten matches the value of the desired prefix, replace it with the prefix.

• EXAMPLE 3.7

Express 0.1 mA as an equivalent number of microamperes.

SOLUTION First we replace the given prefix with the equivalent power of ten as follows:

$$0.1 \text{ mA} = 0.1 \times 10^{-3} \text{ amperes}$$

Next we move the decimal until the power of ten matches the micro prefix (i.e., 10^{-6}). In this case, we will have to move the decimal point three places to the right:

$$0.1 \times 10^{-3} \text{ amperes} = 100 \times 10^{-6} \text{ amperes}$$

Finally, we substitute the equivalent prefix:

$$100 \times 10^{-6} \text{ amperes} = 100 \text{ μA}$$

• Practice Problems

1. Express 0.05 A as an equivalent number of mA.
2. How many kilovolts is equivalent to 0.55 MV?
3. Convert 3300 mΩ to an equivalent number of ohms (no prefixes).
4. 10,000,000 kHz is the same as _____ GHz.
5. 2500 cm is equivalent to _____ km.

✓ Exercise Problems 3.1

1. Convert the following to engineering notation (use the nearest acceptable exponent):
   a. 266×10^{-4}
   b. 77.4×10^{8}
   c. 0.007×10^{-5}
   d. 52×10^{4}
2. Perform the following calculations. Express all answers in engineering notation.
   a. $44.5 \times 10^{3} + 0.5 \times 10^{4}$
   b. $0.008 \times 10^{6} - 7,000,000 \times 10^{-4}$
   c. $244 \times 10^{7} \div 0.4 \times 10^{8}$
   d. $0.99 \times 10^{4} \times 338 \times 10^{2}$
3. Write the standard prefix for each of the following powers of ten:
   a. 10^{-3}
   b. 10^{6}
   c. 10^{9}
   d. 10^{-12}
   e. 10^{-2}
4. Write the following quantities using the nearest standard prefix as listed in Table 3–2:
   a. 22.5×10^{-6} farads
   b. 250×10^{5} hertz
   c. 27×10^{-5} henries
   d. 122×10^{-4} seconds
5. Write the following quantities using the nearest standard symbol as listed in Table 3–2:
   a. 220×10^{-5} H
   b. 2500×10^{-4} W
   c. 0.008×10^{-7} F
   d. 5.99×10^{4} Hz
6. Express the following quantities in the forms indicated:
   a. Express 12,500 hertz as kilohertz.
   b. Express 0.0005×10^{-4} farads as microfarads.
   c. Express 0.0025 milliseconds as microseconds.
   d. Express 390 milliamperes as amperes.

3.2 WIRE AND CABLE

Wire and cable play essential roles in most practical electronic and electromechanical systems. This section will introduce a number of characteristics of wire that affect its performance. Knowledge of these characteristics is important whether you are designing a new system, installing a system upgrade, or replacing components in a defective system. Incorrect wire selection can degrade system performance and, in some cases, can present a safety hazard from accidental fires. Finally, the material presented in this section will fill an important requirement as part of your technical vocabulary and basic knowledge base.

Wire Resistance

We generally use wire to electrically connect various components or portions of a system. In other words, we use wire to provide a path for current flow. The fact that wire is a good conductor (i.e., has low resistance) allows it to be used for this purpose. Nonetheless, wire is not a perfect conductor. It does have some resistance.

The amount of resistance that a given wire offers to current flow depends on several factors including

- Type of material
- Temperature
- Length
- Cross-sectional area

The effects of these factors can be seen by examining Table 3–3. If, for example, you compare the 100′ and 1000′ length entries for any given material of a specified size at a given temperature, you can see that the longer wire has more resistance. We say that resistance of a wire is directly proportional to its length.

TABLE 3–3 There are four primary factors that affect the resistance of wire.

Material	Length (feet)	Area (cmil)	Resistance @ 20°C	Resistance @ 75°C
Silver	100	6530	0.14	0.17
		101	9.11	11.01
	1000	6530	1.41	1.70
		101	91.09	110.13
Annealed copper	100	6530	0.16	0.19
		101	10.27	12.49
	1000	6530	1.59	1.93
		101	102.67	124.87
Constantan	100	6530	4.51	4.51
		101	291.78	291.80
	1000	6530	45.13	45.13
		101	2917.82	2917.96
Gold	100	6530	0.22	0.27
		101	14.53	17.25
	1000	6530	2.25	2.67
		101	145.35	172.53
Platinum	100	6530	0.92	1.07
		101	59.55	69.38
	1000	6530	9.21	10.73
		101	595.54	693.81

By contrast, if you compare wires that are identical except for area, you will find the wire with the largest cross-sectional area has the least resistance. Therefore, you can say that the resistance of wire is inversely proportional to its cross-sectional area. By comparing two wires that are identical except for the material, it is evident that the material affects the resistance of the wire. Silver is the best conductor of those listed in Table 3–3.

The unit of measure for cross-sectional area in Table 3–3 is circular mils (cmil). The circular mil area of a wire is calculated with Equation 3–1:

$$A_{\text{CMIL}} = d^2 \tag{3-1}$$

where A_{CMIL} is the circular mil area and d is the diameter of the wire in mils (i.e., 0.001 inch).

EXAMPLE 3.8

What is the circular mil area of a wire that has a diameter of 25 mil (0.025 inch)?

SOLUTION We apply Equation 3–1 as follows:

$$A_{\text{CMIL}} = d^2$$
$$= (25 \text{ mil})^2 = 625 \text{ cmil}$$

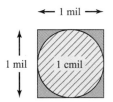

FIGURE 3–3 Circular mils and square mils are both units of measure for area, but they have different values.

It is important to note that area of wire expressed in circular mils is not the same numeric value as the area expressed in square mils. Figure 3–3 illustrates the relationship between these two units of measure. In Figure 3–3, the shaded area represents one circular mil. It is superimposed on a one square mil area. Clearly, one circular mil has less cross-sectional area than one square mil. The use of the circular mil unit of measure simplifies the calculation of wire area ($A = d^2$ instead of $A = \pi r^2$). It can often be estimated with a mental calculation.

Next, by comparing the two temperature columns for a given material, wire size, and length, you can see that temperature also affects the resistance of wire. Resistance increases with increasing temperature for all of those materials listed in Table 3–3. Therefore, you might think that resistance is directly proportional to temperature. In most cases, you would be right, but it is very important to know that the resistance of some materials (e.g., carbon, silicon, and other semiconductors) decreases with higher temperatures. The temperature characteristic (called the **temperature coefficient**) of a particular type of material can be found in many standard reference books. The resistances of materials with positive and negative temperature coefficients increase and decrease, respectively, with increasing temperature. A few materials (e.g., constantan) have temperature coefficients that are very close to zero and are relatively unaffected by moderate temperature changes (refer to Table 3–3).

EXAMPLE 3.9

You are working on a system that uses lengthy wires to allow communication between the various parts of the system (e.g., Ethernet cables). The present system uses 300′ cables, but you need to make a modification that will increase wire length to 800′. If the resistance of the wire was initially 7.7 Ω, what is the resistance of the 800′ wire?

SOLUTION We know that resistance of a wire is directly proportional to its length. Therefore, we can compute the new resistance as follows:

$$R_2 = \frac{l_2}{l_1} \times R_1$$

$$= \frac{800'}{300'} \times 7.7 \; \Omega = 20.53 \; \Omega$$

Now, whether or not this will interfere with the proper operation of the system depends on other system characteristics.

Wire Sizes

You will be working with many different sizes of wire. Some wire is so thin it can barely be seen with the naked eye. Other wire is so large, special tools are required to work with it. The choice of wire size for a given application is largely determined by the current that must pass through the wire. The current-carrying capacity of a wire is sometimes called its **ampacity**.

If a wire is too small for the current being carried, then it will drop excessive voltage across the resistance in the wire ($V = IR$). Additionally, excessive power dissipation in the wire resistance ($P = I^2R$) will cause heating of the wire and may even damage it. This latter event is of particular concern when choosing wire for power distribution in a home or office. If the wire is too small, it may get hot and cause a fire in the walls of the building.

Wire size is specified by gauge numbers. There are at least six different industry standards for assigning a particular wire size to a particular gauge number. They are not the same in all cases. Most electronics applications in the United States rely on the American Wire Gauge (also called Brown and Sharpe [B&S] gauge) numbers to specify wire size. Table 3–4 shows several of the American Wire Gauge (AWG) numbers along with the approximate diameter and cross-sectional area. Table 3–4 also shows the approximate resistance for a 1000-foot length of a given wire gauge for annealed copper wire. The complete AWG table includes both smaller and larger wire sizes than those listed in Table 3–4.

Inspection of Table 3–4 shows that larger gauge numbers correspond to smaller wire sizes. You might also note that the cross-sectional area approximately doubles (or halves) for every three gauge numbers.

The maximum current that a given wire size can carry is determined by several factors including the following:

- Maximum allowable increase in temperature of the wire
- Ambient temperature
- Maximum allowable voltage drop ($V = IR$)
- Type of insulation on the wire
- How the wire is routed (e.g., free air, bundled, conduit, and so on.)

Standards are available for most applications that can be used to determine the correct wire size for a given current. Care should be taken when interpreting manufacturers' wire data. Some manufacturers define maximum current-carrying capacity as the amount of current needed to melt either the wire or the insulation. Neither of these definitions is consistent with the safety-based standards that prevail in the industry. For example, the National Electrical Code cites a maximum current of 20 amperes (less for some types of insulation) for 14-gauge copper wire. By contrast, a leading wire manufacturer lists 27 to 45 amperes (depending on the type of insulation) for 14-gauge wire.

Types of Wire and Cable

To this point in our discussion, we have been focusing on single-conductor wires with or without a layer of insulation. In practice, there are many types of wire to

TABLE 3-4 Wire sizes for standard annealed copper wire.

American Wire Gauge Number	Diameter (mils at 20°C)	Cross-Sectional Area (approximate cmil)	Ohm's per 1000 Feet of Annealed Copper Wire at 20°C
0000	460	211,600	0.04901
000	409.6	167,772	0.06180
00	364.8	133,079	0.07793
0	324.9	105,560	0.09827
1	289.3	83,694	0.1239
2	257.6	66,358	0.1563
3	229.4	52,624	0.1970
4	204.3	41,738	0.2485
5	181.9	33,088	0.3133
6	162.0	26,244	0.3951
7	144.3	20,822	0.4982
8	128.5	16,512	0.6282
9	114.4	13,087	0.7921
10	101.9	10,383	0.9989
11	90.74	8,234	1.26
12	80.81	6,530	1.588
13	71.96	5,178	2.003
14	64.08	4,106	2.525
15	57.07	3,257	3.184
16	50.82	2,583	4.016
17	45.26	2,048	5.064
18	40.3	1,624	6.385
19	35.89	1,288	8.051
20	31.96	1,021	10.15
21	28.45	809	12.8
22	25.35	643	16.14
23	22.57	509	20.36
24	20.1	404	25.67
25	17.9	320	32.37
26	15.94	254	40.81
27	14.2	202	51.47
28	12.64	160	64.9
29	11.26	127	81.83
30	10.03	101	102.67

choose from for a given application. We will discuss some of the major factors that distinguish wire types.

Solid versus Stranded

Figure 3–4 illustrates the difference between solid-conductor and stranded wire. A given wire gauge of stranded wire consists of a number of smaller gauge wires tightly braided together. The overall diameter determines the final gauge size of

FIGURE 3–4 (a) Solid-conductor wire and (b) stranded wire.

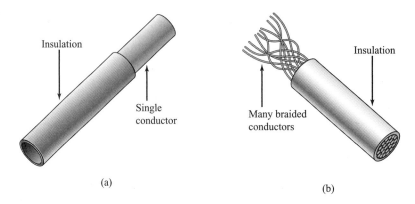

the stranded wire. Stranded wires are less prone to breaking when they are flexed repeatedly. There are many combinations of strand gauge and number of strands to achieve a given overall gauge size. Some common examples are listed in Table 3–5.

TABLE 3–5 Stranded wire consists of several strands of smaller gauge wire.

Overall Gauge	Number of Strands	Strand Gauge
24	10	34
	19	36
	41	40
18	7	26
	16	30
	41	34
14	7	22
	41	30
	105	34

Type of Insulation

Most wires used in electronics are covered with some form of insulation. The insulation prevents shorts to other wires or circuit components and reduces the probability of electrical shock. Many different materials are used for wire insulation. They vary in ways that include the following:

- Breakdown voltage
- Useable temperature range
- Resistance to water, acid, alkali, and petroleum products
- Resistance to flame
- Resistance to weathering
- Cost
- Mechanical flexibility

Breakdown voltage is the potential difference that will cause the insulation to break down. When an insulator breaks down, it loses its ability to stop current flow. During breakdown, a high current can flow and the insulator may be permanently damaged. Breakdown voltage depends on the type and thickness of insulation. Table 3–6 contrasts some of the characteristics of several common materials used to make wire insulation.

It is important for you to realize that not all wire insulation materials are as obvious (visually) as the insulation on standard hook-up wire. Some wires (collectively called **magnet wire**) have a thin, nearly invisible insulation that is applied by

TABLE 3–6 Comparison of characteristics for wire insulating materials.

Material	Breakdown Voltage (per mil thickness)	Resistance to					Flame Retardancy
		Heat	Water	Alkali	Acid	Abrasion	
Butyl rubber	600	Good	Excellent	Excellent	Excellent	Fair	Poor
Kynar®	260	Excellent	Good	Excellent	Excellent	Good	Excellent
Neoprene	600	Good	Excellent	Good	Good	Excellent	Good
Polypropylene	750	Good	Excellent	Excellent	Excellent	Fair	Poor
Polyurethane	500	Good	Good	Fair	Fair	Excellent	Good
PVC	400	Good	Good	Good	Good	Good	Excellent
Silicon rubber	400	Excellent	Excellent	Good	Good	Fair	Good
Teflon®	500	Excellent	Excellent	Excellent	Excellent	Fair	Excellent

dip-coating the conductor in a varnish-like material. Magnet wire is used to wind motors, transformers, speaker windings, and other electromagnetic devices. The thin insulation permits many turns of wire in a relatively compact space. When soldering this type of wire, care must be used to ensure that the insulation has been removed from the portion being soldered.

Single-Conductor versus Cable

Previous discussions have focused on single-conductor wires (either solid or stranded). Figure 3–5 illustrates how two or more *individually insulated* wires may be bundled in a common outer jacket. The bundled assembly is called a **cable**.

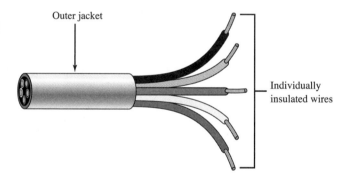

FIGURE 3–5 A cable consists of a number of individually insulated wires in a common sheath.

The number of conductors in a cable varies from as low as two to over one hundred. Most cables use color-coding to distinguish the wires. The individually insulated wires may be either solid or stranded. The outer jacket, which bonds the wires into a single cable, is generally an insulator. Multiconductor cables in which each of the individual wires is actually a coaxial cable are also available.

Coaxial Cable

Figure 3–6 on page 62 shows the construction of **coaxial cable** (called coax for short). It consists of an insulated center conductor surrounded by a braided wire shield. A tough outer jacket encases both the center conductor and the braided shield.

The braided shield, which surrounds the center conductor, is connected to a zero-volt potential in most applications. This provides an electrical barrier, which prevents stray fields from inducing noise voltages into the center conductor. This is important when low-level signals must be routed through an electromagnetically

FIGURE 3–6 Coaxial cable reduces electromagnetic coupling between the center conductor and other nearby circuits.

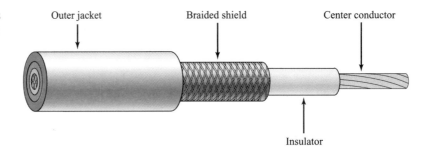

harsh environment. Coaxial cables are also used to prevent high-frequency voltages on the inner conductor from interacting with adjacent circuitry or from causing electromagnetic emissions that would cause interference to nearby receivers.

An alternative to standard coaxial cables that still provides shielding for the center wire(s) consists of a spiral-wrapped metal foil. The foil surrounds the center conductor(s) and is generally connected to a zero-volt potential. An uninsulated wire, called a **drain wire**, runs along the length of the metal foil and provides a means for connecting to it.

Ribbon Cable

Figure 3–7 shows a multiconductor cable where the individually insulated wires are side by side forming a flat strap. The bound wires are collectively called a **ribbon cable**. Ribbon cable is frequently used to interconnect circuit boards and devices in a computer system. Some types have a braided shield that surrounds the entire ribbon cable. The shield and the other conductors are then covered with a protective insulative jacket.

FIGURE 3–7 Ribbon cable is frequently used in computer systems.

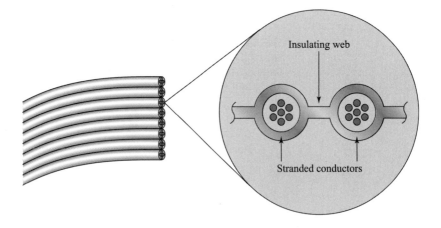

The individual wires in a ribbon cable can be split out and soldered individually if necessary. In most cases, however, crimp-on connectors are applied to the cable ends that make simultaneous connection to all of the wires. The connectors can then be plugged into mating sockets on circuit boards or other devices such as CD and disk drives.

A variation of the basic ribbon cable consists of flat copper conductors (like traces on a printed circuit board) sandwiched between two sheets of plastic laminate. The resulting package provides a flexible, multiconductor connection between two points.

Troubleshooting Wire and Cable

Defects in wires and cables can generally be found with an ohmmeter as long as all power is removed from the system. An **ohmmeter** is a test instrument that measures the resistance between its two test leads. These common wire and cable defects can be identified with an ohmmeter:

- One or more wires open
- Two or more wires shorted together
- One or more intermittent connections

Direct measurements can be used for short wires. In the case of long wires, you can temporarily short two wires together at one end of a long cable and then measure between the wires on the opposite end. A low resistance reading indicates that neither wire is open. If a high resistance reading is obtained, then you must repeat this process with one of the original two wires and a third wire in order to determine which wire is actually open.

Shorts between wires can also be detected with an ohmmeter. A low resistance reading between two wires that are supposed to be insulated from each other indicates a short between the wires. Intermittent opens or shorts can be found in a similar manner, except that you must flex the wire or cable while measuring its resistance. A dramatic change in resistance when the wire or cable is flexed indicates an intermittent connection.

✓ Exercise Problems 3.2

1. The resistance of a wire is unaffected by temperature. (True/False)
2. If the resistance of a material decreases as the temperature increases, the material is said to have a (*positive/negative*) temperature coefficient.
3. The resistance of a wire is (*directly/inversely*) proportional to the length of the wire.
4. The resistance of a wire is (*directly/inversely*) proportional to the cross-sectional area of the wire.
5. The resistance of a wire is not related to the material used to make the wire. (True/False)
6. Compute the circular mil area of a wire that has a diameter of 30 mils.
7. What is the diameter of a wire that has an area of 2025 cmil?
8. Semiconductors often have a (*positive/negative*) temperature coefficient.
9. Which of the following wire gauges would have the greatest ampacity: 0, 14, or 30?
10. What is the resistance of 3000 feet of 18-gauge annealed copper wire at 20° C?
11. What is the cross-sectional area of a 24-gauge wire?
12. Name at least one advantage of stranded wire over solid-conductor wire.
13. If a wire is to be used in a 120-volt circuit, then the insulation must have a breakdown voltage rating of 120 volts or less. (True/False)
14. What is the approximate breakdown voltage for a 35-mil thickness of PVC insulation?
15. What type of insulation is used on magnet wire?
16. Describe the physical construction of coaxial cable.
17. Describe the physical construction of ribbon cable.

3.3 ELECTRONIC COMPONENTS

As you study the material in this and subsequent texts, you will learn the electrical characteristics of many electrical and electronic components. This section is meant to introduce you to the physical appearance of some of the components you will be studying. Similar devices can often have dramatically different appearances. You must learn to recognize a component from its physical appearance.

You will also learn some of the more outstanding or unique electrical characteristics of each component type. Resistors, switches, and circuit protection devices will be discussed extensively in this section. Discussions of the remaining components will serve as an introduction.

Voltage Sources

Every electronic circuit requires some type of voltage source. The choices are extensive and range from automobile batteries to solar cells to complete electronic systems connected to commercial power lines.

You are already familiar with several types of cells and batteries that are used as sources of voltage: flashlight battery (D cell), 9-volt transistor radio battery, AAA cells, 12-volt car battery, watch battery, and so on. In general, a cell is a stand-alone source of electrical energy. Cells develop electrical energy by converting other forms of energy such as chemical energy, solar energy, or thermal energy. A battery consists of two or more interconnected cells. By connecting multiple cells, a battery can be made to deliver higher voltages and/or higher currents than a single cell. Cells and batteries have voltage and ampere-hour (Ah) ratings. The **ampere-hour rating** is an indication of capacity. For example, a 1 Ah battery might be able to deliver 1 A of current for 1 hour, 0.25 A of current for 4 hours, or perhaps 2 A of current for $\frac{1}{2}$ hour.

Figure 3–8 shows several typical cells and batteries. Some batteries are small enough to fit inside of a watch or hearing aid. Others have a volume of several cubic feet and weigh hundreds of pounds.

FIGURE 3–8 (a) Batteries vary in voltage, current capacity, chargeability, and physical appearance. (*Courtesy of Duracell*) (b) Schematic symbol for a cell. (c) Schematic symbol for a battery.

Figure 3–8 also shows the schematic symbols for a cell and a battery. The negative terminal is identified by the short line segment; the positive terminal is identified by the longer line segment.

The label "electronic voltage source" is somewhat misleading, since an electronic voltage source does not really supply electrical energy in the same way a cell or battery does. Rather, an electronic voltage source changes and conditions electrical power that is initially supplied from some other source such as a battery or commercial power line. The electronic voltage source (called a **power supply**) in a computer is a good example. Most computers require 120 V from the commercial power line for their operation. However, the various electronic devices within the computer typically require from 1.5 V to as high as 5 V for proper operation. It is the electronic power supply within the computer that converts the power line voltage into the lower voltages required by the internal components.

Fuses and Circuit Breakers

Electrical and electronic circuits are susceptible to damage by excessive current flow. The increased current normally flows as a result of a defective component or an accidental short circuit. When a very low-resistance path bypasses a normal path for current flow, we call the low-resistance path a **short circuit**. Short circuits are generally characterized by increased current flow. Fuses and circuit breakers provide protection against damage from excessive current flow. Figure 3–9 shows an assortment of fuses and circuit breakers.

Fuses

Fuses are connected such that the current flowing through the protected circuit also flows through the fuse. Figure 3–10 shows the operation of a fuse. There is a resistive link inside the fuse body that heats up when current flows through it. If

FIGURE 3–9 (a) An assortment of fuses and circuit breakers. (b) Schematic symbol for a fuse.

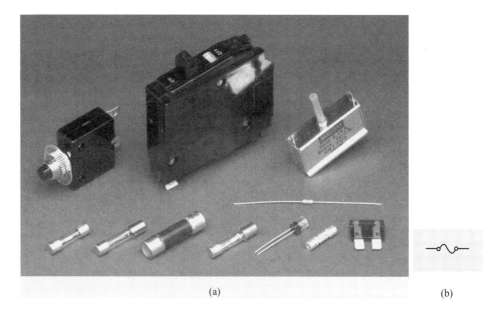

FIGURE 3–10 A fuse protects circuits or devices from excessive current flow.

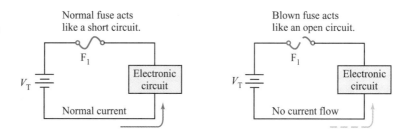

the current is sufficiently high, the resistive link burns open, which stops all current flow in the circuit. When a fuse has burned open, we say it is blown.

Fuses have three electrical ratings, which are particularly important:

- **Current Rating** The current rating of a fuse indicates the maximum *sustained* current that can flow through the fuse without causing the element to burn open. The actual current required to blow the fuse may be less than the rating if the fuse is operated in a high-temperature environment. A fuse normally has a substantially higher current rating than the value of normal operating current in the circuit to be protected.
- **Voltage Rating** The voltage rating on a fuse specifies the minimum amount of voltage required to arc across the fuse *after it has blown*. The voltage rating of the fuse should exceed the highest voltage to be expected in the protected circuit. If the circuit voltage exceeds the fuse rating, then the voltage may arc (like a miniature lightning bolt) across the open fuse link. This effectively reconnects the circuit and defeats the purpose of the fuse.
- **Response Time** When the current through a fuse exceeds its current rating, the fuse link will burn open. However, it takes a certain amount of time for the resistive element to heat up to the point of disintegration. The time required for the link to burn open is called **response time**. Different types of fuses have different response times. The response time of any given fuse decreases (opens faster) as the percentage of overcurrent increases. Typical response times vary from several milliseconds to several seconds.

Manufacturers classify fuses into three general categories based on the approximate response times:

- **Slo-Blo** These fuses are designed to withstand currents which greatly exceed their current rating as long as the overcurrent condition is only momentary.

A short-duration current burst, called a **transient** or **surge**, is characteristic of normal operation for some devices such as motors and high-capacitance circuits. A slo-blo fuse can open quickly if the overcurrent value is high enough. Typical response times for slo-blo fuses vary from one-tenth second to 10 seconds. Slo-blo fuses in glass packages can often be identified by a coiled spring internal to the fuse housing.
- **Fast-Blo** These fast-acting fuses are designed to have very short response times. They are often used to protect sensitive electronic components (such as transistors) which may be damaged by high current transients. Typical response times include the range of submillisecond through hundreds of milliseconds.
- **Normal-Blo** These fuses have response times in between those of slo-blo and fast-acting fuses. Remember, though, the response time of all three types of fuses varies dramatically with the percentage of overload. Figure 3–11 shows several types of fuses.

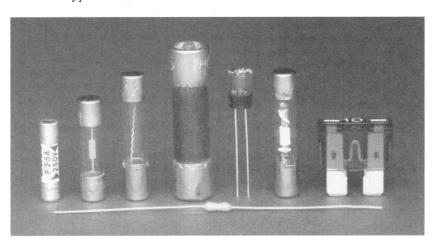

FIGURE 3–11 Some representative types of fuses.

Circuit Breakers

Fuses have a distinct disadvantage of having to be physically replaced once they have blown. Circuit breakers also provide overcurrent protection for circuits and devices, but they do not have to be replaced after they have opened the circuit. When a circuit breaker opens in response to excessive current flow, we say that it has been tripped. Once a circuit breaker has been tripped, it must be reset before the current path can be restored to the circuit. Most circuit breakers require the user to manually reset the breaker, but some are designed to reset automatically after a time delay.

Although circuit breakers have the advantage of reusability, they have a disadvantage for certain applications since they generally have longer response times than fuses. Figure 3–12 shows several examples of circuit breakers.

Resistors

A resistor is one of the most fundamental components used in electronic circuits. A resistor is constructed to have a specific amount of resistance to current flow. Resistors range in value from less than one ohm to well over 20 million ohms. They also vary in size from microscopic through devices that are too large to carry. In general, we can classify resistors into two categories: fixed and variable. Fixed resistors have a single value of resistance. Variable resistors are made to be adjustable so that they can provide different values of resistance. We will examine both classes of resistors in the following paragraphs.

Resistor Power Rating

When current flows through a resistor, it encounters opposition, which creates heat. If a resistor dissipates too much heat, it will be damaged. Every resistor has a

FIGURE 3–12 Some representative examples of circuit breakers. (*Courtesy of Weber US*)

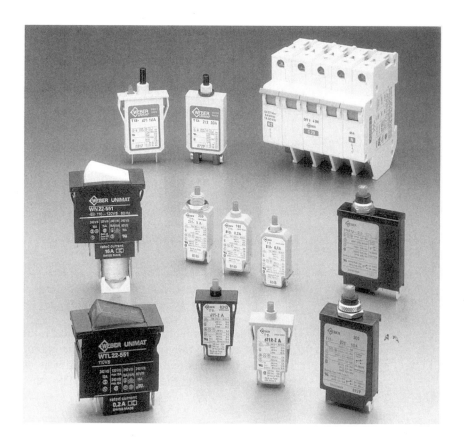

power rating. This indicates the amount of power that can safely be applied to the resistor for an indefinite amount of time without degrading the performance of the resistor.

The power rating of a resistor is largely, but not solely, determined by the physical size of the resistor. The greater the surface area of the resistor, the more power it can dissipate. Thus, as a rule, the physical size of a resistor is an indication of its power rating. Resistor power ratings range from less than one-tenth watt to many hundreds of watts. The most common power ratings are $\frac{1}{16}$, $\frac{1}{10}$, $\frac{1}{8}$, $\frac{1}{4}$, $\frac{1}{2}$ and 1 watt. Figure 3–13 shows resistors with several different power ratings.

FIGURE 3–13 The power rating of a resistor is largely determined by its physical size.

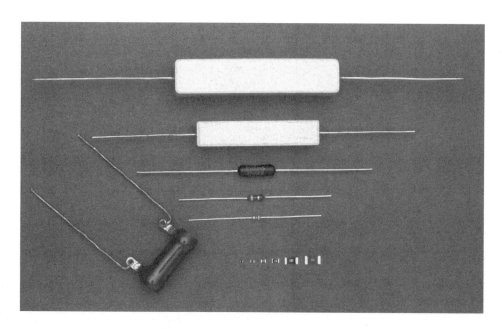

Resistor Tolerance

Manufacturers cannot make resistors with exactly the right value. There is inevitably a certain degree of variation between resistors that are ideally the same value. The manufacturer guarantees that the actual value of the resistor will be within a certain percentage of its nominal or marked value. The allowable variation—expressed as a percent—is called the **resistor tolerance**.

The most common resistor tolerances today are 1% and 5%; several years ago, 10% and 20% tolerances were common. As a rule, tighter tolerance resistors are more expensive. Resistors with tolerances lower than 2% are often called **precision resistors**. The tolerance may be either positive or negative. That is, the actual resistor value may be smaller (negative tolerance) or larger (positive tolerance) than the marked value. The maximum deviation between actual and marked values for a particular resistor is computed by multiplying the tolerance percentage by the marked value of the resistor. This computation is given by Equation 3–2.

$$\text{maximum deviation} = \text{tolerance} \times \text{marked value} \qquad (3\text{–}2)$$

We can determine the highest resistance value that a particular resistor can have and still be within tolerance by adding the maximum deviation to the marked value. The minimum acceptable resistance value is computed by subtracting the maximum deviation from the marked value. Thus the *range* of resistance values is given by Equation 3–3.

$$\text{resistance range} = \text{marked value} \pm \text{maximum deviation} \qquad (3\text{–}3)$$

EXAMPLE 3.10

A certain resistor is marked as 1,000 Ω with a 10% tolerance. Compute the maximum deviation and the range of possible resistance values.

SOLUTION First we apply Equation 3–2 to determine the maximum deviation.

$$\text{maximum deviation} = \text{tolerance} \times \text{marked value}$$
$$= 0.1 \times 1{,}000 \; \Omega = 100 \; \Omega$$

Calculator Application

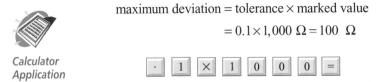

We now apply Equation 3–3 to calculate the range of resistor values that are within the tolerance specification.

$$\text{resistance range} = \text{marked value} \pm \text{maximum deviation}$$
$$\text{lowest value} = 1{,}000 - 100 = 900 \; \Omega$$
$$\text{highest value} = 1{,}000 + 100 = 1{,}100 \; \Omega$$

Calculator Application

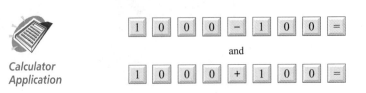

•—Practice Problems

1. A certain resistor is marked as 330 Ω with a 1% tolerance. Compute the maximum deviation and the range of possible resistance values.
2. Determine the highest and lowest resistor value that a 22-kΩ resistor can have, if it has a 5% tolerance.
3. If a resistor that was marked as 39 kΩ, 10% actually measured 37,352 Ω, would it be within tolerance?

Resistor Technology

We shall examine four major classes of fixed resistor technology:

- Carbon-composition resistors
- Film resistors
- Wirewound resistors
- Surface-mount technology

Carbon-composition resistors are one of the oldest types of resistors and are seldom used today. Figure 3–14 shows a cutaway view of a carbon-composition resistor. They are manufactured by making a slurry of finely ground carbon (fairly low-resistance material), a powdered filler (high-resistance material), and a liquid binder. The slurry is pressed into cylindrical forms and the leads are attached. Finally, the resistor is coated with a hard nonconductive coating and then color banded to indicate its value. Altering the ratio of carbon to filler material creates resistors of different values. The higher the percentage of carbon, the lower the resistance of the resistor. The power rating is made higher by increasing the physical size of the resistor.

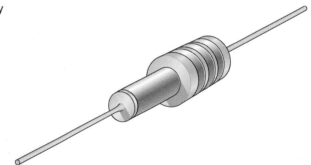

FIGURE 3–14 A cutaway view of a carbon-composition resistor.

Film resistors come in several varieties. A film resistor is made by depositing a thin layer of resistive material onto an insulating tube or rod called the substrate. The general range of resistance is established by the type of the material used. Leads are attached to end caps, which contact the ends of the resistive film.

Once the resistive layer has been deposited, it is trimmed with a high-speed industrial laser. As the laser etches away portions of the deposited film, the resistance of the device increases. Normally, the laser trims the resistor to the correct value by cutting a spiral pattern along the length of the resistor body. This method of trimming is called **spiraling**. The laser, in combination with automatic test fixtures, can produce resistors that are very close to the desired value. Figure 3–15 on page 70 shows a cutaway view of a typical film resistor.

Wirewound resistors are made by winding resistive wire around an insulating rod. The ends of the wire are connected to leads, and the body of the resistor is coated with a hard insulative jacket. The value of a particular wirewound resistor is determined by the type of wire, diameter of the wire, and the length of the wire used to wind the resistor. Wirewound resistors are generally used for their high power ratings, but they are sometimes selected because of their precise values.

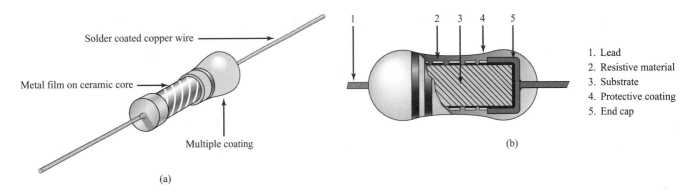

FIGURE 3–15 A cutaway view of a film resistor.

Figure 3–16 shows a cutaway view of a wirewound resistor. Sometimes a wide colored band is painted on the resistor body to identify it as a wirewound type.

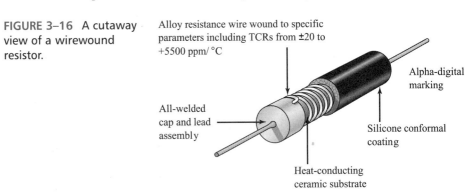

FIGURE 3–16 A cutaway view of a wirewound resistor.

Because there is a continuing demand for smaller electronic products, there is an associated demand for smaller electronic components. One common method for reducing the size of an electronic product is to utilize **surface-mount technology** (SMT). *Surface-mount resistors* have no leads. Their contacts solder directly to pads on the printed circuit board. Surface-mount resistors are manufactured by spreading a resistive layer onto a ceramic substrate. The resistive coating is then covered with a layer of glass for protection. Figure 3–17(a) on page 71 shows a cutaway view of a surface-mount resistor. In Figure 3–17(b) several surface-mount resistors are pictured next to the head of a straight pin to provide a size comparison. This is the most commonly used style of resistor for new products. The physical sizes of these components range from as small as 0.020″ × 0.01″ to as large as 0.18″ × 0.12″.

Resistor Color Code (three- and four-band)

Since many resistors are physically small, it is impractical to print their values with numbers big enough to read. Instead, manufacturers often mark resistors with three to five colored bands to indicate their value. The colors are assigned according to a code standardized by the Electronics Industries Association (EIA). It is essential for you to memorize this code and know how to use it.

Table 3–7 shows the relationship between a particular digit and its corresponding color in the standard color code. Figure 3–18 shows how to interpret the colored bands on a three- or four-band resistor.

Three- or four-band resistors can be interpreted by applying the following procedure:

1. The first two bands represent the first two digits of the resistance value.
2. Multiply the digits obtained in step one by the value of the third (multiplier) band.

3.3 • ELECTRONIC COMPONENTS   71

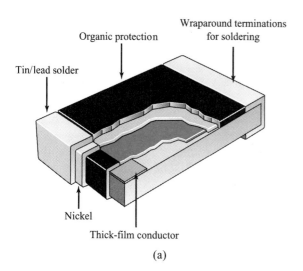
(a)

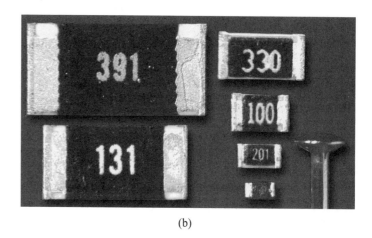
(b)

FIGURE 3–17 (a) A cutaway view of a surface-mount resistor. (b) Typical surface-mount resistors. The head of a pin is shown for perspective.

TABLE 3–7 Resistor color code for three- and four-band resistors.

Band Type	Color	Digit Value	Multiplier Value	Tolerance Value
Digit/ Multiplier Bands	Black	0	10^0	—
	Brown	1	10^1	—
	Red	2	10^2	—
	Orange	3	10^3	—
	Yellow	4	10^4	—
	Green	5	10^5	—
	Blue	6	10^6	—
	Violet	7	10^7	—
	Gray	8	10^8	—
	White	9	10^9	—
Multiplier/ Tolerance Bands	Gold	—	10^{-1}	±5%
	Silver	—	10^{-2}	±10%
	No band	—	—	±20%

FIGURE 3–18
Interpretation of the resistor color-code bands.

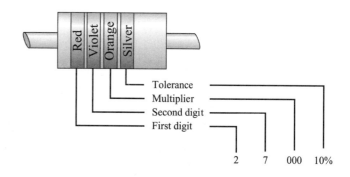

3. If there is a fourth band, it indicates the tolerance, otherwise tolerance is ±20%. It should be noted that 20% resistors (no tolerance band) are obsolete for most practical purposes.

•—EXAMPLE 3.11

Determine the value of each of the resistors pictured in Figure 3–19.

SOLUTION

a. The first band is the one nearest to an end of the resistor. In this case, the resistor value is decoded as shown in Table 3–8. The final decoded value is 27,000 Ω, ±10%.

TABLE 3–8 Decoding the value of the resistor in Figure 3–19(a).

Band	Color	Value
First	Red	2
Second	Violet	7
Third (multiplier)	Orange	000 (×10³)
Fourth (tolerance)	Silver	10%

b. The first two bands (orange and white) give us the first two digits of 39. The next band (red) is the multiplier and tells us to add two zeros. This gives us a base value of 3900 Ω. Finally, the fourth band is gold, which identifies the tolerance as 5%. Thus, we have a 3.9 kΩ, ±5% resistor.

c. The first two bands (green and blue) give us the digits 56. The multiplier band is yellow so we add four zeros to produce our base resistance value of 560,000 Ω. There is no tolerance band, which means that the resistor has a 20% tolerance. The final decoded value is 560 kΩ, ±20%.

d. The first two bands (brown and gray) give us the digits 18. The multiplier band is black so we add zero 0s (i.e., add nothing). The silver multiplier band indicates a 10% tolerance rating. The final converted value is 18 Ω, ±10%.

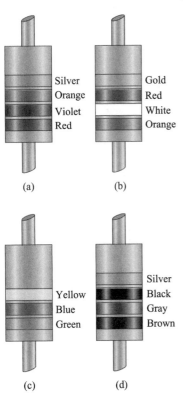

FIGURE 3–19 Determine the value of these resistors.

•—Practice Problems

1. Determine the values of the resistors pictured in Figure 3–20.

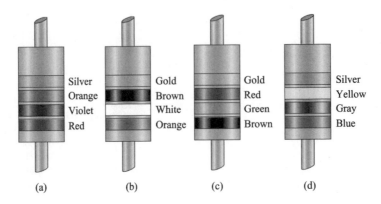

FIGURE 3–20 Determine the value of these resistors.

Resistor Color Code (five-band)

Figure 3–21 shows a resistor with five coding bands. The bands are interpreted as follows:

1. The first three bands determine the first three digits of the resistance value.
2. The fourth band is the multiplier.
3. The fifth band indicates tolerance.

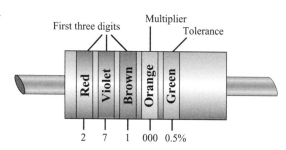

FIGURE 3–21 A resistor that uses a five-band color code.

Precision resistors often use this coding scheme. The color values shown in Table 3–7 can be used to decode the first four bands. The tolerance band is decoded as shown in Table 3–9.

It should be noted that the color interpretation listed in Table 3–9 reflects the EIA standard number RS-196-A. Some resistors use different five-band color code schemes. One such alternative code is shown in Figure 3–22. Here the first four bands are interpreted according to the four-band color code standard (Table 3–7). The fifth band is used to indicate the reliability (actually, the failure rate) of the resistor. Reliability is expressed as a percentage of failures in 1,000 hours of operation. Thus, a value of 0.01% means that if one million resistors were operated for 1,000 hours, then no more than 100 resistors would go out of tolerance. The fifth-band colors and their interpretation are shown in Table 3–10.

TABLE 3–9 Interpretation of tolerance band values for a five-band resistor code.

Tolerance Band Color	Tolerance Value
Brown	1%
Red	2%
Gold	5%
Silver	10%
Green	0.5%
Blue	0.25%
Violet	0.1%
Gray	0.05%

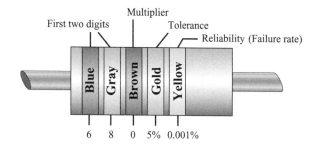

FIGURE 3–22 An alternative five-band color code scheme.

TABLE 3–10 Interpretation of the reliability band on a five-band resistor code.

Reliability Band Color	Failures per 1000 Hours
Brown	1.0%
Red	0.1%
Orange	0.01%
Yellow	0.001%

Surface-Mount Resistor Markings

Many surface-mount components, including some resistors, have no visible markings to indicate their resistance value. Rather, you must use test equipment (ohmmeter) to measure the resistance. Some manufacturers indicate the resistance value of surface-mount resistors with a numerical code in which the first digit or digits in the code represent the digits in the resistance value, and the last digit in the code indicates the number of trailing zeros. For example, a surface-mount resistor that is coded as 103 would be a 10,000-ohm resistor.

Schematic Symbol

As you know, schematic diagrams are used to represent the electrical connections in a circuit. Every component has a symbol that is used to represent it on a schematic diagram. The schematic shows how the various components are interconnected. Figure 3–23 shows the schematic symbol for a fixed resistor.

FIGURE 3–23 Schematic symbol for a fixed resistor.

Variable Resistors

Sometimes it is desirable to change the value of a resistor once it has been installed in a circuit. For these applications, we can use a variable resistor. The resistance of a variable resistor can be adjusted by turning a knob, rotating a screw, or moving a slider. There are two major classes of variable resistors: **rheostats** and **potentiometers**. Often these two classes are actually the same physical device; it is the electrical connection that distinguishes the two types of variable resistors.

Figure 3–24 shows cutaway views for both a rheostat and a potentiometer. In both devices, a sliding contact is moved along a strip of resistive material to adjust the value of the component. In the case of a rheostat, the resistance between its two external terminals changes as the slider is moved. By contrast, the total resistance between two of the three external terminals on a potentiometer remains constant as the slider is moved. It is equal to the resistance of the internal strip of resistive material. The third external terminal on a potentiometer is connected to the moveable slider or wiper arm. The resistance between this terminal and the other two external connections changes as the slider is moved. As the slider moves closer to one end of the resistive strip, there is less resistive material between the wiper arm and the contact connected to this end. Similarly, there is greater resistance between the wiper arm contact and the connection to the opposite end of the resistive strip.

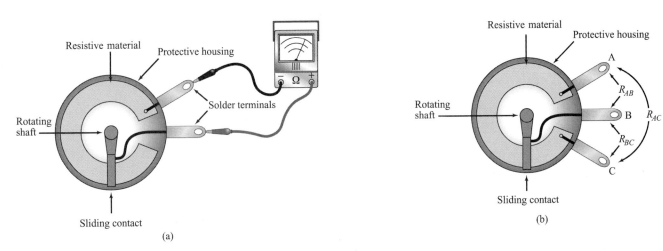

FIGURE 3–24 Cutaway views of (a) a rheostat and (b) a potentiometer.

Figure 3–25 on page 75 shows the schematic symbols that are used to represent variable resistors. The symbols for a rheostat and a potentiometer are shown in Figures 3–25(a) and 3–25(b), respectively. In Figure 3–25(c), two of the three terminals of a potentiometer have been connected together to form a two-terminal rheostat. Similarly, Figure 3–25(d), shows a potentiometer with only two of its three terminals used, which means it will function as a rheostat. Figure 3–26 on page 75 shows an assortment of potentiometers. Some have internal gear mechanisms. This allows multiple turns of the adjusting screw to have smaller effects on the wiper arm position, which, in turn, allows the resistor to be set to more precise values.

Capacitors

A capacitor is a basic component that is used to store electrical energy in the form of an electrostatic field. The operation of every electronic system depends on capacitors. A personal computer system, for example, uses hundreds of capacitors throughout the overall circuit. Capacitors vary widely in value, appearance, and physical size. Figure 3–27 shows several typical capacitors and the schematic symbol for a capacitor.

3.3 • ELECTRONIC COMPONENTS 75

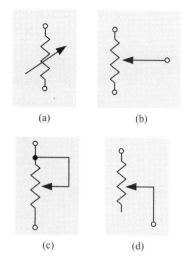

FIGURE 3–25 Schematic symbols for (a) rheostats, (b) potentiometers, (c) a potentiometer connected as a rheostat, and (d) a potentiometer connected as a rheostat.

FIGURE 3–26 An assortment of potentiometer types.

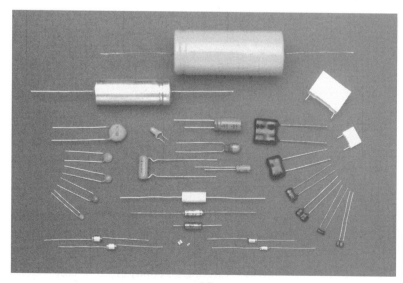

FIGURE 3–27 (a) Typical capacitors. (b) Schematic symbol for a capacitor.

Inductors

The inductor is a fundamental electrical component—also called a coil—that consists of a spiraled or coiled wire. The inductor is used to store electrical energy in the form of an electromagnetic field. Figure 3–28 shows several representative inductors and the schematic symbol for an inductor.

FIGURE 3–28 (a) Representative inductors. (b) Schematic symbol for an inductor.

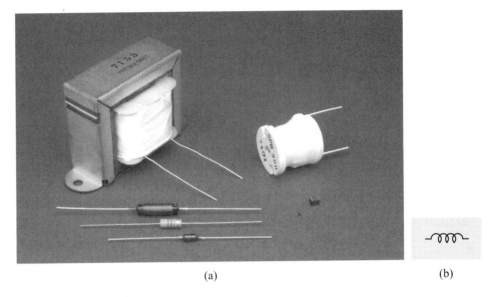

Transformers

A transformer is basically two or more inductors whose electromagnetic fields interact. Transformers are used to transform (i.e., increase or decrease) alternating voltages like the 120-volt, 60-hertz voltage that is supplied to your house by the power company. Figure 3–29 shows some sample transformers. Transformers range in size from less than 0.2 inch on a side to more than 10 feet on a side. There are many uses for transformers in computer equipment including reduction of power line voltage to a lower value useable by the various components, development of the high voltages required by display panels, power line isolation, and interfacing the computer circuits to communication lines such as Ethernet.

FIGURE 3–29 (a) Typical transformers. (b) Schematic symbol for a transformer.

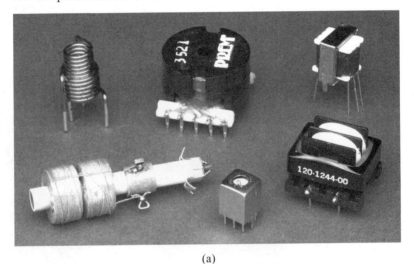

Switches

A switch is another basic, but important, component used in electronic circuits. Its purpose is to break (open) or make (close) circuit connections. You are familiar with the switches in your home that are used to control the lighting. When you operate the switch to turn the lights on, you are closing the circuit and connecting the light bulb to the power line. When the switch is off, the light bulb is isolated from the power line.

Basic Switch Operation

Figure 3–30 shows a basic switch circuit. The type of switch shown in the picture is called a knife switch. Although it is not often used in electronic circuits, this type of switch is easy to understand, and its operation is representative of other, less obvious switch types. When the switch is in the upper position, as in Figure 3–30(a), the circuit is open and the lamp remains dark. When the switch closes the circuit, as in Figure 3–30(b), electrons can travel from the battery, through the lamp, through the switch, and return to the positive side of the battery. This simple switch represents a general class of switches called single-pole single-throw (SPST). The schematic symbol for an SPST switch is shown in Figure 3–31 along with the symbols for several other types of switches.

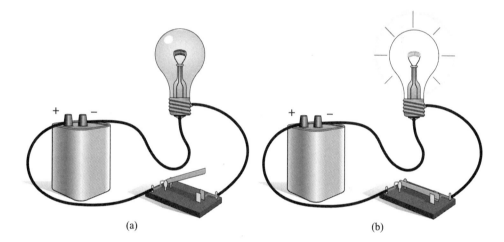

FIGURE 3–30 A basic switch circuit in the (a) open and (b) closed conditions.

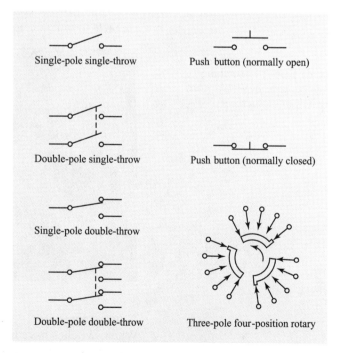

FIGURE 3–31 Schematic symbols for several types of switches.

The term **pole** refers to the moveable portion of the switch. For example, a double-pole switch has two moveable arms; it acts as two electrically separate switches that are mechanically linked so they operate simultaneously. The pole of a

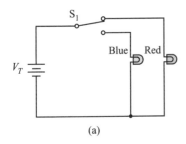

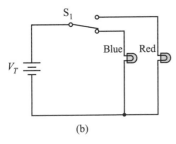

FIGURE 3–32 An SPDT switch can alternately connect power to two different circuits.

FIGURE 3–33 There are many types of switching mechanisms available from switch manufacturers.

switch is often identified with an arrowhead on the switch symbol. The term **throw** identifies the number of circuits that are opened or closed by each pole when the switch is operated. For example, a double-throw switch opens or closes two circuits on each pole.

Figure 3–32 illustrates how to connect a single-pole double-throw (SPDT) switch into a lamp circuit. In Figure 3–32(a), the red lamp is lit while the blue lamp is off. In the switch position shown in Figure 3–32(b), the circuit for the blue lamp is closed and the red lamp is extinguished.

Switching Mechanism

Any of the basic switch types (e.g., SPST, SPDT, DPST, DPDT, and so on) can be made with different mechanical means for operating the switch. Some of the more common means for switch operation are listed:

- Toggle (e.g., light switches in many homes)
- Rocker (e.g., power switches on many printers)
- Push-button (e.g., power switches on many desktop computers)
- Rotary (e.g., selection switch to choose between multiple printers)
- Slide (e.g., power switches on many notebook computers)

There is a tremendous range of switch forms based on these basic classes of mechanisms. Figure 3–33 shows a sampling of switch types.

Momentary-Contact Switches

Some switches open and/or close a circuit when you operate the switch, but return to the original state as soon as the switch button is released. This type of switch is called a **momentary-contact switch**. Doorbell and car horn switches are two examples of momentary-contact switches. Additionally, each alphanumeric key on a computer keyboard is mechanically linked to a momentary-contact switch.

It is important to know which contacts are closed and which ones are open when the switch is in its normal state. There are two ways this is indicated on schematic diagrams. First, the poles for all switch contacts are drawn in the *normal* (not operated) position. Second, the letters NC and NO are sometimes written near the contacts to indicate normally closed and normally open contacts, respectively.

Ganged Switches

When switches have multiple poles, the individual poles are mechanically linked, or ganged, together so that they operate simultaneously. This mechanical link is shown on schematic diagrams as a dotted line. The multiple-pole switches shown in Figure 3–31 have dotted lines to show the mechanical link between sections of the switch.

Switch Specifications

Switch manufacturers provide extensive catalogs (nearly always available on the Internet) that detail the physical and electrical characteristics of their switches. You will often need this information to select a suitable replacement for a defective switch. Some of the factors to be considered in switch selection are:

- Contact form (i.e., SPDT, DPDT, and so on)
- Switching mechanism (e.g., push-button, toggle, rocker, and so on)
- Voltage rating
- Current rating
- Environmental performance (e.g., moisture immunity, ruggedness, and so on)

Relays

Relays are essentially electromagnetically operated switches. They are used extensively in modern electronic systems, especially in industrial systems. Applications for relays include motor control circuits for industrial machines, circuits to protect workers, switching circuits in automobiles, and power switching circuits in notebook computers.

Relay Construction

Figure 3–34 shows the basic component parts of a relay. Although relays vary greatly in size and appearance, they all have the components called out in Figure 3–34. It is important to remember that the relay is essentially a switch—a simple on-off/open-closed switch. The relay differs from a standard switch in the way the switch is moved from one position to the other. A basic switch changes state when its mechanism is moved by the force from a human finger. The switching mechanism of a relay, by contrast, is moved between states by the force of attraction provided by an electromagnetic field.

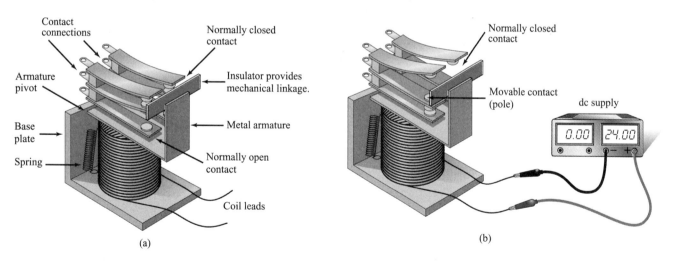

FIGURE 3–34 The basic component parts of a relay.

First, locate the switch contacts on the relay illustration in Figure 3–34(a). The particular relay illustrated has double-pole double-throw (DPDT) contacts. Effectively, these are two electrically separate switches that change states at the same time. The relay in Figure 3–34(a) is shown in its *normal* or *deenergized* state. In this state, the normally closed switch contacts are closed (shorted together), while the normally open contacts are open (physically separated). Spring tension on the moveable contact forces the contacts to stay in the normal position.

Figure 3–34(b) shows the relay in its energized position. Here, current has been applied to the relay coil, producing a magnetic field near the metal **armature**. The moveable metal armature is attracted by the magnetic field and moves into the position shown in Figure 3–34(b). The "pole" (moveable contact) portion of the switches is mechanically linked to the metal armature. When the armature is attracted to the coil, the switch contacts change state; the normally closed contacts open and the normally open contacts close.

As long as sufficient current flows through the coil of the relay, the contacts will remain in the energized position. If the coil current is interrupted or allowed to fall below a certain value, then the force of the spring will snap the switch contacts back into the normal state shown in 3–34(a). It is important to understand that the coil circuit and the switch circuit are electrically isolated from each other. Thus, for example, we might use a 5-V control circuit to activate the coil of a relay whose contacts switch a 2000-V circuit.

Figure 3–35 shows four alternative schematic symbols for a relay. Many companies identify relays on their schematic diagrams with a "K" or a "CR" label. One of the most confusing aspects of the symbols shown in Figures 3–35(a) through 3–35(c) is that the contacts are not necessarily drawn in the same vicinity as the relay coil. If the coil is labeled K_5 for example, then the contacts might be labeled K_{5A}, K_{5B}, and K_{5C} to identify them as being associated with the K_5 coil, but they could be drawn on any part of the schematic. When analyzing the circuit, you must remember that *all* contacts associated with a particular relay coil will change states simultaneously.

FIGURE 3–35 Alternative schematic symbols for a relay.

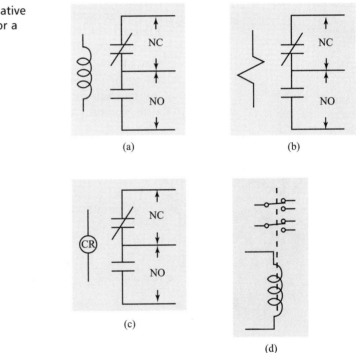

Relays and Computers

Although physical relays are commonly used in many computer applications, there is another important application of relays used with industrial controllers. A **programmable logic controller** or PLC is a small computer system used to monitor and control industrial machines and processes. In many cases, PLCs provide the programmer/operator with a graphical representation of the overall system. Relays are widely used as the controlling or switching elements on the graphical screen. The actual physical circuits may consist of transistors, integrated circuits, or other components (including actual relays), but regardless of the physical implementation, relay coils and contacts are used to represent the behavior of the system. Thus, a knowledge of relay operation is necessary to operate or program an industrial control system based on a PLC.

Transistors and Diodes

Transistors and diodes are called solid-state components or semiconductors. They are used in many electronic circuits including amplifiers, computers, and industrial controls. Diodes are used to alter information signals, to convert alternating current (like that used in your home) into direct current (such as provided by a battery), and as protective devices against voltages surges. Transistors are used as amplifiers to increase the strength of an electrical signal. They are also used as switching devices. The operation of a computer or other digital system is based on transistors operated as switches. Figure 3–36 shows several transistors and diodes.

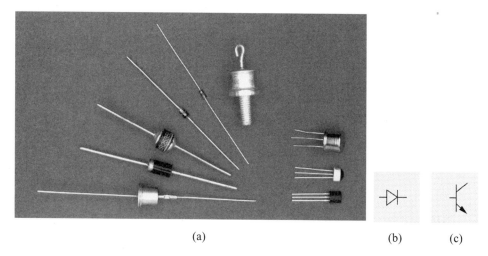

FIGURE 3–36 (a) Assortment of transistors (three leads) and diodes (two leads). (b) Schematic symbol for a diode. (c) Schematic symbol for a transistor.

Integrated Circuits

These incredibly complex devices are actually miniature versions of complete circuits as briefly described in chapter 1. Some integrated circuits have over a million equivalent components (resistors, transistors, and so on). Figure 3–37 shows several types of integrated circuits. A typical computer system will have tens or even hundreds of these devices.

FIGURE 3–37 Several types of integrated circuits. (*Courtesy of Altera Corporation*)

✓ Exercise Problems 3.3

1. Solar cells and flashlight batteries are examples of _____ _____.
2. Two or more cells can be combined to form a _____.
3. What is the primary advantage of a circuit breaker over a fuse?
4. The voltage rating of a fuse tells how much voltage it takes to blow the fuse. (True/False)

5. The physical size of a resistor is an indication of its resistance value. (True/False)
6. What is the range of values that a 3.9-kΩ, 1% resistor could have and still be within its allowable tolerance?
7. If a resistor that was marked as 470 kΩ, 5% actually measured 441,800 Ω, would it be within tolerance?
8. Wirewound resistors are sometimes used because of their high power ratings. (True/False)
9. Most modern electronic systems utilize surface-mount components. (True/False)
10. What is the expected value of a resistor that has the following color stripes: Red, Red, Orange, Gold?
11. What is the range of acceptable resistance values for a resistor with the following color stripes: Blue, Gray, Brown, Silver?
12. What is the expected value of a resistor that has the following color stripes: Yellow, Violet, Green, Orange, Brown?
13. What is the nominal value of a surface-mount resistor that has the following numbers printed on it: 271?
14. A potentiometer can be wired to work as a rheostat. (True/False)
15. A rheostat can be wired to work as a potentiometer. (True/False)
16. Capacitors and inductors both store energy in fields. (True/False)
17. How many terminals or connections would you expect to find on an SPST switch?
18. How many terminals or connections would you expect to find on a DPDT switch?
19. A _____ is an electromagnetically-operated switch.
20. How many leads does a diode have?

•—SUMMARY

Technical notation eases the burden of expressing very large and very small numbers used in electronics. Engineering notation is based on powers of ten and requires the exponent to be zero or evenly divisible by three. Metric prefixes also simplify the expression of large and small numbers. Here, a prefix such as milli or micro is added to the basic unit of measure to express multiple or submultiple units. Each metric prefix has a corresponding power of ten representation.

Many of the components that make up an electronic system are connected together with wire and cable. Since wire is not a perfect conductor, it has some resistance that is determined by the type of material, the temperature, and the length of the wire. Wire size (i.e., diameter) is expressed as gauge. Higher gauge numbers correspond to wire with smaller diameters. Wire may be stranded or solid. Multiple wires may be formed into a cable. Ribbon cables are widely used in computer applications. Coaxial cables shield signals from external interference, and they prevent interference to and from adjacent circuits.

All electronic systems are composed of a few basic types of electronic components. Voltage sources supply power to electronic circuits. Fuses and circuit breakers protect other components from damaging current. Resistors limit the current in a circuit. Their value is sometimes indicated by colored bands. The physical size of a resistor is an indication of its power-handling ability. Most fixed-value resistors used in computer circuitry today utilize surface-mount packages. Variable resistors called rheostats or potentiometers have two or three terminals, respectively. Capacitors and inductors store electrical energy in electrostatic and electromagnetic fields, respectively. A transformer is used to increase or decrease certain types of voltages. It consists of multiple inductors with interacting fields.

Switches are used to open (break) and close (make) a path for current. There are many types of actuators such as push-button, toggle, slide, and rotary. There are also many different arrangements of electrical contacts such as SPST, SPDT, and DPDT. A relay is essentially an electromagnetically-operated switch. Current in one circuit energizes an electromagnet that causes switch contacts to change in another circuit.

Transistors and diodes are solid-state components used for many applications. Integrated circuits are complete functional systems in a single, multi-pin package.

•—REVIEW QUESTIONS

Section 3.1: Technical Notation

1. The expression $\times 10^{-3}$ is the same as multiplying by 1,000. (True/False)
2. Arrange the following powers of ten in order of magnitudes, beginning with the smallest: $\times 10^{-9}$, $\times 10^{6}$, $\times 10^{3}$, and $\times 10^{-3}$.

3. The number 54,700 can also be written as 54.7 × 10$^?$.
4. The number 0.0056 can also be written as _____ × 10^{-4}.
5. Express 3,400 in engineering notation (with base number between 1 and 999).
6. Express 0.00000091 in engineering notation (with base number between 1 and 999).
7. Express 1,250,000,000 in engineering notation (with base number between 1 and 999).
8. Express the number 0.551 × 10^3 in engineering notation (with base number between 1 and 999).
9. Convert 3.65 × 10^{-2} to engineering notation (with base number between 1 and 999).
10. Convert 100 × 10^5 to engineering notation (with base number between 1 and 999).
11. A capacitance of 25 × 10^{-12} F can also be written as _____ using a metric prefix.
12. A frequency of 1,750,000,000 Hz can be expressed as _____ using a metric prefix.
13. Express an inductance 390 × 10^{-9} H using a metric prefix.
14. Express a resistance of 4 million ohms using a metric prefix.
15. Convert 0.05 µA to _____ nA.
16. A voltage of 3,750 millivolts is the same as _____ volts.
17. Arrange the following resistances in order based on their values, beginning with the smallest: 2.5 MΩ, 75 mΩ, 1,200 kΩ.
18. A current of 1,200 amperes could be expressed in either kiloamperes or milliamperes. Which would produce the most readable result?
19. A resistance of 56,000,000 ohms could be expressed in either kilohms or megohms. Which would produce the most readable result?
20. Which metric prefix provides the most readable choice for a voltage value of 0.05 volts?

Section 3.2: Wire and Cable

21. An ideal conductor or wire has _____ resistance.
22. Practical wires have (*low/high*) resistance.
23. The resistance of a wire is partially determined by the type of material used to make the wire. (True/False)
24. The resistance of copper wire (*increases/decreases*) as temperature rises.
25. As a certain type of wire is made longer, its resistance (*increases/decreases*).
26. If all other factors are the same, then a wire with a larger cross-sectional area will have (*less/more*) resistance.
27. Assuming all other factors are the same, arrange the following wires in order of their resistance, beginning with the lowest resistance: 10 ga, 30 ga, 12 ga, and 18 ga.
28. Gold is a better conductor than annealed copper. (True/False)
29. What is the circular mil area of a wire that has a diameter of 0.035 inches?
30. If the resistance of each wire in a 50′ communication cable that is used to connect two computers is 1,500 milliohms, what is the value of resistance if the cable length is reduced to 20′?
31. Which of the following copper wires has the greatest ampacity: 10 ga, 12 ga, 24 ga, or 30 ga?
32. Stranded wires are less easily damaged by repeated flexing than solid-conductor wires. (True/False)
33. When two or more individually insulated wires are bundled in a common outer jacket, the bundled assembly is called a _____.
34. Coaxial cables are useful for protecting low-level signals from external electromagnetic interference. (True/False)
35. What is the name of the cable that consists of several individually insulated conductors molded into a flat web or strap?

Section 3.3: Electronic Components

36. Two or more interconnected cells are called a _____.
37. A fuse will be destroyed and must be replaced if an overcurrent condition occurs. (True/False)
38. What type of fuse is designed to withstand short-term current surges that exceed their current ratings?
39. Power ratings of $\frac{1}{10}$, $\frac{1}{8}$, and $\frac{1}{4}$ watt are common for resistors. (True/False)
40. A certain resistor is marked as 1,000 ohms with a 1% tolerance. Compute the maximum deviation and the range of possible resistance values.
41. Determine the highest and lowest resistor value that a 560-Ω resistor can have if it has a 5% tolerance.
42. If a resistor that was marked as 2.7 MΩ, 10% actually measured 2.916 MΩ, would it be within tolerance?
43. What is the preferred resistor technology if small physical size is the primary concern?
44. What is the preferred resistor technology if a high power rating is required?
45. What are the resistance value and tolerance of a resistor with the following color stripes: Orange, White, Red, Gold?
46. What are the resistance value and tolerance of a resistor with the following color stripes: Gray, Red, Brown, Silver?
47. What are the resistance value and tolerance of a resistor with the following color stripes: Red, Red, Green, Red, Violet?

48. What is the value of a surface-mount resistor with the following markings: 104?
49. A potentiometer and a rheostat have _____ and _____ electrical connections, respectively.
50. How many electrical connections would you expect to find on an SPDT, push-button switch?

◆ CIRCUIT EXPLORATION

Relays are widely used in today's electronic systems. Although they are simple in principle, understanding relay circuits can be challenging. The following exercises provide you with an opportunity to practice analyzing and explaining the operation of unfamiliar relay circuits.

The material presented in this chapter and the following list of pointers are all you really need to successfully analyze relay circuits:

- Whenever current flows through a relay coil, the contacts change state from their normal position (i.e., the position shown on the schematic).
- Current will flow in an electrical circuit whenever there is a complete and unbroken (e.g., no open switches in line with the current) path that includes a voltage source.
- Current can flow through closed-switch or relay contacts.
- Current cannot flow through opened-switch or relay contacts.
- A lamp will illuminate if it has current through it.

Problem 1

Construct the circuit shown in Figure 3–38 (either in the lab or in a circuit simulation environment), and then answer the following questions:

1. What is the condition of the relay (i.e., energized or deenergized) when S_1 is in its normal position?
2. What is the condition of the two lamps (i.e., On or Off) when S_1 is in its normal position?
3. What is the state of the relay (i.e., energized or deenergized) when S_1 is depressed?
4. What is the condition of the two lamps (i.e., On or Off) when S_1 is depressed?

Problem 2

Construct the relay circuit shown in Figure 3–39 (either in the lab or in a circuit simulation environment). Assume that S_2 has just been pressed and released, then respond to the following questions:

5. What is the state of the relay?

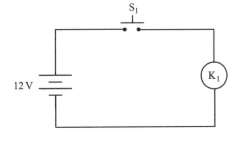

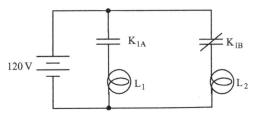

FIGURE 3–38 Analyze this relay circuit.

Fig03_38.msm

6. What is the condition of the lamp (i.e., On or Off) when S_1 is in its normal position?

Now, assume S_1 is pressed and held.

7. What is the condition of the lamp (i.e., On or Off)?
8. Are the K_{1A} contacts open or closed at this time?

Now, assume S_1 is released.

9. What is the condition of the lamp (i.e., On or Off)?
10. Are the K_{1A} contacts open or closed at this time?

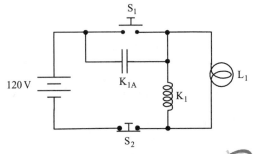

FIGURE 3–39 Analyze this relay circuit.

Fig03_39.msm

ANSWERS TO PRACTICE PROBLEMS

Page 51

1. 5000×10^0
2. 10060×10^{-6}
3. $0.00000001006 \times 10^6$
4. -2.47×10^5

Page 52
1. 25×10^{-6} converts to 0.025×10^{-3}.
2. 0.001×10^{-6} converts to 1000×10^{-12}.
3. 105×10^3 converts to $105{,}000 \times 10^0$.
4. 0.045×10^9 converts to 45×10^6.
5. 0.2×10^6 converts to 200×10^3.

Page 54
1. 56 kΩ
2. 2.2 mA
3. 5000 kW
4. 12 μs
5. 0.175 GHz

Page 55
1. 50 mA
2. 550 kV
3. 3.3 ohms
4. 10 GHz
5. 0.025 km

Page 69
1. maximum deviation: 3.3 Ω
   range: 326.7 Ω to 333.3 Ω
2. 20.9 kΩ and 23.1 kΩ
3. Yes.

Page 72
1. a. 27 kΩ, ±10%
   b. 390 Ω, ±5%
   c. 1.5 kΩ, ±5%
   d. 680 kΩ, ±10%

CHAPTER 4
Circuits

•—KEY TERMS
branch
complex circuit
ground
ground plane
Kirchhoff's Current Law
Kirchhoff's Voltage Law
open circuit
parallel circuit
series circuit
series-parallel circuit
voltage divider equation
voltage drop

•—OBJECTIVES
After studying this chapter, you should be able to:
- State the requirements for current flow.
- Positively identify each of the following types of circuit configurations: series, parallel, series-parallel, and complex.
- State intuitive relationships regarding the circuit values of each of the following types of circuits: series, parallel, and series-parallel.
- Calculate circuit values of resistance, voltage, current, and power for each of the following circuit types: series, parallel, and series-parallel.
- Describe the effects of series and parallel connections of voltage sources.
- Discuss and utilize ground and other reference points.

•—INTRODUCTION
Just as all electrical circuits are composed of a few basic component types, there are only four basic ways to interconnect the various components in an electrical circuit. You must be able to analyze electrical and electronic circuits to determine the expected values of circuit quantities such as voltage, current, and resistance. You must also be able to measure the actual circuit quantities with electronic test equipment. Finally, you must be able to apply the principles of basic electronics theory to account for any differences between the expected and the actual circuit values.

In this chapter we will examine the four fundamental configurations of electrical circuits and learn how to identify each form. Additionally, we will study relationships between voltage, current, resistance, and power, both intuitively and mathematically.

Finally, we will discuss the important topic of circuit references, including the concept of ground. The material presented in this chapter forms an essential part of the foundation for your career in computers and electronics in general.

4.1 BASIC REQUIREMENTS FOR CURRENT

Before you can trace current or measure circuit values, it is important to be able to identify a circuit capable of having current flow. The term current flow, as used in this chapter, refers to *sustained* current flow. Momentary or transient currents are considered in a later chapter. There are two basic requirements for sustained current in a circuit: electromotive force and a complete path for current.

Voltage Source

You will recall from chapter 2 that electromotive force is a difference in potential that does not decay as charges are transferred. In most cases, we are interested in the electrons that leave the negative side of the electromotive force (power source) and flow through the external circuit. As the electrons return from their trip through the circuit, they enter the positive side of the power source. Internal to the power source, electrons are moved from the positive side of the source to the negative side. This requires an expenditure of energy. This energy may be provided chemically, as in the case of a battery, or perhaps from energy supplied by the power company.

The electromotive force maintains a potential difference and provides the electrical pressure that causes electrons to flow through the circuit. Ohm's Law describes the relationship of applied voltage and current flow as $I = V/R$. If there were no electromotive force in a circuit, then the V in the Ohm's Law equation would be zero. This would produce a resulting current flow of zero regardless of the value of resistance in the circuit. Figure 4–1 illustrates the requirement for electromotive force. This figure also provides a hydraulic analogy.

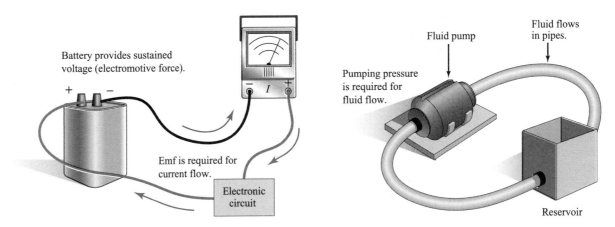

FIGURE 4–1 A circuit must have an electromotive force (emf) to have current. Emf in an electrical circuit is similar to pressure in a hydraulic circuit.

Complete Path for Current

The second requirement for current flow is a complete path or closed loop between the two terminals of the electromotive force. For most purposes, you can consider an **open circuit** to have infinite resistance. In the hydraulic analogy, this is equivalent to a blockage in the pipe. Thus, if there is any open in a circuit, the current flow will be essentially zero, since no amount of electromotive force can cause current to flow through an infinite opposition.

•EXAMPLE 4.1

Which of the circuits shown in Figure 4–2 have complete paths for current flow?

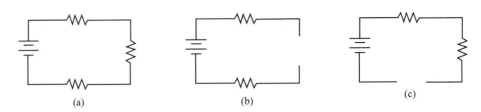

FIGURE 4–2 Which circuits have complete current paths?

SOLUTION If we begin at the negative battery terminal and proceed toward the positive terminal in Figure 4–2(a), we find that we can complete the entire journey. Therefore, Figure 4–2(a) has a complete path for current.

As we progress around the circuits shown in Figures 4–2(b) and (c), we encounter an open circuit. No current can flow through an open circuit; there is no complete path for current.

•—Practice Problem

1. Which of the circuits in Figure 4–3 have complete paths for current flow?

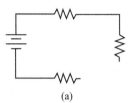 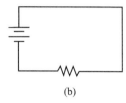 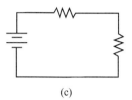

(a)   (b)   (c)

FIGURE 4–3 Which circuits have complete paths for current flow?

Direction of Current

As discussed in chapter 2, electron current moves from negative to positive since electrons are negatively charged particles. We are now ready to expand this concept to include *relative* differences in potential.

Figure 4–4 shows an electrical circuit consisting of a voltage source and a resistor. We know that current leaves the negative side of the voltage source, flows through the resistor (from point B to point A), and returns to the positive side of the voltage source. In this figure, we can see that point A is positive and point B is negative since they are connected to the positive and negative terminals, respectively, of the voltage source. We can say that point A is positive with respect to point B. We could just as well say that point B is negative with respect to point A. Either of these points can be positive or negative with respect to some other point, but as long as they have the indicated polarity, the current will flow from point B to point A.

Figure 4–5(a) shows an electromotive source with four different voltage terminals. All of the voltages are measured with reference to the common line. If we connect a resistor between the +5-volt output and common, as shown in Figure 4–5(b), then we expect current to flow from point B to point A. Similarly, if

FIGURE 4–4 A simple electrical circuit.

FIGURE 4–5 Current flows from a *relatively* negative potential to a point that is *relatively* positive.

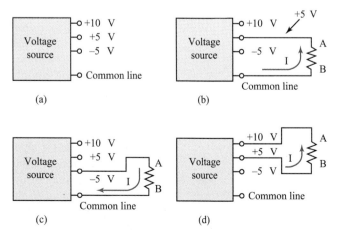

we connect a resistor between the −5-volt output and common, as shown in Figure 4–5(c), we expect current to flow from point A to point B since point A is now the most negative. Finally, Figure 4–5(d) illustrates a less obvious possibility. If we connect a resistor between the +10- and +5-volt outputs, as shown in Figure 4–5(d), then current will flow from point B to point A. In this case, point A is five volts more positive than point B. The current through the resistor will be the same as if it had a simple 5-volt source across it. This is a very important concept. Be sure you understand it.

Figure 4–6 further illustrates this idea. Here, a scale of electrical potentials is compared to the rungs on a ladder. For example, +45 volts is negative with respect to +50 volts and positive with respect to +30 volts, just as eight feet above ground is low with respect to a point which is twelve feet above ground, but high with respect to a point which is two feet above ground.

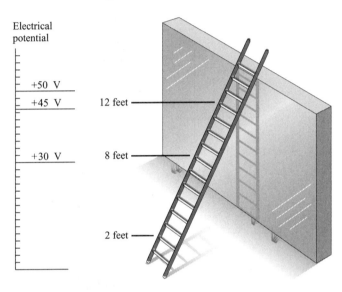

FIGURE 4–6 An analogy to explain *relative* polarities.

✓ Exercise Problems 4.1

1. What are the two primary requirements in order to have sustained current flow in a circuit?
2. No current can flow through an open circuit. (True/False)
3. If a current path exists between two points having potentials of −2 V and +5 V with respect to a third point, will current flow between the two points? If yes, which way will current flow?

4.2 SERIES CIRCUITS

It is essential that you develop a high level of expertise in the analysis and intuitive understanding of series circuits. Not only are these skills important for the solution of practical problems, they also form the basis for understanding and analyzing more complex circuit configurations that are more common in practical applications.

Positive Identification

A **series circuit** is characterized as having only a single path for current flow. Individual components may be connected in series with each other, or the entire circuit may be a series circuit. In order for a circuit to be classified as a series circuit, every component in the circuit must be connected such that there is only a single path for current in the entire circuit.

Series Components

In order for two components to be in series, there must be one and only one current path that includes both components. Current must enter the first component, flow through both components, and then exit the last component without encountering any branches in the circuit. A **branch** is a point where the current can divide into two or more paths.

•—EXAMPLE 4.2

Which of the components shown in Figure 4–7 are in series?

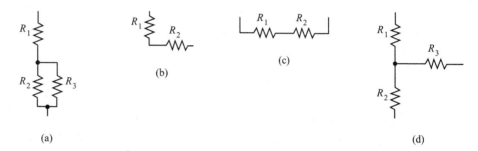

FIGURE 4–7 Which of these components are connected in series with each other?

SOLUTION If we trace through R_1 in Figure 4–7(a), we encounter a branch at the junction of R_1, R_2, and R_3. Therefore, these components are not in series. There are no branches in Figures 4–7(b) or (c), which means that the components in these circuits are in series. Finally, we can see that the circuit in Figure 4–7(d) has a branch where R_1, R_2, and R_3 join. Thus, the components in Figure 4–7(d) are not in series.

•—Practice Problem

1. Which of the components shown in Figure 4–8 are connected in series?

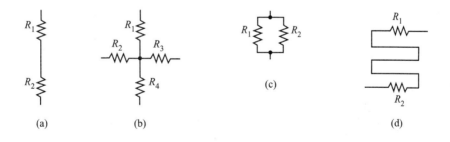

FIGURE 4–8 Which of these components are in series with each other?

Series Circuits

If every electron that leaves the negative terminal of the power source has only one path for current and flows through every component before returning to the positive side of the power source, then the circuit is a series circuit. Every component in a series circuit is in series with every other component in the sense that they are part of the common, single current path.

•EXAMPLE 4.3

Which of the circuits shown in Figure 4-9 can be classified as series circuits?

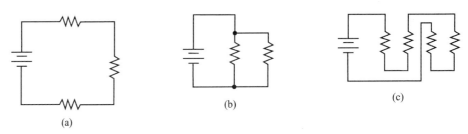

FIGURE 4-9 Which of these circuits can be classified as series circuits?

SOLUTION If we begin at the negative side of the power source in Figure 4-9(a) and move through the circuit toward the positive side of the battery, we find that there are no alternate routes. That is, there is only one path for current, and it includes every component. Therefore, the circuit in Figure 4-9(a) is a series circuit.

If we trace the current path for the circuit shown in Figure 4-9(b), we find we have a branch where the two resistors connect. We could complete our path to the positive side of the battery by passing through either resistor. Since the circuit has more than one path for current, it does not qualify as a series circuit.

If we begin at the negative battery terminal in Figure 4-9(c), we find that we can trace through the entire circuit before we arrive at the positive supply terminal. Since there was only a single path for current flow, we can classify Figure 4-9(c) as a series circuit.

•—Practice Problem

1. Which of the circuits shown in Figure 4-10 can be classified as series circuits?

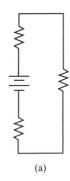
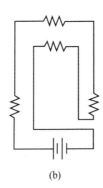
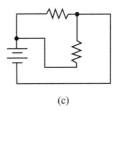

FIGURE 4-10 Which of these circuits can be classified as series circuits?

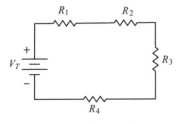

FIGURE 4-11 A basic series circuit.

Intuitive Relationships

Let us examine the series circuit in Figure 4-11 and make some observations that will help us later during mathematical analyses of series circuits. In many technical positions, you will use your intuitive understanding of electronics more often than your ability to calculate circuit values.

All Currents Are Equal

First, observe that there is only one possible path for current. This, of course, is a basic requirement for a series circuit. However, since there is only one current path, it follows that the value of current must be the same at all points in the circuit. For example, if we know how much current is flowing through one resistor in a series circuit, we immediately know how much flows through every other resistor in the same circuit. Likewise, we know how much current must be flowing through the voltage source.

Total Resistance Is Greater than Any One Resistance
We know that resistance offers opposition to current. It follows that each time we add another resistance to a series circuit, the total opposition to current in the circuit must increase. Therefore, we know intuitively that the total resistance in the circuit is always greater than any individual resistance. Of course, in the special case where there is only one resistor in the circuit, you could reason that total resistance is the same as the value of the single resistor.

Total Power Is Greater than Any One Component Dissipation
Recall that electrical power in a resistor is dissipated as heat. All of the power that is dissipated in any circuit must be supplied by the voltage source, since no individual component can create energy on its own. So, if each resistor dissipates a certain amount of power and all of the power must come from the source, it follows that total power (the power supplied by the source) must be greater than the dissipation of any individual component.

Larger Resistances Have Higher Voltage Drops
We know that it takes a certain amount of voltage to cause a particular current to flow through a given resistance (Ohm's Law). In the case of a series circuit, all components have the same current. It follows that the voltages across the various components may be different depending on their individual values of resistance. Those with more resistance must have more voltage in order to have the same current as the other components. We refer to the voltage across the terminals of a component as a **voltage drop**.

Total Voltage Is Greater than Any One Component Voltage
In a previous paragraph, we saw that total resistance in a series circuit is greater than any individual resistance in the circuit. We also noted that voltage drops are proportional to resistance values in a series circuit. Since the total resistance in a series circuit is greater than any individual resistance, and since current is the same, it follows that the total voltage must be greater than any one component voltage drop.

Mathematical Relationships

The intuitive relationships discussed in the preceding paragraphs provide the foundation for more detailed mathematical analyses of circuits. Additionally, the intuitive concepts allow you to catch calculation errors. Say, for example, you analyzed a series circuit mathematically and determined (wrongly) that the voltage drop across R_1 was greater than the voltage drop across R_2 even though R_2 was the larger resistor. You would know intuitively that this could not be correct.

Current
We have already seen that the current is the same in all parts of a series circuit. We can express this mathematically with Equation 4–1:

$$I_A = I_1 = I_2 \ldots = I_N \tag{4-1}$$

where I_A is the applied or total current and I_1, I_2, and I_N are individual component currents. The "N" subscript on the final current (I_N) implies that this general expression can be extended indefinitely.

Resistance

We know that each resistor in a series circuit contributes to total resistance. We express this mathematically with Equation 4–2:

$$R_T = R_1 + R_2 \ldots + R_N \qquad (4\text{–}2)$$

●─ **EXAMPLE 4.4**

Calculate the total resistance for the circuit shown in Figure 4–12.

SOLUTION We apply Equation 4–2 as follows:

$$R_T = R_1 + R_2 + R_3$$
$$= 2.7 \text{ k}\Omega + 3.9 \text{ k}\Omega + 4.7 \text{ k}\Omega = 11.3 \text{ k}\Omega$$

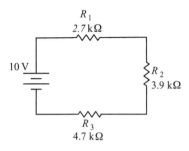

FIGURE 4–12 Calculate total circuit resistance.

Fig04_12.msm

Calculator Application

●─ **EXAMPLE 4.5**

What is the value of R_2 in Figure 4–13?

SOLUTION We apply Equation 4–2 as follows:

$$R_T = R_1 + R_2 + R_3, \text{ so}$$
$$R_2 = R_T - R_1 - R_3$$
$$= 810 \text{ }\Omega - 150 \text{ }\Omega - 270 \text{ }\Omega = 390 \text{ }\Omega$$

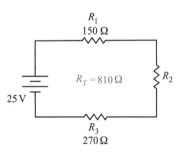

FIGURE 4–13 Find the value of R_2.

Voltage

We know that total or applied voltage is greater than any individual component voltage drop in a series circuit. We also know that the various component voltage drops are proportional to the resistance values (i.e., larger resistances have larger voltage drops). These lead to a mathematical relationship for the voltages in a series circuit that is expressed with Equation 4–3.

$$V_A = V_1 + V_2 \ldots + V_N \qquad (4\text{–}3)$$

This can be expressed verbally as

> ▶ **KIRCHHOFF'S VOLTAGE LAW** The applied voltage in a closed loop is equal to the sum of the component voltage drops.

Of course, a series circuit is a closed loop as referenced in the above rule, but application of this rule extends to include *any* circuit configuration. This important relationship is called **Kirchhoff's Voltage Law**, and it represents a fundamental principle of electronics. It is *essential* for you to understand this law in order to be proficient at analyzing and understanding unfamiliar circuits. Kirchhoff's Voltage Law can also be a powerful troubleshooting aid. Let's explore it further.

Polarity of Voltage Drops When current flows through a resistor, there is a corresponding voltage drop across that resistor. Figure 4–14 shows the polarity of voltage drops in a series circuit. As shown in Figure 4–14, the resistor voltage is negative on the end where electron current enters the resistor. The end where electron current exits the resistor will be the more positive of the two ends.

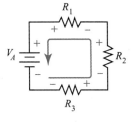

FIGURE 4–14 The polarity of voltage drops in a series circuit.

•–EXAMPLE 4.6

Label the polarities of the voltage drops on the resistors shown in Figure 4–15.

FIGURE 4–15 Label the resistor polarities.

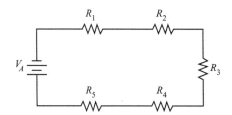

SOLUTION Figure 4–16 shows the polarity for each of the resistors originally shown in Figure 4–15. First, we determine the direction of current flow. It leaves the negative side of the power source, flows through the circuit, and returns to the positive side of the power source. Next, we label each resistor such that the negative side of its voltage drop is on the end of the resistor where the current enters the resistor. The end where the current exits is labeled as the positive side of the voltage drop.

FIGURE 4–16 The polarity of each resistor originally shown in Figure 4–15.

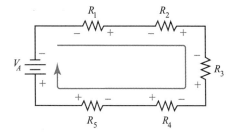

Closed-Loop Equations Although mathematically similar to Equation 4–3, Kirchhoff's Voltage Law is often expressed in another way. Verbally, we can state the relationship as follows:

> ➤ **KIRCHHOFF'S VOLTAGE LAW** The algebraic sum of all the voltage drops and all the voltage sources in *any* closed loop equals zero.

The following sequential procedure can be used to produce a closed-loop equation to describe the circuit:

1. Label the polarity of all voltage sources and all voltage drops.
2. Start at any point in the circuit and write the voltages (including polarity) as you progress around the loop in either direction. Stop when you have completed the loop.

Either the polarity of the voltage sources will be given, or you can tell the polarity from the voltage source symbol (recall that the shortest line on the symbol for a voltage source is the negative terminal). The polarity of the voltage drops is determined by the direction of current flow. Recall that the negative end of a voltage drop will be the point where current enters the component.

Step two is the actual writing of the equation. The recommended method is stated as follows: *Move around the loop in either direction and write the voltages with the polarity that is on the exit end of the component.* Some people prefer to use the polarity of the voltage drop that is on the end where you enter the component. Although this second method will produce the same mathematical answer for the type of problem currently being discussed, it will handicap us for other techniques discussed at a later point in the text.

•—EXAMPLE 4.7

Use Kirchhoff's Voltage Law to determine whether the voltage drops shown on the circuit in Figure 4–17 are correct.

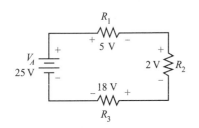

FIGURE 4–17 Are the voltage drops correct?

SOLUTION We begin (arbitrarily) at the negative side of the voltage source and write the Kirchhoff's Voltage Law equation for the closed circuit loop as we move clockwise (arbitrarily):

$$V_A - V_1 - V_2 - V_3 = 0$$
$$25\text{ V} - 5\text{ V} - 2\text{ V} - 18\text{ V} = 0$$
$$25\text{ V} - 25\text{ V} = 0$$

Since this latter statement is true, we can conclude that the labeled voltages are most likely correct. If the equation were invalid (e.g., 5 = 9), then we would know for sure that an error had been made.

•—EXAMPLE 4.8

Apply Kirchhoff's Voltage Law to determine whether the voltages shown in Figure 4–18 are labeled correctly.

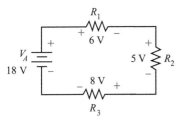

FIGURE 4–18 Are the voltage drops correctly labeled?

SOLUTION Let us arbitrarily begin at the negative side of the voltage source and move in a clockwise direction. Remember, you can start at any point and move in either direction. The closed-loop equation is written as

$$V_A - V_1 - V_2 - V_3 = 0$$
$$18\text{ V} - 6\text{ V} - 5\text{ V} - 8\text{ V} = 0$$
$$18\text{ V} - 19\text{ V} = 0 \quad [\text{error}]$$

Since this last equation is clearly erroneous, we can safely say that the voltages labeled on Figure 4–18 are incorrect. We don't know which one is wrong, but there is *definitely* an error.

Voltage Divider Equation Recall from a previous discussion that the voltage drops across the various resistors in a series resistive circuit are proportional to the resistance values. This allows us to express another useful relationship as Equation 4–4. This is often called the **voltage divider equation**.

$$V_X = \left(\frac{R_X}{R_T}\right) V_A \qquad (4\text{–}4)$$

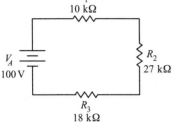

FIGURE 4–19 What is the voltage drop across R_2?

Fig04_19.msm

EXAMPLE 4.9

Calculate the voltage drop across R_2 in Figure 4–19.

SOLUTION We apply Equation 4–4 as follows:

$$V_2 = \left(\frac{R_2}{R_T}\right)V_A$$

$$= \left(\frac{27\text{ k}\Omega}{55\text{ k}\Omega}\right)100\text{ V} = 49.1\text{ V}$$

Calculator Application

SYSTEM PERSPECTIVE: Series Circuits

Oscilloscopes are one of the most valuable and versatile types of test equipment used to diagnose defects in electronic systems. In short, they display a graph of voltage (vertical axis) versus time (horizontal axis). Every oscilloscope has definite limits on the value of voltage that can be measured without risking damage to the instrument. However, many circuit voltages (e.g., the high voltage used inside video displays) exceeds the capability of the oscilloscope. To measure such a voltage, you use a special probe called a 10:1 probe. This effectively multiplies the voltage capability of the scope by ten. As shown in the adjacent figure, the internal resistance of an oscilloscope is typically one megohm. The 10:1 probe has a nine megohm series resistor. This divides the input voltage between the probe and the internal resistance of the scope, so that 90 percent of the measured voltage is dropped across the probe resistance. You may also find a need to use 100:1 or even 1000:1 probes, which work on the same principle but have correspondingly higher probe resistances.

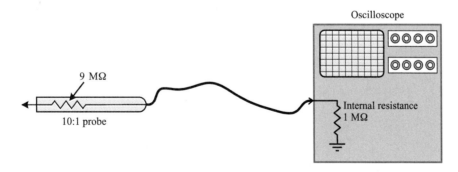

Power

Power in a resistive circuit can be thought of as heat dissipated by the resistors. The more power that is dissipated by a particular resistor, the hotter that resistor will be. All of the power that is dissipated by the various resistors in a circuit must be supplied by the source. A laser printer, for example, has a heating element that gets hot and fuses the toner onto the paper as a page is printed. Clearly, all of the power dissipated by the heating element (and the rest of the printer circuitry) must be supplied by the voltage source (the power company in this case).

To determine the total power supplied by a voltage source, we simply sum the individual power dissipations in the circuit as indicated by Equation 4–5.

$$P_T = P_1 + P_2 \ldots + P_N \tag{4-5}$$

•EXAMPLE 4.10

Calculate the power that must be supplied by the source in Figure 4–20.

SOLUTION There are many ways to solve this problem. Here is one approach: Compute the power dissipated in each resistor and then apply Equation 4–5. When analyzing a series circuit, it is always good to calculate total current since that is common to all components. We might, for example, decide to use Ohm's Law as follows:

$$I_T = \frac{V_A}{R_T}$$

The supply voltage V_A is given in the problem, but total resistance R_T must be computed before we can use this equation. So, let's compute total resistance.

$$R_T = R_1 + R_2$$
$$= 100\ \Omega + 330\ \Omega = 430\ \Omega$$

We can now continue with our calculation of total current.

$$I_T = \frac{V_A}{R_T}$$
$$= \frac{10\ \text{V}}{430\ \Omega} \approx 23.26\ \text{mA}$$

Now we are able to compute the power in each resistor by applying one of the power equations discussed in chapter 2.

$$P_1 = I_1^2 R_1$$
$$= (23.26\ \text{mA})^2 \times 100\ \Omega = 54.1\ \text{mW}$$

Similarly,

$$P_2 = I_2^2 R_2$$
$$= (23.26\ \text{mA})^2 \times 330\ \Omega = 178.54\ \text{mW}$$

Finally, we apply Equation 4–5 as follows:

$$P_T = P_1 + P_2$$
$$= 54.1\ \text{mW} + 178.54\ \text{mW} = 232.64\ \text{mW}$$

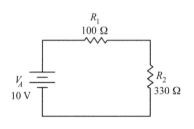

FIGURE 4–20 Compute power supplied by the source.

Fig04_20.msm

Relative Values of Series Resistance

Let us now make an observation about series circuits that will lead to a very practical rule, which will allow you to greatly simplify some problems you are likely to encounter in industry. We will consider two series resistors, but the concept can be extended to include any number of resistors in a series circuit. Since total resistance is equal to the sum of the individual resistances in a series circuit, if one resistor is much larger than the other(s), then it will dominate the calculation for total resistance. That is, total resistance will be nearly the same as the large resistance alone. For example, if a 1.0-MΩ resistor and a 1.0-Ω resistor are in series, the mathematical result for total resistance is 1,000,001 Ω, which is nearly the same as the larger resistor alone. The greater the difference between the two resistor values, the better this rule works. As a general guideline, you can often apply this rule if one resistor is at least 100 times (and sometimes as little as 10 times) the other.

Series Voltage Sources

Many electronic circuits require more than one voltage source for proper operation. A typical computer, for example, might require +3-volt, +5-volt, +12-volt, and −12-volt power sources. Let us now consider the effects of multiple voltage sources that are connected as part of a series circuit.

Series-Aiding Voltage Sources

Figure 4–21 shows two voltage sources that are connected in the same polarity. They are said to be series-aiding, since both voltage sources cause current flow in the same direction. The effective voltage in the circuit is the sum of the series-aiding voltages (Equation 4–6).

$$V_T = V_{A_1} + V_{A_2} \ldots + V_{A_N} \quad (4-6)$$

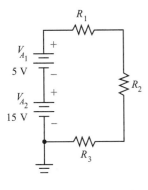

FIGURE 4–21 Series-aiding voltage sources.

•—EXAMPLE 4.11

What is the effective applied voltage for the circuit shown in Figure 4–21?

SOLUTION First we note that both voltage sources are connected in the same polarity; they are series-aiding. The total voltage is the sum of the two individual sources (Equation 4–6):

$$V_T = V_{A_1} + V_{A_2}$$
$$= +5\text{ V} + 15\text{ V} = +20\text{ V}$$

Series-Opposing Voltage Sources

Figure 4–22 shows two voltage sources that are connected in the opposite polarity. That is, they try to cause current to flow in opposite directions. This configuration of voltage sources is called series-opposing. The effective voltage of two series-opposing voltage sources is the difference between the two individual sources (Equation 4–7).

$$V_T = V_{A_1} - V_{A_2} \quad (4-7)$$

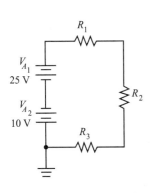

FIGURE 4–22 Series-opposing voltage sources.

The polarity of the net voltage will be the same as the larger of the two opposing sources. The direction of current flow is determined by the relative magnitudes of the series voltage sources. Current flows in the direction determined by the polarity of the net effective voltage.

•—EXAMPLE 4.12

Calculate the effective voltage for the circuit in Figure 4–22 and determine the direction of current flow.

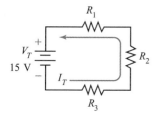

FIGURE 4–23 An equivalent circuit for the circuit shown in Figure 4–22.

SOLUTION The effective voltage is computed with Equation 4–7 as follows:

$$V_T = V_{A_1} - V_{A_2}$$
$$= 25 \text{ V} - 10 \text{ V} = 15 \text{ V}$$

The polarity of the net voltage is the same as the polarity of V_{A_1}. Therefore, the current will flow in the direction determined by V_{A_1}. In the present case, the current will flow in a counterclockwise direction. Figure 4–23 shows an equivalent circuit.

Complex Voltage Sources

It is also important for you to be able to analyze circuits that involve combinations of series-aiding and series-opposing voltage sources. A circuit that has both series-aiding and series-opposing voltage sources is said to have a complex voltage source. Figure 4–24 shows a circuit with a complex voltage source.

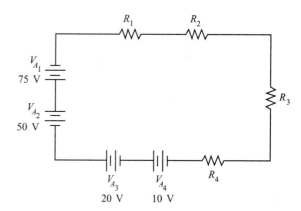

FIGURE 4–24 A complex voltage source has both series-aiding and series-opposing voltage sources.

The effective applied voltage for a complex voltage source configuration is the algebraic sum of the individual voltage sources as indicated by Equation 4–8:

$$V_T = \pm V_{A_1} \pm V_{A_2} ... \pm V_{A_N} \qquad (4\text{–}8)$$

The sign (plus or minus) for a given polarity of voltage source is essentially arbitrary as long as we apply our choice consistently throughout the problem. It should also be noted that the voltage sources may be distributed around the series circuit. That is, the sources do not have to appear immediately adjacent to each other on the schematic diagram.

EXAMPLE 4.13

What is the effective circuit voltage for the circuit in Figure 4–24?

SOLUTION We apply Equation 4–8 as follows, arbitrarily choosing the polarity of V_{A_1} to be positive.

$$V_T = +V_{A_1} - V_{A_2} - V_{A_3} - V_{A_4}$$
$$= +75 \text{ V} - 50 \text{ V} - 20 \text{ V} - 10 \text{ V} = -5 \text{ V}$$

Calculator Application | 7 | 5 | – | 5 | 0 | – | 2 | 0 | – | 1 | 0 | = |

Since the result is negative, it merely means the effective voltage is opposite the polarity of V_{A_1}, which was initially assigned a positive value.

✓ Exercise Problems 4.2

1. How can a series circuit be identified with regard to current flow?
2. If two components are connected in series, what percentage of the current flowing through the first component also flows through the second?
3. The largest resistor in a series circuit drops the most voltage. (True/False)
4. A 10-kΩ resistor and a 22-kΩ resistor are connected in series with a voltage source. The 10-kΩ resistor has 800 µA flowing through it. Calculate the current through the 22-kΩ resistor.
5. A series circuit consists of a 10-V battery and six 5-Ω resistors. What is the total resistance in the circuit?
6. Analyze the circuit shown in Figure 4–25 and determine if the voltage drops (both polarity and value) are correctly labeled on each resistor.

FIGURE 4–25 Are the voltage drops correct?

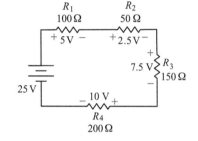

7. Calculate the voltage drop across R_2 in Figure 4–26.
8. Refer to Figure 4–27 and complete the following table.

FIGURE 4–26 What is the voltage drop across R_2?

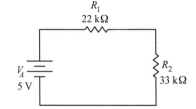

FIGURE 4–27 A series circuit for analysis.

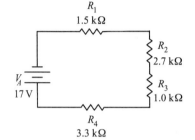

Component	Resistance	Voltage	Current	Power
R_1				
R_2				
R_3				
R_4				
Total				

4.3 PARALLEL CIRCUITS

Parallel components and **parallel circuits** establish another fundamental way to connect electrical and electronic devices where the components are connected across one another. Most of the electrical lights and appliances in your home, for example, are connected in parallel when they are turned on. This is also true for the electrical and electronic devices in an automobile. Parallel circuits have both similarities and differences when compared to series circuits. First, strive to understand and appreciate the similarities and differences intuitively. Then move on to numeric analysis of parallel circuits.

Positive Identification

A parallel circuit is characterized by having the same voltage across every component in the circuit. Sets of components may be connected in parallel, or the entire circuit may be connected as a parallel circuit.

Parallel Components

Two components are in parallel with each other if both ends of both components connect *directly* together. Since the components are connected directly across each other, they will inherently have the same voltage across them.

•—EXAMPLE 4.14

Which of the sets of components shown in Figure 4–28 are connected in parallel?

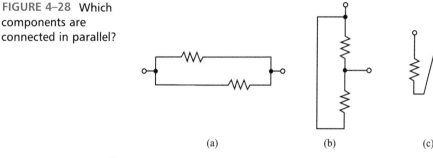

FIGURE 4–28 Which components are connected in parallel?

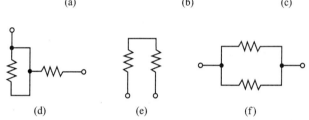

SOLUTION The ends of the resistors in Figure 4–28(a) connect directly together; therefore, they are in parallel. The resistors shown in Figure 4–28(b) are also connected in parallel, since both ends of both components connect directly together. We can see by inspection of Figure 4–28(c), that the resistors do *not* have both ends connected together. Therefore, these resistors are not in parallel. The components in Figures 4–28(d) and (e) do not have both ends connected and are not in parallel. Finally, the resistors shown in Figure 4–28(f) are connected in parallel, since both ends of both components are connected directly together.

Parallel Circuits

If every component in a circuit is connected directly across every other component in the circuit, then the entire circuit may be classed as a parallel circuit. Every component in a parallel circuit will have the same voltage across its terminals as every other component in the circuit.

•—EXAMPLE 4.15

Which of the circuits shown in Figure 4–29 can be classified as parallel circuits?

SOLUTION Since every component in Figure 4–29(a) is connected directly across every other component, we can classify the circuit as a parallel circuit. The circuit shown in Figure 4–29(b) is not a parallel circuit, because not all of the components are connected directly across each other. We can classify the circuit shown in Figure 4–29(c) as a parallel circuit, since all components are connected directly across each other. The circuits in Figures 4–29(d) and (f) are also parallel circuits for the same reason, although the parallel connections are more difficult to identify. Finally, not every component in the circuit shown in Figure 4–29(e) connects directly across every other component, so the circuit cannot be classified as a parallel circuit.

FIGURE 4–29 Which of these circuits are parallel circuits?

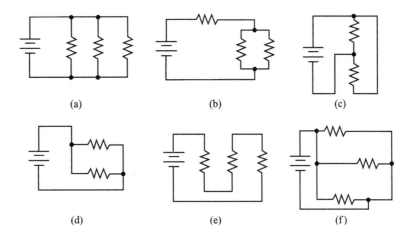

Intuitive Relationships

Let us carefully examine the parallel circuit shown in Figure 4–30 and make some intuitive observations. An intuitive understanding of circuits is a valuable tool for many situations, including its use as an aid to more involved numerical analyses.

FIGURE 4–30 A basic parallel circuit.

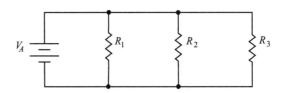

Same Voltage for All Components

Both ends of every component in Figure 4–30 are connected directly together. So, if you tried to measure the voltage across one of the components, you would be simultaneously measuring the voltage across every other component. It follows, then, that the voltage across any component in a parallel circuit is the same as the voltage across every other component.

Total Current Is Greater than Any One Branch Current

Since the voltage is the same across each branch of a parallel circuit, it should seem reasonable that the current through each branch will be determined by that common voltage and by the value of branch resistance (the exact value being determined by Ohm's Law). It is important to realize, however, that the current through any one branch is unaffected by the presence or even the value of any other branch. Figure 4–31 shows the paths for the various branch currents for the circuit originally shown in Figure 4–30.

As you can see in Figure 4–31, the total current that leaves the voltage source divides among the various branches. At the lower end of R_1, part of the total current flows through R_1 and the rest continues on to the R_2 and R_3 branches. Similarly, at the lower end of R_2, the current again divides. Part of it flows through R_2 and the rest continues on to R_3. The current that flows through R_3 merges with the

FIGURE 4–31 Current paths in a parallel circuit.

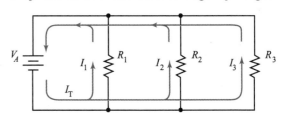

R_2 current at the top of R_2. These combined currents then merge with the current through the R_1 branch to once again form total current that returns to the voltage source.

Clearly, total current is a combination of the various branch currents. So, we know intuitively that total current must be greater than the current through any single branch.

Total Resistance Is Less than Any One Branch Resistance

We know that resistance is the opposition to current flow. We also know that total current in a parallel circuit is greater than any one branch current. Finally, we know that source voltage in a parallel circuit is the same as the voltage across every branch. Since source voltage is the same as branch voltage, but source (i.e., total) current is greater than any branch current, it follows that total resistance has to be less than branch resistance. In short, total resistance and all branch resistances have the same voltage (i.e., electrical pressure) to effectively cause current to flow. Since more current flows through the total resistance than through any one branch, there must be less opposition (resistance) to total current.

Total Power Is Greater than Any One Component Dissipation

We know that electrical power in a resistor is dissipated as heat, and that all power dissipated in any circuit must be supplied by the source. So, since each resistor dissipates a certain amount of power, and all of the power must come from the source, it follows that total power (the power delivered by the source) must be greater than the dissipation of any individual component.

Mathematical Relationships

We are now ready to analyze parallel circuits to calculate the various circuit values. We will rely heavily on Ohm's Law and our intuitive understanding of circuits. We will also utilize Kirchhoff's Current Law. This is a counterpart to Kirchhoff's Voltage Law that we used to analyze series circuits.

Resistance

We can compute the total resistance of several resistors connected in parallel by applying Equation 4–9.

$$R_T = \frac{1}{\frac{1}{R_1} + \frac{1}{R_2} \cdots + \frac{1}{R_N}} \tag{4–9}$$

This is often referred to as the reciprocal equation. It can be used to determine the total resistance for any number of parallel-connected resistors.

•–EXAMPLE 4.16

What is the total resistance if a 10-kΩ resistor and a 22-kΩ resistor are connected in parallel?

SOLUTION We apply Equation 4–9 as follows.

$$R_T = \frac{1}{\frac{1}{R_1} + \frac{1}{R_2}}$$

$$= \frac{1}{\frac{1}{10 \text{ k}\Omega} + \frac{1}{22 \text{ k}\Omega}} = 6.875 \text{ k}\Omega$$

Calculator Application

Note that the calculated total resistance value is smaller than either of the individual resistances. If you ever calculate a value larger than one of the branch resistors, your intuitive understanding of parallel circuits will allow you to catch the error.

•—Practice Problems

1. Calculate the total resistance if the following resistors are connected in parallel: 330 Ω, 470 Ω, and 560 Ω.
2. What is the total resistance if a 220-kΩ resistor is connected in parallel with a 680-kΩ resistor?
3. What is the total resistance of five 1.0-kΩ resistors connected in parallel?

There is a shortcut that can be used to compute the total resistance of several equal-valued resistors connected in parallel. Simply divide the common resistance value by the number of resistors. Problem 3 in the preceding practice problems, for example, could be solved as follows:

$$R_T = \frac{R_N}{N}$$

$$= \frac{1000 \text{ Ω}}{5} = 200 \text{ Ω}$$

Let us state this relationship formally as Equation 4–10.

$$R_T = \frac{R_N}{N} \qquad (4\text{–}10)$$

Equation 4–9 will work for any number of parallel resistors with differing values. Equation 4–10 will work for any number of like-valued resistors. Many times, however, you will have to determine the resistance of two parallel resistors. In this special case, Equation 4–11 can be used.

$$R_T = \frac{R_1 R_2}{R_1 + R_2} \qquad (4\text{–}11)$$

This product-over-the-sum formula is sometimes called a shortcut, but it is misnamed, since it actually takes more calculator keystrokes to apply Equation 4–11 than it does to use the reciprocal formula (Equation 4–9).

Current
Calculating the individual branch currents in a parallel circuit is just a matter of applying Ohm's Law ($I = V/R$), since the full value of applied voltage appears directly across each resistor.

Our intuitive understanding of parallel circuits tells us how the total current first divides into the various branch currents and then ultimately recombines to form total current again. That is, total current is composed of the various branch currents. We can express this relationship more precisely with Equation 4–12.

$$I_T = I_1 + I_2 \ldots + I_N \qquad (4\text{–}12)$$

• EXAMPLE 4.17

Calculate the branch currents and total current for the circuit shown in Figure 4–32.

FIGURE 4–32 Calculate the currents in this circuit.

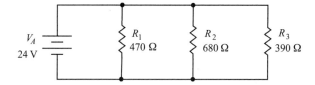

SOLUTION First, we can calculate each of the branch currents by applying Ohm's Law.

$$I_1 = \frac{V_1}{R_1} = \frac{24 \text{ V}}{470 \text{ }\Omega} = 51.06 \text{ mA}$$

$$I_2 = \frac{V_2}{R_2} = \frac{24 \text{ V}}{680 \text{ }\Omega} = 35.29 \text{ mA}$$

$$I_3 = \frac{V_3}{R_3} = \frac{24 \text{ V}}{390 \text{ }\Omega} = 61.54 \text{ mA}$$

Next, we compute total current by summing the branch currents according to Equation 4–12.

$$I_T = I_1 + I_2 + I_3$$
$$= 51.06 \text{ mA} + 35.29 \text{ mA} + 61.54 \text{ mA} = 147.89 \text{ mA}$$

Kirchhoff's Current Law provides another way of viewing the current relationships in a parallel circuit. The following is one way to state this important law:

> ▶ KIRCHHOFF'S CURRENT LAW The total current entering any point in a circuit must be equal to the total current leaving that point.

Figure 4–33 illustrates this principle through some examples. In all cases, the sum of the currents entering a particular point must be equal to the sum of the currents leaving that point.

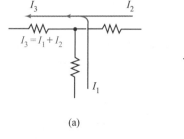
(a)

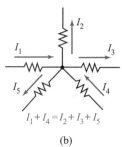
(b)

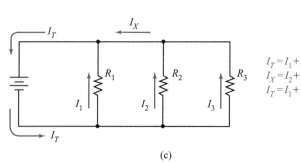
(c)

FIGURE 4–33 Examples to illustrate Kirchhoff's Current Law.

Voltage

The mathematical relationships between the various voltages in a parallel circuit are simple; all voltages are the same in a parallel circuit. Nonetheless, we can state the relationship formally with Equation 4–13.

$$V_A = V_1 = V_2 \ldots = V_N \qquad (4\text{--}13)$$

Power

As you know, power in a resistive circuit can be thought of as heat dissipated by the resistors. All of the power that is dissipated by the various resistors in a circuit is supplied by the source. Therefore, to determine the total power supplied by a voltage source, we simply sum the individual power dissipations in the circuit as indicated by Equation 4–14.

$$P_T = P_1 + P_2 \ldots + P_N \qquad (4\text{--}14)$$

You will note that this is the same way we computed total power in a series circuit. As you will see in later sections, this same method can be used for all resistive circuit configurations regardless of complexity.

Relative Values of Parallel Resistances

If the resistance in one of the branches of a parallel circuit is much smaller than in the other branches, it will dominate the calculation for total resistance. So, as a rule, if one parallel resistance is no more than 1/100 (or even 1/10 for noncritical calculations) of another parallel resistance, then the larger resistance can be ignored without having a dramatic impact on the total resistance calculation. You can also think of this same rule in terms of current. In that case, the very large resistor will draw less than 1% (or 10% if you use the 10:1 rule) of the total current. Therefore, it can be ignored for many calculations without having significant effects on the resulting values.

Parallel Voltage Sources

Sometimes more power is needed for a circuit than can be supplied by a single voltage source. Multiple voltage sources can be connected in parallel to provide additional power. Since the voltage is the same for all components in a parallel circuit, it is essential that each of the parallel-connected sources have the same nominal voltage. If unequal voltage sources are connected in parallel, current will flow in the reverse direction through the lower voltage source. This may or may not damage the voltage sources, but in any case, it does not accomplish the desired goal. If the voltage sources are batteries, then reverse current may cause a violent explosion if it is excessive.

The total voltage for multiple parallel sources is the same as any one of the sources. The current *capacity*, however, is increased. Be careful not to confuse current capacity of a voltage source with total current from the source. Total current is the actual current that flows and can be computed with Ohm's Law. Current capacity of the source, by contrast, is the maximum amount of current that can be supplied by a source. If this value is exceeded, then the supply voltage will decrease, and the voltage source may be damaged.

Theoretically, the total current capacity of parallel voltage sources is the sum of the individual sources. In practice, however, the total current capacity is generally somewhat less than this calculated value, since the voltage sources will not have exactly the same terminal voltage under varying current requirements.

Exercise Problems 4.3

1. Noting that there is only one possible path for current is one way to positively identify a parallel circuit. (True/False)
2. All components in a parallel circuit have the same _____.
3. Total resistance in a parallel circuit is always (*less than/greater than*) any individual branch resistance.
4. Calculate the total resistance of a parallel circuit composed of the following resistors: 1.8 kΩ, 1.1 kΩ, and 2.7 kΩ.
5. What is the total resistance of ten 680-Ω resistors connected in parallel?
6. Compute the combined resistance of a 200-Ω resistor connected in parallel with a 50-Ω resistor.
7. Calculate the branch currents and total current for the parallel circuit shown in Figure 4–34.
8. Find the total resistance of the circuit shown in Figure 4–34.
9. According to Kirchhoff's Current Law, the current entering any point in a circuit must be less than the current leaving that same point. (True/False)

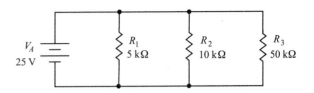

FIGURE 4–34 A parallel circuit.

Fig04_34.msm

10. A wireless joystick for a computer system requires two AA batteries. The batteries are 1.5 V each and are connected in parallel inside the joystick. How much voltage is actually supplied to the joystick by the parallel combination?

4.4 SERIES-PARALLEL CIRCUITS

Series-parallel circuits represent yet another way to interconnect electronic components. As you might expect from the name series-parallel, a **series-parallel circuit** is actually a combination of series and parallel components. A series-parallel circuit is analyzed by applying series circuit techniques to the series portions of the circuit and parallel circuit techniques to the parallel parts of the circuit. Since most practical electronic circuits are series-parallel circuits, it is important for you to be able to confidently analyze and troubleshoot series-parallel circuits.

Positive Identification

As you know, there are four classes of circuits: series, parallel, series-parallel, and complex. You will recall that series circuits have *all* components in series with each other. For components to be in series, they must have one, and only one, end connected together with no other components or wires connected to the common point. A parallel circuit, by contrast, has *both* ends of *all* components connected directly together.

Once you have determined that a circuit is definitely not series and not parallel, then you need only distinguish between series-parallel and complex to positively identify the circuit. To distinguish between series-parallel and complex circuits, apply the following procedure. On a scratch paper, redraw the schematic to be classified, but replace all sets of series resistors with a single resistor. Similarly, replace all truly parallel resistors with a single resistor. Repeat this procedure for the newly drawn figure. Continue until one of the following situations occurs:

- The circuit is reduced to a single resistor.
- The circuit cannot be reduced to a single resistor.

In the first case, you have identified a series-parallel circuit. In the second case, you have identified a complex circuit. It is essential to accurately identify the type of circuit you have, since all subsequent rules, equations, and numerical strategies vary from one circuit type to another.

4.4 • SERIES-PARALLEL CIRCUITS

•—EXAMPLE 4.18

Classify the circuit shown in Figure 4–35 as series-parallel or complex.

FIGURE 4–35 Classify this circuit.

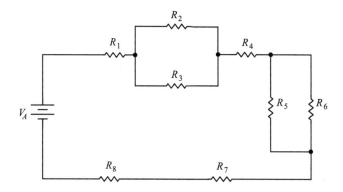

SOLUTION By inspection, we can determine that R_2 and R_3 are in parallel. We can replace these two resistors with a single resistor labeled $R_{2,3}$. Similarly, we replace parallel resistors R_5 and R_6 with $R_{5,6}$. Likewise, the series combination of R_7 and R_8 is replaced with $R_{7,8}$. Figure 4–36(a) shows the simplified sketch up to this point. We now repeat the process with this new sketch.

FIGURE 4–36 Simplification of the circuit in Figure 4–35.

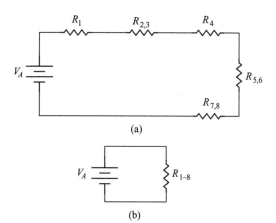

Again, by inspection, we can see that all five resistors are in series, so a single resistor can replace them. We will call this resistor R_{1-8}. This last simplification results in a single resistor as shown in Figure 4–36(b). Because we were able to reduce the circuit to a single resistor by replacing sets of series and parallel components, we can positively identify the original circuit as a series-parallel circuit.

•—EXAMPLE 4.19

Classify the circuit in Figure 4–37 as series-parallel or complex.

FIGURE 4–37 Classify this circuit.

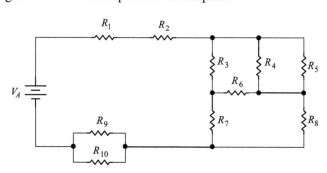

SOLUTION First, we note that R_1 and R_2 are in series; this can be replaced with $R_{1,2}$. Next, we note that R_4 and R_5 are in parallel, so they can be replaced with a single resistor labeled $R_{4,5}$. Similarly, $R_{9,10}$ can replace the parallel combination of R_9 and R_{10}. The simplification to this point is shown in Figure 4–38(a). We now repeat the process for this partially-simplified diagram.

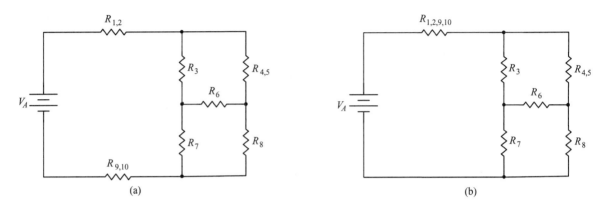

FIGURE 4–38 Steps in the simplification process for the circuit shown in Figure 4–37.

Although not quite as obvious, we can classify resistors $R_{1,2}$ and $R_{9,10}$ as being in series, since there is only a single current path (through the voltage source) for these two resistors. We will replace these two resistors with a single resistor labeled $R_{1,2,9,10}$. The simplification to this point is shown in Figure 4–38(b). We now start to repeat the process, but we soon realize there are no more combinations of pure series or pure parallel components. Since we are unable to reduce the circuit to a single resistor, we can positively identify the original circuit as complex.

Mathematical Relationships

We will now extend our analytical skills to include the numerical analysis of series-parallel circuits. It is very important to remember that Ohm's Law and Kirchhoff's Laws as we discussed in earlier sections are *always* valid. However, you must be careful to apply them correctly. For example, you know the current through a resistor can be found by dividing the voltage across it by its resistance (i.e., $I = V/R$). You also know that it would make no sense to divide the voltage across one resistor by the resistance of a different resistor. For example, the following calculation could not be expected to produce meaningful results:

$$I_4 = \frac{V_3}{R_5}$$

$$= \frac{10 \text{ V}}{500 \text{ }\Omega} = 20 \text{ mA}$$

But, as you will soon see, it is easy to accidentally set up erroneous calculations. To avoid such errors, you are advised to always use subscripts to identify the various currents, voltages, and so forth. If you are trying to compute the current through R_7 using Ohm's Law, then label the current as I_7 and be careful to use V_7 and R_7 (the voltage across R_7 and its resistance) in the calculation.

Resistance

The process for computing the total resistance in a series-parallel circuit is closely related to the simplification procedure previously described for positive identification of a series-parallel circuit. There is only one additional step. Each time you replace a series or parallel network with a single resistor as part of the simplification procedure, you simply compute the value of the equivalent resistance. If you are replacing a series network, then use the equation for series resistances (i.e., sum the individual resistances). On the other hand, use the reciprocal formula (or other equivalent calculation such as product-over-the-sum) to determine the equivalent resistance for simplified parallel networks. These two equations are summarized in Table 4–1.

TABLE 4–1 Select an equation based on the type of circuit section being simplified.

Type of Circuit Section Being Simplified	Equation for Total Resistance
Series	$R_T = R_1 + R_2 \cdots R_N$
Parallel	$R_T = \dfrac{1}{\dfrac{1}{R_1} + \dfrac{1}{R_2} \cdots + \dfrac{1}{R_N}}$

Ultimately, you will reach a point where there is only a single resistor in the simplified circuit. The value of this equivalent resistance is equal to the total resistance of the original series-parallel circuit.

•EXAMPLE 4.20

Determine total resistance of the circuit shown in Figure 4–39.

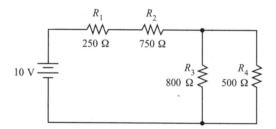

FIGURE 4–39 Find the total resistance in this circuit.

SOLUTION When we first examine Figure 4–39, we see that resistors R_1 and R_2 are in series with each other. We will combine these and label them as $R_{1,2}$. Since they are in series, we must use the series resistance equation to compute their equivalent resistance.

$$R_{1,2} = R_1 + R_2$$
$$= 250 \, \Omega + 750 \, \Omega$$
$$= 1.0 \, k\Omega$$

Similarly, we can see by inspection of Figure 4–39 that resistors R_3 and R_4 are in parallel. Let's combine these and label them as $R_{3,4}$. The equivalent resistance value can be found with the parallel resistance equation.

$$R_{3,4} = \dfrac{1}{\dfrac{1}{R_3} + \dfrac{1}{R_4}}$$
$$= \dfrac{1}{\dfrac{1}{800 \, \Omega} + \dfrac{1}{500 \, \Omega}}$$
$$\approx 308 \, \Omega$$

Figure 4–40(a) shows the results of the simplification process to this point. Now, we examine our partially simplified drawing and see that resistors $R_{1,2}$ and $R_{3,4}$ are in series with each other. We combine these and label them R_{1-4}. Since this is the final step in the simplification process, we know that R_{1-4} is equal to the total resistance (R_T) of the original circuit. Its value is found by applying the series-resistance equation.

$$R_{1-4} = R_T = R_{1,2} + R_{3,4}$$
$$= 1.0 \text{ k}\Omega + 308 \text{ }\Omega$$
$$= 1,308 \text{ }\Omega$$

These last results are shown in Figure 4–40(b).

FIGURE 4–40 Steps in the simplification of the circuit shown in Figure 4–39.

Fig04_40(a).msm

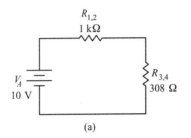
(a)

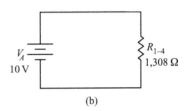
(b)

•—Practice Problems

1. Calculate the total resistance for the circuit shown in Figure 4–41.
2. Find the total resistance of the circuit in Figure 4–42.
3. How much total resistance is in the circuit shown in Figure 4–43?

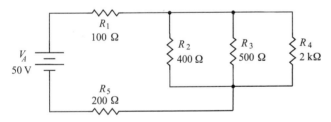

FIGURE 4–41 Calculate total resistance.

Fig04_41.msm

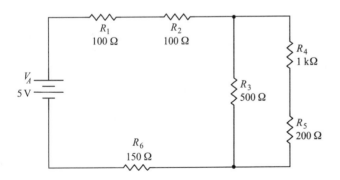

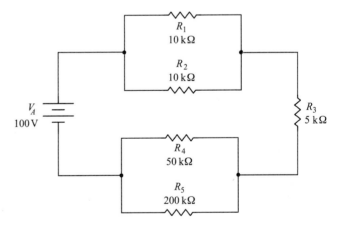

FIGURE 4–42 Find the total resistance.

Fig04_42.msm

FIGURE 4–43 What is the total resistance?

Fig04_43.msm

Current and Voltage

A multistep process is required to determine the voltage and current values throughout a series-parallel circuit. The first step was detailed in the preceding section and consists of finding the total circuit resistance. Once we know total resistance and applied voltage (normally a known quantity in a practical application), we can calculate total current. We then work our way in reverse order through the partially simplified circuit diagrams used to determine total resistance. At each intermediate schematic, we compute all of the current and voltage values possible using Ohm's and Kirchhoff's Laws. When no more values can be determined on a particular diagram, we move to the next schematic and repeat. We continue this process until we have determined all of the voltage and current values throughout the circuit (or at least those of interest to us).

Since we are working backward through the simplification diagrams, we will be expanding equivalent resistances into their component parts. For example, we might expand an equivalent resistance labeled $R_{4,5}$ into R_4 and R_5. When we make this expansion, we must carry the appropriate voltage and current values with us to the next schematic level. If the resistance being expanded represents a parallel network, then the voltage computed for the equivalent resistance will be carried to both (or all) of the resistors that make up the original network. By contrast, if the equivalent resistance represents a series combination of resistors, then the current calculated for the equivalent resistance will be carried forward and assigned to all resistors making up the original series network. These last two procedures should seem reasonable to you, since you know that voltage and current are the same for all resistors in parallel and series networks, respectively.

•―EXAMPLE 4.21

Refer to the circuit shown in Figure 4–44 and complete Table 4–2.

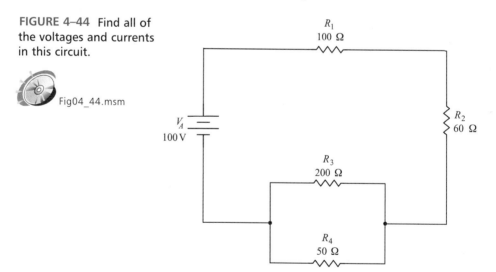

FIGURE 4–44 Find all of the voltages and currents in this circuit.

Fig04_44.msm

TABLE 4–2 Complete this table for the circuit shown in Figure 4–44.

Component	Resistance	Voltage	Current
R_1	100 Ω		
R_2	60 Ω		
R_3	200 Ω		
R_4	50 Ω		
Total		100 V	

SOLUTION First, be aware that there is generally more than one sequence of calculations that will lead you to the correct solutions. Don't try to memorize the exact sequence offered in the following solution. Rather, work to understand the reasons for a particular calculation, the choice of equation, and so forth. You are also encouraged to recognize and explore other sequences.

The first goal must be to find total resistance. By inspection, we see that resistors R_1 and R_2 are in series. We will replace these with a single resistor symbol labeled $R_{1,2}$. We use the series resistance equation to compute the value of $R_{1,2}$.

$$R_{1,2} = R_1 + R_2$$
$$= 100\ \Omega + 60\ \Omega = 160\ \Omega$$

Note that R_3 and R_4 are in parallel. We can replace them with an equivalent resistor symbol labeled $R_{3,4}$. Let's use the product-over-the-sum equation to find the value of $R_{3,4}$.

$$R_{3,4} = \frac{R_3 \times R_4}{R_3 + R_4}$$
$$= \frac{200\ \Omega \times 50\ \Omega}{200\ \Omega + 50\ \Omega}$$
$$= \frac{10{,}000}{250} = 40\ \Omega$$

Calculator Application

We can now sketch an intermediate schematic showing our partially simplified circuit. The simplification at this point is shown in Figure 4–45(a).

We now focus on further simplifying the circuit illustrated in Figure 4–45(a). We can see that $R_{1,2}$ and $R_{3,4}$ are in series. We will combine them using the series-resistance equation and label the result as R_{1-4}.

$$R_{1-4} = R_T = R_{1,2} + R_{3,4}$$
$$= 160\ \Omega + 40\ \Omega$$
$$= 200\ \Omega$$

This completes simplification of our original circuit. The result is shown in Figure 4–45(b).

Now, we are ready to compute all of the voltages and currents we can at each level, beginning with the fully simplified drawing in Figure 4–45(b). Here, we can calculate total current using total voltage and total resistance values along with Ohm's Law.

$$I_T = I_{1-4} = \frac{V_A}{R_T}$$
$$= \frac{100\ \text{V}}{200\ \Omega} = 500\ \text{mA}$$

We now move to the next level of simplification shown in Figure 4–45(a). Since R_{1-4} replaced a series network ($R_{1,2}$ and $R_{3,4}$), we will carry the current value (500 mA) with us. This is the current through both $R_{1,2}$ and $R_{3,4}$.

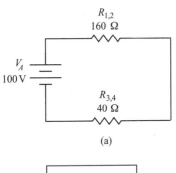

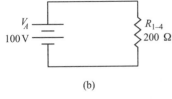

FIGURE 4–45 Simplification steps for the circuit shown in Figure 4–44.

Using Ohm's Law we can now find the voltage across the two series resistances ($R_{1,2}$ and $R_{3,4}$).

$$V_{1,2} = I_{1,2} \times R_{1,2}$$
$$= 500 \text{ mA} \times 160 \text{ } \Omega = 80 \text{ V}$$

And

$$V_{3,4} = I_{3,4} \times R_{3,4}$$
$$= 500 \text{ mA} \times 40 \text{ } \Omega = 20 \text{ V}$$

We can move to the next level which, in this case, is the original circuit shown in Figure 4–44. We will carry the $I_{1,2}$ current and the $R_{3,4}$ voltage values, since these expand to series and parallel networks, respectively.

Since we know the current through R_1 and R_2 (500 mA) and their individual resistances, we can apply Ohm's Law to find their voltage drops.

$$V_1 = I_1 \times R_1$$
$$= 500 \text{ mA} \times 100 \text{ } \Omega = 50 \text{ V}$$

And

$$V_2 = I_2 \times R_2$$
$$= 500 \text{ mA} \times 60 \text{ } \Omega = 30 \text{ V}$$

Now, let's focus on R_3 and R_4. Since $R_{3,4}$ explodes into a parallel network, the voltage across both R_3 and R_4 will be the same as the voltage across $R_{3,4}$ (20 V). This leaves only the current through R_3 and R_4 to be computed, which we can do with Ohm's Law.

$$I_3 = \frac{V_3}{R_3}$$
$$= \frac{20 \text{ V}}{200 \text{ } \Omega} = 100 \text{ mA}$$

And

$$I_4 = \frac{V_4}{R_4}$$
$$= \frac{20 \text{ V}}{50 \text{ } \Omega} = 400 \text{ mA}$$

This completes calculation of the voltages and currents for every component. The results are summarized in Table 4–3. It is always a good practice to check your work by applying Kirchhoff's Voltage and Current Laws to the analyzed circuit to

TABLE 4–3 Calculated values for the circuit in Figure 4–44.

Component	Resistance	Voltage	Current
R_1	100 Ω	50 V	500 mA
R_2	60 Ω	30 V	500 mA
R_3	200 Ω	20 V	100 mA
R_4	50 Ω	20 V	400 mA
Total	200 Ω	100 V	500 mA

be sure the calculated values make sense. For example, we know that $V_{1,2} + V_{3,4}$ should equal V_A. We also know that $I_3 + I_4$ should equal I_1 or I_2 or I_T.

•—Practice Problem

1. Analyze the circuit in Figure 4–46, and complete Table 4–4.

Note: Many additional practice problems have been made available to your instructor in the form of student worksheets.

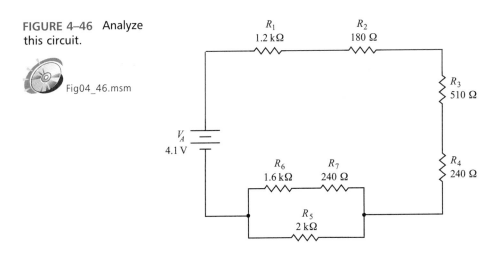

FIGURE 4–46 Analyze this circuit.

Fig04_46.msm

TABLE 4–4 Complete this table with values from Figure 4–46.

	R_1	R_2	R_3	R_4	R_5	R_6	R_7	Total
Resistance	1.2 kΩ	180 Ω	510 Ω	240 Ω	2 kΩ	1.6 kΩ	240 Ω	
Voltage								4.1 V
Current								

Power

You will be pleased to know that power calculations for series-parallel circuits are identical to those in series or parallel circuits. That is, the power dissipation in a resistor can always be found by applying any of the following power equations:

1. $P = VI$
2. $P = I^2 R$
3. $P = V^2/R$
4. $P_T = P_1 + P_2 \ldots + P_N$

Of course, you should always use subscripts to be certain the voltage, current, resistance, and power used in the calculations are all associated with the same physical resistor.

✓ Exercise Problems 4.4

1. Series-parallel circuits are rarely found in industry. (True/False)
2. Examine each of the circuits shown in Figure 4–47 on page 117 and identify the ones that are series-parallel.
3. Refer to Figure 4–48 and complete Table 4–5 on page 117.

4.4 • SERIES-PARALLEL CIRCUITS   117

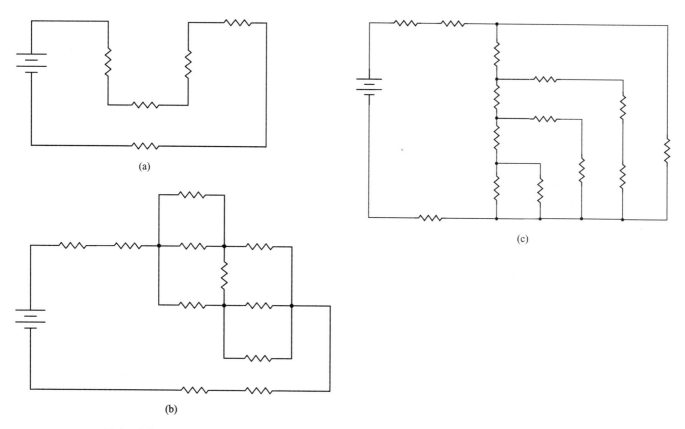

FIGURE 4–47 Which of these circuits are series-parallel?

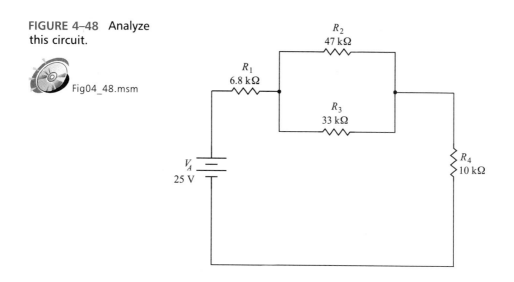

FIGURE 4–48 Analyze this circuit.

Fig04_48.msm

TABLE 4–5 Complete this table with values from Figure 4–48.

	R_1	R_2	R_3	R_4	Total
Resistance	6.8 kΩ	47 kΩ	33 kΩ	10 kΩ	
Voltage					25 V
Current					
Power					

4.5 COMPLEX CIRCUITS

The previous portions of this chapter have discussed three important circuit configurations: series, parallel, and series-parallel. A thorough understanding of these three types will allow you to solve most practical resistive circuit problems you are likely to encounter. There is, however, one additional circuit configuration that you should be aware of. A **complex circuit** is one that contains some components that are neither in series nor in parallel with each other.

Positive Identification

The identification procedure for series-parallel circuits is essentially the same for complex circuits. Identification can be accomplished by performing the following procedure: On a scratch paper, redraw the schematic to be classified, but replace all sets of series resistors with a single resistor. Similarly, replace all truly parallel resistors with a single resistor. Repeat this procedure for the newly drawn figure. Continue until one of the following situations occurs:

- The circuit is reduced to a single resistor.
- The circuit cannot be reduced to a single resistor.

In the first case, you have identified a series-parallel circuit as discussed in the preceding section. In the second case, you have identified a complex circuit. It is important to be able to identify a complex circuit so that you won't try to analyze it using incorrect strategies.

Circuit Analysis Software

Numerical analysis of complex circuits is beyond the scope of this text; however, you should still be able to solve a complex circuit problem if you are ever faced with one. Fortunately, circuit analysis software packages do not distinguish between network types; all are solved with equal ease. So, in the unlikely event that you need to compute circuit values in a complex circuit, simply enter the network into the schematic capture portion of your circuit simulation package. The analysis software can then determine all required currents, voltages, and power dissipations.

Now, you might be thinking to yourself that the same strategy could be applied to other circuit configurations. While that may be true, technically speaking, knowledge of circuit analysis techniques for series, parallel, and series-parallel circuits is an important skill. There are many times that you will want to make a quick estimate of a circuit quantity without necessarily having immediate access to a computer with a circuit analysis package. Perhaps more importantly, the insights gained by solving circuit analysis problems without the aid of computer software provide you with a solid foundation for the study of more advanced topics.

✓ Exercise Problems 4.5

1. Which of the circuits in Figure 4–49 shown here and on page 119 are complex?
2. Most circuit analysis software packages cannot solve complex circuit analysis problems. (True/False)
3. How can you distinguish between a series-parallel circuit and a complex circuit?

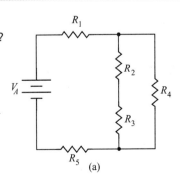

FIGURE 4–49 Which of these are complex circuits?

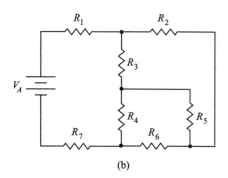
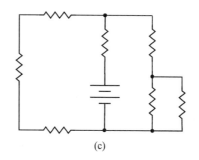
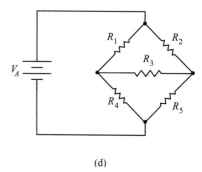

(b) (c) (d)

FIGURE 4-49 (continued)

4.6 GROUND AND OTHER REFERENCE POINTS

Anytime a voltage is measured, the voltmeter is actually indicating the voltage that is present at one point in the circuit *with reference to another point*. We briefly discussed this idea in the first part of this chapter with regard to relative polarities.

Concept of Ground

In most electronic circuits, there is a point from which all circuit voltages are measured. This common point is called the circuit **ground**. Most often, but certainly not always, ground is considered to be one side of the circuit's power source. In all cases, ground is considered to be zero volts. This is sensible, since ground is the reference point for all measurements, and since any voltage is zero with reference to itself (i.e., there is no *difference* in potential). Figure 4–50(a) shows a series circuit where the negative side of the voltage source is considered to be the circuit ground. Note the symbol (⏚) used to represent ground. Figure 4–50(b) shows exactly the same circuit as Figure 4–50(a), but two ground symbols are used. Both ground symbols represent the same electrical point in the circuit. The interconnecting wire is assumed but not shown. The omission of the interconnecting ground wires on schematic diagrams is the normal practice since it removes unnecessary clutter from otherwise complex diagrams.

When a circuit is physically constructed, ground can take a number of different forms. If the circuit has a metal chassis or frame associated with it, then the frame is usually connected to circuit ground. This is especially true for high-frequency devices or high-speed digital circuits. Most circuits are constructed on printed circuit boards (PCBs). The conductors on a PCB are traces of copper bonded to an insulating material. The copper traces connect the various components together just as wire might do in a laboratory circuit. Ground on a printed circuit board is often a very wide trace that wanders throughout the board. If the PCB has more than one layer of traces, then one or more of the layers are frequently dedicated as a ground plane. A **ground plane** is an entire PCB layer (sheet of copper) that is connected to ground potential. Most computer circuit boards have at least two dedicated ground planes.

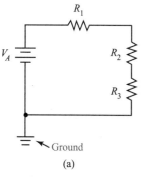

(a)

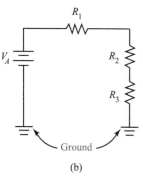

(b)

FIGURE 4–50 Ground is a reference point from which other voltages are measured.

Voltages with Reference to Ground

Figure 4–51 shows a series circuit with one point labeled as the ground connection. The remaining points in the circuit are labeled A through E. When we measure the voltages at points A through E, we connect one lead of the voltmeter to ground and the other lead to the point being measured. For example, suppose we wanted to measure the voltage at point C in Figure 4–51 with reference to ground. We would connect one side of the voltmeter to ground and the other to point C. You can see from inspection of Figure 4–51, that the meter is essentially connected across R_3, so it will indicate 6 V. We label the voltage at point C as V_C. So in the present example, $V_C = 6$ V.

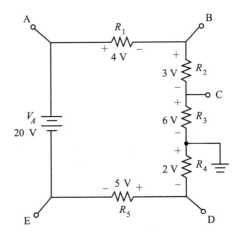

FIGURE 4–51 Voltages can be measured with respect to ground.

The resulting measurement would be less obvious if we were to measure the voltage at point E. There is a simple procedure, however, that will allow you to determine the correct voltage at any point in a circuit.

1. Start at the reference point and write the voltage drops as you move toward the measured point. Use the polarity of the voltage drop that is on the end where you exit a component.
2. When you reach the measured point, the algebraic sum of the accumulated voltage drops will be the correct voltage value.

The polarities of the voltage drops are determined in exactly the same manner as if we were writing the Kirchhoff's Voltage Law equation for the loop. The only differences are that we do not complete the loop, and we do not set the voltages equal to zero.

EXAMPLE 4.22

Determine the voltage at point E in Figure 4–51.

SOLUTION We begin at the reference point (ground) and move toward the measured point (point E). As we pass through R_4, we will write its voltage drop as -2 V. Continuing through R_5, we find a drop of -5 V. Since we have reached our destination, the measured value will be equal to the algebraic sum of our accumulated voltage drops. In this particular case, the measured voltage will be

$$V_E = -2 \text{ V} - 5 \text{ V} = -7 \text{ V}$$

It is important to realize that you may move in either direction as you progress from the reference point toward the measured point. Although the individual voltage drops may be different, the algebraic sums of the two directions will be identical. If, in the present example, we had moved in the opposite direction, we would have accumulated the following sequence of voltages: +6 V, +3 V, +4 V, and −20 V (the voltage source). The algebraic sum of these voltages is still −7 V.

Calculator Application

4.6 • GROUND AND OTHER REFERENCE POINTS

•—Practice Problems

1. Determine the voltage that would be measured between ground and point A in Figure 4–51.

2. Refer to Figure 4–52 and determine the following voltages: V_A, V_B, V_C, and V_D.

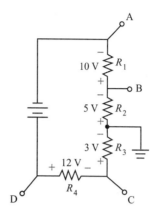

FIGURE 4–52 Determine the voltages at points A, B, C, and D with reference to ground.

Nonground References

In some cases, you must measure a voltage that is referenced to some point other than ground. Suppose, for example, we want to determine the voltage that would be measured at point B with reference to point D in Figure 4–53. We apply the exact same procedure that we used for ground references. In this case, we begin at the reference point (point D) and move toward point B. As we go, we will generate the following sequence of voltages: +10 V and +4 V. The algebraic sum of these voltages is +14 V.

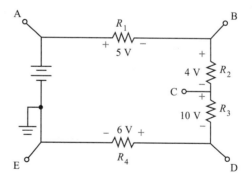

FIGURE 4–53 Nonground references are also common in electronics.

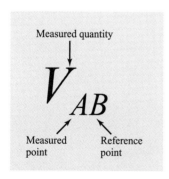

FIGURE 4–54 Method for labeling nonground references.

We must be careful to specify the reference point when writing the voltage at a certain point in a circuit. Figure 4–54 shows the accepted method for indicating voltage references. In the case of a ground reference, the second subscript is omitted and ground is assumed to be the reference point. Exceptions to this labeling method are usually obvious by inspection. In our preceding example, we would label the voltage at point B that was measured with respect to point D as $V_{BD} = +14$ V.

Practice Problems

1. What is the voltage at point E with reference to point C in Figure 4–55?
2. What is the value of V_{AB} in Figure 4–55?
3. What is the value of V_{DA} in Figure 4–55?
4. What is the value of V_{CE} in Figure 4–55?
5. Find the value of V_{AE} in Figure 4–55.

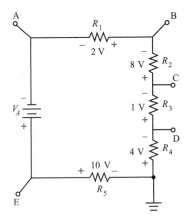

FIGURE 4–55 A series circuit with multiple reference points.

Other Types of Ground References

There are several alternate symbols that are used to represent ground (i.e., 0 V) reference points on schematic diagrams. Some of these are illustrated in Figure 4–56. Although standards committees have documented the "correct" use of a particular symbol, the actual use of the symbols by manufacturers seems to be somewhat arbitrary. The symbol shown in Figure 4–56(a) is generally used to represent earth ground—that is, a point in the circuit that is connected to the safety ground of the 120-Vac power line and ultimately connects to a metal stake driven into the soil. This is a particularly important reference point for several reasons. First, if properly implemented, it provides some degree of increased safety with respect to shock from the 120-Vac power line. In particular, if the metal equipment chassis and other metal parts that are accessible to the equipment user are connected to earth ground, then no shock hazard will exist if an internal wire comes loose and inadvertently touches the chassis.

Many manufacturers use the symbol shown in Figure 4–56(b) to represent circuit ground. In some cases, a given circuit may be designed to have multiple grounds that are essentially isolated from each other. In these cases, the ground symbol shown in Figure 4–56(c) can be labeled to identify the various ground connections. Figure 4–56(c) shows the symbol labeled with the letter "A."

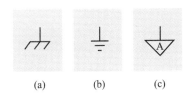

FIGURE 4–56 Several alternative ground symbols are used in the Industry.

✓ Exercise Problems 4.6

1. Refer to Figure 4–57. Determine each of the following voltages:
   a. V_A
   b. V_B
   c. V_C
   d. V_D
   e. V_E
   f. V_F

2. Refer to Figure 4–57. Determine each of the following voltages:
   a. V_{AB}
   b. V_{EA}
   c. V_{DA}
   d. V_{FB}
   e. V_{BF}
   f. V_{DF}

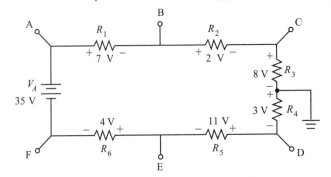

FIGURE 4–57 Determine the voltages in this circuit.

SUMMARY

An electromotive force and a complete path for current are the two requirements for sustained current flow in a circuit. Electron current flows from a given point toward a point with a more positive relative potential.

Series networks are characterized as having only a single path for current. If all components in a given circuit are in series, then the entire circuit is classified as a series circuit. Current is the same in all parts of a series circuit. The sum of the individual voltage drops in a series circuit must be equal to the applied voltage. The individual voltage drops in a series circuit are directly proportional to the resistance values. Total resistance in a series circuit is equal to the sum of the individual resistances. Kirchhoff's Voltage Law states that the algebraic sum of all the voltage drops and all the voltage sources in *any* closed loop equals zero. When voltage sources are connected in series, the net voltage is equal to the algebraic sum of the individual voltage sources.

Parallel networks are identified when both ends of all components connect *directly* together. If all components in a circuit are in parallel, then the entire circuit is classified as a parallel circuit. The voltage is the same for all components in a parallel circuit. Total current in a parallel circuit is equal to the sum of the branch currents. Each branch current is inversely proportional to branch resistance (Ohm's Law). Total resistance in a parallel circuit is always less than the smallest branch resistance. Kirchhoff's Current Law states that the total current entering any point in a circuit must be equal to the total current leaving that point. When voltage sources are connected in parallel, the total voltage is the same as each individual source, but current capacity is increased.

Series-parallel circuits have both series and parallel components or networks within the circuit. Series-parallel circuits are identified by replacing sets of series and parallel components with single equivalent resistances. If the process can be continued until the circuit is reduced to a single equivalent resistor, then the circuit is a series-parallel circuit. Voltage and current calculations are accomplished in a series-parallel circuit by applying series circuit rules to the series components or networks and parallel circuit rules to the parallel components and networks.

Complex circuits are identified in the same way as series-parallel circuits, except the simplified circuit cannot be reduced to a single equivalent resistance by replacing sets of series and parallel components with equivalent resistances. Total power dissipation in any circuit configuration is equal to the total power supplied by the source.

Ground is a reference point from which other circuit voltages are measured. By definition, ground is a zero-volt potential. Nonground references are also used in electronics. These are generally identified by double subscripts (e.g., V_{AB}), where the first subscript is the point measured and the second subscript is the reference point.

REVIEW QUESTIONS

Section 4.1: Basic Requirements for Current

1. One of the basic requirements for current flow is that the circuit must have at least one resistor. (True/False)
2. A voltage source can serve as one of the requirements for sustained current flow in a circuit. (True/False)
3. What is the value of current through an open circuit?
4. One point in a circuit is +25 V, and another is +10 V. If a resistor is connected between these two points, which way will electron current flow?
5. One point in a circuit is −25 V, and another is −10 V. If a resistor is connected between these two points, which way will electron current flow?

Section 4.2: Series Circuits

Refer to Figure 4–58 for questions 6 through 10.

6. Identify the circuit type.
7. What is the total resistance in the circuit?

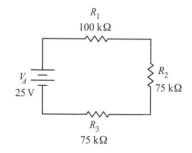

FIGURE 4–58 A circuit for analysis.

Fig04_58.msm

8. How much current flows through resistor R_1?
9. What is the voltage drop across R_3?
10. How much power is dissipated by R_2?
11. Apply Kirchhoff's Voltage Law to determine whether the voltages shown in Figure 4–59 on page 124 are correctly labeled.
12. If a 10-V source, a 3-V source, and a 24-V source are connected in a series-aiding configuration, what is the effective source voltage?

FIGURE 4–59 Are the voltages correctly labeled in this circuit?

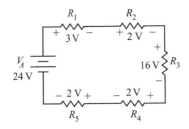

13. Determine the net voltage for the circuit shown in Figure 4–60.

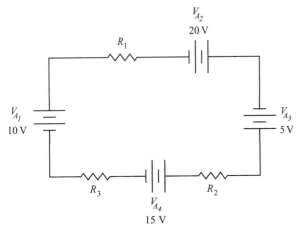

FIGURE 4–60 What is the net voltage in this circuit?

Section 4.3: Parallel Circuits

14. Which of the circuits shown in Figure 4–61 are parallel circuits?
15. The current is the same in all branches of a parallel circuit. (True/False)
16. The voltage is the same across all branches in a parallel circuit. (True/False)
17. Total current in a parallel circuit is (*less than/equal to/greater than*) any branch current.
18. The total resistance in a parallel circuit is always (*less than/equal to/greater than*) the smallest branch resistance.
19. If the total power dissipation in a particular parallel circuit is 100 mW, is it possible that the resistance in one of the branches is dissipating 5 mW?
20. What is the total resistance if a 25-kΩ resistor and a 91-kΩ resistor are connected in parallel?
21. What is the total resistance if ten 5-Ω resistors are connected in parallel?
22. What is the total resistance if all of the following resistors are connected in parallel: 10 kΩ, 12 kΩ, 22 kΩ, and 39 kΩ?
23. Refer to Figure 4–62. What combination of switch positions results in lowest value of total resistance?

FIGURE 4–62 A switchable circuit.

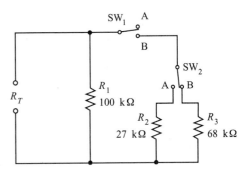

24. Refer to Figure 4–62. What is the value of R_T if SW_1 is in position B and SW_2 is in position B?
25. What combination of switch positions produces the least total current flow for the circuit shown in Figure 4–62?
26. What is the total resistance of the following parallel resistors: 680 Ω, 470 Ω, 910 Ω, and 1.2 kΩ?

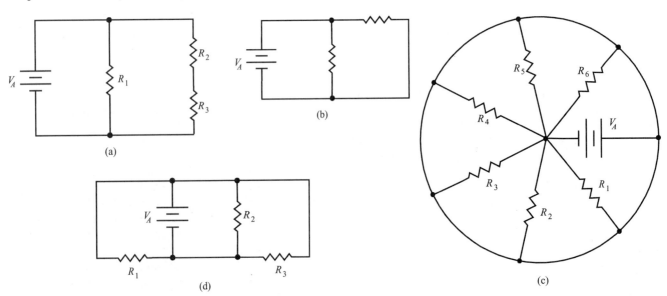

FIGURE 4–61 Identify the parallel circuits.

27. If resistor R_1 in Figure 4–63 is ten ohms, describe the relative change in total resistance when switch SW_1 is moved to the opposite position: (*major change/minor change/can't say without knowing the value of R_2*).

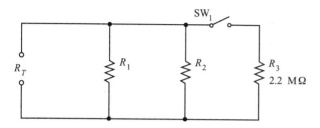

FIGURE 4–63 A switchable circuit.

28. If resistor R_2 in Figure 4–63 is 22 MΩ, describe the relative change in total resistance when switch SW_1 is moved to the opposite position: (*major change/ minor change/can't say without knowing the value of R_1*).
29. Use the reciprocal formula to find the total resistance of the following parallel resistors: 10 kΩ, 22 kΩ, and 18 kΩ.
30. Use the reciprocal formula to find the total resistance of the following parallel resistors: 150 kΩ, 120 kΩ, 68 kΩ, and 220 kΩ.
31. Use the product-over-the-sum formula to calculate the value of the following parallel resistors: 27 kΩ and 39 kΩ.
32. Use the product-over-the-sum formula to calculate the value of the following parallel resistors: 100 Ω and 91 Ω.
33. Use the equal-valued resistor formula to calculate the total resistance produced by paralleling ten 27-kΩ resistors.
34. Use the equal-valued resistor formula to calculate the total resistance produced by paralleling four 100-kΩ resistors.
35. Calculate the total current for the circuit shown in Figure 4–64.
36. How much total power is dissipated by the circuit in Figure 4–64?

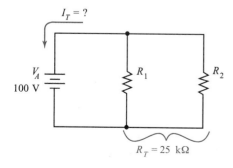

FIGURE 4–64 Example circuit.

37. Calculate the total power dissipation for the circuit shown in Figure 4–65.

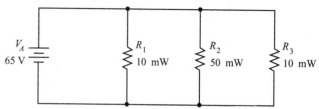

FIGURE 4–65 Example circuit.

38. What is the combined voltage if two 150-V sources are connected in parallel?

Section 4.4: Series-Parallel Circuits

39. Name the four classes of circuit configurations.
40. State the rule for ensuring that two or more components are *definitely* in parallel.
41. State the rule for ensuring that two or more components are *definitely* in series.
42. If a circuit is *not* series and *not* parallel, then describe the procedure for positively identifying its type.
43. Any series-parallel circuit can be reduced to a _____-resistor equivalent circuit having a resistance equal to the _____ resistance of the original circuit.
44. Simplify the circuit shown in Figure 4–66 and compute the value of total resistance.

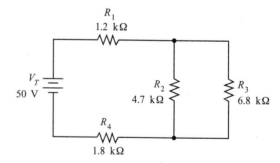

FIGURE 4–66 Circuit for problems 44 through 46.

Fig04_66.msm

45. Refer to Figure 4–66. How much total current flows in the circuit?
46. In Figure 4–66, how much total power must be supplied by the battery?
47. Determine the value of total resistance for the circuit in Figure 4–67 on page 126.
48. What is the value of total current in the circuit shown in Figure 4–67?
49. Compute the amount of power dissipation for R_2 in Figure 4–67.

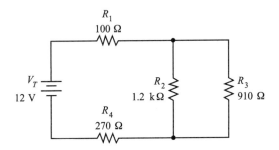

FIGURE 4–67 Circuit for problems 47 through 51.

50. Refer to Figure 4–67. What is the voltage drop across R_3?
51. Which resistor in Figure 4–67 dissipates the most power?
52. Refer to Figure 4–68. How much current flows through resistor R_3?

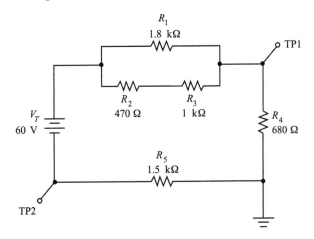

FIGURE 4–68 Circuit for problems 52 through 55.

53. Which resistor dissipates the most power in Figure 4–68?
54. Calculate the voltage drop across R_2 in Figure 4–68.
55. Determine the voltage drop across resistor R_1 in Figure 4–68.

Section 4.5: Complex Circuits

56. Describe how to distinguish between a series-parallel circuit and a complex circuit.
57. Computer-based circuit analysis software packages (circuit simulation) cannot generally solve complex circuit problems. (True/False)

Section 4.6: Ground and Other Reference Points

Refer to Figure 4–69 for questions 58 through 65.

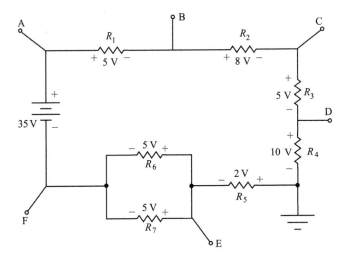

FIGURE 4–69 Circuit for problems 58 through 65.

58. What is the voltage at point D with reference to ground?
59. What is the value of V_E?
60. What is the voltage at point A with respect to ground?
61. What is the value of V_{AB}?
62. What is the value of V_{DC}?
63. What is the value of V_{FA}?
64. Determine the value of V_{EF}.
65. Determine the value of V_{FE}.

•—CIRCUIT EXPLORATION

Here's an opportunity to utilize your circuit simulation software. The following problem is fairly easy to visualize, but it is quite laborious to calculate manually and requires analytical methods beyond those presented in this book. Nonetheless, you can readily analyze the circuit with your circuit simulation software package. Here's the problem.

Visualize a cube made of resistors as pictured in Figure 4–70 on page 127. Each edge of the cube—twelve total—is formed by one of the resistors. Each of the twelve resistors has exactly the same value. Each is 1.0 Ω. As indicated in the figure, you must determine the total resistance as measured between any two diagonal (through the cube) corners.

FIGURE 4–70 Twelve resistors (each 1.0 Ω) form a cube. What is the resistance between diagonal corners?

Fig04_70.msm

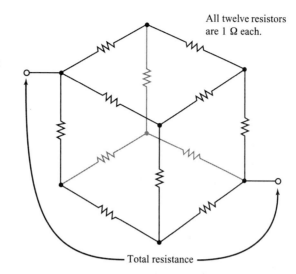
All twelve resistors are 1 Ω each.

Total resistance

ANSWERS TO PRACTICE PROBLEMS

Page 89
1. b and c

Page 91
1. a. R_1 and R_2
   b. None
   c. None
   d. R_1 and R_2

Page 92
1. a and b

Page 105
1. ≈144 Ω
2. 166.22 Ω
3. 200 Ω

Page 112
1. 500 Ω
2. 50 kΩ
3. ≈702.9 Ω

Page 116
1. See table below.

Page 121
1. +13 V
2. $V_A = -15$ V, $V_B = -5$ V, $V_C = +3$ V, and $V_D = +15$ V.

Page 122
1. +15 V
2. −2 V
3. +11 V
4. −15 V
5. −25 V

	R_1	R_2	R_3	R_4	R_5	R_6	R_7	Total
Resistance	1.2 kΩ	180 Ω	510 Ω	240 Ω	2 kΩ	1.6 kΩ	240 Ω	3.09 kΩ
Voltage	1.59 V	238.96 mV	677.06 mV	318.62 mV	1.27 V	1.11 V	165.95 mV	4.1 V
Current	1.33 mA	1.33 mA	1.33 mA	1.33 mA	636.13 μA	691.45 μA	691.45 μA	1.33 mA

CHAPTER 5
Circuit Troubleshooting

•—KEY TERMS
current probe
DMM
sneak path
troubleshooting
voltmeter loading

•—OBJECTIVES
After studying this chapter, you should be able to:

- State the effects of an open circuit on voltage, current, and resistance in any of the following circuit configurations: series, parallel, series-parallel.
- State the effects of a short circuit on voltage, current, and resistance in any of the following circuit configurations: series, parallel, series-parallel.
- State the effects of component tolerance defects on voltage, current, and resistance in any of the following circuit configurations: series, parallel, series-parallel.
- State the effects of power supply defects on voltage, current, and resistance in any of the following circuit configurations: series, parallel, series-parallel.
- Describe and discuss several troubleshooting strategies and practices including all of the following:

Observation	Equipment history
Troubleshooting fences	Equal-probability troubleshooting
Split-half troubleshooting	Signal tracing
Signal injection	Component substitution
Staying on track	

•—INTRODUCTION

Troubleshooting is the process whereby you diagnose a malfunctioning system and identify the specific defect. Once the defect has been positively identified, you must make a judgment to either correct the defect or discard the equipment. This decision can be based on many considerations, but cost of repair, cost of replacement, time required to effect repair, and incidental costs of the defect nearly always play major roles.

This chapter discusses specific troubleshooting strategies that can be used to effectively identify defects in series, parallel, and series-parallel circuits. Several formal troubleshooting techniques are described. These techniques are applicable to all sorts of troubleshooting challenges ranging from basic series circuits to complete electronic systems such as a computer or perhaps an industrial robotics system.

5.1 TROUBLESHOOTING SERIES CIRCUITS

Pure series circuits are relatively rare in practical electronic systems. When they do appear, they are generally quite simple, consisting of perhaps two or three components. But, you may be surprised to learn, the troubleshooting strategies applicable to series circuits play major roles in troubleshooting advanced circuit configurations. So, be sure you fully understand how to effectively troubleshoot a series circuit because you will need to rely on these skills when you troubleshoot more complicated circuits and systems.

Basic Concepts

It is important that you practice only logical troubleshooting procedures because these same procedures will help you diagnose more advanced circuits and systems in the future. When you are first learning to troubleshoot, the circuits are comparatively simple (2 to 10 resistors). Even if you make wild guesses or test every component, you will fix the problem fairly quickly. *But beware!* If you let yourself practice poor habits now, such as guessing or making excessive measurements, then you will probably be unable to effectively diagnose more complex systems. In a practical electronic system, there may be thousands of possible defects. In cases like this, it is unlikely that you can guess where the actual defect is located, and it is physically impractical to measure all of the components.

If you develop a logical and systematic troubleshooting technique early, then not only will you be able to quickly repair simple circuits, but you will also have the skills needed to diagnose complex electronics systems. For now, you should have at least three definite goals as you troubleshoot:

- Make a measurement *only* if you know what a "good" reading would be.
- Make as few measurements as possible.
- Select the best tool for the task at hand.

Know What Is Normal

There is absolutely no sense in measuring the voltage at some point in a circuit if you do not already know what constitutes a "normal" value. Each measurement you take should get you closer to the problem, but if you cannot classify a measurement as "good" or "bad," then it provides no useful information.

Minimize the Number of Measurements

You should make as few measurements as possible. When you are in the laboratory, most of your components are easily accessible and readily measured. This is even truer of a circuit simulation environment. In a real situation, however, it often takes substantial effort just to locate a component that you want to measure. In many cases, portions of the system have to be disassembled to gain access to certain components. So, you should practice making smart, but few measurements.

Select the Best Tool

A voltmeter is your most powerful diagnostic tool for series resistive circuits. You can measure any component without disturbing the circuit (except for possible voltmeter loading), and it is not necessary to turn off power. The term **voltmeter loading** describes a problem (which occurs in mostly older voltmeters) that causes the voltage at the measured point to decrease when the meter is connected. This happens because the meter itself draws excessive current from the circuit. In effect, when you connect the voltmeter to a circuit, you are connecting a parallel resistor. If the voltmeter resistance is very large compared to the other circuit resistances, then the circuit voltage will be relatively unaffected by the presence of the meter. On the other hand, if the voltmeter resistance is comparable to or even smaller than the circuit resistances, then the circuit voltage will drop when the meter is connected, and you will get an incorrect measurement. It is most probable that you will be using a digital voltmeter with an effective resistance of at least 20 MΩ. Thus, voltmeter loading is rarely a problem with today's circuits and test instruments. Should you happen to be using a nonelectronic voltmeter, be aware that voltmeter loading may cause erroneous readings if the circuit resistance is much higher than a few hundred ohms.

When troubleshooting an actual electronic system, opening a circuit for a current measurement or lifting a component for a resistance test requires desoldering components on a delicate and expensive printed circuit board. It should be avoided whenever possible. Once you have identified what you think is the defective component, then you can desolder it, verify your assertion by measuring the resistance of the suspected component, and replace it with a new component.

Intuitive Troubleshooting

The first step toward becoming a skilled troubleshooter of series circuits is to develop a solid intuitive (i.e., nonmathematical) understanding of circuit behavior. For this, you must learn to make mental approximations and be able to classify circuit quantities as high, low, or normal. Let's begin by developing a list of basic rules that will form the basis for all of our troubleshooting efforts. These rules should not only be committed to memory, they must be understood intuitively. You are encouraged to first memorize the rules, and then go on to understand *why* they are valid and be able to visualize them in your mind as you troubleshoot a circuit. We shall consider the effects of open components, shorted components, and components that have changed values. We will also examine the effects of a defective power supply and connectivity problems. In general, an open circuit (infinite resistance) is more common than a short circuit (zero resistance). When a short circuit occurs, however, it can cause extensive damage to other components in the circuit due to the resulting increase in current flow.

Effects of an Open Circuit

An open in an electrical circuit is characterized as having an infinite resistance. In a practical circuit, an open may not literally have infinite resistance, but its resistance will be so high, relative to other circuit resistances, that it can be assumed to be infinite.

Figure 5–1 shows a series resistive circuit that has an open in one of the resistors. If there is an open at any point in a series circuit, then the current will be zero at all points in the circuit. With no current flow, there can be no voltage drop across the various resistors except for the one that is open. The open resistor will have the total applied voltage across it. If this effect is not apparent to you, then think of the open circuit as a very high resistance. This means total current through all of the series components will be extremely small (generally assumed to be zero). Further, recall that the voltage drops in a series circuit are proportional to the resistance

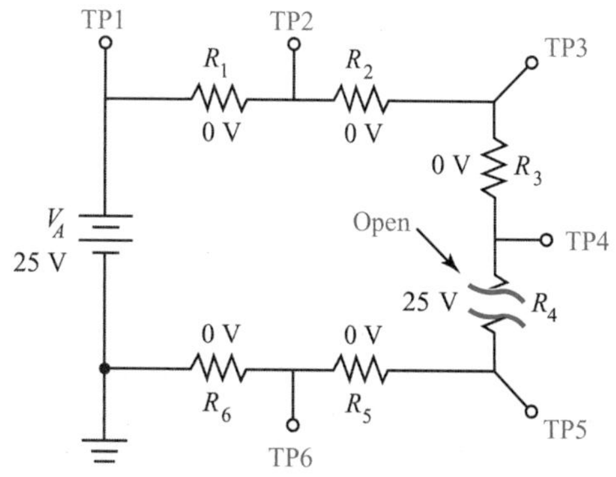

FIGURE 5–1 An open in a series circuit will have the applied voltage across it.

values. Because the open is so much higher in resistance than the other resistors, it will drop nearly all of the voltage (assumed to be 100% for most practical purposes).

When troubleshooting circuits with a voltmeter, it is common practice to connect one side of the voltmeter to ground. The various testpoints are then probed with the remaining voltmeter lead. The following rule tells us how to interpret the measured values:

> ▶ RULE If the open circuit appears between a monitored point and ground, then the meter will indicate the full supply voltage, otherwise the meter will read zero.

We can further refine this rule by considering the *direction* of increase or decrease in voltage at the measured point as a result of the open circuit:

> ▶ RULE When a circuit develops an open, all voltages in the circuit will measure the same as the supply voltage terminal *on the same side of the open.*

•—EXAMPLE 5.1

Refer to the circuit shown in Figure 5–1. What voltage will be measured at each of the indicated testpoints (TP1 through TP6) with respect to ground?

SOLUTION Since testpoints TP1 through TP4 are on the same side of the open as the positive supply terminal, they will all read +25 V. Testpoints TP5 and TP6 are on the same side of the open as the negative terminal of the voltage source. Therefore, they will all read 0 V, because the negative side of the voltage source is connected directly to ground (0 V with respect to ground).

•—EXAMPLE 5.2

Refer to Figure 5–2. Determine the voltage that would be measured at each testpoint with respect to ground.

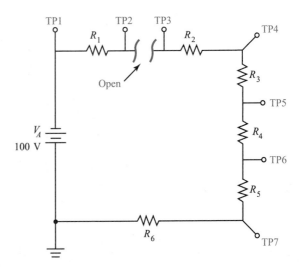

FIGURE 5–2 Determine the voltage at every testpoint.

SOLUTION Since testpoints TP1 and TP2 are on the same side of the open as the negative supply terminal, they will both measure −100 V. All other testpoints (TP3 through TP7) are on the side of the open that is connected to the positive supply terminal (ground). They will all measure 0 V.

•EXAMPLE 5.3

Refer to Figure 5–2. How much voltage would be measured *between* each of the following sets of testpoints?

a. TP1 and TP2
b. TP2 and TP3
c. TP3 and TP4
d. TP4 and TP5
e. TP5 and TP6
f. TP6 and TP7

SOLUTION If there is an open in a series circuit, then there will be no current flow. With no current flow, there will be no voltage drops around the circuit except for across the open. The open will measure the full supply voltage. In the case of Figure 5–2, we would measure the following:

a. TP1 and TP2: 0 V
b. TP2 and TP3: 100 V
c. TP3 and TP4: 0 V
d. TP4 and TP5: 0 V
e. TP5 and TP6: 0 V
f. TP6 and TP7: 0 V

•Practice Problems

Refer to the circuit shown in Figure 5–3.

1. How much voltage would be measured between testpoints TP1 and TP2?
2. How much voltage would be measured between testpoints TP2 and TP3?
3. How much voltage would be measured between testpoints TP3 and TP4?
4. How much voltage would be measured between testpoints TP4 and TP5?
5. How much voltage would be measured between testpoints TP5 and TP6?

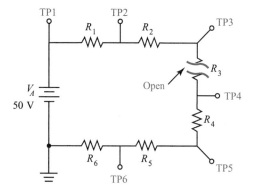

FIGURE 5–3 Determine the voltages between testpoints.

Effects of a Short Circuit

A short in an electrical circuit is characterized as having zero resistance. In a practical circuit, a short may not be literally zero ohms, but its resistance is so much lower than other circuit resistances that it can be assumed to have zero resistance. Technically, the wires used to interconnect the various components can be thought of as short circuits, since they have no resistance for practical purposes. However, these are clearly intentional connections. The short circuit label, however, is generally reserved for cases that have unintentional connections.

Figure 5–4 shows a series resistive circuit with a shorted resistor. When a component shorts in a series circuit, the current will be higher through all other components. This should seem reasonable, because there will be less total

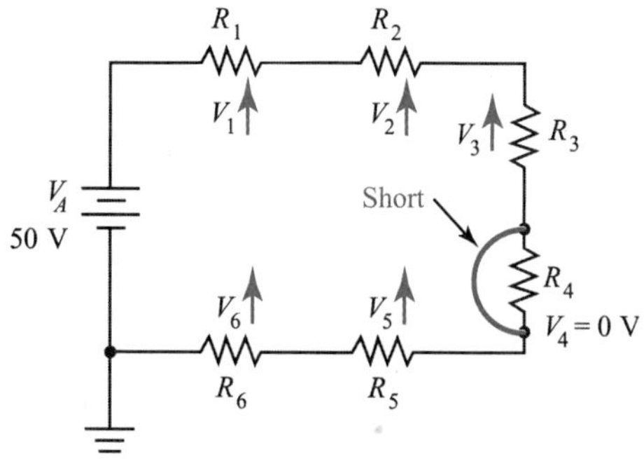

FIGURE 5–4 The voltage across a short circuit is zero. All other component voltage drops are higher than normal.

resistance in the circuit. Since the current is higher, the voltage drops across all but the shorted component will also be higher. The increased voltages are shown in Figure 5–4 as a voltage label followed by an up-arrow symbol (↑). The voltage drop across the shorted component will be zero.

Considering that we have one lead of our voltmeter grounded in accordance with standard practice, we can state the following rule about shorts in series circuits:

> ➤ RULE When a circuit develops a short, all voltages in the circuit (with respect to ground) will measure nearer to the value of the applied voltage terminal on the *opposite* side of the short.

•—EXAMPLE 5.4

Refer to Figure 5–5. Describe the *relative* changes in voltage at each of the indicated testpoints as a result of the short circuit.

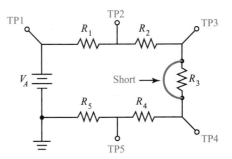

FIGURE 5–5 What relative voltage changes are caused by the short circuit?

SOLUTION Testpoints TP2 and TP3 will measure *nearer* to the negative side of the applied voltage. Since the negative side of the applied voltage is connected directly to ground (0 V), then we can say that the voltage at testpoints TP2 and TP3 will decrease (i.e., get closer to zero). The voltage at TP1 will not be affected by the short circuit, since TP1 connects *directly* to $+V_A$. By contrast, testpoints TP4 and TP5 are on the opposite side of the short and will measure *nearer* to the value of the positive supply terminal. We can say that the voltage at TP4 and TP5 will increase (i.e., become more positive).

•—EXAMPLE 5.5

Refer to Figure 5–6. Describe the *relative* changes in voltage at each of the indicated testpoints as a result of the short circuit.

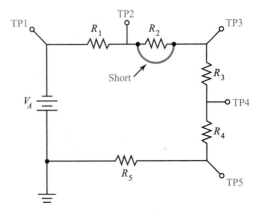

FIGURE 5–6 What relative voltage changes are caused by the short circuit?

SOLUTION Testpoint TP2 will measure nearer to the positive supply terminal. Since the positive supply terminal is grounded, we can say that TP2 becomes less negative (i.e., closer to zero) as a result of the short circuit. The voltage at TP1 will be unaffected by the short circuit, because TP1 is connected *directly* to $-V_A$. Testpoints TP3 through TP5 will measure closer to the value of the negative supply terminal. We can say that TP3 through TP5 become more negative as a result of the short circuit.

• **EXAMPLE 5.6**

Refer to Figure 5–6. Determine the relative *direction* of change caused by the short circuit for each of the voltage measurements that are taken between two testpoints.

a. TP1 and TP2   c. TP3 and TP4
b. TP2 and TP3   d. TP4 and TP5

SOLUTION The short circuit will cause an increased current through all circuit components except R_2. This means that resistors R_1, R_3, R_4, and R_5 will have an increased voltage drop. R_2 will have *no* voltage drop since it is shorted. These observations allow us to determine the following testpoint measurements:

a. TP1 and TP2: Increased voltage   c. TP3 and TP4: Increased voltage
b. TP2 and TP3: 0 V   d. TP4 and TP5: Increased voltage

• **Practice Problems**

Refer to Figure 5–7 for questions 1 through 6.

1. Describe the relative changes in voltage (with reference to ground) at each of the indicated testpoints as a result of the short circuit.
2. Determine the relative *direction* of change caused by the short circuit for the voltage measurement taken *between* testpoints TP1 and TP2.
3. Determine the relative *direction* of change caused by the short circuit for the voltage measurement taken *between* testpoints TP2 and TP3.
4. Determine the relative *direction* of change caused by the short circuit for the voltage measurement taken *between* testpoints TP3 and TP4.
5. Determine the relative *direction* of change caused by the short circuit for the voltage measurement taken *between* testpoints TP4 and TP5.
6. Determine the relative *direction* of change caused by the short circuit for the voltage measurement taken *between* testpoints TP5 and TP6.

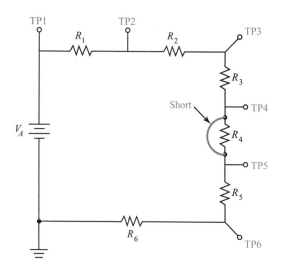

FIGURE 5–7 What relative voltage changes are caused by the short circuit?

Effects of Component Tolerances

In the preceding paragraphs, we considered the effects of opened and shorted components in a series circuit. By definition, an open component has a resistance that approaches infinity while a shorted component has a resistance that approaches zero. We shall now consider the effects when a component changes values (perhaps due to aging or overheating).

The rules are simple. If a component has increased in resistance, then it causes effects similar to those an open component would exhibit, but the symptoms are not as extreme. The voltage across the increased-value component will increase.

All points in the circuit will measure *nearer* to the value of the applied voltage *on the same side of the defect*.

A component that has decreased in value will cause symptoms similar to a shorted component, although the voltage changes will not be as extreme. Thus, the basic rules for locating an incorrectly valued component are identical to the rules for opened and shorted components. These rules for series circuits are summarized in Table 5–1.

TABLE 5–1 Summary of troubleshooting rules for series circuits.

Type of Defect	Voltage		
	With Respect to Ground	Across Defect	Across Other Components
Open	Nearer or equal to supply terminal on *same* side of defect	Full supply	Zero
Increased resistance		Increased	Decreased
Shorted	Nearer or equal to supply terminal on *opposite* side of defect	Zero	Increased
Decreased resistance		Decreased	Increased

If your test equipment is connected *directly* across a voltage source when measuring a particular testpoint, then that voltage will normally be unaffected by either shorts or opens unless the short circuit damages the voltage source. Every time you take a measurement as you troubleshoot a series circuit, you should mentally cross-reference the measured voltage to an entry in Table 5–1. With only a very few measurements, you will have isolated the defect.

Effects of Power Supply Defects

The power supply (i.e., source voltage) can also develop defects that must be identified and corrected. There are two general classes of power supply defects that we will address at this time: no-output voltage and wrong-output voltage.

A power supply with a defect that causes it to produce no-output voltage is fairly easy to identify. In short, there will be no voltage at any point in the circuit including across the power supply terminals. This is such a common defect that many troubleshooters prefer to measure the power supply voltage as a first step. This is largely a matter of personal preference, but it is essential that you recognize this type of defect.

A power supply can also develop defects that cause it to deliver the wrong voltage. If a battery develops this type of defect, then the output voltage will be less than normal. A battery does not develop defects that cause excessive terminal voltage. On the other hand, electronic voltage sources such as those discussed in later chapters can have defects that cause any of these three symptoms: no output, low output, and high output.

Following the troubleshooting strategies previously discussed will automatically lead you to discover any power supply defects. As you measure testpoints and follow the symptoms toward the defect, you will ultimately reach the supply terminals. When you measure an incorrect voltage directly on the supply terminals, you may have located a power supply defect. To be certain, you must disconnect the supply from the rest of the circuit and measure the supply voltage again. If it is still wrong, then you have located a defective supply. If the voltage returns to normal when the supply is disconnected, you probably have a short, or at least a low resistance path, somewhere in your circuit. A short in the circuit can draw excessive current. But, any given power supply can only furnish a certain maximum value of current and still maintain the correct terminal voltage. If it tries to supply more current than its limit, the output voltage will generally decrease. Table 5–2 summarizes power supply symptoms and the probable causes.

TABLE 5–2 Possible power supply symptoms and their causes.

Power Supply Symptom	Cause	Comments
No output voltage	Defect in power supply	No output even when disconnected from rest of circuit
	Short circuit across power supply terminals	Supply voltage returns to normal when disconnected from rest of circuit
High output voltage	Defect in power supply	Few exceptions
Low output voltage	Defect in power supply	Still low with supply disconnected
		Still low when supplying normal or less than normal current
	Short or low resistance in rest of circuit	Supply output returns to normal when disconnected from circuit and/or when supplying a normal current

Circuit Connectivity Problems

In the earlier discussion about open circuits and short circuits, we restricted our focus to open and short components. The wires and/or printed circuit board traces used to interconnect the components were assumed to be normal. In practice, however, wires and circuit board traces can develop defects.

When a wire or trace opens in a series circuit, it acts just like an open resistor. Current in the entire circuit will be zero. No voltage will be dropped across the other components. The full supply voltage will be dropped across the open circuit.

A wire or trace can also develop a short circuit, though not in the same way as previously discussed. A wire or trace is supposed to be a short circuit (i.e., very low resistance) by design. So, shorting end-to-end would have no effect on normal circuit operation. But, a wire or trace can become shorted to (i.e., touch) another nearby wire, trace, or metal chassis. This may happen during the initial soldering process or as a result of repair work on the board. There are no consistent symptoms for a short of this type, since they depend heavily on which two wires are being shorted together. The result may be nearly impossible to detect (i.e., the circuit continues to work normally), or the results may be catastrophic (e.g., components damaged, sparks, the product may catch on fire, and so forth).

✓ **Exercise Problems 5.1**

1. Because pure series circuits are not the most common configuration in real products, there is no reason to be skilled in troubleshooting this type of circuit. (True/False)
2. Explain the reason for the troubleshooting guideline, which says, "Make a measurement *only* if you know what a good reading would be."
3. Describe the effect on current through one component in a series circuit when another component in the circuit develops an open.
4. Describe the effect on voltage across one component in a series circuit when another component in the circuit develops an open.

Refer to Figure 5–8 for questions 5 through 10.

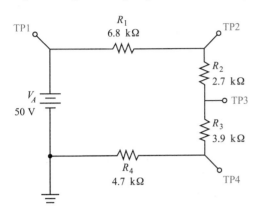

FIGURE 5–8 A defective series circuit.

Fig05_08.msm

5. If the normal voltage at TP2 is 31.2 V, what happens to the voltage at TP2 if resistor R3 is open?
   a. more positive
   b. slightly negative
   c. remains the same
   d. goes to zero
6. If the normal voltage at TP1 is 50 V, what happens to the voltage at TP1 if resistor R_3 is open?
   a. more positive
   b. slightly negative
   c. remains the same
   d. goes to zero
7. What happens to the voltage at TP4 if R_4 becomes shorted?
8. What happens to the voltage at TP3 if R_1 develops a short?
9. What happens to the current through R_4 if R_1 develops a short?
10. If the normal voltage at TP3 is 23.8 V, but actually measures 42.2 V, which of the following could be the possible defects?
    a. R_1 increased in resistance
    b. R_2 open
    c. R_3 increased in resistance
    d. R_4 shorted

5.2 TROUBLESHOOTING PARALLEL CIRCUITS

Parallel circuits are one of the most basic circuit configurations, but they are far from being easy to troubleshoot in a professional manner. It is important to practice only logical, systematic troubleshooting methods while you are learning to troubleshoot. Because the circuits in this text and in your lab experiments have relatively few components, it is very tempting to utilize unprofessional troubleshooting methods. Unprofessional methods may even be faster for simple circuits. However, if you discipline yourself to practice only professional, systematic troubleshooting—even on simple circuits—then you will be rewarded when you begin to troubleshoot more complex systems. A logical procedure will work equally well on a complex circuit. An unprofessional method, by contrast, fails miserably when applied to complex circuits.

Three important rules were discussed in the preceding section on troubleshooting series circuits that are equally applicable to troubleshooting parallel circuits:

- Know what is normal.
- Minimize the number of measurements.
- Use the best tool.

The first two items were adequately discussed in the previous section and will not be repeated here. The third item deserves special consideration with reference to parallel circuits. First, a voltmeter is of only minimal value when troubleshooting a purely parallel circuit, since the voltage across all components (even open or shorted ones) is the same. Second, a current meter is an excellent diagnostic tool in theory, but it is impractical for most real-world troubleshooting situations. The primary limitation of the current meter is that the circuit must be opened to measure the current. This increases the likelihood of damage to an expensive printed circuit board. Finally, an ohmmeter is of minimal value. Unless the circuit is broken, all components will measure the same value (i.e., they are all connected in parallel). The following paragraphs will examine two alternative methods that can help you locate a defect in a parallel circuit.

Troubleshooting with a Current Probe

Many manufacturers of test equipment make digital multimeters that have current probe options. Current probes that work with most standard digital multimeters (**DMMs**) can also be purchased. A representative current probe is shown in Figure 5–9. A **current probe** senses the magnetic field produced by a current-carrying conductor and produces a corresponding voltage. The output voltage is displayed by the DMM and is a direct function of current flow. Some current probes have jaws that open like a pair of pliers. The jaws must be clamped around the conductor where the current is to be measured. This type of probe is excellent for troubleshooting circuits that have discrete interconnecting wires. Unfortunately, a clamp-on probe is of little value for troubleshooting printed circuit boards (PCBs), since the probe cannot be clamped around a PCB trace. Additionally, the clamp-type probe generally responds to alternating currents only, so it is of no value when troubleshooting dc circuits.

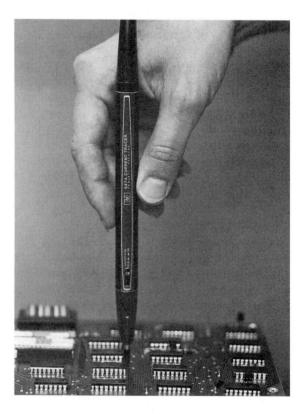

FIGURE 5–9 A representative current probe. (*Courtesy of Hewlett-Packard Company*)

Another type of current probe has the advantage of not having to be clamped around the measured conductor. It can sense the magnetic field associated with the current just by being near the conductor. Its primary disadvantage is that the output is a *relative* indication of current. That is, you cannot measure the exact amount of current, but can usually tell when one branch is more or less than another. This type of current probe also responds to dc.

Both types of current probes have the advantage of not having to break the circuit to get an indication of the current through a conductor. This provides us with a tool for locating defects in a parallel circuit.

•–EXAMPLE 5.7

Show how a current probe could be used to locate an open component in a parallel circuit.

SOLUTION Figure 5–10 illustrates one possible technique. The current probe is used to monitor the currents in the various parallel branches. The good branches will have a normal value of current flow. The open branch has no current and is easily detected. If the physical implementation of the circuit will allow, then you can use the split-half method (described later in this chapter) to reduce the number of measurements needed to locate the defect.

FIGURE 5–10 A current probe is a valuable tool for troubleshooting parallel circuits without having to break the circuit.

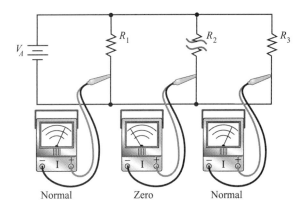

•—EXAMPLE 5.8

Describe how a current probe could be used to locate a resistor whose resistance was substantially lower than normal.

SOLUTION The procedure illustrated in Figure 5–10 also applies to locating resistances that are lower than normal. In this case, however, the defective branch would be identified by its excessive current flow.

Troubleshooting with a Temperature Probe

A temperature probe is another attachment that you can buy for a DMM. It generally has a small, metal surface that is used to touch an object. The temperature of the contacted object is displayed on the DMM readout. Figure 5–11 shows a representative temperature probe.

FIGURE 5–11 A representative temperature probe. (*Courtesy of Hewlett-Packard Company*)

Since power dissipation in an electronic component produces heat, we can measure the *relative* temperatures of the components in the branches of a parallel circuit and locate a defect. The exact temperature of a component is usually not an important troubleshooting consideration. But its temperature relative to normal and relative to the other branches is a valuable diagnostic tool for locating defects in parallel circuits.

If a component in a parallel circuit becomes open, it will dissipate no power. Consequently, the temperature probe will indicate a temperature much lower than normal. Similarly, if the resistance of a particular branch becomes lower than normal, it will dissipate more power. The temperature probe can sense the resulting increase in temperature.

Neither the current probe nor the temperature probe is the ideal tool for every application, but they are very helpful for many types of circuits that you may encounter.

Effects of a Short in a Parallel Circuit

If a branch in a parallel circuit develops a short, extremely high currents will flow from the source through the short circuit. In most cases, the source will be damaged by the high current unless a fuse, a circuit breaker, or some other form of current limiting protects it. In any case, the voltage across all branches of the parallel circuit will be zero—initially because of the zero resistance of the short, and ultimately because of the action of the protection device.

✓ Exercise Problems 5.2

1. A voltmeter is generally the best tool to use when locating defects in a parallel circuit. (True/False)
2. Explain why a standard current meter is not usually a practical tool for troubleshooting parallel circuits.
3. What common characteristic makes current probes and temperature probes good choices for locating defects in parallel circuits?

5.3 TROUBLESHOOTING SERIES-PARALLEL CIRCUITS

Locating defects in a series-parallel circuit can be as exciting as it is challenging. It is important that you practice only logical troubleshooting procedures, because these same procedures will help you diagnose more complex circuits and systems in the future.

Basic Concepts

It is important for you to develop a logical and systematic troubleshooting technique early in your career so that you will be able to quickly repair both simple circuits and complex electronics systems. We will discuss other techniques later in the chapter, but for now, you should rely on three definite troubleshooting guidelines for series-parallel circuits:

- Don't make a measurement unless you know what a "good" reading would be.
- Make as few measurements as possible.
- Rely mostly on voltage measurements. Use ohmmeter checks as the final proof of a defective component. Use current measurements only in very special cases.

The first two guidelines have been discussed previously. We will focus on the third guideline regarding choice of instruments. A voltmeter is your most powerful diagnostic tool for series-parallel resistive circuits. You can measure any component without disturbing the circuit, and it is not necessary to turn off the power. When troubleshooting an actual electronic system, opening a circuit for a current measurement or lifting a component for a resistance test requires desoldering components on a delicate (and expensive) printed circuit board. It should be avoided whenever possible. Once you have identified what you think is the defective component by making voltage checks, then you can desolder it, verify your assertion by measuring the resistance of the suspected component, and then replace it with a new component.

Visualizing an Overall Series Circuit

The first step toward becoming a skilled troubleshooter of series-parallel circuits is to develop a solid intuitive (i.e., nonmathematical) understanding of circuit behavior. For this you must learn to make mental approximations, and be able to classify circuit quantities as high, low, or normal.

Let's begin by briefly reviewing the troubleshooting chart presented in Table 5–1 with reference to series circuits. It is repeated in Table 5–3.

TABLE 5–3 Troubleshooting chart indicating relative voltage values in a series circuit with a defective component.

Type of Defect	Voltage		
	With Respect to Ground	Across Defect	Across Other Components
Open	Nearer or equal to supply terminal on *same* side of defect	Full supply	Zero
Increased resistance		Increased	Decreased
Shorted	Nearer or equal to supply terminal on *opposite* side of defect	Zero	Increased
Decreased resistance		Decreased	Increased

As you troubleshoot series-parallel circuits, the methods presented in Section 5.1 and the use of Table 5–3 are still applicable. The only significant difference is that you must mentally view parallel combinations of resistors as a single resistor. This is essentially the same process used to simplify a parallel portion of a series-parallel circuit when you are determining total resistance. Once you have made the mental transformation of the circuit, you proceed exactly as you would if you were troubleshooting a series circuit.

If the defect is isolated to one of the true series components of the circuit, then you are finished. On the other hand, if the defect is isolated to an equivalent resistor that represents a bank of parallel resistors, you can then apply the troubleshooting methods discussed for parallel circuits to identify the specific defect within that bank of components.

✓ Exercise Problems 5.3

Refer to Figure 5–12 on page 143 for questions 1 through 3.

1. When the circuit is mentally viewed as a series circuit for troubleshooting purposes, how many resistors are there in the resulting series circuit?
2. If the voltage at TP4 measures zero, name at least three possible defects.
3. If the voltage at TP2 is higher than normal but less than 20 V, name at least three possible defects.

Refer to Figure 5–13 on page 143 for questions 4 through 6.

4. Name at least two defects that would cause the voltage at TP3 to measure zero.
5. Name at least two defects that would cause the voltage at TP5 to measure higher than normal but less than 150 V.
6. Name at least two defects that would cause the voltage at TP4 to be lower than normal but greater than zero.

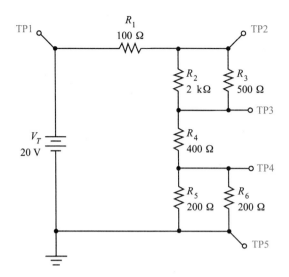

FIGURE 5–12 A series-parallel circuit.

Fig05_12.msm

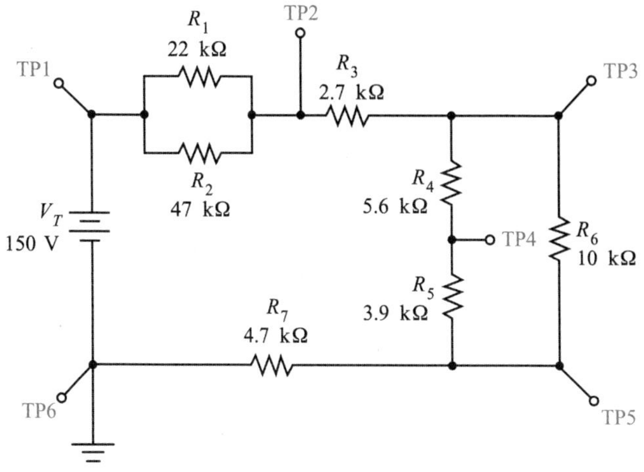

FIGURE 5–13 A series-parallel circuit.

Fig05_13.msm

5.4 TROUBLESHOOTING STRATEGIES

This section will introduce you to several troubleshooting techniques that are applicable to all kinds of troubleshooting situations. The concepts work equally well for simple circuits, complex systems, and even nonelectronic systems.

Observation and Equipment History

We discussed the importance of keen observation skills and the value of historical troubleshooting records in chapter 1. These critical issues will not be repeated here except as a reminder of their place in your overall array of troubleshooting tools.

Troubleshooting Fences

A fence is typically thought of as something that defines boundaries. In the case of troubleshooting, we will use troubleshooting fences to define the boundary between those components that may be defective and those that are definitely not defective. The fences will enclose and identify those components that are still suspect. When you first begin troubleshooting a particular problem, the fences may enclose the whole system. As you take measurements and begin to localize the defect, you move the fences so they continue to surround only the suspected components. Eventually, they will surround only a single component, and the defect will have been positively identified.

Figure 5–14 on page 144 briefly illustrates how troubleshooting fences might be applied to the diagnosis of a home computer system. In Figure 5–14(a), the fences enclose the whole system because we don't have any clues yet.

After milking the front panel and making a series of observations, let's suppose we determine that the only problem seems to be that we cannot print any documents. In that case, we might move our troubleshooting fences to enclose only the computer and the printer as shown in Figure 5–14(b). In other words, based on our understanding of the system and the behavior that we have observed, we are

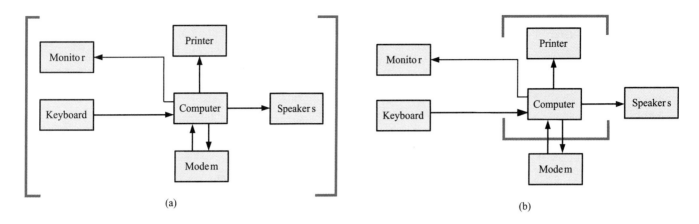

FIGURE 5–14 Troubleshooting fences are used to identify suspected components.

willing to say that the problem must lie in either the printer or the computer. We do not believe that defects in the monitor, keyboard, speakers, or modem could cause the symptoms we have.

This process of reducing the number of components within the fences continues to whatever level is required to fix the system. In the present case, for example, we might be satisfied to localize the problem to the printer and then replace the printer. Alternatively, if we had access to schematics for the printer circuitry, we might resume our fencing operation within the printer. Initially, the fences would enclose the whole printer. They would then shrink to include only a single circuit board. More measurements would reduce the fence so that only a few components were enclosed. As further measurements excluded certain components, the fences would ultimately enclose only a single electronic component, which, of course, is the defect that caused the initial symptoms for the overall system.

Equal-Probability Troubleshooting

Equal-probability troubleshooting is a technique that allows you to make a few initial observations and then immediately move your troubleshooting fences to include only one or two suspected components. In other words, it provides an extreme shortcut that allows you to identify the defective component(s) after only brief observation and/or a few quick measurements. Its major drawback is that it cannot be utilized for practical troubleshooting exercises until you have gained substantial experience with the particular system being repaired.

Here's how it works. Throughout the lifetime of a given system, a certain number of defects will occur. The actual number of defective components is always a tiny fraction of all the possible defects. Additionally, there will nearly always be a lot of repetition within those defects. That is, while most components in a typical system never develop defects, certain other components have frequent or repeated defects. So, if you have maintained a good equipment history in your logbook (either physically or mentally), then you will begin to recognize certain symptoms as being associated with a specific defect. When you encounter these symptoms, you can immediately move your fences to include only the component(s) known to have caused similar symptoms in the past. In essence, you are saying to yourself, "There is just as much probability that the defect is with this one (or two) components than there is for all of the other components combined."

In most cases, your experience will prevail, and you will quickly identify the defective component. If it turns out that the suspected component is good, you will have to expand your fences and continue with the normal troubleshooting process.

Split-Half Troubleshooting

Split-half troubleshooting is a technique that allows you to locate defects with fewer measurements. It is most applicable to systems that are sequential in nature. In later studies, you will find that some systems perform a series of operations on an electrical signal in a specific sequence. You could start at one end of the sequence and examine each stage of the overall process in order (or reverse order). If there were 100 sequential operations, then you might have to test all 100 operations to locate the defect if the actual defect was the last operation you tested. By contrast, if the defect was at the first location you tested, you could locate it with only a single measurement or observation. If we applied this procedure to a large number of systems with random defects, we would find that it would take an average of 50 measurements to locate a defect.

Split-half troubleshooting offers an alternative to the sequential measurement technique just described. It is illustrated in Figure 5–15.

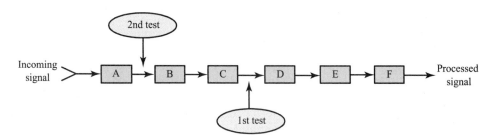

FIGURE 5–15 Split-half troubleshooting reduces the number of measurements needed to locate a defect in a sequence of operations.

In order to apply split-half troubleshooting, you choose the midpoint of the system for your first measurement or observation. If the signal is normal at that point, then the defect must lie in some later stage of the overall process. On the other hand, if the signal is not normal at the midpoint of the system, then the defect must be somewhere in the first half of the overall process. So, with one measurement, you have essentially excluded half of the system from the range of possible defects.

Suppose, for example, the defect was in process A in Figure 5–15. The first measurement between stages C and D would reveal an abnormal signal, since the defect has already occurred. But, you would then know that the defect had to lie within the first half of the system. Your next testpoint should be selected at the midpoint of the suspected range of components. In this case, we would make our second test at either the input or output of stage B. In the present example, let us measure the input of stage B. It will show an abnormal signal, which means the defect has to be in stage A. In two measurements, we have located the defective process.

Let's apply this procedure to the 100-stage process that we discussed earlier. Recall that it took an average of 50 measurements (and sometimes as many as 100) to positively locate the defective stage. For this example, we will assume the defect is in stage 27. Table 5–4 on page 146 shows how the troubleshooting exercise would go. As you can see, with only seven measurements we localized the defect. Regardless of where the defect is located, it would take no more than seven measurements to isolate the defective stage in a 100-stage system.

Signal Tracing and Signal Injection

Signal tracing and signal injection were discussed in chapter 1. At this point, we need only clarify their relationship to the split-half troubleshooting procedure just described. The discussion of split-half troubleshooting in the preceding paragraphs focused on the use of signal tracing. That is, it was assumed that a normal signal was applied to the input of the overall sequence of processes. The

TABLE 5–4 Split-half troubleshooting applied to a 100-stage system with a defect located in stage 27.

Measurement	Between Stages	Result	Possible Defects
0	—	—	1–100
1	50–51	Abnormal	1–50
2	25–26	Normal	26–50
3	37–38	Abnormal	26–37
4	31–32	Abnormal	26–31
5	28–29	Abnormal	26–28
6	27–28	Abnormal	26–27
7	26–27	Normal	27

partially-processed signal was then monitored at various intermediate points in the system to determine signal integrity.

Split-half troubleshooting can also be used in conjunction with signal injection. In this case, we continuously monitor the final output of a series of sequential processes. We then inject a good signal at some intermediate stage and observe the output indicator. If the output signal is normal, then all of the stages between the injection point and the output can be assumed to be normal, in which case the defect would lie ahead of the injection point. In any case, after each test, the injection point is moved to a location that is halfway through the range of remaining suspects.

Signal tracing and signal injection can also be applied to troubleshooting resistive circuits. If, for example, you use a voltmeter and measure the various testpoints on an energized circuit, you are using signal tracing. The "signal" is simply a dc voltage, and the normal versus abnormal decision is based strictly on its value. In the next chapter, you will encounter signals that have other distinguishing characteristics besides voltage.

If you so desired, you could also troubleshoot a resistive circuit by using signal injection. In this case, you would leave the normal voltage source disconnected. You would then inject a signal at the various testpoints by connecting another voltage source with its value set for the normal testpoint voltage. Additionally, you would continuously monitor the testpoint that was closest to ground. If the defect lies between the monitored testpoint and the injection point, the monitored value will be abnormal. On the other hand, if the defect lies outside of this range, the monitored output will be normal when a normal voltage is forced onto a given testpoint.

Signal injection is a very useful strategy for many types of electronic systems, including many types of computer circuits. But, it is not recommended as the preferred method for troubleshooting resistive circuits. The resistive circuit example was given here to illustrate the concept.

Component Substitutions

All troubleshooting exercises hopefully end when you have identified what you believe is the defective component. Once you have reduced the range of possible defects down to a single device, you must confirm your suspicions. Confirmation generally requires removal of the component from the circuit and testing its condition. In the case of a resistor, its condition can be tested with an ohmmeter. There are test instruments available for most components such as inductors, capacitors, transistors, and many integrated circuits. However, there will be many times that you do not have access to these specialized instruments. In these instances, substitution of a good replacement part is the best action.

If the system operation returns to normal when the good replacement part is installed, you have successfully diagnosed the problem. But, what do you do with the bad part? Unless there is some overriding reason for keeping it (such as a warranty claim), you should be certain it is discarded in a place where it cannot find its way back into your stock of parts. If a defective part is inadvertently used as a substitute during a troubleshooting exercise, you will be misled by the symptoms. Even if you have correctly identified the defect, you won't realize it, because your substitute part (which was also defective) failed to repair the system. Always dispose of parts known to be defective.

Now, consider the case in which you remove a component from a circuit board, substitute it with a new part, and find that the system displays identical symptoms. In other words, the suspected (and removed) component was not actually defective. What should you do with the removed component that is still good? If the component is either very rare (locally) or quite expensive, then you may want to keep it with your good parts if you are absolutely certain the removed part is good. In the vast majority of cases, however, your best strategy is to discard the removed component regardless of its condition. This apparently wasteful action is recommended on the observation that most common electronic components cost only a few cents (many resistors and capacitors cost only fractions of a cent). Second, tiny surface-mount components can be damaged in the process of removing them from the circuit board, and the damage is not always visible to the eye. Do not risk introducing confusion to an already challenging troubleshooting task by allowing bad parts to mingle with good parts.

In-Circuit Testing of Components

Up to this point, we have always suggested that a component be removed from the circuit in order to verify its condition. In most cases, this is the best course of action. However, there are times when you can verify the condition of a component without first having to unsolder it. Some of these situations are described here.

Figure 5–16(a) shows a series-parallel resistive circuit. Your troubleshooting to this point has led you to suspect that one of the resistors in the R_3–R_4 network is defective. If you open S_1 (or simply disconnect the voltage source), you can then use an ohmmeter in the circuit without risk of damaging the ohmmeter. Now, if you connect an ohmmeter across the R_3–R_4 network, what will you actually measure? Figure 5–16(b) on page 148 shows the effective paths that will be seen by the ohmmeter.

As you can see, the ohmmeter will be measuring the effective resistance of three paths. The total resistance indicated by the meter (if there is no defect) is about 915 Ω. Now, suppose you are trying to determine if R_4 is open. Well, if it is open, the ohmmeter will measure 964 Ω. This is not much of a change. If your meter indicates 964 Ω, you cannot reliably determine whether R_4 is actually open, or

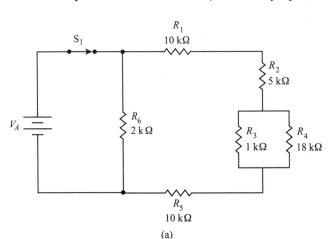

FIGURE 5–16
In-circuit resistance measurements.

(a)

FIGURE 5–16 (continued)

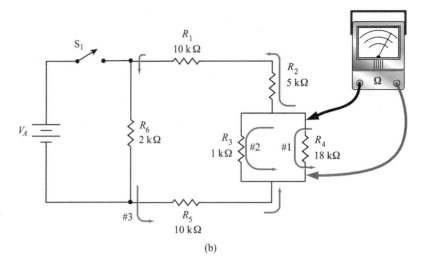

(b)

whether some of the other resistors are simply near the limits of their tolerance values.

On the other hand, suppose your troubleshooting steps have led you to believe that R_4 is shorted. If it is shorted, then an ohmmeter across it will indicate zero regardless of what else is also in parallel with it. In this case, the in-circuit ohmmeter check could provide some useful information. It still doesn't positively prove that R_4 is shorted, but confirms that it might be. On the other hand, if the measured value is over 900 ohms, then you would know for certain that R_4 is not shorted without ever having removed it from the circuit.

As a general rule, in-circuit resistance measurements are good from the standpoint that they avoid possible damage to the circuit board during component removal. But, in order for the measurements to be useful, you have to be able to account for the effects of parallel current paths, which are often called **sneak paths**.

Staying on Track

The ability to stay focused on the trail of a defect while troubleshooting a complex system is an important skill, but it is one that may be hard to appreciate until you gain personal experience with larger circuits. In short, if you take a measurement that indicates a defect, then you must follow that path through to completion. It is not uncommon for a malfunctioning system to have more than one physical defect at the same time (generally one is the result of another). A common pitfall is to follow one trail for a while and then be distracted by a new symptom that is indicative of a different malfunction. If you come across a new symptom that indicates a defect (but not the defect you are currently pursuing), note it for future reference, but don't follow it until you exhaust your current path. Failure to follow this simple guideline can result in your wandering in circles without ever localizing the actual defects.

✓ Exercise Problems 5.4

1. Troubleshooting fences separate what two classes of components?
2. At the end of a troubleshooting exercise, how many components are inside the troubleshooting fences?
3. Equal-probability troubleshooting is a strategy that is best suited to beginning troubleshooters. (True/False)
4. Suppose you applied split-half troubleshooting to a 16-stage sequential system. How many measurements would you have to make to positively localize the defective stage?
5. Why is it often a good idea to discard a component removed from a circuit board even if the component is thought to be good?

SUMMARY

When troubleshooting circuits, you should make a measurement only if you know what a normal value would be. You should try to minimize the number of measurements that you take. Voltmeter tests are a good choice for troubleshooting series circuits and series-parallel circuits. Defects in parallel circuits are easier to locate by using a temperature probe or a current probe. These generally provide only relative indications but can often isolate a defective branch.

An open in a series circuit results in zero current for all components. The voltage across the open is equal to the applied voltage, while the voltage across all other components is zero. A short in a series circuit causes higher current and higher voltage drops in all other components. The voltage across the shorted component will be zero.

An open branch in a parallel circuit has no effect on the current or voltage of the other branches, but it does reduce the value of total current. A shorted branch in a parallel circuit will normally activate a protection device such as a fuse or circuit breaker. Otherwise, the voltage source is likely to be damaged.

During troubleshooting, a series-parallel circuit can be viewed as a series circuit by mentally collapsing all parallel networks. Series troubleshooting strategies are applied to the effective series circuit. If the defect is localized to an equivalent resistor that represents a parallel network, then parallel troubleshooting techniques can further localize the defect to a specific component.

Observation skills are essential for effective troubleshooting. The use of troubleshooting fences helps maintain focus and reduce distractions. Applying equal-probability and split-half troubleshooting strategies can dramatically reduce the number of measurements required to positively identify a defect.

Some components can be tested in the circuit, provided you account for the effects of sneak paths. Substitution of a known good component is a very effective technique, but it is essential to keep good and defective parts well isolated from each other at all times.

REVIEW QUESTIONS

Section 5.1: Troubleshooting Series Circuits

1. When troubleshooting simple circuits, it is best to guess at the defect, since that is often faster. (True/False)
2. Why is it important to know what constitutes a normal value before deciding to take a measurement during a troubleshooting exercise?
3. Why should you strive to minimize the number of measurements taken during a troubleshooting task?
4. A current meter is the ideal tool for troubleshooting a series circuit. (True/False)

Refer to Figure 5–17 for questions 5 through 15.

5. If R_1 opens, what is the value of current through R_3?
6. What is the relative effect on current through R_2 if resistor R_4 develops a short?
7. If the voltage at TP3 measures zero, circle all of the following that could be possible defects:
   a. R_1 open
   b. R_1 shorted
   c. R_2 decreased in value
   d. R_3 shorted
   e. defective voltage source
   f. R_4 open
8. If R_3 developed a short, then the voltage at TP4 would be higher than normal. (True/False)
9. What happens to the relative voltage at TP2 if resistor R_3 increases in value?

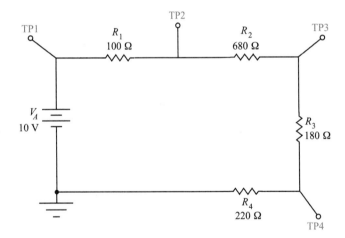

FIGURE 5–17 A series circuit.

 Fig05_17.msm

10. What is the relative effect on the voltage at TP3 if the supply voltage is lower than normal?
11. What is the relative effect on voltage measured between TP2 and TP3 if R_4 develops a short?
12. If R_1 increases in resistance, what happens to the voltage measured between TP3 and TP4?
13. What is the effect on the voltage measured between TP1 and TP2 if resistor R_1 shorts?

14. What is the effect on the voltage measured between TP1 and TP2 if resistor R_4 opens?
15. What is the effect on the voltage measured between TP2 and TP4 if R_3 develops a short?

Section 5.2: Troubleshooting Parallel Circuits

16. A voltmeter is the ideal test instrument for diagnosing defects in parallel circuits. (True/False)
17. Give an advantage and a disadvantage of using a current meter for troubleshooting parallel circuits.
18. What is the most outstanding difference between a meter with a current probe and a standard current meter?
19. How could a temperature probe be used to identify a branch of a parallel circuit that had increased in resistance?

Refer to Figure 5–18 for questions 20 through 23.

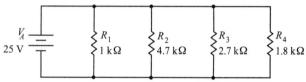

FIGURE 5–18 A parallel circuit.

Fig05_18.msm

20. What is the effect on the voltage across R_3, if resistor R_1 opens?
21. If R_2 develops an open, what happens to the current through R_4?
22. If R_4 increases in resistance, how does this affect the current through R_3 and the current through the voltage source?
23. If resistance of R_1 is reduced to 50% of its correct value, what happens to the voltage drop across R_2?

Section 5.3: Troubleshooting Series-Parallel Circuits

24. Series-parallel circuits can be visualized as simple parallel circuits for the purpose of localizing defects. (True/False)
25. A voltmeter is a valuable troubleshooting tool for series-parallel resistive circuits. (True/False)
26. When would a temperature probe be of value for troubleshooting a series-parallel circuit?

Refer to Figure 5–19 for questions 27 through 30.

27. The voltage at TP3 is higher than normal. Circle all of the following that could be possible defects:
    a. R_1 open   d. R_4 shorted
    b. R_2 open   e. R_6 open
    c. R_3 shorted

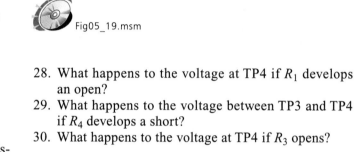

FIGURE 5–19 Circuit for questions 27 through 30.

Fig05_19.msm

28. What happens to the voltage at TP4 if R_1 develops an open?
29. What happens to the voltage between TP3 and TP4 if R_4 develops a short?
30. What happens to the voltage at TP4 if R_3 opens?

Section 5.4: Troubleshooting Strategies

31. Why is observation an important part of troubleshooting?
32. What percentage of the total system components is contained within the troubleshooting fences at the very beginning of a troubleshooting exercise?
33. How many components are contained within the troubleshooting fences at the very end of a troubleshooting exercise?
34. The equal-probability troubleshooting technique can only be effectively utilized after you have gained experience on the particular system being repaired. (True/False)
35. If a system consisting of 25 sequential processing stages is diagnosed using the split-half troubleshooting method, it will take 12 measurements to localize the defect to a specific stage. (True/False)
36. What is the primary reason for utilizing the split-half troubleshooting method?
37. Which of the following techniques are compatible with the split-half troubleshooting method?
    a. signal injection   c. both
    b. signal tracing   d. neither

38. In most cases, it is best to discard components that are removed from a printed circuit board even if they are thought to be good. (True/False)

Refer to Figure 5–20 for questions 39 through 41.

39. If you suspect R_2 of being shorted, can you test it with an ohmmeter without unsoldering it from the circuit?
40. If you suspect R_4 of being open, can you test it with an ohmmeter without unsoldering it from the circuit?
41. If you suspect R_3 of being open, can you test it with an ohmmeter without unsoldering it from the circuit?

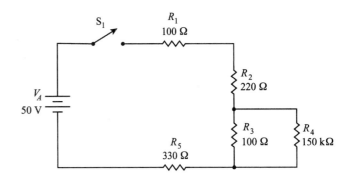

FIGURE 5–20 Circuits for questions 39 through 41.

Fig05_20.msm

●—CIRCUIT EXPLORATION

A circuit simulation software package can be a valuable tool to help you develop practical troubleshooting skills. First, you construct the circuits in the schematic capture portion of the program. Next, you simulate the circuit under normal conditions and document the normal voltages at each testpoint (node) in the circuit. You will need these in order to contrast the voltages measured with a defect in the circuit. Finally, you insert a defect into a component and troubleshoot it by measuring the various testpoints. Even though you can make any number of measurements in the simulator environment, you will benefit the most if you strive to minimize the number of tests required to positively identify the defect.

It is generally more fun and more challenging when you work with a partner. Insert a defect without your partner's knowledge and let your partner try to locate the defect. Once the problem has been identified, correct it immediately so that you won't forget. Now have your partner insert a defect and give you a chance to troubleshoot the circuit. Since you have no way of knowing where the defect is located, you will have to employ your troubleshooting skills to find it.

Any circuit can be used for this purpose, but the circuit shown in Figure 5–21 will provide you with a good starting point.

Enter the circuit into the schematic capture portion of your simulator package and complete the two "Normal" columns in Table 5–5 on page 152. Use these normal values for contrasting with values measured in the presence of a defect to make judgments concerning the nature and location of the defect.

You will need a new set of "Actual" columns for each new defect. One way to avoid having to reproduce multiple copies is to lay a piece of transparency material over

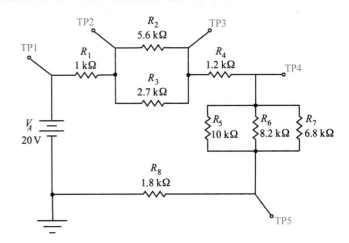

FIGURE 5–21 A circuit for troubleshooting practice.

Fig05_21.msm

the printed table. If you make your entries with water-based markers or grease pencils, you can then easily erase them and be ready for a new problem. The down side of this method is that it robs you of the chance to keep historical data on a particular circuit, which means you won't be able to streamline your troubleshooting of a particular circuit based on your experience as noted in your logbook.

Some circuit simulation packages, such as MultiSIM, allow you to insert defects directly into the component models. On other packages that do not provide this convenience, you can simulate defects in resistors by entering the largest possible value (at least 1.0 GΩ) to represent an open, the smallest value (e.g., 0 Ω or 1 mΩ) to simulate a short, and of course a wrong value to represent an out-of-tolerance component. Clearly, these values should not be made visible as you troubleshoot or you will be giving out unreasonable clues.

You can measure improvements in your skill level based on how long it takes you to locate a defect and how many measurements are required.

TABLE 5-5 A measurement table for the circuit in Figure 5–21.

Testpoint	Normal		Actual	
	Voltage	Resistance	Voltage	Resistance
TP1				
TP2				
TP3				
TP4				
TP5				
TP1–TP2				
TP2–TP3				
TP3–TP4				
TP4–TP5				
Total number of measurements required to locate the defect				

ANSWERS TO PRACTICE PROBLEMS
Page 133
1. 0 V
2. 0 V
3. 50 V
4. 0 V
5. 0 V

Page 135
1. TP1 is unaffected since it is connected directly to the voltage source. TP2 decreases. TP3 decreases. TP4 decreases. TP5 increases. TP6 increases.

2. Increase
3. Increase
4. Increase
5. Decrease (0 V)
6. Increase

CHAPTER 6
Alternating Current

•—KEY TERMS

alternation
asymmetrical waveform
cycle
digital signal
doppler effect
fundamental frequency
half-cycle
harmonic frequency
lagging
leading
peak value
peak-to-peak
phase
piconet
root-mean-square (rms)
sine wave
slip rings
symmetrical waveform

•—OBJECTIVES

After studying this chapter, you should be able to:

- Explain how a sinusoidal waveform can be generated.
- Describe the relationship between frequency and period of a sine wave.
- Compute any of the following voltage or current values for a given sine wave: peak, peak-to-peak, average, rms, and instantaneous at a specified angle.
- Discuss the meaning of phase when referencing sinusoidal waveforms.
- Analyze resistive ac circuits.

•—INTRODUCTION

Alternating current and voltage is critical to the operation of nearly all practical electronic devices. The 120 volts supplied to your home by the power company, the signal voltages that cause music from your stereo, the radio waves that allow radio and television transmission, the radar waves used in speed detectors, the ultrasonic waves used in burglar alarms, and the telemetry signals used to communicate with satellites are all examples of alternating current and voltage. Many computers are interconnected with a communications link known as Ethernet. The signals on an Ethernet cable are also examples of alternating current and voltage.

6.1 GENERATION OF ALTERNATING VOLTAGE

A direct voltage or current (dc) such as might be supplied by a battery has a certain value and a certain polarity. It remains relatively constant indefinitely. By contrast, an alternating voltage or current (ac) varies in both amplitude and polarity. To better appreciate these important characteristics of an alternating voltage or current, let us see how these voltages may be produced.

Rectangular Waveshapes

Figure 6–1 illustrates a very simple way to produce an alternating voltage. When the switch in Figure 6–1(a) is in position A, a positive voltage (with reference to ground) is applied to the resistor. V_{A_1} is essentially connected across the resistor. These times are labeled in Figure 6–1(b). When the switch is moved to position B, V_{A_2} is connected across the resistor, and a negative voltage (with reference

154 CHAPTER 6 • ALTERNATING CURRENT

FIGURE 6-1 A simple way to produce alternating voltage.

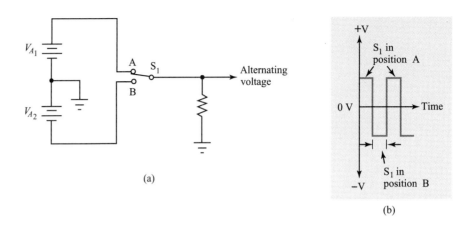

to ground) can be measured. These times are also labeled in Figure 6–1(b). The horizontal axis of the graph is measured in time (e.g., seconds). Clearly, the rate at which the switch is moved between positions determines the duration of each voltage polarity interval on the graph.

Figure 6–2 shows the graph (called a waveform) of a voltage (or current, since $I = V/R$ and R is constant) that might be produced by the circuit shown in Figure 6–1. Several important characteristics of alternating voltages are labeled.

FIGURE 6-2 Some important characteristics of alternating waveforms.

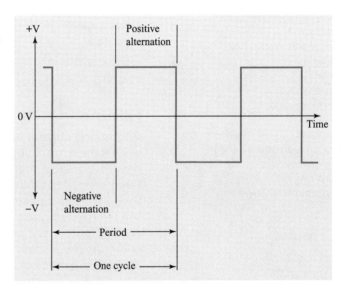

Note that the waveform in Figure 6–2 is repetitious and consists of a continuing series of **cycles**. The time required for one complete cycle is called the period of the waveform. A period consists of two, opposite polarity **alternations**. The lengths of the positive and negative alternations are not necessarily the same, but the combined length is always equal to the period of the waveform. If the positive and negative alternations are equal, it is said to be a **symmetrical waveform**. Waveforms with unequal alternations are called **asymmetrical waveforms**. If the cycles of the waveform are continuous (regardless of symmetry), the waveform is said to be periodic.

A waveform like the one shown in Figure 6–2, which makes abrupt changes in polarity, is called a **digital signal**, a rectangular waveform, a square wave (if symmetrical), a pulse, or other similar classification. You will learn additional characteristics of this type of waveform when you study digital electronics. The Circuit Exploration exercise at the end of this chapter also explores rectangular waveforms.

Sinusoidal Waveshapes

Figure 6–3 shows a loop of wire being rotated in a magnetic field. This is the basis of an ac generator or alternator. The **slip rings** and brushes in Figure 6–3 perform the task of making a continuous electrical connection to the rotating conductor. As shown in Figure 6–3, each end of the rotating loop is connected to a slip ring. The slip rings are mounted on the shaft that supports the rotating loop. Figure 6–3 shows that each of the slip rings is a continuous ring of smooth metal. The spring-loaded brushes slide along the surface of the slip rings as the conductor assembly rotates.

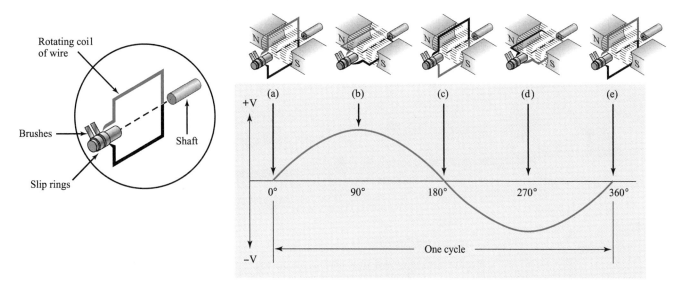

FIGURE 6–3   Construction of a basic alternator.

Figure 6–3 also shows a graph that plots the magnitude and polarity of the voltage that is produced as the conductor loop rotates. The first position shown in Figure 6–3(a) is labeled as 0°. Here the conductors are moving parallel to the flux lines, so no voltage is generated. The graph shows 0 V at 0°. The sketch in Figure 6–3(b) shows the conductor as it passes directly across the pole faces of the magnet. In this case, the conductor is cutting a maximum number of flux lines and produces a maximum voltage. Since the conductor has moved one-fourth of a complete rotation, we label this the 90° point. The graph shows that we have a maximum induced voltage at this point.

In Figure 6–3(c), the coil has reached the halfway point, and the conductors are once again moving parallel to the magnetic flux lines. No voltage is induced. This is shown graphically as the 180° point. Figure 6–3(d) shows the conductors moving across the pole faces and generating maximum voltage. Realize, however, that each of the conductors is moving across the opposite pole face from that shown in Figure 6–3(b). Thus, we would expect an opposite polarity of induced voltage. This expectation is labeled as the 270° point on the graph.

Finally, Figure 6–3(e) shows the conductor back in its original position. The relative movement of the conductor and magnetic field is again zero, so no voltage is induced. This is identified as the 360° point on the graph. The graph clearly shows that the voltage varies continuously in amplitude and periodically changes polarity. Although the horizontal axis of the graph in Figure 6–3 is labeled in degrees of rotation, it could just as easily be labeled in units of time. That is, if the conductor is rotated at a given speed, then it will take a definite amount of time to make a full 360° rotation. A practical alternator requires the combined output from many rotating loops of wire to produce useable levels of voltage.

Figure 6–4 shows two full cycles of alternating voltage with several important points labeled. A period is the time required for one full cycle. As with a rectangular wave, a period consists of two opposite polarity alternations. Each alternation is also called a **half-cycle**. A waveform like that shown in Figure 6–4 is called a **sine wave**, or more generally a sinusoidal waveform. The two half-cycles of a sinusoidal waveform are identical except for their opposite polarities. Each half-cycle has a maximum value called the **peak value**.

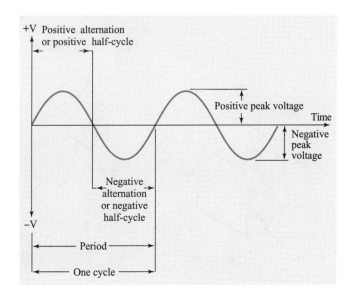

FIGURE 6–4 A basic sinusoidal waveform.

Function Generators

Figure 6–5 shows a function generator (generically called a signal generator). This is a common electronic test instrument. Nearly all function generators are capable of producing fundamental waveforms such as sine waves and rectangular waveforms. Some function generators can produce many other waveshapes, including arbitrary waveforms that can be programmed by the technician or operator. In addition to selecting a particular waveshape, the voltage, symmetry, and time characteristics of the waveforms can be adjusted.

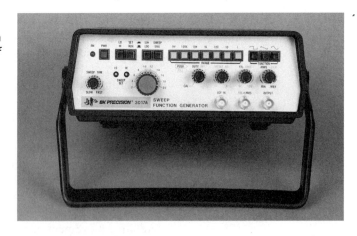

FIGURE 6–5 A representative function generator. (*Courtesy of B&K Precision*)

✓ Exercise Problems 6.1

1. Match each of the labeled points on the waveform in Figure 6–6 to the following list of waveform characteristics:
   a. positive alternation _____
   b. negative alternation _____
   c. period _____
   d. time axis _____
   e. voltage or current axis _____

2. Match each of the labeled points on the waveform in Figure 6–7 to the following list of waveform characteristics:

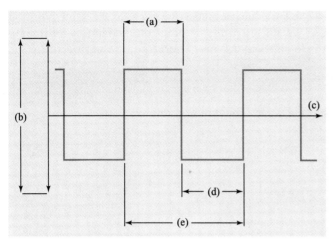

FIGURE 6–6 Identify the labeled areas.

a. positive half-cycle _____
b. positive peak _____

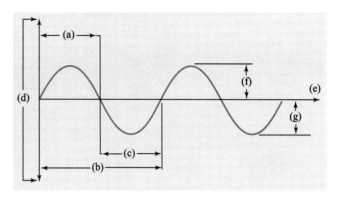

FIGURE 6–7 Identify the labeled areas.

c. negative peak _____
d. negative half-cycle _____
e. period _____
f. time axis _____
g. voltage or current axis _____

6.2 SINE WAVE CHARACTERISTICS

We examined some of the basic characteristics of sine waves in the previous section. We are now ready to examine these basic characteristics more closely and to identify other important characteristics that distinguish one sine wave from another. It is important to understand that the following characteristics apply to the sine waveshape itself. The sine wave may represent voltage, current, or some other parameter.

Period

We have seen that the period of a sine wave is the time required for one full cycle of the waveform to occur. The measurement of a period does not have to begin at zero degrees. That is,

> ▶ DEFINITION The period of a sine wave is the time from any given point on the cycle to the same point on the following cycle.

The period is measured in time (t), and in most cases, it is measured in seconds (i.e., s, ms, µs, ns, ps).

•—EXAMPLE 6.1

Determine the period for the sine wave shown in Figure 6–8.

FIGURE 6–8 What is the period for this sine wave?

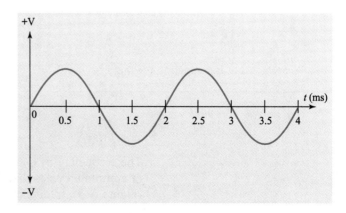

SOLUTION We measure the time from any point on one cycle to the same point on the following cycle. Suppose we started at the 0° point on the first cycle ($t = 0$). The 0° point on the next cycle starts at 2.0 ms. Therefore, the period is 2.0 ms − 0 ms = 2.0 ms.

EXAMPLE 6.2

Determine the period for the sine wave shown in Figure 6–9.

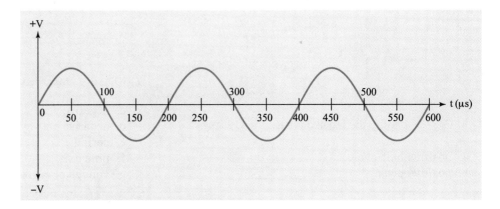

FIGURE 6–9 What is the period for this sine wave?

SOLUTION Let's measure the time between two consecutive 90° points. The first 90° point occurs at $t = 50$ μs. The 90° point on the next cycle occurs at $t = 250$ μs. The period is the difference between these two times, or $t = 250$ μs − 50 μs = 200 μs. Always subtract the smaller time value from the larger. A negative period has no meaning.

Practice Problems

1. What is the period for the sine wave shown in Figure 6–10?

2. Determine the period of the sine wave shown in Figure 6–11.

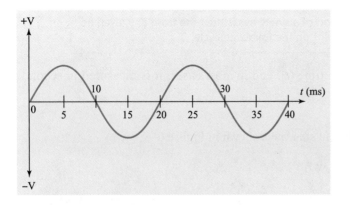

FIGURE 6–10 What is the period for this sine wave?

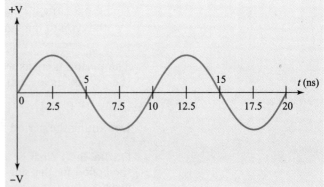

FIGURE 6–11 Determine the period of this sine wave.

Frequency

The frequency of a sine wave (or any other periodic waveform) is the number of complete cycles that occur in one second. In the case of the simple alternator (Figure 6–3), frequency (number of cycles per second) is equivalent to the speed of the alternator (revolutions per second). This is because each complete revolution produces one complete sine wave cycle.

Frequency (f) is measured in hertz (Hz). One hertz corresponds to one cycle per second. In fact, frequency used to be expressed in units of cycles per second. Useable frequencies range from less than one hertz (although 15 to 20 Hz is a more common low-frequency end) to frequencies over 100 billion hertz (100 GHz).

Frequency and period have an inverse relationship. We use Equation 6–1 to determine the period of a known-frequency sine wave.

$$t = \frac{1}{f} \tag{6-1}$$

Similarly, we use Equation 6–2 to find the frequency of a sine wave with a known period.

$$f = \frac{1}{t} \tag{6-2}$$

•—EXAMPLE 6.3

If the period of a sine wave is 100 ms, what is its frequency?

SOLUTION We apply Equation 6–2 as follows:

$$f = \frac{1}{t}$$

$$= \frac{1}{100 \text{ ms}} = 10 \text{ Hz}$$

Calculator Application

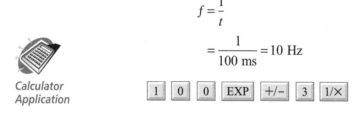

•—EXAMPLE 6.4

What is the period of a 250-MHz sine wave?

SOLUTION We apply Equation 6–1 as follows:

$$t = \frac{1}{f}$$

$$= \frac{1}{250 \text{ MHz}} = 4 \text{ ns}$$

Frequency-to-period and period-to-frequency conversions are especially easy with an electronic calculator. Simply enter the known quantity and press the reciprocal button (1/x).

•—Practice Problems

1. What is the period of a 10-kHz sine wave?
2. What is the frequency of a sine wave that has a period of 80 μs?
3. Find the period of a 4-GHz sine wave.
4. What frequency corresponds to a period of 175 ms?
5. One sine wave has a period of 3 ms, and another has a period of 6 ms. Which one has the higher frequency?

Peak Value

We briefly discussed the peak sine wave value in an earlier section. It is the maximum voltage on a sine wave. By inspection of a sine wave graph, we can conclude several things about the peak value of a sine wave:

- Peak voltage occurs at two different points in the cycle.
- One peak point is positive, while the other is negative.
- The positive peak occurs at 90°.
- The negative peak occurs at 270°.
- The positive and negative peaks have equal magnitudes (opposite polarities).

•—EXAMPLE 6.5

If a sine wave has a value of +10 volts (v_1) as it passes through the 90° point on the cycle, what voltage (v_2) will it have as it passes through the 270° point?

SOLUTION The 90° and 270° points correspond to the positive and negative peaks, respectively. These two points have equal magnitudes, but opposite polarities. Thus, we would expect the sine wave to have a value of $v_2 = -v_1 = -10$ volts as it passes through the 270° point.

Note the use of lowercase letters to represent instantaneous voltages. Instantaneous voltages are values that occur at a specific instant in time. This measurement is discussed more completely in a later section. Throughout the remainder of this text we will use lowercase letters to represent instantaneous values of alternating voltage and current, and uppercase letters for other values. This is an accepted practice in industry.

•—Practice Problems

1. If a certain sine wave of current has a negative peak of 1.5 amperes, what is its positive peak current value?

2. If the positive peak of a sine wave of voltage is 170 V, what is the value of its negative peak?

Average Value

The average value of any measured quantity is determined by summing all of the intermediate values and then dividing by the number of intermediate values. In the case of a sine wave, we need to consider two conditions: the average value for a full cycle and the average value of a half-cycle.

Average Value of a Full Sine Wave

The sine wave shown in Figure 6–12 has seventeen labeled points. If we measured the voltage at these seventeen points, added the values together, and then divided by 17, we could get an estimate of the average value for the full sine wave. It would only be an estimate, since we did not use an infinite number of points.

FIGURE 6–12
Determination of the average value of a sine wave.

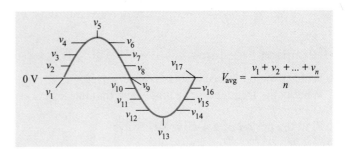

$$V_{avg} = \frac{v_1 + v_2 + \ldots + v_n}{n}$$

Nevertheless, by inspection of Figure 6–12, we can see that no matter how many equally distributed points we choose, there will be as many positive values as there are negative values. Further, every positive value will have a corresponding negative value. For example, the positive values v_3 and v_5 in Figure 6–12 correspond to the negative values v_{11} and v_{13}, respectively. If we add equal positive and negative values together, we will always get zero. Therefore, we can conclude the following about the average value of a sine wave:

> ▶ CONCLUSION   A full cycle of any sine wave has an average value of zero.

This is an important observation when troubleshooting some types of circuits (e.g., low-distortion amplifiers).

●—Practice Problems

1. What is the average value of a full sine wave that has a +100-V peak value?
2. What is the average value of a full sine wave with a negative peak of 3 V and a frequency of 10 MHz?
3. If the period of a certain sine wave is 5 ms, what is the average value for a full cycle?
4. What is the average value of a full sine wave if the peak value is 2.4 A?

Average Value for One Half-Cycle

If we sum the voltage at a number (ideally an infinite number) of points in a single half-cycle of a sine wave and then divide by the number of sampled points, we can find the average value for one half-cycle. You can perform this calculation in the laboratory by measuring a large number of points on a sine wave and then dividing by the number of points. However, since a sine wave has a consistent shape, higher mathematics can be used to prove the following relationships (Equations 6–3 and 6–4):

$$V_{avg} = 0.637\, V_P \qquad (6\text{–}3)$$

And

$$I_{avg} = 0.637\, I_P \qquad (6\text{–}4)$$

where V_P and I_P are the peak voltage and current values. Of course, current is still directly proportional to voltage in a given circuit as described by Ohm's Law.

●—EXAMPLE 6.6

What is the average value for one half-cycle of a sine wave that has a peak voltage of 100 V?

SOLUTION   We apply Equation 6–3 as follows:

$$V_{avg} = 0.637\, V_P$$
$$= 100 \text{ V} \times 0.637 = 63.7 \text{ V}$$

Since the average value of a single half-cycle provides a more meaningful result in most cases than the average value of a full sine wave, it is customary to interpret the expression "average value of a sine wave" to mean average value of a half-cycle. While this may be technically incorrect, it is very common throughout the industry. It is important that you understand this interpretation. For the remainder of this text, we shall interpret V_{avg}, I_{avg}, and so on to mean the average of one half-cycle, unless otherwise stated.

• Practice Problems

1. Compute the average value (V_{avg}) for a sine wave with a peak value of 20 V.
2. Find V_{avg} for a sine wave with a peak value of 150 mV.
3. What is the average voltage (V_{avg}) for a 1000-V peak sine wave?
4. Compute I_{avg} for a sine wave with a 150-µA peak current.

rms Value

One of the most important characteristics of a sine wave is its rms or effective value. The rms value describes the sine wave in terms that can be compared to an equivalent dc voltage. Figure 6–13(a) shows a 10-V_{rms} sine wave voltage source connected to a resistor. The resistor dissipates power in the form of heat. Figure 6–13(b) illustrates that a 10-V dc source will produce the same amount of heat in the resistor.

> ► CONCLUSION The rms value of a sine wave produces the same heating effect in a resistance as an equal value of dc.

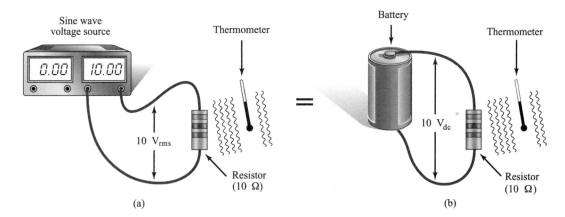

FIGURE 6–13 The rms voltage of a sine wave has the same heating effect as a similar dc voltage source.

The abbreviation rms stands for **root-mean-square**. For a given sine wave, we can take a large number (ideally an infinite number) of voltage values, square each value, sum our squared values together, divide by the number of points to get the average or mean squared value, and then extract the square root to get the rms value. Fortunately, higher mathematics can be used to derive Equations 6–5 and 6–6, which describe the rms value of a sine wave in terms of its peak values.

$$V_{rms} = 0.707\ V_P \tag{6-5}$$

And

$$I_{rms} = 0.707\ I_P \tag{6-6}$$

EXAMPLE 6.7

What is the rms value of a sine wave that has a 25-V peak?

SOLUTION We apply Equation 6–5 as follows:

$$V_{rms} = 0.707\, V_P$$
$$= 0.707 \times 25\text{ V} = 17.68\text{ V}$$

EXAMPLE 6.8

What peak value of a sine wave is required to produce the same heating effect as a 48-V battery pack?

SOLUTION Since the rms value of a sine wave produces the same heating effect as dc, we know that our sine wave will have an rms value of 48 V. We need only compute the peak voltage by applying Equation 6–5.

$$V_{rms} = 0.707\, V_P$$

Or

$$V_P = \frac{V_{rms}}{0.707} = \frac{48\text{ V}}{0.707} = 67.89\text{ V}$$

This latter calculation is used so often that it is generally remembered as separate equations as shown in Equations 6–7 and 6–8.

$$V_P = \frac{V_{rms}}{0.707} = 1.414\, V_{rms} \qquad (6\text{–}7)$$

And

$$I_P = \frac{I_{rms}}{0.707} = 1.414\, I_{rms} \qquad (6\text{–}8)$$

It is important to note that most ac voltmeters—both digital and analog—are calibrated to display the rms value of the measured ac voltage. Most ac voltmeters only respond accurately to sinusoidal waveforms. Additionally, standard voltmeters are severely limited to relatively low frequencies; they will not read correctly for waveforms with frequencies that exceed their capabilities.

Peak-to-Peak Value

Figure 6–14 on page 164 illustrates another measurement that can be used to describe a sine wave. The **peak-to-peak** voltage or current value of a sine wave is the difference between the two peak values. We can express this mathematically with Equations 6–9 and 6–10.

$$V_{PP} = 2\, V_P \qquad (6\text{–}9)$$

And

$$I_{PP} = 2\, I_P \qquad (6\text{–}10)$$

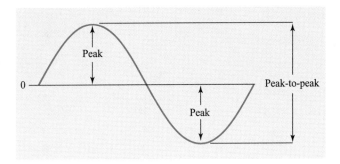

FIGURE 6–14 The peak-to-peak value of a sine wave is equal to the difference between the two peak values.

•EXAMPLE 6.9

What is the peak-to-peak current of a sine wave that has a 750-mA peak current?

SOLUTION We apply Equation 6–10 as follows:

$$I_{PP} = 2\, I_P$$
$$= 2 \times 750 \text{ mA} = 1.5 \text{ A}$$

•Practice Problems

1. If a sine wave has a peak voltage of 500 mV, what is its peak-to-peak value?
2. What is the peak-to-peak voltage of a sine wave that has a peak value of 12 V?
3. If the peak-to-peak voltage of a sine wave is 150 V, what is its peak voltage?

It is important to note that the value of peak-to-peak voltage never really exists as a measurable voltage at any particular point in time. Recall that the horizontal axis of a sine wave graph can be marked off in units of time. This means that the positive and negative peaks of the sine wave do not occur at the same exact time. Therefore, as stated in an earlier section, the highest voltage or current of a sine wave is its peak value. The peak-to-peak value is mathematically twice the peak value, *but it does not exist as a measurable voltage at some instant in time.* It is, however, an important way to describe a sine wave. It is also the easiest value to interpret with an oscilloscope, since the oscilloscope display can show voltages that occur at different times. You will use an oscilloscope throughout your career.

Phase

Phase is a relative term that is used to compare two or more sine waves that have the same frequency. When two sine waves are in phase, the various points on one sine wave (e.g., 0°, 90°, 180°, and so on) occur at exactly the same time as the corresponding points on the second sine wave. When two sine waves are out of phase, the corresponding points of the two sine waves occur at different times.

Figure 6–15(a) shows one method to help you understand phase relationships. Here, two identical alternators are rotating at exactly the same speed; they will produce two identical sine waves. However, one of the alternators is one-quarter turn ahead of the other. Therefore, the corresponding points of the two resulting sine waves will occur at different times. This is illustrated in Figure 6–15(b). We say that the two sine waves are out of phase. In the case shown in Figure 6–15(b), we can say the two sine waves are 90° out of phase.

FIGURE 6–15 Corresponding points on out-of-phase sine waves occur at different times.

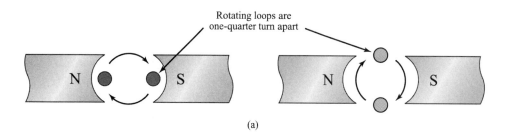

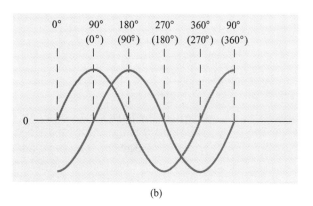

In order to describe the relative order of two out-of-phase waveforms, we use the terms **leading** and **lagging**. A waveform that has a leading phase of 45° is simply 45° ahead of another waveform. That is, any given point on the leading waveform occurs one-eighth (45°/360°) of a cycle before the corresponding point of the second waveform. If we prefer, we can refer to the second waveform as lagging by 45°. We will further examine the use of the terms leading and lagging in later chapters.

Phase angles can be expressed in either degrees or radians. Both of these are units for angular measurements. Throughout the remainder of this textbook, we will utilize degrees for all angular measurements including phase. If you prefer to use radians or if you encounter a problem with angles expressed as radians, simply make the necessary conversion on your calculator. Even the most basic engineering calculator provides simple conversions between degrees and radians. Refer to your calculator manual for details. Alternatively, you can convert degrees to radians by dividing the angle in degrees by 57.3. Similarly, an angle expressed in radians can be converted to degrees by multiplying by 57.3.

Some alternators (e.g., commercial power stations and alternators used in cars) have three sets of loops or windings on a single rotating shaft. Each coil of wire connects to its own set of slip rings. The coils are physically positioned 120° apart on the shaft. As the shaft rotates, three separate sine waves are generated, which are 120° out of phase with each other. This type of device is called a three-phase alternator. In general, three-phase systems can provide more power than comparable single-phase systems.

•—Practice Problems

1. 45° is the same as _____ radians.
2. 350° is the same as _____ radians.
3. 270° is the same as _____ radians.
4. 90° is the same as _____ radians.
5. 180° is the same as _____ radians.
6. 2.5 rad is equivalent to _____ degrees.
7. 6.2827 rad is equivalent to _____ degrees.
8. 4 rad is equivalent to _____ degrees.
9. 1.75 rad is equivalent to _____ degrees.
10. 5.25 rad is equivalent to _____ degrees.

Instantaneous Value

As a sine wave progresses through its cycle, we can express the voltage (or current) at any instant in time as the instantaneous voltage (or current). For example, the instantaneous voltage of a sine wave is zero at 0°. Similarly, a sine wave with a 10-V peak amplitude will have an instantaneous voltage of +10 V at 90° and −10 V at 270°. Equations 6–11 and 6–12 give us a way to determine the instantaneous value of a sine wave at any given angle.

$$v = V_P \sin \theta \quad (6\text{--}11)$$

And

$$i = I_P \sin \theta \quad (6\text{--}12)$$

where the lowercase v and i are used to represent instantaneous voltage and current values, θ is the angle, and sin is the trigonometric sine function.

The value for the sin of an angle is easily found on a scientific calculator by keying in the given angle (θ) and pressing the SIN button. Most scientific calculators can be configured to work in either degrees or radians. They will have a key labeled DRG, RAD, DEG or other similar identification. Be sure your calculator is in the correct mode for the units of angular measurement you prefer to use.

•–EXAMPLE 6.10

What is the instantaneous voltage at the 45° point of a sine wave with a peak voltage of 10 V?

SOLUTION We apply Equation 6–11 as follows:

$$\begin{aligned} v &= V_P \sin \theta \\ &= 10 \text{ V} \times \sin 45° \\ &= 10 \text{ V} \times 0.7071 = 7.071 \text{ V} \end{aligned}$$

Calculator Application

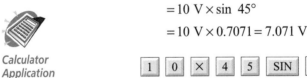

•–EXAMPLE 6.11

What is the peak voltage of a sine wave that has an instantaneous voltage of −25 volts at 225°?

SOLUTION We apply Equation 6–11 as follows:

$$v = V_P \sin \theta$$

Or

$$V_P = \frac{v}{\sin \theta}$$

$$= \frac{-25 \text{ V}}{\sin 225°} = \frac{-25 \text{ V}}{-0.707} = 35.36 \text{ V}$$

Practice Problems

1. What is the instantaneous voltage of a sine wave at 30° if its peak voltage is 500 V?

2. What is the peak current in a circuit that has a sinusoidal current wave with an instantaneous current of 1.2 A at 50°?

Time and Frequency Domains

All electronic signals can be viewed from either of two perspectives: time domain and frequency domain. It is useful to have a conceptual view of how these two domains interact. A French mathematician named Fourier made some important observations that led to the conclusion that any nonsinusoidal waveform can be created by adding sinusoidal waveforms of the correct frequencies, phases, and amplitudes. The work of Fourier forged a mathematical link between the time and frequency domains.

Time Domain

We think of a time-domain signal as one whose instantaneous voltage changes over time. An oscilloscope, for example, displays the instantaneous values of a waveform as they change with time. We say an oscilloscope is a time-domain instrument. To this point in the text, all discussions concerning sine waves have viewed the waveforms from a time-domain perspective.

Frequency Domain

We can also analyze an electronic signal in terms of its frequency content. That is, any given waveform can be shown to be composed of one or more sinusoidal signals at specific frequencies, amplitudes, and phases. A spectrum analyzer can

SYSTEM PERSPECTIVE: Time and Frequency Domains

Computer circuits operate with square waves or pulses. The voltage waveform for a typical computer signal might switch between 0 V and 3.3 V; the transition time between these two states is made as short as possible for functionality reasons. Many computer signals switch states in less than 1 ns. Computers typically connect to some sort of video display device. The signals that are required to be on the cable for functional reasons have short periods and fast transition times. Unfortunately, these desirable time-domain characteristics are necessarily associated with high-frequency harmonic energy in the frequency domain. The sometimes lengthy video cable can inadvertently serve as a radiating antenna for these high frequencies. Thus, the computer signals actually radiate (i.e., transmit) energy into the air just like a radio transmitter might, causing interference to nearby electronic systems. This is the reason you are normally required to turn off your notebook computer during airplane takeoffs and landing. The FCC and other agencies regulate the levels of high-frequency radiation that can legally be emitted from a computer. The shorter the transition times, the stronger the radiated emissions at higher harmonic frequencies.

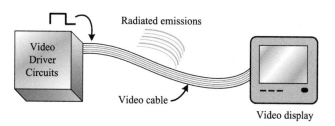

be used to measure the frequency content of a waveform. We say that a spectrum analyzer is a frequency-domain instrument.

Converting between the time and frequency domains is a well-defined process, but it requires advanced mathematics (Fourier analysis). There are also many computer programs available that can make these conversions.

Harmonics

Any repetitive, nonsinusoidal waveform in the time domain can be shown to be composed of a fundamental frequency and some combination of harmonic frequencies. The **fundamental frequency** is the basic frequency of the waveform as determined by the period of the waveform. **Harmonic frequencies** are exact multiples of the fundamental frequency. Both fundamental and harmonic frequency components of the original waveform have sinusoidal characteristics in the time domain.

A frequency component of a waveform that is exactly twice the fundamental frequency is called the second harmonic. The third and fourth harmonic frequencies occur at three and four times the frequency of the fundamental, respectively. The first, third, fifth, and seventh harmonics (and so on) are collectively called odd harmonics. Similarly, the second, fourth, sixth, and eighth harmonics (and so on) are collectively called the even harmonics.

Now let us consider a specific example. Figure 6–16 illustrates that a square wave in the time domain corresponds to a mixture of the fundamental and the odd harmonic frequencies. Figure 6–16(a) shows the fundamental and the third and fifth harmonics. Figure 6–16(b) shows the algebraic sum of the three sine waves shown in Figure 6–16(a). Clearly, a rectangular waveform is starting to form. Figures 6–16(c) and (d) illustrate the effects of adding additional harmonic frequencies. The more harmonics we include, the more nearly the result resembles a square wave. A perfect square wave would have an infinite number of odd harmonics. The lower frequencies are primarily responsible for flat, horizontal portions of the square wave. The higher harmonic frequencies contribute to the steepness of the rising and falling edges. Although the square wave is a selected example, any periodic nonsinusoidal waveform in the time domain can be decomposed into the fundamental and some combination of harmonics in the frequency domain.

FIGURE 6–16 A square wave in the time domain corresponds to a mixture of the fundamental and odd harmonic frequencies in the frequency domain.

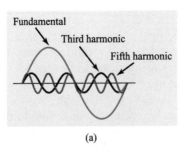

(a)

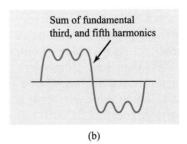

(b)

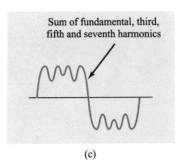

(c)

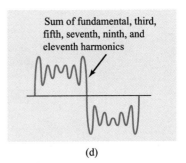

(d)

EXAMPLE 6.12

What is the third harmonic of a 1-kHz sine wave?

SOLUTION The third harmonic is three times the frequency of the fundamental, or

$$f_3 = 3 \times 1.0 \text{ kHz} = 3.0 \text{ kHz}$$

Practice Problems

1. What is the second harmonic frequency if the fundamental frequency is 2.5 MHz?
2. List the first four harmonics of a 25-MHz waveform.
3. List the first three odd harmonics of a 100-MHz sine wave.
4. List the first three even harmonics of a 250-MHz waveform.

Combined AC and DC Waveforms

All sinusoidal waveforms illustrated so far have varied equally above and below the 0-V reference. In many practical circuits (e.g., transistor amplifiers), both ac and dc voltages appear at the same point at the same time. In this case, the actual instantaneous voltage of the given point is simply the algebraic sum of the ac and dc values.

EXAMPLE 6.13

What are the maximum excursions of a point in a circuit that has a 10-V peak sine wave and a 5-V dc level at the same time?

SOLUTION The positive peak would extend to +15 V, and the negative peak would go to −5 V. These values are computed as follows.

$$\text{positive excursion} = \text{dc} + \text{peak voltage} = +5 \text{ V} + 10 \text{ V} = 15 \text{ V}$$

And

$$\text{negative excursion} = \text{dc} - \text{peak voltage} = +5 \text{ V} - 10 \text{ V} = -5 \text{ V}$$

✓ Exercise Problems 6.2

1. The time required for the complete period of a sine wave is called its _____.
2. There are _____ degrees in a full cycle of a sine wave.
3. The time required for a sine wave to pass from the 90° point on one cycle to the 90° point on the next cycle is (*shorter than/longer than/same as*) the period of the sine wave.
4. The frequency of the signals on an Ethernet transmission cable is 100 MHz. What is the period of these signals?
5. How long is the period of a 2.5-GHz sinusoidal waveform?
6. What is the frequency of a sine wave whose period is 5 ns?
7. If the period of a sine wave is 4 ms, what is its frequency?
8. The positive peak value of a certain sine wave is +75 V and occurs at 90° on the waveform. The negative peak of the same sine wave is _____ and occurs at _____° on the waveform.
9. If the peak value of a sine wave is 5 mA, what is its average value?
10. The rms value of the wall voltage in your home is 120 V. What is the peak voltage?
11. If a sine wave has a peak voltage of 150 mV, what is its instantaneous voltage at 25°?
12. What is the fifteenth harmonic of a 66-MHz computer clock?

6.3 WORKING WITH PHASE ANGLES

Computer technicians sometimes have to solve problems that involve phase relationships of one or more sinusoidal waveforms. It is very difficult to sketch sinusoidal curves accurately, but we need an accurate method of analyzing sinusoidal waveforms. Phasors provide us with a convenient way to represent sinusoidal waveforms and to simplify many calculations.

Phasor Diagrams

Figure 6–17 shows a phasor diagram. The phasor itself consists of a vector or "arrow" that can be rotated around a central point. The length of the phasor corresponds to the peak value of the sine wave. The angle of the phasor (θ) relative to the right-most horizontal axis corresponds to the instantaneous phase angle of the sine wave. Finally, the distance from the point of the phasor to the horizontal axis represents the instantaneous voltage at the specified angle.

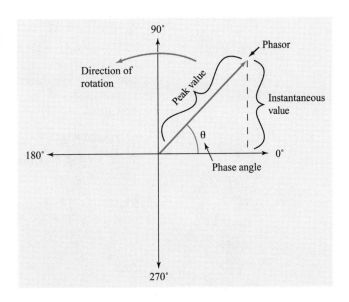

FIGURE 6–17 A phasor can be used to represent a sinusoidal quantity.

•—EXAMPLE 6.14

Figure 6–18(a) shows a phasor diagram for a particular sine wave. What is the peak value of the sine wave? What is the instantaneous voltage?

FIGURE 6–18 An example phasor diagram.

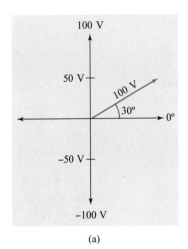

(a)

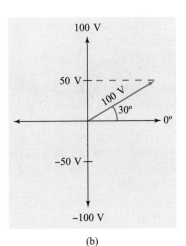

(b)

SOLUTION The peak voltage is represented on a phasor diagram by the length of the phasor. In the case shown in Figure 6–18, the peak voltage is 100 V. In Figure 6–18(b), we have drawn a horizontal dotted line from the point of the phasor to the vertical axis. The point of intersection is the instantaneous voltage. In this case, the instantaneous voltage is 50 V.

If the angle were not given, we could find it with a protractor. However, our accuracy would be limited by the preciseness of the phasor sketch. This would be tedious and inaccurate. Thus, we need yet a better way to calculate phasor values.

Right-Triangle Relationships

Figure 6–19 shows a phasor diagram. We have drawn a dotted line from the point of the phasor to the horizontal axis. This forms a triangle. Side "a" is a portion of the horizontal axis. Side "b" corresponds to the instantaneous voltage of the sine wave. Side "c" represents the peak voltage of the sine wave. The angle of the sine wave is represented by θ. A triangle drawn in this way will always have a right angle (90°) where the dotted line from the phasor point intersects the horizontal axis. Solving problems involving triangles requires the application of a branch of mathematics called trigonometry. Fortunately, the right-triangle calculations required to solve phasor problems are a well-defined subset of trigonometry. You do not have to understand all there is to know about trigonometry to solve these problems.

Figure 6–20(a) shows a right triangle with the sides labeled according to standard practice. The hypotenuse is always the longest side and is across from the right angle. Angle θ is used to represent the phase of a sine wave. The other two sides are named according to their relationship to θ. That is, the adjacent side is adjacent (i.e., forms one side of the angle) to θ, and the opposite side is opposite (i.e., across the triangle from) θ.

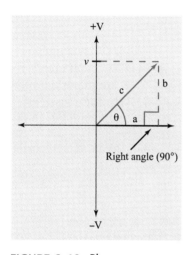

FIGURE 6–19 Phasor calculations are based on right-triangle calculations.

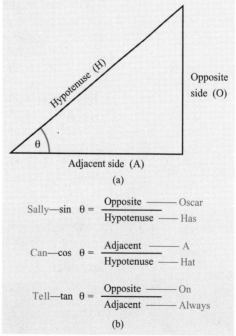

FIGURE 6–20 A right triangle with (a) standard labels and (b) a memory aid.

There are three equations that we must know in order to solve basic phasor problems.

$$\sin \theta = \frac{\text{opposite side}}{\text{hypotenuse}} \qquad (6\text{–}13)$$

$$\cos \theta = \frac{\text{adjacent side}}{\text{hypotenuse}} \qquad (6\text{–}14)$$

$$\tan \theta = \frac{\text{opposite side}}{\text{adjacent side}} \qquad (6\text{–}15)$$

Sin, cos, and tan are basic functions (sine, cosine, and tangent) that are easily computed by your calculator. Equations 6–13, 6–14, and 6–15 are important relationships. They should be committed to memory. Figure 6–20(b) shows a popular memory aid. Some people use different words, but basically the first letter in the expression "**S**ally **C**an **T**ell **O**scar **H**as **A** **H**at **O**n **A**lways" corresponds to the words **S**ine, **C**osine, **T**angent, **O**pposite, **H**ypotenuse, and **A**djacent. If you remember the phrase, then you can quickly construct the three equations.

Calculator Operation

There are several functions on your calculator that have not been required in the preceding chapters. We will, however, use them in the following chapters, and you will certainly need to use them in your career. The trigonometric functions are of immediate interest to us. You might want to view the trigonometric function values as a code. Every angle has a corresponding numeric code called the sine (sin) of the angle. Each angle also has a different code called the cosine (cos) of the angle. Finally, for every angle, there is yet a different coded numerical value called the tangent (tan) of the angle. Scientific calculators have built-in look-up tables that allow you to determine the function values (or codes) for any given angle. Think of your calculator as your decoding device. You key in an angle. Your calculator gives you a corresponding code. Alternatively, you can key in one of the codes (sin, cos, or tan), and the calculator will produce the value of the corresponding angle.

Trigonometric Functions

A scientific calculator will compute the sin, cos, and tan functions by simply keying in the value and pressing the appropriate button. The sine of an angle is computed when the SIN button is pressed. The cosine is found by pressing the COS button. And, finally, the tangent of an angle can be calculated by using the TAN button.

Many scientific calculators can work with angles expressed in either degrees or radians. Be certain that your calculator is in the degree mode if your angles are expressed in degrees. Similarly, you must put your calculator in the radian mode if you want to enter the angle values as radians.

•—**EXAMPLE 6.15**

Find the value of $\cos \theta$, when $\theta = 55°$.

SOLUTION The key sequence required to obtain the value of a trigonometric function is the same for a standard engineering calculator or an RPN calculator. It consists of keying in the angle and pressing the appropriate function key. We must also ensure that the calculator is in the correct mode. Figure 6–21 shows a typical key sequence for the given problem. You should get a result of 0.5736. If your

FIGURE 6–21 A typical calculator key sequence to compute cos 55°.

6.3 • WORKING WITH PHASE ANGLES

calculator produces an answer of 0.0221, then it is in the radian mode instead of the degree mode. Some calculators have the sin, cos, and tan functions as second, or shifted, functions. In these cases, you will need to press the "2nd" function key first to invoke the cos function.

FIGURE 6–22 A typical calculator key sequence to compute tan 2.75 rad.

•EXAMPLE 6.16

Find the value of tan θ, when θ = 2.75 rad.

SOLUTION Put your calculator into the radian mode. Look for a DRG or RAD button or consult your calculator manual. Figure 6–22 shows a typical key sequence. Your calculator should produce a result of −0.4129. If you get 0.048 for an answer, then your calculator is in the degree mode.

•Practice Problems

Find the value of each of the following functions:

1. cos 28° = _____
2. sin 3.6 rad = _____
3. tan 150° = _____
4. sin 20° = _____
5. tan 175° = _____
6. tan 5 rad = _____
7. cos 1.8 rad = _____
8. sin 3.75 rad = _____
9. cos 200° = _____
10. sin 90° = _____

Inverse Trigonometric Functions

There will be many times that we know the sine, cosine, or tangent of an angle, but we do not know the angle itself. That is, we know the "code," but we don't know the original angle. The sine, cosine, and tangent operations are called functions. The corresponding opposite (or decoding) operations are the arcsin, arccos, and arctan. These are called the inverse trigonometric functions. You can think of "arc" as meaning "angle whose." Thus, the expression arcsin 0.2 can be interpreted as "angle whose sine is 0.2."

•EXAMPLE 6.17

We know the sine of an angle is 0.5. What is the size of the angle in degrees?

SOLUTION We need to find the angle whose sine is 0.5. That is, we need to find arcsin 0.5. First, we check to be sure our calculator is in the degree mode. Then, we key in the value of 0.5 and press the inverse sine button. In many cases, the inverse sine function is a second or shifted function of the sine key. You may have to press the "2nd" then the sine key to invoke the arcsin function. Some calculators label the inverse trigonometric functions as sin^{-1}, cos^{-1}, and tan^{-1}. Again, they are probably shifted functions. Another alternative labeling used on some calculators is ASIN, ACOS, and ATAN.

FIGURE 6–23 A typical calculator key sequence to compute arcsin 0.5.

Figure 6–23 shows a typical key sequence. You should get a result of 30°. If you get an answer of 0.524, then your calculator is in the radian mode.

•Practice Problems

Find the value of the following inverse functions.

1. arcsin 0.7 = _____
2. arccos −0.633 = _____
3. arctan 0.466 = _____
4. arcsin 0.0872 = _____
5. arccos −1 = _____
6. arctan −0.577 = _____
7. arcsin −0.174 = _____
8. arccos 0.259 = _____
9. arctan −0.7 = _____
10. arcsin 0.2 = _____

174 CHAPTER 6 • ALTERNATING CURRENT

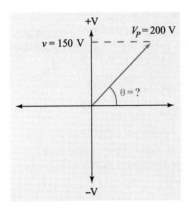

FIGURE 6–24 Find the angle θ.

Phasor Calculations

Phasor calculations are used extensively to analyze circuits that contain inductors and/or capacitors. For now, however, let us concentrate on using the sine, cosine, and tangent functions and their inverse functions to solve right-triangle problems and basic sine wave phase problems.

•—EXAMPLE 6.18

Figure 6–24 shows a phasor diagram with known values for the peak and instantaneous voltages. Calculate the angle θ.

SOLUTION In Figure 6–25, we have extended a vertical line from the phasor tip to the horizontal axis. This makes a right triangle. As long as we know the value of any two sides or at least one side and one of the smaller angles, we can find the rest by using Equations 6–13 through 6–15. In the present case, we know the hypotenuse and the opposite side. We can use Equation 6–13 to find the sine of the angle.

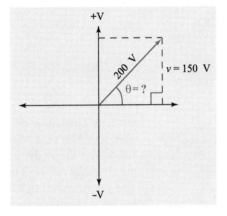

FIGURE 6–25 Form a right triangle and solve for θ.

$$\sin \theta = \frac{\text{opposite side}}{\text{hypotenuse}}$$

$$= \frac{150 \text{ V}}{200 \text{ V}} = 0.75$$

If we know the sine of an angle, then the arcsin function will give us the value of the angle in degrees or radians.

$$\theta = \arcsin 0.75$$
$$= 48.59° \text{ or } 0.848 \text{ rad}$$

•—EXAMPLE 6.19

Figure 6–26 shows a right triangle with the length of two sides given. Find the value of the angle θ.

FIGURE 6–26 Find the angle θ.

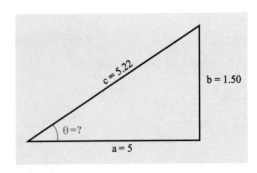

SOLUTION Recall that if we know any two sides, we can find all other missing values. In this case, we know the adjacent side (a), and we know the opposite side (b). We can use Equation 6–15 to find the tangent of angle θ.

$$\tan \theta = \frac{\text{opposite side}}{\text{adjacent side}}$$

$$= \frac{1.5}{5} = 0.3$$

The arctan function will give us the actual value of the angle.

$$\theta = \arctan 0.3$$

$$= 16.7° \text{ or } 0.29 \text{ rad}$$

• Practice Problems

1. Find the length of side c and the value of angle θ in Figure 6–27.
2. Find the length of sides b and c in Figure 6–28.
3. Determine the length of sides a and c in Figure 6–29.
4. Calculate the value of θ in Figure 6–30.
5. Find the value of θ in Figure 6–31.

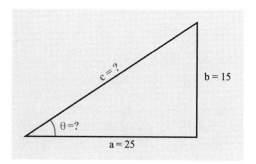

FIGURE 6–27 Find the missing values.

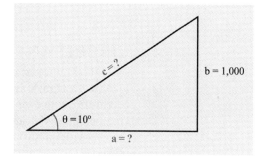

FIGURE 6–29 Find the missing values.

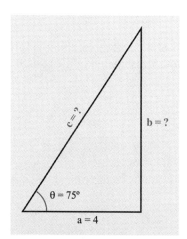

FIGURE 6–28 Find the missing values.

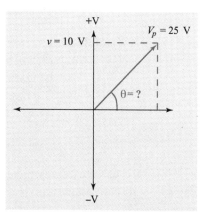

FIGURE 6–30 What is the value of θ?

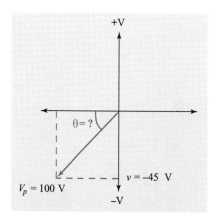

FIGURE 6–31 What is the value of θ?

Exercise Problems 6.3

1. The angle of a phasor is measured with reference to the left-most horizontal axis of the phasor graph. (True/False)
2. The length of a phasor represents _____ voltage of the sine wave.
3. The distance from the point of the phasor to the horizontal axis represents the _____ voltage of the sine wave.
4. What does the expression arctan 35° mean?
5. Figure 6–32 shows a phasor diagram for a sine wave. What is the peak value of the sine wave?
6. What is the instantaneous voltage of the sine wave represented by the phasor diagram in Figure 6–32?
7. What is the value of θ (in degrees) in Figure 6–33?
8. Find the length of side a in Figure 6–33.

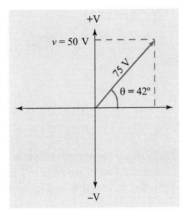

FIGURE 6–32 A phasor diagram.

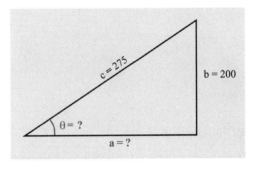

FIGURE 6–33 Find the missing values.

6.4 CIRCUIT ANALYSIS OF AC RESISTIVE CIRCUITS

Circuits composed of resistors and supplied by an ac voltage source can be analyzed with Ohm's and Kirchhoff's Laws just as we did with dc sources. You must be careful, however, that you use the correct values. For example, if you want to find peak current, then use peak voltage in your calculations. If you want to find the rms voltage drop across a resistor, then use the rms value of current. For most purposes, power dissipation is computed using rms values of voltage and/or current, although peak and average power are also useful computations for some applications (e.g., power supply circuits, antenna calculations, speaker design). Unless specifically noted, *power calculations in the remainder of the text will refer to rms values.*

•EXAMPLE 6.20

Figure 6–34 shows a simple series circuit composed of an ac source and a single resistor. Find the peak current in the circuit and the power dissipated in the resistor.

FIGURE 6–34 An ac circuit for analysis.

$V_p = 10$ V
$f = 1$ kHz
$R = 2$ kΩ

SOLUTION Since the value of the voltage source is given as a peak voltage, we can apply Ohm's Law directly to determine peak current.

$$I_P = \frac{V_P}{R}$$

$$= \frac{10 \text{ V}}{2 \text{ k}\Omega} = 5 \text{ mA}$$

We can use any of the power formulas to compute power, but we will need to use rms values of voltage and/or current. Let's use the $P = V^2/R$ formula. First, we will need to find the rms value of voltage.

$$V_{rms} = 0.707\, V_P$$
$$= 0.707 \times 10 \text{ V} = 7.07 \text{ V}$$

We can now compute power dissipation.

$$P = \frac{V^2}{R}$$

$$= \frac{(7.07 \text{ V})^2}{2 \text{ k}\Omega} = 24.99 \text{ mW}$$

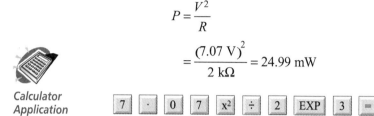

Calculator Application

• **EXAMPLE 6.21**

Figure 6–35 shows a series-parallel circuit with an ac source. Complete Table 6–1 with reference to this circuit.

TABLE 6–1 Values for Figure 6–35.

	Voltage				Current			
	V_P	V_{PP}	V_{rms}	V_{avg}	I_P	I_{PP}	I_{rms}	I_{avg}
R_1								
R_2								
R_3								
Total								

Fig06_35.msm

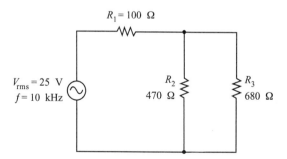

FIGURE 6–35 Analyze this circuit.

SOLUTION As with most circuit analysis problems, there are many ways to solve the problem. Let us begin by simplifying the circuit and sketching intermediate schematics. First, we note that R_2 and R_3 are in parallel. We can replace them with an equivalent resistance having a value of

$$R_{2,3} = \frac{R_2 R_3}{R_2 + R_3}$$

$$= \frac{470 \text{ }\Omega \times 680 \text{ }\Omega}{470 \text{ }\Omega + 680 \text{ }\Omega} = 277.9 \text{ }\Omega$$

This step is shown in Figure 6–36.

FIGURE 6–36
Simplification of the circuit shown in Figure 6–35.

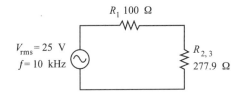

We can now find the total circuit resistance by combining the series resistances of R_1 and $R_{2,3}$.

$$R_T = R_1 + R_{2,3}$$
$$= 100 \text{ } \Omega + 277.9 \text{ } \Omega = 377.9 \text{ } \Omega$$

The result at this point is shown in Figure 6–37.

FIGURE 6–37 An equivalent circuit for Figure 6–35.

As a next step, let's compute the various voltage values for the applied voltage. V_{rms} is given. From this, we can find the others. We find V_P by using Equation 6–7.

$$V_{T(P)} = 1.414 \text{ } V_{T(rms)}$$
$$= 1.414 \times 25 \text{ V} = 35.35 \text{ V}$$

We use Equation 6–9 to compute V_{PP}.

$$V_{T(PP)} = 2 \text{ } V_{T(P)}$$
$$= 2 \times 35.35 \text{ V} = 70.7 \text{ V}$$

Finally, we can compute the average value by applying Equation 6–3.

$$V_{T(avg)} = 0.637 \text{ } V_{T(P)}$$
$$= 0.637 \times 35.35 \text{ V} = 22.52 \text{ V}$$

This completes the first four boxes on the bottom row of Table 6–1. Since we know all forms of the applied voltage, and we know the value of total resistance, we can complete the lower row of Table 6–1 by using Ohm's Law. We find I_P as follows:

$$I_{T(P)} = \frac{V_{T(P)}}{R_T}$$
$$= \frac{35.35 \text{ V}}{377.9 \text{ } \Omega} = 93.54 \text{ mA}$$

Similarly, we compute I_{PP} as

$$I_{T(PP)} = \frac{V_{T(PP)}}{R_T}$$
$$= \frac{70.7 \text{ V}}{377.9 \text{ } \Omega} = 187.1 \text{ mA}$$

I_rms is computed as

$$I_{T(\text{rms})} = \frac{V_{T(\text{rms})}}{R_T}$$

$$= \frac{25 \text{ V}}{377.9 \text{ }\Omega} = 66.16 \text{ mA}$$

And we compute the total average value as

$$I_{T(\text{avg})} = \frac{V_{T(\text{avg})}}{R_T}$$

$$= \frac{22.52 \text{ V}}{377.9 \text{ }\Omega} = 59.59 \text{ mA}$$

If we wanted to compute total power dissipation, we could use any of the basic power formulas along with the relevant values of total current, voltage, or resistance.

We have computed everything we can with Figure 6–37, so let us move back to Figure 6–36 and continue. First, we note that total current also flows through R_1 since it is in series with the source. This observation immediately gives us the values for all the R_1 currents in Table 6–1. We can find the voltages across R_1 as follows:

$$V_{1(P)} = I_{1(P)} \, R_1 = 93.54 \text{ mA} \times 100 \text{ }\Omega = 9.354 \text{ V}$$

$$V_{1(PP)} = I_{1(PP)} \, R_1 = 187.1 \text{ mA} \times 100 \text{ }\Omega = 18.71 \text{ V}$$

$$V_{1(\text{rms})} = I_{1(\text{rms})} \, R_1 = 66.16 \text{ mA} \times 100 \text{ }\Omega = 6.616 \text{ V}$$

$$V_{1(\text{avg})} = I_{1(\text{avg})} \, R_1 = 59.59 \text{ mA} \times 100 \text{ }\Omega = 5.959 \text{ V}$$

This completes the first row of answers in Table 6–1.

Kirchhoff's Law can be used to determine the various voltages across the $R_{2,3}$ resistance as follows:

$$V_{2,3(P)} = V_{T(P)} - V_{1(P)} = 35.35 \text{ V} - 9.354 \text{ V} = 26 \text{ V}$$

$$V_{2,3(PP)} = V_{T(PP)} - V_{1(PP)} = 70.7 \text{ V} - 18.71 \text{ V} = 51.99 \text{ V}$$

$$V_{2,3(\text{rms})} = V_{T(\text{rms})} - V_{1(\text{rms})} = 25 \text{ V} - 6.616 \text{ V} = 18.38 \text{ V}$$

$$V_{2,3(\text{avg})} = V_{T(\text{avg})} - V_{1(\text{avg})} = 22.52 \text{ V} - 5.959 \text{ V} = 16.56 \text{ V}$$

Because resistors R_2 and R_3 are in parallel, they will have equal voltages across them. Therefore, we can use the values for $V_{2,3}$ in both R_2 and R_3 rows of Table 6–2. This leaves only the current for R_2 and R_3 to be calculated. For this, we must move to Figure 6–35. There are many ways to calculate these currents. Let us choose to use Ohm's Law for the R_2 calculations and Kirchhoff's Current Law for the R_3 calculations. The calculation for the R_2 currents are

$$I_{2(P)} = \frac{V_{2(P)}}{R_2} = \frac{26 \text{ V}}{470 \text{ }\Omega} = 55.32 \text{ mA}$$

$$I_{2(PP)} = \frac{V_{2(PP)}}{R_2} = \frac{51.99 \text{ V}}{470 \text{ }\Omega} = 110.6 \text{ mA}$$

$$I_{2(\text{rms})} = \frac{V_{2(\text{rms})}}{R_2} = \frac{18.38 \text{ V}}{470 \text{ }\Omega} = 39.11 \text{ mA}$$

$$I_{2(\text{avg})} = \frac{V_{2(\text{avg})}}{R_2} = \frac{16.56 \text{ V}}{470 \text{ }\Omega} = 35.23 \text{ mA}$$

Using Kirchhoff's Current Law for the R_3 current gives us the following results:

$$I_{3(P)} = I_{T(P)} - I_{2(P)} = 93.54 \text{ mA} - 55.32 \text{ mA} = 38.22 \text{ mA}$$

$$I_{3(PP)} = I_{T(PP)} - I_{2(PP)} = 187.1 \text{ mA} - 110.6 \text{ mA} = 76.5 \text{ mA}$$

$$I_{3(\text{rms})} = I_{T(\text{rms})} - I_{2(\text{rms})} = 66.16 \text{ mA} - 39.11 \text{ mA} = 27.05 \text{ mA}$$

$$I_{3(\text{avg})} = I_{T(\text{avg})} - I_{2(\text{avg})} = 59.59 \text{ mA} - 35.23 \text{ mA} = 24.36 \text{ mA}$$

The completed solution matrix is shown in Table 6–2.

TABLE 6–2 Completed solution matrix for Figure 6–35.

	Voltage (V)				Current (mA)			
	V_P	V_{PP}	V_{rms}	V_{avg}	I_P	I_{PP}	I_{rms}	I_{avg}
R_1	9.354	18.71	6.616	5.959	93.54	187.1	66.16	59.59
R_2	26	51.99	18.38	16.56	55.32	110.6	39.11	35.23
R_3	26	51.99	18.38	16.56	38.22	76.5	27.05	24.36
Total	35.35	70.7	25	22.52	93.54	187.1	66.16	59.59

✓ Exercise Problems 6.4

1. Complete a solution matrix similar to Table 6–2 for the circuit shown in Figure 6–38.

2. Refer to Figure 6–39 and complete a solution matrix similar to Table 6–2.

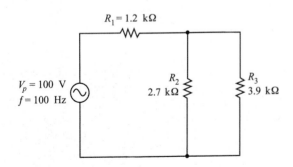

FIGURE 6–38 Complete a solution matrix for this circuit.

Fig06_38.msm

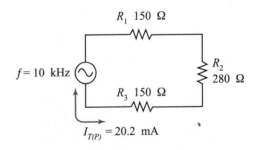

FIGURE 6–39 Complete a solution matrix for this circuit.

Fig06_39.msm

6.5 ALTERNATING VOLTAGE APPLICATIONS

There is an endless array of applications that rely on alternating voltage and current. The following paragraphs provide a brief sampling of applications in several frequency ranges.

60-Hz Power Distribution

Undoubtedly, the most common application of alternating voltage is the 60-Hz power distribution that is supplied by power companies throughout the United States. Alternators at such places as Hoover Dam near Las Vegas, Nevada initially produce the voltage. In the case of Hoover Dam, the force of water rotates the alternators as the Colorado River flows past the dam. Other power-generating plants use coal or nuclear power to produce steam, which ultimately provides the mechanical energy needed to spin the alternators. Figure 6–40 shows an alternator used to generate commercial power.

FIGURE 6–40 A 60-Hz alternator used for generation of commercial power. (*Courtesy of Tampa Electric Company*)

The voltage is initially generated at a level of several thousand volts. Near the power station, it is increased to several hundred thousand volts for long-distance transmission. Once the power has reached your neighborhood, it is again reduced to several thousand volts for local distribution. Finally, at a point just outside your home, the voltage is lowered to 220 V and routed into your house. At the service entrance (fuse or breaker box) in your house, the voltage is tapped off at a 120-V level for most appliances and lights, while the full 220 V is available for larger equipment such as air conditioners and electric dryers.

It may surprise you to learn that 120 V is an ideal rms value. (The peak voltage is nearly 170 V.) The actual rms voltage may change significantly throughout the day. It may be less than 100 V or higher than 130 V at certain times depending on the load demands and the capabilities of your power company.

The frequency is nominally 60 Hz and is established by the speed of the alternators. The frequency does vary slightly throughout the day. However, it is carefully compensated so that the long-term average is very nearly 60 Hz. Many electrical and electronic devices (e.g., clocks and turntables) use the 60-Hz frequency as a time reference.

In most cases, transmission of 60-Hz energy is restricted to conductive wires and cables. No substantial energy is radiated through the air at 60 Hz.

Sound Waves

Sound waves are made up of sinusoidal pressure changes in the air. We use sound waves in electronics for things like microphones and speakers. Sound also plays an important role in computers. Nearly all computers emit a "beep" to alert the user to an error. Computers can also create and play music. Many industrial applications utilize sound as an audible annunciator to attract the operator's attention.

The frequency range for audible sound is from about 15 to 20 Hz on the low-frequency end to about 15 to 20 kHz on the high-frequency end of the range. Sound waves travel through air at about 1,130 feet per second. This is roughly 1 ms per foot.

Ultrasonic Waves

Ultrasonic waves, like sound waves, are pressure changes in the air or other material. Ultrasonic applications include burglar alarms, range finders, and nondestructive inspection (NDI) equipment. In these cases, the ultrasonic sound is emitted by a transducer (the equivalent of a speaker). It travels outward, reflects off objects (e.g., a burglar) and returns to a receiving transducer (equivalent of a microphone). Figure 6–41 illustrates this echo principle.

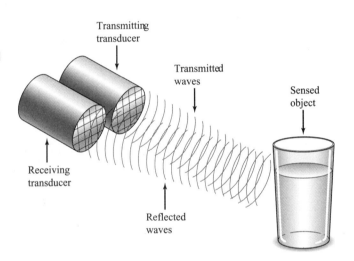

FIGURE 6–41 An ultrasonic receiver can detect an echo bounced off a sensed object.

Some applications (e.g., a range finder on a camera) simply measure the amount of time that it takes the ultrasonic wave to travel from the "speaker" to the object and back to the "microphone." Since the waves travel at 1,130 feet per second, we can easily compute the distance to the object.

Other applications rely on a detected frequency difference between the transmitted and received signal. When the reflecting object is stationary, the received frequency is the same as the transmitted frequency. If the object is moving toward or away from the transmitter/receiver, then the received signal will be higher or lower, respectively, than the transmitted signal. This shift in frequency is known as the **doppler effect**. This is the same phenomenon that causes the sound of a rapidly approaching train or racecar to sound like a higher frequency. As the train or car passes you, the frequency drops to a much lower value.

Many medical electronics applications use ultrasonic frequencies. Ultrasound imaging systems can effectively "see" inside the body without the dangers associated with X rays. Oral hygienists use ultrasonics to clean your teeth. Figure 6–42 shows a medical application for ultrasonics.

Ultrasonic energy travels through air at about 1,130 feet per second. The frequency range is from about 25 kHz to several hundreds of kilohertz.

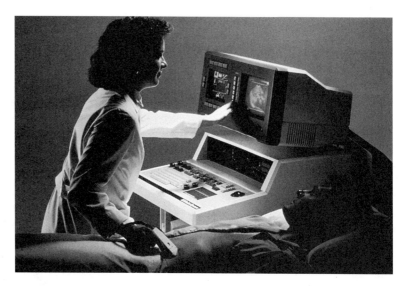

FIGURE 6–42 An ultrasonic machine used to view inside a human body. (*Courtesy of GE Medical Systems*)

Radio Waves

Radio, or electromagnetic, waves are used by a wide array of electronic devices. The most obvious applications include radio and television broadcasts. Other common applications include wireless Internet, microwave communications links, satellite communications, and a variety of radar applications. Radio frequencies are also used for navigation. A small handheld receiver can intercept signals from a satellite and locate any spot in the world within a few feet.

Microwave ovens use radio frequencies. Other radio-frequency (rf) devices are used in industry to heat metal, melt plastic, and dry plywood. A similar device can be used to induce an artificial fever in a human.

Radio waves are transmitted from an antenna and travel through the air at very nearly the speed of light. The waves travel at about 300×10^6 meters per second, or 186,000 miles per second. Unlike sound waves, which are high- and low-pressure regions moving through the air, rf waves are electromagnetic. The moving wave has characteristics of both magnetic fields and electrostatic fields. The frequency range for electromagnetic radio waves begins below 20 Hz and extends into hundreds of gigahertz. Figure 6–43 shows an antenna used for transmission of electromagnetic waves.

Bluetooth™

Bluetooth is a global specification that defines a wireless communication technology that ultimately is intended to replace the rat's nest of cables generally associated with a computer system. There are other similar standards such as IEEE 802.11 maintained by the Institute of Electrical and Electronic Engineers. Devices that conform to the Bluetooth specification have low-cost transmitters and receivers, which operate over relatively short distances (10 centimeters to 100 meters). The radio communication links are designed to be transparent to the user. A given product can automatically detect and communicate with any other Bluetooth-compliant device within its operating range.

Applications for Bluetooth-compliant products are seemingly endless, but here are a few examples of how this wireless technology can be utilized:

- Your home computer could communicate with the mouse, the keyboard, and your printer with no interconnecting cables.

FIGURE 6–43 A radio-frequency antenna used to transmit electromagnetic energy. (*Courtesy of Andrew Corporation*)

- You could automatically transfer notes and diagrams from an electronic whiteboard used by your instructor in a classroom directly into your notebook computer. Again, no wires are necessary.
- You go to a meeting with other computer experts. As you enter the room, your notebook computer automatically detects the others and exchanges business cards. As you start your presentation, your notes are automatically displayed on each participant's computer screen.

Bluetooth devices operate in the 2.4-GHz to 2.4835-GHz frequency range. Bluetooth devices that are within range of each other automatically (i.e., under software control) form themselves into small groups. Each group is called a **piconet** and has one master and as many as seven active slaves (although there can be many more inactive slaves that have been identified but are not actively communicating). All active devices within a piconet can communicate with each other simultaneously. Additionally, any one device can be a member of multiple piconets. For example, a given device can be a master in one piconet while simultaneously participating in another piconet as a slave. Bluetooth technology is discussed further in chapter 17.

Wireless Internet Links

At the start of the millennium, the vast majority of people who connected to the Internet did so via phone wires, cables, and fiber-optic links. Today, there are hundreds of millions of people accessing the Internet via wireless radio links.

Wireless Internet access has many advantages. The most obvious advantage is the absence of the tethers represented by physical connections to cables or fiber-optic lines. With wireless devices, you can be fully mobile and still maintain your link to the Internet for web access and e-mail. Wireless Internet equipment also provides very high transfer rates. Communication to the Internet via wireless links can be as much as 350 times as fast as a standard in-home connection.

FIGURE 6–44 Photo of a product with wireless Internet access capability. (*Courtesy of Palm, Inc.*)

✓ Exercise Problems 6.5

1. A standard flashlight battery is a common source of alternating voltage. (True/False)
2. The voltage supplied to your home by the power company is alternating voltage. (True/False)
3. What is the approximate speed that sound waves travel through air?
4. Both sound waves and radio waves travel through the air as pressure changes. (True/False)
5. Sound waves and ultrasonic waves travel at approximately the same speed. (True/False)
6. At what approximate speed would a 10-GHz radio wave travel through the air?
7. What is the maximum range for a Bluetooth communication link?
8. Bluetooth technology eliminates the need for many computer cables. (True/False)
9. Several nearby Bluetooth-compliant devices can form into a group called a _____.
10. A wireless Internet connection is slower than most phone line Internet connections. (True/False)

•─SUMMARY

Alternating voltage and current is at the heart of a vast majority of electronic applications. A sinusoidal voltage continuously varies in amplitude and periodically reverses polarity. A sinusoidal current continuously varies in amplitude and periodically changes direction. Sinusoidal waveforms consist of a series of repetitious cycles. The time for one cycle is called the period of the waveform. Frequency describes how many full cycles occur in a one-second interval. Each cycle is composed of two half-cycles with opposite polarities. Sine waves can be generated by an alternator or by an electronic circuit. If the durations of the two half-cycles are equal (as in a sine wave), the waveform is said to be a *symmetrical waveform*. An asymmetrical waveform has unequal half-cycles.

The highest instantaneous voltage or current point on a sine wave is called the peak. Each sine wave has a positive and a negative peak. The average value of a full sine wave is zero, since it is positive as long as it is negative. As a general practice, the term *average*, when used to describe sine waves, refers to the average of one half-cycle. The average voltage in this case is $0.637 \times V_P$. The root-mean-square (rms) voltage of a sine wave describes the heating effect of the waveform. It takes an amount of dc equal to the rms value of a sine wave to produce the same amount of heat in a resistance. The rms voltage is equal to $0.707 \times V_P$. The peak-to-peak voltage does not actually exist as an instantaneous value, but it is useful to describe the sine wave. It is always equal to twice the peak value.

The phase of a sine wave describes an angular point on the wave at a specific time. Zero and 180 degrees coincide to the zero-crossing points of the sine wave. The peak values occur at 90° and 270°. If two sine waves are in phase, then the corresponding points of the two waveforms occur at the same time. Out-of-phase sine waves have corresponding points occurring at different times.

Phase angle can be expressed in degrees or radians. One radian is equal to 57.3°. Either unit of measure is easy to use with a scientific calculator.

The instantaneous voltage of a sine wave varies with each instant in time. For a given peak amplitude, the instantaneous voltage depends on the instantaneous phase of the sine wave. More specifically, $v = V_P \sin \theta$.

A phasor diagram is a pictorial representation of the instantaneous phase in a sine wave. It can also show the phase relationships between multiple sine waves. The length of a phasor arrow represents the peak voltage of a sine wave. The distance of the arrowhead from the horizontal axis represents the instantaneous voltage of the sine wave. Finally, the angle formed by the phasor and the horizontal axis corresponds to the instantaneous phase of the sine wave.

Phasor calculations are based on right-triangle calculations. You should be able to compute the sine, cosine, and tangent functions of an angle. Each of these functions also has a corresponding inverse function (arcsin, arccos, and arctan). Table 6–3 on page 186 provides a brief summary of the more common trigonometric functions.

Resistive circuits with an ac source are analyzed in a manner consistent with the dc analysis presented in previous chapters. Care must be exercised to ensure that corresponding voltages and currents are used (e.g., rms voltage with rms current).

Applications that utilize alternating voltage and current are endless. New applications are being created every day. The applications can be grossly categorized according to frequency of operation. The frequency range of alternating voltages and currents used in electronics extends from subaudio to well over 100 GHz.

TABLE 6–3 A summary of common trigonometric functions.

Function	Value	Example
sine (sin)	opposite/hypotenuse	sin 45° = 0.7071
cosine (cos)	adjacent/hypotenuse	cos 25° = 0.9063
tangent (tan)	opposite/adjacent	tan 80° = 5.67

•—REVIEW QUESTIONS

Section 6.1: Generation of Alternating Voltage

1. The time required to complete a full cycle of alternating voltage is called the _____.
2. How many alternations are required to make one full cycle?
3. If the positive and negative half-cycles of an alternating voltage are equal, the waveform is (*symmetrical/asymmetrical*).
4. If the cycles of a waveform are continuous, the waveform is said to be _____.
5. What is another name for an ac generator?
6. Slip rings provide sliding contact to the brushes in an alternator. (True/False)
7. Why are the brushes on an alternator spring loaded?
8. The frequency of a sine wave produced by an alternator is affected by the speed of rotation of the alternator. (True/False)
9. The maximum instantaneous value of a sine wave is called the _____ voltage.
10. What type of test equipment can generate sine waves, rectangular waves, and waves of other shapes?

Section 6.2: Sine Wave Characteristics

11. What is the basic unit of measurement for the period of a sine wave?
12. What sine wave quantity describes how many cycles occur in a one-second interval?
13. What is the basic unit of measurement for the frequency of a sine wave?
14. If the period of a certain sine wave is 2.5 ms, what is its frequency?
15. What is the frequency of a sine wave that has a period of 75 ns?
16. A sine wave with a period of 100 ns has a higher frequency than a sine wave with a period of 100 ms. (True/False)
17. What is the period of a 5-MHz sine wave?
18. If the frequency of a certain sine wave is 2.5 GHz, what is its period?
19. What is the period of a 60-Hz sine wave?
20. What is the period of a 550-kHz sine wave?
21. How many times does a peak occur during one cycle of a sine wave?
22. The positive peak of a sine wave occurs at 180°. (True/False)
23. The negative peak of a sine wave occurs at 270°. (True/False)
24. If a sine wave has a positive peak current of 2.75 amperes, what is its negative peak value?
25. What is the average value of a *full cycle of a sine wave* if the peak voltage is 10 V?
26. What is the average value of a *full sine wave* if the peak voltage is 1000 V?
27. What is the average value (V_{avg}) of a sine wave with a peak voltage of 25 V?
28. If the average value of a sine wave is 108 volts, what is its peak value?
29. It takes an amount of dc equal to the average value of a sine wave to cause the same heat in a resistance. (True/False)
30. If a sine wave has a peak-to-peak current value of 10 A, what is its rms value?
31. What is the rms value of a sine wave that has a 150-mV average value?

32. What is the peak voltage of a sine wave that has a 200-µV rms value?
33. What is the peak-to-peak voltage of a 10-V peak sine wave?
34. Which sine wave would cause the most heat in a resistor?
    a. 10 V_{PP} c. 10 V_{avg}
    b. 10 V_{rms} d. 10 V_P
35. Angular measurement of a sine wave can be expressed in _____ or _____.
36. When the corresponding points on two similar-frequency sine waves occur at different times, we say the sine waves are (in/out) of _____.
37. If the instantaneous phase of a certain sine wave is 45° and the peak voltage is 100 V, what is the instantaneous voltage?
38. What is the instantaneous voltage of a sine wave at 25° if its peak voltage is 30 V?
39. What is the instantaneous voltage of a sine wave at 105° if the rms voltage is 100 V?
40. What is the peak voltage of a sine wave that has an instantaneous voltage of 10 V at 55°?
41. If a sine wave at 100° has an instantaneous voltage of 120 V, what is its peak-to-peak value?
42. What is the frequency of the fifth harmonic of a 75-MHz waveform?
43. Seventy-five megahertz is the _____ harmonic frequency of a 25-MHz fundamental.

Section 6.3: Working with Phase Angles

44. The length of a phasor represents the _____ voltage of a sine wave.
45. The phase of a sine wave is represented on a phasor diagram by the angle between the phasor and the vertical axis. (True/False)
46. The distance of the phasor tip from the horizontal axis represents the _____ voltage of a sine wave.
47. The longest side of a right triangle is called the _____.
48. The cosine of an angle in a right triangle is obtained by dividing the _____ by the _____.
49. If the _____ in a right triangle is divided by the _____, you will get a number that is the tangent of the angle.
50. If you know the tangent of an angle, how do you find the angle itself?
51. If you know the cosine of an angle, how do you find the angle itself?
52. What is the sine of a 35° angle?
53. What is the sine of a 3.2-rad angle?
54. What is the tangent of a 23° angle?
55. What is the cosine of a 1.6-rad angle?
56. If the tangent of an angle is 5.6, what is the angle in degrees?
57. If the tangent of an angle is 4.25, what is the size of the angle in radians?

58. Refer to Figure 6–45. What is the length of side a?

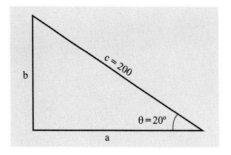

FIGURE 6–45 A right-triangle problem.

59. What is the length of side b in Figure 6–45?
60. What is the value of θ in Figure 6–46?

FIGURE 6–46 A right-triangle problem.

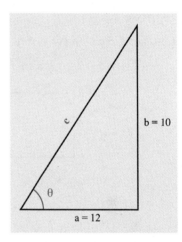

61. What is the length of side c in Figure 6–46?
62. What is the peak voltage of the sine wave represented in Figure 6–47?

FIGURE 6–47 A phasor diagram.

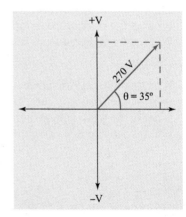

63. What is the instantaneous voltage of the sine wave represented in Figure 6–47?
64. What is the instantaneous voltage of the sine wave represented in Figure 6–48 on page 188?
65. What is the phase angle of the sine wave represented in Figure 6–48?
66. Compute the value of arccos 0.6 and express in radians.

FIGURE 6–48 A phasor diagram.

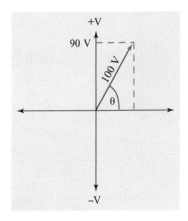

FIGURE 6–50 Analyze this circuit.

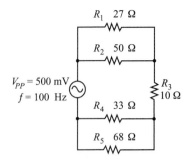

78. What is the peak-to-peak voltage across R_2?
79. What is the rms current through R_3?
80. What is the total average current?
81. What is the total average current over a full cycle?
82. What is the total power dissipation?
83. What is the rms current through R_4?

67. Express the value of arcsin 0.8 in degrees.
68. The value of arctan 20 corresponds to _____ degrees.

Section 6.4: Circuit Analysis of AC Resistive Circuits

Refer to Figure 6–49 for questions 69 through 73.

FIGURE 6–49 Analyze this circuit.

69. What is the peak current?
70. What is the power dissipation in the resistor?
71. What is the average value of input voltage?
72. What is the peak-to-peak input voltage?
73. What is the instantaneous current at 50°?

Refer to Figure 6–50 for questions 74 through 83.

74. What is the total resistance?
75. What is the total rms current?
76. What is the peak voltage across R_1?
77. What is the average current through R_4?

Section 6.5: Alternating Voltage Applications

84. What is the frequency of the power distribution network in the United States?
85. When the power used in your home is first generated, it is a much lower voltage. (True/False)
86. The nominal value of voltage available at the wall sockets in your home is about 108 V average. (True/False)
87. Sound and ultrasonic waves both travel through air at about 1,130 feet per second. (True/False)
88. What is the name of the effect that explains the frequency shift in an ultrasonic system, when the reflecting object is moving?
89. An ultrasonic transmitter transducer is like a _____ in an audio system.
90. An ultrasonic receiver transducer is like a _____ in an audio system.
91. Radio frequency waves travel through air at approximately _____ meters per second.
92. Radio waves move through the air as high- and low-pressure changes. (True/False)
93. Radio waves are called electromagnetic waves. (True/False)
94. It is possible to have radio waves at least as high as 2000 MHz. (True/False)

•—CIRCUIT EXPLORATION

In this Circuit Exploration exercise, you will become familiar with several alternating voltage sources that are commonly provided in circuit simulation programs. An understanding of these signal sources increases your ability to simulate circuits presented in subsequent chapters and practical circuits that you encounter in industry.

For this project, we will focus on the following basic signal sources:

1. *Sinusoidal voltage source*—A sinusoidal voltage generator with adjustable values that include a minimum of amplitude and frequency. Requirements: 1.5-MHz and 500-mV peak voltage.
2. *Digital clock source*—A simple square wave generator with adjustable frequency, duty cycle, and amplitude. Requirements: 33 MHz and 40% duty cycle with voltage levels of 0 V and 5 V.

3. *Pulsed voltage source*—A more complex rectangular wave generator with adjustable values that include a minimum of amplitude, rise time, fall time, period, and pulse width. Requirements: Voltage levels of 0 V and 3.3 V, rise time of 1.5 ns, fall time of 3 ns, period of 200 ns, and a pulse width of 50 ns.

For each of the above sources, locate the generator in your circuit simulation package, select it into the schematic capture environment, and then configure it according to the requirements listed previously. Once configured, use a virtual oscilloscope to verify the various waveforms.

Figure 6–51 clarifies the various measurements for a rectangular waveform such as is used in this exercise.

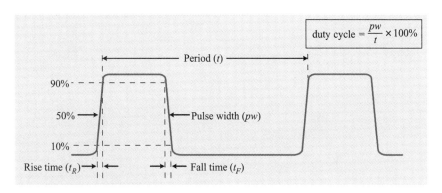

FIGURE 6–51 Retangular waveform measurements.

ANSWERS TO PRACTICE PROBLEMS

Page 158
1. 20 ms
2. 10 ns

Page 159
1. 100 µs
2. 12.5 kHz
3. 250 ps
4. 5.7 Hz
5. The shorter period (3 ms) has the higher frequency (333.3 Hz).

Page 160
1. 1.5 amperes
2. −170 V

Page 161
1. Zero
2. Zero
3. Zero
4. Zero

Page 162
1. 12.74 V
2. 95.55 mV
3. 637 V
4. 95.55 µA

Page 164
1. 1.0 V
2. 24 V

3. 75 V

Page 165
1. 0.785
2. 6.11
3. 4.7
4. 1.57
5. 3.14
6. 143.2
7. 360
8. 229.2
9. 100.3
10. 300.8

Page 167
1. 250 V
2. 1.57 A

Page 169
1. 5 MHz
2. 50 MHz, 75 MHz, 100 MHz, 125 MHz
3. 300 MHz, 500 MHz, 700 MHz
4. 500 MHz, 1000 MHz, 1500 MHz

Page 173 (top)
1. 0.883
2. −0.443
3. −0.577

4. 0.342
5. −0.087
6. −3.381
7. −0.227
8. −0.572
9. −0.94
10. 1

Page 173 (bottom)
1. 44.4° or 0.78 rad
2. 129.27° or 2.26 rad
3. 24.99° or 0.44 rad
4. 5° or 0.087 rad
5. 180° or 3.14 rad
6. −29.98° or −0.52 rad
7. −10.02° or −0.175 rad
8. 74.99° or 1.31 rad
9. −34.99° or −0.611 rad
10. 11.54° or 0.2 rad

Page 175
1. θ = 30.96° or 0.54 rad; c = 29.15
2. b = 14.93; c = 15.45
3. a = 5,671.28; c = 5,758.77
4. 23.58° or 0.412 rad
5. −26.74° or −0.467 rad

CHAPTER 7
Inductors, Capacitors, and Transformers

•—KEY TERMS
bandpass
bandwidth
capacitive reactance
coefficient of coupling
cutoff frequency
dielectric
dielectric constant
differentiator
dot notation
ESL
ESR
half-power point
henry
impedance
inductance
inductive reactance
integrator
long time constant
pass band
permittivity
primary
Q
reflected impedance
resonant frequency
secondary
selectivity
self-resonant frequency
short time constant
step-down transformer
step-up transformer
time constant
turns ratio

•—OBJECTIVES
After studying this chapter, you should be able to:

- List the factors that affect electromagnetic induction, inductance, capacitance, capacitive reactance, and inductive reactance.
- Name at least three types of inductors, three types of transformers, and five types of capacitors and describe an application for each.
- Determine the total inductance and total inductive reactance of inductors connected in series or parallel.
- Determine the total capacitance and total capacitive reactance of capacitors connected in series or parallel.
- State the phase relationships between voltage and current in an inductor and in a capacitor.
- Explain how to identify defects in an inductor, a capacitor, and a transformer.
- Calculate RL and RC time constants and classify them as short, medium, or long for a given input frequency.
- State the general requirements and characteristics of RC integrator and differentiator circuits.
- State the following characteristics of series and parallel RLC circuits at resonance, above resonance, and below resonance: (a) phase relationships and (b) relative component voltages and currents.
- Analyze transformer circuits by applying the relationships between turns ratio, voltage ratio, current ratio, and impedance ratio between the primary and secondary windings.
- Name four general classes of frequency-selective filters.

•—INTRODUCTION
This chapter introduces three more basic electronic components: inductors, capacitors, and transformers. We will focus on learning the physical and electrical characteristics that distinguish these important building blocks. Additionally, we will perform circuit calculations to illustrate how these components affect circuit behavior. Several applications that show how each of these components is used for practical applications are presented. Finally, we discuss procedures for detecting and identifying defects in each of these basic building blocks.

7.1 INDUCTORS

Inductors, like resistors, are fundamental building blocks of electronic circuits. Like resistors, inductors will tend to impede current flow. Unlike resistors, however, the opposition offered by inductors is different for dc and ac circuits. Inductors are also called coils or chokes. The latter term comes from the use of some inductors to suppress (i.e., choke out) undesired frequencies. The basis for understanding the effects of an inductor in an electronic circuit lies in understanding electromagnetic induction.

Electromagnetic Induction

Figure 7–1 illustrates the basic principle of electromagnetic induction. Here, two coils of wire are wound on a common core. The core is a material with a high permeability. Permeability is a measure of how easily a magnetic field can be set up in a material. High permeability indicates that the flux lines are more concentrated in the core. One of the coils is connected to an alternating current source. As the current changes continuously, so does the magnetic flux within the core. Since a significant percentage of this flux is common to the second coil, the changing flux lines will intercept the turns of the second coil. When moving magnetic flux cuts a conductor, a voltage is induced. Figure 7–1 shows that a voltage can be measured in the second coil, even though there is only a magnetic connection to the first coil. The process of inducing a voltage into a conductor with a changing magnetic field is called electromagnetic induction.

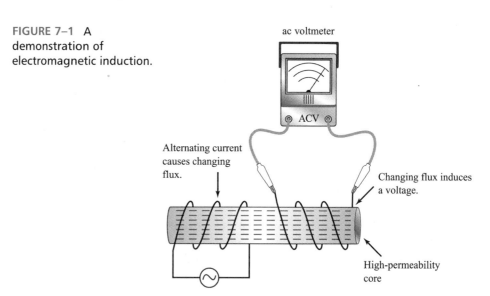

FIGURE 7–1 A demonstration of electromagnetic induction.

Factors Affecting Induction

There are several factors that determine the amount and polarity of induced voltage. The factors are:

- Magnetic field strength
- Rate of relative motion between flux and conductor
- Angle of relative motion between flux and conductor
- Number of turns of wire in the coil
- Direction of relative motion between flux and conductor
- Polarity of the magnetic field

These factors are illustrated in Figure 7–2 for a simple case.

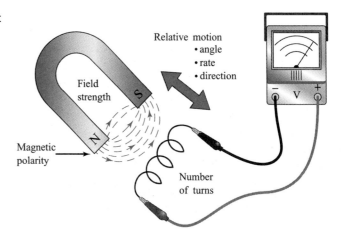

FIGURE 7–2 Factors that affect electromagnetic induction.

Effects of Induced Current

When current flows in a wire, it causes a magnetic field to be formed around the wire. If the wire is formed into a coil, then the magnetic field becomes concentrated, and magnetic poles are formed. The polarity of induced voltage (or current in a closed circuit) is determined by the polarity of the magnetic field and the relative direction of motion. Lenz's Law describes the relationship of induced current polarity to the polarity of the moving magnetic field that caused the induced current. Lenz's Law can be summarized as follows:

> ► **LENZ'S LAW** When a current is induced in a coil by a changing magnetic field, the current creates a second magnetic field. The magnetic field produced by the induced current has a polarity that opposes the changes in the original magnetic field.

This important principle is illustrated in Figure 7–3. In Figure 7–3(a), the north pole of a magnet is approaching a coil. An induced current flows in the coil such that a magnetic field is produced that opposes the motion of the moving field. In the case shown in Figure 7–3(a), the coil forms a north pole on the end nearest the approaching north pole. The two fields now oppose the relative motion.

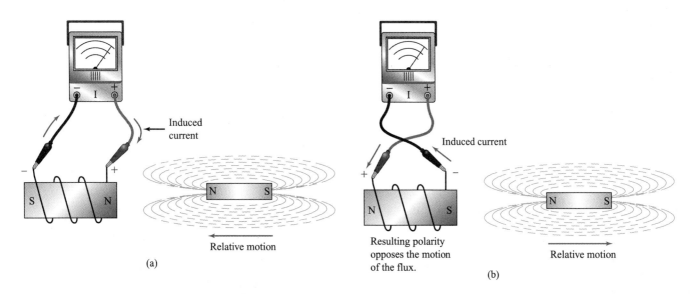

FIGURE 7–3 An illustration of Lenz's Law.

Figure 7–3(b) illustrates the effects when the magnet is moved away from the coil. Since the field is moving in the opposite direction, we would expect an opposite polarity to be induced in the coil. As shown in Figure 7–3(b), the resulting current flow causes a magnetic field that opposes the motion of the original magnetic field. In the case shown in Figure 7–3(b), the induced current causes a south pole to form nearest the departing pole. The unlike poles tend to attract, which retards the movement of the departing field.

Inductance and Its Factors

An inductor is essentially a coil of wire. When current flows through the turns of an inductor, a magnetic field is produced around each turn. If the current is changing, then the magnetic field will change. Any change in the magnetic field associated with one turn in the coil will produce a voltage (induced voltage) in the other turns of the coil as the flux lines intersect the wire. Thus, any change in current produces a change in flux that causes an induced voltage in each turn of the coil. Further, Lenz's Law tells us that the polarity of induced voltage in each turn will be such that it opposes the original change in current. This is a *very* important concept that describes the basic behavior of an inductor.

> ➤ CONCEPT   When current changes in an inductor, a voltage is induced in the turns of the inductor that opposes the initial change in current.

The key word here is change. If, for example, a certain inductor had a very high—but steady—current, then there would be a very strong magnetic field around the coil, but there would be no induced voltage, since the field is stationary (i.e., not changing). If, by contrast, a current changes in an inductor, then there can be a very high value of induced voltage. The value of induced voltage is determined by the number of turns and by how quickly the current (and therefore the flux) changes.

This behavior of an inductor is often stated another way:

> ➤ CONCEPT   An inductor opposes changes in current.

This is a simple but important concept. An inductor does not oppose the actual flow of current, but it opposes any changes in current. Thus, if the current tries to increase or decrease, the inductor will generate a self-induced voltage that opposes the changing current and will try to keep it constant. This is also why inductors tend to limit current in ac circuits but have no effect on steady dc.

The property of an inductor, which allows it to exhibit the characteristics just described, is called **inductance** (L). The unit of measure for inductance is the **henry** (H). More specifically, if the current through an inductor changes at the rate of one ampere per second and causes one volt of self-induced voltage, then the inductor has a value of one henry. Table 7–1 clarifies the physical factors that affect inductance.

TABLE 7–1   Physical factors that affect inductance.

Factor	Proportionality	Example
Length of the coil	Inverse	Length ↑ Inductance ↓
Cross-sectional area of the coil	Direct	Area ↑ Inductance ↑
Number of turns (N)	Direct (N^2)	Turns ↑ (×3) Inductance ↑ (×9)
Permeability of the core (μ)	Direct	μ ↑ Inductance ↑

Inductive Reactance

When an inductor is used in a circuit that has a sinusoidal current, the self-induced voltage is also sinusoidal. This means the self-induced voltage is constantly changing. However, the polarity of the induced voltage always opposes the changing current. This opposition results in a lower value of current than would otherwise flow. The opposition to a sinusoidal alternating current flow presented by an inductor is called **inductive reactance**. It is measured in ohms. The generic symbol for reactance is X. Inductive reactance is represented by the symbol X_L.

Anything that affects the value of self-induced voltage in the inductor will affect its inductive reactance. Thus, we would expect the number of turns on the coil, permeability of the core, rate of change of current, and other such factors to be related to inductive reactance. They are. However, we can group all the factors such as physical coil characteristics and magnetic core characteristics together, since they collectively determine the value of inductance. That is, by referring to the value of inductance, we are inherently referring to all of the physical and magnetic factors. We can account for the rate of change of current factor by referring to the frequency of the current or voltage in the inductor. Equation 7–1 provides us with a direct method for computing the inductive reactance of an inductor:

$$X_L = 2\pi f L \qquad (7\text{–}1)$$

•─EXAMPLE 7.1

How much inductive reactance is provided by a 10-mH inductor when operated at a frequency of 25 kHz?

SOLUTION We apply Equation 7–1 as follows:

$$X_L = 2\pi f L$$
$$= 2 \times 3.14 \times 25 \text{ kHz} \times 10 \text{ mH}$$
$$= 1.57 \text{ k}\Omega$$

Calculator Application

Phase Relationships

You will recall from a previous discussion that the voltage across an inductor is proportional to the rate of change of current through the inductor. If the current waveform is a sine wave, then it follows that the points of highest self-induced voltage will coincide with the 0° and 180° points of the current waveform, since these points have the highest rates of change. Similarly, there will be no self-induced voltage as the current wave passes through the 90° and 270° points, since the rate of change of current at these points is zero. These relationships lead to the waveforms shown in Figure 7–4 on page 196.

As you can see from the waveforms in Figure 7–4(b), the voltage peaks coincide with the points of maximum rate of change of current. Similarly, the zero-crossing points of the voltage waveform coincide with the peaks (minimum rate of change) of the current waveform.

Both current and voltage waveforms are sinusoidal, but the voltage waveform is 90° out of phase with the current waveform. That is, the current and voltage peaks

FIGURE 7-4 The current in an inductor lags the voltage by 90°.

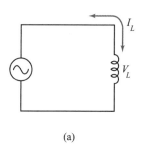

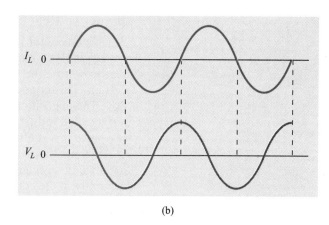

(a)

(b)

do not occur at the same time. More specifically, the voltage waveform leads (i.e., is ahead of) the current waveform by 90°.

Alternatively, we may say the current waveform lags (i.e., is behind) the voltage waveform. Let's state this important characteristic of an inductor as a rule.

> **RULE** The sinusoidal voltage across an inductor will always lead the sinusoidal current by 90°.

Inductor Types and Applications

All inductors are similar in certain respects. They all possess inductance, they all generate a self-induced voltage in direct proportion to the rate of change of current, and the polarity of self-induced voltage will always oppose any changes in current. There are many different types of inductors, however, and they vary dramatically in value, primary application, and physical appearance.

Inductors can be grossly categorized into two classes: fixed and adjustable (variable). Figure 7–5 shows the schematic symbols for these two general classes of inductors.

Inductors can also be classified by the type of core material used. Some, but not all, manufacturers indicate the type of core material by using a slightly modified schematic symbol. Table 7–2 summarizes a number of inductor types including variations to the schematic symbol and a representative application.

Ferrite beads are also widely used to add inductance (and resistance) to circuits. They are available in a wide array of sizes and material types. A ferrite bead has a hole (or holes) through which a wire or leaded component may be inserted. (Note: Some beads are sold with the wire already inserted.) The presence of the bead adds

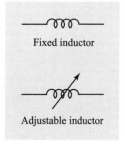

FIGURE 7-5 Generic schematic symbols for fixed and adjustable inductors.

TABLE 7-2 A summary of inductor types.

Inductor Type	Picture	Schematic Symbol	Applications
Air: Coil is wound on a nonmagnetic form.		—	Radio-frequency circuits

TABLE 7-2 (continued)

Inductor Type	Picture	Schematic Symbol	Applications
Iron: Laminated core consists of several thin iron sheets.			Audio circuits and power supplies
Powdered-iron: Core of individually-insulated iron granules is pressed into a solid pellet.			Radio-frequency circuits
Ferrite: Core is high permeability ceramic material called ferrite.			Used from low kilohertz frequencies to hundreds of megahertz. Popular in computers as high-frequency filters.
Molded: Similar to air core but encapsulated.			Radio-frequency circuits
Surface-mount			Used in applications ranging from audio to several gigahertz

series inductance and series resistance to the wire at high frequencies. The bead has no effect at low frequencies.

It is important to note that although the bead is slipped over a wire (even an insulated wire in many cases), the bead's effective inductance and resistance characteristics appear as if they were in series with the inserted wire. Ferrite beads find extensive use in high-frequency circuits. They are also widely used in personal and handheld computers to suppress unwanted emissions that might otherwise interfere with nearby radio and television reception. They are sometimes inserted on external computer cables (e.g., video monitors and keyboards) to reduce high-frequency radiation. Figure 7–6 on page 198 shows a selection of ferrite components.

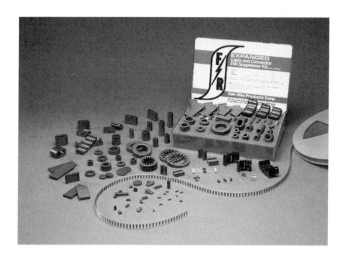

FIGURE 7-6 Ferrite components come in a wide range of shapes and sizes. (*Courtesy of Fair-Rite Products Corporation*)

Series and Parallel Inductors

Multiple inductors can be connected together much like resistors. With inductors, however, we must consider the effect on both inductance and inductive reactance. We will restrict our formal discussion to series and parallel connections, but the concepts can be directly extended to series-parallel connections as discussed with resistive circuits. The following analyses assume there is no interaction between the magnetic fields of the various inductors. This is normally a valid assumption for individual inductor packages.

Series Inductances

Figure 7-7 shows three inductors connected in series. Each inductor responds to changes in the common series current. The self-induced voltages all act to oppose any changes in current. The combined inductance of series-connected inductors is the simple sum of the individual inductors. This is expressed in Equation 7-2.

$$L_T = L_1 + L_2 + L_3 + \ldots + L_N \qquad (7\text{-}2)$$

This equation is not valid if any of the coils are magnetically linked.

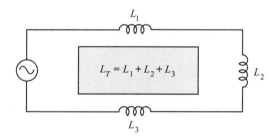

FIGURE 7-7 Series-connected inductors combine like series resistors.

•—EXAMPLE 7.2

If $L_1 = 10$ µH, $L_2 = 50$ µH, and $L_3 = 25$ µH in Figure 7-7, what is the total circuit inductance?

SOLUTION We apply Equation 7-2 as follows:

$$\begin{aligned} L_T &= L_1 + L_2 + L_3 \\ &= 10 \text{ µH} + 50 \text{ µH} + 25 \text{ µH} \\ &= 85 \text{ µH} \end{aligned}$$

Series Inductive Reactances

When the individual inductive reactances of two or more series-connected inductors are known, the total inductive reactance can be found by summing the individual reactances. Equation 7–3 expresses this relationship.

$$X_{L_T} = X_{L_1} + X_{L_2} + X_{L_3} + \ldots + X_{L_N} \qquad (7\text{–}3)$$

•—EXAMPLE 7.3

Determine the total inductive reactance for the circuit shown in Figure 7–8.

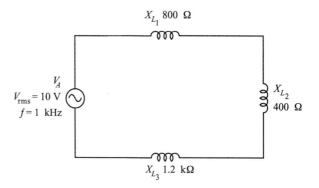

FIGURE 7–8 What is the total inductive reactance in this circuit?

SOLUTION We apply Equation 7–3 as follows:

$$X_{L_T} = X_{L_1} + X_{L_2} + X_{L_3}$$
$$= 800\ \Omega + 400\ \Omega + 1.2\ k\Omega$$
$$= 2.4\ k\Omega$$

Parallel Inductances

Figure 7–9 shows three parallel-connected inductors. Since the total current divides among the various inductors, any given inductor experiences less than the total change in current. The self-induced voltage for a given inductor will be correspondingly smaller than if the total current change were applied to the coil. As with parallel-connected resistors, the total inductance is found by using a reciprocal formula shown in Equation 7–4.

$$L_T = \dfrac{1}{\dfrac{1}{L_1} + \dfrac{1}{L_2} + \dfrac{1}{L_3} + \cdots + \dfrac{1}{L_N}} \qquad (7\text{–}4)$$

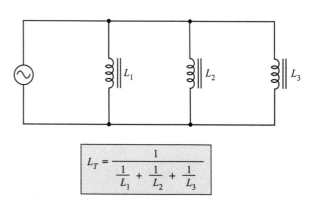

FIGURE 7–9 Parallel-connected inductors combine like parallel resistors.

As you might expect, the combined inductance of parallel-connected inductors is always less than the smallest individual inductance.

•EXAMPLE 7.4

If $L_1 = 80$ mH, $L_2 = 50$ mH, and $L_3 = 35$ mH in Figure 7–9, what is the total circuit inductance?

SOLUTION We apply Equation 7–4 as follows:

$$L_T = \frac{1}{\dfrac{1}{L_1} + \dfrac{1}{L_2} + \dfrac{1}{L_3}}$$

$$= \frac{1}{\dfrac{1}{80 \text{ mH}} + \dfrac{1}{50 \text{ mH}} + \dfrac{1}{35 \text{ mH}}} = 16.37 \text{ mH}$$

Calculator Application

Parallel Inductive Reactances

When inductive reactances are connected in parallel, the total inductive reactance is less than the smallest individual reactance. The exact value can be computed with the reciprocal formula shown in Equation 7–5.

$$X_{L_T} = \frac{1}{\dfrac{1}{X_{L_1}} + \dfrac{1}{X_{L_2}} + \dfrac{1}{X_{L_3}} + \cdots + \dfrac{1}{X_{L_N}}} \qquad (7\text{–}5)$$

•EXAMPLE 7.5

What is the total inductive reactance for the circuit shown in Figure 7–10?

FIGURE 7–10 Find the total inductive reactance in this circuit.

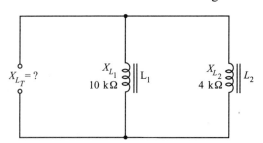

SOLUTION We apply Equation 7–5 as follows:

$$X_{L_T} = \frac{1}{\dfrac{1}{X_{L_1}} + \dfrac{1}{X_{L_2}}}$$

$$= \frac{1}{\dfrac{1}{10 \text{ k}\Omega} + \dfrac{1}{4 \text{ k}\Omega}} = 2.86 \text{ k}\Omega$$

Circuit Analysis of Inductive Circuits

Previous sections have shown how to compute the total inductance and total inductive reactance for series and parallel combinations of inductances. Once you have determined the inductive reactance for each inductor in a multiple inductor circuit, you can then determine all component voltages and currents using the same techniques previously outlined for resistive circuit analysis. This approach works for series, parallel, or series-parallel networks. For example, you could utilize any of the following variations of Ohm's Law:

$$I_L = \frac{V_L}{X_L}$$

$$V_L = I_L X_L$$

$$X_L = \frac{V_L}{I_L}$$

As with resistive circuits, care must be exercised to ensure that consistent values are used in the equations (e.g., rms, peak, average, and so forth). If you are solving for an unknown component value, then all other values used in the equation must be associated with the same component. Practice on this type of problem is provided in the review exercises.

Troubleshooting Inductors

Since an inductor is nothing more than a wire wrapped around a core, there are only three defects that are probable. An open in the coil winding is probably the most common malfunction. Figure 7–11 illustrates the use of an ohmmeter to detect the open winding. If the winding is open, the ohmmeter will indicate infinite (∞) resistance.

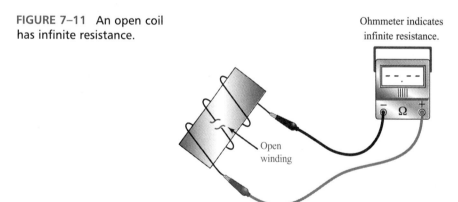

FIGURE 7–11 An open coil has infinite resistance.

When an ohmmeter is used to check the condition of a good coil, the meter will indicate the resistance of the wire. This varies with the type of coil being considered and can be less than one ohm or as high as several thousand ohms. You should measure and record the normal resistances of coils used in circuits you are expected to troubleshoot. That way, you can contrast the resistance values of suspected coils with those of known good coils.

Figure 7–12 on page 202 shows another possible defect that can occur in a multilayer coil. Here, the insulation between two adjacent turns has disintegrated allowing the windings to short together. This effectively shorts out or bypasses a portion of the coil. Depending on the location of the short and the method of winding the coil, the short may only bypass a single turn, or it may bypass a significant

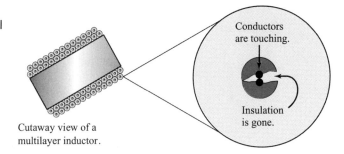

FIGURE 7–12 Adjacent turns on a multilayer coil can become shorted if the insulation breaks down.

portion of the inductor. A shorted coil is sometimes very difficult to detect with an ohmmeter. The ohmmeter will read a value of coil resistance that is less than the normal value for the coil. However, since the normal value may only be one or two ohms, a typical ohmmeter may not absolutely identify shorted turns.

Windings usually become shorted as a direct result of overheating or internal arcing. In either case, the short may be accompanied by physical clues such as discoloration of the coil, visible melting of the insulation, or a characteristic, pungent odor.

Figure 7–13 shows a test instrument that you can use to measure the value of inductance in a coil. If a significant number of turns in the coil are shorted, or if the coil is open, then the inductance meter will easily and quickly identify the defect. For accurate measurements in most circuits, at least one lead of the inductor must be removed from the circuit while it is being measured.

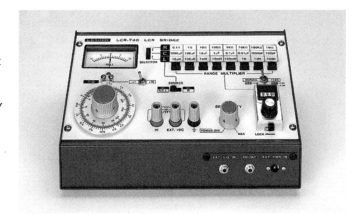

FIGURE 7–13 An inductance meter measures the actual value of an inductor. It can be used to detect opens or shorts in the coil winding. (*Courtesy of Leader Instruments Corporation*)

The final defect that is likely to occur in coils is similar to the case of shorted windings. If the coil overheats, the insulation that separates the winding from the metallic core can break down. This allows the coil winding to contact the conductive core material. In the case of iron-core coils, this is generally a catastrophic failure, since the core is often connected directly to the chassis ground of the equipment. This means that the current that would ordinarily flow through the coil is bypassed directly to ground. Additionally, the short to ground effectively bypasses the inductive reactance of the coil, which may allow a substantial current to flow.

An ohmmeter can be used to detect a winding that is shorted to the core. Figure 7–14 illustrates this technique. Figure 7–14(a) shows that a good coil will have an infinite resistance (at least higher than a typical ohmmeter will read) between the coil winding and the metallic core. If the winding shorts to the core, as shown in Figure 7–14(b), then an ohmmeter check will easily detect this condition by indicating zero ohms. See Table 7–3 for a quick reference.

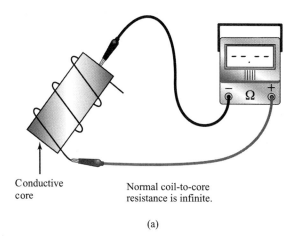
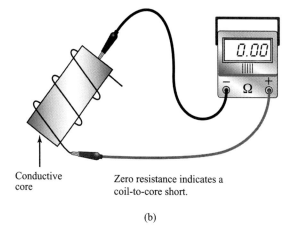

FIGURE 7–14 (a) An ohmmeter can distinguish between a normal coil winding and (b) a winding that is shorted to the core.

TABLE 7–3 A summary of inductor defects and their indications when using an ohmmeter.

Defect	Ohmmeter Indication
No defect	Resistance of the coil winding. Infinite resistance between core and coil.
Open winding	Infinite
Shorted turns	Lower than normal. May require the use of an inductance meter.
Coil-to-core short	Low resistance value between core and coil

✓ Exercise Problems 7.1

1. The behavior of inductors is based on the principles of electromagnetic _____.
2. The strength of the magnetic field affects the magnitude of the induced voltage when the field intercepts a coil of wire. (True/False)
3. The polarity of the magnetic field affects the magnitude of the induced voltage when the field intercepts a coil of wire. (True/False)
4. The relative speed of the magnetic field affects the magnitude of the induced voltage when the field intercepts a coil of wire. (True/False)
5. An inductor _____ changes in current.
6. A current of 10 A supplied by a battery will produce more self-induced voltage across an inductor than a current, which changes from 5 μA to 10 μA in one nanosecond. (True/False)
7. What is the inductive reactance of a 100-μH coil operated at 300 MHz?
8. How much inductance must a coil have in order to provide 50 Ω of inductive reactance to a 150-MHz sine wave?
9. Sinusoidal current *leads/lags* voltage by ____° in an inductor.
10. What is the total inductance if five 100-mH coils are connected in series?
11. What is the total inductance if five 100-mH coils are connected in parallel?
12. Which would have the greatest total inductive reactance, three series-connected 50-μH coils or four parallel-connected 500-μH coils?
13. Calculate the voltage drop across L_2 in Figure 7–15.

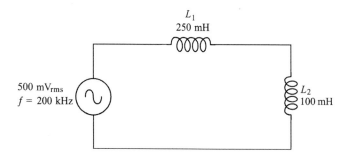

FIGURE 7–15 Calculate the voltage drop across L_2.

Fig07_15.msm

14. How does an ohmmeter indicate an open in a coil winding?

15. What is the normal ohmmeter reading between a coil and its core?

7.2 TRANSFORMERS

Transformers are basic circuit components like resistors and inductors. Their operation is based on the same electromagnetic principles discussed with respect to inductors. Transformers are used in nearly every electronic system that operates from the ac power line and in most battery-operated circuits as well. Most computer systems use multiple transformers in applications that include power supplies, multimedia circuits, modem circuits, LCD display panels, and Ethernet drivers. You will need to understand the theory of operation of transformers, know how to analyze transformer circuits, and be able to troubleshoot transformers to locate defects.

Basic Transformer Action

Before we begin to analyze and troubleshoot transformer circuits, it is important to understand the basic nature of transformer operation. It is an application of electromagnetic induction.

Magnetic Coupling

You already know that when current passes through an inductor, a magnetic field is produced around the inductor. The strength of the field is proportional to current and the physical properties of the coil (e.g., number of turns, type of core, and so on). You will also recall that a changing magnetic field can induce a voltage into the windings of an inductor. The amount of voltage induced is proportional to such things as the relative rate of movement of the field, strength of the field, and physical properties of the coil.

Figure 7–16 shows two coils that are physically close to each other. One coil is connected to an alternating voltage source. The other coil is connected to an ac voltmeter. The alternating current in L_1 will cause the surrounding flux (shown in Figure 7–16 as dotted lines) to continuously expand and contract. Because L_2

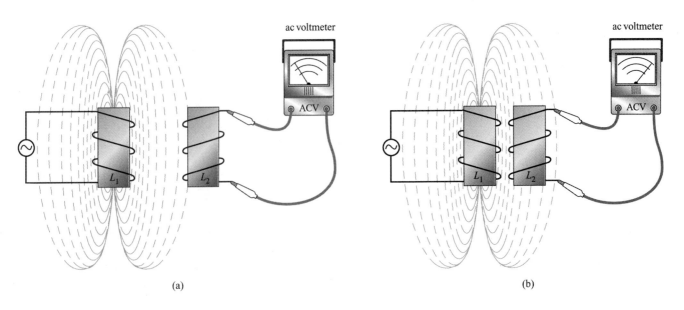

FIGURE 7–16 Two nearby coils can be magnetically linked.

is nearby, some of the flux produced by L_1 cuts the windings of L_2 and induces a voltage into the L_2 coil. This is a measurable, useful voltage as indicated by the voltmeter in Figure 7–16. As indicated in Figure 7–16(a), not all of the flux lines produced by L_1 pass through L_2.

Now, if the two coils are brought closer together as shown in Figure 7–16(b), a greater percentage of the flux is shared by the two coils. Because more flux lines are cutting the L_2 turns, a higher voltage will be produced in L_2 as indicated by the voltmeter in Figure 7–16(b).

The percentage of flux from one coil that passes through a second coil is called the **coefficient of coupling**. The coefficient of coupling can range from zero (i.e., the two coils share no flux lines) to 100 (i.e., the two coils share 100% of the flux lines). Generally, the coefficient of coupling is expressed as a decimal, so the range is from zero to one.

A transformer is a basic component whose operation depends on the magnetic linkage between two or more coils. The coils in the transformer are wound on a common core to increase the amount of flux linkage. In many cases, the coils are wound on overlapping layers to obtain the highest degree of coupling.

Transformer Terminology

Figure 7–17 shows two coils wound on a common magnetic core. A sinusoidal voltage is applied to one of the windings. This winding is called the **primary** of the transformer. The inductive reactance of the primary winding limits the primary current. The magnetic flux in the core varies as the current in the primary varies. Because the core is permeable, much of the flux created by the primary passes through the other coil called the **secondary**. The changing flux induces a voltage in the secondary winding. As illustrated in Figure 7–17, the secondary voltage is a measurable, usable voltage. As you might expect, the greater the coefficient of coupling, the higher the value of secondary voltage for a given primary voltage.

FIGURE 7–17 Basic transformer action.

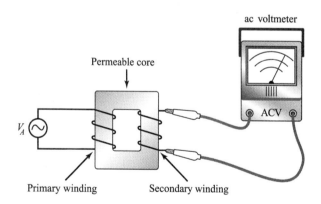

The polarity of the voltage in the secondary depends on the direction in which the secondary winding is wrapped. In certain applications, the primary and secondary windings must be connected into the circuit with proper phase relationships. Schematic diagrams often indicate the phase relationships in the transformer with **dot notation**. Figure 7–18 on page 206 shows the schematic symbol for one type of transformer and illustrates the use of dot notation.

As shown in Figure 7–18, dots are added to both primary and secondary windings. The locations of the dots indicate similar instantaneous polarity. As indicated by the sine wave outputs in Figure 7–18, the relative positioning of the primary and secondary phase dots indicates the input/output phase relationship.

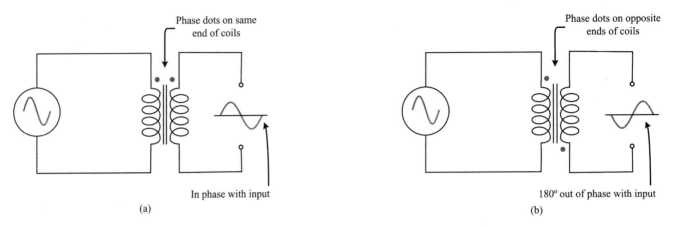

FIGURE 7–18 Schematic symbol for a basic transformer including the use of dots to indicate transformer phase.

Transformers can be constructed to provide secondary voltages that are different from the primary voltage. If the secondary voltage of a transformer is higher than its primary voltage, we call it a **step-up transformer**. Similarly, if the secondary voltage is lower than the primary voltage, we call it a **step-down transformer**. Figure 7–19 illustrates step-up and step-down transformers.

FIGURE 7–19 Illustration of step-up and step-down transformers.

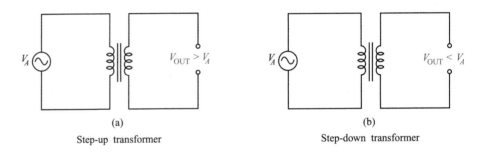

Transformer Types and Applications

There are several ways to categorize transformers. They can be classified according to their intended application, the type of core material, or the way their windings are configured.

Classification by Application

Let us briefly examine how transformers can be categorized based on their primary application. Some transformers are designed for connection to the power line and used to connect power to electronic circuits. This type of transformer is generally called a power transformer. Power transformers can be used to increase or decrease the voltage applied to the primary. If the transformer does *not* increase or decrease the voltage, but is used strictly to provide isolation between the primary and secondary circuits, it is often called an isolation transformer.

Some transformers are specifically designed for operation throughout the audio range. They are, for example, used to connect audio amplifiers to speakers. In any case, these transformers are commonly called audio transformers.

Transformers that are constructed for operation at high radio frequencies are classified as r.f. (radio frequency) transformers. The internal circuits of a radio receiver utilize signals that are called intermediate frequencies. The transformers used in these circuits are called i.f. (intermediate frequency) transformers.

Many pulse (e.g., rectangular waveform) circuits utilize transformers for coupling signals from one point to another. These transformers are called pulse transformers and are optimized for coupling nonsinusoidal waveforms.

Ethernet cables are commonly used to link computers together so they can share data on an office or factory network. An Ethernet transformer is used inside each computer to provide the interface between the computer circuits and the Ethernet cable.

SYSTEM PERSPECTIVE: Transformers and Inductors

Nearly all business computers and many home computers now have Ethernet connections. Ethernet is a method of connecting two or more computers together. Once connected via an Ethernet cable, the various computers can pass information freely back and forth. They can also share system resources such as printers, fax machines, and modems. High-frequency (30 MHz to several hundred MHz) noise generated within the computer can find its way to the Ethernet cable. The long cable can serve as a radiating antenna structure for the high-frequency noise. This can very easily cause illegal interference to other nearby electronic systems. Additionally, external electrical noise can be inadvertently coupled onto the long Ethernet cable as it winds through an office or factory. If this noise is allowed to enter the computer, errors and/or damage may result. To reduce both of these problems, many computer manufacturers include an isolation transformer and a common-mode choke on their Ethernet connections. Both of these devices offer very high impedance to signals (called common-mode noise) that are common to both signal lines. This is the nature of most interference. By contrast, these devices offer very little opposition to signals that appear between the two signal lines (called differential voltages or signals). This is the nature of the actual Ethernet signals. Noise suppression is a very common application for transformers and inductors in a computer.

Classification by Core Material
Another common way to categorize transformers is according to the type of core material used in the transformer. There are three primary types of core material used in transformers: air, iron, and ferrite.

Air-Core Transformers Air-core transformers, like air-core coils, are wound around a nonmagnetic coil form. The form may be plastic, cardboard, or any other material with a relative permeability near unity. Since the core material has such a low permeability, it follows that much of the flux escapes the core. The flux that is external to the core and does not cut both primary and secondary windings is called leakage flux. Figure 7–20 shows an air-core transformer and its generic schematic symbol. These devices are generally used at high radio frequencies.

FIGURE 7–20 A representative air-core transformer and its schematic symbol.

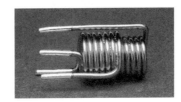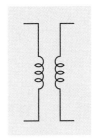

Iron-Core Transformers The primary and secondary windings in an iron-core transformer are wound on a common, high-permeability core. Because of the high permeability of the core, the coefficient of coupling approaches unity (i.e., there is minimal leakage flux). Iron-core transformers are generally limited to frequencies in the audio range. One of the most common applications for iron-core transformers is as power transformers used to couple power from the ac power line. Figure 7–21 shows a typical iron-core transformer and its generic schematic symbol.

FIGURE 7–21 A typical iron-core transformer and its schematic symbol.

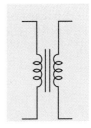

Ferrite-Core Transformers The third category of transformer cores is ferrite. Ferrite is a high-permeability ceramic material that is relatively brittle. Ferrite-core transformers are used for applications extending from audio frequencies to well into the high-megahertz range. The cores used in ferrite-core transformers come in many different shapes and sizes. The torroid core has a very low level of leakage flux. That is, nearly all of the flux is contained within the continuous core material. Figure 7–22 shows a representative ferrite-core transformer and the generic schematic symbol.

FIGURE 7–22 A representative ferrite-core transformer and its schematic symbol.

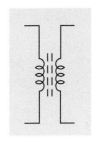

Classification by Winding Connection

Figure 7–23 shows a pictorial sketch and the schematic diagram for a transformer with multiple secondary windings. The changing (i.e., sinusoidal) primary current causes corresponding changes in the magnetic flux. The high-permeability core material causes most of this changing flux to pass through the two secondary windings. Each of the secondary windings will have an induced voltage that can be used by circuits or devices connected to the secondary winding. The two (or more) secondary windings are essentially independent of each other.

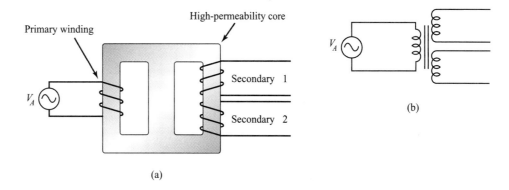

FIGURE 7–23 A transformer can have multiple secondary windings. Each secondary acts like a separate transformer.

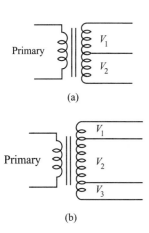

FIGURE 7–24 Tapped transformers provide multiple access points to the transformer winding.

Figure 7–24 shows another way to alter the connection of transformer windings. In the case shown in Figure 7–24(a), the secondary winding is tapped in the center. During manufacture, we can position the tap at any convenient point along the length of the winding. Although the winding is continuous, the tap allows us to access two different voltages (V_1 and V_2). As you would expect, the sum of V_1 and V_2 is equal to the voltage of the entire secondary. Figure 7–24(b) shows a secondary with multiple taps. We can obtain a useful voltage between any two of the connection points. Although not specifically pictured in Figure 7–24, we can also provide taps on the primary of the transformer. This option is often used when constructing transformers that can be operated from two different voltages (e.g., 120 Vac and 240 Vac). The voltage selection can be done by soldering to a different tap on the primary or, more commonly, by using a switch to change the primary connection.

Figure 7–25 shows yet another way that the windings of a transformer can be constructed. Here, a single, tapped winding serves as both the primary and the secondary of the transformer. One end of the single coil serves as a common line for both primary and secondary windings. This configuration is called an autotransformer.

If the input voltage is applied between the two ends of the transformer winding, as illustrated in Figure 7–25(a), then the output voltage will be lower than the input voltage. By contrast, a step-up autotransformer can be constructed, as shown in Figure 7–25(b), by connecting the primary voltage between the common and the

FIGURE 7–25 An autotransformer has a common primary and secondary winding.

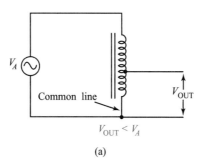
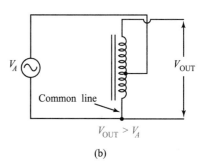

tap. In either case, the amount of increase or decrease in primary voltage is determined by the relative location of the tap. Some manufacturers make autotransformers with the tap connected to a moveable slider. The slider is connected to an external knob that can be used to adjust the position of the tap and, therefore, the amount of step-up or step-down provided by the transformer.

Autotransformers are also frequently used in high-voltage power supplies. A common example is the power supply in your television set or computer monitor, which generates tens of thousands of volts.

Analysis of Transformer Circuits

We shall limit our discussion to the analysis of iron-core transformer circuits. Furthermore, we will assume that our transformers are ideal and have unity coupling. This not only simplifies calculations, but it provides adequate accuracy for many practical applications.

Turns Ratio

A given transformer has a certain number of turns in the primary and a certain number of turns in the secondary. The ratio of primary turns to secondary turns is called the **turns ratio** of the transformer. This ratio is the heart of many transformer calculations. We can express it formally with Equation 7–6.

$$\text{Turns ratio} = \frac{N_P}{N_S} \qquad (7\text{–}6)$$

It should be noted that some textbooks define turns ratio as N_S/N_P. This is acceptable, provided the reciprocals of all other transformer ratios are also used.

•–EXAMPLE 7.6

What is the turns ratio for a transformer that has 400 turns in the primary winding and 100 turns in the secondary winding?

SOLUTION The given transformer is illustrated in Figure 7–26. We compute the turns ratio by dividing the primary turns by the secondary turns (Equation 7–6). Rather than express the turns ratio as a single number (4 in this case), it is customary to express it as a reduced fraction (4/1 or simply 4:1 in this case).

FIGURE 7–26 A transformer turns ratio example.

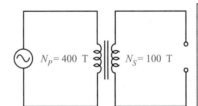

Voltage Ratio

For purposes of this discussion, we are assuming unity coupling. This means that exactly the same flux cuts both primary and secondary windings. Whatever voltage is induced into a single turn of the primary will also be induced into a single turn in the secondary. That is, the volts per turn are the same for both primary and secondary windings. This should seem reasonable, since the coils are wound on a common core and are cut by the same flux lines.

Now, we know from inspection (or Kirchhoff's Voltage Law) that the sum of the voltages induced into the various turns in the primary winding must equal the applied voltage. The voltage in the secondary, however, depends on the number of turns in the secondary (remember the volts per turn are the same as the primary). Therefore, the ratio between primary and secondary voltages is equal to the turns ratio of the transformer. This is an important relationship and is formally stated as Equation 7–7.

$$\frac{N_P}{N_S} = \frac{V_P}{V_S} \tag{7-7}$$

EXAMPLE 7.7

A transformer with a 4:3 turns ratio is connected to a 120-Vac source. Calculate the secondary voltage.

SOLUTION We apply Equation 7–7 as follows:

$$\frac{N_P}{N_S} = \frac{V_P}{V_S} \text{ or}$$

$$N_P V_S = N_S V_P \text{ or}$$

$$V_S = \frac{N_S V_P}{N_P}$$

$$= \frac{3\,\text{T} \times 120\,\text{V}}{4\,\text{T}} = 90\,\text{V}$$

Calculator Application

Power Ratio

The power ratio for an ideal transformer is unity. That is, with 100% coupling and no transformer losses, the output power must be the same as the input power. In particular, power in the primary (i.e., power taken from the source) must always be equal to the power in the secondary (i.e., power delivered to devices or circuits connected to the secondary winding).

Practical transformers have losses and less than 100% coupling. Nevertheless, these ideal assumptions will satisfy our immediate needs and the needs for many practical analyses.

Current Ratio

We know that power is the product of voltage and current (i.e., $P = VI$). We also know that the primary power must equal the secondary power. It follows, therefore, that if the secondary voltage is higher than the primary voltage, then the secondary current must be correspondingly smaller than the primary current in order for the two powers to be equal. This means that if the voltage is stepped up in a transformer, then the current will always be stepped down proportionally—the current ratio is equal to the inverse of the voltage and turns ratios. We can express this relationship with Equation 7–8.

$$\frac{N_P}{N_S} = \frac{I_S}{I_P} \qquad (7\text{--}8)$$

•—EXAMPLE 7.8

If the turns ratio of a transformer is 5:1 and the primary current is 100 mA, what is the value of secondary current?

SOLUTION We apply Equation 7–8 as follows:

$$\frac{N_P}{N_S} = \frac{I_S}{I_P} \text{ or}$$

$$I_S = \frac{N_P I_P}{N_S}$$

$$= \frac{5\text{ T} \times 100\text{ mA}}{1\text{ T}} = 500\text{ mA}$$

Impedance Ratio

Resistance and reactance are both measured in ohms and both tend to limit current flow. The term **impedance** is often used to include the effects of both resistance and reactance. We know that a transformer can alter both voltage and current levels between the primary and secondary. It should seem reasonable, then, that a transformer can also alter the impedance between primary and secondary.

Reflected Impedance Consider the circuit shown in Figure 7–27. In Figure 7–27(a), the secondary is connected to a high-resistance load. The flux produced by the secondary tends to cancel (i.e., it is moving in the opposite relative direction) some of the induced voltage in the primary. This allows more current to flow in the primary.

Now, if the resistance in the secondary is made smaller, as in Figure 7–27(b), then more secondary current flows. This causes more secondary flux to induce an opposite polarity voltage in the primary, which again allows even more primary current to flow. Finally, Figure 7–27(c) shows that if the secondary load resistance is made even smaller, more secondary current, and therefore more primary current, will flow.

Since changes in the resistance or, more universally, the impedance of the secondary cause current changes in the primary (with the source voltage held constant), it follows that the impedance changes in the secondary are transferred to the primary. The impedance in the primary that results from the secondary load is called the **reflected impedance**. Changes in secondary impedance are reflected to the primary.

Impedance Calculations Now, let us consider the magnitude of the reflected impedance. Equation 7–9 expresses the relationship between the turns ratio and the impedance ratio.

$$\frac{N_P}{N_S} = \sqrt{\frac{Z_P}{Z_S}} \qquad (7\text{--}9)$$

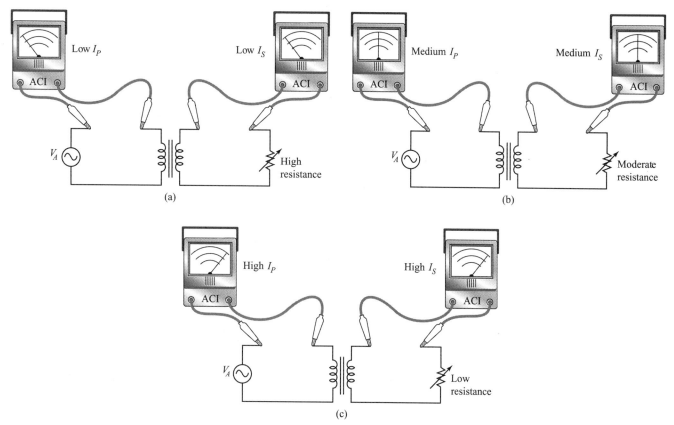

FIGURE 7–27 The impedance in the secondary of a transformer is reflected into the primary.

●—EXAMPLE 7.9

The secondary of a transformer with a 5:1 turns ratio is connected to a 100-Ω resistor. What is the value of impedance reflected into the primary?

SOLUTION We apply Equation 7–9 as follows:

$$\frac{N_P}{N_S} = \sqrt{\frac{Z_P}{Z_S}} \quad \text{or}$$

$$\left(\frac{N_P}{N_S}\right)^2 = \frac{Z_P}{Z_S} \quad \text{so}$$

$$Z_P = Z_S \left(\frac{N_P}{N_S}\right)^2$$

$$= 100 \ \Omega \times \left(\frac{5 \text{ T}}{1 \text{ T}}\right)^2 = 100 \ \Omega \times 25 = 2.5 \text{ k}\Omega$$

Calculator Application

This means that the source sees the 100-Ω resistance reflected back as a 2.5-kΩ resistance. That is, the current drawn from the primary source will be the same as if a 2.5-kΩ resistor were connected directly across the source.

The ability of a transformer to change impedances is an important property and is the sole purpose of the transformer in many applications. For example, maximum power will be transferred between a signal source and a load when the output impedance of the source is the same as the impedance of the load.

Troubleshooting Transformers

In general, transformers can develop all of the same defects discussed with reference to inductors. The following is a list of common transformer defects:

- Open primary
- Open secondary(s)
- Shorted turns or shorted winding
- Winding shorted to the core
- Short between primary and secondary

Most defective transformers can be detected through effective use of three things: observation, voltmeter checks, and ohmmeter checks.

Observation

When a transformer develops a short circuit, it generally results in increased current flow and increased internal power dissipation. If the primary of the transformer is fused, then a shorted winding will likely cause the fuse to blow as soon as power is applied. If you disconnect one side of the secondary (one side of each secondary in the case of multiple windings) and the fuse still blows, then you can be sure the transformer is shorted. Use your powers of observation before you begin an extensive troubleshooting process.

Iron-core transformers typically use varnish or shellac as insulation on the wires and between adjacent laminations in the core. If some of the turns in one winding of the transformer become shorted (e.g., the insulation breaks down), then the excessive heat will cause the shellac or varnish to emit an odor and possibly visible smoke. A similar symptom occurs if the transformer is not properly fused and is subjected to an excessive load in its secondary (e.g., a shorted component). In any case, the smell of an overheated transformer is very distinct. Once you have smelled the odor, you will remember it forever. Be alert when troubleshooting. By detecting this aroma, you may be able to quickly locate a defective transformer and save a lot of troubleshooting time.

There may also be visible evidence indicating that a transformer has been overheated. The visible paper insulation may be discolored or the visible varnish may have a bubbled appearance. Use your powers of observation to help you quickly locate transformer defects.

Voltmeter Tests

If a winding on a transformer is open (either primary or secondary), there will be no voltage developed across the secondary winding. In the case of multiple secondary windings, an open primary results in no voltage in any secondaries. An open secondary, on the other hand, results in no voltage across the defective secondary but relatively normal voltages across all other secondaries.

Ohmmeter Tests

An ohmmeter can be used to absolutely confirm that a particular winding is open. Desolder one end of the suspected winding and measure its resistance. An open winding will read infinite ohms. In many cases, the normal resistance of the transformer windings is so much lower than any parallel sneak paths that you can test for an open winding without desoldering one lead first. If it checks open, then it definitely is. If it checks good, then check carefully to be sure you are not measuring the resistance of a parallel sneak path.

An ohmmeter is also useful for detecting a winding-to-core short or a primary-to-secondary short. In either of these cases, the normal resistance is infinite ohms. If the measured resistance is very low or even zero, then you have located a short circuit.

Ohmmeters do not usually provide adequate resolution to reliably detect shorted turns in a transformer winding. That is, if the resistance of a normal winding were 2.8 ohms, then a few shorted turns might cause it to have a resistance of 2.75 ohms. Not only is this small change difficult to detect reliably, but also it is well within the normal variation of different transformers (of the same type) and different ohmmeters (i.e., measurement error).

✓ Exercise Problems 7.2

1. Transformers are rarely used in computer circuits. (True/False)
2. Alternating current in the primary winding of a transformer causes a voltage to be induced in the secondary winding with no electrical connection between the two windings. (True/False)
3. The greater the coefficient of coupling between primary and secondary windings in a transformer, the (*lower/higher*) the secondary voltage for a given primary voltage.
4. When a transformer is used in a power supply application, the 120-Vac line voltage is applied to the (*primary/secondary*) winding.
5. What characteristic of a transformer is communicated with dot notation?
6. If the secondary voltage is lower than the primary voltage, the transformer is called a _____ transformer.
7. Name three types of core materials used in transformers.
8. The sum of the voltages in the secondary windings of a multiple-secondary transformer must equal the value of the primary voltage. (True/False)
9. What type of transformer has common primary and secondary windings?

Refer to Figure 7–28 for questions 10 through 14.

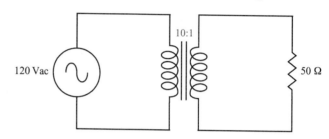

FIGURE 7–28 A transformer circuit.

Fig07_28.msm

10. What is the turns ratio?
11. What is the value of secondary voltage?
12. What is the value of secondary current?
13. What is the value of primary current?
14. What is the reflected impedance seen by the primary?
15. Name at least three defects that can occur in a transformer.

7.3 CAPACITORS

A capacitor is another fundamental electronic component like resistors, inductors, and transformers. Certain characteristics of the capacitor are similar to characteristics of resistors or inductors. Other characteristics are unique to capacitors. Since capacitors are used in nearly every practical electronic system, you will need to learn all you can about how they work.

Characteristics

Capacitance can be defined as the ability to store electrical energy in an electrostatic field. This is somewhat comparable to inductance, which is the ability to store electrical energy in an electromagnetic field. A capacitor is simply a physical device designed to have a certain amount of capacitance.

Basic Construction

Figure 7–29(a) shows the essential parts of a simple parallel-plate capacitor. It consists of two conductors—generally called the plates of the capacitor—separated by an insulator. The insulator material is called the **dielectric**. Figure 7–29(b) shows the generic schematic symbol for a capacitor. Although capacitors come in a wide range of shapes and sizes, the simple sketch in Figure 7–29 represents the heart of all capacitors: two conductors separated by an insulator.

FIGURE 7–29 (a) Basic construction of a capacitor and (b) the generic symbol for a capacitor.

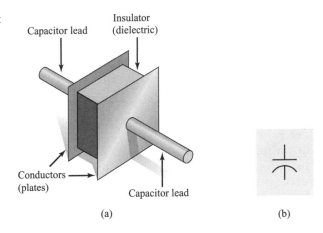

Charges and Electric Fields

Figure 7–30 shows a capacitor connected through a switch to a battery and a series resistor. In Figure 7–30(a), the switch is open and no current is flowing in the circuit. The capacitor has zero volts across it, as you might anticipate.

In Figure 7–30(b), the switch has been closed. The potential on the negative battery terminal causes electrons to leave the battery and move to the lower plate of the capacitor. Since the dielectric of the capacitor is an insulator, the electrons cannot continue through the capacitor. However, a similar process is occurring on the upper capacitor plate. Here, the positive potential of the battery terminal attracts electrons from the upper capacitor plate. As electrons leave the upper plate and travel to the battery terminal, the upper plate of the capacitor is left with a deficiency of electrons (i.e., a positive charge).

You will recall from chapter 2 that an electric field exists between charged bodies. In the case of the capacitor in Figure 7–30(b), there is an electrostatic field set up within the dielectric that extends between the positive and negative plates of the capacitor. We refer to the accumulation of charges on the capacitor plates as "charging the capacitor." A voltmeter connected across the capacitor, as shown in Figure 7–30(b), will show increasing voltage as the capacitor is charged.

If you closely examine the polarity of the charge that is accumulating on the capacitor plates in Figure 7–30(b), you will see that the increasing capacitor voltage is opposing the battery voltage. Each time an electron moves to the lower capacitor plate and/or leaves the upper plate, the charge on the capacitor and the opposition to the battery voltage increases. The increased opposition results in reduced current flow.

Eventually, enough electrons will have moved around the circuit to produce a capacitor charge that is equal (but opposite) to the battery voltage. This condition is pictured in Figure 7–30(c). Since the battery and capacitor voltages are equal and opposite, there will be no current flow in the circuit. This is a stable condition and will remain as long as the battery voltage is available.

In Figure 7–30(d), the switch has been opened. The electrons that have accumulated on the lower plate of the capacitor cannot move around the circuit to neutralize the positive charge on the upper plate. The charge is trapped on the capacitor. In theory, the voltage (charge) will stay on the capacitor indefinitely. In practice, it will eventually leak off primarily due to imperfections in the dielectric.

FIGURE 7–30 Charging a capacitor.

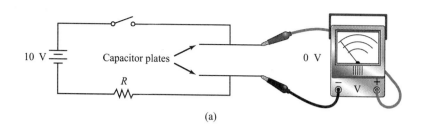

(a)

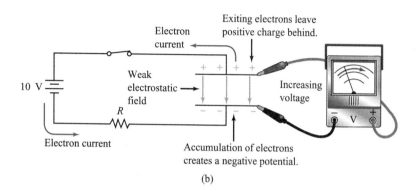

(b)

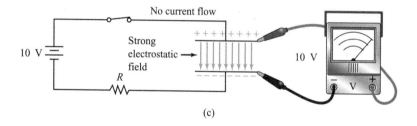

(c)

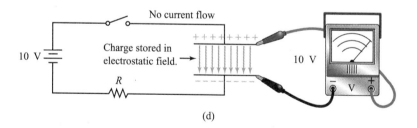

(d)

Figure 7–31 on page 218 shows the results of connecting a charged capacitor across a circuit. In Figure 7–31(a), the switch is open, so the charge remains on the capacitor. No current flows, and there is no voltage dropped across the resistor.

In Figure 7–31(b), the switch has closed, thus providing a path for the accumulated electrons on the lower plate to move around the circuit to the positive charge on the upper plate. As the electrons move around the circuit, they represent current through the resistor. The electron movement (amount of current) is highest when the switch is first closed. As each electron makes its trip around the circuit, both positive and negative plates become more neutral. As the plates lose their charge, there is less potential available to cause current flow. The voltage across the capacitor and across the resistor will decay as the capacitor discharges. The time required to fully discharge the capacitor depends on several circuit variables, but can range from fractions of a picosecond to literally months.

FIGURE 7–31 Discharging a capacitor.

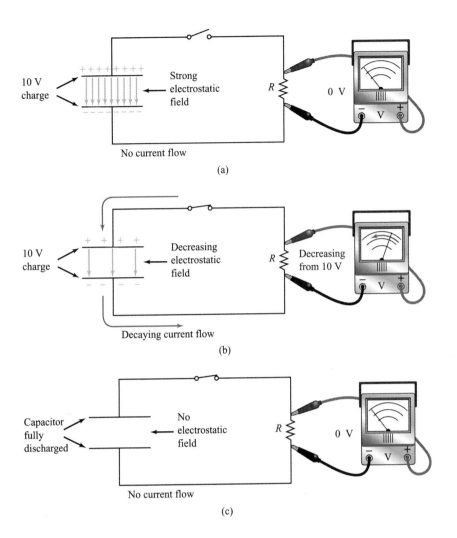

Unit of Measure

The unit of measure for capacitance is the farad. Values for practical capacitors are generally in the sub-farad range. Values of 27 pF, 0.05 µF, and 10,000 µF are representative values for typical capacitors. Although use of the nano prefix is not technically incorrect, it is common to avoid this usage for capacitance values. So, a 1.0-nF capacitor would most likely be expressed as 0.001 µF or 1000 pF.

Factors Affecting Capacitance

The physical characteristics of a capacitor determine its capacitance. There are three primary factors to consider: area of the plates, distance between the plates, and the type of material used for the dielectric. Figure 7–32 illustrates these factors.

FIGURE 7–32 The physical characteristics of a capacitor determine its value.

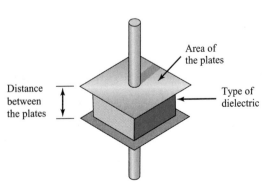

Plate Area The greater the plate area, the more charge the capacitor can hold. If all other factors remain constant and we double the area of the plates, then we can store twice as much charge for a given voltage. The value of capacitance is directly proportional to the area of the plates. Manufacturers often use multiple sets of capacitor plates to obtain higher values of capacitance.

Plate Separation By contrast, if all other factors remain the same and we double the distance between the plates of a capacitor, then we will have only half as much electric field intensity for a given voltage. The value of capacitance is inversely proportional to the distance between the plates.

Dielectric Material **Permittivity** is a measure of a material's ability to concentrate an electrostatic field. This is similar to permeability with reference to magnetic fields. Relative permittivity (ϵ_r) of a material is the ratio of its absolute permittivity (ϵ_o) to the absolute permittivity of a vacuum (ϵ_v). The relative permittivity of a dielectric is generally called its **dielectric constant** (k). The value of capacitance is directly proportional to the permittivity of the dielectric material. Practical values for dielectric constants range from 1.0 (air) to greater than 1000 (ceramic).

Table 7–4 summarizes the physical factors that determine the capacitance of a given capacitor.

TABLE 7–4   Relationship of physical factors affecting capacitance.

Factor	Proportionality	Example
Plate area	Direct	Area ↑ Capacitance ↑
Distance between plates	Inverse	Distance ↑ Capacitance ↓
Dielectric constant (k)	Direct	k ↑ Capacitance ↑

Current Flow in a Capacitive Circuit

It is interesting that a capacitor has an internal insulator (open circuit), and yet current can flow through the external circuit. It is important for you to be able to visualize this action. The discussion relevant to Figure 7–30 illustrated current flow in a dc capacitive circuit. In the circuit external to the capacitor, there will be current flow (i.e., movement of electrons) as long as the capacitor is either charging or discharging. Once the capacitor has fully charged or fully discharged, there is no current in the circuit.

One interesting application of capacitance—although not intentional—is illustrated in Figure 7–33. You already know that an open in a series circuit will drop the entire applied voltage. An open circuit is essentially a capacitor, since it is two conductors (the wires on either side of the open) separated by an insulator (the air between the two conductors). As you might expect, the value of capacitance is very small, so it charges almost instantly to the applied voltage. Nevertheless, it is a capacitor, and now you have yet another way to view the effects of an open circuit.

FIGURE 7–33 An open circuit forms an accidental capacitance that charges to the value of the applied voltage.

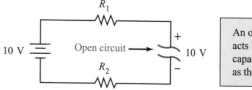

An open circuit acts like a capacitor with air as the dielectric.

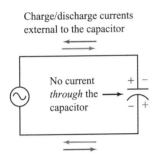

FIGURE 7-34 Current flow in an ac circuit.

Figure 7-34 illustrates current flow in an ac circuit. In this case, the source voltage is continuously changing—in both polarity and amplitude—so the capacitor is continuously charging or discharging. This means there will always be current in the external circuit. It appears as if the current were actually flowing through the capacitor, but it is important to remember that the dielectric is an insulator, so no current can literally flow through the capacitor.

SYSTEM PERSPECTIVE: Capacitors

High-speed computer circuits have outputs that are at one of two voltage levels. An equivalent circuit for such an output is shown in the accompanying illustration. The two switches are always in the opposite states. Now, when the output first switches to the state shown, the input capacitance (C_i) for all driven circuits must be charged via the output pin of the driving gate. This results in a brief surge current called a transient. If this current is allowed to flow through the parasitic inductance (inductance that results from the nonideal characteristics of components and conductors) of the power supply and connecting circuit board traces, a significant voltage will be dropped across the inductance. This causes a drop in voltage across the integrated circuit, which may cause it to malfunction. To reduce this effect, a local decoupling capacitor is connected very close to the driving integrated circuit. The transient current is now supplied by the charge on the decoupling capacitor, so negligible surge current flows through the parasitic inductance. Therefore, the voltage across the integrated circuit is held more constant. In a practical computer, every integrated circuit will have at least one, and sometimes many, decoupling capacitors.

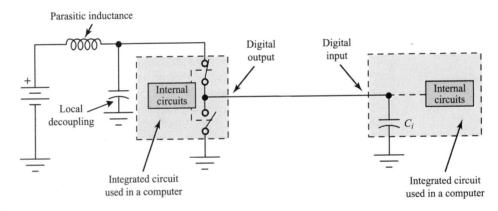

Capacitive Reactance

The value of current in a capacitive circuit with an alternating source is directly proportional to the value of the capacitor. Greater capacitance corresponds to increased current flow. Current is also directly proportional to frequency. That is, if the frequency is higher, then the capacitor will have to charge and discharge more times per second. This results in a higher current (coulombs per second) in the external circuit.

Since the current in a capacitive circuit can be altered without changing the applied voltage, it follows that the opposition to current must be changing. This opposition, called **capacitive reactance**, is measured in ohms. As we have seen, it must be inversely proportional to capacitance and frequency. We can compute capacitive reactance with Equation 7–10.

$$X_C = \frac{1}{2\pi f C} \qquad (7\text{--}10)$$

EXAMPLE 7.10

What is the capacitive reactance offered by a 0.05-µF capacitor when operated at a frequency of 60 Hz?

SOLUTION   We apply Equation 7–10 as follows:

$$X_C = \frac{1}{2\pi f C}$$

$$= \frac{1}{2 \times 3.14 \times 60 \text{ Hz} \times 0.05 \text{ µF}} = 53.08 \text{ k}\Omega$$

Calculator Application

Capacitor Types

There are many ways to categorize capacitors and to distinguish between them. This section will consider several of these ways.

Lead Styles

There are four major lead styles that are used with capacitors, although there are also many variations on these basic designs plus a number of specialized lead styles. The four basic lead styles are axial, radial, surface mount, and integrated. Figure 7–35 shows some representative capacitors with axial leads (protruding from opposite ends of the body). Figure 7–36 on page 222 shows several capacitors with radial leads (protruding from the same side of the body). Figure 7–37 shows two surface-mount capacitors. These are essentially leadless. The end contacts are soldered directly to pads on a printed circuit board. Finally, Figure 7–38 shows some integrated capacitors in which a single package houses several capacitors.

FIGURE 7–35
Capacitors with axial leads.

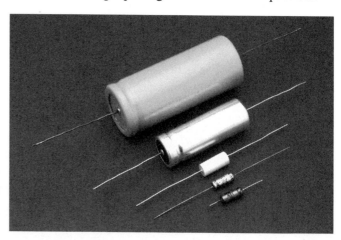

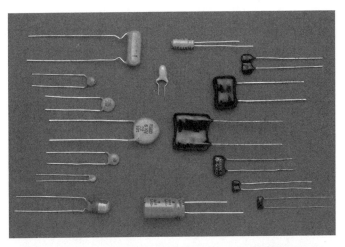

FIGURE 7–36
Capacitors with radial leads.

FIGURE 7–37 Surface-mount capacitors. The head of a common straight pin is shown for a size comparison.

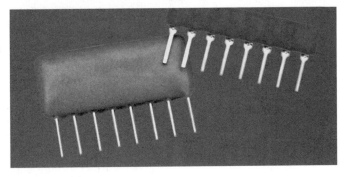

FIGURE 7–38
Integrated capacitors have multiple capacitors inside a single, multipin package.

Fixed and Variable Capacitors

Capacitors can be either fixed or variable. All of the capacitors pictured in Figures 7–35 through 7–38 are fixed capacitors. Their value is determined at the time of manufacture and cannot be altered. Figure 7–39 shows several variable capacitors. The range of adjustment is established during manufacture, but the exact value can be adjusted in the field.

FIGURE 7–39 The value of variable capacitors can be adjusted.

Polarized and Nonpolarized Capacitors

Most capacitors are nonpolarized, which means that they can be inserted into the circuit in either direction. Some types of capacitors, however, are polarized. Polarized capacitors have markings to indicate the polarity of the capacitor. One common marking method is to put a plus (+) sign near the positive terminal. Alternatively, some manufacturers place a series of minus signs (−) along one side of the capacitor with an arrow pointing to the negative terminal. When you insert a polarized capacitor into a circuit, it is *very* important that the positive side of the capacitor be connected to the more positive terminal in the circuit. If a polarized capacitor is inserted backward, then it is very likely to explode violently. At a minimum, you will be startled when the capacitor explodes with the loudness of a firecracker. Worse, you may receive eye damage or skin cuts from flying debris and have chemicals splattered in your face. You should not be afraid of polarized capacitors, but it certainly makes sense to be very certain they are connected properly before applying power. Since a polarized capacitor must always have the correct voltage polarity across it, you should never connect a polarized capacitor to an ac source.

Capacitor Ratings

Capacitors have three ratings that are of primary interest to persons working with computer circuits: capacitance, voltage, and temperature coefficient. The capacitance value indicates the nominal value of the capacitor (e.g., 100 pF). There will be a tolerance associated with this value much as there was a tolerance for resistor values. Tolerances range from less than 5% to as much as 80% depending on the capacitor type and grade.

The voltage rating of a capacitor indicates the maximum voltage that can safely be applied to the capacitor without risking damage to the dielectric. In other words, if the voltage rating of a capacitor is exceeded, the dielectric may break down. In most cases this destroys the capacitor, although there are certain types of dielectrics that are considered self-healing.

The temperature coefficient of a capacitor describes how the value of capacitance varies with changes in temperature. The temperature coefficient specification consists of two parts: polarity and magnitude.

The polarity of the temperature coefficient tells whether the capacitance increases or decreases with increasing temperature. The capacitance of a capacitor with a positive temperature coefficient increases as temperature increases. The capacitance of a capacitor with a negative temperature coefficient decreases as the temperature increases. Some capacitors are designed to have a zero temperature coefficient, which means their capacitance value is relatively unaffected by changes in temperature.

The magnitude portion of the temperature coefficient specification tells how much the capacitance value changes for a given change in temperature. It is specified in parts per million per degree Centigrade (ppm/°C).

The complete temperature coefficient is always specified in the manufacturer's specification sheets, but is not always printed on the physical capacitor. When it is printed on the capacitor, the polarity is listed first as N, P, or NP for negative, positive and zero temperature coefficients, respectively. The magnitude is listed after the letter designation.

EXAMPLE 7.11

Three capacitors have the markings N750, P350, and NPO. What do these marks mean?

SOLUTION These are the temperature coefficients for the capacitors. The N750 mark indicates a capacitor with a negative temperature coefficient of 750 ppm/°C. The P350 mark identifies a positive temperature coefficient of 350 ppm/°C.

The NPO marking specifies a capacitor with a zero temperature coefficient. Practical capacitors with NPO ratings still vary slightly with temperature, but their coefficients are generally less than 30 ppm/°C.

Capacitor Technologies

The technologies used to manufacture capacitors can also be used to classify them. Table 7–5 lists some of the more common capacitor technologies and a brief description of each. Figures 7–40 through 7–44 show representative samples of each capacitor type.

TABLE 7–5 Capacitor technologies and characteristics.

Capacitor Technology	Polarized	Characteristics
Aluminum 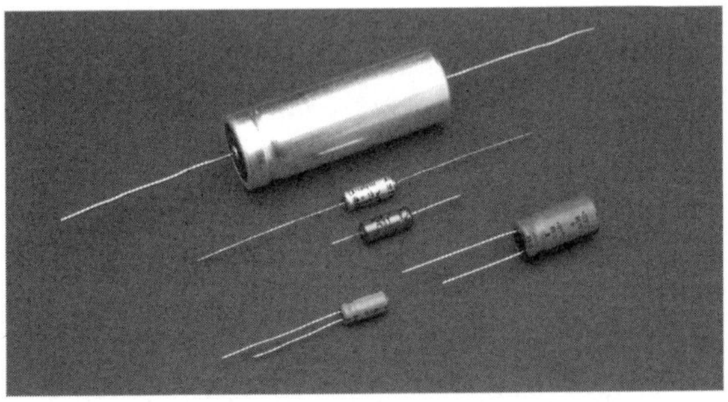 FIGURE 7–40 Some representative aluminum electrolytic capacitors.	Yes	These capacitors are called electrolytics. They have high capacitance values (e.g., 1.0 µF to > 100,000 µF). Primarily used for low frequencies. A thin (e.g., 10^{-8} inch) film of aluminum oxide serves as the dielectric.
Tantalum 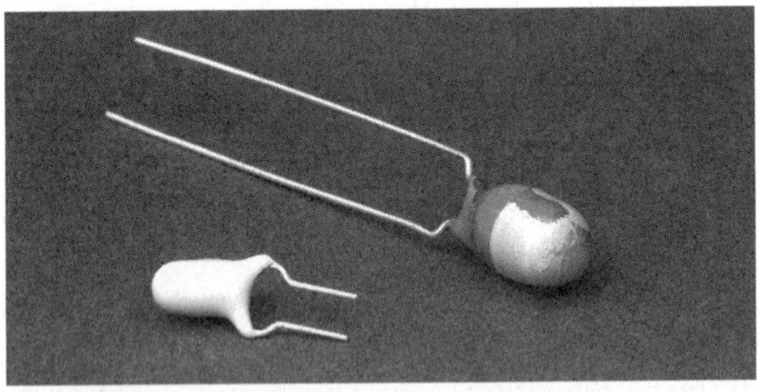 FIGURE 7–41 Tantalum capacitors offer a very high capacitance-to-volume ratio.	Yes	These are electrolytic capacitors. They are much smaller than aluminum capacitors for a given capacitance. Dielectric is tantalum oxide. Generally limited to low-voltage circuits (i.e., < 100 V). Values range from 0.1 µF to at least 2,200 µF. Common as surface mount.
Ceramic  FIGURE 7–42 Ceramic capacitors. (*Courtesy of AVX Corporation*)	No	Uses ceramic as the dielectric. Most common type of capacitor in computers and other digital products. Values range from 1 pF to at least 10 µF. Well-suited for high-frequency applications (e.g., wireless communication). Common as surface mount. Available with negative, positive, and zero temperature coefficients. Voltage ratings as high as tens of thousands of volts.

Capacitor Technology	Polarized	Characteristics
Film	No	Made with alternate layers of plastic film (dielectric) and metal foil (or metal deposit). Values range from 47 pF to 75 μF. Voltage ratings as high as 1000 V.
FIGURE 7–43 Plastic film capacitors.		
Mica	No	Uses thin sheet of mica as the dielectric and deposited silver for the plates. Also called silver mica capacitors. Values range from 1 pF to tens of thousands of picofarads. Voltage ratings as high as several thousand volts.
FIGURE 7–44 Representative mica capacitors.		

Series and Parallel Connections

Two or more capacitors may be connected in series, parallel, or other configuration to produce an equivalent value of capacitance. The equations for combining combinations of capacitors are different from the equations for resistors, but they are very easy to remember if you learn them in a logical way (i.e., avoid casual memorization).

Series Capacitances

Figure 7–45(a) shows two series-connected capacitors. Figure 7–45(b) illustrates a way to help you remember the relative (i.e., larger or smaller) value of the equivalent capacitance. As shown in Figure 7–45(b), when the capacitors are connected in series, we are essentially increasing the thickness of the dielectric. That is, we are increasing the distance between the plates. You already know that increased plate separation results in decreased capacitance. So, we can already make the following conclusion:

> **RULE** When capacitors are connected in series, the total capacitance is less than the value of any of the individual capacitances.

FIGURE 7–45 (a) Series-connected capacitors. (b) The distance between the plates is effectively increased when capacitors are connected in series.

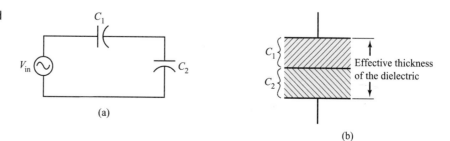

Although the reasons are totally different, you should recognize this general statement as being similar to the rule for parallel-connected resistors. Equation 7–11 can be used to compute the equivalent capacitance for series-connected capacitors. Note that it is the same basic form as the reciprocal equation used to compute the equivalent resistance of parallel-connected resistors.

$$C_T = \frac{1}{\frac{1}{C_1} + \frac{1}{C_2} + \ldots + \frac{1}{C_N}} \qquad (7\text{–}11)$$

•—EXAMPLE 7.12

A 4.7-µF and a 6.8-µF capacitor are connected in series. What is the total capacitance?

SOLUTION We apply Equation 7–11 as follows:

$$C_T = \frac{1}{\frac{1}{C_1} + \frac{1}{C_2}}$$

$$= \frac{1}{\frac{1}{4.7\ \mu F} + \frac{1}{6.8\ \mu F}} = 2.78\ \mu F$$

Series Capacitive Reactances

Capacitive reactance is the opposition to alternating current offered by a capacitor. This is similar to the opposition offered by inductors (inductive reactance) and resistors (resistance). We know, intuitively, that adding more opposition in series will increase the total opposition. We express this notion more formally with Equation 7–12.

$$X_{C_T} = X_{C_1} + X_{C_2} + \ldots + X_{C_N} \qquad (7\text{–}12)$$

•—EXAMPLE 7.13

If three capacitors having capacitive reactances of 100 Ω, 275 Ω, and 150 Ω are connected in series, what is the total capacitive reactance?

SOLUTION We apply Equation 7–12 as follows:

$$X_{C_T} = X_{C_1} + X_{C_2} + X_{C_3}$$
$$= 100\ \Omega + 275\ \Omega + 150\ \Omega = 525\ \Omega$$

Parallel Capacitances

Figure 7–46(a) shows two capacitors connected in parallel. Figure 7–46(b) illustrates a way to help you remember the relative (i.e., larger or smaller) value of the equivalent capacitance. As capacitors are added in parallel, the effective plate area, and therefore the total capacitance, increases.

FIGURE 7–46 (a) Parallel-connected capacitors. (b) The effective area of the plates is increased when capacitors are connected in parallel.

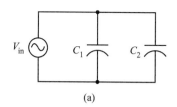

(a)

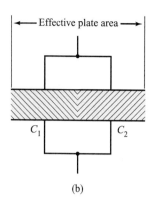

(b)

We can already make the following conclusion:

> RULE When capacitors are connected in parallel, the total capacitance is equal to the sum of the individual capacitances.

Although the reasons are totally different, you should recognize this general statement as being similar to the rule for series-connected resistors. Equation 7–13 can be used to compute the equivalent capacitance for parallel-connected capacitors.

$$C_T = C_1 + C_2 + \ldots + C_N \tag{7–13}$$

• EXAMPLE 7.14

A 33-µF and a 10-µF capacitor are connected in parallel. What is the total capacitance?

SOLUTION We apply Equation 7–13 as follows:

$$C_T = C_1 + C_2 = 33 \; \mu F + 10 \; \mu F = 43 \; \mu F$$

Parallel Capacitive Reactances

We know from our studies of resistance and inductive reactance that connecting resistances or reactances in parallel reduces the total opposition to current. In the case of capacitive reactance, we express this relationship with Equation 7–14.

$$X_{C_T} = \frac{1}{\dfrac{1}{X_{C_1}} + \dfrac{1}{X_{C_2}} + \ldots + \dfrac{1}{X_{C_N}}} \tag{7–14}$$

• EXAMPLE 7.15

Compute the total reactance for the parallel circuit shown in Figure 7–47.

FIGURE 7–47 What is the total capacitive reactance in this circuit?

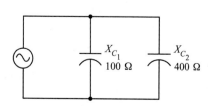

SOLUTION We apply Equation 7–14 as follows:

$$X_{C_T} = \frac{1}{\frac{1}{X_{C_1}} + \frac{1}{X_{C_2}}}$$

$$= \frac{1}{\frac{1}{100\ \Omega} + \frac{1}{400\ \Omega}} = 80\ \Omega$$

Circuit Analysis of Capacitive Circuits

Once you compute the capacitive reactance values for all capacitors in a given network, you can solve for all of the component voltages and currents in the circuit by applying the identical strategies used for resistive and inductive networks. This analytical approach is applicable to all configurations including series, parallel, and series-parallel.

Troubleshooting Capacitors

Capacitor malfunctions may be divided into three general classes of failures: open, short, and increased leakage. You must be able to test capacitors and determine their condition. The following paragraphs discuss three methods that can be used to diagnose capacitor malfunctions.

Substitution

The condition of a suspected capacitor can easily be determined by replacing it in the circuit with another capacitor known to be good. At first this appears to be a poor technique, but it is actually wise in many cases. First, most capacitors are very inexpensive and readily available. You can remove and replace a capacitor faster than you can remove, test, and replace it. Generally, the labor savings far outweighs the cost of the capacitor. Second, the act of removing the capacitor for testing and the subsequent resoldering can stress an otherwise good capacitor, which may introduce even more trouble. In the case of surface-mount capacitors (the most popular type), substitution is the recommended method. You should always keep a supply of common capacitor values on hand for this purpose.

Substitution for purposes of troubleshooting is of no value unless the substituted component is known to be good. Many troubleshooters and repair shops eventually accumulate a wide assortment of new and used parts mixed together in a box or drawer. Beware! If you accidentally substitute a defective capacitor, then your results will be misleading.

Capacitance Testers

Second only to substitution, the use of a capacitor tester is the preferred way to determine the condition of a capacitor. Although some capacitor testers are restricted to the testing of capacitors only, many technicians use an LCR tester which can be used to test inductors, capacitors, and resistors. A representative LCR tester is shown in Figure 7–48.

The use of the capacitor (or LCR) tester is very straightforward. You simply connect the capacitor to be tested, and the display indicates the capacitance value. If the capacitor is open or shorted, the tester indicates its condition.

Ohmmeter Tests

In some cases, you can get an estimate of the condition of a capacitor by using an analog ohmmeter (VOM). The method is illustrated in Figure 7–49. First, the capacitor to be tested is fully discharged by shorting its leads as shown in Figure 7–49(a).

FIGURE 7–48 A representative LCR tester. (*Courtesy of B&K Precision*)

FIGURE 7–49 Testing the condition of a capacitor with an ohmmeter.

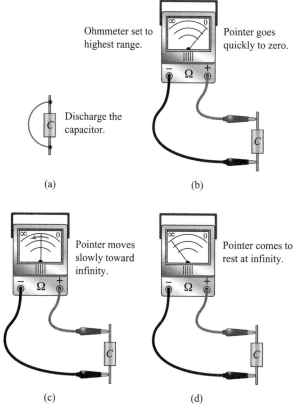

Next, the ohmmeter is switched to one of the higher ranges (e.g., R × 10 k, R × 100 k, or R × 1 M) and connected across the capacitor (observe the polarity of electrolytics). As soon as the leads make contact, the meter pointer will swing to near 0 Ω on the scale, as indicated in Figure 7–49(b). It will then move slowly toward infinity on the scale, as shown in Figure 7–49(c), as the capacitor charges to the internal battery voltage of the ohmmeter. Finally, the pointer will come to rest near infinite ohms as shown in Figure 7–49(d). Now that we have discussed the general method, let us be more specific.

First, the time required for the meter to move between zero and infinite ohms (Figure 7–49[b] through Figure 7–49[d]) varies dramatically from several minutes to a few microseconds. The exact time depends on the meter (and range) used as well as the size of the capacitor. For very small capacitances, the time span may be so short that it is difficult (or even impossible) to detect the pointer movement. In this case, use the highest range on the ohmmeter. For larger capacitor sizes (e.g., 0.1 µF and up), the pointer movement is easily detected. Generally, the larger the value of capacitance, the lower the range on the ohmmeter that gives a useable indication.

Regardless of the capacitor size, a shorted capacitor will cause the ohmmeter to deflect to zero ohms and remain there. This means the dielectric has been damaged, and the capacitor plates are electrically connected. The capacitor must be replaced.

If the capacitor is open, there will be no ohmmeter indication. This is a useful test for large capacitors but provides no detectable information for smaller capacitors.

Finally, a capacitor with low dielectric leakage resistance (called a leaky capacitor) will cause the pointer to come to rest at some resistance lower than infinity. This means that the capacitor dielectric is not a very good insulator. You must classify the measured value as good or bad based on a comparison with a good capacitor of the same type. Ceramic capacitors, for example, have normal leakage resistances so high that most ohmmeters indicate infinite ohms. Aluminum electrolytics, by contrast, may have normal leakage resistances as low as 1.0 MΩ.

In general, the ohmmeter test for capacitors should be interpreted as follows. If the ohmmeter test indicates a defective capacitor (i.e., open, shorted, or leaky), then it is definitely defective. If, however, no defective indication can be observed on the meter, the capacitor *might* be good. You cannot positively prove that a capacitor is good with an ohmmeter test.

✓ Exercise Problems 7.3

1. Capacitors store energy in an electromagnetic field. (True/False)
2. If the area of the plates in a capacitor is decreased during the manufacturing process, the value of capacitance will (*increase/decrease*).
3. What is the value of current in a series dc capacitive circuit once the capacitor is fully charged?
4. Capacitance (*increases/decreases*) as the distance between the plates decreases.
5. As frequency is increased in a capacitive circuit, the capacitive reactance (*increases/decreases*).
6. If you wanted to increase the current in a capacitive ac circuit, you could (*increase/decrease*) the capacitance.
7. What is the capacitive reactance offered by a 1000-pF capacitor when operated at a frequency of 40 Hz?
8. A 0.1-µF capacitor is used in a multimedia circuit of a notebook computer to couple audio signals from one circuit to another. How much capacitive reactance does the capacitor have at 10 kHz?
9. What type of lead configuration would be selected for a capacitor in an application where small size was an important consideration?
10. When replacing a defective electrolytic capacitor in a circuit, why is it essential to be sure it is installed properly?
11. If you replaced a defective 100-pF capacitor that was marked with an NPO label in a sensitive computer circuit with a 100-pF capacitor marked as N750, what might happen?
12. When two capacitors are connected in series, the total capacitance is (*less/more*) than any one of the capacitors.
13. You need to replace a 10,000-pF capacitor in a microprocessor circuit, but all you have available are 0.005-µF devices. Explain how you could connect multiple 0.005-µF capacitors to replace the defective capacitor.
14. Capacitive reactances can be thought of as resistances with respect to the effects of series and parallel connections. (True/False)
15. How could an ohmmeter be used to detect a shorted capacitor?
16. Name three defects that can occur within a capacitor.

7.4 *RC* AND *RL* CIRCUITS

We will consider *RC* and *RL* circuits to be nothing more than applications for resistors, inductors, and capacitors. We will briefly examine their combined effects and then examine a number of practical applications. Emphasis will be placed on *RC* circuits, since you will be much more likely to encounter them in computer applications. However, the analytical strategies for *RL* circuits will also be presented in abbreviated form.

Characteristics with Sine Wave Inputs

Let us first consider the behavior of an ac circuit having both resistance and capacitance. You would expect the characteristics of the circuit to lie somewhere between the characteristics of a pure resistive and a pure capacitive circuit. We will concentrate on understanding the operation of series and parallel *RC* circuits. A solid intuitive understanding of these circuits will serve you well for any practical *RC* circuits you are likely to encounter.

Series *RC* Circuits

Many of the characteristics of series *RC* circuits should seem logical to you, since they are really just extensions of characteristics you already know.

Current Figure 7–50(a) shows a simple series *RC* circuit. We know that the current is the same in all parts of a series circuit. Therefore, the resistor, capacitor, and source currents are identical.

FIGURE 7–50 A series *RC* circuit and its phase relationships.

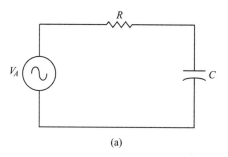

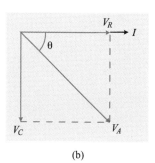

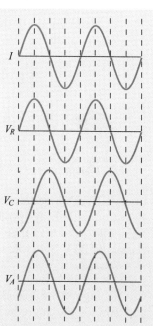

Voltage Drops and Phase Relationships The voltage drops across each component can still be found with Ohm's Law (e.g., $V_R = IR$ and $V_C = IX_C$.) But, can we use Kirchhoff's Voltage Law to sum the component voltages? The answer is a qualified "yes." Let's examine this a little closer. First, we know the currents are identical for both components. We also know there is a 90° phase relationship between current and voltage in the capacitor, but current and voltage are in phase in a resistor. This means the voltage waveforms across the two series components must be out of phase with each other by 90°. These observations are illustrated in Figures 7–50(b) and 7–50(c).

Figure 7–50(b) shows the phasor diagram for the *RC* circuit in Figure 7–50(a). Since current is the same everywhere in a series circuit, we draw the current phasor in the reference position. We know that current and voltage are in phase in a resistor, so we sketch the phasor for resistor voltage (V_R) at the same angle as current. Next, the capacitor voltage phasor is added 90° behind the current, since we know current leads voltage by 90° in a capacitor. We can now complete a parallelogram from the points of the V_C and V_R phasors, as shown with dashed lines. The diagonal of this parallelogram represents the applied voltage (V_A). This procedure is called phasor addition. That is to say, we can add or sum two quantities that are 90° out of phase by using the parallelogram method. We can also express the value of V_A numerically by applying Equation 7–15.

$$V_A = \sqrt{V_R^2 + V_C^2} \quad (7\text{–}15)$$

So, yes, Kirchhoff's Voltage Law still applies. However, when adding quantities that are 90° out of phase, you must add them using phasor addition, since the voltages do not occur at the same time (i.e., they are out of phase).

•—EXAMPLE 7.16

The component voltage drops in a series *RC* circuit are 10 V and 20 V for the resistor and capacitor, respectively. What is the value of applied voltage?

SOLUTION According to Kirchhoff's Voltage Law, we add the individual voltage drops to compute the applied voltage. We do that here as well, but we must use phasor addition.

$$V_A = \sqrt{V_R^2 + V_C^2} = \sqrt{(10 \text{ V})^2 + (20 \text{ V})^2} = 22.36 \text{ V}$$

Calculator Application

Here's a good way to check for errors. The result of a phasor addition calculation will always produce a value that is greater than either of the original numbers but is less than their arithmetic sum. In this case, 22.36 V is greater than either 10 V or 20 V but is less than their sum (30 V).

Figure 7–50(b) also identifies the overall phase angle (θ) of the circuit. This is always the angle between total current and total voltage. Figure 7–50(c) shows a waveform representation of the phase relationships between the various circuit quantities.

Impedance In a series circuit, we can add resistances or reactances to compute total impedance (Z). But, in the case of an *RC* circuit, we must sum the resistances and reactances using phasor addition as reflected in Equation 7–16.

$$Z = \sqrt{R^2 + X_C^2} \qquad (7\text{–}16)$$

EXAMPLE 7.17

What is the total impedance of a series circuit consisting of 250 Ω of capacitive reactance and 100 Ω of resistance?

SOLUTION We add the reactance and resistance using phasor addition.

$$Z = \sqrt{R^2 + X_C^2} = \sqrt{(100\ \Omega)^2 + (250\ \Omega)^2} = 269.26\ \Omega$$

Parallel *RC* Circuits

As with series *RC* circuits, many of the previously discussed techniques for parallel circuit analysis can be extended to include parallel *RC* circuits.

Voltage Drops All of the components in a parallel *RC* circuit and every other parallel circuit you will ever encounter have equal voltage drops. So, in a parallel *RC* circuit, the resistor voltage, capacitor voltage, and source voltage are identical.

Currents Figure 7–51(a) shows a parallel *RC* circuit. You already know how to determine the individual branch currents in a parallel circuit using Ohm's Law. You also know that the current in one branch is unaffected by other branches. Therefore, you can safely conclude that the currents through the resistive and capacitive branches for the circuit in Figure 7–51(a) are determined using the same methods

FIGURE 7–51 A parallel *RC* circuit and its phase relationships.

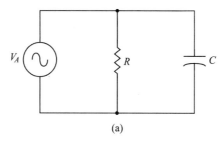

(a)

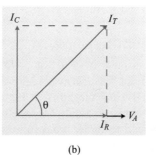

(b)

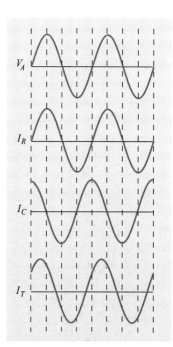

(c)

discussed for parallel resistive circuits and parallel capacitive circuits, respectively. But, how do we find total current?

First, let's examine the phasor diagram in Figure 7–51(b). The applied voltage (V_A) is used as a reference because it is common to all components. Resistor current (I_R) is shown in phase with applied voltage because there is no phase difference between voltage and current in a resistor. As you know, current leads voltage by 90° in a capacitor. This is also reflected in the phasor diagram in Figure 7–51(b). Now, according to Kirchhoff's Current Law, we should be able to sum the branch currents to find total current. That remains true here, but since the resistive and capacitive currents are 90° out of phase, we must add them using phasor addition as indicated by Equation 7–17.

$$I_T = \sqrt{I_R^2 + I_C^2} \qquad (7\text{–}17)$$

•—EXAMPLE 7.18

The branch currents for a parallel RC circuit are 5 µA and 4 µA for the resistive and capacitive branches, respectively. What is the total current provided by the source?

SOLUTION We add the branch currents to find total current, but we must use phasor addition since the currents are 90° out of phase with each other.

$$I_T = \sqrt{I_R^2 + I_C^2} = \sqrt{(5 \text{ µA})^2 + (4 \text{ µA})^2} = 6.4 \text{ µA}$$

Impedance The total impedance of a parallel circuit is most easily found by direct application of Ohm's Law using total voltage and total current values.

•—EXAMPLE 7.19

What is the impedance of a parallel RC circuit that has 250 mA of current when a source of 100 V is connected?

SOLUTION We apply Ohm's Law as follows:

$$Z = \frac{V_A}{I_T} = \frac{100 \text{ V}}{250 \text{ mA}} = 400 \text{ Ω}$$

RL Circuit Characteristics

The behavior of a either a series or parallel RL circuit is very similar to the characteristics discussed for series and parallel RC circuits. The important characteristics are summarized in Table 7–6.

Pulse Circuit Response

We have discussed the analysis of RC and RL circuits with sinusoidal inputs. Nonsinusoidal input waveforms cause a radically different circuit response and require different analytical methods. It is not necessary for you to be able to thoroughly analyze the nonsinusoidal response of an RC or RL circuit at this time, but it is important for you to understand time constants and how RC and RL circuits respond to pulse or rectangular waveforms such as those found in digital circuits.

TABLE 7–6 A summary of RL circuit characteristics.

Configuration	Characteristic	Description
Series RL	Current	Same through all components
	Voltage drops	Ohm's Law: $V_L = IX_L$ and $V_R = IR$
	Applied voltage	Kirchhoff's Voltage Law (phasor sum): $V_A = \sqrt{V_R^2 + V_L^2}$
	Impedance	Phasor sum: $Z = \sqrt{R^2 + X_L^2}$
	Phase	Resistor current and voltage: 0° Inductor current and voltage: 90° (current lags) Total current and voltage: $0 < \theta < 90°$ (current lags)
Parallel RL	Voltage	Same across all components
	Branch currents	Ohm's Law: $I_L = \dfrac{V}{X_L}$ and $I_R = \dfrac{V}{R}$
	Total current	Kirchhoff's Current Law (phasor sum): $I_T = \sqrt{I_R^2 + I_L^2}$
	Impedance	Ohm's Law: $Z = \dfrac{V}{I_T}$
	Phase	Resistor current and voltage: 0° Inductor current and voltage: 90° (current lags) Total current and voltage: $0 < \theta < 90°$ (current lags)

RC Time Constants

Figure 7–52 shows a simple circuit that will manually produce pulse waveforms to an RC circuit as the switch is moved between positions A and B. When the switch is first moved to position A (Figure 7–52a), current will flow through the RC circuit and begin to charge the capacitor. The initial value of current is limited only by the resistor. Charging current continues until the capacitor voltage is equal to the battery voltage.

For a given size capacitor to charge to a particular voltage requires a definite amount of charge (i.e., a definite number of electrons). It takes time for the electrons to move around the circuit and accumulate on the capacitor. How long does it

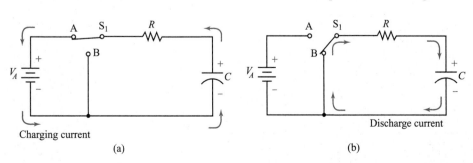

FIGURE 7–52 An RC circuit with a pulse input.

take for enough electrons to move around the circuit to charge the capacitor to a value equal to the supply voltage? Well, that depends on the value of charging current. Remember, current is a measure of how many electrons move past a given point in one second. As stated previously, the resistor limits the value of current in the circuit. The more resistance we have, the lower the current and the longer it takes the capacitor to reach full charge.

It is also important to note that the value of charging current decays as the capacitor becomes charged. In essence, the accumulating charge on the capacitor provides more and more opposition to the applied voltage (like series-opposing voltage sources). When the capacitor is fully charged, the opposing voltages are equal, and no more current can flow.

The total time required for the capacitor to fully charge is divided into five time periods called **time constants**. The Greek letter tau (τ) is used to represent one time constant. In the case of RC circuits, τ is computed with Equation 7–18.

$$\tau = RC \qquad (7\text{–}18)$$

where τ is measured in seconds, R in ohms, and C in farads.

Figure 7–53(a) illustrates the timing of the capacitor voltage and the charging current. The graph indicates that after a time interval of $t = \tau$, the capacitor voltage has increased to 63.2% of its full-charge value. At that same time, charging current has decreased to 36.8% of its initial value. During each additional time period of τ seconds, the capacitor voltage increases by 63.2% of the remaining voltage. For example, at the end of the first time interval (τ), the capacitor voltage has another 36.8% to go. During the time between τ and 2τ it will increase 63.2% of this 36.8% or 23.3% for a total of 86.5% at time $t = 2\tau$.

FIGURE 7–53 An RC time constant graph: (a) charging and (b) discharging.

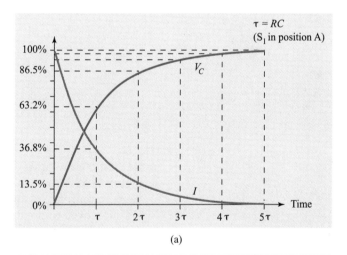

(a)

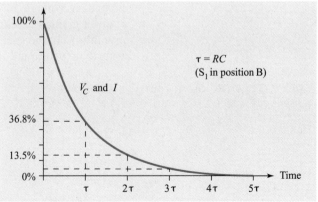

(b)

Once 5τ seconds have elapsed, the capacitor voltage will have increased to more than 99% of its full-charge value. Likewise, the charging current will have decreased to less than 1% of its initial value. For practical purposes, we consider the capacitor to have reached full charge in five time constants (5τ).

If the capacitor in Figure 7–52(a) has reached full charge with the switch in position A, there will be no additional current flow. The charge on the capacitor is equal, but opposite, to the battery potential. Now, if we move the switch to position B, the charge on the capacitor acts like a voltage source and causes a discharge current to flow. This action is illustrated in Figure 7–52(b). Again, the value of initial current is limited by the resistor. The higher the discharge current (i.e., the lower the value of the resistor), the quicker the capacitor can discharge its accumulated voltage. The discharge current will continue until the charge on the capacitor has decreased to zero.

Figure 7–53(b) illustrates the timing of the current and capacitor voltage during the discharging period. At time $t = 0$, the capacitor is at full charge, and the discharge current is maximum. At time $t = τ$, the capacitor voltage and discharge current have decayed to 36.8% of their initial values. The current and voltage continue to decrease by 63.2% of the remaining voltage (or current) each time constant. Again, for practical purposes, we consider the capacitor to be fully discharged after five time constants (5τ).

It is important to realize that the time required to fully charge or discharge a capacitor (5τ) is determined only by the value of resistance and capacitance. It is unaffected by the value of voltage in the circuit. If either the resistor or the capacitor is made larger, then it takes longer to charge or discharge the capacitor.

•—EXAMPLE 7.20

If a 10-kΩ resistor and a 2.5-μF capacitor are connected in series across a 10-V battery, how long does it take for the capacitor to have a 10-V charge?

SOLUTION   We know it takes five time constants to charge or discharge a capacitor. So, let's compute the time constant with Equation 7–18.

$$τ = RC = 10 \text{ kΩ} \times 2.5 \text{ μF} = 25 \text{ ms}$$

We multiply by five to find the total charge (or discharge) time.

$$\text{charge time} = 5τ = 5 \times 25 \text{ ms} = 125 \text{ ms}$$

•—Practice Problems

1. What is the RC time constant for the circuit shown in Figure 7–54?
2. How many time constants are required for the capacitor in Figure 7–54 to reach full charge once the switch has been closed?
3. What will be the approximate value of current in the circuit shown in Figure 7–54, after five time constants?
4. Repeat questions 1 through 3 for a battery voltage of 1,000 V.

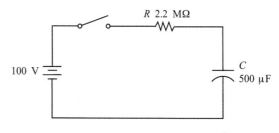

FIGURE 7–54   An RC circuit.

Fig07_54.msm

RL Time Constants

RL circuits also have a time constant associated with them. An RL time constant is computed with Equation 7–19.

$$\tau = \frac{L}{R} \qquad (7\text{–}19)$$

When a voltage change is first applied to an RL circuit, the current remains at its initial value (recall that inductors oppose changes in current). In one time constant, the current builds to 63.2% of its final value. It takes five time constants for the RL circuit to stabilize.

Circuit Waveforms

Charging and discharging of a capacitor in an RC circuit requires a definite amount of time (determined by the values of resistance and capacitance). If the input voltage changes levels too quickly, the capacitor may not have enough time to fully charge or discharge. Figure 7–55 illustrates the response of an RC circuit to a pulse of three different frequencies.

In Figure 7–55(b), each polarity of the input pulse is present for a period of time equal to one time constant. Thus, the capacitor does not have sufficient time to reach full charge before the input voltage is removed. Similarly, it is not given time to decay to zero between pulses.

Figure 7–55(c) illustrates a **short time constant**. Here, the input pulse duration is present for an amount of time equal to ten time constants. Clearly, the capacitor

FIGURE 7–55 (a) An RC circuit and the effect of (b) moderate, (c) short, and (d) long time constants on RC circuit response.

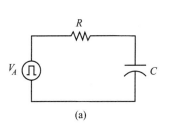
(a)

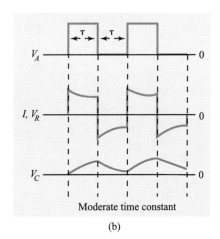
Moderate time constant
(b)

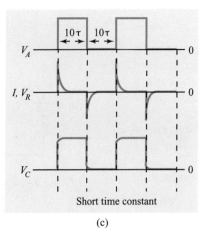
Short time constant
(c)

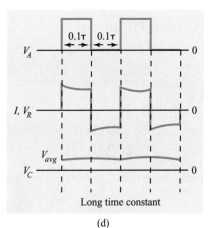
Long time constant
(d)

voltage is allowed time to reach maximum and to decay to zero. The resistor voltage appears as short-duration voltage spikes. Notice that the resistor waveform has a dual polarity caused by the opposite direction of current flow during the charge and discharge periods. The waveform across the resistor is called a differentiated waveform. The entire circuit is often called a **differentiator** circuit. To qualify as a differentiator, the circuit must have a short time constant and the output must be taken across the resistor. An *RL* circuit can also serve as a differentiator if it has a short time constant and the output is taken across the inductor.

Finally, Figure 7–55(d) illustrates the effects of a **long time constant**, where the input pulse duration is one-tenth of one time constant. Here, the capacitor voltage is barely allowed time to increase or decrease. The resistor voltage has a waveform that is nearly identical to the input waveform. This effect is useful for coupling signals between subsequent stages in an amplifier circuit. When the output is taken across the resistor, the circuit is called an *RC* coupling circuit. If the output is taken across the capacitor, we call the circuit an **integrator** circuit. Both coupling and integrator circuits require long time constants. An *RL* circuit can also serve as an integrator if it has a long time constant and the output is taken across the resistor.

Applications

There are many applications of *RC* and *RL* circuits in computers as well as in nearly every practical electronic device. We will explore a few of them in the following paragraphs.

Power Supply Circuits

Figure 7–56 illustrates the basic operation of a filter circuit for an electronic power supply. Most electronic circuits require dc voltage for operation, but it is often desirable to operate the system from the standard 120-Vac power line. An electronic power supply circuit is used to convert the 120-Vac, 60-Hz power distributed by the power company to a pulsating dc voltage. This pulsating or surging voltage can then be applied to an *RC* circuit as shown in Figure 7–56. The *RC* network is designed to have a long time constant. This means the voltage across the capacitor cannot follow the quick increases and decreases in the pulsating voltage from the electronic power supply. Rather, the capacitor voltage will be a relatively smooth dc. The voltage across the capacitor is used to supply power to the subsequent electronic circuitry. The overall result is a smooth dc voltage similar to that which would be produced by a battery.

FIGURE 7–56 An *RC* circuit can smooth the output of an electronic power supply.

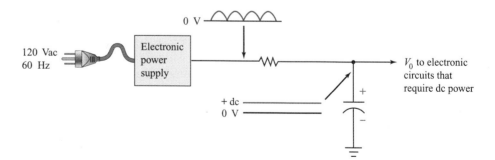

Passive Filter Circuits

RC and *RL* circuits can be used for frequency-selective filter circuits that pass or reject certain signals based on their frequencies. There are four basic types of filter configurations: low-pass, high-pass, bandpass, and band-reject. The low-pass and high-pass filters allow relatively low and high frequencies, respectively, to pass through with little or no reduction. Other frequencies, however, are greatly reduced or attenuated. Bandpass and band-reject filters pass or reject, respectively, a well-defined range of frequencies. Figure 7–57 shows a low-pass filter that is typically used on mouse ports and printer ports of computers.

In Figure 7–57, movements of the computer mouse send pulses to the computer. The values of R and C are selected such that the capacitive reactance of the capacitor is very high for all frequencies that make up the pulses coming from the mouse. Recall that a pulse waveform consists of a fundamental and a number of harmonics. In the case of mouse signals, the frequencies of these components are relatively low. Therefore, the reactance of the capacitor can be quite high (ideally it acts as an open circuit to the mouse signals).

The resistor in Figure 7–57 is selected so that its resistance is too small to have any significant effect on the mouse pulses. From the standpoint of the mouse pulses that represent mouse movement, the RC circuit has no effect.

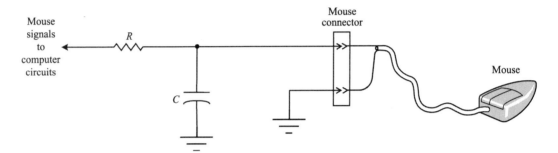

FIGURE 7–57 A low-pass filter circuit often used on computer ports.

There are many high-frequency waveforms in a typical computer system (e.g., 1-GHz microprocessor signals, high-frequency video signals, 100-MHz Ethernet signals, and many more). Most of these waveforms are pulses (i.e., digital), so they are composed of many high-frequency harmonics. Although not intended or desired, these signals (or at least reduced versions of them) can inadvertently find their way onto external computer cables such as the one represented in Figure 7–57. If this is allowed to happen, the high frequencies can actually radiate from the external cable and cause interference to nearby radios, televisions, and other wireless equipment. The RC filter shown in Figure 7–57 prevents the unwanted radio frequencies from escaping. At these high frequencies, the reactance of the capacitor is quite low relative to the resistance value. Therefore, according to Ohm's Law and Kirchhoff's Voltage Law, most of the high-frequency voltage will be dropped across the resistor leaving very little to be dropped across the capacitor. Since the mouse cable is connected in parallel with the capacitor, it too will have very little high-frequency voltage.

Time Delay Circuits

RC circuits are widely used to provide time delay action in electronic circuits. Figure 7–58 illustrates one particular method.

When the reset button in Figure 7–58 is pressed, the capacitor voltage is quickly discharged via the low resistance of the switch contacts. When the button is

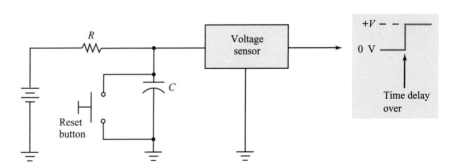

FIGURE 7–58 RC circuits can be used to provide electronic time delays.

released, the capacitor begins to charge via resistor R. As the capacitor charges, its voltage is monitored by an electronic circuit called a voltage sensor. As long as the capacitor voltage is below a certain voltage level (established by the voltage sensor circuit), then the output of the voltage sensor will be zero volts. When the capacitor reaches the trigger-point voltage established by the voltage sensor circuit, the output of the voltage sensor quickly switches to a different voltage level. This change in output voltage can be used to drive an indicator lamp, sound an audible alarm, or activate a computer circuit. This type of circuit is often used in industrial computers as part of a safety circuit called a watchdog timer. If the computer is operating normally, then it electronically presses the reset button every few milliseconds. If something occurs that causes the computer to stop executing the intended program, then the reset button does not get pressed, the capacitor charges, and the output of the voltage sensor changes state. This changing voltage can be used to provide a master reset to the entire computer, which forces it back into the correct program execution.

So what determines the length of the time delay? The time delay is determined by how long it takes the capacitor to charge to the trigger voltage. If we increase the value of either R or C, the time delay will increase. Of course, if we use a variable resistor for R, then we can have an adjustable time delay.

✓ Exercise Problems 7.4

1. In a series RC circuit, the voltage is always identical across all components. (True/False)
2. In a series RL circuit, the current and voltage phase relationship in the inductor is _____.
3. In a series RC circuit, the capacitor and resistor voltages are each 25 V. What is the value of applied voltage?
4. In a series RL circuit, the resistance and inductive reactance are 250 Ω and 500 Ω, respectively. What is the total impedance of the circuit?
5. In a parallel RC circuit, all component _____ are identical.
6. Express the value of source current in a parallel RL circuit relative to the values of resistive and inductive branch currents.
7. A parallel RL circuit has 50 Ω of resistance and 100 Ω of inductive reactance with an applied voltage of 100 V. What is the impedance of the circuit?
8. How long does it take to fully charge a 1000-pF capacitor through a 1.5-kΩ resistor using a 5-V source?
9. An RC differentiator circuit has a (*short/medium/long*) time constant.
10. The output from an RC integrator circuit is taken across the (*resistor/capacitor*).

7.5 *RLC* CIRCUITS

Much of the behavior of RLC circuits will seem natural to you, since these circuits are merely extensions of RC and RL circuits. But there are one or two characteristics of RLC circuits that will likely surprise you. This surprise behavior is associated with a condition called resonance, which plays a critical role in all forms of radio communications.

Characteristics

We will consider RLC circuit characteristics under both resonant and nonresonant conditions. We will first examine nonresonant circuits, which are essentially just a mixture of the characteristics discussed for RC and RL circuits.

Nonresonant *RLC* Circuits
We will examine series and parallel circuits separately.

Series RLC Circuits Figure 7–59(a) shows a series *RLC* circuit. What do we know about this circuit already? First, it is a series circuit, so we know the current is the same through every component. Second, we know that the individual voltage drops can be determined with Ohm's Law (e.g., $V_R = IR$, $V_L = IX_L$, and $V_C = IX_C$). Third, we know the voltage across the resistor is in phase with the common current. Similarly, we know that the voltage leads the common current by 90° in the inductor, while the voltage across the capacitor lags the common current by 90°. This means the inductor voltage and capacitor voltage are 180° out of phase with each other. The phasor diagram in Figure 7–59(b) illustrates this relationship. When two waveforms are 180° out of phase, they oppose or subtract from each other.

FIGURE 7–59 A series *RLC* circuit.

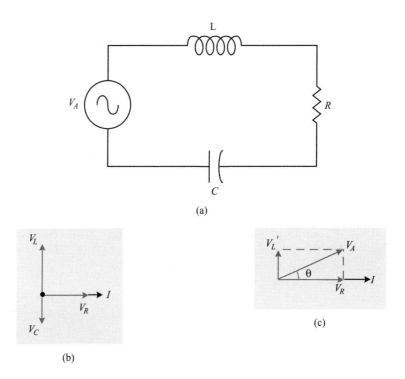

Figure 7–59(c) shows how the phasors are simplified. First, the phasors representing capacitor voltage and inductor voltage are subtracted from each other. In the case of Figure 7–59(b), the inductor phasor is assumed to be larger. So, after subtracting the capacitor phasor, we end up with a smaller phasor in the direction of the original inductor voltage. This shorter phasor is shown in Figure 7–59(c) as V_L'. If the capacitor voltage were greater, then the net reactive voltage phasor would be labeled V_C' and would appear 90° behind the current phasor. The remainder of the circuit can be analyzed as we would a pure *RC* or pure *RL* circuit. As shown in Figure 7–59(c), the net reactive voltage is combined with the resistive voltage by phasor addition to equal the source voltage. In summary, the two reactive components tend to cancel each other's effects with the larger of the two dominating. The portion of dominating reactance left after cancellation combines with the resistive effects to produce the overall characteristics of the circuit. We can make a slight modification to our phasor addition equation, as shown in Equation 7–20, to include the effects of both inductive and capacitive components in a series circuit.

$$V_A = \sqrt{V_R^2 + (V_L - V_C)^2} \qquad (7\text{--}20)$$

This same cancellation effect also occurs with reactance as illustrated in the modified impedance equation for series *RLC* circuits shown in Equation 7–21.

$$Z = \sqrt{R^2 + (X_L - X_C)^2} \quad (7\text{–}21)$$

•EXAMPLE 7.21

Find the source voltage and the total circuit impedance for the series *RLC* circuit shown in Figure 7–60.

FIGURE 7–60 Find the source voltage and impedance in this circuit.

Fig07_60.msm

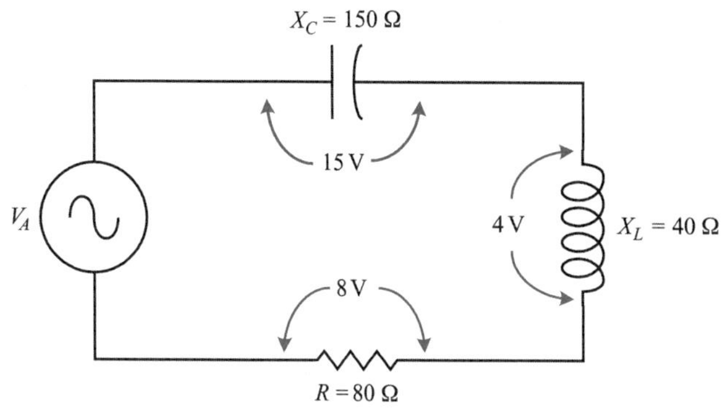

SOLUTION We apply Equation 7–20 and Equation 7–21 as follows:

$$V_A = \sqrt{V_R^2 + (V_L - V_C)^2} = \sqrt{(8 \text{ V})^2 + (4 \text{ V} - 15 \text{ V})^2} = 13.6 \text{ V}$$

And

$$Z = \sqrt{R^2 + (X_L - X_C)^2} = \sqrt{(80 \text{ }\Omega)^2 + (40 \text{ }\Omega - 150 \text{ }\Omega)^2} = 136 \text{ }\Omega$$

Now, if you are alert, this result may concern you. It clearly shows that the voltage drop across one of the series components (the capacitor in this case) is actually larger than the source voltage! Equally alarming is the observation that the total circuit impedance of this circuit is less than one of the series components (X_C in this case). Be curious, but don't be confused at this point. We will examine this more in our discussion of resonant *RLC* circuits.

Parallel RLC Circuits Figure 7–61(a) on page 244 shows a parallel *RLC* circuit. What do we know about this circuit already? First, it is a parallel circuit, so we know the voltage is the same across every component. Second, we know that the individual branch currents can be determined with Ohm's Law (e.g., $I_R = V/R$, $I_L = V/X_L$, and $I_C = V/X_C$). Third, we know the current through the resistor is in phase with the common voltage. Similarly, we know that the current leads the common voltage by 90° in the capacitor, while the current through the inductor lags the common voltage by 90°. This means the inductor current and capacitor current are 180° out of phase with each other. The phasor diagram in Figure 7–61(b) illustrates this relationship. When two waveforms are 180° out of phase, they oppose or subtract from each other.

Figure 7–61(c) shows how the phasors are simplified. First, the phasors representing capacitor current and inductor current are subtracted from each other. In the case of Figure 7–61(b), the capacitive current phasor is assumed to be larger. So, after subtracting the inductor current phasor, we end up with a smaller phasor

FIGURE 7–61 A parallel *RLC* circuit.

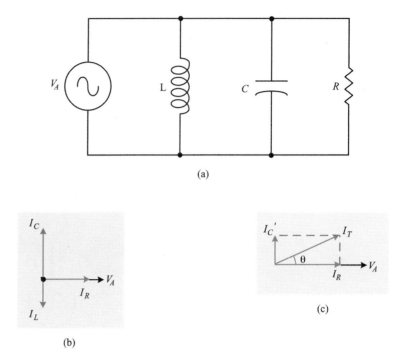

in the direction of the original capacitor current. This shorter phasor is shown in Figure 7–61(c) as I_C'. If the inductor current were greater, then the net reactive current phasor would be labeled I_L' and would appear 90° behind the voltage phasor. The remainder of the circuit can be analyzed as we would a pure *RC* or pure *RL* circuit. As shown in Figure 7–61(c), the net reactive current is combined with the resistive current by phasor addition to equal the source current. In summary, the two reactive components tend to cancel each other's effects with the larger of the two dominating. The portion of dominating current left after cancellation combines with the resistive current to determine the overall characteristics of the circuit. We can make a slight modification to our phasor addition equation, as shown in Equation 7–22, to include the effects of both inductive and capacitive components in a parallel circuit.

$$I_T = \sqrt{I_R^2 + (I_L - I_C)^2} \qquad (7\text{–}22)$$

•EXAMPLE 7.22

Find the source current and the total circuit impedance for the parallel *RLC* circuit shown in Figure 7–62.

FIGURE 7–62 Find the source current and impedance in this circuit.

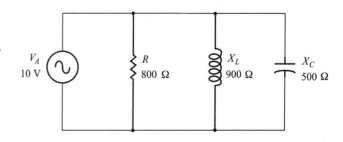

SOLUTION   We first apply Ohm's Law to determine the branch currents.

$$I_R = \frac{V_R}{R} = \frac{10 \text{ V}}{800 \text{ }\Omega} = 12.5 \text{ mA}$$

$$I_L = \frac{V_L}{X_L} = \frac{10 \text{ V}}{900 \text{ }\Omega} = 11.11 \text{ mA}$$

$$I_C = \frac{V_C}{X_C} = \frac{10 \text{ V}}{500 \text{ }\Omega} = 20 \text{ mA}$$

Next, we apply Equation 7–22 to find the total source current.

$$I_T = \sqrt{I_R^2 + (I_L - I_C)^2} = \sqrt{(12.5 \text{ mA})^2 + (11.11 \text{ mA} - 20 \text{ mA})^2} = 15.34 \text{ mA}$$

Finally, we apply Ohm's Law to compute circuit impedance.

$$Z = \frac{V_A}{I_T} = \frac{10 \text{ V}}{15.34 \text{ mA}} = 651.9 \text{ }\Omega$$

Again, the results from this example might spark your curiosity. Note that the total current is less than one of the branch currents (I_C in this case). Also, note that one of the branches (capacitor branch) has a lower reactance than the total circuit impedance. This apparent magic will be even more astounding when we evaluate resonant *RLC* circuits in the next section.

Resonant *RLC* Circuits

The concept of resonance is very important since it plays such a central role to the operation of many electronic devices. Although we will address the subject of resonance as a separate topic, it is very important for you to realize that resonance is nothing more than a description of the circuit characteristics of an *RLC* circuit at a particular operating frequency. That is, every *RLC* circuit that we have previously discussed and analyzed has a resonant frequency and exhibits the characteristics of resonance that are detailed in this section. A resonant *RLC* circuit is exactly like any other *RLC* circuit, except it is operating at a particular frequency called the **resonant frequency**.

Mechanical Resonance Before we begin our discussion of resonant *RLC* circuits, let's briefly consider a simple example of mechanical resonance. Many readers will be familiar with the phenomenon that is illustrated in Figure 7–63. Here, we have a crystal goblet containing only a small amount of liquid. If a finger (a moist finger works best) is rubbed smoothly and lightly around the rim of the glass, the glass will begin to vibrate at a very definite frequency. Once it has started to oscillate, you can sustain the oscillation with only the slightest movement of your finger along the rim. The crystal goblet is exhibiting mechanical resonance. If we add energy at the right time during each cycle of its oscillation period, we need only add enough energy to compensate for losses. In Figure 7–63, we are adding the energy from a sliding finger. Some singers can add the energy with their voice (excessive energy can shatter the glass). The frequency of oscillation is determined by the physical characteristics of the glass (e.g., size, thickness, amount of liquid). Unless we change these physical characteristics, we cannot easily alter the frequency of oscillation. This natural frequency of oscillation that is exhibited by the crystal goblet is called its resonant frequency.

FIGURE 7–63 An interesting demonstration of resonance.

Rub moistened finger gently and smoothly around rim.

Glass vibrates at its resonant frequency.

Parallel Resonance The similarities between electrical and mechanical resonance are easier to understand when considering a parallel *RLC* circuit, so let's begin by studying the characteristics of a parallel resonant circuit.

Figure 7–64 shows a parallel *RLC* circuit being operated at three frequencies: (a) below resonance, (b) above resonance, and (c) at resonance. The first two conditions were discussed in preceding sections, but are included here for contrast. The interesting condition called resonance occurs when the inductive and capacitive currents are equal. There is only one frequency for a given circuit where this condition exists, and that frequency is called the resonant frequency.

If the input frequency is adjusted such that the inductive reactance and the capacitive reactance are equal, this will be the resonant frequency of the circuit as illustrated in Figure 7–64(c). Because the inductive and capacitive branches have equal reactances and equal voltages, we know the two branch currents will be equal in magnitude. The phasor diagram in Figure 7–64(c) illustrates that the two equal reactive currents (regardless of their exact values) completely cancel each other (i.e., net reactive current is zero). This leaves only resistive current to form I_T. We observe the following characteristics of a parallel *RLC* circuit operating at its resonant frequency:

1. Inductive and capacitive reactances are equal.
2. Inductive and capacitive currents are equal.
3. Total current is equal to the resistive branch current.
4. Total current is minimum (only resistive current).
5. Circuit impedance is maximum (sometimes called a resonant rise of impedance).
6. Phase angle between total current and total voltage is zero (i.e., circuit acts resistive).

The reactive currents are determined by the reactance and the applied voltage as with any *RLC* circuit. However, at resonance they may be substantially higher than the line current (I_T). The smaller line current only has to supply losses in the circuit (e.g., power dissipation in the resistor). This is akin to the small amount of energy that must be supplied to a mechanically resonant circuit to keep it oscillating. As with a mechanical system, the external energy must be timed correctly. In the case of an *RLC* circuit, the external energy must be added at the resonant frequency. If we shift the frequency either higher or lower, we start to minimize the resonance effect as illustrated in Figures 7–64(a) and (b).

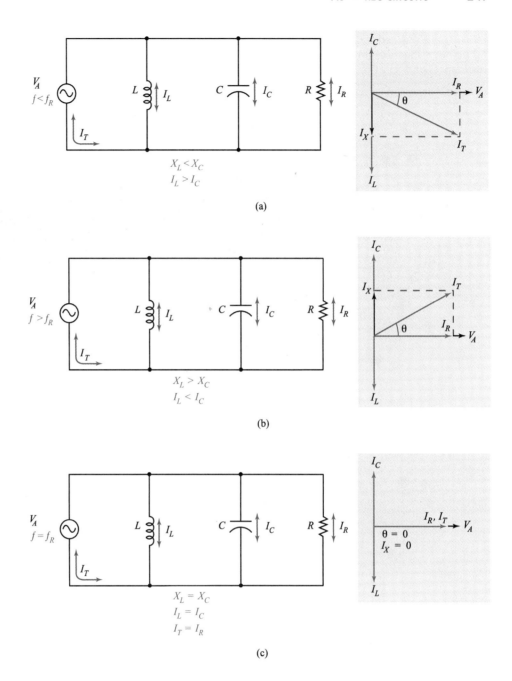

FIGURE 7-64 A parallel RLC circuit (a) below resonance, (b) above resonance, and (c) at resonance.

Series Resonance Figure 7-65 on page 248 shows a series *RLC* circuit being operated at three frequencies: (a) below resonance, (b) above resonance, and (c) at resonance. The first two conditions were discussed in earlier sections, but are included here for contrast. Resonance occurs when the inductive reactance and capacitive reactance are equal. There is only one frequency for a given circuit where this condition exists, and that is the resonant frequency.

If the input frequency is adjusted such that the inductive reactance and the capacitive reactance are equal, this will be the resonant frequency of the circuit as illustrated in Figure 7-65(c). Because the inductive and capacitive components have equal reactances and equal currents, we know the two voltage drops will be equal in magnitude. The phasor diagram in Figure 7-65(c) illustrates that the two equal reactive voltages (regardless of their exact values) completely cancel each other (i.e., net reactive voltage is zero). This leaves only resistive voltage to equal

FIGURE 7–65 A series *RLC* circuit (a) below resonance, (b) above resonance, and (c) at resonance.

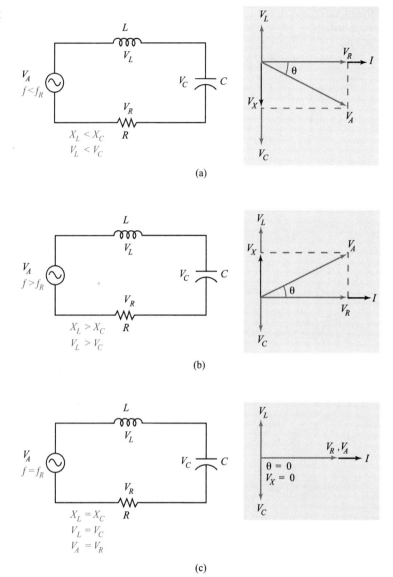

V_A. We observe the following characteristics of a series *RLC* circuit operating at its resonant frequency:

1. Inductive and capacitive reactances are equal.
2. Inductive and capacitive voltage drops are equal.
3. The resistive voltage drop is equal to the applied voltage.
4. Circuit impedance is minimum (reactances cancel).
5. Current is maximum (limited only by the resistance).
6. Phase angle between total current and total voltage is zero (i.e., circuit acts resistive).

The reactive voltages are determined by the reactance and the current as with any *RLC* circuit. However, at resonance they may be substantially higher than the applied voltage (V_A). This increase in voltage is often called a resonant rise in voltage.

Computing Resonant Frequency The frequency that causes the inductive and capacitive reactances to be equal in an *RLC* circuit (either series or parallel) is called the resonant frequency. You can determine the resonant frequency with Equation 7–23.

$$f_R = \frac{1}{2\pi\sqrt{LC}} \qquad (7\text{-}23)$$

• EXAMPLE 7.23

A series *RLC* circuit consists of the following components: 10-Ω resistance, 2.5-µH coil, and a 100-pF capacitor. What is the resonant frequency of the circuit?

SOLUTION   We apply Equation 7–23 as follows:

$$f_R = \frac{1}{2\pi\sqrt{LC}} = \frac{1}{2 \times 3.14 \times \sqrt{2.5\ \mu\text{H} \times 100\ \text{pF}}} = 10.07\ \text{MHz}$$

Calculator Application

Resonant Circuit Measurements   The degrees to which the resonance effects are evident in an *RLC* circuit are determined by the quality, or *Q*, of the circuit. *Q* has no units. The higher the value of *Q*, the more marked the effects of resonance. Figure 7–66 shows graphs of current as functions of frequency for (a) series and (b) parallel *RLC* circuits. There are three cases plotted on each graph representing low-, moderate-, and high-*Q RLC* circuits. The highest *Q* produces a very sharp response at resonance. The lower the *Q*, the less pronounced the resonance effect.

Values of *Q* range from less than 10 to over 100 for standard *RLC* circuits. In general, we classify circuits whose *Q* factor is less than 10 as low-*Q* circuits. By contrast, circuits with *Q* factors over 100 are considered to be high-*Q* circuits.

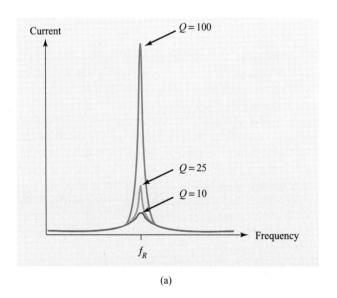

(a)

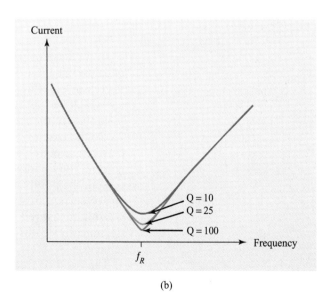

(b)

FIGURE 7–66   Current as functions of frequency for (a) series and (b) parallel *RLC* circuits. The *Q* of the circuit determines the sharpness of the resonance response.

Later in your electronics education, you will study circuits and devices that have Q factors greater than 1,000. In general, higher resistance in a series circuit decreases the Q of the circuit, whereas higher resistance in a parallel circuit increases Q.

Selectivity of an *RLC* circuit describes how sharply the circuit distinguishes between the resonant frequency and frequencies on either side of the resonant frequency. Frequencies that cause a response that is 70.7% of the maximum (voltage, current, or impedance) response or greater are considered to be passed or selected by the circuit. The frequencies that cause a response of less than 70.7% of the maximum (voltage, current, or impedance) response are considered to be rejected (i.e., not selected) by the circuit.

Figure 7–67 shows the response of three *RLC* circuits with different Q factors. Their 100% responses have been adjusted to be equal. Regardless of the Q factor, all frequencies with responses above 70.7% of the maximum voltage, current, or impedance are considered to be selected or passed by the circuit. Clearly, there is a band or range of frequencies that is above the 70.7% level. The greater the selectivity (i.e., the higher the Q) of the circuit, the narrower the range of frequencies that produces a response of greater than the 70.7% of the maximum.

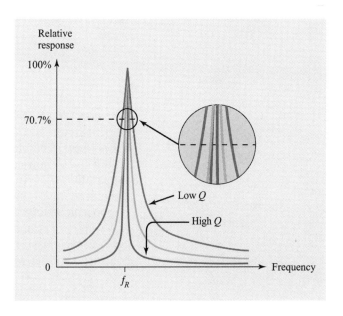

FIGURE 7–67 The Q factor determines the selectivity and bandwidth of an *RLC* circuit.

In the case of parallel *RLC* circuits, we measure the response in terms of impedance. That is, the impedance is maximum at the resonant frequency (f_R) and decreases on either side of f_R. In the case of series *RLC* circuits, it is common to use resistive voltage, power, or circuit current as the indicator. All of these are maximum at the resonant frequency and decrease for frequencies on either side of f_R. In the case of current or voltage responses, we use 70.7% as the dividing line between selection and rejection. Since power is equal to the product of voltage and current, the dividing line for a power response is 70.7% × 70.7% or 50%. For this reason, the points at which the response falls below 70.7% of the maximum voltage or current in a series *RLC* circuit are also called the **half-power points**. Power is often measured in decibels, so these same points are also called 3-dB points (technically −3-dB points), since a −3-dB power loss corresponds to a 50% power reduction.

The band of frequencies that causes a response above the half-power points is called the **bandpass** of the *RLC* circuit. Bandpass (also called **pass band**) is expressed as a range of frequencies. The lowest frequency that still produces a response above the half-power point is called the lower **cutoff frequency** (f_L). The upper cutoff frequency (f_H) is the highest frequency that still produces a response above the half-power point. The bandpass of an *RLC* circuit is the range of frequencies between the lower and upper cutoff frequencies.

The width (measured in hertz) of the bandpass of an *RLC* circuit is called the **bandwidth** of the circuit. The bandwidth is not a measure of specific frequencies like the bandpass, but rather it is a measure of the width of the bandpass. We can compute bandwidth with Equation 7–24.

$$BW = f_H - f_L \qquad (7\text{–}24)$$

where f_L is the lower cutoff frequency and f_H is the upper cutoff frequency of the *RLC* circuit.

•—EXAMPLE 7.24

What is the bandwidth of an *RLC* circuit that has 70.7% or higher of the maximum response for frequencies higher than 150 kHz and lower than 250 kHz?

SOLUTION We apply Equation 7–24.

$$BW = f_H - f_L = 250 \text{ kHz} - 150 \text{ kHz} = 100 \text{ kHz}$$

Figure 7–68 further clarifies the definitions and relationships affecting the bandwidth of an *RLC* circuit.

FIGURE 7–68 The relationship of bandpass and bandwidth to the 70.7% response level.

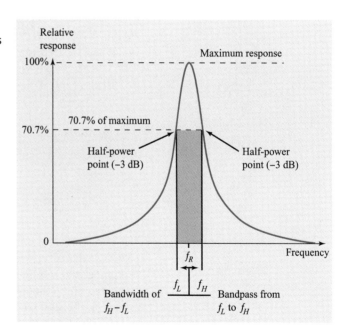

Applications

RLC circuits (often called *LC* circuits because the *R* portion is parasitic) are used in many electronic products because of their ability to discriminate signals based on frequency. We will briefly highlight a few examples.

Wireless Communication
All forms of radio communication such as Bluetooth devices, wireless Internet products, AM/FM radios, televisions, and satellite communication links rely on *RLC* circuits for their operation. Figure 7–69 on page 252 shows a simplified block diagram of a radio receiver.

FIGURE 7-69 An *LC* circuit can select a particular radio station and reject all other frequencies.

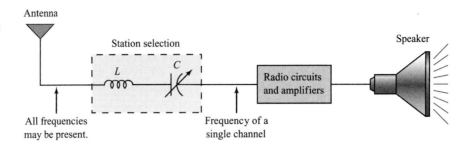

As shown in Figure 7-69, frequencies from all radio wave sources can strike the receiving antenna. However, there is only a very narrow range of these frequencies that represents the signal from a specific radio station. An *LC* circuit can be tuned (with the station-selector knob) such that the desired station falls within the bandpass of the *LC* circuit. All other frequencies fall outside this bandpass and are rejected.

Rejection of Interference

Figure 7-70 shows how an *RLC* circuit can be used to separate a desired range of frequencies from another specific frequency that would otherwise cause interference. As shown in Figure 7-70, the circuit is essentially a two-part voltage divider. The first part is the series resistance (*R*). The second part of the voltage divider is the series *LC* circuit. The *LC* circuit is tuned to the frequency of the interference.

FIGURE 7-70 An *RLC* circuit can be used to eliminate interference.

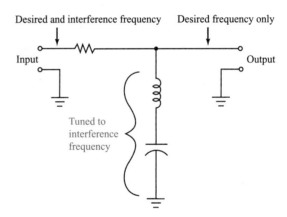

Recall that a series *LC* circuit has minimum impedance at resonance (i.e., the net reactance is zero ohms). In the case of the circuit shown in Figure 7-70, there will be a zero ohm (theoretically) impedance through the *LC* portion of the circuit to ground at the frequency of the interference. This means all of the interference voltage will be dropped across the series resistance and will not pass on to subsequent circuitry.

The desired frequency, on the other hand, is either higher or lower than the resonant frequency of the *LC* circuit. Recall that a series *LC* combination can have substantial impedance at frequencies either side of the resonant frequency. In the case shown in Figure 7-70, the impedance of the *LC* circuit at the desired frequency is much greater than the value of series resistance. This means that nearly all of the desired signal voltage will be felt across the *LC* portion of the circuit and will be available for use in subsequent circuitry connected in parallel with the *LC* circuit.

Parasitic Circuit Elements

For many practical applications in computer technology, you can consider all circuit components to be ideal. That is, they behave as you would expect them to behave. However, in circuits with high frequencies (usually tens of megahertz and higher), all components (even wires) begin to exhibit nonideal characteristics. Since many computers have frequencies in excess of 1.0 GHz (and nearly all have frequencies of several hundred megahertz), you need to be alert to nonideal characteristics of common components. We will highlight two of the most important nonideal components in the following sections.

Nonideal Inductors

Figure 7–71 shows that a practical coil has resistive and capacitive characteristics in addition to the intended inductive characteristics. The resistance results from nonideal wire. The capacitance appears between adjacent turns on the coil. (Remember that a capacitor is simply two conductors separated by an insulator.)

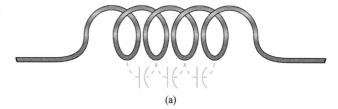

FIGURE 7–71 A physical inductor also has resistance and capacitance.

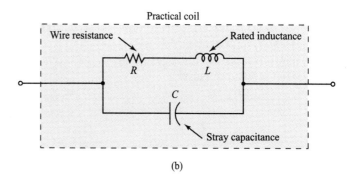

Because the practical coil is actually an *RLC* circuit, it will exhibit resonance. This frequency is called the **self-resonant frequency** of the coil. At frequencies above the self-resonant frequency, the capacitive reactance is less than the inductive reactance, so the coil is effectively bypassed.

Nonideal Capacitors

Practical capacitors also exhibit nonideal properties. They have equivalent series inductance (**ESL**) and equivalent series resistance (**ESR**), which represent losses in the wires and within the dielectric. Figure 7–72 shows the equivalent circuit for a practical capacitor. The rated capacitance and the parasitic resistance and inductance all appear in series. Thus, a capacitor alone becomes a series resonant circuit at some high frequency.

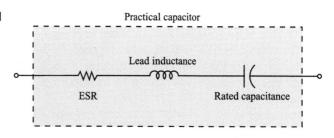

FIGURE 7–72 A practical capacitor acts like a series *RLC* circuit at high frequencies.

Exercise Problems 7.5

1. The voltage is the same across all components in a series *RLC* circuit. (True/False)
2. In a series *RLC* circuit, the capacitor voltage is _____° out of phase with the inductor voltage.
3. Compute the source voltage in a series *RLC* circuit if it has the following component voltage drops: $V_R = 3$ V, $V_C = 25$ V, and $V_L = 21$ V.
4. What is the impedance of a series *RLC* circuit with a resistance of 5 Ω, a capacitive reactance of 25 Ω, and an inductive reactance of 18 Ω?
5. The voltage is the same across all components in a parallel *RLC* circuit. (True/False)
6. In a parallel *RLC* circuit, the capacitor voltage is _____° out of phase with the inductor voltage.
7. A series *RLC* circuit at resonance has (*minimum/maximum*) impedance.
8. A parallel *RLC* circuit at resonance has (*minimum/maximum*) source current.
9. A series *RLC* circuit has a 2-μH coil, an 82-pF capacitor, and 2 Ω of resistance. What is the resonant frequency of the circuit?
10. A parallel *RLC* circuit is being operated below its resonant frequency. What happens to source current as the operating frequency is increased toward resonance?
11. A series *RLC* circuit is being operated above its resonant frequency. What happens to the voltage across the resistor as the operating frequency is decreased toward the resonant frequency?
12. What is the phase difference between source voltage and source current in a series *RLC* circuit that is being operated at resonance? Repeat for a parallel *RLC* circuit.

●─ SUMMARY

Inductors store energy in an electromagnetic field and oppose changes in current. Inductance is a property of an inductor that allows it to oppose changes in current. Inductance is measured in henrys. For sinusoidal currents, the opposition to changing currents is called inductive reactance and is measured in ohms. Inductive reactance is directly proportional to inductance and frequency. Inductors offer no opposition to dc. Multiple inductors and inductive reactance in series or parallel can be combined in the same fashion as series and parallel resistances. Current through an inductor always lags the voltage by 90°. A coil can have several defects including open windings, shorted turns, and coil-to-core shorts.

Transformers are used to increase (step up) or decrease (step down) alternating voltages. The supply voltage is connected to the primary winding and the load circuitry is connected to the secondary. Transformers are also used to transform or alter impedances in order to make a certain load appear as a specific resistance to a source. Transformers can develop defects that include open windings, shorted turns or windings, and winding-to-core shorts.

Capacitors store a charge in an electrostatic field. Capacitance is measured in farads. The opposition to sinusoidal currents is called capacitive reactance and is measured in ohms. Capacitive reactance is inversely proportional to capacitance and frequency. Capacitors offer infinite opposition to dc (i.e., they act as open circuits). Series and parallel capacitances can be combined in the same fashion as parallel and series resistances, respectively. Series and parallel capacitive reactances combine in the manner as series and parallel resistances, respectively. A capacitor can have several defects including opens, shorts, and leaky dielectrics. Current through a capacitor always leads the voltage by 90°. Many students use the mnemonic phrase "ELI the ICE man" to help remember the phase relationships in capacitors and inductors. In this phrase, E represents voltage (i.e., electromotive force), L identifies an inductor, I is for current, and C identifies a capacitor. So, in the word ELI, voltage comes before current, so you can remember voltage leads current in an inductor. Similarly, current comes before voltage in the word ICE, helping you recall that current leads voltage in a capacitor. Table 7–7 serves to summarize and contrast inductor and capacitor behavior.

RC and *RL* circuits exhibit characteristics that lie between pure resistive and pure capacitive (or inductive)

TABLE 7-7 Inductor and capacitor comparisons.

Item	Inductor	Capacitor
Energy storage	Magnetic Field	Electric Field
Opposes changes in	Current	Voltage
Series combinations	$L_T = L_1 + L_2$	$C_T = \dfrac{1}{\dfrac{1}{C_1} + \dfrac{1}{C_2}}$
Parallel combinations	$L_T = \dfrac{1}{\dfrac{1}{L_1} + \dfrac{1}{L_2}}$	$C_T = C_1 + C_2$
ac current	Voltage Leads Current	Current Leads Voltage

when driven with sinusoidal voltages. The opposition to sinusoidal current offered by RC or RL circuits is called impedance and is measured in ohms. When used with pulse waveforms or switched dc, RC and RL circuits introduce delays. A time constant is the time required for a capacitor to charge or discharge (or the time for inductor current to build or decay) to 63.2% of the full charge or discharge level. For practical purposes, it takes five time constants for an RC or RL circuit to stabilize following a change in applied voltage. RC circuits with short time constants can serve as differentiator circuits if the output is taken across the resistor. RC circuits with long time constants can be integrators or coupling circuits when the output is taken across the capacitor or resistor, respectively. RL circuits can also serve as differentiators and integrators when the output is taken across the inductor or resistor, respectively.

RLC circuits can exhibit the properties of RC circuits, RL circuits, or purely resistive circuits depending on the frequency of operation relative to the resonant frequency of the circuit. At resonance, series RLC circuits have maximum current and impedance that is equal to the resistance in the circuit only. Parallel RLC circuits have maximum impedance at resonance, which is also equal to the circuit resistance. The phase relationships in an RLC circuit can range from 90° lagging to 90° leading depending on the frequency of operation relative to the resonant frequency. Voltage and current are in phase at resonance. Higher Q circuits exhibit greater resonance effects. They also have greater selectivity and narrower bandwidths. Practical inductors and capacitors act as parallel and series RLC circuits, respectively, due to parasitic elements.

REVIEW QUESTIONS

Section 7.1: Inductors

1. Inductors are also called _____.
2. Inductors offer essentially no opposition to dc. (True/False)
3. Inductors oppose _____ in current.
4. What is the unit of measure for inductance?
5. What is the unit of measure for inductive reactance?
6. What is the inductive reactance of a 3-mH coil operated at 250 kHz?
7. Current _____ the voltage in an inductor by 90°.
8. Air-core inductors are generally used for audio applications. (True/False)
9. When a ferrite bead is slipped over a wire, both resistance and inductance are added in series with the wire. (True/False)
10. What is the total inductance of three series-connected coils if each coil is 25 mH?
11. What is the total inductance of four parallel-connected coils if each coil has an inductance of 5 µH?
12. What is the total reactance of a series circuit having individual inductive reactances of 100 Ω, 300 Ω, and 250 Ω?
13. What is the total reactance of a parallel circuit having individual inductive reactances of 25 kΩ, 100 kΩ, and 75 kΩ?
14. What happens to the total inductive reactance in a series inductive circuit if the frequency is increased?
15. What happens to the total inductive reactance in a parallel inductive circuit if the frequency is increased?

Section 7.2: Transformers

16. There is no direct electrical connection between primary and secondary windings in a standard transformer. (True/False)

17. The percentage of flux from the primary that passes through the secondary of a transformer is called the _____ __ _____.
18. How is the phase relationship between the primary and secondary windings of a transformer indicated on a schematic diagram?
19. If the secondary voltage in a transformer is lower than the primary voltage, it is called a _____-_____ transformer.
20. What is the turns ratio of a transformer whose sole purpose is to provide isolation between primary and secondary circuits?
21. What class of transformers is wound on a nonmagnetic coil form?
22. Iron-core transformers are generally used for low-frequency (power line and audio frequency) applications. (True/False)
23. It is possible to have a single transformer that provides both step-up and step-down windings. (True/False)
24. A transformer with a turns ratio of 5:1 has its primary supplied by a 120-Vac source. What is the value of secondary voltage?
25. The secondary current of a transformer with a 6:1 turns ratio is 250 mA. What is the value of primary current?
26. The impedance in the primary that results from a secondary load is called the _____ _____.
27. The secondary of an audio transformer with a 4:1 turns ratio is connected to an 8-Ω speaker. What is the value of impedance reflected into the primary?
28. While troubleshooting a transformer, you find that the primary winding measures infinite (∞) ohms with an ohmmeter. This means the transformer is ____.
    a. definitely defective
    b. definitely good
    c. possibly defective
29. While troubleshooting a transformer, you find that the primary winding measures 0.6 Ω with an ohmmeter. This means the transformer is ____.
    a. definitely defective
    b. definitely good
    c. possibly defective
30. The primary of a transformer is connected to 120 Vac, and the secondary is disconnected from the load. The secondary voltage measures zero. This means the transformer is ____.
    a. definitely defective
    b. definitely good
    c. possibly defective

Section 7.3: Capacitors

31. Capacitors store energy in an _____ field.
32. Once a practical capacitor is fully charged, it can be removed from the voltage source and it will remain charged forever. (True/False)
33. If the total surface area of the plates in a capacitor is reduced during manufacture, the value of the capacitor will (*decrease/increase*).
34. If the distance between the plates of a capacitor is increased, the value of capacitance will (*decrease/increase*).
35. What is another name for the relative permittivity of a material?
36. If the frequency in a capacitive circuit is increased, what happens to the value of capacitive reactance?
37. A capacitor has 10 mA of current when connected to a 100-V source operating at 100 kHz. What value of current will flow if the capacitor is connected to a 10-Vdc source?
38. What is the capacitive reactance offered by a 1000-pF capacitor when operated at a frequency of 125 MHz?
39. Calculate the capacitive reactance for a 25-pF capacitor that is operated at 850 kHz.
40. What type of capacitor package has no protruding leads?
41. What does the voltage rating on a capacitor indicate?
42. Which of the following capacitors are polarized (circle all that apply)?
    a. mylar   d. film
    b. tantalum   e. mica
    c. ceramic   f. aluminum
43. What is the total capacitance if a 100-pF, a 270-pF, and a 150-pF capacitor are connected in series?
44. Which of the following combinations results in the greatest capacitance?
    a. 1000 pF, 200 µF, and 250 µF in series
    b. 470 pF, 10 µF, and 1000 pF in parallel
45. A capacitive circuit has 100 Ω, 250 Ω, and 330 Ω of capacitive reactance in parallel. What is the total capacitive reactance?
46. A capacitive circuit has 1.2 kΩ, 2.5 kΩ, and 4.2 kΩ of capacitive reactance in series. What is the total capacitive reactance?
47. If you have three fixed-value capacitors that will be operated at a specific frequency, how would you connect them (series or parallel) in order to get the greatest total capacitive reactance?
48. When troubleshooting a circuit with surface-mounted components, what is the best action to take if you suspect that a capacitor is defective?
    a. remove it and replace it
    b. measure it with an ohmmeter
    c. measure it with a capacitance tester
    d. measure it with an LCR tester

49. If a capacitor is shorted, how will its condition be indicated by an ohmmeter?

Section 7.4: *RC* and *RL* Circuits

50. An *RL* circuit consists of 10 mH and 4.7 kΩ in series across a 100-kHz source. If the resistor has 3.1 mA of current, how much current flows through the inductor?
51. What is the phase relationship between resistor voltage and resistor current in a series *RC* circuit? Repeat for a parallel *RC* circuit.
52. What is the phase relationship between capacitor voltage and capacitor current in a series *RC* circuit? Repeat for a parallel *RC* circuit.
53. What is the phase relationship between inductor voltage and inductor current in a series *RL* circuit? Repeat for a parallel *RL* circuit.
54. The voltage drops in a series *RC* circuit are 10 V for the capacitor and 15 V for the resistor. What is the value of source voltage?
55. Determine the source voltage in a series *RC* circuit if the capacitor and resistor have voltage drops of 500 mV and 750 mV, respectively.
56. What is the total impedance of a series *RC* circuit consisting of 2.5 kΩ of capacitive reactance and 5 kΩ of resistance?
57. An *RL* circuit consists of 25 µH and 2.2 kΩ in parallel across a 12-MHz source. If the resistor has a voltage drop of 5 V, what is the voltage drop across the inductor?
58. The branch currents in a parallel *RC* circuit are 25 mA for the capacitor and 20 mA for the resistor. What is the value of source current?
59. Determine the source current in a parallel *RC* circuit if the capacitor and resistor have branch currents of 50 µA and 10 µA, respectively.
60. What is the total impedance of a parallel *RC* circuit consisting of 400 Ω of capacitive reactance and 400 Ω of resistance connected to a 3.2-V source?
61. What is the total impedance of a series *RL* circuit consisting of 35 kΩ of inductive reactance and 50 kΩ of resistance?
62. What is the *RC* time constant of a circuit that has a 4.7-kΩ resistor in series with a 2700-pF capacitor?
63. How long does it take an 1800-pF capacitor to reach full charge when connected across a 25-V source if it must charge through a 1.2-kΩ resistor?
64. Repeat the previous question for a 1.5-V source voltage.
65. An *RC* differentiator circuit has a (*short/long*) time constant, and the output is taken across the (*resistor/capacitor*).
66. What name is given to an *RC* circuit with a long time constant if the output is taken across the resistor?
67. What is the name given to a passive filter circuit that allows low frequencies to go through with minimal reduction but greatly attenuates higher frequencies?

Section 7.5: *RLC* Circuits

68. Describe the phase relationship between resistor current and inductor current in a series *RLC* circuit that is operating below its resonant frequency.
69. Describe the phase relationship between resistor current and inductor voltage in a series *RLC* circuit that is operating above its resonant frequency.
70. A series *RLC* circuit has (*minimum/maximum*) impedance at resonance.
71. A parallel *RLC* circuit has (*minimum/maximum*) source current at resonance.
72. The voltage drop across the resistor in a series *RLC* circuit (*decreases/increases*) as the operating frequency is brought closer to resonance.
73. Describe the phase relationship between inductor voltage and capacitor voltage in a parallel *RLC* circuit as the operating frequency is adjusted from lower than the resonant frequency to some frequency above the resonant frequency.
74. A parallel *RLC* circuit is operating at its resonant frequency. The capacitive reactance is 5 kΩ and the resistance is 100 kΩ. What is the value of inductive reactance?
75. A series *RLC* circuit consists of the following components: 20-Ω resistance, 2.5-µH coil, and a 27-pF capacitor. What is the resonant frequency of the circuit?
76. A parallel *RLC* circuit consists of the following components: 1-MΩ resistance, 5-µH coil, and a 100-pF capacitor. What is the resonant frequency of the circuit?
77. If the *RLC* circuit used to dial the stations in your radio had poor selectivity, describe what might be the audible symptoms.
78. A nonideal capacitor acts like a (*series/parallel*) *RLC* circuit.
79. A nonideal inductor acts like a (*series/parallel*) *RLC* circuit.
80. If an *RLC* circuit has a very sharp change in its characteristics when the operating frequency approaches the resonant frequency of the circuit, then the circuit has a (*low/high*) *Q*.

CIRCUIT EXPLORATION

This learning exercise will give you the opportunity to use your circuit simulation tool to further explore the behavior of *RC*, *RL*, and *RLC* circuits. You will also learn more about a very important application: passive filter circuits.

For this exercise, you will need to construct (in the schematic capture environment) the five circuits that are presented in Figures 7–73 through 7–77. Once each circuit is constructed, perform the following tasks:

1. Plot the frequency response curve for the circuit.
2. Determine the resonant frequency for *RLC* circuits and the cutoff frequency for *RC* and *RL* circuits.
3. For *RLC* circuits, change the value of series resistance both up and down by a factor of five and observe the effect on Q and bandwidth.
4. Classify each filter circuit into one of the following categories: low-pass, high-pass, bandpass, or band-reject.

How you choose to plot the frequency response curves is somewhat dependent on the specific circuit simulation tool you are using. Some simulators allow you to use a swept frequency source. If you choose this option, be sure to set the limits of the sweep such that the range includes the cutoff frequency ($R = X_C$ or $R = X_L$) for *RC* and *RL* circuits or the resonant frequency ($f_R = \frac{1}{2\pi\sqrt{LC}}$) in the case of *RLC* circuits. MultiSIM provides a Bode plotter as a test instrument. Here you merely connect the +V in jack to the voltage source, the +V out jack to the VOUT point of the circuit, and the two −V jacks to ground. The default settings will allow you to get a basic plot, but you will want to adjust the scales to get better definition in the areas of interest (e.g., near resonance and near cutoff).

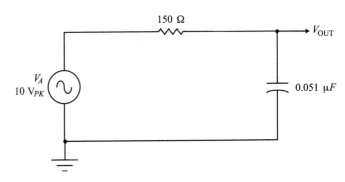

FIGURE 7–75 An *RC* filter circuit.

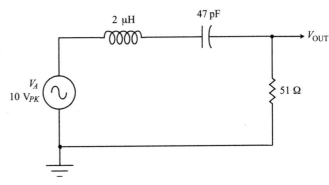

FIGURE 7–76 An *RLC* filter circuit.

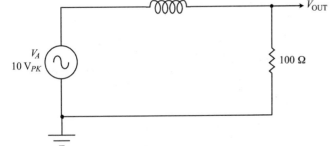

FIGURE 7–73 An *RC* filter circuit.

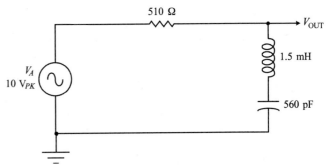

FIGURE 7–74 An *RL* filter circuit.

FIGURE 7–77 An *RLC* filter circuit.

ANSWERS TO PRACTICE PROBLEMS
Page 237
1. 1,100 seconds or 18.33 minutes
2. 5
3. zero
4. None of these values is affected by battery voltage, so the answers are the same.

CHAPTER 8
Semiconductor Technology

•—KEY TERMS
acceptor atom
avalanche current
barrier potential
bipolar
conduction band
covalent bonding
crystal
depletion region
diode
donor atom
doping
electron-hole pairs
extrinsic semiconductor
forward bias
front-to-back ratio
hole
intrinsic semiconductor
junction breakdown
majority carriers
minority carriers
n-type
p-type
pentavalent
pn junction
recombination
reverse bias
reverse current
solid-state
tetravalent
trivalent
unipolar
valence band

•—OBJECTIVES
After studying this chapter, you should be able to:
- Classify materials based on their relative conductivity.
- Make a sketch of a simple atom and label its key parts.
- Describe several important characteristics regarding the behavior and movement of valence and free electrons.
- Describe the basic characteristics of a semiconductor crystal.
- Describe the effects of doping on a semiconductor material, including characteristics of the dopant and characteristics of the resulting semiconductor.
- Name and describe two ways that current can flow through a semiconductor material.
- Contrast and explain the characteristics of a pn junction under conditions of no bias, forward-bias, and reverse-bias.
- Explain what happens when a semiconductor junction experiences reverse-voltage breakdown.
- Describe how a semiconductor junction is affected by changes in operating temperature.
- Describe several factors that must be considered when troubleshooting semiconductor devices including the following: electrical, mechanical, and soldering/desoldering techniques.

•—INTRODUCTION
This chapter introduces you to semiconductor technology. The operation of all computers is strongly based on semiconductor devices. These devices include such things as diodes, transistors, microprocessors, and an endless array of integrated circuits. The material presented in this chapter will help you understand the operation of all semiconductor components.

8.1 BASIC ATOMIC THEORY

It is not necessary for you to have a detailed understanding of the atomic activity that occurs within semiconductor devices (like transistors) to understand their operation and to effectively utilize them in practical computer applications. It is beneficial, however, to develop a simplified, visual picture of the internal activity. This picture can make the operation of the device seem more natural and logical, and it makes learning the material easier.

Classifications of Material

There are many ways that materials can be classified. One obvious way is to categorize materials into solid, liquid, or gaseous states. All of the devices presented in the remainder of this text are constructed from materials that are solid (i.e., not liquid or gaseous). For this reason, they are often called **solid-state** devices. Most of the solid-state devices that we will discuss had vacuum or gas tube ancestors in which electrons were made to flow through a vacuum or gas-filled region.

There are many other ways to categorize materials such as by electrical conductivity, density, color, hardness, strength, resiliency, composition, chemical activity, and many others. For our immediate purpose of understanding the operation of solid-state devices, we shall restrict our comparisons to electrical conductivity. As discussed in chapter 2, all materials can be roughly categorized into four classes based on their conductivity: insulators, conductors, semiconductors, and superconductors. Insulator, conductor, and semiconductor materials are used in the manufacture of practical solid-state devices such as microprocessors for computers.

Recall that the relative conductivity of a semiconductor material falls between the conductivities of conductors and insulators. That is, a semiconductor has less resistance than an insulator material but more resistance than a conductor having similar dimensions. The most common semiconductor materials are germanium, silicon, and carbon. Silicon is the most widely used semiconductor material in the manufacture of solid-state devices.

Review of Basic Atomic Model

Figure 8–1 illustrates a common model for a single atom. We picture the atom as being composed of a central nucleus densely packed with protons and neutrons, which is surrounded by orbiting electrons. The protons have a positive charge and the neutrons are neutral (no charge), which means that the nucleus will have a net positive charge.

As shown in Figure 8–1(a), negatively-charged electrons orbit around the nucleus. The number of orbits and the number of electrons in a given orbit vary with the type of material being considered. Nevertheless, all materials are considered to

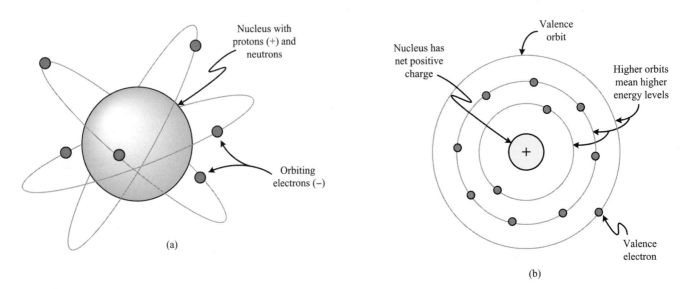

FIGURE 8–1 The structure of a basic atom is (a) three dimensional but is often shown as (b) a two-dimensional sketch.

have orbital electrons. It is convenient to think of the electrons as being held in a stable orbit by the balance between the centrifugal force of the orbiting electrons and the attractive force of the positively-charged nucleus.

The orbits of the electrons are three-dimensional as illustrated in Figure 8–1(a), but for simplicity in drawing, the orbits are often shown like those in Figure 8–1(b) as two-dimensional. In the two-dimensional view, it is clear that the electrons have definite, nonoverlapping orbits. It is the energy content of a particular electron that determines the specific orbit. In general, the higher the energy content of an electron, the higher (i.e., greater altitude from the nucleus) the orbit.

You may recall from earlier studies that if an electron in a given orbit acquires additional energy (perhaps by absorbing external heat or light energy), it will move to a higher orbit. Furthermore, it must either remain in its initial orbit or move completely to a higher orbit. For example, the various orbits pictured in Figure 8–1(b) have regions between them where no electrons can orbit. It takes a definite amount of energy absorption to move to a higher level. Likewise, electrons must give up a definite amount of energy to return to a lower orbit.

You may also recall that the outermost orbit is called the **valence band** or valence orbit of the atom. The electrons in this outer orbit are called valence electrons. These are the electrons that take part in chemical reactions, that bond with other atoms, and that can be most easily freed to produce current flow in a conductor.

Energy Levels

Figure 8–2 shows a diagram that further clarifies the relationship between the altitude (i.e., radius) of the electron orbit and the energy content of the electron. The valence-band electrons are farthest from the nucleus and have higher energy levels than the electrons in the lower orbits. The electrons in all orbits, up to and including the valence-band orbit, are closely associated with a particular atom. All of the electrons will remain in orbit unless they are influenced by an external energy source.

As shown in Figure 8–2, the region beyond the valence band is called the **conduction band**. You can view this region as simply another electron orbit. However,

FIGURE 8–2 The size or altitude of an electron orbit is directly related to the energy content of the electron.

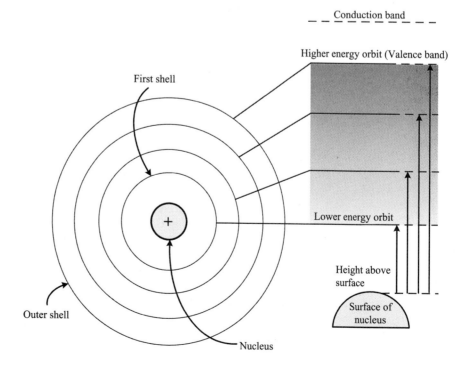

since it is so far away from the influence of the positive nucleus, and since the only electrons in this band have substantial energy, the electrons in the conduction band have little or no association with the parent atom. That is, if an electron in the conduction band acquires even a slight bit of additional energy, then it may well escape the pull of the nucleus and become a free electron. You will recall from your earlier studies that the movement of free electrons is what we call current flow.

As you will see in the next section, current flow in semiconductors occurs in two regions: the conduction band and the valence band. The conduction band current flow in a semiconductor is similar to current flow in a conductor. The valence-band current, though, is an important new concept that is unique to semiconductors.

✓ **Exercise Problems 8.1**

1. Materials with extremely high resistance are called _____.
2. Materials that have resistances that fall between those of conductors and insulators are called _____.
3. Electrons are negatively-charged particles. (True/False)
4. The nucleus of an atom has a net negative charge. (True/False)
5. A valence-band electron has more energy than an inner-orbit electron. (True/False)
6. If a valence-band electron absorbs energy, it can move into the _____ band.
7. In a semiconductor, current flow occurs in both the _____ band and the _____ band.

8.2 SEMICONDUCTOR THEORY

We are now ready to narrow our discussion of materials to the behavior of semiconductor materials. We will briefly examine the atomic structure of a semiconductor to see how it is altered to achieve the desired electrical properties. We will also see how current flow occurs in a semiconductor material. This is an important step toward understanding the operation of transistors and other solid-state devices.

Intrinsic Semiconductors

Silicon, germanium, and gallium arsenide are the primary materials most often used in the manufacture of transistors and other solid-state devices. Silicon and germanium are elements. The pure forms of these elements are called **intrinsic semiconductors**. In this pure form, the materials do not exhibit the electrical characteristics needed for practical solid-state devices. Nevertheless, an understanding of their structure and behavior is an important building block.

Isolated Semiconductor Atoms

Figure 8–3 shows the atomic structure of both silicon and germanium atoms. Both of the atoms illustrated in Figure 8–3 are electrically neutral. That is, each atom has the same number of protons (+) as orbiting electrons (−). Each silicon atom has fourteen protons in the nucleus and fourteen orbiting electrons (two in the first orbit, eight in the second orbit, and four in the valence-band orbit). By contrast, each germanium atom has thirty-two protons in the nucleus that are balanced by thirty-two orbiting electrons (two in the lowest orbit, eight in the second orbit, eighteen in the third orbit, and four in the valence-band orbit). Note that both silicon and germanium have four valence-band electrons. For this reason they are called **tetravalent** atoms. This is an important characteristic of semiconductor atoms.

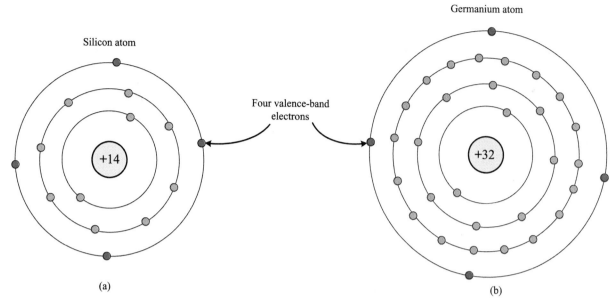

FIGURE 8–3 Both (a) silicon atoms and (b) germanium atoms have four valence-band electrons in their pure form.

Semiconductor Crystals

Tetravalent atoms in materials like silicon, gallium arsenide, and germanium bond together in a special way to form a geometric pattern of atoms called a **crystal** or crystal lattice. You may be familiar with diamond crystals that are formed by carbon (another semiconductor material) atoms. Figure 8–4 shows a two-dimensional sketch of a semiconductor crystal.

Figure 8–4 shows a central atom surrounded by four neighboring atoms. Each of the five atoms shown has four valence-band electrons. For simplicity, the single

FIGURE 8–4 Semiconductor atoms combine to form a regular geometric pattern called a crystal lattice.

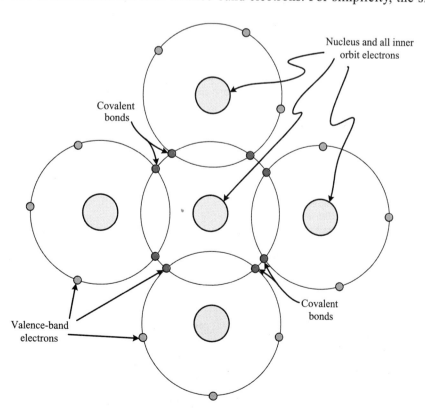

central region drawn for each atom represents all lower-orbit electrons and its nucleus. Valence electrons in the orbit of one semiconductor atom have very nearly the same energy level as valence electrons in a neighboring atom. Thus, with minimal changes in energy, a valence-band electron could begin orbiting in the valence band of a neighboring atom. For reasons beyond the scope of this text, each semiconductor atom positions itself between four other atoms as pictured in Figure 8–4. Each valence-band electron is shared with one of the neighboring atoms. Thus, the central atom shown in Figure 8–4 effectively has eight valence-band electrons (i.e., four of its own and four from neighboring atoms). This sharing of valence electrons is called **covalent bonding** and results in a very tightly bound structure. You may recall from earlier studies that atoms with eight valence-band electrons are very stable and chemically inactive. (Inert gases are examples of this condition.) In the case of semiconductor materials, the covalent bonding makes it more difficult to get conduction-band electrons. That is, pure semiconductor crystal is a good insulator (ignoring the effects of thermal energy, which are present even at room temperature).

Now, for a moment, try to visualize a three-dimensional view of the crystal structure represented in Figure 8–4. It may help to visualize each atom as a cube with a valence electron at each of its four corners. As you mentally progress through the crystal, you will see a very definite and repetitive geometric pattern of atoms. Effectively, each atom has eight valence-band electrons. Any given valence electron is being shared with another atom (covalent bonding). When you reach the edges of the crystalline solid, the outermost row of atoms will have fewer than eight valence-band electrons, since there are no neighboring atoms to establish a covalent bond. This imperfection in the crystal structure contributes to nonideal characteristics of solid-state devices. It is very important to realize the individual atoms in the crystal are immobile. The covalent bonds hold the atoms in a fixed location within the crystal. We will refer to this point in a later section.

Electron Distribution

Now let us briefly examine the distribution of electrons in a semiconductor crystal at two temperatures: absolute zero and room temperature. Figure 8–5 shows energy-band drawings for each of these cases.

In Figure 8–5(a), all electrons are at their lowest energy level, since there is no external energy added at absolute zero. Notice that the valence band is full and the conduction band is empty. This means the material will act like an insulator, since there are no free electrons to participate in current flow.

Now consider the room temperature case illustrated in Figure 8–5(b). As shown, two of the valence-band electrons have absorbed enough thermal energy to break their covalent bonds and move to the conduction band (as essentially free electrons). Each of the broken covalent bonds represents an orbital vacancy where

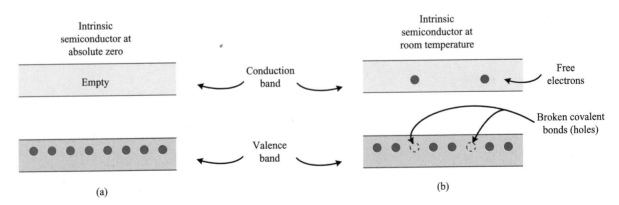

FIGURE 8–5 Electron distribution into valence and conduction bands at (a) absolute zero and (b) room temperature.

an electron could reside. But since no electron is there, the broken covalent bond is called a **hole**. It is important to note that there is a one-to-one relationship between the conduction-band electrons and the valence-band holes in Figure 8–5(b). This is always true of intrinsic semiconductor material; for every electron in the conduction band, there is a corresponding hole in the valence band. We call them **electron-hole pairs** (EHPs).

Comparison of Figures 8–5(a) and 8–5(b) will also reveal another important characteristic of intrinsic semiconductors. In Figure 8–5(a), there are no conduction-band electrons, since the crystal is void of external energy. As the temperature increases, however, more and more valence-band electrons will gain enough energy to break their covalent bonds and move to the conduction band. The greater the number of conduction-band electrons, the easier current will flow through the material (i.e., more free electrons). Thus, we can conclude that the resistance of intrinsic semiconductor material decreases with increasing temperature. We say it has a negative temperature coefficient. Figure 8–6 further illustrates this important characteristic.

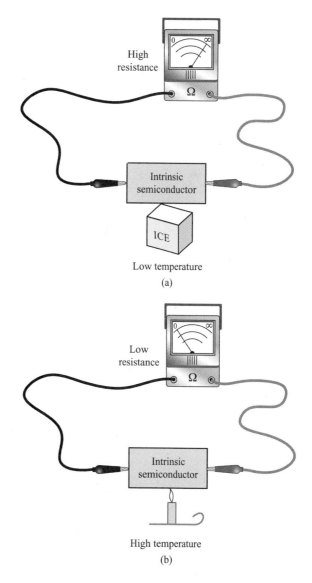

FIGURE 8–6 Intrinsic semiconductor materials have a negative temperature coefficient.

Semiconductor Doping

To produce the electrical characteristics that are required for a practical transistor or other solid-state device, we must add impurities to the intrinsic semiconductor material. The process of adding impurities is called **doping**. The resulting material is called **extrinsic semiconductor** material since it is no longer pure.

There are two primary classes of impurities that are used to dope semiconductors. The atoms of one class have three valence electrons and are called **trivalent** atoms. Examples include aluminum, gallium, and boron. The other class is called **pentavalent** since each atom has five outer-orbit electrons. Examples of pentavalent impurities include antimony, arsenic, and phosphorous. These two types of atoms are represented in Figure 8–7. Both examples shown in Figure 8–7 have essentially the same atomic structure as silicon, but have one fewer (trivalent) or one more (pentavalent) valence electron.

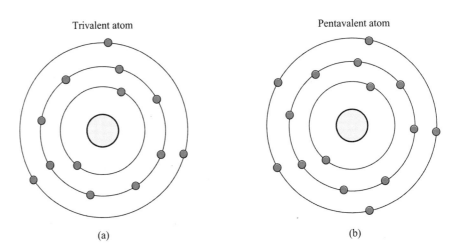

FIGURE 8–7 (a) Trivalent and (b) pentavalent atoms are used as dopants in the manufacture of practical semiconductor materials.

In either case, the impurity atoms displace some of the semiconductor atoms within the crystal structure. Each impurity atom technically represents a "defect" in the crystal structure, but the presence of the crystal defect produces desirable characteristics. The amount of doping varies with the type of semiconductor device being manufactured; levels on the order of one impurity atom for every ten million silicon atoms might be representative. Now let's examine how the two different types of impurities affect the crystal structure and the electrical characteristics of the semiconductor.

Trivalent Doping
Silicon is currently the most widely used semiconductor material. For this and subsequent discussions throughout this text, we shall focus our discussions on silicon unless specifically noted. Figure 8–8 illustrates a crystal lattice composed mostly of silicon atoms. However, a few trivalent atoms have been inserted into the crystal. You will recall that each silicon atom in an intrinsic crystal forms covalent bonds with four neighboring atoms. Figure 8–8 shows that the trivalent impurity atom is only able to form covalent bonds with three neighboring atoms. The missing bond effectively creates a hole in the valence band. As you will see in a later section, these broken covalent bonds or holes in the valence band provide a mechanism for current to flow through the crystal structure. Every trivalent impurity atom contributes one valence-band hole to the crystal. It is particularly important to understand, however, that each silicon atom and each trivalent atom have just as many protons as orbiting electrons. That is, none of the atoms have any net charge.

Since a hole represents a broken covalent bond, it is possible that a conduction-band electron will recombine with a given hole. The recombined electron then orbits the impurity atom. Because each trivalent impurity atom has the ability to accept an additional electron (via recombination), it is sometimes called an **acceptor atom**.

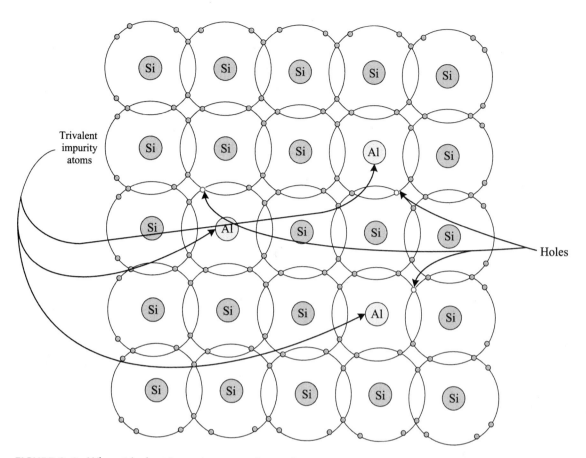

FIGURE 8–8 When trivalent impurity atoms (e.g., aluminum) are added to a silicon crystal, holes are produced in the valence band.

Once a silicon crystal has been doped with trivalent impurity atoms, it is called **p-type** semiconductor material. Figure 8–9 illustrates the conversion of intrinsic silicon into p-type silicon.

FIGURE 8–9 Intrinsic (pure) silicon is converted into p-type semiconductor material by doping it with trivalent impurity atoms.

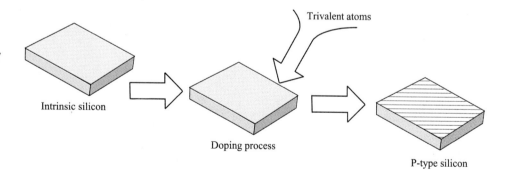

Pentavalent Doping

Figure 8–10 on page 270 shows the results of doping a silicon crystal with pentavalent atoms (atoms with five valence-band electrons). Here, each pentavalent atom forms covalent bonds with four neighboring silicon atoms just as in an intrinsic crystal. This effectively fills the outer orbit of the impurity atom (maximum of eight electrons permitted). The remaining electron is excluded from its valence-band orbit and must reside in the conduction band. Each pentavalent atom contributes one electron to the conduction band. For this reason, pentavalent atoms are sometimes called **donor atoms**. Be sure to remember, though, that neither the

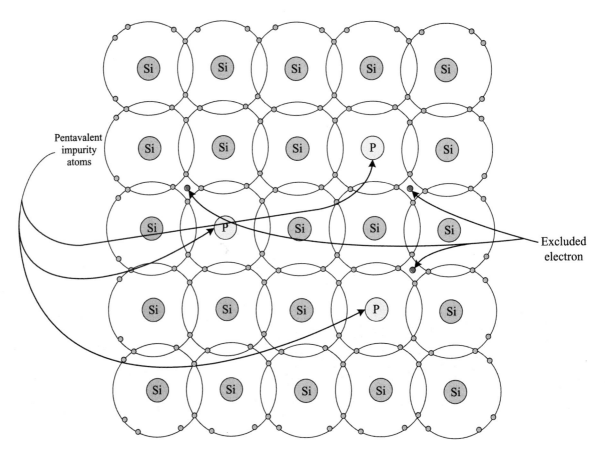

FIGURE 8–10 When pentavalent atoms are added to a silicon crystal, electrons are produced in the conduction band.

silicon nor the pentavalent atoms have any net charge. That is, they both have equal numbers of protons and electrons.

Once a silicon crystal has been doped with pentavalent impurity atoms, it is called **n-type** semiconductor material. Figure 8–11 illustrates the conversion of intrinsic silicon into n-type silicon.

FIGURE 8–11 Intrinsic (pure) silicon is converted into n-type semiconductor material by doping it with pentavalent impurity atoms.

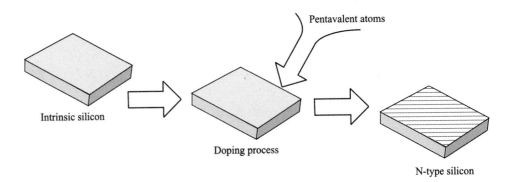

Energy Levels

Many people find the discussion of energy levels to be too abstract or detailed to have any practical value. While it is unlikely that you will ever have the opportunity to apply your knowledge of energy levels to the solution of a practical problem, the knowledge is still valuable to you. This is especially true while you are learning how solid-state devices operate. We will minimize our discussion of energy levels, but you should realize that they will provide you with substantially greater insight into semiconductor operation.

We previously examined the energy levels of an intrinsic semiconductor crystal. You will recall that at absolute zero there were no conduction-band electrons and no valence-band holes. Further, at higher temperatures there were the same number of conduction-band electrons as there were valence-band holes (i.e., EHPs only). Now let's compare these to the energy levels in a doped crystal.

Figure 8–12(a) illustrates the energy distribution of a p-type semiconductor at absolute zero. All electrons are at their lowest possible energy levels. As you can see, there are no conduction-band electrons, but there are some holes in the valence band. The holes exist because there are simply not enough electrons available to form complete covalent bonds. Recall that each trivalent impurity atom produces a shortage of one electron.

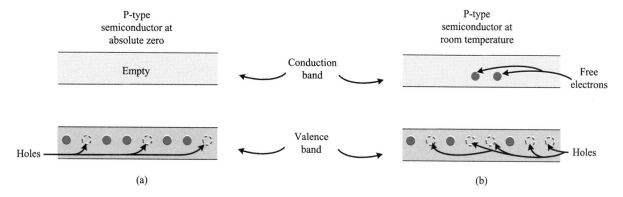

FIGURE 8–12 Energy levels in a p-type crystal at (a) absolute zero and at (b) room temperature.

Figure 8–12(b) shows the energy levels in a p-type crystal at room temperature. As the electrons absorb external (thermal) energy, some of them break their covalent bonds and move to higher conduction-band orbits. This, of course, creates a hole in the valence band where the covalent bond was broken. Due to the presence of the trivalent impurity atoms, however, there will always be more holes in the valence band than electrons in the conduction band.

Figure 8–13(a) shows the energy distribution of an n-type semiconductor at absolute zero. All electrons are at their lowest possible energy levels. As you can see, there are no valence-band holes, but there are still some conduction-band electrons. There will be one excluded electron for each impurity atom present in the crystal. The conduction-band electrons exist because they were excluded from covalent bonding and had to take a higher (conduction-band) orbit.

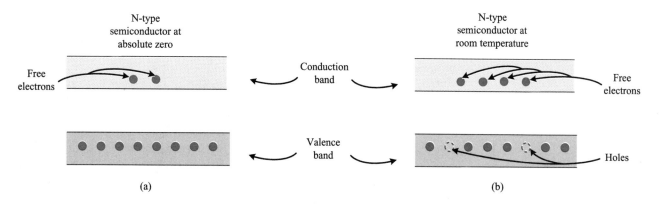

FIGURE 8–13 Energy levels in an n-type crystal at (a) absolute zero and (b) at room temperature.

Figure 8–13(b) shows the energy levels in an n-type crystal at room temperature. As the electrons absorb external (thermal) energy, some of them break their covalent bonds and move to higher conduction-band orbits. This, of course, creates a hole in the valence band where the covalent bond was broken. Due to the presence of the pentavalent impurity atoms, however, there will always be more electrons in the conduction band than holes in the valence band.

Current Flow in a Semiconductor

We are now in a position to discuss how current flows through a doped semiconductor crystal. We shall connect a voltage across the crystal and study the mechanism of current flow within the crystal structure. External to the crystal, it makes no difference whether we are discussing p- or n-type material. That is, electrons leave the negative side of the voltage source, flow into and through the crystal, exit on the opposite end, and continue to the positive terminal of the external voltage source. This simple process is illustrated in Figure 8–14.

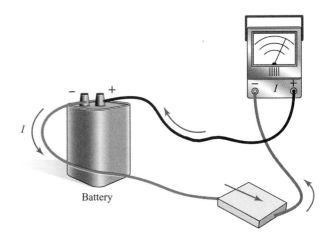

FIGURE 8–14 Electron current can flow through a doped semiconductor crystal regardless of its type (p or n).

Although electrons can enter, pass through, and leave both p- and n-type materials, the mechanism for current flow inside of the semiconductor material is radically different in the two types of materials.

Current Flow through N-Type Material

You will recall that n-type material has many conduction-band electrons as a result of the pentavalent dopant. A conduction-band electron is essentially a free electron, since it is so loosely bound by the atom. If a voltage is connected across the crystal, then the free electrons will move toward the positive potential. Every time an electron leaves its conduction-band orbit and moves toward the positive battery potential, the parent atom takes on a positive charge. Remember, the atoms were initially neutral since they had equal numbers of electrons and protons. Free electrons entering the crystal from the negative terminal of the voltage source are attracted to the positively-charged atoms in the crystal.

To summarize, let's imagine what it would be like to be inside of an n-type crystal before and after an external voltage source was connected across it. With no external voltage applied, we would see a very regular pattern of atoms. They are fixed in their relative positions. There are many free electrons whizzing by, but their motion is generally random. They will orbit an impurity atom for a while, fly out of orbit, and eventually begin orbiting another impurity atom. For the most part, all of the other electrons are participating in covalent bonds and do not wander throughout the crystal. Occasionally, we see a valence-band electron break its covalent bond (due to thermal energy) and join the free electrons in the crystal. Similarly, on occasion the path of a free electron will encounter one of the valence-band holes that were created due to thermal energy. When this happens, we see recombination occurring. The hole disappears, and the free electron becomes part of the covalent bond.

Now, when a voltage is connected across the crystal, it begins to look like we are in a wind tunnel with a bunch of Ping-Pong balls. All the free electrons that were previously moving in random directions are now rushing toward the positive end of the crystal. When they reach the positive end, they move into the connecting wire and head for the positive source terminal. At the negative end of the crystal, we see an inrush of electrons coming into the crystal from the wire leading to the negative terminal of the voltage source.

Current Flow through P-Type Material

Current flow in a p-type semiconductor is substantially different from that described for n-type crystals. Recall that in p-type material there are more holes in the valence band than there are electrons in the conduction band. An electron in the valence band can move to another location in the valence band with only minimal changes of energy required. That is, if a hole exists at a particular point in the crystal, then a nearby valence electron can break its covalent bond and move into the hole without significant energy changes.

If a voltage is connected across the crystal, the following activity occurs. Electrons enter the crystal from the negative voltage terminal. Upon entering the crystal, the free electrons encounter an abundance of holes. In most cases, the electrons recombine with the holes and become valence-band electrons. As valence-band electrons, they can readily move from one atom to the next (i.e., from one hole to the next as a new hole becomes available).

Clearly, the primary current flow in a p-type crystal consists of electrons moving in the valence band. Nevertheless, it is customary to speak in terms of hole flow. Figure 8–15 clarifies the concept of hole flow.

Figure 8–15(a) shows a row of blocks. The positions are labeled 1 through 8. The block at position 2 is missing. We shall refer to a position with no block as a hole. In the case of Figure 8–15(a), there is a hole in position 2. Now, if the block in position 3 is moved to position 2, there will be a hole in position 3 (shown in

FIGURE 8–15 When electrons move from hole to hole in the valence band of a crystal, the holes effectively move in the opposite direction.

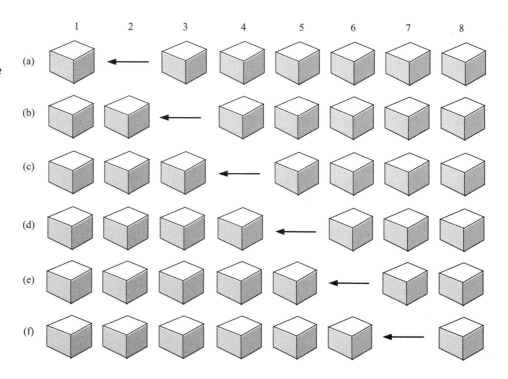

Figure 8–15[b]). We could say that the hole has moved from position 2 to position 3. In Figures 8–15(c) through 8–15(f), blocks are moved to the left. As each block is moved, the hole appears to be moving to the right.

The demonstration illustrated in Figure 8–15 is similar to the concept of hole flow in a semiconductor. Holes in the valence band are not physical things. They are simply the absence of a complete covalent bond. As valence-band electrons move from one covalent bond to another, the holes move in the opposite direction. So, even though it is common to say that current in a p-type semiconductor consists of hole flow that moves from positive to negative, never forget that the current flow can also be thought of as electron flow. The electron flow in p-type material occurs in the valence band, whereas electron movement in n-type material occurs in the conduction band.

Majority versus Minority Current Carriers

We have seen that the primary mechanism for current flow is conduction-band electrons in an n-type crystal and valence-band holes in a p-type crystal. For this reason, we say that conduction-band electrons are the **majority carriers** in n-type semiconductors, while holes are the majority carriers in p-type material. Valence-band holes and conduction-band electrons are **minority carriers** for n- and p-type material, respectively. A device whose operation depends upon only one type of current carrier is called a **unipolar** device. If a component's operation relies on both holes and conduction-band electrons, it is called a **bipolar** device.

You will recall that at temperatures above absolute zero n-type material also had holes in the valence band due to thermally-produced electron-hole pairs, but the number of holes was small compared to the number of conduction-band electrons. A small part of the total n-type semiconductor current flows via holes in the valence band. Therefore, we say that holes are the minority current carriers in an n-type semiconductor. Similarly, although p-type material has an abundance of holes, there are some thermally-produced conduction-band electrons that can contribute to the overall current flow. We call the conduction-band electrons minority current carriers when they are in p-type material.

Since minority current carriers exist largely due to the absorption of thermal energy (i.e., more broken bonds at higher temperatures), we would expect minority current flow to be greater at higher temperatures. This is one of the reasons that semiconductor devices are sensitive to temperature.

✓ Exercise Problems 8.2

1. Another name for pure silicon is extrinsic silicon. (True/False)
2. At absolute zero, all of the atoms in the central region of an intrinsic silicon crystal have complete outer orbits (i.e., eight valence electrons). (True/False)
3. At room temperature, an intrinsic silicon crystal has the same number of conduction-band electrons as valence-band holes. (True/False)
4. When two electrons are effectively shared between two atoms, a _____ _____ is formed.
5. A broken covalent bond is called a _____.
6. Pentavalent atoms are sometimes called donor atoms. (True/False)
7. When acceptor atoms are added to intrinsic silicon, p-type semiconductor material is formed. (True/False)
8. The primary path for current flow through an n-type semiconductor is in the valence band. (True/False)
9. The primary path for current flow through a p-type semiconductor is in the valence band. (True/False)
10. The minority current carriers in n-type semiconductor material are _____ in the _____ band.
11. The minority current carriers in p-type semiconductor material are _____ in the _____ band.

8.3 SEMICONDUCTOR JUNCTIONS

The preceding section introduced many new terms and concepts that will be important to our overall understanding of semiconductor operation. However, the intrinsic semiconductor and the p- and n-type materials as described have limited use in the manufacture of practical solid-state devices. In fact, doped silicon (as discussed so far) is little more than an expensive resistor. That is, the heavier the doping level, the more current carriers there are and the lower the resistance of the material. Now, let us expand the use of these materials into more practical semiconductor devices.

The operation of practical semiconductor devices like transistors is made possible when one region of a *single* silicon crystal is doped to form p-type material and another region is doped to form n-type material. The point in the crystal where the p- and n-type materials join is called a **pn junction**. A device having only a single pn junction is called a **diode**. Figure 8–16 shows the construction of a pn junction.

Although pn junctions are often discussed (and illustrated) as though they had abrupt transitions from p- to n-type material, there is actually a more gradual transition. We are now ready to examine the region near the pn junction more closely.

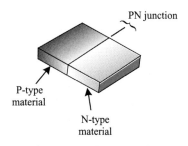

FIGURE 8–16 When p-type material meets n-type material within a single silicon crystal, a pn junction is formed.

Unbiased Junction

In practice, the pn junction is automatically formed in the process of creating the semiconductor device. For our immediate purposes, however, let us assume that a pn junction has somehow been created as an abrupt line. The material is purely n-type on one side of the junction and purely p-type on the other side. Figure 8–17 illustrates such a crystal.

FIGURE 8–17 An abrupt junction of p- and n-type semiconductor material within a single crystal.

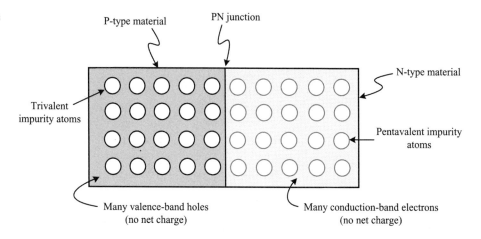

For simplicity in Figure 8–17, only the impurity atoms are represented. Remember, however, that there are many more silicon atoms than impurity atoms.

Before Carrier Migration

On the p-side of the pn junction shown in Figure 8–17, the trivalent impurity atoms provide an abundance of holes in the valence band. On the n-side of the junction, the pentavalent doping creates an abundance of conduction-band electrons. On both sides of the junction, however, there is no net charge, since there are just as many protons (+) as electrons (−).

Carrier Migration

At any temperature above absolute zero, some of the conduction-band electrons on the n-side of the junction will have enough energy to migrate (diffuse) across the pn junction. When the conduction-band electrons enter the p-type region, they encounter numerous holes. The probability that a migrating electron will encounter a hole is very high since the p-region has so many holes. When their paths cross, **recombination** occurs. Recombination of a migrated conduction-band electron (originally from the n-side) and a valence-band hole (on the p-side) produces several effects:

- The conduction-band electron disappears.
- The valence-band hole disappears.
- The trivalent impurity atoms in the p-type crystal take on a net negative charge after recombination of the EHP.
- The pentavalent impurity atoms in the n-type crystal take on a net positive charge after recombination of the EHP.

Figure 8–18 illustrates carrier migration near the pn junction.

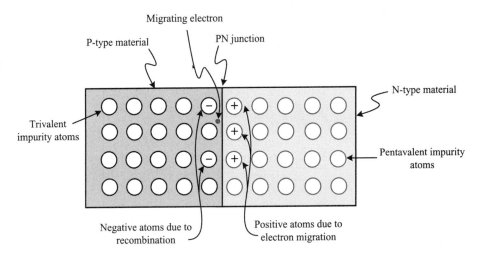

FIGURE 8–18 Conduction-band electrons on the n-type side migrate into the p-type material and recombine with the abundant holes.

Figure 8–18 shows that the trivalent atoms whose holes have been filled by the migrating electrons take on a negative charge. Recall that the trivalent atoms were initially neutral. Although each trivalent impurity atom had one fewer valence electron, it also had correspondingly fewer protons. After recombination, however, there will be more electrons than protons, giving the trivalent atom an overall negative charge. In a similar manner, the pentavalent atoms in the n-type material take on a net positive charge when the excluded electrons migrate across the junction.

At first, it may seem that the migration process would continue until recombination had eliminated all holes and all conduction-band electrons. However, there is a very important point to remember: The impurity atoms are not mobile; their positions within the crystal are rigidly fixed in the lattice structure. As the electrons migrate and the impurity atoms take on net charges, an electrostatic field begins to form between the fixed charges on the impurity atoms. Each electron that crosses the junction from the n-side and recombines on the p-side increases the strength of the electrostatic field.

Figure 8–19 shows the semiconductor crystal after electron migration has been allowed to continue for some time. The polarity of the electrostatic field is such that it opposes any further electron migration. That is, every electron that crosses the junction and recombines makes it more difficult for other electrons to make similar trips. Eventually, the field becomes strong enough to essentially stop the electron migration.

In the vicinity of the junction, there are neither conduction-band electrons nor valence-band holes. They have both been eliminated through recombination. Of

FIGURE 8–19 The diffusion of electrons from the n-type material and elimination of holes in the p-type material creates a barrier potential.

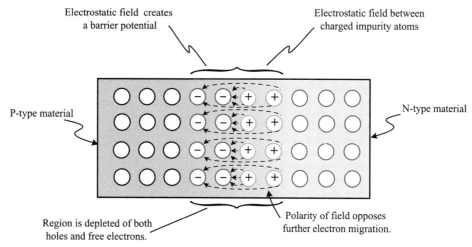

course, there may be a few thermally-produced EHPs, but these are of minor concern in our present discussion. Since the area has been depleted of current carriers (holes and electrons), we label it the **depletion region**. Can you see that the portion of the crystal that contains the depletion region has been effectively converted into *intrinsic* silicon? That is, the only current carriers that exist in the depletion region are EHPs produced by absorption of thermal energy. The difference, of course, is that the atoms in the depletion region have a net charge, whereas intrinsic silicon atoms are neutral.

The charges on the fixed-impurity atoms produce an effective difference of potential across the junction. It is this difference of potential that stops further electron migration. We label this charge difference the **barrier potential**. Barrier potentials in pn junctions range from 0.1 V to 1.5 V depending on the materials used. In the case of silicon devices, the barrier potential is on the order of 0.6 V to 0.7 V (0.2 V to 0.3 V for germanium). These are important values, and they provide us with powerful troubleshooting tools.

Forward-Biased Junction

If we connect an external voltage source to a pn junction, we can either aid or oppose the barrier potential. If the positive side of the external voltage source is connected to the p-type material and the negative side is connected to the n-type material, then it will be in opposition to the barrier potential. We call this polarity **forward bias**. A forward-biased pn junction is illustrated in Figure 8–20.

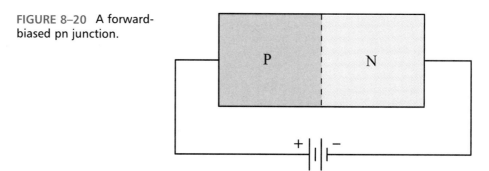

FIGURE 8–20 A forward-biased pn junction.

Forward Bias Overcomes Barrier Potential

You will recall that it was the electrostatic field associated with the barrier potential that stopped electron migration through the depletion region. As shown in Figure 8–21, the electrostatic field produced by the external voltage source has a polarity opposite the field of the barrier potential. If the externally-produced field exceeds 0.7 volts (nominal barrier potential), then the effects of the barrier potential will be overcome, and the electrons can once again migrate across the junction.

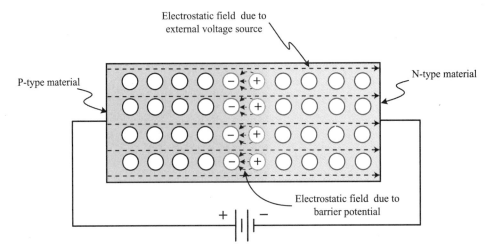

FIGURE 8–21 Forward bias sets up an electrostatic field within the crystal that opposes and ultimately overcomes the barrier potential.

Forward Bias Narrows the Depletion Region

When forward bias is applied to a pn junction, the majority carriers in both types of materials are pushed toward each other (i.e., toward the junction). A conduction-band electron crossing the junction under these conditions has a very short life-time. That is, since the valence-band holes in the p-type material are concentrated near the junction, there is a good chance that a migrating electron will recombine shortly after crossing the pn junction. This means that the effective width of the depletion layer will be less with forward bias than when the junction was unbiased, since the migrating electrons don't have to travel as far into the p-region to encounter a hole.

Maximum Current Flow with Forward Bias

Recall that it was the electrostatic field of the depletion layer that ultimately stopped the diffusion across the junction in an unbiased crystal. Since forward bias opposes and overcomes the barrier potential, diffusion can continue without significant opposition. With forward bias, an ideal pn junction may be thought of as a short circuit.

The activity within the crystal under conditions of forward bias is illustrated in Figure 8–22 and is explained as follows. First, there is an abundance of electrons on the negative terminal of the external voltage source. This negative potential repels many of the conduction-band electrons in the n-type material toward the pn junction. As the conduction-band electrons move toward the junction, they are replaced by other electrons that enter the crystal from the wire connected to the negative supply terminal.

Since the effects of the barrier potential have been overcome by the forward bias, the conduction band electrons that are being pushed toward the junction can continue their path across the pn junction. As they enter the "p" region of the crystal, they encounter many holes and, in most cases, recombine to become valence electrons. As valence electrons, they can continue toward the positive supply terminal by moving from hole-to-hole as valence-band current. When the electrons reach the positive end of the crystal, they are swept out of the crystal by the positive potential of the external voltage source.

In summary, current flows easily through a forward-biased pn junction. It flows as conduction-band current on the n-side of the junction and as valence-band current on the p-side of the junction.

FIGURE 8–22 Current flows easily through a forward-biased pn junction.

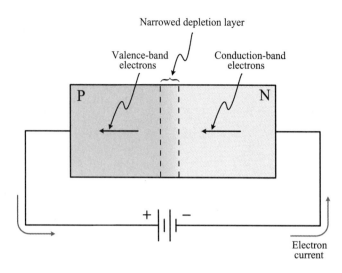

Reverse-Biased Junction

If the positive side of an external voltage source is connected to the n-type material and the negative side is connected to the p-type material in a pn junction, then the external voltage will aid the barrier potential. We call this polarity **reverse bias**. A reverse-biased pn junction is illustrated in Figure 8–23.

FIGURE 8–23 A reverse-biased pn junction.

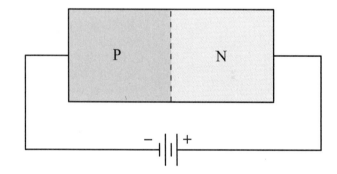

Reverse Bias Strengthens the Barrier Potential

As you know, it is the electrostatic field associated with the barrier potential that stops electron migration through the depletion region. As shown in Figure 8–24, the electrostatic field produced by the external voltage source has a polarity that aids the field of the barrier potential. The barrier potential will increase until it equals the value of reverse bias supplied by the external voltage source.

FIGURE 8–24 Reverse bias sets up an electrostatic field within the crystal that aids the barrier potential.

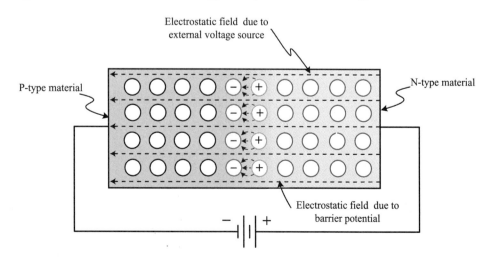

Reverse Bias Widens the Depletion Region

When reverse bias is applied to a pn junction, the majority carriers in both types of materials are pulled away from each other (i.e., away from the junction). This means that if a conduction-band electron does gain enough energy to cross the junction, it will have to travel farther into the p-type crystal before it can recombine. Thus, the effective width of the depletion layer is widened with reverse bias.

Minimum Current Flow with Reverse Bias

Since reverse bias causes increased separation of the current carriers on either side of the junction, we should expect minimal recombination to occur. Recombination, you will recall, was required for a particular electron to pass through the crystal. The net effect of reverse bias is to produce a condition within the crystal that allows very little current flow. With reverse bias, an ideal pn junction can be thought of as an open circuit.

Reverse Current and Junction Breakdown

We know that an ideal pn junction has zero current when it is reverse biased. We also know that a practical pn junction does permit some current to flow in the reverse direction, although the current that flows is very small. However, if the reverse voltage is increased beyond a certain point, then the pn junction loses its ability to stop current flow and permits very large currents to flow. Let's examine each of these **reverse currents** in more detail.

Reverse Current is Temperature Dependent

At temperatures above absolute zero, thermal energy continually generates electron-hole pairs in the crystal. Once the EHP appears, the hole and electron are free to contribute to the total current flow in the crystal. That is, the conduction-band electron will be swept toward the positive terminal of the external voltage source, and the hole will move toward the negative terminal.

Since the EHPs are generated as a function of thermal energy, we should not be surprised to learn that the value of this reverse current is strongly dependent on the temperature of the crystal. Higher temperatures produce more EHPs and consequently higher reverse current.

Avalanche Current

If we increase the reverse bias enough, we will reach a point where the reverse current in the diode increases dramatically. We refer to this as reverse **junction breakdown**. The voltage required to cause junction breakdown is called the reverse breakdown voltage.

Figure 8–25 illustrates what is happening inside the crystal during reverse breakdown. First, an EHP is created due to thermal energy. Because the electrostatic field through the crystal is so strong (i.e., relatively high reverse voltages),

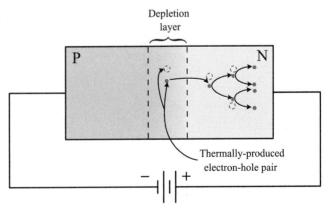

FIGURE 8–25 Accelerating and colliding electrons cause avalanche currents during junction breakdown.

the newly freed electron is rapidly accelerated toward the positive supply terminal. If the accelerating electron collides with a valence-band electron, the valence-band electron can absorb energy and break its covalent bond. As shown in Figure 8–25, there are now two electrons being accelerated toward the positive potential. Each of these electrons can collide with still other valence-band electrons and break their covalent bonds before reaching the positive supply terminal. The net effect is that the current is effectively multiplied to high values. We call this current **avalanche current**.

It is important to realize that reverse junction breakdown does not inherently damage the semiconductor device. However, the resulting avalanche current can cause destructive power dissipation. If the avalanche current is controlled, then reverse breakdown does not destroy the crystal. In chapter 9, we will study a type of diode that is specifically designed to operate in the reverse breakdown region.

Energy Levels in a PN Junction

Let's examine the behavior of a pn junction from the perspective of energy levels. We will consider three conditions: unbiased, forward-biased, and reverse-biased.

Unbiased PN Junction

Figure 8–26 shows the energy band diagram for an unbiased pn junction. Note that the energy levels on the p-side of the junction are slightly higher than the energy levels on the n-side of the junction. This means that conduction-band electrons on the n-side will not be able to enter the p-region unless they acquire additional energy. This is a stable condition and will remain so unless we change the temperature of the crystal or apply an external voltage.

FIGURE 8–26 The energy level diagram for an unbiased semiconductor junction.

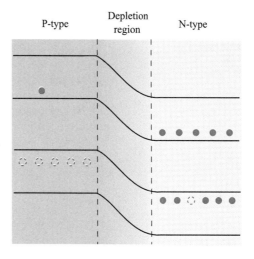

Forward-Biased PN Junction

Figure 8–27 on page 282 illustrates the energy levels within a crystal under conditions of forward bias. When the negative terminal of an external voltage source is connected to the n-type material, the energy levels in the n-type crystal are raised. You might think of this as meaning the electron orbits are higher due to the abundance of electrons (negative charges) in the crystal.

In a similar manner, the connection of a positive potential to the p-type side of the junction causes electrons to exit the crystal. This has the effect of reducing the energy levels within the p-type material.

By inspection of Figure 8–27, it is apparent that the conduction-band electrons on the n-side of the junction will have sufficient energy to cross the pn junction. Once they enter the p-region, they will probably encounter a hole and recombine to

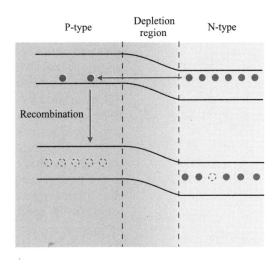

FIGURE 8–27 The energy level diagram for a forward-biased semiconductor junction.

become valence-band electrons. As electrons leave the n-type material, they are replaced by electrons pouring in from the negative supply terminal. Once an electron has crossed the junction, recombined, and made its way through the p-type crystal, it is swept out of the crystal by the attractive force of the positive supply terminal. So, under conditions of forward bias, we see that the energy levels within the crystal permit the flow of current.

Reverse-Biased PN Junction

Figure 8–28 illustrates the energy levels within a crystal during conditions of reverse bias. Clearly, the energy levels in the p-type material (relative to the n-type material) are even higher than in an unbiased junction. Under these conditions, the conduction-band electrons in the n-type material do not have enough energy to cross the depletion region and enter the p-type region. This means there can be no substantial current flow through the pn junction under conditions of reverse bias (unless the breakdown voltage is exceeded).

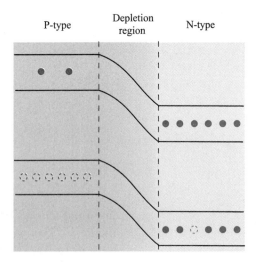

FIGURE 8–28 The energy level diagram for a reverse-biased semiconductor junction.

✓ Exercise Problems 8.3

1. The point in a semiconductor crystal where p- and n-type materials meet is called a _____.
2. A semiconductor device having only a single pn junction is called a _____.
3. The impurity atoms in a semiconductor crystal are highly mobile. (True/False)
4. The area near a pn junction that has neither valence-band holes nor conduction-band electrons is called the _____ _____.

5. The barrier potential for a silicon diode is approximately _____.
6. Current flows easily through a (*forward-/reverse-*) biased pn junction.
7. Forward bias narrows the depletion region. (True/False)
8. An ideal forward-biased pn junction acts like an open. (True/False)
9. Reverse bias on a pn junction aids the barrier potential. (True/False)
10. Under normal conditions, there is very little current through a reverse-biased pn junction. (True/False)
11. Reverse bias causes the depletion layer to become wider. (True/False)
12. If the _____ _____ _____ is exceeded on a reverse-biased pn junction, large (possibly damaging) currents will flow.

8.4 TROUBLESHOOTING SEMICONDUCTORS

Throughout this book we will present techniques and procedures that will help you effectively troubleshoot and repair circuits that use semiconductor devices. In this section, we will examine some general procedures that apply to troubleshooting all types of semiconductor devices.

Mechanical Considerations

Some semiconductor devices (e.g., small diodes) have glass packages. You must use care when removing or replacing this type of device to prevent damage to the component. In particular, avoid stressing the leads near the component body. Figure 8–29 shows how a pair of needle-nose pliers placed near the body can prevent damage while you are forming the component leads. As shown in Figure 8–29, the pliers are used to support the lead. You can then bend the lead over the jaw of the pliers. Use your finger to bend the lead. Don't twist the pliers to make the bend.

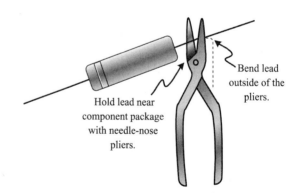

FIGURE 8–29 Needle-nose pliers should be used when bending a component lead.

You must remember that the leads on many solid-state devices are made from relatively small wire. If you bend it repeatedly or make very sharp bends, then the lead itself can break. This can render the component useless, if the broken lead is too short. Therefore, you should minimize the number of times a component lead must be bent. When you do shape a lead, try to make gradual bends. Finally, try to avoid bending the leads immediately adjacent to the package. Bending too near the package tends to make sharper bends that can weaken the lead. Additionally, if the bending is near the package, mechanical stress may damage the package seal.

Figure 8–30 on page 284 illustrates proper and improper lead formation.

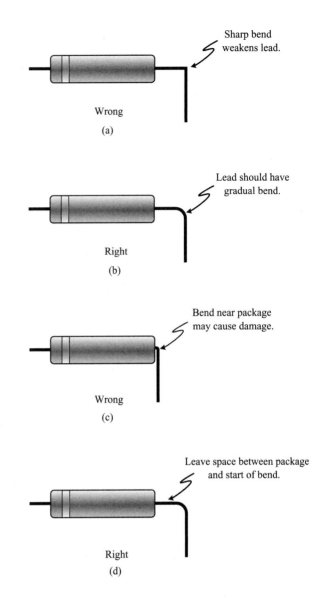

FIGURE 8–30 Do's and don'ts when forming component leads.

Soldering/Desoldering Precautions

Many troubleshooting exercises ultimately require desoldering and soldering of components. Particular care must be used when soldering or desoldering semiconductor devices, since they are temperature sensitive and can be permanently damaged by excessive heat.

Probably the most fundamental precaution is to choose a soldering iron that has the correct wattage rating. Although different powers are required for different tasks, you should generally use as low a power as you can and still obtain good performance.

Figure 8–31 illustrates a method often used to protect sensitive devices from the heat required for proper soldering. Here, a pair of needle-nose pliers is clamped on the component lead and held in place by snugly looping a rubber band over the handles. It is essential to clamp the pliers between the component and the solder joint being heated. As the heat travels up the lead, it is stopped or slowed by the metal jaws of the pliers. The metal pliers act as a heat sink. This protects the heat-sensitive component. Some people use hemostats (available from dental and medical supply houses as well as from electronic distributors) as a heat sink. These are especially handy, since they have a built-in locking mechanism.

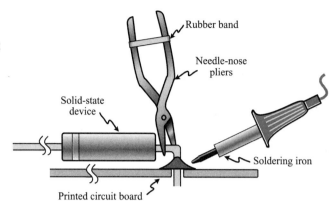

FIGURE 8-31 Pliers clamped between the component and the heat source can act as a heat sink to protect the component.

As an entry-level employee, it is important for you to be aware of another potential problem: bad advice from an experienced employee. For example, there are many technicians and engineers in industry who never use a heat sink when soldering semiconductor devices. These people are likely to profess that heat sinks are no longer needed with modern semiconductors. To be sure, some semiconductor devices are quite rugged and are unlikely to be damaged by soldering. But just as certain, there are other semiconductor components that are extremely sensitive and are almost guaranteed to be damaged if proper precautions are not taken. Further, even if damage does occur, it is not necessarily immediately apparent. It is good practice to employ the methods described previously to ensure that you do not damage semiconductor components as you solder and desolder them.

Electrical Considerations

We will learn additional techniques in later chapters, but we already know enough to understand two important techniques that can be used to locate defective semiconductor devices.

Junction Voltage Measurements

You will recall that a forward-biased silicon pn junction will have approximately 0.7 V across it. A forward-biased germanium pn junction will have about 0.3 V across it. This can be a crucial key for troubleshooting semiconductor devices.

In short, if you measure the voltage across a pn junction and it measures any more than a few tenths of a volt forward bias, then that device is defective. If the junction is reverse biased, it can have almost any value of voltage. But, in the forward-biased condition, if it exceeds a few tenths of a volt, then it is no good. This important characteristic is illustrated in Figure 8-32.

FIGURE 8-32 The voltage drop across a forward-biased pn junction can be used to locate a defective device.

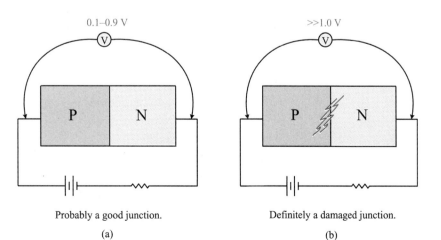

Interpretation of the voltage drop across a forward-biased pn junction can be summarized as follows:

1. If the voltage drop across a forward-biased pn junction is less than about one volt, it may be good or bad, but
2. If the voltage drop across a forward-biased junction is much greater than one volt, it is definitely defective.

These guidelines apply to standard pn junctions. As we will see in later chapters, there are some specialized devices (such as a light-emitting diode) that have greater voltage drops when forward biased.

Ohmmeter Checks

Since a pn junction readily conducts current when it's forward biased but resists current flow when reverse biased, we can use an ohmmeter to determine the condition of a pn junction.

If you are using an analog VOM, simply measure the resistance of the pn junction in both directions. The absolute value of resistance is not as critical as the ratio of the two measurements. The ratio of the resistance in the forward-biased direction to that in the reverse-biased direction is called the **front-to-back ratio**. A good pn junction will have a ratio of at least 1:10 and probably 1:100 or more.

Any one of the following symptoms indicates a defective pn junction:

1. A front-to-back ratio of less than 1:10
2. A short indication (i.e., < 1 ohm) in either direction
3. An open (i.e., infinite ohms) in both directions

Some analog ohmmeters can damage sensitive pn junctions if used on their lowest resistance ranges. It is good practice to start on a higher range and only switch to a lower range if necessary to obtain a useful reading.

A digital ohmmeter (DMM) can also be used to determine the condition of a pn junction, but it requires a different procedure. Most digital meters have a position on their function switch that is specifically designed for testing pn junctions. It is often indicated by the symbol for a diode:

The good/bad indications vary with the specific meter being used. Many meters display a value that is twice the barrier potential. Thus, when forward biased, a good silicon pn junction will indicate about 1.4 on the meter. Consult the operator's manual for your particular meter.

There is one general rule that you should remember when testing pn junctions with an ohmmeter (digital or analog). If the measurements indicate that the pn junction is defective, then it really is defective. On the other hand, if the measurements indicate that the device is good, it might be. You cannot be absolutely certain that a pn junction is good based on tests with an ohmmeter.

Semiconductor Testers

There are several different kinds of test sets that are specifically designed to measure the condition of semiconductor devices. Their operation varies greatly, but it is generally obvious from the labeling of the controls or by reading the user's manual. In general, semiconductor testers provide a more comprehensive evaluation than ohmmeter tests and can generally be relied upon to classify a device as good or bad.

✓ Exercise Problems 8.4

1. When forming semiconductor leads, you should avoid making sharp bends. (True/False)
2. The best way to bend semiconductor leads is to clamp on a pair of pliers and then twist the pliers. (True/False)
3. You should always leave a space between the body of a semiconductor component and the start of a bend in a lead. (True/False)
4. A high-wattage soldering gun is generally recommended for soldering/desoldering most semiconductor devices. (True/False)
5. A _____ _____ can be used between the soldering point and the component body to prevent heat damage during soldering.
6. The voltage across a particular pn junction is 3.5 V when it is reverse biased. Does this mean it is definitely bad?
7. If a pn junction measures 1.6 Ω in both directions, does this mean it is definitely bad?
8. If a forward-biased silicon pn junction measures 2.6 V, does this mean it is definitely bad?

•—SUMMARY

Materials can be classed into four categories based on electrical conductivity: insulators, conductors, semiconductors, and superconductors. Insulators prevent current flow, conductors allow substantial current flow, and superconductors offer no resistance to current flow. Semiconductor materials lie between conductors and insulators with regard to electrical conductivity.

Negatively-charged electrons orbit the dense positively-charged nucleus in an atom. Higher energy electrons have correspondingly higher orbits. Outer-orbit electrons are called valence electrons. Valence-band electrons can acquire energy and escape the influence of the atom. These free electrons are said to be in the conduction band. Current flow in a semiconductor consists of electron movement in both valence and conduction bands.

Silicon is the most widely-used semiconductor and has four valence-band electrons. Covalent bonding between neighboring atoms forms a rigid structure called a crystal lattice. Thermal energy can break some covalent bonds and produce free electrons. Intrinsic (pure) silicon has relatively few free electrons, which gives it a high resistance. For every electron in the conduction band of an intrinsic silicon crystal, there is a corresponding hole (broken covalent bond) in the valence band.

Intrinsic silicon is doped with impurities to form extrinsic silicon. Trivalent dopant atoms make p-type silicon, whereas pentavalent doping results in n-type silicon. N-type material has more conduction-band electrons than valence-band holes. P-type material has more valence-band holes than conduction-band electrons. Electrons and holes are the majority and minority current carriers, respectively, in n-type material. Holes and electrons are the majority and minority current carriers, respectively, for p-type material.

If p- and n-type material are adjacent within a single crystal, a pn junction is formed where the two materials meet. Due to immobile charges within the crystal, a barrier potential forms across the junction. This potential (≈ 0.7 V for silicon) must be overcome by an external voltage before substantial current can flow through the crystal. If an external voltage opposes the barrier potential, it is called forward bias. If an external voltage aids the barrier potential, it is called reverse bias. A pn junction has little opposition to current flow when forward biased but has substantial opposition when reverse biased.

There are no free electrons or holes (except those produced thermally) in the immediate vicinity of an unbiased pn junction. This area is called the depletion region. The width of the depletion region is reduced with forward bias and increased with reverse bias.

The current that flows when a pn junction is reverse biased is very low (ideally zero). A substantial part of this current is a result of thermally-produced current carriers, so it increases with increasing temperature. If the value of reverse-bias voltage is increased enough, the pn junction will eventually break down and lose its ability to resist current flow. If the breakdown voltage of a pn junction is reached, then the current will increase dramatically. This breakdown current is called avalanche current and may be destructive to the semiconductor material.

Care must be exercised when working with semiconductor devices since some are both electrically and mechanically sensitive and easily damaged. They can also be damaged by excessive heat during a soldering operation. The voltage drop across a forward-biased pn junction can be used to judge its electrical condition. If the forward voltage is much more than one volt, then it is defective. An ohmmeter can also help locate a defective pn junction. A good pn junction has a high (> 1:10) front-to-back resistance ratio.

REVIEW QUESTIONS

Section 8.1: Basic Atomic Theory

1. An ideal insulator has _____ resistance.
2. An ideal conductor has _____ resistance.
3. Superconductors have _____ resistance.
4. The resistance of a semiconductor material lies between the resistances of _____ and _____.
5. An atom with an equal number of protons and electrons has no net charge. (True/False)
6. Protons are found in the nucleus of an atom. (True/False)
7. Neutrons are found in orbits around the nucleus of an atom. (True/False)
8. The outermost orbit of an atom is called the valence orbit or valence _____.
9. The electrons of an atom that participate in chemical reactions are located in the _____ band.
10. The altitude of an electron's orbit is directly related to its energy level. (True/False)
11. Electrons in the valence band are considered to be free electrons. (True/False)
12. Electrons in the conduction band are considered to be free electrons. (True/False)
13. Protons have a (*positive/negative*) charge.
14. Electrons have a (*positive/negative*) charge.
15. Neutrons have a negative charge. (True/False)

Section 8.2: Semiconductor Theory

16. A material that contains nothing but silicon atoms is called intrinsic silicon. (True/False)
17. Silicon and germanium atoms are tetravalent atoms. (True/False)
18. Silicon atoms can move easily within a crystal lattice structure. (True/False)
19. What type of atomic bonds are formed between neighboring atoms in a semiconductor crystal?
20. At absolute zero, there are no conduction-band electrons in an intrinsic silicon crystal. (True/False)
21. A broken covalent bond is called a _____.
22. Electron-hole pairs are produced when thermal energy breaks covalent bonds in a semiconductor crystal. (True/False)
23. The process of adding impurities to a pure silicon crystal is called _____.
24. A semiconductor material that has impurity atoms is called an (*intrinsic/extrinsic*) material.
25. An atom with three valence-band electrons is called a (*trivalent/pentavalent*) atom.
26. An atom with five valence-band electrons is called a (*trivalent/pentavalent*) atom.
27. Trivalent atoms are also called (*acceptor/donor*) atoms.
28. Pentavalent atoms are also called (*acceptor/donor*) atoms.
29. If a silicon crystal is doped with trivalent atoms, then the resulting material is called (*p-type/n-type*) silicon.
30. If a silicon crystal is doped with pentavalent atoms, then the resulting material is called (*p-type/n-type*) silicon.
31. Regardless of temperature, p-type material always has more conduction-band electrons than valence-band holes. (True/False)
32. At room temperature, n-type material has more conduction-band electrons than valence-band holes. (True/False)
33. At absolute zero, n-type material has the same number of conduction-band electrons as it has valence-band holes. (True/False)
34. The primary mechanism for current flow in a p-type crystal is hole flow in the valence band. (True/False)
35. Conduction-band electrons are called majority carriers in a (*p-type/n-type*) crystal.
36. Holes are minority carriers in a (*p-type/n-type*) crystal.
37. Operation of a unipolar device depends on both holes and conduction-band electrons to form the total current flow. (True/False)

Section 8.3: Semiconductor Junctions

38. A semiconductor device having only a single pn junction is called a _____.
39. In the depletion region of an unbiased pn junction, there are no conduction-band electrons nor any valence-band holes except those produced by thermal energy. (True/False)
40. If an external voltage is applied to a pn junction that aids the barrier potential, it is called (*forward/reverse*) bias.
41. If an external voltage is applied to a pn junction that opposes the barrier potential, it is called (*forward/reverse*) bias.
42. The depletion layer is (*wider/narrower*) with forward bias.
43. Current flows easily through a forward-biased diode. (True/False)
44. A reverse-biased diode has a wide depletion layer and offers substantial opposition to current flow. (True/False)
45. The current through a reverse-biased diode is temperature dependent. (True/False)
46. Avalanche current can occur in a (*forward-/reverse-*) biased pn junction.
47. Avalanche current, by its very nature, permanently damages the semiconductor material. (True/False)

Section 8.4: Troubleshooting Semiconductors

48. Why do technicians often clamp pliers or other metal tools on the lead between a semiconductor device and the end of the lead being soldered or desoldered?
49. If the voltage across a forward-biased pn junction measures 5.2 V, then the junction is definitely defective. (True/False)
50. If the voltage across a reverse-biased pn junction measures 5.2 V, then the junction is definitely defective. (True/False)
51. A diode with a 1:1 front-to-back resistance ratio is definitely defective. (True/False)
52. A diode with a 275:1 front-to-back resistance ratio is probably good. (True/False)

•—CIRCUIT EXPLORATION

Although this chapter had no emphasis on new circuits, it did introduce many new terms and concepts. The following crossword puzzle is offered in lieu of a circuit analysis problem. Your knowledge of the terminology presented in this chapter and your basic knowledge of electronics will permit you to solve the puzzle.

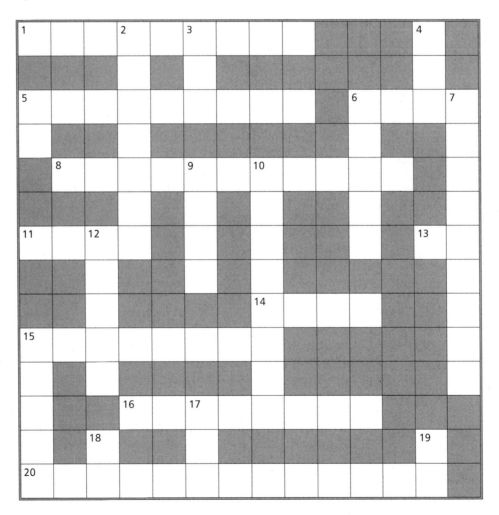

Across:

1. The type of impurity atoms needed to make a p-type semiconductor.
5. The name of a region where there are no current carriers.
6. If you _____ an intrinsic crystal, you will produce an extrinsic crystal.
8. An atom with five valence electrons.
11. A broken covalent bond.
13. A prefix meaning two.

14. Recombination destroys an electron-hole ____.
15. The type of atomic bond that holds a silicon crystal together.
16. Conduction-band electrons in n-type material are called ____ carriers.
20. What generally occurs when a conduction-band electron leaves the n-side of a pn junction and enters the "p" material?

Down:

2. The band where covalent bonds occur.
3. A forward-biased pn junction will ____ current flow.
4. There is an energy ____ between the valence and conduction bands.
5. A diode can convert ac into ____.
6. Pentavalent impurities are also called ____ atoms.
7. A silicon crystal after impurities have been added.
9. An impurity _____ is immobile because of the crystal structure.
10. A trivalent impurity atom is also called an ____.
12. What electrons do to make a positively-charged atom.
15. A code used to mark some electronic components.
17. A solid knowledge of semiconductor devices can help you get a good ____.
18. An abbreviation for integrated circuit.
19. The type of junction formed where "p" and "n" material meet.

CHAPTER 9
Diodes and Diode Circuits

●─KEY TERMS

anode
cathode
hot-carrier diode
knee current
knee voltage
load resistor
negative resistance
pulsating dc
rectify
ripple voltage
voltage regulation
zener voltage

●─OBJECTIVES

After studying this chapter, you should be able to:

- Contrast the characteristics of an ideal diode with those of a practical diode.
- Explain the operation of each of the following diode circuits:

  Biphase half-wave rectifier   Clamper
  Full-wave bridge rectifier   Half-wave rectifier
  Series clipper   Shunt clipper

- State the purpose and describe the operation of the following power supply filter circuits:

  Capacitor   LC

- Describe techniques and procedures that can be used to troubleshoot diode circuits.
- Describe the operational characteristics and state at least one application for each of the following special-purpose diodes:

  Current regulator   PIN   Schottky
  Step-recovery   Tunnel   Varactor
  Zener

●─INTRODUCTION

We are now ready to focus our attention on real diodes and practical diode applications. A semiconductor diode is simply a pn junction, but its applications are quite extensive. We will examine a wide range of diode applications in this chapter. In addition to basic diode circuits, we will also examine the operation and application of several special-purpose diodes, many of which are commonly used in computer circuits. Finally, we will discuss troubleshooting techniques and strategies that are applicable to diode circuits.

9.1 DIODE CHARACTERISTICS

There are three diode characteristics that will be of primary interest to us with regard to computer applications: forward voltage drop, reverse current, and reverse breakdown voltage. We will first examine these characteristics with respect to an ideal diode and then contrast the same characteristics for a practical diode.

Diode Elements

Since a diode is a pn junction, it follows that it has two leads or connections to the external circuit. Furthermore, since a diode behaves differently under conditions of forward and reverse bias, it is essential that we be able to distinguish between the two leads so that we can correctly insert the diode into the circuit. Figure 9–1 illustrates how the leads are named and identified.

FIGURE 9–1 Identification of diode elements.

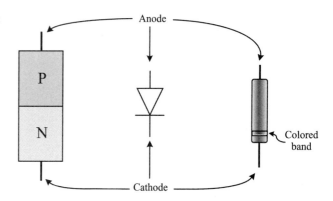

The **anode** lead is connected to the p-type portion of the silicon crystal. The n-type material is connected to the **cathode** lead. Figure 9–1 also shows how a colored band is used on the diode body to identify the cathode connection. There are other packaging styles, but the cathode can always be identified in some way. If the diode package is large enough, for example, it is common to print the schematic symbol of a diode on the side of the body to indicate the polarity of the device. Surface-mounted diodes are sometimes too small for physical identification. In this case, you must rely on an ohmmeter to identify the terminals. Diode measurements with an ohmmeter will be discussed in a later section.

Ideal Diodes

When troubleshooting a defective computer circuit, you can generally consider a diode to be ideal, but you can gain additional insights by understanding how practical diodes differ from ideal diodes.

Forward Bias

When an ideal diode is forward biased, it allows current to flow freely with no opposition. That is to say, a forward-biased ideal diode has no resistance. If it has no resistance, then it follows that there will be no voltage drop across the diode regardless of how much current passes through it. When an ideal diode is conducting, you can think of it as a short circuit or simply a piece of wire. The behavior of a forward-biased diode is illustrated in Figure 9–2.

FIGURE 9–2 An ideal forward-biased diode.

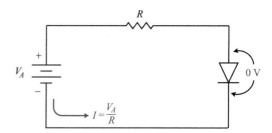

We identify a forward-bias condition by noting that the anode is more positive than the cathode. Notice that the forward voltage drop across the diode is 0 V. This means that all of the applied voltage will be dropped across the series resistance (R). Therefore, the current in the circuit is V_A/R, according to Ohm's Law.

Reverse Bias

When a pn junction is reverse biased, the depletion layer widens, and it becomes more difficult for current to flow through the device. In the case of an ideal diode with reverse bias, we assume that no current can pass through the diode regardless of the voltage applied. This means the diode has infinite resistance under conditions of reverse bias. You can think of a reverse-biased diode as an open circuit. This behavior is illustrated in Figure 9–3.

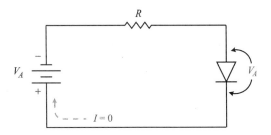

FIGURE 9–3 An ideal reverse-biased diode.

We know the diode is reverse biased because the anode is negative with respect to the cathode. Since the reverse-biased diode acts as an open circuit, there will be no current, and we will expect the full applied voltage to be across the diode as illustrated in Figure 9–3.

Current-versus-Voltage Curve

Figure 9–4 illustrates the current-versus-voltage curve for an ideal diode. The horizontal axis represents the voltage across the diode. The right portion of the horizontal scale is forward bias, while the left half is reverse bias. The vertical axis indicates current through the diode. Forward current is represented on the top portion of the vertical scale, while reverse current is represented on the lower half.

As we can see from Figure 9–4, there is no voltage across the diode when it is forward biased, even though the current can go to high values. In the reverse-bias polarity, there can be any amount of voltage across the diode, but no current is allowed to flow.

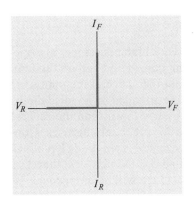

FIGURE 9–4 Current versus voltage for an ideal diode.

Practical Diodes

Let's briefly examine the characteristics of a practical diode and compare its behavior to an ideal diode.

Forward Bias

A practical diode readily conducts current when it is forward biased. However, it does offer some opposition to current. The most significant opposition is a result of the barrier potential across the pn junction. You will recall that the barrier potential acts as a small internal voltage (≈ 0.7 V for silicon). The external forward-bias voltage must oppose and overcome the barrier potential before any significant current can pass through the device. So, as long as the forward bias is less than the barrier potential, a practical diode continues to act like a high resistance. Once the forward voltage exceeds the internal barrier potential, a diode then acts like a low-value resistor.

Since a practical diode does have a voltage drop (≈ 0.7 V) and some small resistance when it is forward biased, it follows that it will dissipate power (e.g., $P = I^2R$ or $P = VI$). Therefore, there is a practical limit to the amount of forward current a diode can conduct without damage. Additionally, as the current through a diode increases, the voltage across it also increases slightly as a result of the voltage drop across its resistance. This latter effect is fairly small. Even with high currents, the forward voltage across a diode will rarely exceed 1.0 V. Figure 9–5 illustrates a practical diode under three forward-biased conditions.

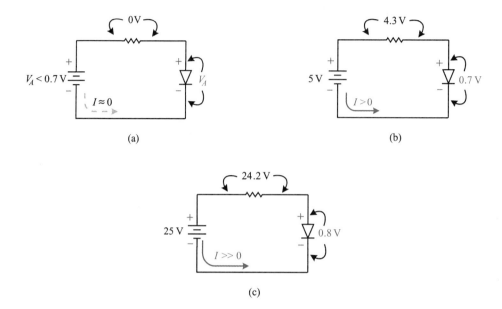

FIGURE 9–5 A forward-biased diode with (a) very low bias, (b) moderate bias, and (c) relatively high bias.

As you can see, for most practical applications (including troubleshooting), the assumption of ideal diode characteristics does not lead to any significant errors.

Reverse Bias

When a practical diode is reverse biased, it acts as a very high resistance. The specific resistance depends on the particular diode being considered and on the temperature, but values of at least several hundred kilohms to several megohms are representative. Clearly, this is not the infinite resistance provided by an ideal diode. However, if the other circuit resistances are substantially lower than the reverse resistance of the diode, then it is effectively infinite for practical purposes. Consider, for example, a 100-Ω resistor in series with a reverse-biased diode having an effective resistance of 1 MΩ. If 10 V are connected across the network, then less than 1 mV will be across the resistor. The rest of the voltage will be dropped across the diode that is, for practical purposes, an open circuit.

Because the reverse resistance of a diode is so high, it follows that the reverse current will be quite low. Of course, the exact value depends on applied voltage and the temperature of the pn junction. For many computer circuits, reverse diode currents range from hundreds of picoamps to tens of microamps.

Unlike an ideal diode that can withstand any amount of reverse voltage and still maintain infinite resistance, a practical diode can only block current up to a certain voltage called the reverse breakdown voltage (V_{BR}). If the reverse bias exceeds the breakdown voltage value, then avalanche occurs in the pn junction as described in the preceding chapter. Once the junction breaks down, it maintains a relatively constant voltage drop, but the current can increase greatly. Unless provisions are made to limit the current to a safe value, the power dissipation in the diode quickly exceeds its ratings, and the diode is destroyed. Figure 9–6 illustrates a reverse-biased diode under three conditions.

Again, for most practical computer applications (including troubleshooting), you can safely assume that a diode has ideal characteristics. Of course, if you locate a defective diode, you must be sure to replace it with one that has equivalent ratings. Even though we may assume all diodes are ideal for many purposes, we cannot assume that they are all equal with regard to ratings.

FIGURE 9–6 A reverse-biased diode with (a) low bias, (b) high bias, and (c) bias that exceeds the reverse breakdown voltage.

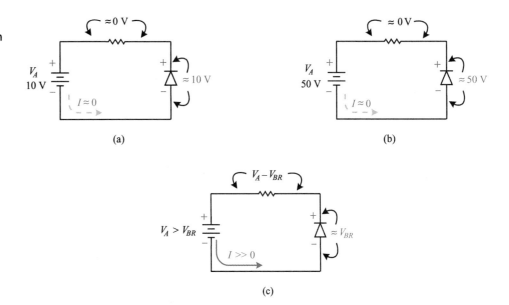

Current-versus-Voltage Curve

Figure 9–7 shows the current-versus-voltage curve for a representative practical diode. The nonideal characteristics are evident in several ways. First, notice that there is no significant current in the forward-bias polarity until the barrier voltage is reached. Beyond this point, current can increase substantially without dramatic increases in forward voltage drop. However, as the forward current increases, there is a slight increase in the forward voltage drop (i.e., the curve has a slight angle to the right). This is a result of the increasing voltage drop across the internal resistance of the diode.

In the reverse-bias direction, we see that only a small current flows as long as the voltage is less than the reverse breakdown voltage (V_{BR}) of the diode. Beyond this point, the current can increase greatly with only modest increases in reverse voltage. For standard diodes, we always try to avoid applying reverse voltages high enough to cause the diode to break down. In most cases, reverse breakdown will cause instant diode damage due to the high power dissipation (i.e., $P = VI$, where both voltage and current are relatively high).

The part numbers for practical diodes often, but not always, begin with the prefix "1N." Examples include 1N4005, 1N914A, and 1N3064.

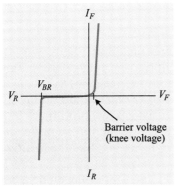

FIGURE 9–7 A current-versus-voltage curve for a practical diode.

✓ Exercise Problems 9.1

1. A diode is a ____-terminal device.
2. What names are used to identify the terminals of a diode?
3. Sometimes a diode has a colored band or marking adjacent to one lead. What does this indicate?
4. A diode is connected into a resistive circuit with 5 V applied. If the diode is forward biased, what is the voltage drop across the diode if it is an ideal diode?
5. A reverse-biased ideal diode acts as (*a short/an open*).
6. If the anode of a diode is more negative than the cathode, the diode is (*forward-/reverse-*) biased.
7. If a practical diode has +10.5 V on the anode and +9.8 V on the cathode, it is
   a. forward biased
   b. reverse biased
   c. defective
8. The forward-bias portion of the current-versus-voltage curve for an ideal diode is vertical. Explain why.
9. Name two characteristics that distinguish a practical diode from an ideal diode under conditions of forward bias.
10. Name two characteristics that distinguish a practical diode from an ideal diode under conditions of reverse bias.

9.2 POWER SUPPLY APPLICATIONS

With minor exceptions, all computers have some sort of power supply circuitry. This same statement applies to all electronic products with the exception of certain single-voltage, battery-operated systems. In the case of systems that operate from the ac power line, the power supply circuits must accomplish the following:

- Convert the ac power line voltage into the dc voltage required by the electronic circuits
- Reduce the voltage from 120 Vac to a lower value
- Continuously adjust the dc output voltage to keep it constant even though the current drawn may change and the input voltage may vary

Our primary focus at this point is the conversion from ac to dc. You have already learned about how transformers can be used to reduce an ac voltage to a lower value. The maintenance of a constant output voltage under varying conditions (called **voltage regulation**) will be discussed in chapter 14.

Half-Wave Rectifier

Figure 9-8 shows the schematic diagram for a half-wave rectifier circuit. The term **rectify** is used to describe the conversion of ac (bidirectional current) into dc (unidirectional current). As you will soon see, only one half-cycle of the input sine wave is allowed to pass through to the output. This is why it is called a half-wave rectifier circuit.

Circuit Operation

During the positive alternation of the input, as represented in Figure 9-8(a), the diode is forward biased and acts like a short circuit (i.e., no voltage drop). This means the full applied voltage will be dropped across the resistor (R_L). This is called the **load resistor**, because it represents the circuitry that would ordinarily serve as the load for the supply. For example, the resistor shown in Figure 9-8 might represent all of the circuitry in a computer. Because the full supply voltage is dropped across the load, the output waveform looks just like the positive half-cycle of the input voltage.

During the negative alternation, as represented in Figure 9-8(b), the diode is reverse biased and acts like an open circuit. Since no significant current can flow through a reverse-biased diode, there can be no voltage drop across the load. That is, all of the voltage will be dropped across the diode and none across the load. For this period of time, the output voltage (voltage across the load) is zero.

Figure 9-8(c) shows that the overall output voltage is actually **pulsating dc**. That means that the voltage is dc, since it has only a single polarity, but it is pulsating rather than a smooth, constant-valued voltage. Most electronic circuits require smooth, constant-valued dc similar to the voltage of a battery. So, in most cases, we follow the rectifier circuit with additional circuitry (called filters) to smooth

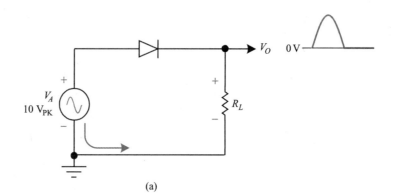

FIGURE 9-8 A positive half-wave rectifier circuit showing circuit conditions for (a) positive and (b) negative alternations, as well as (c) the overall voltage output.

FIGURE 9-8 (continued)

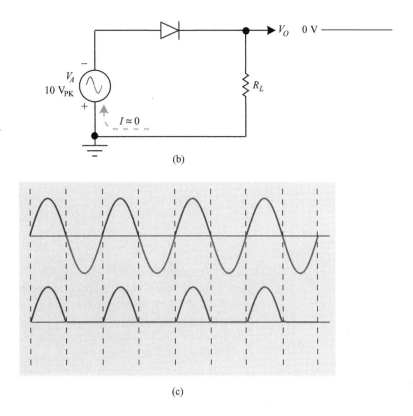

(b)

(c)

the pulsating voltage into steady dc. We will discuss filters in a later section. The circuit in Figure 9–8 is called a positive half-wave rectifier, since the output voltage is positive with respect to ground.

The amplitude of the pulsating voltage in the output of a half-wave rectifier is equal to the peak amplitude of the input voltage (ideally). If we consider the effects of a nonideal diode, then the peak output voltage will be slightly (≈ 0.7 V) lower than the input voltage.

Figure 9–9 shows how simply reversing the diode polarity can make a negative half-wave rectifier.

FIGURE 9-9 A negative half-wave rectifier circuit showing circuit conditions for (a) positive and (b) negative alternations, as well as (c) the overall voltage output.

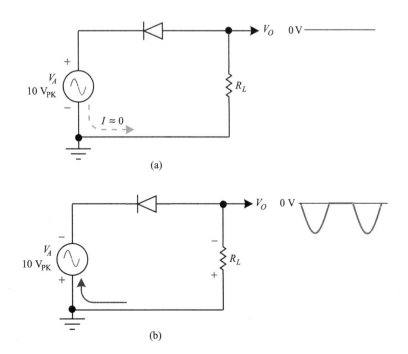

FIGURE 9–9 (continued)

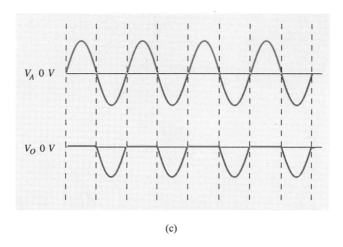

(c)

Industry Application

Figure 9–10 shows a simple application for a half-wave rectifier circuit. Here, a transformer is used to reduce the line voltage to a lower value. Additionally, the primary has a power switch and a fuse. The secondary voltage provides the input to the half-wave rectifier circuit. A rechargeable cell is the load.

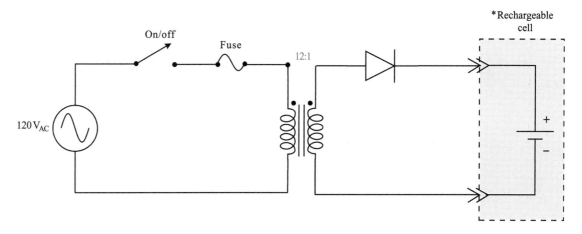

*Safety Note: Some types of rechargeable cells require additional circuitry to prevent overheating, cell damage, and possible injuries resulting from cell explosion.

FIGURE 9–10 A half-wave rectifier used to charge a cell.

During the positive (when the top is positive with respect to the bottom in Figure 9–10) alternations of the secondary winding, the diode conducts. This allows electron current to flow in a counterclockwise direction in the secondary circuit of Figure 9–10. Note that this is "backward" from normal cell or battery current, which leaves the negative terminal and returns to the positive terminal. This reverse current charges the battery. In this application, there is no requirement for a smooth, constant dc voltage. The cell itself provides a certain degree of filtration.

It is very important for you to realize that the circuit in Figure 9–10 has been simplified and is not directly applicable for all types of rechargeable cells. Many types of cells have strict requirements on the amount of charging current. If this is not carefully monitored and controlled, the cell can be damaged. In some cases, the heat caused by improper charging current can cause a violent explosion or fire. Therefore, practical chargers often require additional electronic circuitry.

Full-Wave Rectifier

Figure 9–11 shows the schematic diagram for a full-wave rectifier circuit. It is also called a biphase half-wave rectifier and a center-tapped rectifier circuit. As indicated in Figure 9–11, both alternations of the input cycle appear as positive voltages in the output of the rectifier.

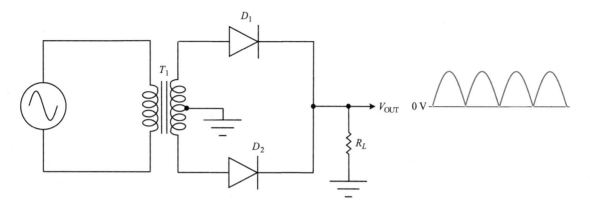

FIGURE 9–11 A full-wave rectifier circuit.

Let's begin by assuming the alternation that causes the top of the secondary to be positive with respect to the bottom, as illustrated in Figure 9–12(a). Electron current will leave the negative side of the top half of the winding (i.e., the center-tapped point) and flow through ground to the lower end of the load. It then progresses up through the load, giving the polarity of voltage drop indicated in Figure 9–12(a). Note that the output is positive with respect to ground. From the

FIGURE 9–12 A full-wave rectifier circuit on (a) one alternation and (b) the opposite alternation.

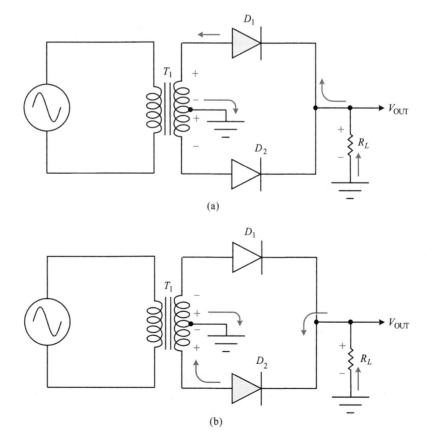

top of the load resistor, current continues through forward-biased D_1 and returns to the positive side of the secondary winding. If we visualize diode D_1 as being a short circuit during this half-cycle, we can see two important points. First, the load is connected directly across the upper half of the secondary during this time interval. This means the output voltage will be the same amplitude and shape as the voltage across one-half of the secondary. Next, note that diode D_2 is definitely reverse biased since its anode is connected to the negative end of the secondary and its cathode is connected to the positive end of the secondary (through forward-biased D_1).

The next alternation is shown in Figure 9–12(b). Here the top of the secondary is negative with respect to the bottom of the winding. This forward biases D_2 and reverse biases D_1. Throughout this alternation, electron current leaves the center tap, flows through ground to the lower end of the load, and then flows through D_2 to the positive side of the secondary. Note that the polarity of voltage drop across the load is still positive on V_{OUT} with respect to ground. If we consider forward-biased D_2 to be a short and reverse-biased D_1 to be open, we can readily see that the output is effectively connected across the lower half of the secondary during this alternation. Again, the output amplitude and shape will be the same as one-half of the secondary winding. With D_2 effectively shorted, you can see how D_1 is essentially connected across the entire secondary such that its anode is negative with respect to its cathode (i.e., it is reverse biased.)

If we combine the two alternations just described, we obtain the full-wave pulsating output voltage that was illustrated in Figure 9–11. This waveform still requires filtering and regulation for most practical applications as previously stated for half-wave rectifiers. However, the output of a full-wave circuit is easier to filter. If the diodes are both reversed, then a negative full-wave output is produced.

Bridge Rectifier

Figure 9–13 shows the schematic diagram of another type of full-wave rectifier circuit called a bridge rectifier. This circuit is more widely used than the center-tapped rectifier since it does not require a center-tapped transformer. This greatly reduces the overall size, weight, and cost of the product in most cases.

FIGURE 9–13 A full-wave bridge rectifier circuit.

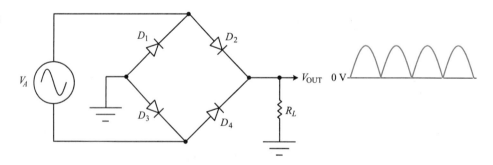

Circuit Operation

Let's examine the operation of a full-wave bridge circuit by considering each alternation of the input separately. Figure 9–14(a) shows one alternation. This polarity of input voltage causes diodes D_2 and D_3 to be forward biased throughout this half-cycle.

Electron current starts at the negative side of the input voltage source and flows to the junction of D_3 and D_4. It then flows through forward-biased D_3 to ground. Current continues through ground to the bottom of R_L. As it moves upward through R_L, it develops the voltage drop shown (i.e., positive on V_{OUT} with respect to

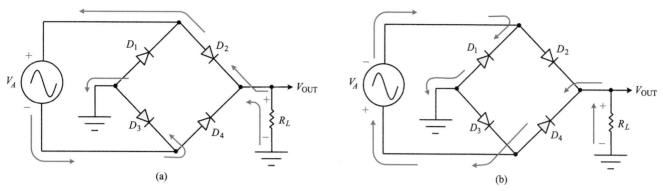

FIGURE 9–14 A full-wave bridge rectifier circuit on (a) one alternation and (b) the opposite alternation.

ground). From there, electron current goes through forward-biased diode D_2 and returns to the positive side of the source. If you mentally replace diodes D_2 and D_3 with wires (recalling that forward-biased diodes act as short circuits), you can see that D_1 and D_4 are in parallel. Further, the parallel combination is connected directly across the source in a reverse-biased polarity. Now, if you mentally remove D_1 and D_4 (recalling that reverse-biased diodes act as opens), you can readily see that the load is connected directly across the source with V_{OUT} connected to the positive side of the source. Therefore, throughout this particular alternation, V_{OUT} will track V_A.

The next alternation is illustrated in Figure 9–14(b) where the top side of V_A is negative. During this alternation, diodes D_1 and D_4 are forward biased while diodes D_2 and D_3 are reverse biased. Again, if you mentally replace the forward-biased diodes with wire and remove the reverse-biased diodes, you will find the load is still connected directly across the source. More importantly, the V_{OUT} end of the load is still connected to the positive side of the source even though the polarity has changed. The diode bridge network effectively switches the load such that V_{OUT} is always connected to the positive side of the source.

Industry Application

Figure 9–15 shows how a bridge rectifier can be used to allow a product to operate from ac or dc. The user merely connects the Power-In jacks to any convenient source of voltage with the correct amplitude. It can be either an ac or dc source, but a dc source can be connected without regard to polarity. In all of these cases, the output from the bridge circuit will be the same polarity (as marked). The output of the bridge is then passed through some type of filtration to smooth the voltage (since an ac input produces a pulsating dc from the bridge) and through regulation circuitry to maintain a constant voltage for the electronic circuits to follow.

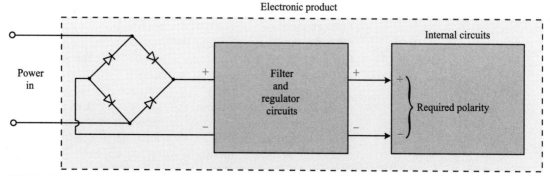

FIGURE 9–15 A bridge circuit allows a product to operate from ac or either polarity dc.

Filter Networks

A few applications (e.g., some types of dc motors) can operate normally with pulsating dc from a half-wave or full-wave rectifier circuit. However, the vast majority of electronic applications require smooth dc such as that provided by a battery.

Half-Wave Rectifier with Filter

Figure 9–16 shows how a capacitor can be added to a basic rectifier circuit to filter the pulsating dc into smooth dc.

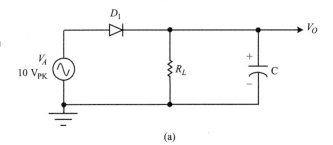
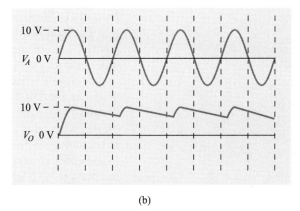

FIGURE 9–16 A filter can be used to smooth the pulsating dc from a rectifier circuit.

On the positive half-cycle of the input voltage, diode D_1 conducts as we discussed previously. Current not only flows up through the load resistor, but it also charges filter capacitor C. You will recall that the peak input voltage (less the forward voltage drop of the diode) appears across the load. Capacitor C charges to this peak voltage. The time constant for the capacitor charging path is quite small, since the resistance is only that of the forward-biased diode. This initial charging action is illustrated in Figure 9–16(b). Notice that the output voltage essentially tracks the input voltage for the first quarter of a cycle.

Now, as the input voltage passes 90° and begins to decrease in amplitude, the diode is reverse biased. The reverse-bias condition results as the input voltage drops relatively fast, but the charge on the capacitor decays more slowly. Recall that the capacitor charged quickly through the diode, but it cannot discharge through the diode because of the unidirectional nature of the diode. So, the capacitor must discharge through the load. This is intentionally designed to be a long time constant. The discharge current from the capacitor keeps current going through the load in the same direction.

As shown in Figure 9–16(b), the capacitor continues to decay throughout the negative alternation and well into the next positive half-cycle. Eventually, the input voltage rises above the decaying capacitor voltage. This causes the diode to again be forward biased, which recharges the capacitor to the peak voltage. This process continues as long as the input voltage is applied. The longer the time constant for the capacitor's discharge, the smoother the dc output voltage will be. The remaining variations in the output voltage are called **ripple voltage**. The amplitude of the ripple voltage is a measure of the quality of a power source. The smaller the ripple voltage, the better the supply (i.e., the closer it is to the ideal dc supply, which has no ripple voltage).

The output waveform shown in Figure 9–16 assumes ideal diode behavior. The peak voltage would be slightly less (e.g., 9.3 V) if a practical silicon diode was illustrated.

Full-Wave Rectifier with Filter

Figure 9–17 illustrates a capacitive filter added to the output of a full-wave bridge rectifier circuit. The waveforms in Figure 9–17(b) illustrate the operation of the circuit (assuming ideal diodes). The capacitor charges to the peak of the input voltage (less two diode drops for practical diodes). As with the half-wave circuit, the filter capacitor begins a slow discharge as soon as the input voltage passes the peak and starts to decay. However, it will have less time for discharge in the full-wave circuit because the pulsating dc occurs during both alternations of input voltage. Consequently, the ripple voltage on a full-wave circuit will be less for a given load and a given value filter capacitor. The diodes in Figure 9–17 are considered to be ideal, so they have no voltage drop. A practical silicon diode will drop about 0.7 V, so the peak output voltage for the circuit in Figure 9–17 will actually be about 8.6 V instead of 10 V. This is true because, on any given half-cycle, two diodes are effectively in series and drop 0.7 V each.

FIGURE 9–17 A full-wave bridge rectifier with a capacitor filter.

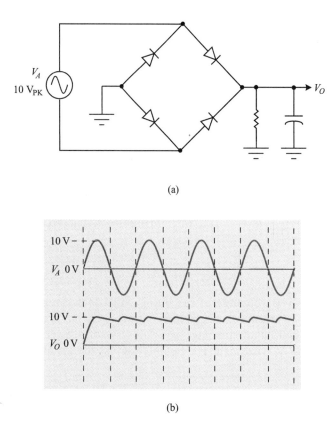

(a)

(b)

Other Types of Filter Networks

The simple capacitor filters discussed in the preceding section are adequate for many applications, but sometimes a voltage source with even smoother dc is required. In these cases, more complex filter networks can be used. Figure 9–18 on page 304 shows two of the more common filter networks. The network in Figure 9–18(a) is called an L filter, since its schematic representation is generally drawn in an L shape. For similar reasons, the network in Figure 9–18(b) is called a

FIGURE 9–18 (a) An L-type filter network and (b) a pi-type filter network.

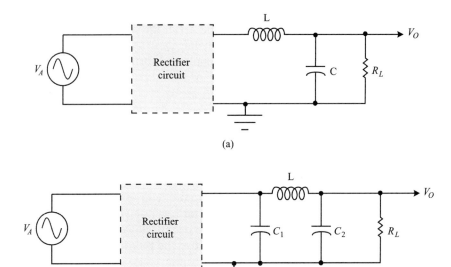

pi network. By inspection, you can see that a pi filter is really a simple capacitor filter followed by an L-type filter.

In an earlier chapter, we discussed passive filter networks. The filters presented here are all low-pass filters. The goal of the filter is to pass the low-frequency components of the waveform (specifically dc) while greatly attenuating higher-frequency components (specifically ripple voltage).

✓ Exercise Problems 9.2

1. The output of a half-wave rectifier circuit is _____.
   a. ac
   b. dc
   c. pulsating dc
2. The output of a full-wave bridge rectifier circuit is _____.
   a. ac
   b. dc
   c. pulsating dc
3. What percentage of the input cycle does the diode pass current in a negative half-wave rectifier circuit?
4. If the rms input voltage on a full-wave bridge rectifier is 12 V, what is the peak voltage across the load?
5. If the full (end-to-end) secondary winding voltage in a full-wave center-tapped rectifier circuit is 25-V peak, what is the amplitude of the pulsating dc output voltage (assume ideal diodes)?
6. What is the primary impact of a filter network on the output of a rectifier circuit?
7. A capacitor filter on the output of a rectifier circuit should be designed to have a (*short/long*) charge time constant and a (*short/long*) discharge time constant.
8. What class of passive filter is used to lessen the amplitude of the ripple in a rectifier circuit?

9.3 MISCELLANEOUS DIODE APPLICATIONS

This section will examine several practical circuit applications that rely on diodes for their operation. As you study each example, be sure you can accomplish the following:

1. Be able to recognize and name the circuit
2. Determine output waveforms where applicable
3. Know the purpose of each component

Clipper Circuits

A clipper (also called a limiter) restricts the maximum voltage that a waveform can have in a given polarity. In many cases, a clipper will serve as a protection circuit to prevent damage from excessive voltage.

Clipper Operation

Figure 9–19 shows the schematic diagram for a clipper circuit. As we explore its operation, recall that a diode will conduct when it is forward biased (i.e., anode more positive than the cathode). It acts as a short when it is conducting. By contrast, it acts as an open when it is reverse biased (i.e., anode less positive than the cathode).

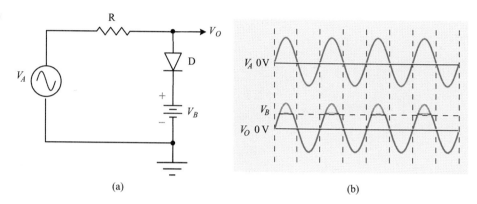

FIGURE 9–19 A clipper (or limiter) circuit.

Let's determine what must occur for the diode to become forward biased. The dc voltage source V_B holds the cathode at a constant potential of $+V_B$ volts with respect to ground. So, in order to forward bias the diode, the anode side of the diode must be returned to a voltage that is more positive than $+V_B$.

Consider how the circuit behaves if the peak input voltage is less than V_B. Can you see that the anode will never be returned to a voltage greater than V_B, so the diode will never be forward biased? This means it will continue to act as an open circuit. If the diode is an open circuit, then no substantial current will flow through R, and no voltage will be dropped across R. Therefore, whatever potential is on one end of R will also be on the other end (i.e., 0 V across it). The left end of R is connected directly to V_A, which means that the other end (V_O) will be at the same potential as V_A. As long as the input voltage never exceeds the value of V_B, the clipper circuit has no effect on the input waveform. That is, the output is the same as the input.

Now, consider what happens when the peak voltage of V_A exceeds the value of V_B. We know the diode will be forward biased during these positive excursions. If the diode is forward biased and acting as a short, can you see that V_B is essentially connected directly to the output? So, during all times that the input voltage is more positive than V_B, the output will remain at the V_B voltage. This clearly distorts the waveform, as shown in Figure 9–19(b), but it limits the maximum voltage that can appear at the V_O terminal.

The orientation of the diode determines which polarity of the input voltage is potentially clipped. The value and polarity of V_B determine the level beyond which clipping occurs. Figure 9–20 on page 306 shows several clipper circuits with different diode orientations and biasing polarities.

Clippers in Computers

Limiters are quite common in computer circuits. Figure 9–21 on page 306 shows how a clipper can be used to protect an industrial computer from excessive input voltage on a transducer input.

FIGURE 9–20 Limiter circuit variations.

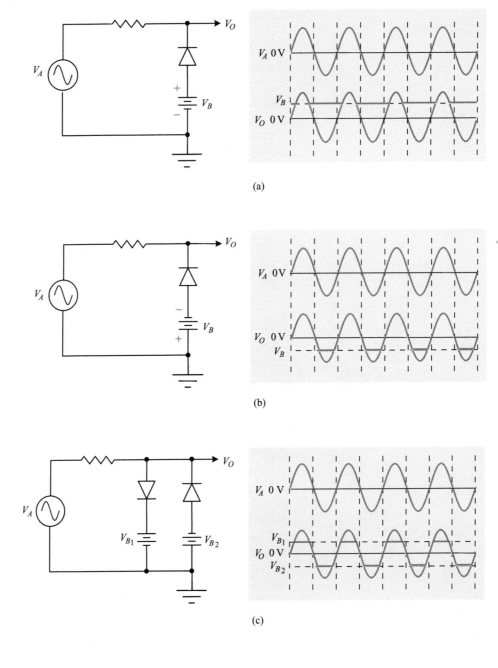

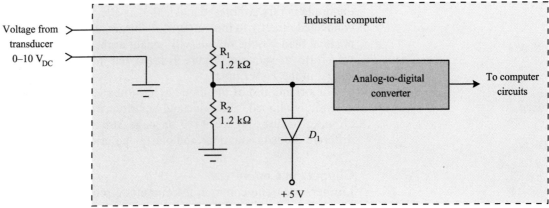

FIGURE 9–21 A clipper circuit used to protect an industrial computer.

Industrial computers often have to monitor and control manufacturing processes by measuring conditions such as pressure, temperature, pH value, density, depth, velocity, and many others. These quantities are sensed by transducers. There are many types of transducers, but basically each is designed to vary some electrical characteristic (e.g., voltage, current, resistance, frequency, and so on) on its output in proportion to changes in the sensed physical value.

The computer illustrated in Figure 9–21 is designed to accept transducer signals in the range of 0 V to +10 V. As you can see, the transducer voltage is reduced to a lower level (0 V to 5 V) through the voltage divider action of R_1 and R_2. This reduced voltage range is then applied to the input of an analog-to-digital (A/D) converter. This is usually a single, integrated circuit that accepts a range of voltages (or currents in some cases) on its input. A signal of this type that can assume any one of an infinite number of possible values is called an analog signal. Digital computers, however, operate on digital signals that are characterized as having only two possible levels. In short, an A/D converter transforms the varying voltage from a transducer into digital signals that can be interpreted and manipulated by a digital computer.

The transducer input to the computer illustrated in Figure 9–21 must necessarily be routed a long distance from the computer through a harsh industrial environment. It is possible that voltage in excess of the expected 0 V to 5 V range will appear at the input of the A/D converter through human errors in wiring the transducers, from electrical defects, or for other reasons. To protect the sensitive and often expensive A/D converter, a clipper can be used as illustrated in Figure 9–21. Here, the cathode of diode D_1 is connected to the +5-V power source used by the rest of the computer circuits. This means the anode of the diode (which is also the input to the A/D converter) can never go more positive than +5 V (+5.7 V for a practical diode). Thus, the A/D converter is protected from excessive input voltages.

Isolation Diodes

There are many computer circuit applications that rely on diodes to selectively isolate one circuit from another. Consider, for example, a computer memory that has a battery backup such as the one illustrated in Figure 9–22. As long as the computer is turned on, system power from the computer is provided to the memory circuits. However, when the main computer power is turned off, a small battery continues to provide power to the memory circuits. The diodes illustrated in Figure 9–22 isolate the battery from the computer power when the power is on. They also isolate the rest of the computer circuits from the battery and memory when system power is off. This isolation is necessary because the small battery cannot supply enough power for the whole computer; it would quickly discharge if it were not isolated from the bulk of the system circuitry.

Let's begin by assuming system power is applied to the computer illustrated in Figure 9–22. The +5 V system power forward biases D_1 and provides power to the memory circuits. D_2, however, is reverse biased since its anode is less positive than

FIGURE 9–22 Isolation diodes for a computer memory circuit with battery backup.

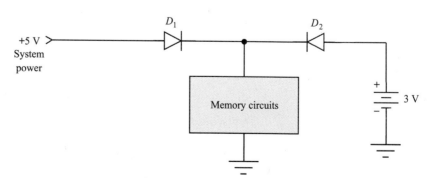

its cathode. That is, its anode is +3 V with respect to ground, while the cathode is +5 V with respect to ground (assuming ideal diodes).

Now, if system power is removed, then the battery forward biases diode D_2 and provides backup power to the memory circuits. But, D_1 is reverse biased since its anode is near zero (system power is off) and its cathode is near +3 V. Reverse-biased D_1 acts as an open to prevent the rest of the computer circuits from draining energy from the backup battery.

Battery backup is frequently used in personal and office computers to keep the real-time clock circuits operating. In other words, you would not want to reset the clock and calendar in the computer every time you turned it on. Industrial computers also use battery backup for memory circuits that store critical values. When the system is powered down, the critical values are stored in the battery-powered portion of the memory circuits. When main system power is reapplied, these values are retrieved and operation can resume where it left off.

Figure 9–23 shows another simple application of a diode that isolates a circuit under certain conditions. In this case, the diode is used to protect a product in the event its power supply wires are connected to the wrong polarity of voltage.

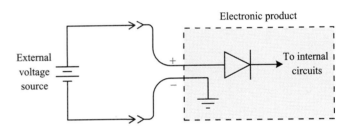

FIGURE 9–23 A diode can protect a product from reversed-polarity power supply lines.

If the external dc power is properly connected as illustrated in Figure 9–23, then the diode is forward biased and acts as a short. For practical purposes, the diode has no effect on circuit operation when external power is properly connected. But, if the power input wires are inadvertently reversed, then the diode will be reverse biased and will act as an open circuit. The open diode prevents any current flow and possible damage to the circuit under these conditions.

Asymmetrical Time Constants

We can use diodes to create an RC circuit that has different time constants for charge and discharge. Figure 9–24 shows an example of how this can be done.

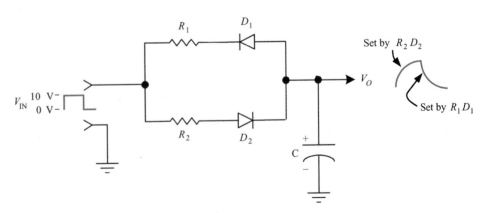

FIGURE 9–24 A circuit with asymmetrical time constants for charge and discharge.

To discuss the operation of the circuit in Figure 9–24, let's assume that the input (V_{IN}) is a pulse waveform. When V_{IN} is positive, diode D_2 will be forward biased. This will allow C to charge through R_2 with the polarity shown. The RC time constant is determined by $R_2 C$ (ignoring the small forward resistance of diode D_2).

When the input voltage returns to a 0-V level, capacitor C will discharge. The discharge current path is from the bottom of C, through ground to the source, through the source, through R_1, and through D_1 to the positive side of the capacitor. The time constant in this case is established by $R_1 C$. By selecting the correct values for R_1 and R_2, we can achieve any desired relationship between charge and discharge times. One application for this type of circuit uses a fast charge and a slow discharge time. It is called a fast-attack and slow-decay circuit. It is used in music synthesizers, industrial computer interfaces, and test equipment (e.g., spectrum analyzers). In a later chapter, we will also utilize this basic concept to create pulse waveforms with arbitrary duty cycles. (A duty cycle describes the ratio of high time to total cycle time in a pulse waveform.)

AM Detector Circuit

Modulation is a process whereby a low-frequency information signal (such as voice, music, video, or computer-data signals) is used to alter some characteristic (such as amplitude, frequency, phase, and so on) of a higher-frequency signal called the carrier. The higher-frequency signal can be transmitted much more efficiently over distances. Amplitude modulation (AM) is one of several ways that information is carried via electromagnetic waves. Your AM radio, for example, is designed to receive amplitude-modulated signals, as is the picture portion of your television receiver.

Overview

Figure 9–25 introduces the basic concepts of an amplitude-modulated transmitter/receiver system. Figure 9–25(a) shows the transmitter portion of the system. First, there must be an information source such as sound waves detected by a microphone, video signals produced by a television camera, or perhaps data signals from a computer. It is this information, called the modulating signal, that is to be recovered at the receiver.

The transmitter circuitry generates a high-frequency signal called the carrier. It is the carrier wave that allows the information to be transmitted over great distances with reasonably small antennas. The carrier wave and the modulating (information) signal are combined in a circuit called the modulator. In essence, the instantaneous amplitude of the high-frequency carrier wave is determined by the instantaneous amplitude of the lower-frequency modulating signal. The modulated signal consists of a high-frequency component whose amplitude varies at a rate

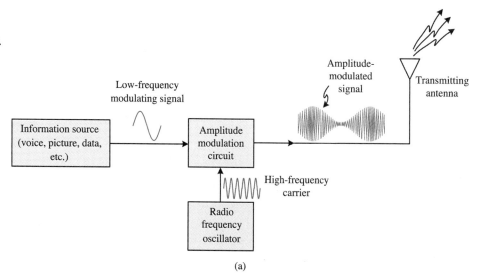

FIGURE 9–25 An amplitude modulation system consists of (a) a transmitter and (b) a receiver. Information is carried from the transmitter to the receiver by a high-frequency radio wave called a carrier.

FIGURE 9–25 (continued)

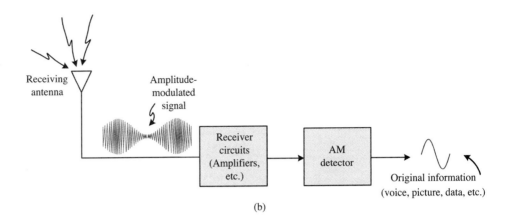

(b)

determined by the frequency of the modulating signal. This composite wave is sent to the transmitting antenna as shown in Figure 9–25(a).

Figure 9–25(b) shows a simplified representation of an AM receiver. First, the receiving antenna intercepts the radio wave. The microvolt-level signal is then amplified and processed before reaching the AM detector portion of the receiver circuitry. It is the job of the AM detector to recover the original modulating signal. The recovery process is often called demodulation or detection. The detector is essentially the reverse of the modulator at the transmitter end. The modulator mixes the information signal and the carrier wave, while the AM detector separates the information signal from the carrier wave. The output of the AM detector, then, is the original information signal (voice, picture, data, and so on).

Industry Example
Figure 9–26 shows an AM detector circuit capable of detecting an amplitude-modulated signal similar to those found in an AM radio receiver. Diode D_1 rectifies the modulated signal. Capacitor C_1 filters the rectified signal. R_2 and C_2 provide additional filtration. The time constants for the two-stage filter network must be short relative to the original modulating signal but long relative to the high-frequency carrier signal. When this condition is met, the high-frequency rectified pulses are eliminated, and only the lower-frequency variations remain. The lower-frequency changes, of course, correspond to the original information that was used to modulate the carrier wave.

FIGURE 9–26 A simple AM detector circuit.

Fig09_26.msm

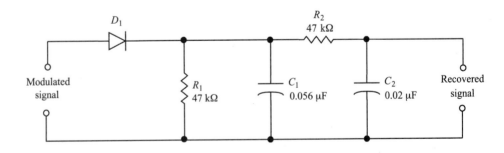

Figure 9–27 shows a circuit simulation of the AM detector circuit shown in Figure 9–26. The upper and lower waveforms are the modulated carrier wave and the original information signal, respectively. The rf carrier and the information signal were combined by the modulator at the transmitter to form the modulated signal (upper waveform in Figure 9–27) that is actually transmitted through the atmosphere as a radio wave. Finally, the modulated waveform passes through the detector circuit (Figure 9–26), where the original information is recovered. The recovered information waveform is shown as the bottom waveform in Figure 9–27.

FIGURE 9–27 Circuit simulation result for the AM detector shown in Figure 9–26.

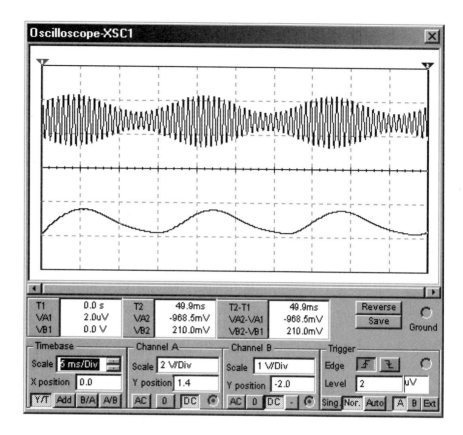

✓ Exercise Problems 9.3

Refer to Figure 9–28 for questions 1 through 5.

FIGURE 9–28 Circuit for Exercise Problems 1 through 5.

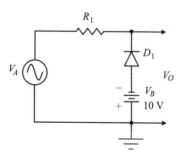

1. If the peak input voltage is 5 V, describe the output waveshape (V_O).
2. If the peak input voltage is 25 V, which peak will be clipped?
3. If the diode is reversed, then V_O will never go more positive than $-V_B$. (True/False)
4. If diode D_1 becomes open, what are the symptoms in terms of V_O?
5. If the value of V_B is reduced to 0 V, the output waveform will be approximately the same as the output of a positive half-wave rectifier circuit. (True/False)
6. Refer to Figure 9–26. If C_1 develops a short, describe the symptoms in terms of the output waveform.

9.4 SPECIAL DIODES

The diodes presented in this section have special characteristics that make them well-suited to certain applications. Nevertheless, they are still diodes and share many of the characteristics normally associated with diodes: unidirectional current flow, sensitivity to temperature, and limitations such as current, power, and reverse voltage. The characteristics of some of the diodes that are presented here (e.g., zener diodes) represent only slight deviations from the rectifier diodes previously discussed. Others, however, possess additional characteristics that make them radically different. But as we proceed, don't forget that they are all diodes.

Zener Diodes

A zener diode is very much like a standard rectifier diode in many respects. It is designed, however, for normal operation in the reverse-breakdown portion of its operating curve. When operated in this region, the voltage across the zener remains relatively constant even though the current through the diode may change substantially. This provides us with a constant voltage source. Figure 9–29(a) shows the schematic symbol for a zener diode. Its characteristic curve is shown in Figure 9–29(b). You should recognize the similarity between this curve and the comparable curve for a standard rectifier diode.

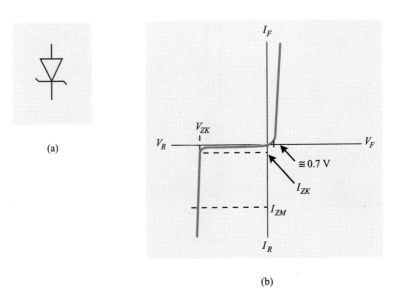

FIGURE 9–29 (a) The symbol for a zener diode and (b) its current-versus-voltage curve.

Basic Zener Characteristics

Although the shape of the basic current-versus-voltage curve is essentially the same as a standard rectifier diode, let's study the reverse-bias portion of the curve more closely. As you can see from Figure 9–29(b), no significant current is allowed to flow in the reverse-bias direction as long as the bias is below the breakdown voltage. Beyond this point, current can change dramatically, but the voltage across the zener remains fairly constant. This is where the zener operation occurs.

The point where the reverse current begins to increase sharply while the voltage remains constant is called the **knee voltage** (V_{ZK}). The small current that flows at this same time is called the **knee current** (I_{ZK}). Now, if the reverse bias is increased beyond the knee voltage, the current increases quickly. As indicated on the curve in Figure 9–29(b), there is a maximum allowable current (I_{ZM}). If the current goes higher than this, the resulting power dissipation will damage the zener.

Therefore, normal operation for a zener diode is in the reverse-bias mode, beyond the knee voltage, and with a reverse current between I_{ZK} and I_{ZM}. Under these conditions, the voltage across the zener is relatively constant, and the power dissipation is not excessive.

Zener Voltage Regulator Circuit

One of the primary applications for zener diodes is use as a voltage regulator. The purpose of a voltage regulator is to maintain a constant voltage across a circuit even though the current drawn by the load changes or the input supply voltage changes. To illustrate this application, we will examine the simple circuit shown in Figure 9–30(a).

FIGURE 9–30 (a) Circuit that allows input voltage changes to cause load voltage changes as shown in (b) the circuit simulation.

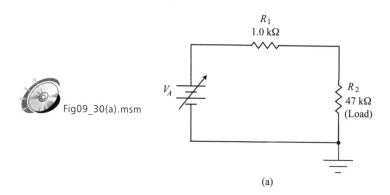

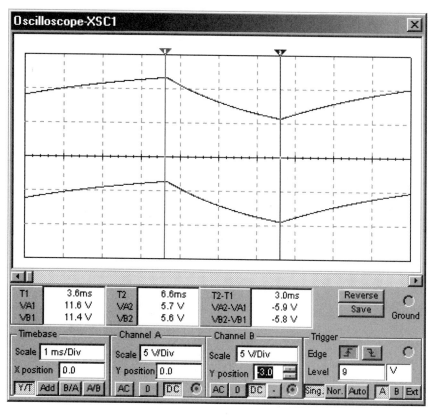

The variable voltage source in Figure 9–30(a) will introduce intentional variations in the supply voltage. As you would expect, variations in the input voltage cause corresponding variations in the load voltage (voltage across R_2 in this example). Figure 9–30(b) shows the results of a circuit simulation. The upper trace in Figure 9–30(b) represents the input voltage, which varies from 11.6 V to 5.7 V. The lower trace shows the voltage across the load (R_2). As you can see, the load voltage varies from 11.4 V to 5.6 V. Clearly, this is not a constant-load voltage.

Now, let's contrast this performance with that of a similar circuit that utilizes a zener diode as a voltage regulator. Figure 9–31(a) on page 314 shows the same circuit, but a zener diode has been connected in parallel with the load. First, notice that the diode is connected in such a way as to be reverse biased. The particular diode utilized in this circuit is a 1N4733 zener diode. It has a **zener voltage** (i.e., reverse-breakdown voltage) of approximately 5 V. As long as we keep the zener in a reverse-bias condition with reverse currents that lie between the knee current and the maximum allowable current, we can expect to have a voltage across the zener that is very close to its rated voltage.

FIGURE 9–31 (a) Circuit with a variable input voltage but a constant output voltage as shown in (b) the circuit simulation.

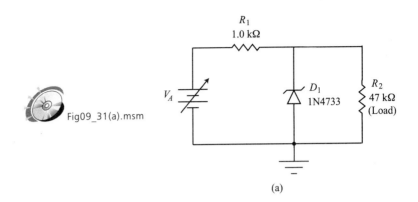

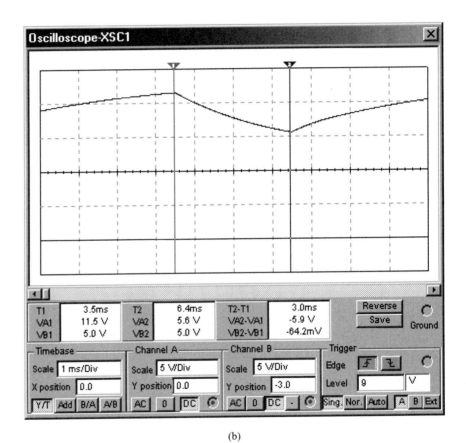

(b)

Figure 9–31(b) shows the results of a circuit simulation for this circuit. The upper trace shows the input voltage variations. Here V_A is changing from 11.5 V to 5.6 V. But, look at the lower trace, which represents load voltage. It remains constant at 5 V. So, the load (R_2) now has a constant voltage in spite of the fact that the input voltage is making rather dramatic changes. According to Kirchhoff's Law, the common voltage drop across the parallel combination of the load and the zener plus the voltage drop across the series resistor R_1 must equal the supply voltage at any given time. In essence, current through the zener changes as necessary to increase or decrease the voltage drop across R_1 such that the voltage across the zener itself (and the load) remains nearly constant. Of course, this operation can only be successful as long as we maintain sufficient reverse bias on the zener. If the input voltage drops too low, then the zener will not be able to maintain a constant voltage.

Varactor Diodes

You will recall that a reverse-biased diode has a depletion region that is essentially free from current carriers. This means the depletion region is effectively an insulator. The p- and n-type materials on either side of the depletion layer have many current carriers (either holes or electrons) and are therefore conductive. With two conductors separated by an insulator we have a capacitance. This junction capacitance is present in all reverse-biased diodes and is generally an undesired quantity that interferes with operation at high frequencies and fast switching speeds. However, the junction capacitance in a varactor diode is optimized and is the primary reason for choosing the device in a particular application.

When the reverse bias on a varactor diode is increased, the depletion layer widens (as with any reverse-biased diode). This causes the junction capacitance to decrease due to the increased distance between the capacitor's plates (p and n regions). By contrast, a decrease in reverse bias increases the junction capacitance. Varactors are available with capacitance values of less than 2 pF to at least 1000 pF. The ratio of maximum-to-minimum capacitance for a specific diode is typically on the order of 5:1 to 6:1, but higher ratios are available. Figure 9–32 shows several schematic symbols used by manufacturers to represent varactor diodes.

FIGURE 9–32
Schematic symbols used to represent a varactor diode.

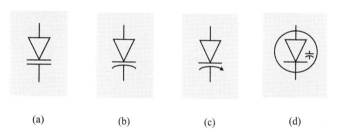

(a)　　(b)　　(c)　　(d)

Electronic tuning is probably the biggest application for varactor diodes. Figure 9–33 illustrates how the variable capacitance of a varactor allows a dc voltage to control the resonant frequency of an LC tank circuit. Capacitor C_2 and L_1 form an LC tank circuit that is part of a radio frequency (rf) circuit (typically an oscillator circuit). An oscillator is a circuit that converts dc into ac. Capacitor C_1 is selected to have a low reactance at rf frequencies. The low reactance of C_1 means that D_1 is effectively in parallel with C_2 at rf frequencies. Thus, the total capacitance of the tank circuit is determined by C_2 in parallel with C_{D_1}, where C_{D_1} is the effective capacitance of the varactor diode. C_{D_1} varies with the amount of reverse bias provided by the dc control voltage. Capacitor C_1 acts as a high-pass filter. It provides a low-impedance path for the rf energy, but it acts like an open circuit to the dc control voltage. This prevents L_1 from shorting out the dc source. Circuits like the one shown in Figure 9–33 are widely used in the frequency-selection circuits for FM radios, televisions, and other radio-frequency receivers. They are also called voltage-variable capacitance diodes.

FIGURE 9–33 A varactor diode can serve as the variable element in a resonant LC tank circuit.

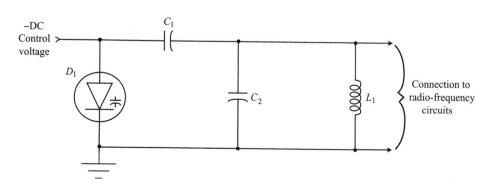

Schottky Diodes

You will recall that a varactor diode is designed to optimize the effect of the junction capacitance. A Schottky diode, by contrast, is designed to minimize the junction capacitance. With minimal junction capacitance, a Schottky diode is able to rectify much higher frequencies and can switch between conducting and nonconducting states much faster than standard diodes. The most significant identifying characteristic of a Schottky diode is high switching speeds.

Surprisingly, a Schottky diode has no pn junction. Rather, it is made by forming a junction between metal (gold or aluminum) and n-type silicon. The junction still has the ability to rectify a signal (i.e., current can only flow in one direction), but the capacitance normally associated with a pn junction is eliminated. Conduction-band electrons in the n-type silicon have higher energy levels than the free electrons in the metal. When the anode (metal) is positive with respect to the cathode (n-type silicon), electrons in the n-type material can move easily into the metal due to their higher energy level. Once in the metal, they are just like any other free electron and flow through the external circuit as current. So, under conditions of forward bias, a Schottky diode will easily pass current. The metal-silicon junction causes a forward voltage drop of 0.3 V. Because the electrons enter the metal at an elevated energy level, Schottky diodes are sometimes called **hot-carrier diodes**.

When the diode is reverse biased, free electrons in the metal do not have sufficient energy to achieve conduction-band orbits in the n-type silicon material. Therefore, current flow is negligible in the reverse-bias condition. It is important to note, however, that the current-blocking property of a reverse-biased Schottky diode does not rely on the formation of a depletion layer. Therefore, the time required for a Schottky diode to switch from on to off (reverse recovery time) is very short (in the low picoseconds). The maximum reverse breakdown voltage for a Schottky diode is usually relatively low (< 50 V). Since electrons are the only current carriers required in a Schottky diode, it is called a unipolar device.

Figure 9-34 shows the schematic symbol for a Schottky diode. Care must be taken not to confuse it with the symbol for a zener diode. The primary use for Schottky diodes is in high-speed digital circuits. Several families of digital devices are manufactured with internal Schottky diodes. The use of Schottky technology allows the circuits to change states much faster. Nearly all state-of-the-art computers utilize Schottky diodes in one or more circuits.

FIGURE 9-34 The schematic symbol for a Schottky diode.

Current-Regulator Diodes

Current-regulator diodes are designed to provide a relatively constant forward current over a wide range of voltages. The diode functions as a constant-current source. An ideal constant-current source has infinite internal impedance. Therefore, it is not surprising that the forward resistance of a current-regulator diode is very high. Typical impedances range from as low as 250 kΩ to well over 20 MΩ. This impedance range corresponds to a typical current range of as high as 5 mA to as low as 200 μA.

Figure 9-35 shows the characteristic curve and schematic for a typical current-regulator diode. Note that it offers minimal opposition to reverse current. Also note that the forward voltage drop can be much higher than the 0.7 V normally associated with forward-biased diodes. Remember, the current-regulator diode is designed for maximum forward resistance, whereas low resistance is a goal for standard rectifier diodes. Of particular importance in Figure 9-35 is the portion of the forward-biased curve that is essentially flat. This shows that current is fairly constant over a wide range of forward voltages.

Two or more current-regulator diodes can be connected in parallel to obtain higher current ratings than would be available from a single diode. For technical

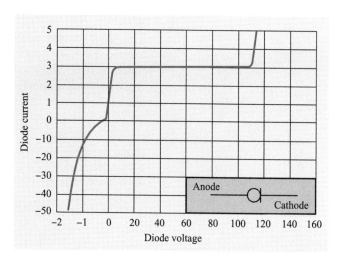

FIGURE 9–35 The characteristic curve and schematic symbol for a representative current-regulator diode.

clarity, we should also point out that even though this device has two leads and is called a diode, it is in fact a field-effect transistor (discussed in chapter 10) in disguise.

Step-Recovery Diodes

A step-recovery diode is characterized by extremely fast switching times. For this reason, its primary use is in communication circuits operating above 1 GHz. Although the step-recovery diode does have a pn junction, the method of doping the crystal is what produces the characteristic switching rates.

The p- and n-type regions of the crystal are lightly doped in the immediate vicinity of the pn junction. The doping level is higher as the distance from the junction increases. This nonuniform doping gives rise to its unique performance. Consider, for example, what occurs as the bias on the diode is changed from forward to reverse. First, the depletion layer begins to reform. During this brief interval, minority-carrier current continues to flow through the diode. This behavior is identical to a standard diode. But, as the depletion layer reforms, the diode suddenly becomes an open circuit. That is, the current through the device drops to zero almost instantly (low picosecond range). This sudden step in current during the reverse recovery time leads to the diode's name.

Now, according to basic electronics theory, high edge rates on a waveform in the time domain correspond to high harmonics in the frequency domain. For example, a square wave with faster rise and fall times has more energy in the higher-frequency harmonics than a square wave with slower rise and fall times. So, since the reverse current in a step-recovery diode makes a *very* sharp transition, we know that a lot of energy will be generated at harmonic frequencies. This characteristic of the step-recovery diode makes it well-suited for use in frequency-multiplier circuits at microwave frequencies. For example, a 500-MHz waveform can be applied to the step-recovery diode. The snap action of the diode previously described distorts the waveform and generates harmonic energy. Now, if the waveform is fed into a resonant circuit that is tuned to a harmonic frequency such as 4 GHz, this frequency will be passed on to subsequent circuits. The original 500 MHz and harmonics outside the passband of the resonant circuit will be severely attenuated. It should also be noted that some step-recovery diodes can operate at fairly high rf power levels (tens of watts).

Figure 9–36 shows two schematic symbols used to represent step-recovery diodes. Note that one of the symbols (used by major manufacturers) is a standard diode symbol. In this case, you would have to refer to a manufacturer's data sheet to know that the diode was a step-recovery device.

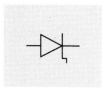

(a)

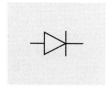

(b)

FIGURE 9–36 Schematic symbols for a step-recovery diode.

Tunnel Diodes

The tunnel diode is another device designed for operation at microwave frequencies. It has a pn junction like a standard rectifier diode, but the doping level is much heavier. The tunnel diode is generally used in its forward-biased state where it exhibits an important characteristic called negative resistance. Figure 9–37 shows the current-versus-voltage curve for a forward-biased tunnel diode.

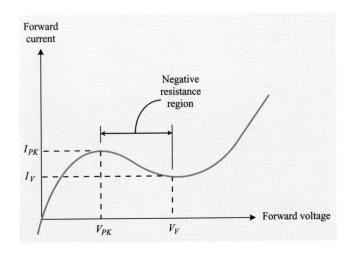

FIGURE 9–37 The characteristic curve of current versus voltage for a tunnel diode.

According to Ohm's Law, current and voltage are directly proportional. As voltage increases, so does current. But, in the region between peak voltage (V_{PK}) and valley voltage (V_V) shown in Figure 9–37, the voltage is increasing while the current is decreasing. This apparent violation of Ohm's Law defines a property we call **negative resistance**. It is the negative-resistance portion of the characteristic curve that allows a tunnel diode to be used as the active element in high-frequency oscillator and amplifier circuits. When used as an oscillator, for example, a dc bias is used to set the normal operating point to the middle of the negative-resistance region of the characteristic curve. However, since the device has negative resistance in this area, the operating point is unstable. The instantaneous operating point will swing between the peak and valley points of the characteristic curve. The resulting ac energy is coupled to a resonant circuit, which constrains the oscillations to a particular frequency. The rf power level in tunnel diode circuits is generally quite low (a few milliwatts).

Figure 9–38 shows two schematic symbols that are used to represent tunnel diodes.

(a)

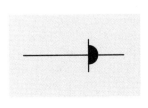

(b)

FIGURE 9–38 Schematic symbols used to represent tunnel diodes.

PIN Diodes

A PIN diode is yet another device intended for operation at extreme frequencies (from 100 MHz to at least 100 GHz). Figure 9–39 shows how the PIN diode is constructed. A layer of p-type material is separated from a layer of n-type material by a layer of intrinsic or *very* lightly doped silicon. This semiconductor sandwich of **p**-type, **i**ntrinsic, and **n**-type materials leads to the name of the diode (PIN).

The level of doping determines the depth that the depletion layer penetrates into the crystal for a given reverse voltage. Higher concentrations of doping result in a shallow depletion layer. By contrast, the depletion layer extends farther into a lightly doped material for similar values of reverse bias. Since the intrinsic silicon layer in the PIN diode has few carriers, the depletion layer will extend far into this region. This has the effect of increasing the separation between the "plates" of the junction capacitance. Therefore, we expect the junction capacitance of a PIN diode to be quite low. The junction capacitance of a Motorola

FIGURE 9–39 A PIN diode consists of a three-layer semiconductor sandwich. A layer of intrinsic silicon separates the p- and n-type materials.

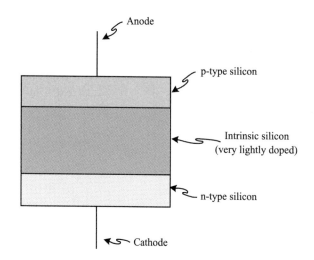

MPN3404 PIN diode, for example, is typically 1.3 pF with a reverse voltage of 15 V. Further, this capacitance remains fairly constant over a wide range of reverse voltages. This is one of the reasons a PIN diode is well-suited for operation at high frequencies.

Now, let's examine the switching characteristics of a PIN diode. If the diode is forward biased, then electrons pour into the intrinsic material from the n-type material. They pass through the intrinsic material, enter the p-type material and exit the crystal via the anode terminal. Since the intrinsic material is lightly doped, recombination is very limited. This means that at any given instant in time (under conditions of forward bias) there will be many free electrons in the intrinsic material. In effect, the intrinsic silicon has been temporarily converted to n-type material. We know the resistance of semiconductor material decreases as more current carriers (holes or electrons) are made available. In the case of the PIN diode, the resistance of the intrinsic region decreases (due to more free electrons) as the forward current increases.

Let us suppose the bias on the PIN diode is suddenly changed from forward to reverse. At the first instant, there will still be large numbers of free electrons in the intrinsic material. They will have to be removed (i.e., the depletion layer must reform) before the diode will be capable of blocking reverse-current flow. Due to the thickness of the intrinsic material, a relatively long time is required to clear the free electrons out of the intrinsic material. So, visualize what would happen if the bias voltage suddenly switched back to forward bias before the electrons were removed. Further, imagine this process continuing. Can you see that the diode would no longer be acting like a rectifier? It would be acting like a resistor. This is what happens when a PIN diode is operated at frequencies above a few hundred megahertz. But, changing the amount of dc forward-bias current can vary the value of rf resistance. Recall that higher currents result in more free electrons in the intrinsic material, which corresponds to lower resistance. Figure 9–40 on page 320 shows how the resistance of a PIN diode decreases with increasing forward currents.

The rf resistance of a typical PIN diode can range from less than one ohm with a high forward current to well over one thousand ohms for lesser currents. One application that uses the two extremes of device resistance is PIN crystal switching in a scanning communications receiver. These receivers use several crystals that must be selectively enabled. By using a PIN diode in conjunction with each crystal, we can connect or isolate a particular crystal by applying forward or reverse bias, respectively, to the PIN diode.

Below about 100 MHz, the PIN diode behaves similar to a standard rectifier diode, although the forward voltage drop is slightly higher and the forward

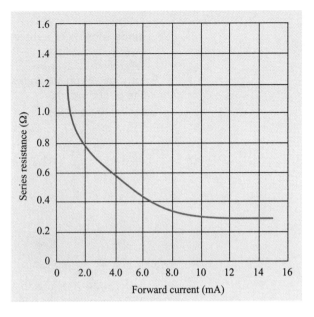

FIGURE 9–40 The effective rf resistance of a PIN diode decreases with higher forward currents. It can be used as a current-controlled resistor.

current-versus-voltage curve has a much less pronounced (more rounded) knee voltage. Most manufacturers use a standard rectifier diode symbol to represent the PIN diode on schematic diagrams.

✓ Exercise Problems 9.4

1. A _____ diode is commonly used as part of the capacitance in a resonant circuit.
2. What do the letters PIN represent in a PIN diode?
3. The capacitance of a varactor diode (*increases/ decreases*) as the reverse bias is reduced.
4. What type of diode is used as a variable capacitor whose value can be changed with a dc voltage?
5. What type of diode uses metal for one of the two junction materials?
6. A Schottky diode has very short switching times. (True/False)
7. A Schottky diode is also called a _____-_____ diode.
8. What type of diode has a very high resistance?
9. What is the name of the diode that is often used above 1 GHz as a key element in a frequency-multiplier circuit?
10. What type of diode discussed in this chapter has a negative-resistance region?
11. Voltage and current are directly proportional in a device that has negative resistance. (True/False)
12. A PIN diode is generally used as an audio-frequency oscillator due to its negative resistance. (True/False)
13. A PIN diode acts like a current-controlled resistor for high rf frequencies. (True/False)

9.5 TROUBLESHOOTING DIODE CIRCUITS

Diodes are extremely common in the electronics industry. It is essential that you be able to troubleshoot and repair systems that employ diodes. This means that you must be able to test a diode and ultimately classify it as good or bad.

In chapter 8, we discussed methods that could be used to test and classify a pn junction. Since a junction diode consists of a pn junction, it should seem reasonable that all of the troubleshooting methods discussed with reference to pn junctions also apply to troubleshooting diodes. They will be briefly summarized here. Refer to chapter 8 for greater detail. Let's first look at the possible defects that can occur in a diode. Then we will examine ways to locate these defects.

Diode Defects

There are several defects that can occur in a diode. They include:

1. Anode-to-cathode short
2. Anode-to-cathode open
3. Low front-to-back ratio
4. Out-of-tolerance parameters (e.g., reverse current)

The first three defects cited can be identified with the methods listed in the following paragraphs and detailed in chapter 8. The fourth defect (out-of-tolerance parameters) can be very difficult to locate. It often requires an in-depth knowledge of the operation of the system being repaired. Then, based on careful observation of the symptoms, you can identify the diode as a suspect component. In this case, the diode may test good on available test equipment. Correct diagnosis can only be confirmed by substituting the suspect diode with a known good one. If the symptoms are corrected with substitution, then the diode was indeed defective.

Now, let us examine specific tests that can be used to locate a defective diode. We will categorize our diode troubleshooting methods into three classes: voltage measurements, ohmmeter tests, and diode testers.

Voltage Measurements

Probably the most informative voltage measurement for testing the condition of a diode is measurement of the forward voltage drop. If the diode is forward biased with a dc voltage, then simple monitoring of the anode-to-cathode voltage with a voltmeter will provide the necessary data. Recall that if the forward voltage drop is substantially greater than one volt (except for some special-purpose diodes), the diode is definitely defective. If, on the other hand, the forward voltage drop measures a few tenths of a volt, then it is *probably* good.

If the diode is in an ac circuit, the same theory applies, but you will need to use an oscilloscope to measure the forward voltage drop. That is, since the diode is forward- and reverse-biased on alternate half-cycles of the input waveform, we need to be able to distinguish between half-cycles. The oscilloscope provides this capability.

Ohmmeter Tests

A VOM can be used to measure the forward- (front) and reverse- (back) resistance of a diode. Good diodes generally have front-to-back resistance ratios of 100 or more.

Many digital ohmmeters cannot measure the front-to-back resistance ratio directly, but they often have a special provision for testing diodes and other pn junctions. Consult the operator's manual for the particular DVM being used.

Diode Testers

A diode tester is a specialized test set designed to evaluate the condition of a diode. These testers provide a more reliable means of determining a diode's condition. Unfortunately, you may not have access to a diode tester for routine troubleshooting tasks. Thus, the other methods described must be utilized.

Rectifier Circuit Defects

As a first step toward troubleshooting rectifier circuits, let us begin by classifying the symptoms of the defect into one of two classes:

- The power supply is defective, but there is no visible damage and no fuses are blown.
- The rectifier circuit (or the circuit board) shows visible signs of damage or a fuse is blown.

Each of these symptoms dictates a different approach to trouble isolation. Let's examine each one separately.

No Visible Damage/No Blown Fuses

This is the easiest type of rectifier malfunction to diagnose. Input power can be safely applied to the system and voltage measurements taken with a voltmeter or, preferably, an oscilloscope. The first step is to determine whether the rectifier circuit itself is defective or whether the external load circuit contains the defect. Figure 9–41 illustrates how these two cases can have similar symptoms.

FIGURE 9–41 A defect in the external load circuitry can cause symptoms (i.e., zero volts output) that are similar to a defective rectifier circuit.

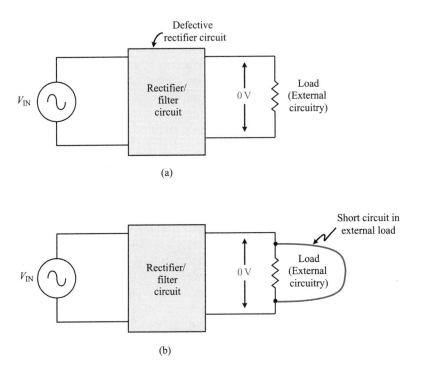

In Figure 9–41(a), the rectifier circuit is actually defective (perhaps an open diode in a half-wave rectifier circuit). The result is that there is no dc output voltage. In Figure 9–41(b), the symptoms are identical, but they are caused by a short in the external load circuit.

The simplest way to determine the general location of the defect (i.e., internal or external to the power supply) is to replace the external load circuitry with a simple resistor. The value of the resistor should be selected (Ohm's Law) so that the expected amount of current will be drawn from the supply. If the output voltage returns to normal after substituting the load circuit, then the defect is in the external circuitry. If the power supply still has no output, even with a resistive load, then the defect is internal to the rectifier/filter circuit. Following are some of the possible internal defects:

- Open rectifier diode(s)
- Open wire or trace
- Open series element of the filter

In any case, the defect is likely an open circuit. If we had a short circuit, then the symptoms would generally include a blown fuse or circuit breaker, which would place the defect into one of the categories discussed in the following sections.

Another nondestructive symptom that can occur is excessive ripple voltage. Given that the circuit worked correctly in the past (i.e., the design is correct), then the most likely problem is a defective capacitor in the filter circuit. It may be open, leaky, or have a high internal resistance, but in any case, it needs to be replaced.

Visible Damage/Blown Fuses

Before extensive troubleshooting can begin three tasks must be accomplished:
- Repair damage to circuit board traces.
- Replace any visibly damaged components.
- Check for shorted components with an ohmmeter.

It is important to understand why each of these steps is needed. The first two steps are self-evident. The defective traces and components must be repaired or replaced if we expect the circuit to operate. The third step is very important, but it is easily overlooked. Figure 9–42 illustrates the importance of ohmmeter checks after repair of visible damage and *before* reapplication of ac power.

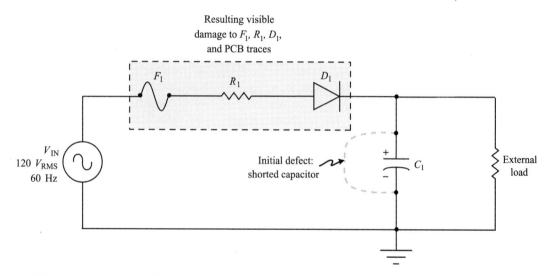

FIGURE 9–42 A shorted filter capacitor (or other component) can cause visible damage to other parts of the circuit masking the location of the original defect.

The original defect in the circuit shown in Figure 9–42 was a shorted filter capacitor. This resulted in extremely high currents that visibly damaged other components: fuse open, printed circuit board (PCB) traces vaporized, R_1 scorched, and D_1 disintegrated. If we simply repair the PCB trace and replace the components with visible damage (F_1, R_1, and D_1), then the new components will be damaged as soon as we reapply power since the initial defect (C_1 shorted) still exists. So, after replacing components that show visible damage, we must carefully check for shorted components with an ohmmeter before reapplying power to the circuit.

There is one more recommended step that can be taken if you have access to a variable ac power source (e.g., a variable autotransformer). After you have cleared all obvious defects and checked for dc shorts, you can reapply ac power gradually while monitoring the current. If the current increases too rapidly, then you know the circuit still has problems, but you can stop before damage occurs. If you reapply full power instantly, the newly-replaced components may be damaged.

The most valuable troubleshooting tool for diagnosing defects in diode circuits is a solid understanding of basic circuit theory. Additionally, the more you understand about the normal operation of the specific circuit being repaired, the faster you can focus in on the actual defective component(s). In short, rely on your knowledge of theory to guide your measurements and to interpret your results.

In-Circuit Diode Resistance Checks

Except in a few unique applications (e.g., white noise generators and logarithmic amplifiers), diodes are used for their unidirectional characteristics. Because of this, their failure is often easy to detect by observing the symptoms. The two most commonly occurring defects in diodes are:

- Open
- Short

Although other defects are possible, it seems these occur most frequently. When diagnosing a circuit whose operation is based on a diode, ask yourself if the observed symptoms could be caused by a particular open or shorted diode. If the answer is yes, then confirm your suspicions with an ohmmeter. In many circuits, both open and shorted diodes can be detected with an ohmmeter *without* removing them from the circuit. This assertion relies on the fact that the forward resistance of a diode is generally less than the resistance of any parallel sneak-current paths.

Defects in Special-Purpose Diodes

The special-purpose diodes presented in this chapter are used in many types of electronic equipment. You must be able to troubleshoot circuits that use these diodes, and you must be able to classify a given diode as good or bad.

For the most part, the specialized diodes discussed in this chapter can be tested in much the same way as standard rectifier diodes. For example, defects in zener, Schottky, step-recovery, PIN, and varactor diodes can often be located with simple ohmmeter tests as described previously. As with rectifier diodes, a more conclusive test can be made with a semiconductor tester.

Working with low-power microwave devices requires particular caution. They are often sensitive and can be very expensive. Some ohmmeters (particularly analog meters) can provide sufficient power to destroy a sensitive diode. When in doubt, check the manufacturer's datasheet ratings (reverse breakdown voltage and maximum forward current) for the diode in question and compare them to the voltage and current available from the ohmmeter on a particular range. You can even measure the open-circuit voltage and the short-circuit current of the ohmmeter with a DMM or VOM.

If you routinely troubleshoot circuits that use a particular special-purpose device such as a tunnel diode, current-regulator diode or step-recovery diode that may produce inconclusive results with ohmmeter measurements, you can build a simple test circuit (e.g., a microwave oscillator) that uses the diode. When you suspect a defective diode, you can remove it from the actual system and insert it into the test circuit. If the test circuit functions normally, then the diode is good. Otherwise, the diode is defective. In most cases, the test circuit can be quite simple and may even be a duplicated portion of the real circuit.

✓ Exercise Problems 9.5

1. Which of the following diode defects can generally be identified with an ohmmeter? Circle all that apply.
   a. anode-to-cathode short
   b. anode-to-cathode open
   c. low front-to-back ratio
2. It is possible for a diode to be defective and still test as a good device with an ohmmeter test. (True/False)
3. Explain how measuring the forward voltage drop of a rectifier diode can provide a good indication of the condition (good or defective) of the diode.
4. A rectifier diode measures 125 Ω and 250 kΩ in the forward and reverse directions, respectively. This indicates that the diode is probably good. (True/False)

5. An ohmmeter generally gives a better indication of a diode's condition than a diode tester. (True/False)
6. If you find that a printed circuit board trace has been burned open in a power supply circuit, it is not a good practice to simply repair the trace and immediately reapply power. Why not?
7. If the primary symptom of a defective power supply is that the output is low or zero, explain why replacing the real load with a simple resistor is a good troubleshooting step.
8. Defects in zener diodes can be identified using the same strategies as those used for troubleshooting rectifier diodes. (True/False)

●—SUMMARY

A pn junction diode is a two-terminal device that passes current easily in one direction (forward bias) and essentially stops current in the opposite direction (reverse bias). A diode is forward biased when its anode (p-type material) is more positive than its cathode (n-type material). An ideal diode acts as a short and an open when forward and reverse biased, respectively. A practical diode has a small voltage drop (≈ 0.7 V) when forward biased and a small leakage current when reverse biased.

A rectifier circuit is used to convert ac into pulsating dc. In most cases, the pulsating dc must be filtered to provide a smooth dc for electronic circuits, which require minimal ripple voltage. A half-wave rectifier produces output pulses at the same frequency and essentially the same peak voltage as the input. A full-wave center-tapped rectifier and a bridge rectifier produce output pulses at twice the frequency of the input. The peak output voltage of a bridge is about the same as the input peak. The peak output voltage of a center-tapped rectifier (also called a biphase half-wave rectifier) circuit is equal to one-half of the full secondary voltage.

A rectifier filter circuit consists of a low-pass filter. Common choices include simple capacitor filters, L-type filters consisting of a series inductor followed by a shunt capacitor, and a Pi-type (π-type) filter consisting of a capacitor filter followed by an L-type filter. The better the filtration in a rectifier filter network, the lower the ripple voltage from the power supply circuit.

A limiter (also called a clipper) circuit has no effect on the input waveform as long as the peak excursions are below the clipping level of the circuit. If the peak voltage exceeds the clipping level of the limiter circuit, then the output waveform is clipped at the clipping level. Clipping may occur on either or both alternations of the waveform depending on the design of the limiter circuit.

Isolation diodes can selectively connect or isolate a circuit. The operation of an isolation diode relies on the unidirectional property of a diode. Diodes can be used to provide asymmetrical time constants for the charging and discharging of a capacitor in an RC circuit. This chapter also illustrated how a diode can be used to recover the information from an amplitude-modulated rf signal.

There are a number of special-purpose diodes whose characteristics are particularly well-suited for specific applications. Table 9–1 provides a brief summary of the special diodes discussed in this chapter.

TABLE 9–1 Summary of various diodes.

Diode Type	Characteristics	Application	Symbol
Zener	Fairly constant reverse voltage drop beyond breakdown. Operated between knee current and maximum current.	Voltage regulators	
Varactor	Junction capacitance varies inversely with reverse bias	Electronically-controlled resonant tanks in communication circuits	

TABLE 9–1 (continued)

Diode Type	Characteristics	Application
Schottky	Metal-silicon junction, low junction capacitance, fast switching times (low picoseconds)	High-speed digital circuits
Current-Regulator	High forward resistance and relatively constant current.	Constant-current source
Step-Recovery	High-frequency (> 1 GHz), graduated doping level, reverse recovery has snap action	Microwave frequency multipliers
Tunnel	Heavily doped, has negative resistance region, low power	High-frequency oscillators and amplifiers
PIN	P-type/intrinsic/n-type sandwich, low capacitance, current-controlled rf resistance	Rf resistance or switch

REVIEW QUESTIONS

Section 9.1: Diode Characteristics

1. The anode lead of a junction diode is connected to (*p-type/n-type*) material.
2. When a rectifier diode is forward biased, the _____ terminal is more positive than the _____ terminal.
3. The anode of a diode is −10 V with reference to ground. If the diode is forward biased, what is the voltage on the cathode with respect to ground?
4. If one end of a diode has a colored band, what does this indicate?
5. How does an ideal diode act when it is forward biased?
6. How does an ideal diode act when it is reverse biased?
7. Is it possible for a practical diode to have a large current flow in the reverse-bias direction?
8. A rectifier diode has +10 V on its cathode and +5 V on its anode. The diode is (*forward/reverse*) biased.
9. What is the primary difference between an ideal diode and a practical diode in the forward-bias polarity?

Section 9.2: Power Supply Applications

10. One of the main purposes of the power supply in an electronic product such as a computer is to convert ac into dc. (True/False)
11. The input voltage to a half-wave rectifier circuit is 18 V_P at 60 Hz. What is the approximate peak output voltage?
12. The input voltage to a half-wave rectifier circuit is 24 V_{rms} at 60 Hz. What is the frequency of the ripple voltage?
13. If the full secondary voltage of a biphase half-wave rectifier circuit is 100 V_P, what is the approximate peak output voltage?
14. What is the relationship between input frequency and output ripple frequency in a center-tapped full-wave rectifier circuit?

15. The diodes in a center-tapped rectifier circuit conduct at the same time. (True/False)
16. If the input frequency to a bridge rectifier circuit is 400 Hz, what is the ripple frequency in the output?
17. How many diodes are required to build a full-wave bridge rectifier?
18. In a bridge rectifier circuit, two diodes conduct at the same time. (True/False)
19. If the peak input voltage to a bridge rectifier is 25 V_P, what is the approximate peak output voltage?
20. The larger the capacitance in a rectifier filter circuit, the (*smaller/larger*) the ripple voltage.

Section 9.3: Miscellaneous Diode Applications

Assume ideal diodes for questions 21 through 28.

21. Refer to Figure 9–43. If the peak input voltage is 2 V, draw the output waveform.
22. Refer to Figure 9–43. If the peak input voltage is 20 V, draw the output waveform.
23. Refer to Figure 9–44. If the peak input voltage is 2 V, draw the output waveform.
24. Refer to Figure 9–44. If the peak input voltage is 20 V, draw the output waveform.
25. Refer to Figure 9–45. If the peak input voltage is 2 V, draw the output waveform.
26. Refer to Figure 9–45. If the peak input voltage is 20 V, draw the output waveform.
27. Refer to Figure 9–46. Draw the proper battery polarity for B_1 and B_2 to produce the output waveform shown.
28. Refer to Figure 9–47 and assume that all lamps are designed to be operated from 120 Vac. Complete

FIGURE 9–44 Circuit for questions 23 and 24.

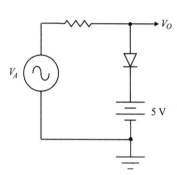

FIGURE 9–43 Circuit for questions 21 and 22.

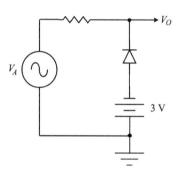

FIGURE 9–45 Circuit for questions 25 and 26.

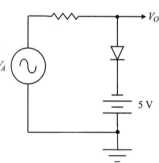

FIGURE 9–46 Circuit for question 27.

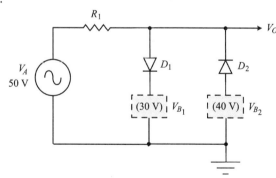

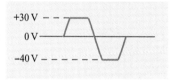

FIGURE 9–47 Circuit for question 28.

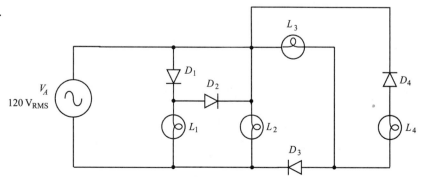

the following table to indicate the relative lamp brightness in terms of off/dim/bright.

Lamp	Brightness
L_1	
L_2	
L_3	
L_4	

Section 9.4: Special Diodes

29. What type of diode is generally selected for use as a voltage regulator?
30. What is the most outstanding electrical characteristic of a Schottky diode?
31. A zener diode is generally operated in the (forward-/reverse-) biased portion of its operating curve.
32. Refer to Figure 9–48. Which of the following statements is correct? Circle all that apply.
    a. This is a voltage regulator circuit.
    b. The diode must be reversed in order for this circuit to work properly.
    c. The breakdown voltage rating of D_1 must be greater than 15 V.
    d. The breakdown voltage of D_1 must be less than 8 V.

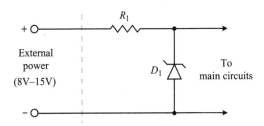

FIGURE 9–48 Circuit for question 32.

33. What type of diode is generally used as a variable capacitance in an rf tuning circuit?
34. What type of diode serves as a constant-current source?
35. What type of diode has a layer of intrinsic silicon between the p- and n-type layers?
36. What type of diode has a region of negative resistance?
37. What type of diode has a snap action when it goes from the conducting to the nonconducting state?
38. What is the approximate knee voltage of a 10-V zener diode?
39. What name is given to the process of maintaining a constant output voltage even though the input voltage has variations?
40. A PIN diode is also called a hot-carrier diode. (True/False)

Section 9.5: Troubleshooting Diode Circuits

41. An ohmmeter can be used to detect an anode-to-cathode short in a diode. (True/False)
42. If an ohmmeter shows no apparent indication in either polarity while measuring a diode, the diode is probably shorted. (True/False)
43. If the forward resistance of a diode measures 35 Ω with an ohmmeter, and the reverse resistance measures 75 Ω, the diode is probably defective. (True/False)
44. If you suspect that a diode has excessive reverse-leakage current, what is the best choice of troubleshooting strategies to verify your suspicion?
45. If a forward-biased rectifier diode has a voltage drop of 18.7 V, what does this indicate?
46. If a reverse-biased rectifier diode has a voltage drop of 50 V, it is definitely defective. (True/False)
47. If there is evidence of physical damage in a power supply circuit (such as scorched components and open circuit board traces), then there may be a shorted component in the circuit. (True/False)
48. Why is it not a good practice to test the condition of a low-power microwave diode with an ohmmeter?

•—CIRCUIT EXPLORATION

Since it is literally impossible to study the operation of every possible type of circuit you will encounter in industry, it is important for you to be able to analyze unfamiliar circuits to discover how they work. This Circuit Exploration will give you an opportunity to do just that.

We have not formally studied the behavior of the circuit configuration shown in Figure 9–49 on page 329, but it is a signal-conditioning application for a diode.

Study the circuit and accomplish the following tasks:

1. Determine the waveform across each component.
2. Construct the circuit in the laboratory or in a simulation environment.
3. Use an oscilloscope (real or virtual) to determine the actual waveforms across each component.
4. Take the time to understand any discrepancies between the waveforms you anticipated in step 1 and the actual waveforms measured in step 3.

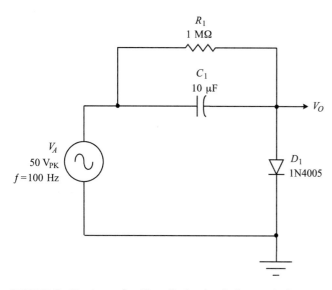

5. Write a brief theory of operation that explains how the circuit operates.
6. Based on your knowledge of circuit operation, answer the following questions:
    a. What is the effect on output waveshape if the input voltage amplitude is doubled?
    b. How would the circuit respond to a square wave input instead of a sine wave?
    c. This circuit is sometimes called a clamper circuit. Why?

FIGURE 9–49 An unfamiliar diode circuit for analysis.

 Fig09_49.msm

CHAPTER 10
Transistors and Transistor Circuits

•—KEY TERMS
alpha
base
beta
bipolar transistor
channel
collector
cutoff
depletion mode
drain
emitter
enhancement mode
field-effect transistor
gate
inversion layer
quiescent operating point
saturation
source
totem-pole output

•—OBJECTIVES
After studying this chapter, you should be able to:

- Describe the construction of and state the bias requirements for bipolar junction transistors, JFETs, and MOSFETs.
- Trace all current paths through a properly biased transistor and indicate the relative magnitudes of all currents.
- Describe how a transistor can be used like a switch.
- Test a transistor and classify it as good or defective.
- Draw and label the schematic symbol for both types of bipolar transistors, both n- and p-channel JFETs, and both n- and p-channel MOSFETS (including both depletion- and enhancement-mode devices).
- Contrast the general characteristics of an amplifier that are required for classification into the following amplifier classes or configurations:

class A	class C	difference
class AB	current	power
class B	Darlington	voltage

- Explain the operation of each of the following amplifier circuits:

| cascode | common-collector | common-drain |
| common-emitter | common-gate | common-source |

- Explain the operation of a basic CMOS inverter.

•—INTRODUCTION
During the first part of the 1960s, most electronic applications made the transition from vacuum tube amplifiers and computer logic to transistor equivalents. Transistors performed the same operations, but they were less expensive, more reliable, and required less power. Now, the practical reality is that transistors as discrete components are quickly disappearing except for specific isolated applications. Most circuit designs previously accomplished with many individually-packaged transistors now use integrated circuits with hundreds or even millions of transistors in a single package. In many circuits (and computers are no exception), there are usually one or two discrete transistors that are included for some special function, but the days of a product based entirely on individually-packaged (i.e., discrete) transistors are gone.

Indeed, if discrete transistors are quickly vanishing from new designs, then why should you devote any effort to learning how they work? There are some very good reasons. First, the integrated circuits that are replacing the discrete transistors in most applications are comprised of hundreds, thousands, or even millions of transistors integrated into a single silicon crystal. By understanding the behavior of a single transistor, you are in a much better position to comprehend the operation of a complex integrated circuit that utilizes thousands or millions of transistors. Second, although the use of discrete transistors now represents only a tiny portion of a typical new design, many (if not most) systems still utilize one or more discrete transistors. And that is not likely to change in the immediate future. Consequently, you must understand their operation.

The material in this chapter focuses heavily on the operational characteristics of various types of discrete transistors. This knowledge will allow you to appreciate their behavior, analyze unfamiliar applications, and diagnose defective circuits. However, we do not devote extensive time to mathematical analysis of transistor circuits. While this is a most enjoyable activity, it is simply inconsistent with the role that transistors play in current designs.

10.1 BIPOLAR TRANSISTORS

The pn junction diodes discussed in previous chapters were bipolar devices, since their operation relied on two types of current carriers: holes and electrons. In this section, we explore the construction and operation of two additional bipolar devices: npn and pnp **bipolar transistors**.

Construction

Let's briefly examine the internal construction of bipolar transistors. You will need to keep these images in mind as we progress into transistor operation.

NPN

Figure 10–1(a) shows a layer diagram for an npn transistor. As you can see, it consists of three layers of semiconductor material. A very thin layer of p-type material is sandwiched between two regions of n-type material. As illustrated in Figure 10–1(a), two pn junctions are formed.

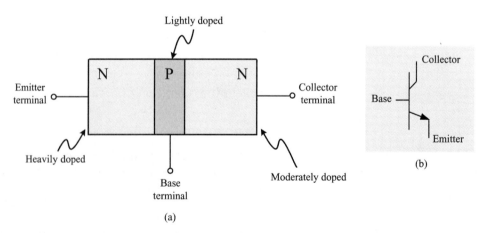

FIGURE 10–1 (a) Construction and (b) the schematic symbol for an npn transistor.

A lead is attached to each of the three semiconductor regions. The thin, lightly-doped p-type region is called the **base** of the transistor. The heavily-doped and moderately-doped n-type regions are called the **emitter** and **collector**, respectively. Figure 10–1(b) shows the schematic symbol for an npn transistor.

PNP

Figure 10–2(a) shows the construction of a pnp transistor. This is essentially the complement of an npn from the standpoint that the p- and n-type regions are interchanged. However, it is important to note that the emitter terminal is still connected to the heavily-doped region on one end of the semiconductor sandwich. Similarly, the base still consists of a thin, lightly-doped region in the center of the semiconductor sandwich. Figure 10–2(b) shows the schematic symbol for a pnp transistor. Note that the only difference between npn and pnp transistor symbols is the direction of the arrow on the emitter terminal.

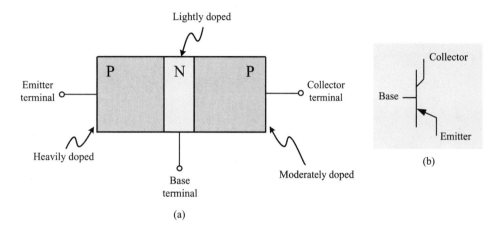

FIGURE 10–2 (a) Construction and (b) the schematic symbol for a pnp transistor.

For the remainder of the section, we will discuss npn transistors in greater detail, and then briefly examine the same characteristics in a pnp device. Both are important, but as you will soon see, the behaviors are very similar and their primary differences are well-defined.

Biasing

In order for a transistor to serve a useful purpose, the two pn junctions must be properly biased. Except for specific discussions, which are obvious by their context, the term bias implies dc voltages and currents.

Base-Emitter Junction

You will be pleased to know that the base-emitter junction of a transistor behaves like any other pn junction when viewed alone. Figure 10–3(a) and Figure 10–3(b) on page 334 illustrate forward and reverse bias, respectively, on the base-emitter junction of an npn transistor.

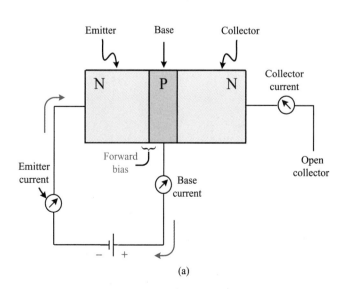

FIGURE 10–3 (a) Forward and (b) reverse bias on the base-emitter junction of an npn transistor. (c) Forward and (d) reverse bias on the base-emitter junction of a pnp transistor.

FIGURE 10–3 (*continued*)

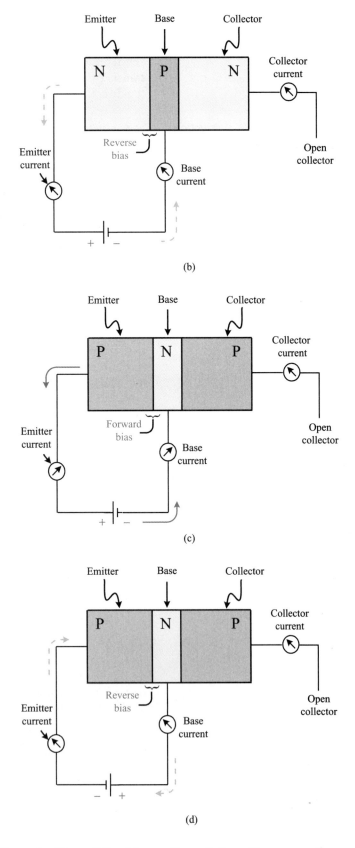

First, notice that the polarities of the bias voltage follow the conventions discussed for diodes. That is, if the p-type material is more positive than the n-type material, the junction is forward biased. Second, observe that a forward-biased transistor junction readily permits current flow. There will be only a small

(≈ 0.7 V) forward voltage drop across a forward-biased junction, just as we found with diodes. According to Kirchhoff's Current Law, we know that the base and emitter currents will be equal under these conditions.

Figure 10–3(c) and Figure 10–3(d) illustrate forward and reverse bias, respectively, on the base-emitter junction of a pnp transistor. As you would expect, the polarities are necessarily opposite those in an npn device, but the behavior is consistent. So, up to this point, there are no surprises in the behavior of the semiconductor sandwich.

Base-Collector Junction

When biased independent of the emitter-base junction, the collector-base junction behaves just as you would expect. It has negligible current when reverse biased (i.e., it acts as an open) and readily passes current (i.e., it acts as a short) when forward biased. Additionally, it drops approximately 0.7 V across the forward-biased junction. Again, Kirchhoff's Current Law tells us that the base and collector currents will be identical under these conditions. Figure 10–4 summarizes the behavior of an isolated collector-base junction.

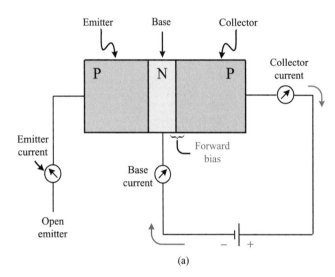

FIGURE 10–4 The behavior of (a) a forward- and (b) a reverse-biased collector-base junction.

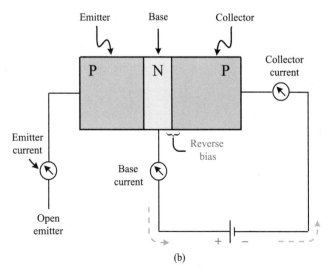

Effects of Simultaneous Biasing

Now, let's examine the current flow in a transistor when both of its junctions are biased at the same time. There are four possible combinations of forward and reverse bias for the two junctions. Only three of these play key roles in practical applications, so we will focus on these.

Reverse-Biased BE and Reverse-Biased CB Junctions Figure 10–5 illustrates the case where both junctions are reverse biased at the same time. There are no surprises here. No significant current flows in any lead of the transistor. This is an important biasing mode that is essential to the operation of digital systems including computers. The state of the transistor is called **cutoff**. During cutoff, a transistor essentially acts as an open circuit. This is one of two states used to represent information and control activity in a digital computer. This state is not generally used for linear applications such as amplifiers except for specially-designed amplifier circuits that use multiple transistors to amplify a single waveform.

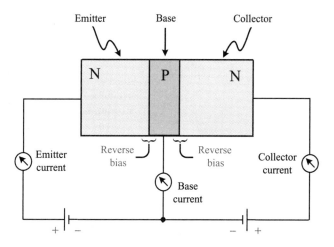

FIGURE 10–5 When both junctions are reverse biased, the transistor is cut off.

Forward-Biased BE and Reverse-Biased CB Junctions Figure 10–6 illustrates a transistor with a forward-biased emitter-base junction and a reverse-biased collector-base junction. At first glance, you might expect the emitter and base current to be high and equal in value, while the collector current remains near zero (since the collector-base junction is reverse biased.) But, as illustrated in Figure 10–6, the emitter and collector currents are nearly equal, and the base current is very small. Here is where your understanding of the internal operation of semiconductor material is important. Let's go inside of the crystal to see what is causing this unexpected behavior.

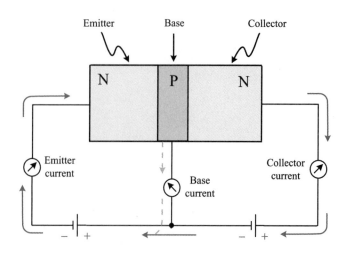

FIGURE 10–6 A transistor with a forward-biased base-emitter junction and a reverse-biased base-collector junction.

Figure 10–7 further illustrates the internal activity that is occurring. The depletion layer associated with the collector-base junction is quite wide. You will recall that the depletion layer gets wider as the reverse bias is increased. Also, remember that the depletion layer extends farther into a lightly-doped region. Since the base region has less doping than the collector, the depletion layer of the collector-base junction will reach farther into the base as indicated in Figure 10–7. Essentially, the depletion layer has no current carriers. However, if an electron ever found itself in the depletion layer (as when a covalent bond breaks due to thermal energy), it would be quickly swept out of the depletion layer by the strong electrostatic field. It would move through the n-type crystal material and exit as collector current.

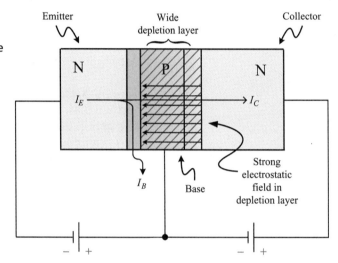

FIGURE 10–7 Electrons pour into the base from the emitter, but most are swept into the collector by the electrostatic field of the base collector.

Now, let's examine another aspect of the condition illustrated in Figure 10–7. The emitter-base junction is forward biased. This means that conduction-band electrons in the emitter can easily move into the p-type base region. Further, since the emitter is heavily doped, electrons crossing the emitter-base junction will flood the base.

Once electrons have entered the base region, they are free to drift or diffuse throughout the base region until one of two events occurs. If a given electron encounters a hole in the p-type base material, then recombination will occur and the electron will exit the base lead as a base current. This is exactly the same behavior as that which occurs in a standard rectifier diode. But, in the case of a transistor, another possibility exists. If a particular electron encounters the depletion layer of the base-collector junction before it finds a hole for recombination, then the electron will be swept into the collector region and will exit the transistor as collector current.

The transistor is constructed in a way that encourages electrons to make the trip through the base and emerge as collector current. First, the base is lightly doped, which makes it more difficult for recombination to occur. Second, the base is very thin, which increases the chances that an electron will encounter the depletion layer before it encounters a hole. In a practical transistor, over 95% and often over 99% of the emitter current succeeds in passing on through to the collector. This is sometimes called the **alpha** (α) of the transistor. It is simply the ratio of collector current to emitter current (i.e., $\alpha = I_C/I_E$).

The discussions in this section can be extended to include pnp transistor operation, but the voltage polarities and electron current directions are opposite. Note that the operation of npn and pnp transistors depends upon two types of current carriers: holes and electrons. This is why they are called bipolar transistors.

This biasing condition (BE junction forward biased and CB junction reverse biased) is required for linear amplification using a single transistor. The output waveshape of a linear amplifier is identical to the input waveshape.

One extremely important characteristic of transistor operation is this: as long as a transistor is operated in its linear range of operation, there will be a relatively constant relationship between the three transistor currents. That is, emitter and collector currents will be nearly equal and base current will be quite small. Now, if we controlled or varied the base current while the transistor is biased into its linear range, then it follows that the emitter and collector currents would vary proportionally. This is the basis for amplification; changes in a small signal cause corresponding changes in a larger signal. We will explore this important concept more in a subsequent section.

We have seen that base current is relatively small and is related to the higher value of collector current. The ratio between collector and base current is often called **beta** (β). That is, $\beta = I_C/I_B$. This same value is called h_{FE} on a transistor datasheet.

Forward-Biased BE and Forward-Biased CB Junctions When both pn junctions are forward biased, current is relatively high in all three transistor leads. Each pn junction acts like a low resistance and permits a relatively high current flow. This set of bias conditions is used in digital circuits. It is called **saturation**. During saturation, a transistor readily conducts current. In effect, it acts like a short circuit. Saturation and cutoff are the two conditions used by the millions of transistors in a typical computer or microprocessor system. For reasons discussed in the previous section (e.g., thin and lightly-doped base), most of the current still flows from emitter to collector (or collector to emitter in the case of a PNP transistor). That is, base current is still small relative to the other two currents. However, it is substantially higher than it is when the transistor is operated in its linear mode (i.e., BE and CB junctions forward and reverse biased, respectively). The following voltages are typical for an npn transistor in saturation: $V_E = 0$ V, $V_B = +0.7$ V, and $V_C = +0.2$ V. Since the base terminal is more positive than the emitter and collector terminals, both junctions are forward biased.

Biasing Requirements

Table 10–1 summarizes the biasing requirements based on the application. These conditions must be satisfied in order for the circuit to operate properly. Recognition of these conditions (or their absence) can provide valuable insights when troubleshooting defective transistor circuits.

TABLE 10–1 Biasing requirements for transistor applications.

Application		Transistor Junction Bias Requirement	
		Base-Emitter	Collector-Base
Digital circuits	Saturation	Forward	Forward
	Cutoff	Reverse	Reverse
Amplifier circuits	Linear	Forward	Reverse

Bipolar Transistor Amplifiers

Bipolar transistors (both pnp and npn) can be used as amplifiers. Amplifiers are used to make a small signal (current, voltage, or power) into a corresponding signal with a larger amplitude. A public address system is a good example of an amplifier. Here, an amplifier is used to increase the weak microphone output to a level large enough to drive many speakers in an auditorium.

Amplifier Basics

We will now examine the general characteristics of amplifiers. These characteristics are applicable whether the amplifier circuit is constructed around discrete transistors or integrated circuits.

Gain Gain is a term used to indicate the amount of increase or amplification given to a signal by an amplifier. Figure 10–8 shows an amplifier with a 1-V input and a 10-V output. This amplifier can be described as having a voltage gain of 10, since the output is ten times as large as the input.

FIGURE 10–8 Gain describes the ratio of output to input for an amplifier.

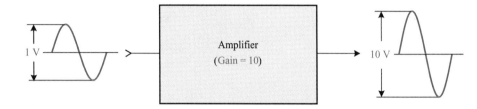

We can express gain mathematically with Equation 10–1.

$$\text{gain} = \frac{\text{output}}{\text{input}} \qquad (10\text{–}1)$$

An amplifier can be designed to amplify current, voltage, or power. The example in Figure 10–8 is called a voltage amplifier, since the voltage is increased in amplitude.

• EXAMPLE 10.1

Calculate the current gain of an amplifier with an input current of 1 µA and a corresponding output current of 2 mA.

SOLUTION We apply the generic gain equation (Equation 10–1) as follows:

$$\text{gain} = \frac{\text{output}}{\text{input}}$$

$$\text{Current gain} = \frac{\text{output current}}{\text{input current}} = \frac{2 \text{ mA}}{1 \text{ µA}} = 2000$$

Observe that gain has no units of measure. It is simply a ratio.

---

dB Gain Gain (or loss) can also be expressed in terms of decibels. Decibels are based on logarithms, and they provide a convenient way to express very large ratios (e.g., voltage gain is a ratio of output voltage to input voltage). Like the trigonometric functions discussed in chapter 6, logarithm values can be found easily with an engineering calculator.

• EXAMPLE 10.2

Find the logarithm of each of these numbers: (a) 100, (b) 25,000, and (c) 146.

SOLUTION Simply enter the original number into your calculator and press the log button. (Note that this is not the same as the ln button.) Following this procedure, we get the following results: (a) 2, (b) 4.398, and (c) 2.164.

Equations 10–2 through 10–4 provide the conversion formulas for voltage, current, and power gains, respectively.

$$\text{Voltage gain (dB)} = 20 \log \frac{\text{output voltage}}{\text{input voltage}} \qquad (10\text{--}2)$$

$$\text{Current gain (dB)} = 20 \log \frac{\text{output current}}{\text{input current}} \qquad (10\text{--}3)$$

$$\text{Power gain (dB)} = 10 \log \frac{\text{output power}}{\text{input power}} \qquad (10\text{--}4)$$

Notice in each case, the term to the right of the log operator is simply the gain ratio of the amplifier.

EXAMPLE 10.3

The output power from a certain amplifier is 10 W with only 0.1 W of input power. Express the gain of this amplifier in decibels.

SOLUTION   We apply Equation 10–4 as follows:

$$\text{Power gain (dB)} = 10 \log \frac{\text{output power}}{\text{input power}}$$
$$= 10 \log \frac{10 \text{ W}}{0.1 \text{ W}} = 10 \log 100$$
$$= 10 \times 2 = 20 \text{ dB}$$

Calculator Application

Phase Relationships   In the process of being amplified, a signal may or may not experience a phase shift. Varying amounts of phase shift are possible, but we often generalize amplifier behavior into one of two classes based on phase shift: inverting or noninverting. Figure 10–9(a) shows that the output of an inverting amplifier is 180° out of phase with its input. By contrast, Figure 10–9(b) shows that the input and output for a noninverting amplifier are in phase. The input/output phase relationship is one way to characterize the performance of a particular amplifier circuit.

Input and Output Resistance   Figure 10–10 illustrates an amplifier driven by a signal source and driving a load. This might represent, for example, a microphone

FIGURE 10-9 An amplifier may cause a phase inversion.

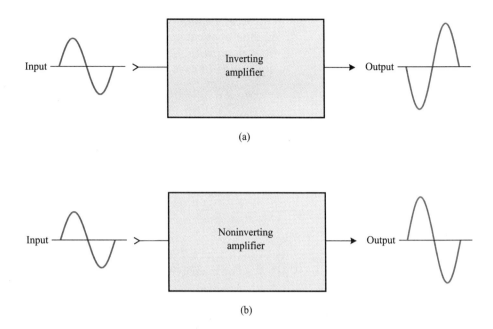

(a)

(b)

(source), an amplifier, and a speaker (load). The diagram could equally represent a series of three amplifiers where the output of one amplifier provides the input to the next one and so on.

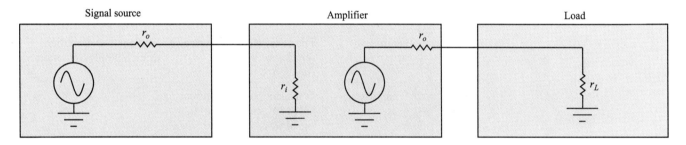

FIGURE 10-10 Practical amplifiers have input and output resistances.

The magnitude of the effective input signal to the amplifier in Figure 10–10 is whatever voltage is dropped across its input resistance (r_i). We know from basic circuit theory (i.e., Ohm's Law) that this depends on three things: the amplitude of the signal source, the value of the source output resistance (r_o), and the value of the input resistance of the amplifier (r_i). For minimal loading effects, we want the source to have a low output resistance while the amplifier has a high input resistance. It can be shown mathematically that for maximum transfer of power, the output resistance of the driving circuit should be identical to the input resistance of the driven circuit. So, for example, you would connect an 8-Ω speaker to an amplifier whose output resistance was also 8 Ω if you wanted maximum power to be transferred to the load.

The terms input resistance and output resistance are more generally and correctly called input and output impedance, respectively. That is, a practical amplifier may have reactive components in its input and output circuits. In most cases, we will not distinguish between resistance and impedance for our discussion of transistor amplifiers. The input impedance of a practical amplifier can range from tens of ohms to several megohms depending on the type of device and its configuration. The output impedance of a practical amplifier can range from fractional ohms to thousands of ohms.

Classes of Amplifiers Amplifiers can be classified into different categories based on their application or their operation. Table 10–2 lists several categories of amplifiers along with their distinguishing characteristics and at least one representative application of each.

TABLE 10–2 Classifications of amplifiers.

Classification	Characteristics	Typical Applications
Voltage amplifier	Increases the amplitude of the input voltage.	Audio amps, video amps, communication circuits, test equipment
Current amplifier	Output current is proportional to but greater than input current.	Transducer amplifiers, communication circuits, industrial controls (motor and relay drivers), computers
Power amplifier	Output signal power is greater than input signal power. Also amplifies voltage and/or current.	Audio amps, video amps, communication circuits, test equipment, industrial controls, computers
Impedance matching	Input impedance same as source. Output impedance same as load.	Transducer interface
Class A amplifier	Transistor conducts current for 360° of the signal waveform.	Low-level audio amplifiers
Class B amplifier	Transistor conducts current for 180° of the signal waveform.	Audio power amplifiers using multiple transistors
Class AB amplifier	Transistor conducts current for >180° and <360° of the signal waveform.	
Class C amplifier	Transistor conducts current for <180° of the signal waveform.	Communication circuits
Class D amplifier	Transistor operates in only two states of conduction: cutoff and saturation.	High-power amplifiers
Logarithmic amplifier	Output voltage (or current) is proportional to the logarithm of the input voltage (or current).	Optical equipment, audio circuit, communication circuits

Amplifier Configurations

There are many different ways that a transistor can be connected into a circuit to function as an amplifier. We will briefly examine three basic circuit configurations.

Common-Emitter Amplifier Figure 10–11 shows a basic common-emitter amplifier circuit. The $+V_{CC}$ symbol represents a dc voltage source. The emitter is connected to ground through R_4. The base is at some positive voltage established by the voltage divider action of R_1 and R_2. Therefore, the base-emitter junction is forward biased. The collector is returned through R_3 to the $+V_{CC}$ source (the most positive point in the circuit). As long as the voltage drop across R_3 is not excessive, the collector will be more positive than the base voltage. Therefore, the collector-base junction is reverse biased. You will recall that these biasing conditions are required for linear amplification.

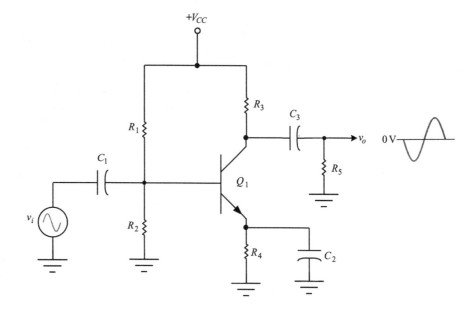

FIGURE 10–11 A common-emitter amplifier circuit.

Capacitor C_1 acts as an open to the dc biasing voltages and thus isolates the source from the effects of the dc voltage on the base. But, C_1 has a low reactance to the input signal, so the source is effectively connected to the base from an ac standpoint. C_1 is called a coupling capacitor, but you will recognize this basic behavior as a high-pass filter.

As the input signal goes through its positive and negative alternations, it adds and subtracts, respectively, from the dc bias on the base-emitter junction. In other words, during the positive alternation of the input, additional forward bias (i.e. increased base current) is provided by the input signal (actually a result of charging C_1). By contrast, the negative alternation reduces the net forward bias on the base-emitter junction. Therefore, base current is increasing and decreasing from the dc or quiescent (i.e., no signal) level at a rate and magnitude determined by the input signal.

Now, recall that the ratios of the various transistor currents remain relatively constant as long as the transistor is biased in its linear range of operation (i.e., not saturated and not cutoff). These ratios are determined by the construction of the transistor (e.g., thickness of the base, relative doping levels, and so on). This means that as the base current changes in response to the input signal, the emitter and collector currents will show similar changes. However, due to transistor operation, the corresponding changes in the emitter and collector currents will be substantially larger. Typical ratios between collector current and base current (β) range from less than 35 to over 1000.

In the common-emitter circuit shown in Figure 10–11, the varying collector current flows through R_3. The voltage drop across R_3, then, varies as the input signal varies. According to Kirchhoff's Voltage Law, the voltage measured at the collector of the transistor (with respect to ground) will be equal to $+V_{CC}$ minus the voltage drop across R_3. Since $+V_{CC}$ is constant, we know the voltage at the collector will be sinusoidal (i.e., the same as the input waveform). However, the positive alternation on the input causes more base current and therefore more collector current. This results in a greater voltage drop across R_3, leaving less voltage on the collector of the transistor. That is, an increasing input causes a decrease in collector voltage.

This means the waveform on the collector of the transistor is 180° out of phase with the input waveform. A common-emitter amplifier is an inverting-amplifier configuration.

Finally, the amplified waveform at the collector of Q_1 is passed through coupling capacitor C_3 and appears across R_5. R_5 represents the actual load or the input resistance of a subsequent circuit.

Resistor R_4 (in conjunction with R_1 and R_2) helps to establish the **quiescent operating point** (e.g., the amount of emitter and collector current that flows with no input signal). Capacitor C_2 bypasses R_4 for ac purposes, which allows the circuit to have more amplification.

The circuit shown in Figure 10–11 is called a common-emitter amplifier because the emitter is common to both input and output circuits. More specifically, the emitter is connected to ground by the R_4–C_2 network. Since C_2 has a low reactance at the operating frequency, the emitter is effectively connected to ground for ac purposes. The input is applied between base and ground (i.e., base and emitter). The output is taken between the collector and ground (i.e., collector and emitter).

SYSTEM PERSPECTIVE: Bipolar Transistors

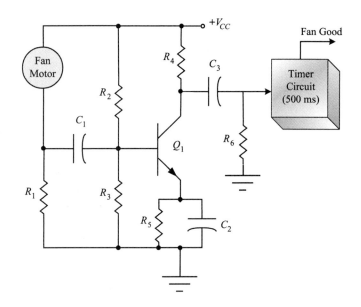

The microprocessors in many high-speed computers dissipate substantial power and will destroy themselves unless external cooling is provided. One common means of external cooling is forced airflow from a small fan physically mounted on top of the microprocessor itself. The circuit shown is used to verify that the fan is actually operating. As the fan runs, it draws current in pulses. The current pulses are converted to voltage pulses by resistor R_1. Transistor Q_1 is configured as a common-emitter amplifier. The pulses from the fan are amplified and then coupled through C_3 to a timer circuit. Each pulse applied to the timer circuit causes it to reset. So, as long as the fan is running, the timer never completes its timing cycle. If for some reason the fan stops turning, pulses will no longer be applied to the timer circuit, and it will time out after 500 ms. This causes the "Fan Good" signal to change logic states. The microprocessor senses this and shuts down or takes other corrective action before the excessive heat causes any damage. Here, a basic amplifier configuration using a discrete npn transistor plays a key role in a high-speed digital computer.

Emitter-Follower Amplifier Figure 10–12 shows the schematic diagram of an emitter-follower amplifier. This configuration is also called a common-collector amplifier. Resistor R_1 provides a path for base current (i.e., forward bias for the base-emitter junction). The complete path is from ground to the emitter, out the base, through R_1 to $+V_{CC}$, through the power supply, and back to ground. R_1 and R_2 are sized such that the transistor operates in its linear range between cutoff and saturation.

FIGURE 10-12 An emitter-follower or common-collector amplifier.

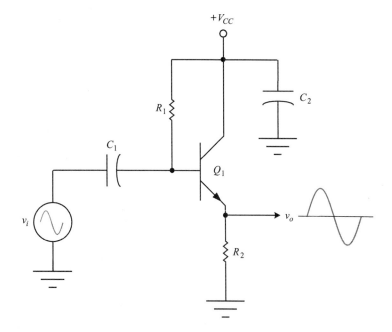

The changing input voltage is coupled through C_1 to the base-emitter circuit. The positive alternation of the input signal increases the base current (more forward bias), while the negative alternation reduces the forward bias. As the base current increases and decreases in response to the input signal, the emitter current will have corresponding, but substantially larger, current changes. Since the emitter current flows through R_2, it follows that a sinusoidal (in this case) waveform will appear across R_2. This serves as the output of the amplifier.

Capacitor C_2 is called a decoupling capacitor. It places the collector at ground potential for ac purposes. The input signal is applied between the base and ground (i.e., collector), and the output is taken between the emitter and ground (i.e., collector). Hence the circuit is often called a common-collector amplifier. A positive-going input signal increases forward bias, increases base and emitter currents, and increases the voltage drop across R_2 (the output signal). Therefore, the output signal is in phase with the input signal. Further, the difference in potential between the base and emitter terminals of a forward-biased transistor is very low and nearly constant (≈ 0.7 V). Consequently, ac variations on the base are also felt on the emitter. This means the output voltage is nearly identical to the input voltage in terms of amplitude, phase, and shape. However, the output signal can be capable of supplying substantial current to a load or subsequent stage. This type of circuit is generally used to buffer a high-impedance source from a relatively low-impedance load or circuit.

Common-Base Amplifier Figure 10-13 on page 346 shows a common-base amplifier circuit. Resistors R_1 and R_2 act as a voltage divider to establish the dc base voltage. This voltage in conjunction with R_3 determines the amount of forward bias on the base-emitter junction. These values are chosen such that the quiescent operating point is midway between cutoff and saturation.

The base is at ground potential for purposes of ac as a result of the low reactance of decoupling capacitor C_1. The input signal is applied between emitter and ground (i.e., base) via coupling capacitor C_2. Holding the base voltage constant and changing the emitter voltage has the same effect as holding the emitter voltage constant and changing the base voltage. Thus, the input signal causes increases and decreases in forward bias on the base-emitter junction. This in turn causes increases and decreases in emitter and collector currents. As the varying collector current flows through R_4, a corresponding voltage drop will be developed. This

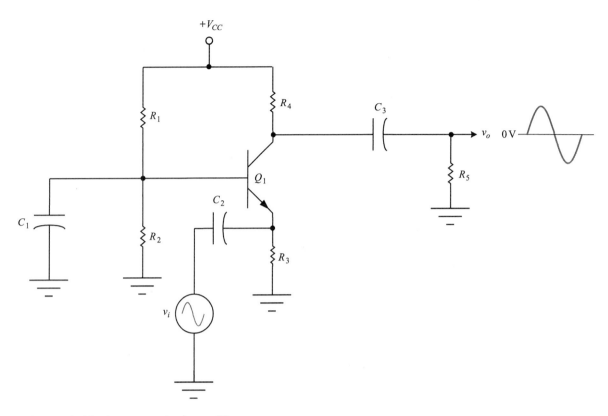

FIGURE 10–13 A common-base amplifier.

voltage drop subtracts from the steady $+V_{CC}$ potential (per Kirchhoff's Voltage Law) to produce a varying voltage on the collector of the transistor with respect to ground. This is the output signal of the amplifier, which is coupled through C_3 to the load (represented by R_5.)

The positive alternation of the input signal causes a reduction in forward bias (more positive potential on n-type emitter material). This in turn causes less collector current, less voltage drop across R_4, and a higher output voltage. Thus, the input and output waveforms are in phase. The common-base is a noninverting amplifier configuration.

Comparison of Amplifier Configurations Table 10–3 serves as a summary and comparison of the major characteristics of the three basic amplifier configurations.

TABLE 10–3 Relative performance characteristics of the three basic amplifier configurations.

Characteristic	Amplifier Configuration		
	Common-Emitter	Common-Base	Common-Collector
Voltage gain	High	High	Low (≈ 1)
Current gain	High	Low (<1)	High
Power gain	Very high	Moderate	Low
Input resistance	Moderate to high	Low	Very high
Output resistance	Moderate	Moderate	Low
Input terminals	Base and emitter	Emitter and base	Base and collector
Output terminals	Collector and emitter	Collector and base	Emitter and collector
Phase shift	180°	0°	0°

Summary of Bipolar Transistor Characteristics

The preceding sections in this chapter have presented a lot of important material with only short discussions. This section provides you with a bulleted list of important bipolar transistor characteristics. Most of the listed items were presented in earlier topics. Others will be new to you.

- The base of a transistor is thin and lightly doped.
- The emitter is heavily doped.
- The collector region is moderately doped.
- According to Kirchhoff's Current Law, the sum of base and collector currents is equal to emitter current (i.e., $I_E = I_C + I_B$).
- Base current is normally less than 5% of emitter current.
- Emitter and collector currents are nearly equal.
- The voltage across a forward-biased base-emitter junction will be near 0.7 V for a typical silicon transistor.
- Cutoff is a condition where there is no substantial transistor current. This condition occurs when the collector-base junction is reverse biased and the base-emitter junction is either reverse biased or has no bias at all.
- Saturation is a condition where the transistor is conducting its maximum current. During saturation, both junctions are forward biased.
- Saturation and cutoff represent the two extremes of transistor operation. These two states are used by digital systems to represent information and to control circuits.
- Between saturation and cutoff lies the linear region of transistor operation. Here, changes in the input signal cause corresponding changes in the output.
- The quiescent operating point is established by dc biasing components. The ac input signal causes the transistor currents to vary above and below the quiescent currents.
- There are many ways to properly bias a transistor for linear operation, but all must provide forward bias for the base-emitter junction and reverse bias for the collector-base junction. The quiescent operating point is normally midway between cutoff and saturation for a linear amplifier.

The part numbers for practical bipolar transistors often, but not always, begin with the prefix "2N." Examples include transistors such as 2N2905, 2N2222, 2N3055, and 2N3904.

✓ Exercise Problems 10.1

1. A bipolar transistor relies on two types of current carriers. (True/False)
2. The base region of a pnp transistor is made from ____-type material.
3. Which element of a bipolar transistor has the lightest level of doping?
4. The majority of the electrons entering the base from the emitter in a properly biased npn transistor exit the transistor as base current. (True/False)
5. The base region of a pnp transistor is heavily doped relative to the other two regions. (True/False)
6. In order to bias a transistor in its linear operating range, the base-emitter junction must be (*forward/reverse*) biased, and the collector-base junction must be (*forward/reverse*) biased.
7. What state is the transistor in if both junctions are reverse biased?
8. If a transistor is properly biased for linear operation, what is the approximate base-to-emitter voltage?
9. If the base voltage of an npn transistor operating as a linear amplifier measures +10 V with a dc voltmeter, what would you expect the emitter voltage to measure?
10. The depletion layer on the collector-base junction extends equally into the collector and base regions. (True/False)
11. A 10-mV input signal to a voltage amplifier causes a 5-V output. What is the voltage gain of the amplifier?

12. A transistor amplifier has a power gain (i.e., output power/input power) of 250. Express this gain in decibels.
13. An inverting amplifier causes a ____-degree phase shift to the input signal.
14. A noninverting amplifier causes a ____-degree phase shift to the input signal.
15. What characteristic of a transistor amplifier determines how much of a load it will present to a voltage source?
16. A class A amplifier conducts for the full 360° of the input waveform. (True/False)
17. A common-emitter amplifier has a relatively low power gain (typically <1). (True/False)
18. What type of amplifier configuration provides a voltage gain of less than one with no inversion of the signal and has the output taken from the emitter?
19. Which basic amplifier configurations can serve as inverting amplifiers?

10.2 JUNCTION FIELD-EFFECT TRANSISTORS

Field-effect devices are another whole class of electronic components that are used to solve application problems. In some ways, a **field-effect transistor** (FET) is similar to a bipolar transistor (npn or pnp). For example, they both can be used as amplifiers and electronic switches. On the other hand, many FET characteristics and even the basic construction and operation of FETs are dramatically different from those of a bipolar device.

Construction

Our first step in understanding JFET operation is to examine the internal semiconductor structure. This, coupled with your knowledge of basic semiconductor technology, will provide a good introduction into device behavior.

Figure 10–14 shows that the internal semiconductor structure of a JFET consists of both p- and n-type materials, but the layout is different from a bipolar transistor. The fabrication begins with a block (substrate) of p-type material as shown in Figure 10–14(a). Next, a pocket of n-type material is diffused into the p-type substrate as shown in Figure 10–14(b). Finally, as indicated in Figure 10–14(c), a smaller p-type region is diffused into the n-type layer. Contacts and lead wires are added such that one end of the central region makes contact with two of the external leads called the **source** and **drain** terminals. The two p-type regions are electrically joined and then connect to the third transistor lead: the **gate** terminal. As you will soon see, current will flow through the n-type material between the source and drain connections. The narrow n-type region between the two p-type regions is called the **channel**. Therefore, this type of transistor is called an n-channel JFET.

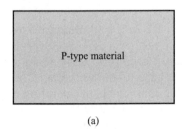
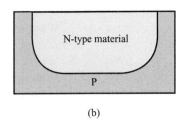
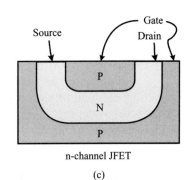

FIGURE 10–14 Fabrication of a junction field-effect transistor (JFET).

Fabrication of a p-channel JFET is identical to that of an n-channel device except the positions of the material types are reversed. Schematic symbols for both types of JFETs are shown in Figure 10–15.

10.2 • JUNCTION FIELD-EFFECT TRANSISTORS 349

FIGURE 10-15 Schematic symbol for (a) an n-channel JFET and (b) a p-channel JFET.

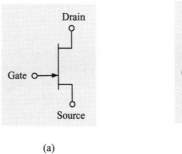
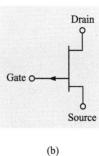

Biasing

Figure 10–16 shows how we bias an n-channel JFET for linear operation (such as for an amplifier). Note that the two parallel pn junctions in the JFET are reverse biased with voltage source V_{GG}. This causes a depletion layer to extend into the channel. In Figure 10–16(a), the reverse bias is low, so the depletion layers are small. This leaves most of the channel unrestricted. V_{DD} is connected across the n-type material that forms the channel. Source-to-drain current (I_D) can readily flow through this region. In effect, the n-type material merely acts as a resistor whose initial value is set by the level of doping.

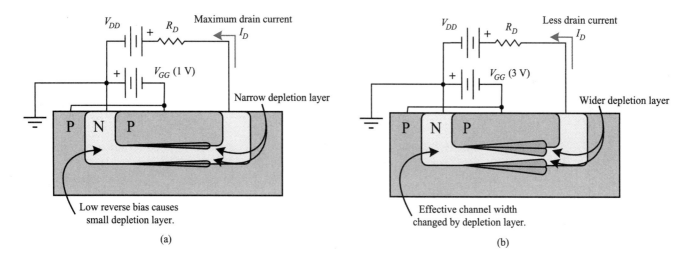

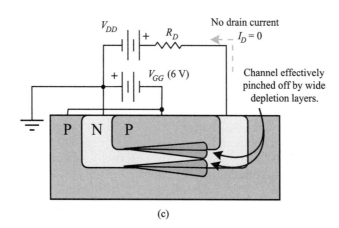

FIGURE 10-16 The depletion layers of the reverse-biased pn junctions affect the area and effective resistance of the channel.

The reverse bias of the gate-source pn junctions has been increased in Figure 10–16(b). Note the expansion of the depletion layers into the channel region. Recall that a depletion layer is relatively void of current carriers. Thus, as the depletion layer expands, the conductive area of the channel is reduced. This has the effect of increasing channel resistance. Therefore, we expect a decrease in the amount of drain current as indicated in Figure 10–16(b).

Finally, if we apply a high level of reverse bias to the gate-source junction as illustrated in Figure 10–16(c), we can expand the depletion layers until the conductive portion of the channel is fully consumed. When this occurs, no source-to-drain current can flow, and we say the device is cutoff. This corresponds to cutoff in a bipolar transistor.

Based solely on our present understanding of a JFET, you could easily conclude that the source and drain terminals are interchangeable. Current should be able to flow either way through the channel. To some degree, this is true. In fact, there are applications that actually pass a bidirectional current through the channel circuit. However, in order to obtain the performance characteristics described in the manufacturer's data sheet and to maintain reverse bias on the gate junction, we must distinguish between the source and drain terminals. In the case of an n-channel device, the drain should be more positive than the source. P-channel JFETs, by contrast, require the drain to be more negative than the source.

One significant difference between JFETs and bipolar transistors is that bipolar devices require forward bias on the base-emitter junction in order to have current flow. In other words, they are off with no bias. A JFET, by contrast, is fully on with no bias. We have to apply reverse bias to the gate-source junction in order to move the device toward cutoff. Also note that the operation of a JFET depends on only one type of majority current carrier (electrons for an n-channel device and holes for p-channel JFETs). Thus, a JFET is considered a unipolar transistor.

JFET Amplifiers

The common-emitter, common-base, and common-collector amplifier configurations for bipolar transistors have equivalent JFET configurations. The JFET equivalents are called common-source, common-gate, and common-drain or source-follower. Examples of these basic amplifier configurations are shown in Figure 10–17.

Let's examine the operation of a common-source amplifier in greater detail. Figure 10–17(a) shows the schematic for a common-source JFET amplifier. This configuration is comparable to a common-emitter amplifier. Capacitor C_S in Figure 10–17(a) is a bypass capacitor, which effectively grounds the source for ac purposes. The input voltage is applied between the gate and source (ground) terminals. The output is taken between the drain and source (ground) terminals. Thus, the source is common to both input and output circuits.

There is a 180° phase shift between the input and output voltage waveforms in a common-source amplifier. This is similar to the behavior found in a common-emitter circuit. The input circuit (gate-to-source) of the JFET is essentially a reverse-biased pn junction. Current flow is negligible—typically measured in picoamperes. This means that the input resistance of the JFET is extremely high. This is probably the most notable characteristic of the JFET amplifier. The output resistance is, for practical purposes, equal to the value of R_D.

A common-source amplifier can provide voltage gain and power gain. The voltage gain available from a JFET amplifier is generally less than that from a comparable bipolar circuit. Technically, it provides an extremely high current gain, but this is not a useful measurement due to the extremely small and unstable value of gate current. The voltage gain of the common-source amplifier illustrated in Figure 10–17(a) is dependent on the JFET characteristics, which vary between

FIGURE 10–17 (a) Common-source, (b) common-gate, and (c) common-drain amplifier configurations.

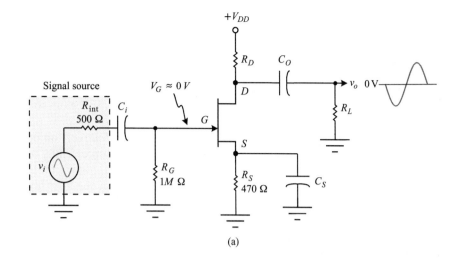

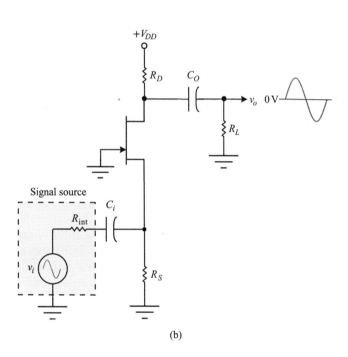

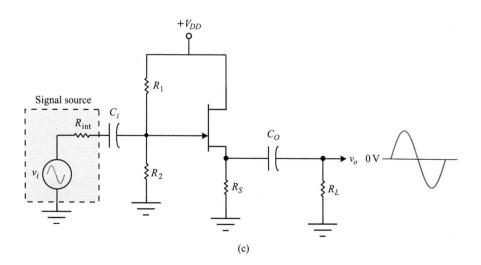

devices. Resistor R_S is called a source-swamping resistor, and it is used to reduce the dependency on transistor characteristics.

The gate of the JFET is returned to ground through R_G. Since no meaningful current flows out of the reverse-biased gate terminal, there is no dc voltage drop across R_G. As current flows through R_S and the resistance of the internal channel, it causes a voltage drop. This places the internal channel at a more positive potential than the gate (ground). Thus, the gate-source junction is reverse biased, as is required for proper operation.

Since the gate terminal is effectively an open circuit (reverse-biased junction), it has no loading effect on the source. Thus, the input signal couples through C_i and appears across R_G. Part of the input signal is lost across the internal impedance of the source (R_{int}). However, because of the high input resistance of the JFET and the allowable high value of R_G, most of the input signal is coupled to the gate of the transistor. As the voltage on the gate (voltage drop across R_G) varies in response to the input signal, the reverse bias of the gate-source junction also varies. This, in turn, causes corresponding variations in drain current, and the changing drain current causes a varying voltage drop across R_D. The output signal is taken from the drain terminal and is 180° out of phase with the input signal.

The operation of the other amplifier configurations is similar to that of their bipolar counterparts. In all cases, the input signal causes variations in the gate-source reverse bias, which causes corresponding variations in the source-to-drain current.

Summary of JFET Characteristics

The preceding topics have introduced the construction and behavior of JFETs. This section provides you with a bulleted list of important JFET characteristics. Most of the listed items were presented in earlier topics. Others will be new to you.

- The source and drain terminals connect to opposite ends of the channel material in a JFET.
- The gate-source junctions of a JFET are reverse biased for normal operation.
- Increased reverse bias on the gate-source junction of a JFET causes the depletion layers to expand farther into the channel. This restricts source-to-drain current flow and, in effect, increases the resistance of the channel.
- A JFET has maximum source-to-drain current with no gate-source bias.
- A JFET can be used in digital applications by restricting its operation to only two conditions: fully on (saturated) and fully off (cutoff).
- The voltage gain of a JFET amplifier is generally less than can be obtained from a bipolar amplifier.
- Common-source, common-gate, and common-drain (source-follower) are JFET amplifier configurations that are comparable to the bipolar configurations of common-emitter, common-base, and common-collector, respectively.
- Common-source amplifiers have a 180° phase inversion between input and output. Common-gate and common-drain amplifiers have no phase inversion.
- Common-source amplifiers are characterized as having very high input impedance, since the gate is part of a reverse-biased pn junction. This is often the reason for choosing JFETs over bipolar transistors in a given application.
- For normal operation, the source-drain circuit must be biased such that majority current flows from source to drain (i.e., electrons in n-channel devices and holes in p-channel devices).
- JFETs have practical applications ranging from dc circuits to high-frequency radio circuits. They are also used for switching circuits (i.e., digital operation) such as computer applications.
- JFETs are often used in low-noise amplifier applications.

✓ Exercise Problems 10.2

1. The gate material in a p-channel JFET is made from ____-type material.
2. The source and drain connections of an n-channel JFET are connected to ____-type material.
3. In order to properly bias an n-channel JFET for use as a linear amplifier, the gate must be (*positive/negative*) with respect to the source.
4. The source of a p-channel JFET must be (*positive/negative*) with respect to the gate.
5. Electron current flows from source-to-drain in a ____-channel JFET.
6. Electron current flows from drain-to-source in a ____-channel JFET.
7. Increased reverse bias on the gate-source junction of a JFET (*reduces/increases*) the effective resistance of the channel.
8. What happens in a JFET if the depletion layer associated with the gate reaches entirely across the channel?
9. Zero bias on the gate-source junction of a JFET causes it to be in (*saturation/cutoff*).
10. The input impedance of a common-emitter amplifier is generally much higher than the input impedance of a common-source amplifier. (True/False)

10.3 MOS FIELD-EFFECT TRANSISTORS

The name MOSFET is an abbreviation for metal-oxide-semiconductor-field-effect-transistor. As you will see, the name describes the basic internal construction of the device. Although the fabrication and operation of a MOSFET is different from a JFET, there are many similarities.

Construction

There are two basic types of bipolar transistors (npn and pnp) and two basic types of JFETs (n-channel and p-channel). By contrast, there are two classes of MOSFETs (**depletion mode** and **enhancement mode**) with two types of devices (n-channel and p-channel) in each class. So, there are four basic types of MOSFETs.

Depletion Mode

Figure 10–18(a) shows the internal structure for an n-channel depletion-mode MOSFET. There is a continuous channel of n-type material extending between the source and drain terminals. If a voltage were applied between the source and drain terminals, a current could readily flow. In essence, without considering the effect of the gate, the channel simply acts like a resistor.

As shown in Figure 10–18(a), the gate of the MOSFET consists of a layer of metal (deposited aluminum). The metalized gate is electrically insulated from the channel by a thin layer of silicon dioxide (SiO_2). The three layers consisting of metal (gate), oxide, and semiconductor (channel) lead to the abbreviation used to describe this technology (MOS). The schematic symbol for an n-channel depletion-mode MOSFET is shown in Figure 10–18(b).

FIGURE 10–18 (a) The basic internal structure of an n-channel depletion-mode MOSFET and (b) its schematic symbol.

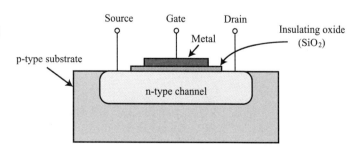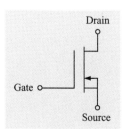

Figure 10–19(a) shows the construction of a p-channel depletion-mode MOSFET. As you might expect, a p-channel depletion-mode MOSFET is constructed like its n-channel complement except the relative positions of the p- and n-type materials are reversed. The schematic symbol for a p-channel depletion-mode MOSFET is shown in Figure 10–19(b). The direction of the arrow distinguishes between n- and p-channel devices. The line with the arrow represents a connection to the substrate of the transistor. As shown in Figures 10–18(b) and 10–19(b), the substrate is internally connected to the source resulting in a three-lead component. This is the normal case. In some cases, however, the substrate connection is brought out as a fourth lead.

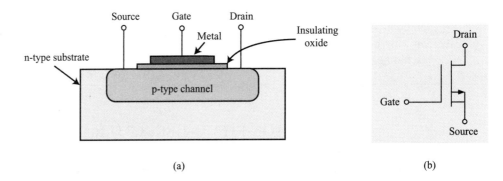

FIGURE 10–19 (a) The symbol for a p-channel depletion-mode MOSFET and (b) its schematic symbol.

Enhancement Mode

Figure 10–20(a) shows the internal construction of an n-channel enhancement-mode MOSFET. The source and drain terminals connect to regions of heavily-doped n-type silicon, but the source and drain regions are not connected. The area between the source and drain diffusions is lightly-doped p-type material. To a rough approximation, the construction resembles an npn transistor with no dc connection to the base lead. As you would expect, there can be no current flow through the device under these conditions. In order to get current to flow between the source and drain terminals, we must apply a gate voltage of the correct polarity. We will explore this in the following section. The schematic symbol for an n-channel enhancement-mode MOSFET is shown in Figure 10–20(b). The broken line serves as a reminder that no inherent current path exists between the source and drain terminals.

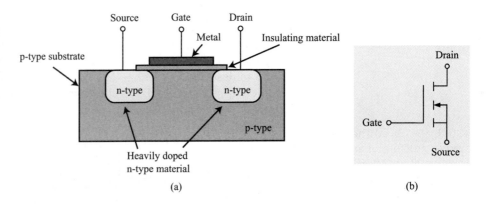

FIGURE 10–20 (a) The basic internal structure of an n-channel enhancement-mode MOSFET and (b) its schematic symbol.

Figure 10–21(a) shows the internal construction of a p-channel enhancement-mode MOSFET. Again, it is similar to its n-channel complement except the material types are reversed. Figure 10–21(b) shows the schematic symbol for a p-channel enhancement-mode MOSFET. As with previously-studied transistors, the direction of the arrow identifies the material type (p-channel or n-channel). The use of a broken line between the source and drain terminals identifies the device as enhancement mode and reminds us that no current path exists without intervention.

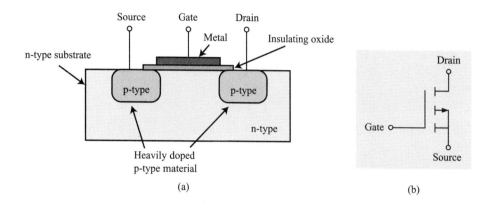

FIGURE 10–21 (a) The basic internal structure of a p-channel enhancement-mode MOSFET and (b) its schematic symbol.

MOSFET Biasing and Operation

The resistance of the channel in a field-effect transistor depends on the level of doping and the physical dimensions of the material. In the case of a JFET, we effectively altered the physical dimensions of the channel by the expanding depletion layers near the gate junction. In the case of a MOSFET, we will alter the effective doping level of the crystal in the channel.

Before we examine how the gate voltage is able to control the drain current in a MOSFET, let's review and emphasize some key points. First, remember that a block of extrinsic (doped) semiconductor acts like a simple resistor. If the doping level is heavier, then the resistance is lower. Second, recall what distinguishes between n- and p-type material. A semiconductor material that has more electrons in the conduction band (essentially free electrons) than corresponding holes in the valence band is classified as n-type. By contrast, p-type material is characterized by having more holes (vacancies) in the valence band than corresponding electrons in the conduction band. Finally, be reminded that a capacitor consists of two conductors separated by an insulator. Anytime we separate two conductors with an insulating material, we form a capacitor. Let's apply these ideas to understand the operation of a MOSFET.

Depletion-Mode Biasing and Operation

If we connected an external dc voltage source between the source and drain of a depletion-mode MOSFET like the one pictured in Figure 10–22(a) on page 356, we would expect to have a measurable current. Since the n-type semiconductor is acting like a resistor, we know the amount of current flow will depend on the resistance of the channel and the value of applied voltage.

Figure 10–22(a) illustrates what happens when an external voltage is connected in a manner that makes the gate more positive than the source. The metal gate and conductive channel act like the plates of a capacitor. The layer of silicon dioxide between the gate and channel forms the dielectric for the capacitor. When the gate voltage is applied, the gate "capacitor" charges to the value of applied voltage. The charging action causes a deficiency of electrons on the gate and an accumulation of electrons in the channel directly beneath the gate. The channel area directly below the gate now has many more electrons in the conduction band. This is equivalent to increasing the n-type doping level. Of course, higher doping levels correspond to lower resistance. Thus, with a positive gate voltage on the MOSFET, we will see an increase in source-to-drain current through the channel.

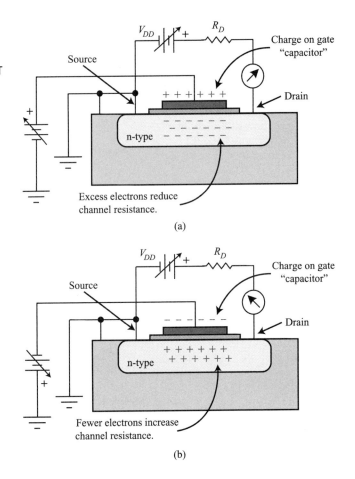

FIGURE 10–22 (a) The effects of making the gate of an n-channel depletion-mode MOSFET positive or (b) negative.

Figure 10–22(b) illustrates the effects of making the gate terminal negative. Again, the gate capacitance charges to the value of gate supply voltage. This time the gate terminal is negative (excess electrons) and the channel below the gate is positive (deficiency of electrons). Since many of the free electrons in the n-type material have been drawn out of the channel area, we have effectively reduced the doping level of the channel. This produces an increase in resistance, so we will see a reduction in channel current through the device.

Sometimes a depletion-mode device is said to be operating in the enhancement mode or depletion mode when the drain current is increased or decreased, respectively, from the no-bias value. Do not confuse this terminology with a true enhancement-mode (only) MOSFET discussed in the next section.

The preceding discussion can be used to reveal some important characteristics of a MOSFET. First, the gate terminal is essentially one side of a capacitor. Thus, for dc purposes, the gate terminal is an open circuit. This gives the MOSFET an extremely high input resistance, which is one of its primary advantages. Gate resistances measured in teraohms (10^{12} ohms) are not uncommon. We should also note from the preceding discussion that there is drain current through a depletion-mode MOSFET with no bias on the gate. Further, this no-bias drain current can be both increased and decreased by the gate voltage. Contrast this behavior with a bipolar transistor, which must have forward bias on the base-emitter to have collector current. Also contrast it with a JFET where the gate-source junction must be reverse biased at all times, so it can only decrease the drain current from a no-bias condition.

Enhancement-Mode Biasing and Operation

Figure 10–23(a) illustrates the internal action of an n-channel enhancement-mode MOSFET when a positive gate voltage is applied. The gate capacitance charges to the value of the gate-source voltage. Since the positive terminal of the source connects to the gate, there will be a deficiency of electrons on the metal gate. Directly below the gate in the channel area, there will be an accumulation of electrons resulting from the charge on the gate capacitance. If we draw enough electrons into the channel region, then the lightly-doped p-type material below the gate will be effectively converted to n-type material. Recall that n-type material is characterized by having more electrons in the conduction band than holes in the valence band. The portion of semiconductor that is temporarily converted from p- to n-type material is called an **inversion layer**. As indicated in Figure 10–23(a), the inversion layer completes the path between the source and drain terminals and allows current to flow. As we make the gate more positive, the depth of the inversion layer increases, which lowers the resistance of the channel and allows more drain current to flow. Because the newly-formed current path is n-type material, this type of device is called an n-channel enhancement-mode MOSFET.

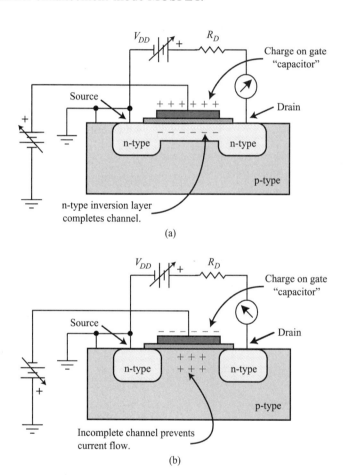

FIGURE 10–23 (a) The effects of a positive gate voltage and (b) a negative gate voltage on an n-channel enhancement-mode MOSFET.

Figure 10–23(b) shows the effect of applying a negative gate voltage to an n-channel enhancement-mode MOSFET. The p-type region immediately below the gate takes on a positive charge. The original p-type material had more holes than conduction-band electrons. The charge on the gate capacitor effectively increases this ratio and has an effect similar to increasing the doping level of the p-type material. Nevertheless, it still acts like an npn transistor with no base connection, so there can be no current flow from source to drain.

If we reverse the positions of the p- and n-type materials, then we will have a complementary device called a p-channel enhancement-mode MOSFET. A negative gate voltage is required to enhance or turn on a p-type enhancement-mode MOSFET. It should be noted that enhancement-mode MOSFETs are much more widely used in industry than depletion-mode devices. They are cutoff with no gate-source bias. Many integrated circuits use enhancement-mode MOSFETs to build complex systems (such as microprocessors) and various support devices (such as video controllers and Ethernet interfaces).

MOSFET Amplifiers

The three basic amplifier configurations that were discussed for bipolar transistors and JFETs have similar MOSFET configurations. They will not be repeated in detail, but an overview will be offered.

Amplifier Biasing

Any of the MOSFET types can be used as a linear amplifier. First, a dc supply voltage is used to bias the channel such that majority current carriers tend to flow from source to drain. Second, the gate-source capacitance is biased such that the transistor is operating midway between cutoff and saturation. In the case of a depletion-mode MOSFET, this often means no gate bias at all. Enhancement-mode devices, by contrast, must be intentionally biased to their midpoint of operation.

Signal Response

Once the transistor is properly biased, we can apply an ac signal between the gate and source. This signal adds to and subtracts from the dc bias, so the source-to-drain current varies in response to the input signal. Variations in source current and drain current cause voltage drops across any resistance in series with the source and/or drain terminals. This causes an amplified version of the input signal to be available at the source and/or drain.

As with their bipolar and JFET counterparts, common-source MOSFET amplifiers have a 180° phase inversion between input and output. Common-gate and common-drain configurations have no phase inversion. The most outstanding characteristic of MOSFET amplifiers is their incredibly high input impedance. Figure 10–24 shows the schematic diagram of a common-source amplifier using an n-channel depletion-mode MOSFET.

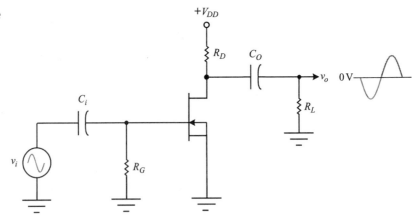

FIGURE 10–24 A common-source amplifier using an n-channel depletion-mode MOSFET.

Figure 10–25 on page 359 shows the schematic diagram of a common-source amplifier using a p-channel enhancement-mode MOSFET.

Low-Noise Amplifiers

Solid-state devices generate internal electrical noise, which mixes with the desired signal and degrades its purity. There are several sources or mechanisms that create noise in a semiconductor device. For example, in bipolar transistors, noise is generated as recombination occurs and emitter current splits into base and collector currents. The random movements and collisions of electrons cause additional

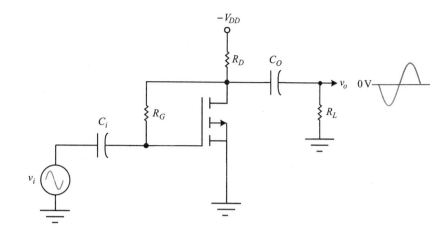

FIGURE 10–25 A common-source amplifier using a p-channel enhancement-mode MOSFET.

noise as the electrons make their way through the semiconductor crystal. Since a FET (either JFET or MOSFET) is a unipolar device, the amount of recombination that occurs in the device is much less than in a bipolar transistor. There are still noise sources within a JFET or MOSFET, but when driven by a high impedance source, the FET device contributes substantially less noise than a bipolar transistor. For this reason, FET amplifiers are often used in the early stages of amplifier circuits. In a multistage amplifier system, all stages after a particular stage will amplify both the desired signal and the noise contributed by the given stage. It is important to use low-noise devices in the early stages of a system where the signal level is smallest. FET devices can be used in amplifiers ranging from dc through the gigahertz range.

MOSFET Handling Precautions

As shown in preceding discussions, the gate of a MOSFET is separated from the channel by a thin insulating layer of silicon dioxide. This insulating layer forms the dielectric of a capacitance. As you know, capacitors have a breakdown voltage rating that, if exceeded, may permanently damage the dielectric of the capacitor. In the case of a MOSFET, the insulation that separates the metal gate from the channel can be damaged by excessive voltage. This may seem obvious to you, but what may not be so apparent is how the transistor may be subjected to damaging voltages.

If we assume that the circuit using the transistor was designed correctly, then there is minimal chance of damaging voltages originating from within the circuit. Damaging voltages could come from nearby lightning strikes and other forms of power line transients, but these types of transients can cause damage to any electronic component. The source of damaging voltages that we must specifically address when discussing MOSFET devices is electrostatic discharge or ESD. You have probably walked across a rug or slid across a car seat in cold, dry weather and then seen a spark jump to your finger as you reached for a metal object or another person. As a general rule, it takes 10 to 20 kV to arc across a distance of one inch. This amount of voltage is easily generated and carried by the human body. If you have an accumulated charge and touch the leads on a MOSFET, the ESD may puncture the gate insulation and damage the transistor. Many transistors and nearly all integrated circuits that utilize MOS technology have protection diodes that help reduce the risk of damage by voltage transients, but even these are relatively limited. Accordingly, manufacturers recommend the following precautions when working with MOS devices:

1. Handle the devices by the case and avoid touching the leads when practical.
2. Always make contact with a grounded object while handling an MOS device. Special grounding bracelets are available if extensive handling is required.

3. Never remove or insert an MOS device while the circuit power is applied.
4. Always store MOS devices in a conductive material to prevent accumulation of charges.
5. Use a soldering iron that has a grounded tip.

Summary of MOSFET Characteristics

We have introduced the construction and behavior of MOSFETs. The following provides you with a bulleted list of important MOSFET characteristics. Most of the listed items were presented in the preceding sections. Others will be new to you.

- The source and drain terminals connect to opposite ends of the channel material in a MOSFET.
- A depletion-mode MOSFET is on with no gate-to-source voltage.
- Channel current in a depletion-mode MOSFET can be increased or decreased from the no-bias condition.
- An enhancement-mode MOSFET is cutoff with no gate-to-source voltage.
- MOSFETs have extremely high input resistances (considered an open circuit for most practical purposes). This is often the reason for choosing MOSFETs for a given application.
- A MOSFET can be used in digital applications by restricting its operation to only two conditions: fully on (saturated) and fully off (cutoff).
- The voltage gain of a MOSFET amplifier is generally less than can be obtained from a bipolar amplifier.
- Common-source, common-gate, and common-drain (source-follower) are MOSFET amplifier configurations that are comparable to the bipolar configurations of common-emitter, common-base, and common-collector, respectively.
- Common-source amplifiers have a 180° phase inversion between input and output, whereas common-gate and common-drain amplifiers have no phase inversion.
- For normal operation, the source-drain circuit must be biased such that majority current flows from source to drain (i.e., electrons in n-channel devices and holes in p-channel devices).
- MOSFETs have practical applications ranging from dc circuits to high-frequency radio circuits in the GHz range.
- MOSFETs are often used in low-noise amplifier applications.
- MOSFETs can be easily damaged by ESD, so care must be used when handling them.

✓ Exercise Problems 10.3

1. What does the "O" in MOSFET represent?
2. Match each of the following MOSFET types with its symbol in Figure 10–26.
   a. n-channel, depletion mode
   b. p-channel, enhancement mode
   c. p-channel, depletion mode
   d. n-channel, enhancement mode
3. Classify the resistance between the gate and drain terminals of a MOSFET as very low, low, moderate, high, or very high.

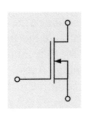   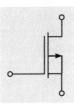   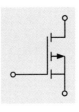

(a)   (b)   (c)   (d)

FIGURE 10–26 Match the MOSFET types to the correct symbol.

4. With no gate-source voltage, the semiconductor material immediately below the gate in a p-channel enhancement-mode MOSFET is effectively (*n-type/p-type*).
5. A _____-mode MOSFET is on, even without gate-source bias.
6. In a properly biased p-channel depletion-mode MOSFET, electrons flow from (*source/drain*) to (*source/drain*).
7. What polarity of bias must be put on the gate (with respect to the source) on an n-channel depletion-mode MOSFET to cause an increase in drain current?
8. What polarity of bias must be put on the gate (with respect to the source) on an n-channel enhancement-mode MOSFET to cause an increase in drain current?
9. What polarity of bias must be put on the gate (with respect to the source) on a p-channel enhancement-mode MOSFET to cause an increase in drain current?
10. MOSFETs are not practical for use in linear amplifiers. (True/False)
11. MOSFETs are not practical for use in digital applications. (True/False)
12. Which of the basic MOSFET amplifier configurations causes a 180° phase inversion between input and output?
13. Why must you use special precautions when handling a MOSFET?
14. Name two types of MOSFETs that have no significant drain current when there is no gate-source bias.

10.4 TRANSISTOR APPLICATIONS

We are now ready to examine a variety of practical transistor applications. The selected applications are representative of current design practices. We have avoided applications that no longer have widespread application in new designs. For example, discrete transistor circuits can be used as logarithmic amplifiers, oscillators to generate square waves, active filters to selectively amplify signals based on their frequency, squelch circuits that selectively pass signals based on their amplitudes, and so on. But, in most cases, these functions are now implemented with integrated circuits, which we will explore throughout the remainder of the text.

Digital Applications

We will first study how transistors are used in digital circuits. Digital circuits are characterized as having only two states. In the case of a transistor, these two states correspond to cutoff and saturation.

Transistor Switch

When a transistor is used as a switch, it is always in one of two states: fully on or fully off. Transistor switches are used in digital circuits, industrial control circuits, burglar alarm circuits, computer circuits, and many other applications.

Figure 10–27 on page 362 shows a typical application for a transistor switch. Here, the transistor is being used as a lamp driver. Many electronics assembly plants have computer-controlled machines (robots) that insert components into circuit boards. When the machine needs human intervention (e.g., the parts supply is empty), it turns on a bright light to get the attention of an operator who is monitoring several machines. A transistor lamp driver is often used for this purpose as illustrated in Figure 10–27.

When the transistor is cutoff, the lamp is extinguished. When the transistor is saturated, the lamp is lit with full brightness. The state of the transistor is determined by the bias condition of the emitter-base junction. If the industrial computer outputs a 0-V level to the transistor base, then the base-emitter junction has no bias (i.e., no base current path). Therefore, there will be no collector current and the lamp will be off. Now, if the computer outputs a positive voltage (e.g., +5 V), then this will provide forward bias for the emitter-base junction. R_B limits the amount of base current. With base current flowing, there will be a corresponding collector

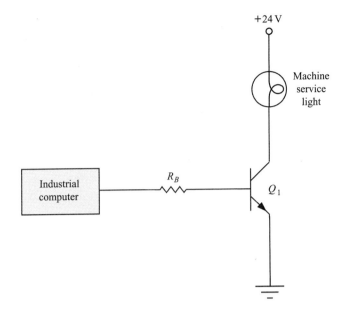

FIGURE 10–27 A transistor lamp driver.

current. If the circuit is properly designed, the transistor will have sufficient forward bias to enter saturation; the lamp will be on.

Now, it would be reasonable to wonder what purpose is served by transistor Q_1. Since the computer output is 0 V and +5 V anyway, why not eliminate the transistor and simply connect the lamp to the computer output? For some applications (such as the disk drive active LED on your computer), this would be a practical solution. But a typical computer output may only be rated for a few tens of milliamperes. A bright lamp, by contrast, may require substantial current flow (e.g., >1 A). As shown in Figure 10–27, the computer output only has to provide base current, which can be very small (a few milliamperes at most). The transistor switches the higher lamp current as its collector current turns on (saturation) and off (cutoff). Additionally, a bright lamp may require higher voltages than are readily available from a computer output, which is usually no more than +5 V. Suppose the lamp in Figure 10–27 requires 24 V at 1.2 A. Recall that the ratio of collector current to base current in a bipolar transistor ranges from less than 35 to over 1000. If we used a transistor with a ratio of 500, the computer output would only have to supply 1.2A/500 or 2.4 mA to the base of the transistor. Further, the 0-V to 5-V change at the computer output would cause a 0-V to 24-V change across the lamp.

The technique illustrated in Figure 10–27 has an extensive range of related applications beyond the basic lamp driver. This same strategy is used to allow a computer to control such devices as relays, electric latches, gas and fluid valves, dc motors, heating elements, audible alarms, and many others.

SYSTEM PERSPECTIVE: Transistors and Relays

A common computer application for relays is to serve as the interface device between a relatively low-current, low-voltage driver in the computer and a high-power industrial device that requires substantial current and/or higher voltage (ac or dc). The adjacent figure shows an application where a computer is controlling a welder used by a robot in an automotive plant. As shown in the figure, a standard high-level logic signal from the computer provides base current for the switching transistor (Q_1), causing it to saturate. A low logic signal (e.g., 0 V) removes the base-emitter bias, placing the transistor in cutoff. When the transistor is on, its collector current energizes a relay coil (K_1). For example, a large relay coil may require 265 mA, but an ACT logic gate can only supply 24 mA. In that case, Q_1 must have a beta value of at *least* 11. D_1 protects the transistor from transients caused by inductance in the relay coil.

When the relay energizes, its contacts close the 240-Vac circuit to the industrial welder. A high current (e.g., 75 A) will flow through the relay contacts. Finally, a step-down transformer provides 1 to 10 V to the load with potentially thousands of amperes of current. The resulting heat ($P = I^2R$) in the resistance of the material fuses the metal together.

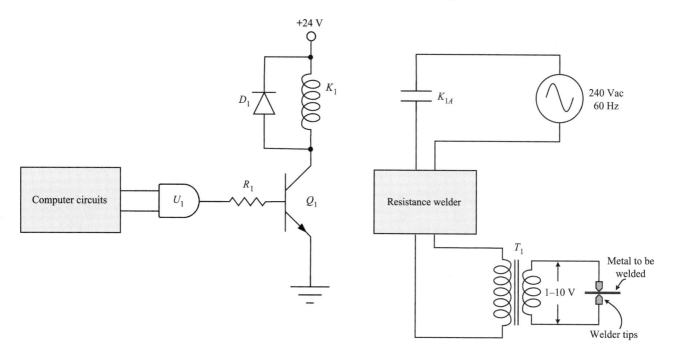

Integrated Logic Circuits

All modern digital computers are built around integrated circuits (ICs), which are composed of tens, hundreds, thousands, or even millions of interconnected transistors. For example, Intel's Pentium® microprocessor contains about six million transistors. Although you don't have access to most of the individual transistors in an integrated circuit, the pins on the IC do provide connections to some of the interface transistors. For instance, the input pins of a microprocessor may be connected to the base of an npn transistor or perhaps to the gate of a MOSFET internal to the IC. Similarly, output pins connect to the emitter, collector, source, or drain of one or more internal transistors. We will now take a brief look at three transistor technologies that are used to make integrated circuits for digital computers and other digital systems.

Bipolar Integrated circuits based on bipolar transistors do not play the central role they did when integration of digital circuits first began. Nonetheless, even newly-designed ICs using MOSFET devices frequently contrast their performance and/or electrical characteristics to those of bipolar ICs.

Figure 10–28(a) on page 364 shows a representative input circuit for a bipolar logic device. It is never necessary to understand detailed current flow internal to an IC unless you are the IC designer. However, by knowing the type of circuitry that connects to the input and output pins, you can better appreciate the electrical characteristics of the device. As you can see from Figure 10–28(a), the input circuit is basically a common-collector configuration using a pnp transistor. Don't be confused by the fact the transistor is drawn upside down from others we have seen.

FIGURE 10–28 A representative (a) input circuit and (b) output circuit for a bipolar integrated circuit.

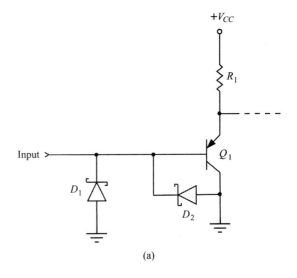

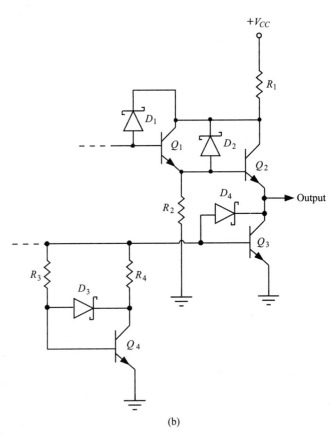

Diode D_1 serves as protection for Q_1 against negative input voltages that might damage Q_1. D_2 is called a collector-clamping diode. It keeps Q_1 from going into deep saturation. This allows the circuit to switch from saturation to cutoff in a shorter period of time. Faster operation is also the reason Schottky diodes are used in the circuit.

A representative bipolar output stage is shown in Figure 10–28(b). Again, Schottky diodes are used as collector clamps to increase switching speed. Two npn transistors are stacked in the output. This common configuration is called a **totem-pole output**. The two totem-pole transistors are always in opposite states. If Q_2 is saturated, then Q_3 is cutoff. This causes the output to go to a positive voltage level. This is one of two logic states in the output. The other logic state occurs when Q_2 is cutoff and Q_3 is saturated. In this state, the output is close to 0 V, since Q_3 is acting as a near short circuit.

MOSFET MOSFETs are used in the construction of the vast majority of integrated circuits used in digital products today. Both p-channel and n-channel enhancement-mode devices are commonly used. Figure 10–29(a) shows the internal transistor configuration for the simplest of digital logic devices: an inverter. The purpose of an inverter is to provide an output that is at the opposite logic state from the input. Although this is a simple configuration, the input and output characteristics are representative of many integrated circuits.

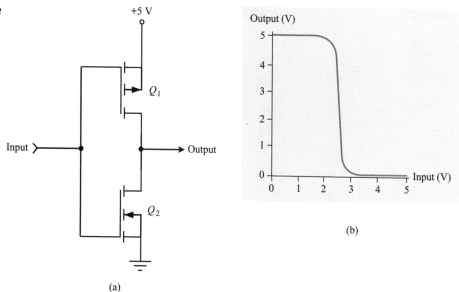

FIGURE 10–29 (a) A representative CMOS inverter and (b) its transfer curve.

This type of structure uses both n-channel and p-channel devices. Since the devices are complementary, this technology is called **c**omplementary **m**etal **o**xide **s**emiconductor or simply CMOS. The input is essentially an open circuit; it connects to two MOSFET gates. The output totem-pole structure is similar to the bipolar totem-pole structure in that the two transistors are always in opposite logic states.

If the input voltage is at 0 V (one of the two allowed logic levels for this particular example), then Q_2 has no gate-source bias and is cutoff. Q_1, by contrast, is saturated since the gate is 5 V less positive than the source. With the upper transistor essentially shorted and the lower transistor effectively open, the output terminal measures approximately +5 V. Note that this is the opposite logic state from the input.

Now, if the input logic state is +5 V, Q_1 will have no gate-source bias (both elements will be at +5 V with respect to ground). It will be in cutoff. Q_2 will have +5 V on the gate with respect to the source and will enter saturation. This circuit condition causes the output terminal to measure near 0 V. Again, this is the opposite logic state from the input. Thus, the inverter circuit is accomplishing its purpose.

Figure 10–29(b) shows the transfer curve for the basic CMOS inverter. A transfer curve shows how the output responds to changes on the input. As you can see, the output stays near +5 V as long as the input is well below +2.5 V. Similarly, if the input is well above +2.5 V, the output is near 0 V. As the input moves through the region near +2.5 V, the output quickly transitions between the two extreme states.

Hybrid Bipolar technologies are generally faster than MOSFET technologies, but they consume substantially more power. Also, bipolar devices can generally drive greater loads at high speeds than MOSFET equivalents. MOSFET technology, though, can achieve greater densities (i.e., more functions integrated into a single package). To get the best of both worlds, some integrated circuit manufacturers build hybrid ICs that have both bipolar and MOSFET devices integrated into the same silicon wafer. The major portion of these circuits is built around CMOS devices. This allows for high levels of functional complexity on a single chip. The input circuits are also CMOS devices, which means they require very little input current. The voltage characteristics of the inputs, however, are adjusted for compatibility with standard bipolar output circuits.

The output driver of a hybrid circuit generally uses bipolar transistors to achieve a higher speed when driving long traces on circuit boards and/or multiple inputs. Texas Instruments builds a line of integrated circuits that use both CMOS and bipolar technology. It is called advanced BiCMOS technology (ABT).

Memory Devices

All computers make extensive use of semiconductor memories. Transistors (both bipolars and MOSFETs, although MOSFETs are more common) are used as memory elements. The state of each piece of digital information (called a binary digit or bit) is represented in the memory IC by a saturated or cutoff transistor.

Linear Applications

Transistors also continue to be designed into linear applications. Several years ago entire systems would be built around transistors; today most transistor applications provide interface functions. That is, they are often used to make the current or voltage levels in one part of the system compatible with those in a subsequent circuit. Most electronic systems are built primarily around functional blocks provided by integrated circuits. But, as with digital circuits, linear ICs consist of many interconnected transistors that perform some desired circuit function (such as a voltage amplifier). Discrete transistors also continue to be widely used in power supply circuits. Here again, it is their relatively high power handling capability that makes them irreplaceable at this time. We will now examine several linear transistor applications.

Current Amplifier for Industrial Computer

Figure 10–30 shows how a transistor can be used as a current amplifier in an industrial computer circuit. Industrial computers are required to respond to both digital (on/off) and analog (continuous) inputs. Similarly, they must control both digital and analog devices within the industrial environment. As shown in Figure 10–30, the computer uses a circuit called a digital-to-analog converter (DAC) to create a varying or continuous voltage or current in response to digital signals from a computer circuit. In other words, all internal computations and data manipulation are accomplished using digital representations of physical quantities such as motor speed, temperature, pressure, and so forth. A varying voltage or current is required to control an analog device such as a heating element or variable-speed motor. These are not on/off (i.e., digital) type devices, so the DAC converts the digital values into corresponding analog voltage or current levels.

The transistor in Figure 10–30 is configured as an emitter-follower (common-collector) amplifier. It is biased by the output of the DAC. To understand its operation, recall that the base-emitter voltage of an npn transistor is nearly zero (≈ 0.7 V) and fairly constant. Therefore, whatever voltage appears on the base with respect to ground will also appear on the emitter with respect to ground (less the small base-emitter drop). From a voltage perspective, the external analog device

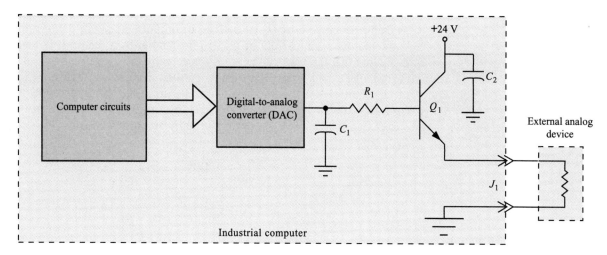

FIGURE 10-30 A current amplifier for an industrial computer.

sees approximately the same voltage as the output of the DAC. You will also recall that an emitter-follower circuit has a voltage gain of about unity.

But the current in the circuit is a different story. The DAC only has to provide base current for Q_1. Depending on the specific transistor used in the circuit, the actual load current may be several hundred or even 1000 times higher since load current is really the emitter current of the transistor.

Capacitors C_1 and C_2 do not directly impact the function of the circuit, but they are included to suppress unwanted high-frequency noise. As you will learn later, digital circuits generate a lot of radio-frequency noise. If this energy is allowed to reach external wiring, it will radiate and cause interference to other products. C_1 and C_2 help attenuate this radio-frequency noise before it can reach the external wiring.

Special Amplifier Configurations

In the next few paragraphs, we will briefly examine several additional amplifier configurations that you may encounter. Each of them involves the interconnection of multiple transistors.

Darlington Pair Figure 10-31(a) shows a transistor configuration called a Darlington amplifier. The two transistors are often referred to as a Darlington pair. This is such a common configuration that many manufacturers provide Darlington pairs in a single package (e.g., MJE1103) as indicated in Figure 10-31(b).

FIGURE 10-31
Darlington pairs can be formed with (a) two discrete transistors or (b) as a single transistor package.

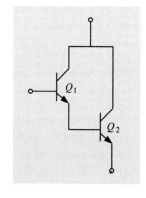
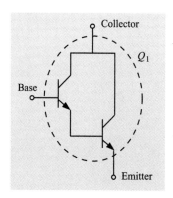

(a)   (b)

Figure 10–32 shows a Darlington pair being used in an emitter-follower circuit. The emitter current of Q_2 is the load or output current for the circuit shown. We know that the base current of Q_2 will be smaller by a factor of β_2 (current gain of Q_2). Note that the base current of Q_2 is actually the emitter current of Q_1. The base current of Q_1 will be smaller than its emitter current by a factor of β_1 (current gain of Q_1). This means that the total current gain of the two interconnected transistors (ignoring the current through the bias resistors) is the product of the individual current gains. We can express this formally as Equation 10–5.

$$A_I \text{ (total)} = \beta_1 \beta_2 \qquad (10\text{–}5)$$

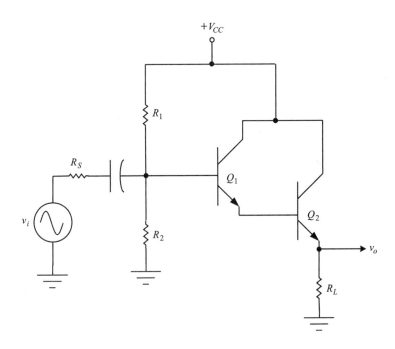

FIGURE 10–32 An emitter-follower circuit based on a Darlington amplifier.

In practice, the resulting current gain or effective overall β is very high. This is the primary advantage or reason for using the Darlington configuration. As an example, the effective current gain for a Motorola TIP101 transistor can be as high as 20,000 with 2,500 being typical.

There is another way to view the effects of the high current gain. If you consider that the emitter of Q_2 and the base of Q_1 have nearly equal voltages (within 1.4 V of each other and fairly constant), but dramatically different currents, then it follows that they have dramatically different resistances. In effect, the resistance seen by the emitter (load resistance in this case) is magnified by the overall current gain of the transistor pair. The result is that the Darlington pair configuration provides an exceptionally high input resistance.

Differential Amplifiers All of the amplifiers discussed to this point have a single input and a single output. The amplitude of the output is proportional to the amplitude of the input. A differential amplifier has two inputs and either one or two outputs. The output voltage amplitude is relatively unaffected by the absolute voltage on either of the input terminals. Rather, the output voltage is proportional to the difference between the two input terminals. This leads to the name-difference amplifier or differential amplifier.

Figure 10–33 shows the schematic diagram of a basic differential amplifier circuit. Note that the two transistors share a common emitter circuit. More specifically, a constant current source supplies the emitter current for the two transistors.

A constant current source is an electronic circuit (e.g., a current regulator diode as discussed in chapter 9) that maintains a constant current regardless of the resistance (within practical limits) of the circuit. The base and collector circuits for the two transistors are not shared. Since the emitter currents are provided by a constant current source, the combined or total emitter currents must remain constant. The current through a given transistor, however, can increase or decrease, but the current through the remaining transistor will change in the opposite direction to maintain a constant total.

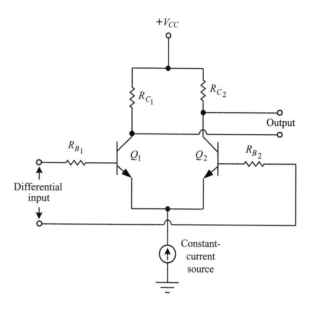

FIGURE 10-33 A basic differential amplifier circuit.

Suppose, for example, that the two input lines have the same dc voltage relative to ground. That is, there is no difference between the two input terminals. Since the two emitters are tied together and the two base connections have the same voltage, we can conclude that the two transistors must have equal emitter-collector currents. If the collector resistors are equal, they will have equal voltage drops, and the two collector voltages will also be equal. This means that the output voltage (taken between the two collectors) will be zero. The preceding discussion assumed that the two input terminals had the same voltage. The results do not depend on the absolute voltage on either terminal. Rather, only the difference between them affects the output, which in our example was zero.

Now, if the voltage on the base of Q_1 becomes more positive and the voltage on the base of Q_2 remains the same, then Q_1 will conduct more due to increased forward bias. The emitter voltage of Q_1 will increase slightly because the base voltage increased and the base-emitter voltage drop of a transistor is relatively constant. Since the two emitters are tied together, we know the emitter voltage of Q_2 must also increase slightly. The base voltage of Q_2, however, has not changed. Therefore, the increased emitter voltage reduces the forward bias on Q_2 and decreases the emitter-collector current. Note that the emitter current of Q_1 increased, but there was a corresponding decrease in the emitter current of Q_2. The total emitter current stays the same. Due to the increased conduction, the collector voltage of Q_1 will decrease. By contrast, the collector voltage of Q_2 will increase due to the reduction of current flow (i.e., less voltage drop across R_{C_2}). Under these conditions, there will be a difference in potential between the two collectors (Q_2 collector is more positive). This difference in potential is the output voltage of the amplifier.

The differential amplifier is a fundamental building block for linear integrated circuits. The two-terminal output of one stage can directly drive the differential inputs of another stage without the need for capacitive coupling. This is ideal for integrated circuits since it is difficult to fabricate practical-valued capacitors in integrated form. We will discuss this important application of differential amplifiers in a later chapter.

Differential amplifiers are also used in industry for amplifying low-level signals that are subject to noise. Figure 10–34 illustrates how a differential amplifier can reject noise that is coupled into both input lines but can still respond to low-level signals that represent a difference between the two input lines. A signal voltage that appears between the input lines is called a differential-mode signal. A signal that is induced into both input lines is called a common-mode signal. Two of the most prevalent common-mode noise sources are 60-Hz power lines and nearby lightning strikes. Both of these sources couple similar voltages into both of the input lines. Since the differential amplifier responds to the difference between its input lines (not the absolute voltage of the lines), the noise voltage is effectively suppressed. It is possible to amplify signals in the microvolt range while in the presence of noise voltages as high as several volts.

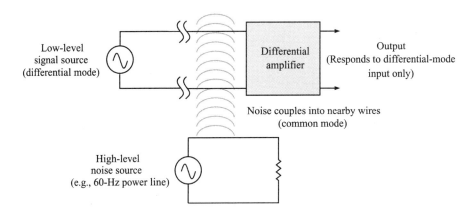

FIGURE 10–34 A differential amplifier can respond to low-level differential-mode signals while suppressing larger common-mode signals (noise).

Cascode Amplifiers Figure 10–35 shows how a common-source and a common-gate amplifier can be connected to form a cascode amplifier. A cascode amplifier configuration provides very high input impedance but a comparatively low input capacitance. The input capacitance of an amplifier tends to shunt high-frequency signals to ground, thus reducing the gain of the amplifier at these frequencies. The lower input capacitance of the cascode connection enables the circuit to operate at higher frequencies.

By carefully examining the cascode amplifier shown in Figure 10–35, we can see that Q_1 uses a voltage divider (R_1 and R_2) for bias and is configured as a common-gate amplifier. Capacitor C_G provides the ac ground for the gate of Q_1. Further examination will show that Q_2 is configured as a common-source amplifier; the input resistance of Q_1 serves as the load for Q_2. You may recall that the input resistance of a common-gate amplifier is relatively low. Therefore, the effective drain resistance for Q_2 is small. This causes the voltage gain of Q_2 to be very low; however, the advantages provided by the extremely high input impedance of Q_2 are still available to us. Most of the voltage gain in the circuit is provided by Q_1.

As mentioned previously, the input capacitance of an amplifier tends to limit high-frequency operation. Input capacitance is affected by the voltage gain of the amplifier. Since the gain of Q_2 is low, the effective shunt input capacitance is reduced, and high-frequency performance is extended.

We can replace both JFETs in Figure 10–35 with MOSFETs and still have a cascode configuration. In many cases, however, we take this transformation one step farther. When the MOSFET is fabricated, the metallized gate layer can be separated into two sections with an external lead attached to each gate section. A third diffusion of heavily-doped n-type material (in the case of an n-channel device) is

FIGURE 10–35 A cascode JFET amplifier circuit.

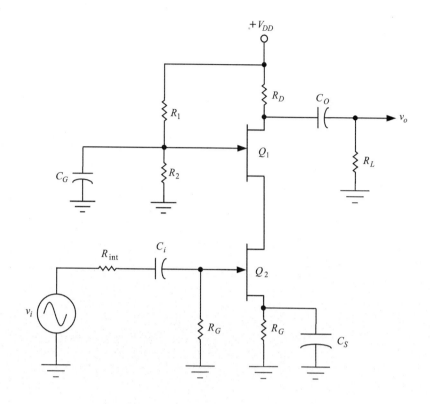

placed between the source and drain diffusions. The resulting device is called a dual-gate MOSFET. The internal structure and schematic symbol for a dual-gate MOSFET are shown in Figure 10–36.

FIGURE 10–36 (a) The internal structure and (b) the schematic symbol for a dual-gate, n-channel, depletion-mode MOSFET.

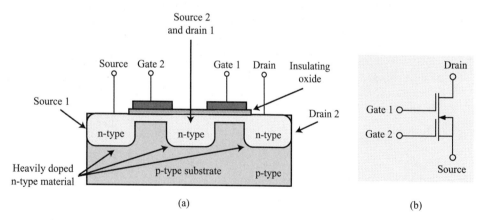

As you can see from the internal structure, it is essentially like having two MOSFETs in series. The center n-type diffusion serves as the drain for one transistor and the source for the other. Either one or both of the gates on a dual-gate MOSFET can be used to control drain current, but the two gates are electrically insulated from each other for dc purposes. We can essentially build a cascode amplifier configuration by using a dual-gate MOSFET to replace both of the transistors in the standard cascode circuit.

Integrated Linear Amplifiers Today, an integrated amplifier circuit can satisfy most amplifier requirements. In most cases, a designer will consider using an integrated amplifier first and resort to a discrete transistor amplifier only when adequate performance cannot be achieved with an integrated device. In many cases, however, the performance of integrated amplifiers is superior to discrete circuits for several reasons.

First, the nonideal or parasitic circuit elements have less effect in an integrated circuit. There are no long leads or copper traces between the components to add resistance, capacitance, and inductance to the circuit. Additionally, due to the small size of the entire circuit, it is less susceptible to interference from external noise. This is particularly true for integrated amplifiers that have metal packages. Another factor that often makes the performance of integrated devices better than that of discrete circuits is that all of the internal components are fabricated within the same silicon wafer. This means that the electrical characteristics of similar components will track very closely over changing conditions. Also, there is very good thermal coupling between the various components.

There is an impressive array of amplifier types that are available as integrated circuits. We will discuss other types later in this text. For now, let us briefly examine the performance of a representative amplifier that might be used as a low-to-medium audio power amplifier. The OPA502 integrated amplifier is manufactured by Burr-Brown. It is designed to operate with supply voltages as high as ±45 V. It can deliver output currents as high as 10 A. These two extremes cannot occur simultaneously, however, since the maximum internal power dissipation is 125 W. The device has a differential input circuit, but in most cases the complete circuit is configured to have a single input terminal. The input impedance of the amplifier is one trillion ohms at dc and decreases at higher frequencies as a result of a 4 to 5 pF shunt capacitance at each input terminal (actually caused by the input transistors). It has a voltage gain of more than 30 throughout the audio range although the gain is normally less when the effects of external components are considered.

The details of this device will be more understandable to you after completing the next few chapters of this book. Nevertheless, an understanding of the circuits presented in this chapter will be important.

✓ Exercise Problems 10.4

1. A transistor being used as a digital switching device is usually biased at midpoint of its operating range between cutoff and saturation. (True/False)
2. Refer to Figure 10–37 and complete the following table.

Computer Output	Relay State (energized/ de-energized)
0 V	
+5 V	

3. Refer to Figure 10–37. When the relay is energized, what is the approximate base voltage with respect to ground?
4. What is the name used to describe the output structure of a digital IC that uses a stacked (one above the other) transistor arrangement?
5. When an emitter-follower amplifier is used as a current amplifier, what is the approximate voltage gain of the circuit?

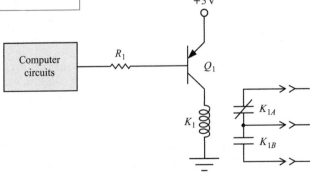

FIGURE 10–37 Circuit for Exercise Problems 2 and 3.

6. A Darlington amplifier is generally used to provide a low input impedance and an exceptionally high voltage gain. (True/False)
7. A difference amplifier has _____ inputs and _____ or _____ output connections.
8. The sum of the collector currents for the two transistors in a difference amplifier is fairly constant. (True/False)
9. A cascode amplifier configuration is most advantageous at high frequencies. (True/False)

10.5 TROUBLESHOOTING TRANSISTOR CIRCUITS

The general troubleshooting strategies discussed in preceding chapters also apply to transistor circuits. However, once you have localized the malfunction to a transistor circuit, you need to be able to determine the condition of the transistor. This section describes methods for testing a transistor to determine its condition.

Identifying Defects in Bipolar Transistors

Today, most electronic systems rely heavily on the use of integrated circuits. One integrated circuit can contain thousands of transistors, but you cannot gain access to a particular transistor to test or repair it. The entire integrated circuit must be replaced. So, do you need to know how to diagnose individual transistor problems? Absolutely! Discrete transistors are still found in most electronic systems. Even state-of-the-art notebook and handheld computers have discrete transistors to perform some of the functions. We will discuss five general strategies for identifying defects in transistors: voltage checks, resistance checks, transistor testers, substitution, and temperature variations.

Voltage Checks

One of the key indicators to a bipolar transistor's condition is the base-emitter voltage. If the transistor is forward biased (which you can determine by measuring the polarity of the base-emitter voltage), then the junction voltage should not measure much more than 0.7 V. If it is significantly higher than 0.7 V, the transistor is definitely defective.

If you determine that the base-emitter junction is forward biased, then the transistor should have current flow through the emitter and collector circuit. If there is no voltage drop across the collector resistance or the full supply voltage appears across the emitter-collector terminals of the transistor, the transistor is defective.

Now, if you find that the base-emitter junction is not forward biased, then the transistor should be cutoff. The emitter-collector circuit should be acting as an open. Therefore, there should be no voltage drop across the emitter resistor or the collector resistor unless the circuit provides some parallel path besides the transistor itself. Additionally, the full available supply voltage should be felt between the emitter and collector of the transistor. If you detect a significant voltage drop across the emitter and collector resistors and/or a very low voltage drop across the transistor (emitter-to-collector) when the transistor should be cutoff, the transistor is definitely defective.

Resistance Checks

The base-emitter junction and the base-collector junction are pn junctions and can be checked with an ohmmeter just like a diode. The absolute resistance is not as important as the front-to-back resistance ratio, which should be at least 1:10 and generally much higher. Additionally, the resistance between the emitter and collector should not be either open or shorted. The actual resistance value depends on the transistor being tested.

There is one other ohmmeter test that works well with bipolar transistors, provided you have access to a VOM (analog meter). Connect the ohmmeter to the emitter and collector terminals such that the emitter-collector circuit is properly biased. That is, the collector should be the more positive terminal for an npn transistor and the more negative terminal for a pnp transistor. Observe the resistance value. With the ohmmeter still connected, short the base terminal to the collector terminal. This will use the voltage in the ohmmeter to forward bias the base-emitter junction. If the transistor is good, you will see a significant drop in the measured resistance (less than the resistance of the forward-biased base-emitter junction alone). Table 10–4 provides a summary of resistance checks on a bipolar transistor.

The tests summarized in Table 10–4 have value, since they are capable of *definitely* identifying a defective transistor. In some cases, however, the results of the tests are inconclusive. When this situation occurs, you can perform similar measurements on a similar transistor that is known to be good. If the results are drastically different, the suspected transistor is definitely defective. It should also be reassuring to note that many transistor defects are catastrophic and can be easily detected with an ohmmeter test.

The same precautions cited for testing diodes with an ohmmeter apply to the testing of transistors. Do not use the lowest or the highest range of an analog meter unless you are certain the voltage and current produced by the meter will not exceed the ratings of the transistor. This is especially important when dealing with low-power microwave transistors.

Resistance measurements are not usually meaningful when using a digital meter because the voltage in the meter won't necessarily be sufficient to forward bias the junctions. However, most digital ohmmeters have a special function that can be used to test diodes (and therefore bipolar transistor junctions). The function is often indicated on the selector switch with a diode symbol. In any case, once the function is selected, connect the meter across the pn junction, checking first one polarity and then the other. In one direction, the junction should read open. The other polarity will generally read a value that is twice the forward voltage drop of a forward-biased junction (frequently 1.4 on the display). Read the operator's manual for your instrument to fully understand its operation.

TABLE 10–4 Interpretation of ohmmeter test on a bipolar transistor.

Measurement	Indication	Conclusion
Front-to-back resistance of base-emitter junction	Less than 10:1	Transistor is defective.
	Between 10:1 and 100:1	Transistor *may* be defective.
	Greater than 100:1	Junction is *probably* good.
Front-to-back resistance of base-collector junction	Less than 10:1	Transistor is defective.
	Between 10:1 and 100:1	Transistor *may* be defective.
	Greater than 100:1	Junction is *probably* good.
Resistance between emitter and collector	Measures very low resistance in both directions	Transistor is defective.
	Measures infinite in both directions	Transistor *may* be defective.
	Measures substantial resistance in both directions (not necessarily the same resistance)	Transistor is *probably* good.
	Decreases when base is shorted to collector while emitter-collector is properly biased by ohmmeter	Transistor is *probably* good.

Transistor Testers
There are many types of transistor testers that can quickly test a transistor and classify it as good or bad. Some are inexpensive and perform only simple tests of the junctions. Others are quite sophisticated and can test the transistor throughout its operational range. In either case, operation of these testers is fairly straightforward and needs no specific explanation at this point.

Substitution
Many technicians believe that the only real way to test a transistor is in the circuit where it will be used. To some extent this is correct. It is certain that the ohmmeter tests described previously will not locate transistor defects such as low gain, reduced breakdown, or deteriorated high-frequency response. Fortunately, many transistor defects are more serious than a slight degradation of some characteristic and are therefore more easily detected. However, the most conclusive test of all is to substitute an exact replacement transistor that is known to be good. Be advised, however, that a competent technician will not use wholesale substitution of components as a replacement for real troubleshooting. Use the techniques presented throughout this book to help you develop a reasonable level of suspicion about a particular component. Then make the substitution only as the final step of confirmation.

Temperature Variations
One type of transistor defect that can be difficult to diagnose involves sensitivity to temperature. The circuit may work normally for a while but then stop working unexpectedly or simply degrade. This type of symptom often indicates a thermal problem in the transistor. It is difficult to diagnose, because when you turn off the power to test the transistor, it cools off and tests normal.

One very effective way to diagnose this type of problem is to use an aerosol can that sprays a freezing mist. There are numerous brands available from any electronics supply house. The can is provided with a long nozzle that can be used to direct the cold spray to a specific component. The basic procedure is to monitor the symptoms and spray the suspected component while the defect is observable. The freezing spray will instantly cool the component and, if that is the problem, clear the defect momentarily. The symptoms will go away. If the symptoms clear up when the spray is applied, you have located the defective component.

Identifying Defects in JFETs
There are basically two regions of a JFET that can be defective: the channel and the pn junction. In either of these cases, the symptoms are (1) the transistor is either completely cutoff (acts like an open), or (2) it is fully saturated (acts like a short). A quick measurement of the dc drain voltage (source voltage in a source-follower circuit) will reveal either of these conditions. If the transistor is cutoff, the drain voltage will be equal to V_{DD} and the source will be equal to V_{SS} (often ground). On the other hand, if the transistor is acting like a short, then the drain voltage will be low (if $R_D > 0$), the source voltage will be high (if $R_S > 0$), and the source and drain voltages will be close to the same value.

If the symptoms indicate that the transistor is cutoff (or internally open), momentarily short the gate and source terminals together while monitoring the dc drain voltage (all but source-follower circuits). If the dc voltage does not change when the gate is shorted to the source, then you have located a defective JFET. If the dc drain voltage does change (goes lower) when the short is applied, the transistor is probably good, and the defect is in the bias components.

If the initial symptoms indicate a shorted or saturated transistor, then remove power and check the JFET with a low-power ohmmeter. Since you are looking for a short, you can generally do the tests without removing the transistor. The gate-to-source and gate-to-drain measurements should indicate the presence of a pn junction. There should be substantial resistance between the source and drain (i.e., not shorted). If the transistor checks "good," the problem must be caused by other defective components.

Identifying Defects in MOSFETs

When a MOSFET develops a defect, it is generally catastrophic and easy to detect. The device is left in a full on (shorted) or full off (open) state. Measurement and comparison of the dc terminal voltages will normally provide enough information to classify the condition of the MOSFET. First, measure the drain (or source), as described in the preceding section for JFETs, to determine whether the MOSFET is fully on or fully off. Next, compare the polarity and value of the gate-to-source voltage with the known state of the device (on or off). If the two disagree, the MOSFET is defective. Otherwise, the problem may be caused by bias components. For example, if the drain voltage indicates that the MOSFET is saturated but the gate-to-source voltage is zero on an enhancement-mode MOSFET, then the transistor is probably defective. That is, $V_{GS} = 0$ V should cause the transistor to be cutoff, but it is fully on (or shorted).

When working with MOSFETs, you must always remember that they can be damaged by static discharge. You can easily damage a replacement MOSFET before you even put it in the circuit by simply touching the leads with your finger. MOSFETs are normally shipped with a wire ring or black conductive foam on the leads. This prevents a charge from accumulating between the leads. Leave the leads shorted together anytime the MOSFET is not in the circuit. After the replacement transistor has been soldered into the circuit, remove the shorting wire from its leads.

✓ Exercise Problems 10.5

Refer to Figure 10–38 for questions 1 through 5. Table 10–5 gives the normal dc and ac voltages at the various testpoints.

1. TP2 measures 4.8 V, TP3 measures 24 V, and TP4 measures 0 V. What is the most probable defect?

FIGURE 10–38 Circuit for Exercise Problems 1 through 5.

Fig10_38.msm

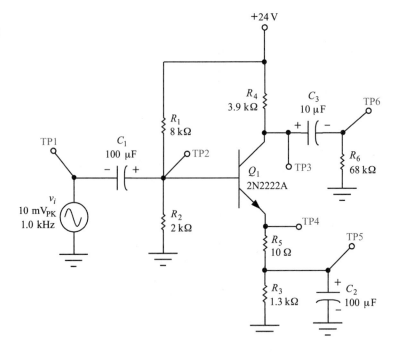

TABLE 10–5 Normal voltage values for the circuit shown in Figure 10–38.

Voltage	Testpoint					
	TP1	TP2	TP3	TP4	TP5	TP6
dc (V)	0	4.777	11.786	4.121	4.090	0
ac (peak)	10 mV	10 mV	1.8 V	5.3 mV	815 µV	1.8 V

2. The dc voltages are normal at all testpoints in the circuit, but there is no ac signal anywhere except at TP1. What is the most likely defect?
3. TP2 measures 4.8 V, TP3 measures 6.034 V, and TP4 measures 6.034 V. What is the most probable defect?
4. TP2 measures 4.777 V, TP3 measures 11.786 V, and TP4 measures 4.121 V. All ac voltages are normal except there is no ac signal across R_6. What is the most probable defect?
5. If resistor R_1 opened, what changes would you expect to see in the dc voltage measurement on TP3?
6. Why must you use care when testing the condition of a MOSFET with an ohmmeter?

•—SUMMARY

Bipolar transistors consist of three-layer sandwiches of semiconductor material: npn and pnp. In both cases, the center region (base) is thin and lightly doped and the emitter is heavily doped. The emitter and collector currents are nearly equal in a properly biased transistor. The ratio of collector current to emitter current is called alpha and is near unity. Base current is small compared to the other currents in a transistor. The ratio of collector current to base current is called beta. Typical values of beta range from 35 to over 1000.

The base-emitter junction is forward biased and the collector-base junction is reverse biased in a bipolar transistor biased for linear operation. The voltage across the forward-biased base-emitter junction is quite small (≈ 0.7 V). For digital operation, the base-emitter and collector-base junctions are both forward biased (saturation) or reverse biased (cutoff).

Amplifiers can amplify the current, voltage, or power of an input signal. Amplification or gain can be expressed as a simple ratio or in decibels (dB). Amplifiers may or may not invert the input signal. Common-emitter (or source) amplifiers have a 180° phase inversion between input and output. Common-base (gate) and common-collector (drain) amplifiers have no inversion.

JFETs are unipolar devices that can be used in either digital or linear applications. A JFET will conduct source-to-drain current with no gate bias. Reverse bias on the gate-source junction is required to decrease conduction and/or to cutoff the transistor. They have a relatively high input impedance due to the reverse-biased gate junction. JFET amplifiers generally have less voltage gain than bipolar equivalents.

There are two general classes of MOSFETs: depletion mode and enhancement mode. Depletion-mode MOSFETs are on with no gate bias. Gate bias (or an ac signal) can increase or decrease the source-to-drain current. Enhancement-mode MOSFETs are off with no gate bias. Gate bias creates an inversion layer below the gate and allows channel current to flow. The gate is insulated from the channel by a thin oxide layer. This is why the input resistance of MOSFET amplifiers can be extremely high. It is also why MOSFETs require special handling to avoid puncturing the oxide layer with ESD that may accumulate in the body. MOSFETs are used in applications ranging from dc to GHz. They are often used for low-noise amplifier applications.

Although most new electronic products are largely built around integrated circuits, discrete transistors are still utilized for both digital and linear applications. In most cases, transistors are selected for their greater current-, voltage-, or power-handling capability.

Darlington amplifiers use two transistors and have high current gain and high input impedance. They can also be purchased as a single three-terminal package. Difference amplifiers have two inputs and either one or two outputs. The amplifier circuit only responds to the difference in voltage (or current in some type of amplifiers) between the two inputs. The circuit is relatively unaffected by the absolute potential on either input. Bipolar, JFET, and MOSFET pairs can be configured as a cascode amplifier. This arrangement reduces the input capacitance of the amplifier and allows it to amplify higher frequencies. A dual-gate MOSFET can also be configured as a cascode amplifier.

REVIEW QUESTIONS

Section 10.1: Bipolar Transistors

1. Why are npn and pnp transistors called bipolar transistors?
2. When viewed independently, each pn junction in an npn transistor acts like a resistor. (True/False)
3. When viewed independently, each pn junction in a pnp transistor acts like a diode. (True/False)
4. The emitter of a pnp transistor is heavily-doped p-type material. (True/False)
5. The base of an npn transistor is a thin region made of lightly-doped p-type material. (True/False)
6. It is possible for a transistor to have a relatively high current through the base-collector junction even if the junction is reverse biased. (True/False)
7. What is the name given to a transistor when it is biased such that $I_C = 0$?
8. A transistor switch is always in saturation. (True/False)

Refer to Figure 10–39 for questions 9 through 12.

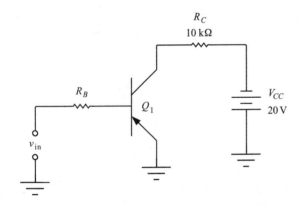

FIGURE 10–39 Circuit for questions 9 through 12.

9. If $V_{IN} = 0$ V, what is the state of the transistor?
10. If $V_{IN} = +5$ V, the transistor may be saturated. (True/False)
11. If $V_{IN} = -5$V, the transistor may be saturated. (True/False)
12. If $V_{IN} = 0$ V, what is the value of collector voltage?
13. What is the approximate value of base-emitter voltage for a silicon npn transistor biased for linear operation?
14. What is the voltage gain of a common-emitter amplifier circuit whose output is 2 V_{PK} when a 25-mV_{PK} input signal is applied?
15. Refer to the previous question. Express this gain in dB.
16. The base current of an emitter-follower amplifier is 200 µA, and the emitter current is 18 mA. What is the current gain of the amplifier circuit?
17. Refer to the previous question. Express this gain in dB.
18. If an audio transistor amplifier delivers 2.5 W to a speaker when 10 mW are applied to its input, what is the power gain of the amplifier circuit?
19. Refer to the previous question. Express this gain in dB.
20. What is the phase relationship between input and output in a common-base amplifier circuit?
21. What is the phase relationship between input and output in a common-collector amplifier circuit?
22. What is the phase relationship between input and output in a common-emitter amplifier circuit?
23. What is the phase relationship between input and output in an emitter-follower amplifier circuit?
24. What class of amplifier conducts for the full 360° of the input waveform?
25. What bipolar amplifier configuration can provide the highest input resistance?
26. Which bipolar amplifier configuration uses the collector and base as output terminals?

Section 10.2: Junction Field-Effect Transistors

27. A JFET can only be used for linear amplifier applications. (True/False)
28. What is the conduction state of a p-channel JFET with no gate-source bias?
29. The gate terminal of a p-channel JFET is connected to n-type material. (True/False)
30. What is the ideal phase relationship between input and output of a common-source amplifier?
31. In order to function as a linear amplifier, the gate-source junction of a JFET must be (*forward/reverse*) biased.
32. What practical use would there be for a JFET circuit that operated only in the cutoff and saturation states?

33. JFETs are often used in low-noise amplifier applications. (True/False)
34. The input resistance of a common-source amplifier is generally much higher than the input resistance of a common-base amplifier. (True/False)

Section 10.3: MOS Field-Effect Transistors

35. Name the four basic classes of MOSFETs.
36. What general type of MOSFET has no significant current flow when the gate-source voltage is 0 V?
37. Are there times when we would want the gate voltage of an n-channel depletion-mode MOSFET to be positive with respect to the source? Negative? Not biased at all?
38. What is an inversion layer in a MOSFET transistor?
39. Which type of MOSFET uses a broken line as part of the schematic symbol?
40. A depletion-mode MOSFET can be operated in an enhancement mode, but an enhancement-mode MOSFET cannot be operated in the depletion mode. (True/False)
41. Name the two most outstanding advantages of a MOSFET transistor over a bipolar transistor for amplifier applications.
42. The input signal is applied between the source and drain in a common-gate MOSFET circuit. (True/False)
43. Why are MOSFET transistors often shipped in conductive bags and/or shipped with their leads shorted together with wire?
44. MOSFETS are limited to applications ranging from dc to no more than 1 to 2 MHz. (True/False)

Section 10.4: Transistor Applications

45. Digital applications are characterized as having _____ states.
46. Industrial computers often use transistor switches to control lamps and relays. Why don't the computer outputs drive the lamps and relays directly?
47. MOSFETs are rarely used to make modern integrated circuits. (True/False)
48. An emitter-follower amplifier using a Darlington pair configuration would provide an exceptionally high voltage gain. (True/False)
49. If an ideal difference amplifier has identical sine waves on its two inputs, describe the output waveshape.
50. A cascode amplifier can be built with either one or two physical transistors. (True/False)

Section 10.5: Troubleshooting Transistor Circuits

51. Regardless of the polarity, if the voltage between the emitter and base of a transistor measures 2.5 V, it is definitely defective. (True/False)
52. If the base-emitter junction of a pnp transistor measures 85 Ω in one direction with an ohmmeter and 375 kΩ in the other direction, it is probably a (good/bad) junction.

Refer to Figure 10–40 for questions 53 through 55. Table 10–6 on page 380 lists the normal voltage values for the circuit shown in Figure 10–40.

53. All dc voltages are normal, and the waveform at TP3 is normal. The dc voltage across R_6 is normal, but

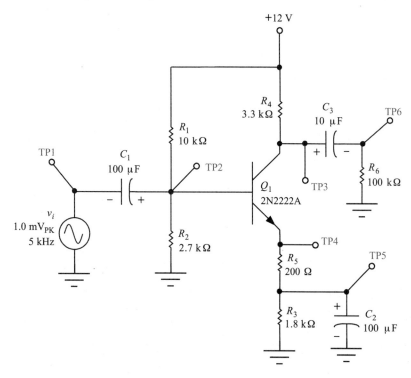

FIGURE 10–40 Circuit for questions 53 through 55.

TABLE 10–6 Normal voltage values for the circuit shown in Figure 10–40.

Voltage	Testpoint					
	TP1	TP2	TP3	TP4	TP5	TP6
dc (V)	0	2.54	8.85	1.92	1.73	0
ac (peak)	1.0 mV	1.0 mV	13.9 mV	873.4 μV	2.5 μV	13.8 mV

there is no ac signal. What is the most probable defect?

54. The dc voltages at TP2, TP3, and TP4 are 0 V, 12 V, and 0 V, respectively. Furthermore, there are no ac waveforms at any of the testpoints TP2 through TP6. What is the most likely defect?

55. The dc voltage at TP2 measures 2.55 V, but the dc voltages at TP3 and TP4 measure 12 V and 0 V, respectively. What is the most probable defect?

—CIRCUIT EXPLORATION

This Circuit Exploration will give you additional opportunities to develop your troubleshooting skills on transistor circuits. Figure 10–41 shows a transistor amplifier circuit. Table 10–7 lists the normal circuit voltages in the first two rows of the table. The remaining rows list the dc and ac voltages that are measured in the presence of a defect. Your task is to associate each list of symptoms with the defect that would cause those symptoms.

Possible defects:
a. C_1 open
b. C_2 open
c. C_3 open
d. C_4 open
e. R_7 open
f. R_2 open
g. Q_2 defective

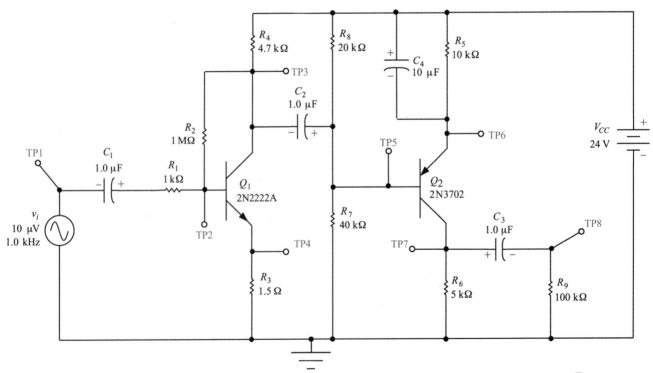

FIGURE 10–41 A transistor amplifier.

Fig10_41.msm

TABLE 10–7 Voltage values for the circuit shown in Figure 10–41.

Voltage		Testpoint							
		TP1	TP2	TP3	TP4	TP5	TP6	TP7	TP8
Normal	dc (V)	0	0.65	11.87	0.004	16.1	16.8	3.58	0
	ac (rms)	7 µV	4.6 µV	.82 mV	0.5 µV	.81 mV	.32 mV	94.9 mV	94.9 mV
Defect 1	dc (V)	0	0.65	11.87	0.004	16.1	16.8	3.58	0
	ac (rms)	7 µV	4.3 µV	1.2 mV	0.5 µV	1.2 mV	1.2 mV	.55 mV	.55 mV
Defect 2	dc (V)	0	0.65	11.87	0.004	24	24	0	0
	ac (rms)	7 µV	4.3 µV	1.3 mV	0.5 µV	1.3 mV	0	0	0
Defect 3	dc (V)	0	0.65	11.87	0.004	16.1	16.8	3.58	0
	ac (rms)	7 µV	0	0	0	0	0	0	0
Defect 4	dc (V)	0	0.65	11.87	0.004	16	24	0	0
	ac (rms)	7 µV	4.3 µV	1.2 mV	0.5 µV	1.2 mV	0	0	0
Defect 5	dc (V)	0	0.65	11.87	0.004	16.1	16.8	3.58	0
	ac (rms)	7 µV	4.1 µV	1.5 mV	.47 µV	0	0	0	0
Defect 6	dc (V)	0	0.017	24	0	16.1	16.8	3.58	0
	ac (rms)	7 µV	7 µV	0	0	0	0	0	0
Defect 7	dc (V)	0	0.65	11.87	0.004	16.1	16.8	3.58	0
	ac (rms)	7 µV	4.6 µV	.82 mV	0.5 µV	.81 mV	.32 mV	99.5 mV	0

CHAPTER 11
Op Amps and Op Amp Circuits

•—KEY TERMS
clipping
closed loop
closed-loop differential input voltage
closed-loop voltage gain
common-mode voltage gain
differential-mode voltage gain
frequency response
input bias current
negative feedback
open loop
open-loop voltage gain
output offset voltage
output saturation voltage
positive feedback
slew rate
virtual ground

•—OBJECTIVES
After studying this chapter, you should be able to:

- Contrast values for each of the following characteristics for ideal and nonideal op amps:

Closed-loop differential input voltage	Frequency response
Input bias current	Input resistance
Open-loop voltage gain	Output offset voltage
Output resistance	Output slew rate

- Calculate the voltage gain for inverting and noninverting amplifier circuits.
- Describe techniques for localizing defects in op amp circuits.
- State the purpose of each component and describe the operation of each of the following op amp circuits:

Active filter	Integrator
Inverting amplifier	Noninverting amplifier
Precision rectifier	Voltage comparator with hysteresis
Voltage follower	

•—INTRODUCTION
An operational amplifier or op amp as discussed in this book refers to an integrated amplifier circuit having well-defined performance characteristics. In practice, a single integrated circuit may contain a single op amp. Alternatively, multiple op amps may be contained in a single integrated circuit. Op amps are widely used in computer circuits for such things as multimedia signal processing in home computers and analog signal conditioning for industrial computer applications.

We will first concentrate on understanding the behavior and characteristics of the individual op amp without regard to how it is ultimately packaged. Next, we will examine two basic amplifier configurations that form the heart of most op amp applications. We will then explore a variety of op amp circuits to gain additional insights into op amp behavior and application. Finally, we will discuss troubleshooting techniques that will help you localize and identify defects in op amp circuits.

11.1 OP AMP CHARACTERISTICS

In this section, we will contrast the characteristics of ideal op amps with those of practical devices. For troubleshooting purposes and for many design purposes, op amps can be viewed as ideal components. That is, when localizing defects in an op amp circuit, we can generally assume the op amp has ideal characteristics.

Schematic Representation

Figure 11–1(a) shows the schematic symbol for an op amp. The symbol includes two input terminals, an output terminal, two power supply connections, and one or more compensation terminals. There are many applications where the compensation pins are not used. In these situations, they are often absent from the schematic symbol. Similarly, not all companies choose to show the power supply connections as part of the schematic symbol. In these cases, the simplified representation shown in Figure 11–1(b) is used. Also note that the orientation of the input pins (i.e., which one goes on top) is generally chosen to eliminate drawing clutter.

FIGURE 11–1 (a) The complete and (b) the simplified schematic symbol for an op amp.

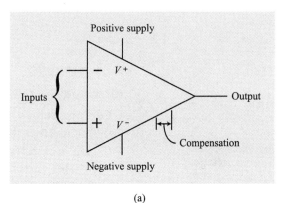

(a)

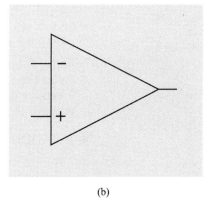

(b)

An op amp is basically a differential amplifier with a single-ended output. Thus, the voltage to be amplified is applied between the two input terminals. As with discrete differential amplifiers, one of the inputs inverts the incoming waveform and the other does not. In the case of an op amp, the inverting input terminal is identified with a minus sign (−). The noninverting input terminal is identified with a plus sign (+).

Many op amp circuits utilize dual power supplies. For example, ±15 V is a popular choice. When a dual power supply system is used to power an op amp, there is no ground connection made directly to the device. Other applications require a single power supply in which one of the power supply terminals of the op amp is grounded, and the single supply voltage is connected to the remaining power supply input terminal.

As shown in Figure 11–1, most op amps have a single output pin. The voltage on this single-ended output is referenced to ground. The output voltage of an ideal op amp can swing between the two power supply levels. For example, the output of an ideal op amp using ±15-V supplies can swing between the limits of +15 V and −15 V. Practical bipolar op amps can usually come within 2 to 3 V of either power supply value, whereas op amps with MOSFET outputs can come much closer (to within a few millivolts).

Some op amps have one or more compensation pins. These pins can be used to fine-tune the behavior of the op amp. For example, with 0 V between the input pins of an ideal op amp, the output will also be at 0 V. In a practical op amp, however, 0 V on the input may result in a slight offset voltage in the output. This is called the **output offset voltage** and can be either polarity. If the op amp is provided with compensation terminals called null offset pins, then the output offset voltage can be nulled or adjusted to zero.

Other op amps include another type of compensation input called frequency compensation inputs. This pin (or pins) can be used to obtain optimum tradeoffs between bandwidth and stability. In short, if an amplifier is designed to be stable at any practical gain, then its bandwidth is necessarily restricted. On the other hand, wider bandwidths are possible if we are willing to restrict the gain of the amplifier to certain portions of its potential range.

Internal Circuitry

It is not necessary for you to understand or analyze the internal circuitry of an op amp in order to troubleshoot or otherwise utilize op amps. The behavior of the overall device is very well defined. Nonetheless, let us briefly examine the schematic shown in Figure 11–2, which shows the internal circuitry of a common op amp. This diagram will remove some of the mystery that might otherwise surround the integrated op amp.

First, note that the op amp is composed of familiar components (e.g., resistors, transistors, diodes, and so on). Next, notice that the input terminals of the overall op amp are actually connected to the bases of a transistor differential amplifier.

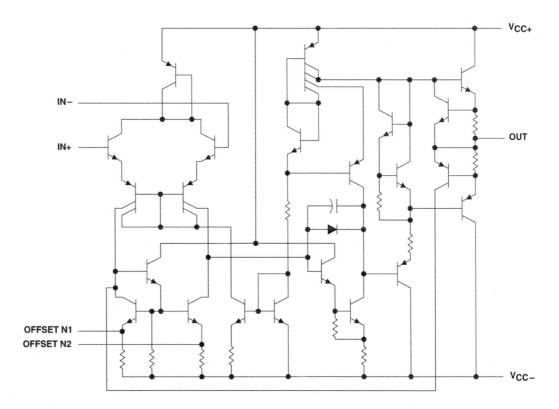

FIGURE 11–2 The internal schematic for a 741 op amp. (*Courtesy of Texas Instruments*)

Now, observe the output circuit. Here, a pair of complementary (npn and pnp) transistors is used to control the voltage on the output terminal of the overall op amp. When the npn transistor conducts harder, the output is pulled toward the positive supply voltage. As the pnp output transistor increases conduction, the output terminal is pulled toward the negative supply voltage. The resistors connected to the output terminal are relatively small, so we can expect the output resistance of the overall op amp to be fairly low.

This particular op amp includes two offset null inputs that can be used to cancel or null the output offset voltage of the op amp. Note that these inputs connect to the emitters of two transistors that are part of the emitter circuits for the input difference amplifier. The offset null inputs essentially cause an intentional imbalance in the difference amplifier to compensate for imperfections in the devices that would otherwise cause a small dc offset voltage to appear at the output when 0 V is present between the input terminals.

The 741 op amp is a relatively old device, but it is still widely used in industry. Other practical op amps include MC1747CD, MC3401, LM301AN, and OPA445BM.

Input Bias Current and Input Resistance

Figure 11–3 shows an op amp with its inputs connected together. If you will recall the internal schematic diagram for a typical op amp (Figure 11–2), you can readily see that the current meter in Figure 11–3 is simply measuring base current for the input difference-amplifier transistors. This current is called the **input bias current**. More specifically, input bias current is considered to be one-half of the sum of the two input currents as measured in Figure 11–3. In other words, the two bias currents are considered to be equal.

FIGURE 11–3 Measuring input bias current.

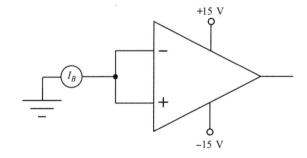

Op amps using bipolar transistor-input circuits typically have bias currents in the tens to hundreds of nanoamperes. Op amps are also available with FET transistors on the input stage. As you might expect, the input bias current is significantly lower for FET devices. Typical values for input bias current on a JFET op amp are generally in the tens of picoamperes. Devices with MOSFET inputs can have input bias current values as low as a few femtoamperes (10^{-15}).

An ideal op amp will be considered to have zero input bias current. For many practical purposes including troubleshooting, we can assume ideal values of input bias current.

When a signal source is connected to the input of an op amp, a certain amount of current is drawn from the source. Input bias current is part of the total input current that must be supplied by the signal source. Additionally, it must provide any current drawn by parasitic (i.e., nonideal) elements in the input circuit of the op amp. The total input resistance of a typical bipolar op amp is several megohms. JFET and MOSFET devices have much higher resistances on the order of 10^{12} to 10^{15} ohms.

An ideal op amp has infinite input resistance. That is, when it is connected to a signal source, it draws no current. For many practical purposes such as troubleshooting, we will assume that an op amp has infinite input resistance.

Negative Feedback Is Required

As you will soon see, for most practical circuit applications we need to modify the performance of a given op amp by providing it with **negative feedback**. We will discuss the effects of negative feedback in greater detail in Section 11.2. In short, this amounts to returning a portion of the output back to the input to be summed (out of phase) with the actual input signal. Figure 11–4 shows two amplifier configurations that utilize negative feedback.

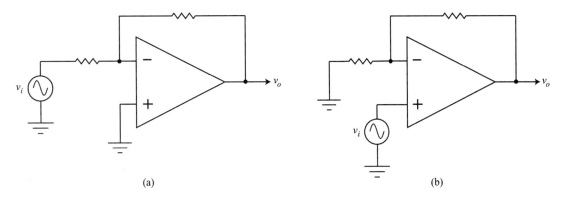

FIGURE 11–4 Negative feedback is required for practical (a) inverting and (b) noninverting amplifiers.

Figure 11–4(a) shows a basic inverting amplifier with negative feedback. Figure 11–4(b) shows a similar noninverting amplifier. We will discuss the operation of these circuits in Section 11.2, but you should begin to recognize these basic circuit configurations as we progress through the remainder of this section. When an op amp is operated with negative feedback, its operation is described as **closed loop**. If there is no feedback path, operation is called **open loop**.

Output Resistance

Figure 11–5 shows one way to view the output of an op amp. We know that in any amplifier circuit a signal at the input terminals causes a corresponding (and generally larger) signal at the output terminal. In Figure 11–5, the output voltage is represented as coming from a source internal to the op amp. This internal source has

FIGURE 11–5 A way to view the output resistance of an op amp.

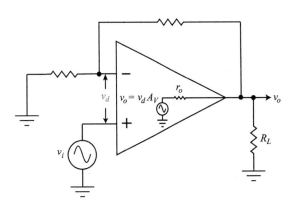

an associated series resistance (r_o). The effective resistance r_o is called the output resistance of the op amp.

As you can see by examining Figure 11–5, the internal source voltage is equal to the effective input voltage (v_d) times the voltage gain (A_v) of the op amp. In an ideal op amp, this is the actual output of the op amp, regardless of what value load is connected to the device. In a practical op amp, we know intuitively that we cannot draw an infinite amount of current. We would expect the output to decrease as we draw more and more current. Resistor r_o in Figure 11–5 drops part of the ideal voltage that would otherwise reach the load. The higher the current drawn from the output of the op amp, the greater the voltage drop across r_o, and the lower the output voltage. An ideal op amp, then, has zero output resistance. The open-loop output resistance of a practical op amp depends on the specific device, but values from 25 ohms to several kilohms are representative.

Negative feedback has many effects, but one of them is to lower the effective output resistance of an op amp. In fact, the effective (i.e., closed-loop) output resistance of an op amp using negative feedback (which includes the majority of op amp applications) is so small as to be insignificant for most practical applications. For purposes other than the initial circuit design (e.g., troubleshooting), the output resistance of a practical op with negative feedback can be considered as ideal (zero ohms).

Output Voltage Swing

As the input signal to an op amp is made larger (or the voltage gain is made higher), we expect the output voltage to increase as well. We know intuitively that there must be practical limits to how large the output voltage swing can be. As stated previously, bipolar op amps can normally have output voltage swings that are within 2 to 3 V of the supply voltages. If the output tries to exceed these limits, the output waveform will be distorted. Figure 11–6 shows an amplifier circuit with four different input signals.

In Figure 11–6(a), the input signal is small, so the output waveform is faithfully reproduced. In Figure 11–6(b), the input signal is larger, so the output is also larger. The input signal is further increased in Figure 11–6(c), but this causes the output

FIGURE 11–6 The output voltage swing of an op amp has limits.

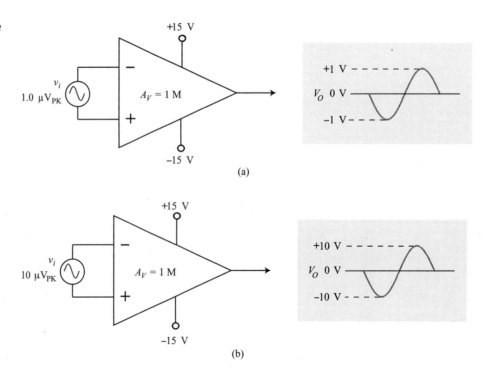

FIGURE 11-6 (continued)

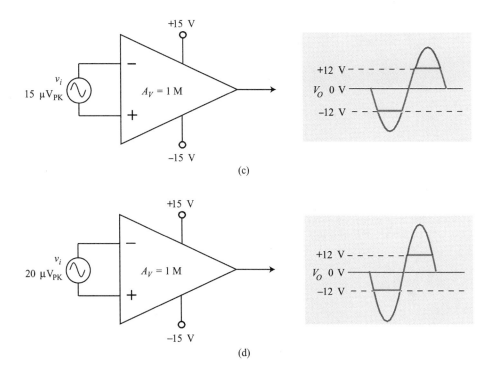

to attempt to swing beyond its limits. Therefore, we begin to see distortion in the output waveform. This particular type of waveform distortion is called **clipping**. Technicians and engineers also refer to this type of operation as "driving the output to the rails." If we continue to increase the input signal as in Figure 11-6(d), the clipping level remains the same (since the op amp internal design and the power supply voltages determine the maximum output excursion), but the percentage of distortion increases. The maximum voltage excursion on the output of an op amp is called the **output saturation voltage**.

Open-Loop versus Closed-Loop Voltage Gain

For reasons that will be more understandable as we progress, the internal voltage gain of an op amp (called the **open-loop voltage gain**) must be much higher than the voltage gain of the overall amplifier circuit (called **closed-loop voltage gain**) that includes the op amp. In fact, an ideal op amp has infinite voltage gain at all frequencies. Practical op amps have typical voltage gains ranging from 100,000 to over 1,000,000 at low frequencies. The internal voltage gain of a practical op amp decreases at higher frequencies, as we will see when we discuss frequency response in the next section.

For most practical op amp applications, the internal op amp voltage gain is so much higher than the overall amplifier voltage gain that the internal gain can be considered infinite. That is, for practical applications that don't exceed the voltage and frequency limits intended by the circuit designer, we can generally consider practical op amps to be ideal with regard to internal voltage gain.

Common-Mode versus Differential-Mode Voltage Gain

The voltage gains of an op amp as discussed so far are more completely described as differential-mode voltage gains. **Differential-mode voltage gain** is the gain given to a voltage that appears between the two input terminals. The ideal value is infinite. By contrast, **common-mode voltage gain** is the gain given to a voltage that appears on both input terminals with respect to ground. Ideally, the op amp will reject voltages that appear on both input terminals, so the ideal common-mode voltage gain is zero.

Frequency Response

The open-loop **frequency response** (i.e., voltage gain versus frequency) curve of an ideal op amp reveals infinite voltage gain over an infinite range of frequencies. Figure 11–7 shows the open-loop frequency response of a practical op amp. At very low frequencies, it has a very high voltage gain (100 dB = 100,000). This should seem quite impressive relative to the discrete transistor amplifiers that we discussed. Another interesting characteristic of an op amp—evident from the frequency response curve—is that it will respond to extremely low frequencies. Although the graph in Figure 11–7 begins at 1 Hz, most op amps respond just as well to dc inputs.

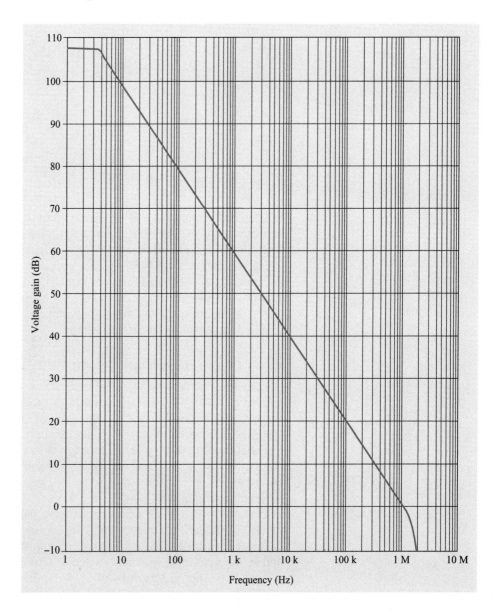

FIGURE 11–7 The frequency response for a practical op amp. (*Courtesy of Texas Instruments*)

Further examination of the frequency response curve in Figure 11–7 shows that for frequencies above 6 or 7 Hz, the voltage gain of the op amp decreases dramatically with frequency. Finally, at 1 MHz the voltage gain has decreased all the way to unity (0 dB = 1). As you will see in the next section, neither the high voltage gain associated with very low frequencies nor the steady decrease in voltage gain with increasing frequency is tolerable for a practical amplifier circuit. By adding some very simple external circuitry, we can bring the closed-loop voltage gain down to a practical value that is constant over a wide frequency range. This will be discussed in Section 11.2.

Output Slew Rate

We previously discussed that the output voltage swing of a practical op amp is limited to maximum values established by the power supply voltages and the internal design of the op amp. The output voltage swing also has a limitation regarding how fast it can change. In other words, there is a limit to how quickly the output voltage can change, regardless of what the input signal is doing. The maximum rate of change of the output voltage is called the **slew rate** of the op amp. Slew rate is specified in terms of volts per second, or more commonly, volts per microsecond (V/μs). Figure 11–8 illustrates the effects of slew rate on an op amp circuit.

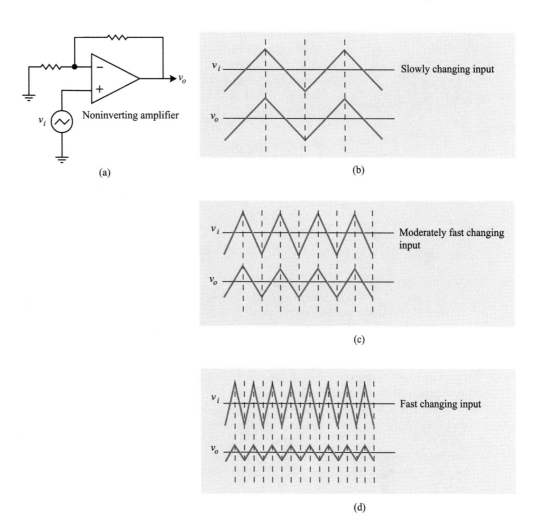

FIGURE 11–8 Op amp slew rate limits the rate of change of output voltage.

Figure 11–8(a) shows a basic noninverting amplifier circuit with a triangle wave input. A triangle wave is simply a voltage source that slowly increases to some maximum value and then steadily decreases to a minimum value. Figure 11–8(b) shows the relationship between input and output when the requested rate of change of output voltage is less than the slew rate of the op amp. The output voltage tracks the input as you would expect from any amplifier circuit.

In Figure 11–8(c) the rate of change of the input voltage has increased. The output tries to match this higher rate, but the requested rate is slightly higher than the slew rate of the op amp. We can make two observations. First, the slope of the output waveform is equal to the slew rate of the op amp under these conditions. In other words, the output is changing as fast as it can, but it cannot quite keep up with the faster changing input. The second observation is that the amplitude of the output waveform will necessarily be reduced. The time allowed for any given alternation is determined by the input waveform. Since the rate of rise and fall on the output is less than required by the input signal, the output voltage will not have time to reach the intended limits.

Figure 11–8(d) shows the input and output relationships for the same amplifier circuit with a very fast input signal. Now the slew rate of the op amp is severely limiting the operation of the circuit. The actual rate of rise and fall at the output is substantially less than that required by the input waveform. Thus, the output waveform has a further reduction in amplitude. Note that the rate of change of the output voltage is equal to the slew rate of the op amp [the same as in Figure 11–8(c)].

Slew-rate limiting of the output can occur regardless of the waveform applied to the input. In the case of ramp waveforms such as those in Figure 11–8, slew-rate limiting is evidenced by reduced amplitude and reduced slope in the output. In the case of a square or rectangular waveform, slew-rate limiting effects show up as lengthened rise and fall times (i.e., the slope of the rising and falling edges of the waveform will be equal to the slew rate of the op amp). If the slew rate is too slow relative to the pulse width, the square wave will be transformed into a triangle wave. In the case of sinusoidal input waveforms, slew-rate limiting causes a reduction in output amplitude and distorts the output waveform.

An ideal op amp has an infinite slew rate (i.e., ∞ V/s). Practical op amps have slew rates that include the range of fractional volts per microsecond (e.g., 0.5 V/μs) to thousands of volts per microsecond (e.g., 5500 V/μs).

✓ Exercise Problems 11.1

1. A typical op amp has _____ inputs and _____ output.
2. An op amp responds to the voltage difference between its two inputs. (True/False)
3. How can you identify the inverting input of an op amp?
4. All op amp circuits require a direct ground connection to the amplifier. (True/False)
5. The output of a typical bipolar op amp can swing to within a few millivolts of the power supply voltages. (True/False)
6. What general type of op amp technology produces devices with the most ideal input bias current characteristics?
7. Most practical op amp circuits require (*positive/negative*) feedback.
8. What is the value of output resistance for an ideal op amp?
9. Most op amps can respond to input frequencies as low as dc. (True/False)
10. Over most of its operational range, the voltage gain of an op amp tends to (*decrease/increase*) with increasing frequency.
11. What op amp parameter defines how fast the output voltage can change?
12. If an op amp circuit is experiencing the effects of slew-rate limiting with a sinusoidal input, what are the symptoms in the output waveform?

11.2 BASIC AMPLIFIER CONFIGURATIONS

We are now ready to thoroughly explore the two most common op amp configurations: inverting and noninverting amplifier circuits. Not only are these two configurations important as standalone applications, but they also form the basis for many other related applications. For each of these circuits, we will learn the purpose of every component, see how to compute the expected voltage gain, predict input/output phase relationships, and discuss the overall operation of the circuit.

Noninverting Amplifier Circuit

Figure 11–9 shows the schematic diagram for a noninverting op amp circuit. This particular schematic shows the power supply connections, but it is common industry practice to omit these connections on the schematic diagram as a way to reduce drawing clutter. In most cases in this book, we will omit the power supply connections from the schematic diagrams. However, it is essential for you to remember that the supply voltages must be connected to the op amp in order for it to operate. Omission of the power supply terminals is merely an industry practice.

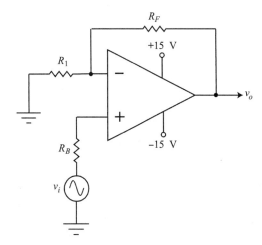

FIGURE 11–9 A noninverting amplifier circuit.

Circuit Identification

It is easy to distinguish between inverting and noninverting op amp circuits. If the input signal is applied to the inverting input (identified by the "−" sign), then the circuit will invert the input signal. By contrast, if the input signal is applied to the noninverting input (identified by the "+" sign), then no inversion will occur. We can readily identify the circuit in Figure 11–9 as a noninverting circuit because the input is applied to the (+) input.

If a portion of the output signal is returned to the (−) input terminal, the circuit is using negative feedback. If the feedback connection is resistive, the circuit is very likely configured to be an amplifier. In the case of the circuit in Figure 11–9, a portion of the output signal is returned to the input via a resistive path (R_F). Therefore, we can be reasonably confident that this circuit is designed to be an amplifier.

Op Amp Analysis Rules

Before we explore the operation of a practical amplifier circuit in more detail, let's evolve a few rules that will be extremely valuable to us throughout our study of op amp circuits. First, remember that the open-loop voltage gain of the op amp itself is extremely high (infinite for an ideal op amp). Second, recall that the output voltage of an op amp responds to the difference in potential between its two input terminals (v_d). Third, recall that the output of an op amp has some very definite limits on voltage swing (typically 1 to 3 V less than the supply voltages). From these three basic characteristics, we can infer a very important rule regarding op amp operation.

We know that the output voltage of any op amp is equal to the open-loop voltage gain times the differential input voltage (v_d). That is, $v_o = A_{\text{VOL}} \times v_d$. Therefore, by transposition, we know that the differential input voltage must be equal to the output voltage divided by the open-loop voltage gain. That is, $v_d = v_o/A_{\text{VOL}}$. This relationship holds true as long as the amplifier is operated within its linear operating range (i.e., not driven to its extremes). So, if the output voltage of an op amp is within its designated limits (i.e., not driven all the way to one of its extremes), then we can infer that the difference in voltage between its two input pins (v_d) is very small. In fact, for the case of an ideal op amp, the **closed-loop differential input voltage** would be infinitesimal, since the voltage gain is infinite. This is an extremely important conclusion, so we will state it as a Rule:

> ▶ RULE The difference in voltage between the input pins of an op amp (v_d) is essentially zero as long as the amplifier output is not driven to one of its extremes.

You will also recall that the input bias current for an op amp is quite small (femtoamperes to low microamperes). These are generally insignificant relative to other currents in the circuit. Therefore, for practical op amp applications such as troubleshooting and functional analysis, we can utilize the following Op Amp Rule:

> ▶ RULE The input pins of an op amp are open circuits and draw no current from the external circuit.

These two op amp analysis rules should be committed to memory. We will rely on them when troubleshooting and when analyzing the operation of unfamiliar op amp circuits.

Circuit Operation

The orientation of the components in the amplifier circuit shown in Figure 11–9 is typical of what you will see in industry. However, for our immediate purposes of understanding circuit operation, let's examine a redrawn version of the circuit. The circuit in Figure 11–10 is electrically the same as Figure 11–9; it is simply redrawn.

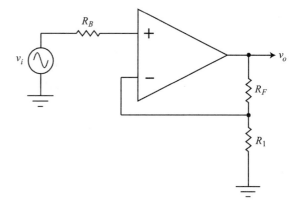

FIGURE 11–10 A noninverting amplifier circuit.

One of our basic op amp rules tells us that the input pins are open circuits. From this we can conclude that there will be no voltage drop across R_B; the full input voltage will be felt at the (+) input of the op amp regardless of the specific value of R_B. We will discuss the purpose of R_B momentarily; but for now, we can ignore it, since it has no significant voltage drop.

Resistors R_F and R_1 form a voltage divider across the output voltage. That is, some percentage of the output voltage will be developed across R_1. The percentage of voltage across R_1 is determined by the ratio of the two resistors. Since the junction of R_F and R_1 is connected to the (−) input, we can conclude that the voltage divider is essentially unloaded. Specifically, the op amp input acts as an open circuit, so it will not affect the voltage division in the voltage divider.

Consider this question. If the instantaneous input voltage on the (+) input is +2 V, what is the voltage on the (−) input pin at that time? According to one of our basic op amp rules, as long as the output is not driven to an extreme, the difference between the input pins (v_d) will be zero. Thus, we would expect to measure +2 V at that same instant in time. Let's look at this important concept another way by means of an example.

•—EXAMPLE 11.1

Suppose the input (v_i) to the circuit in Figure 11–10 is a 350-mV peak-to-peak sine wave. What voltage would be measured on each of the input pins?

SOLUTION Because R_B drops no significant voltage, we expect to measure the full 350-mV peak-to-peak sine wave at the (+) input terminal. Since the voltage between the two inputs (v_d) is always near zero, we will also expect to measure a 350-mV peak-to-peak sine wave at the (−) input terminal.

To appreciate how all this happens, let's slow down the action and follow the circuit operation. We will begin with the input voltage at zero. This means the two input pins will be 0 V. If we multiply the difference between the input pins (0 V) by the open-loop voltage gain of the op amp, we will find that the output voltage is also 0 V at this time. With 0 V at the output, we know the voltage drop across R_1 will also be 0 V. This is the voltage applied to the inverting (−) input terminal, which we initially assumed was at 0 V.

Now, consider what happens as the input voltage begins to increase slightly in the positive direction. This is illustrated in Figure 11–11.

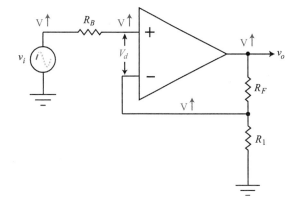

FIGURE 11–11 A rising input voltage on a noninverting amplifier circuit.

As v_i increases in the positive direction (indicated as V↑), the voltage at the (+) input rises at the same time (no voltage drop across R_B since the op amp input draws no current). At this first instant, the output voltage of the op amp, and therefore the (−) input pin, is still at 0 V. This means the (+) input is more positive than the (−) input for this instant in time (i.e., $v_d > 0$). The op amp responds to this difference and creates an output voltage that is larger than this difference by a factor equal to the open-loop voltage gain of the circuit ($v_o = A_{VOL} \times v_d$). Hence, we will see the output voltage begin to rise in a positive direction. We know the output goes positive, since the input is applied to the noninverting pin.

As the output voltage rises, so does the voltage across R_1 due to voltage divider action. This voltage is returned to the (−) input pin, so it too increases in a positive direction. But as the voltage on the (−) pin increases, the difference in voltage between the two inputs (v_d) must be decreasing. Notice that v_d is moving closer to the expected (ideal) value of 0 V. This action will become stable as soon as the output voltage has risen high enough to cause v_d to be equal to v_o/A_V. As a specific example, let us assume the open-loop voltage gain of the op amp itself is 1,000,000. This means that once the amplifier stabilizes, v_d will be equal to v_o divided by 1,000,000. Clearly, this will be a very small voltage, regardless of the exact value of output voltage. Thus, our rule that requires v_d to be essentially zero is upheld. Furthermore, this action happens almost instantaneously, so v_d remains near zero at all times.

Now, how high does the output voltage have to go in order to bring v_d back down to near zero? In short, it will go as high as it needs to go to bring the voltage across R_1 up to the required level. The exact output voltage needed to accomplish this for a given value of input voltage depends on the ratio of the feedback resistors. The smaller the percentage of output voltage fed back to the input, the greater the change in output voltage that will be required in order to reduce v_d to its near zero value. The ratio of the two voltage divider resistors determines the overall voltage gain (closed-loop voltage gain) of the circuit, since that ratio determines how much the output voltage must change for a given input voltage change.

Calculation of Voltage Gain

The closed-loop voltage gain of a noninverting amplifier like the one shown in Figure 11–12 can be computed with Equation 11–1.

$$A_V = \frac{R_F}{R_1} + 1 \qquad (11\text{–}1)$$

FIGURE 11–12 The feedback components determine the overall voltage gain.

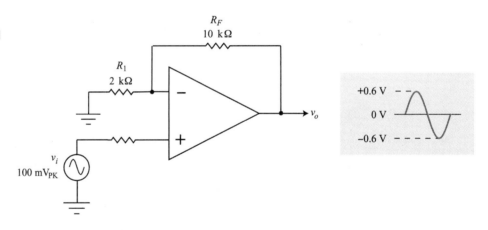

•EXAMPLE 11.2

Compute the voltage gain for the circuit in Figure 11–12.

SOLUTION We apply Equation 11–1 as follows:

$$A_V = \frac{R_F}{R_1} + 1 = \frac{10 \text{ k}\Omega}{2 \text{ k}\Omega} + 1 = 6$$

Calculator Application

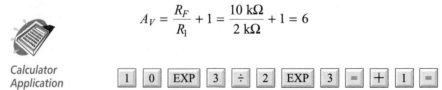

Be careful not to confuse the results of this calculation for closed-loop with the open-loop voltage gain of the op amp itself.

It is particularly important to note the variables in Equation 11–1. The only variables are the values of two external resistors. The closed-loop voltage gain of the circuit does not rely on the specific characteristics of the op amp itself. This is a profound result, since most of the internal characteristics of a given op amp are not especially stable (e.g., they vary between similar devices, and they change with temperature). The relationship expressed in Equation 11–1 is applicable as long as the open-loop gain of the op amp is substantially higher than the closed-loop gain of the circuit (a condition that is true for most practical applications such as functional analysis and troubleshooting).

Effects of Negative Feedback

The noninverting amplifier circuit previously discussed used negative feedback. That is, the portion of the output signal that was returned to the input caused a cancellation of the effects of the input voltage. Recall that input voltage changes attempt to cause corresponding changes in the differential input voltage (v_d) of the op amp. However, the resulting output changes that are fed back to the (−) input reduce the net change in v_d. So, our basic op amp rule, which states that $v_d = 0$ V, is only true because of negative feedback.

We have also seen that negative feedback has a profound effect on the voltage gain of the circuit. In Figure 11–12, for example, the closed-loop voltage gain of the circuit was 6 (≈ 15.6 dB), whereas the open-loop voltage gain of the op amp might be as much as 1,000,000 (120 dB). If we superimpose the closed-loop gain curve onto the open-loop gain curve, we can better appreciate how negative feedback affects the frequency response of the circuit. This is illustrated in Figure 11–13.

Recall from discussions in earlier chapters that the bandwidth of a circuit is defined as the band of frequencies between its two cutoff frequencies. In the case of a

FIGURE 11–13 Closed-loop voltage gain is more stable than open-loop voltage gain and provides a wider bandwidth.

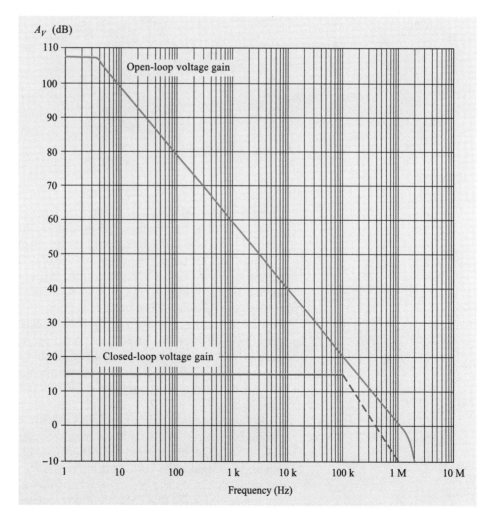

typical op amp circuit, there is no lower cutoff frequency, since the circuit can respond all the way down to dc. The upper cutoff occurs at a frequency where the voltage response decreases by 3 dB. So, in the case of the open-loop gain curve shown in Figure 11–13, the upper cutoff frequency (and the bandwidth) is equal to approximately 5 to 7 Hz. By contrast, the upper cutoff frequency (and bandwidth) for the closed-loop case is higher than 100 kHz. Note that the voltage gain is constant in the closed-loop case until the open-loop voltage gain falls to a value that approaches the closed-loop gain. At this point, our requirement that the open-loop voltage gain be much greater than the closed-loop gain is no longer satisfied.

Negative feedback in an amplifier circuit also reduces waveform distortion. That is, the output waveform more closely approximates the waveform expected from an ideal amplifier circuit. Additionally, negative feedback increases the stability of an amplifier circuit. An unstable amplifier can break into self-oscillations and actually become a signal source. This is an undesirable condition.

Purpose of Components

Refer to the amplifier circuit in Figure 11–14 as we discuss the purpose of each component. The primary purpose of the op amp itself is to provide voltage gain that is significantly higher than the expected closed-loop gain. This is easily accomplished at low frequencies but is limited at higher frequencies as the open-loop voltage gain of the amplifier decreases.

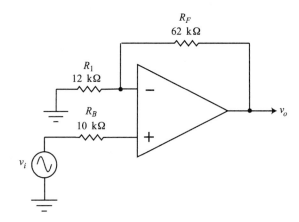

FIGURE 11–14 A noninverting amplifier circuit.

As discussed previously, R_F and R_1 form a voltage divider that determines the percentage of output voltage that is returned to the input as negative feedback. The ratio of these two resistors determines the closed-loop voltage gain of the circuit.

Until now we have been ignoring the effects of resistor R_B, but we can now examine its purpose. You will recall from earlier discussions that while ideal op amps have no input bias currents, practical op amps do require a small input bias current. In the case of the (−) input in Figure 11–14, the bias current flows through R_F and R_1. From this perspective, these two resistors are in parallel. As the bias current flows through this effective resistance, it causes a voltage drop. This voltage drop affects the output voltage of the op amp in much the same way as the normal input voltage does. In other words, even if the input voltage were 0 V, the output voltage might not be exactly 0 V as a result of the effects of input bias current for the (−) input.

Although we can't eliminate bias currents, we can reduce their effects by introducing an equal (but opposite polarity) voltage in the output. This is the purpose of resistor R_B. The bias currents for the two inputs are assumed to be approximately equal. If we make the voltage drop across R_B equal to the voltage drop caused by the bias current on the (−) input terminal, we can cancel the offset voltage in the output. In order for this cancellation to occur, the value of R_B is normally set to a value that is approximately equal to the parallel combination of R_1 and R_F. In practice, cancellation is not perfect because the two currents are not exactly the same and because component values are not precise.

Inverting Amplifier Circuit

Figure 11–15 shows the schematic diagram of an inverting amplifier circuit. We will discuss its operation and the purpose of every component, and we will learn to compute the closed-loop voltage gain of the overall circuit.

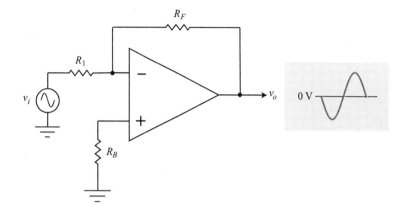

FIGURE 11–15 An inverting amplifier circuit.

Circuit Identification

By inspecting Figure 11–15, we can see that a portion of the output is returned to the input via resistor R_F. That is, the circuit has resistive feedback. The feedback is returned to the (−) input terminal, so we recognize it as negative feedback. This also means there is a strong probability that the circuit is designed to be an amplifier.

The input signal is applied to the inverting (−) input pin. Consequently, we expect the output to be inverted relative to the input waveform. By inspection, we should be able to identify this circuit as an inverting amplifier circuit.

Circuit Operation

Figure 11–16 shows an inverting amplifier circuit and illustrates the important concept of **virtual ground**.

We know that no appreciable current flows through R_B (input bias current only). Therefore, we know that the voltage drop across R_B will be essentially 0 V as labeled in Figure 11–16. Because the amplifier circuit is operated as a closed-loop system and utilizes negative feedback, we know that the differential input voltage (v_d) will be essentially 0 V. So, with respect to ground, we can infer that the

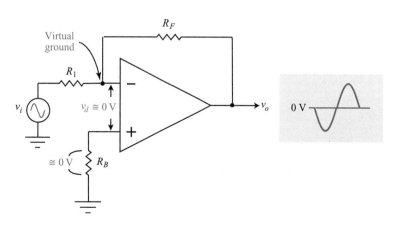

FIGURE 11–16 An inverting amplifier and virtual ground.

(−) input pin must also be at 0 V. Since the voltage drop across R_B and the value of v_d will remain at 0 V regardless of input signals, we can conclude that the voltage on the (−) input will remain at approximately 0 V with respect to ground. Because the (−) input remains at ground potential even though it is not actually connected to ground, it is referred to as a virtual-ground point. It has the characteristics of ground for purposes of circuit analysis, troubleshooting, and explaining the operation of the circuit.

•–EXAMPLE 11.3

If you were troubleshooting the circuit shown in Figure 11–16 with an oscilloscope, what waveform would you expect to see at the (−) input pin with respect to ground?

SOLUTION Since the (−) input pin is a virtual-ground point, you would expect to see 0 V only. That is, you would not see any appreciable sinusoidal waveforms. Now, if you adjust your scope to a very sensitive scale, you will be able to see a very low-level sine wave, but it is insignificant relative to the actual input signal. For most practical purposes, a virtual-ground point can be considered to be 0 V.

Since R_1 is connected between the input voltage and ground (virtual ground), we know that the input voltage will cause a signal current to flow through R_1. We could calculate its value with Ohm's Law. Since the (−) input pin of the op amp is effectively an open circuit, all of the input current through R_1 must continue through R_F according to Kirchhoff's Current Law. The output voltage will be equal to the voltage drop across R_F, since the left end of R_F is connected to ground (virtual ground) and the other end is connected to the output of the op amp.

Now, let us consider the circuit operation as a changing input signal is applied. Changes in the input voltage will make very small changes in the voltage at the (−) input pin. That means that v_d will be *slightly* greater than 0 V at that instant in time. The op amp responds to a nonzero v_d by generating an output voltage that is larger than v_d by an amount equal to the open-loop voltage gain of the op amp. As the output voltage moves away from its former value, the voltage across R_F is changed. This alters the current through R_F correspondingly. The new value of current through R_F, and ultimately R_1, serves to return the value of v_d to near 0 V. So, as with the noninverting amplifier previously discussed, the output of the inverting amplifier circuit will go to whatever level is necessary to maintain v_d at near 0 V.

Calculation of Voltage Gain

Figure 11–17 shows an inverting amplifier with component values. Let's use it to understand voltage gain calculations for this circuit.

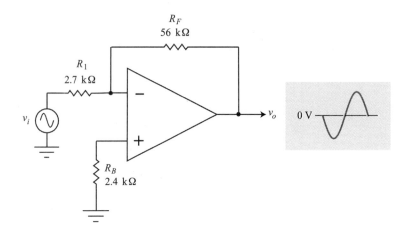

FIGURE 11–17 An inverting amplifier circuit.

The closed-loop voltage gain for this type of circuit can be found by applying Equation 11–2.

$$A_V = -\frac{R_F}{R_1} \qquad (11\text{–}2)$$

The minus sign on the right-hand side of the equation is used to indicate phase inversion. It is not an actual negative gain, which would represent a signal reduction.

•—EXAMPLE 11.4

Calculate the voltage gain for the inverting amplifier in Figure 11–17.

SOLUTION We apply Equation 11–2 as follows:

$$A_V = -\frac{R_F}{R_1} = -\frac{56 \text{ k}\Omega}{2.7 \text{ k}\Omega} = -20.7$$

To further our understanding of circuit operation, let's confirm this calculation with Ohm's and Kirchhoff's Laws. For this analysis, let us stop the input voltage at some specific instant in time and compute the corresponding output voltage. Figure 11–18 shows the same circuit with the input signal replaced with a 0.1-V dc source. We can now use basic circuit analysis techniques to determine the output voltage.

FIGURE 11–18 An inverting amplifier with a dc input voltage.

Fig11_18.msm

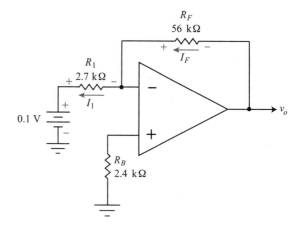

Remembering that the (−) input pin remains at 0 V, we can use Ohm's Law to calculate the current through R_1. Since the input voltage is positive, current will be in the direction labeled in Figure 11–18.

$$I_1 = \frac{V_1}{R_1} = \frac{0.1 \text{ V}}{2.7 \text{ k}\Omega} = 37.04 \text{ μA}$$

Now, according to Kirchhoff's Current Law, all of this current must also flow through R_F, since the (−) input of the op amp is effectively an open circuit. That is, $I_F = I_1$. We can now compute the voltage drop across R_F using Ohm's Law.

$$V_F = I_F \times R_F = 37.04 \text{ μA} \times 56 \text{ k}\Omega = 2.07 \text{ V}$$

The polarity of this voltage drop is labeled in Figure 11–18. Since the left side of R_F is connected to virtual ground, the right side must be −2.07 V with respect to ground. The right side is the output of the op amp, so the output voltage (v_o) is −2.07 V. First, note that the output is negative while the input is positive. That is, this is an inverting amplifier circuit. Second, note that the output voltage is 20.7

times greater than the input voltage. That is to say, the amplifier has a closed-loop voltage gain of 20.7 (inverting). This confirms our previous calculation using Equation 11–2.

Purpose of Components

By now, the purposes of the various components probably seem apparent to you, but let's briefly review them.

- The op amp is necessary to provide the high voltage gain needed for circuit operation. For example, the assumption that the (−) input pin is a virtual-ground point is only true if the op amp has a very high voltage gain.
- Resistors R_F and R_1 set up the feedback path and establish the closed-loop voltage gain of the circuit. Additionally, R_1 determines the input impedance of the circuit since it connects between the source and virtual ground.
- Resistor R_B serves the same purpose as previously discussed for noninverting amplifier configurations. That is, it compensates for the effects of nonzero input bias currents that would otherwise cause a dc offset voltage in the output of the op amp.

✓ Exercise Problems 11.2

1. An inverting amplifier circuit based on an op amp must have _____ feedback to be useful.
2. A noninverting amplifier circuit based on an op amp must have _____ feedback to be useful.
3. If the input signal is applied to the (+) input of an op amp circuit, the circuit is a(n) (*inverting/noninverting*) configuration.
4. Negative feedback is returned to the [(−)/(+)] input terminal.
5. One of the basic op amp rules for analysis states that the differential input voltage of an op amp will always be _____ V as long as the output is not driven to one of its extremes.
6. What is the ideal value of input bias current required by the (−) pin of an op amp?

Refer to Figure 11–19 for questions 7 through 10.

7. Is this an inverting or a noninverting amplifier circuit?
8. What is the voltage gain of the circuit?
9. What is the peak output voltage of the circuit?
10. If you were troubleshooting this circuit with an oscilloscope, what would you expect to see at the (+) input terminal with respect to ground?
11. The closed-loop voltage gain of an op amp circuit is (*less/more*) than the open-loop voltage gain of the op amp.
12. Negative feedback increases the voltage gain of an amplifier. (True/False)
13. Negative feedback increases the useful bandwidth of an amplifier. (True/False)

Refer to Figure 11–20 for questions 14 through 17.

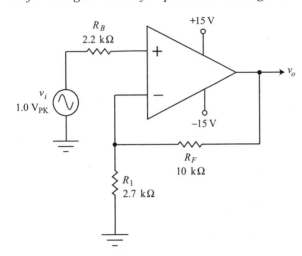

FIGURE 11–19 A circuit for questions 7 through 10.

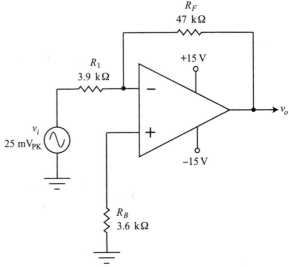

FIGURE 11–20 A circuit for questions 14 through 17.

14. What is the voltage gain of the circuit?
15. What is the peak output voltage?
16. Describe the output voltage waveform and levels if the input voltage is increased to 5 V_{PK}.
17. As long as the output is not driven to its extremes, the (−) input is a virtual-ground point. (True/False)

 SYSTEM PERSPECTIVE: Op Amps

The following schematic illustrates how an op amp is used to regulate the speed of a fractional horsepower dc motor. The motor is in series with an npn transistor connected as an emitter follower. A small spot on the motor shaft is magnetized and induces voltage into an adjacent pickup coil as the shaft rotates. Each pulse from the magnetic pickup is used to trigger a timer. The output of the timer goes positive for a fixed period of time. The time between positive output pulses is determined by the speed of the rotating shaft. Therefore, the average voltage out of the timer is an indication of motor speed. This voltage is continuously compared with a dc voltage set by the output of a computer, thereby allowing the computer to specify the desired motor speed.

Now, suppose the circuit has stabilized, but for some reason the motor slows down (perhaps due to more load). When this happens, the frequency of pulses to the timer will decrease. There will be a corresponding decrease in the average voltage on the output of the timer. This means the inverting input of the op amp is less positive, which means the output of the op amp will become more positive. This condition increases the bias on the transistor, which in turn, increases the current to the motor. As the motor returns to speed, the circuit will again stabilize. Op amps are widely used in industrial applications such as the one shown in the schematic.

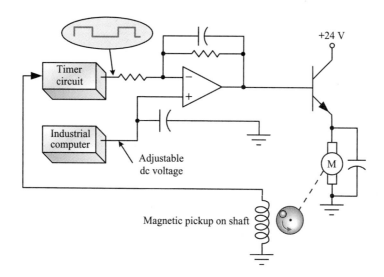

11.3 OP AMP APPLICATIONS

We will now examine the operation of a wide variety of op amp applications. Although it is impractical to review every op amp application that you will likely encounter in industry, the selection in this section is intended to give you a good cross section of applications. As we proceed, try to master the ability to analyze unfamiliar circuits by applying the basic principles that we have already discussed.

Voltage-Follower Circuit

Figure 11–21 shows schematic diagrams for voltage-follower circuits. These are op amp equivalents of an emitter-follower or source-follower circuit. You will recall from these related discussions that a follower circuit has the following characteristics:

- High input impedance
- Low output impedance
- No phase inversion
- Unity voltage gain

These same characteristics apply to the op amp voltage-follower circuits shown in Figure 11–21. These circuits are often used as the first stage of a multi-stage amplifier system because their high input impedance reduces loading of high-impedance sources (e.g., crystal microphones, oscillator circuits, and some industrial transducers).

FIGURE 11–21 A voltage-follower circuit.

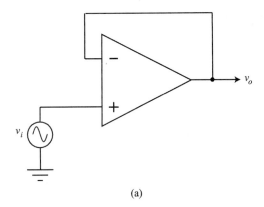

(a)

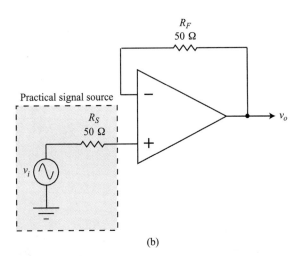

(b)

The high input impedance characteristic is apparent by inspection of the schematic, since we consider the input terminals of an op amp to be effectively open. In practice, the actual input impedance is typically in the tens of megohms.

The low output impedance that characterizes voltage-follower circuits can also be confirmed by examination of the circuits in Figure 11–21. The output is taken directly from the output of the op amp. If we include the effects of negative feedback, a typical op amp might have an output resistance that is less than 1 Ω. In short, it comes close to the ideal output impedance of 0 Ω.

Let us consider the phase relationship between input and output. Since the input signal is applied to the noninverting (+) input of the op amp, we can reason that this is a noninverting circuit. That is, we do not expect an inversion from input to output.

The voltage gain of the circuit can be calculated using the voltage gain equation for noninverting amplifiers (Equation 11–1). By inspecting Figure 11–21(a), we see that $R_F = 0$ and $R_1 = \infty$. We calculate voltage gain as follows:

$$A_V = \frac{R_F}{R_1} + 1 = \frac{0}{\infty} + 1 = 1$$

The circuit in Figure 11–21(b) includes a 50-Ω feedback resistor, but the calculation for voltage gain will still produce a result of unity. Since the (−) input pin of the op amp is open for most practical purposes, there will be no significant current through R_F and, therefore, no voltage drop across it. That is, it does not affect normal circuit behavior. But, you will recall that the op amp inputs do have a small input bias current. In Figure 11–21(b), you can see that the bias current for the (+) input will necessarily flow through the internal resistance (R_S) of the signal source. To compensate for the slight offset voltage that would be produced in the output, we ensure that the bias current for the (−) input has to flow through similar impedance. Therefore, we have two nearly equal, but opposite polarity, voltage offsets in the output, so they tend to cancel.

There is yet another way that you can view the operation of the voltage-follower circuit. Since the (+) input is effectively an open circuit, we expect to see the full input signal voltage between the (+) input and ground. Because the op amp has negative feedback, we know that the differential input voltage between the two input pins of the op amp will be near 0 V. Therefore, by Kirchhoff's Voltage Law, we can conclude that the voltage on the (−) input with respect to ground will be the same as the voltage on the (+) input. Since the (−) input is connected directly to the output terminal, we know that the output must also be the same as the (+) input, which is identical to the original input signal.

Selective Inversion Circuit

There are times (e.g., needing a special feature as part of a test instrument used for production testing of electronic products) when we want to be able to choose to invert a signal or, alternatively, to pass it through without inversion. Figure 11–22 on page 406 shows an op amp circuit that allows selective inversion by means of a two-position switch. The switch in this schematic could be manual as shown, or it could be a solid-state switch controlled by a computer. In either case, the circuit has no effect on any waveform characteristics other than to determine whether the signal should be inverted.

When the switch shown in Figure 11–22 is in the upper position, the circuit acts as a simple inverting amplifier with a gain of unity. That is,

$$A_V = -\frac{R_2}{R_3} = -\frac{10 \text{ k}\Omega}{10 \text{ k}\Omega} = -1$$

Recall that the negative gain merely indicates phase inversion. Note that the original equation (Equation 11–2) uses R_F and R_1 in the equation. Do not let this and other similar references confuse you. Again, don't try to memorize each thing presented, but rather understand the concepts and apply them where applicable. In this case, the feedback resistor happens to be labeled R_2, and the input resistor is labeled R_3. In industry, they might be R305 and R87.

FIGURE 11–22 A selective inversion circuit.

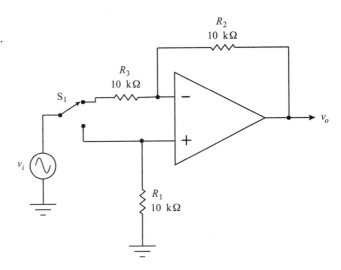

When switch S_1 in Figure 11–22 is in the lower position, resistor R_3 is open. Therefore, it has no effect on the circuit (i.e., no current can flow through it). As you can see by inspection, the circuit is now configured as a simple voltage-follower circuit. Thus, it has a voltage gain of unity, and it does not invert the signal.

Resistors R_3 and R_1 are equal for this application in order to present the same overall input impedance to the source regardless of switch position. You might observe that the effects of input bias current are not resolved, since the net resistances seen by the two input terminals are not identical. This is a necessary consequence of this particular circuit configuration.

Current-to-Voltage Converter

Most of the circuits we have studied in this text respond to signal voltages as their input. However, many devices and circuits produce output current changes that represent the signal. Examples of such devices are phototransistors (transistors whose conduction is controlled by light intensity), some types of temperature and pressure sensors, and certain digital-to-analog converters (circuits that convert digital information into corresponding analog currents or voltages). Suppose, for example, we have a pressure transducer similar to the ones shown in Figure 11–23, whose output current (mA) is proportional to pressure (psi).

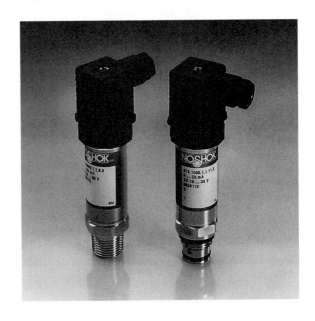

FIGURE 11–23 A close-up photo of pressure transducers with current outputs. (*Courtesy of NOSHOK, Inc.*)

As a specific example, let's assume that when the sensed pressure is 100 psi, the output current is 4 mA. When the pressure increases to 200 psi, the output current is 20 mA. Although the output current is an accurate representation of pressure, there are times that we may prefer to have a voltage that is proportional to pressure. In those situations, we might choose to use a current-to-voltage converter similar to that shown in Figure 11–24.

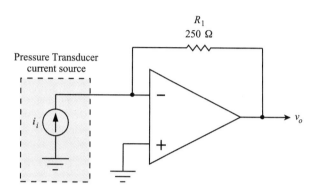

FIGURE 11–24 A current-to-voltage converter.

A current source is designed to have very high internal impedance. As a result, changes in any smaller external resistances have negligible effect on the generated current. Ideally, the current source should be connected to a circuit or device that has an input resistance of 0 Ω. The (−) input on the op amp in Figure 11–24 is a virtual-ground point. Therefore, the input impedance of the op amp circuit is essentially 0 Ω. This satisfies the requirements of the current-source loading.

Any current that is generated by the pressure transducer must flow through R_1, since the (−) terminal of the op amp draws no significant current (only input bias current). In the present case, as a current from 4 mA to 20 mA flows through R_1, the voltage drop across R_1 will be 1 to 5 V according to Ohm's Law. Since the left end of R_1 is ground (virtual ground), it follows that the right end will be 1 to 5 V with respect to ground. This means that a pressure variation from 100 psi to 200 psi in the present assumed case will cause the output to change from 1 V to 5 V. So, the output of the pressure transducer is (20 mA − 4 ma)/(200 psi − 100 psi) or 160 µA per psi. The output of the op amp, by contrast, is (5 V − 1 V)/(200 psi − 100 psi) or 40 mV per psi. Thus, the op amp circuit has converted a signal current into a proportional signal voltage.

If we focus only on the current-to-voltage converter, we can express its conversion relationship as (5 V − 1 V)/(20 mA − 4 mA) or 0.25 V per mA. This is a simple but useful circuit. You should also observe that changes in circuit resistance beyond the op amp have no effect on the transducer. That is, the op amp also provides isolation and buffering for the high-impedance sensor.

Active Rectifier

The half-wave rectifier (and series clipper) circuits discussed in chapter 9 work well for high-amplitude waveforms, but they cannot be used for very small signals. Briefly refer to Figure 11–25 on page 408 to see their shortcomings.

Recall that the diode conducts on one half-cycle and blocks current on the opposite alternation. When the diode conducts, the output is the same as the input waveform except for the small forward voltage drop across the conducting diode. In Figure 11–25(b), we see a peak-rectified output voltage of 9.3 V when 10 V(peak) is applied. The larger the input voltage, the less significant the effects of the diode voltage drop. Figure 11–25(c) shows the results when a 100-mV(peak) signal is applied to the input of the half-wave rectifier circuit. Since 100 mV is not sufficient to overcome the barrier voltage of the diode junction, the diode never conducts,

FIGURE 11-25 A standard half-wave rectifier.

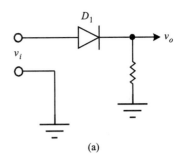

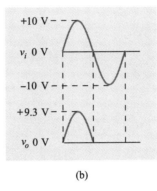

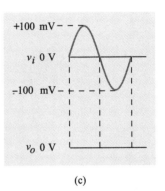

and there is no output. Thus, we cannot use this type of circuit to rectify low-amplitude waveforms.

Figure 11-26 shows the schematic diagram of an active rectifier (also called an ideal rectifier or a precision rectifier). As you will see as we discuss its operation, this circuit acts like a half-wave rectifier with an ideal diode.

FIGURE 11-26 An active rectifier circuit.

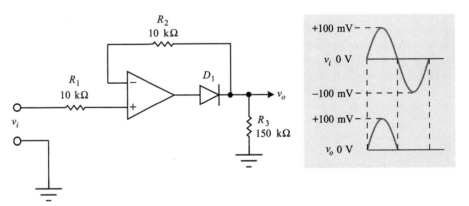

Anytime the input voltage is negative with respect to ground, the output of the op amp goes negative. This reverse biases diode D_1. Under these conditions, the load (resistor R_3) is effectively disconnected from the op amp and has zero volts across it.

Now, if the input voltage is more positive than ground, the output of the op amp will go positive. This forward biases the diode and completes the negative feedback loop through R_2. Under these conditions, the op amp is basically a voltage-follower circuit. Since resistors R_1 and R_2 are each connected directly to an input of an op amp, neither of them will have any significant current flow (input bias current only). Therefore, for purposes of functional analysis or troubleshooting, we can assume that neither resistor will have a voltage drop. This means that whatever voltage is on one end of one of these resistors will also be felt at the other end. In the case of R_1, v_i is on the left end. So, we know that the other end (the noninverting input of the op amp) will also be at the v_i potential.

During periods of positive input voltages when the op amp is operating with negative feedback as outlined previously, there will be 0 V between the two input pins of the op amp (i.e., $v_d = 0$ V). Therefore, the (−) input also has a voltage equal to v_i. Since no voltage is dropped across R_2, we know that the voltage on the right end of R_2, and therefore at v_o, will also be equal to v_i. Note that these relationships hold regardless of the exact forward voltage drop across the diode. That is, $v_o = v_i$ as long as the input voltage is positive.

The circuit will basically work as described even if R_1 and R_2 are both zero ohms. They are included to show how the effects of input bias current could be minimized. In particular, resistor R_1 might represent the source resistance of the input voltage source. Input bias current for the (+) input of the op amp will flow through this resistance, and if not compensated, will cause a dc offset voltage in the output. By making R_2 equal to R_1, we reduce the effects of input bias current.

Integrator

You will recall from an earlier chapter that an integrator can be constructed from an RC circuit with a long time constant. We can also build an integrator circuit around an op amp to gain improved performance. Figure 11–27 shows a simple op amp integrator circuit.

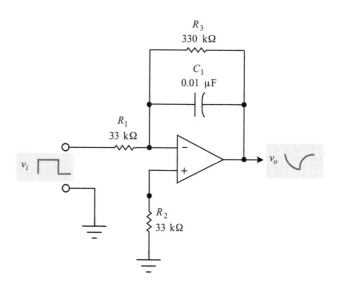

FIGURE 11–27 An integrator circuit.

Because the circuit has negative feedback (R_3 and C_1) and the (+) input is grounded (ignoring the tiny voltage drop across R_2 due to input bias current), we know that the (−) input pin is a virtual-ground point. Thus, the input voltage (v_i) will be dropped across R_1. With Ohm's Law, we could calculate the current through R_1 for any given input voltage. Now, since the (−) input of the op amp is effectively open, none of the current through R_1 enters or leaves the (−) input pin. Rather, it must flow through C_1 and R_3.

We know that C_1 and R_3 are in parallel, so they will always have equal voltage drops. Therefore, the voltage across R_3 (and the current through R_3) can only change as fast as C_1 can accumulate a charge. By design, we will ensure that the time required for C_1 to accumulate a significant voltage is longer than the time allowed. This is equivalent to saying the circuit is designed to have a long time constant relative to the input waveform. So, when the input voltage changes and increases or decreases the current through R_1, nearly all of the current change passes through C_1 as it charges or discharges.

The left end of C_1 is connected to a virtual ground. Therefore, v_o (which is connected to the other end of the capacitor) is equal to the value of capacitor voltage at any instant in time. The voltage across the capacitor can be found by applying Equation 11–3.

$$v_o = \frac{v_i t}{R_1 C} \qquad (11\text{–}3)$$

where v_i is the instantaneous input voltage, t is the time in seconds, R_1 is the input resistor, and C is the value of the integration capacitor. You may never be required to calculate v_o to solve a practical circuit problem, but examination of the expression gives us additional insight into the operation of the integrator circuit. In particular, notice that input voltage (v_i) and time (t) are both in the numerator of the expression. Additionally, the denominator consists of fixed values for a given circuit. This means that the output voltage of the integrator circuit is directly proportional to both time and input voltage. If, for example, the input waveform were rectangular, then v_o would be proportional to both pulse width and pulse amplitude.

Industrial computers frequently use integrators in the circuits that interface to analog devices. Since the output is proportional to time, the integrator tends to ignore short transients. This is desirable when a relatively slow changing, low-level analog signal is monitored in an electrically noisy environment. Integrator circuits are also used to develop linear voltage ramp waveforms (e.g., a triangle waveform). As you can see by examination of Equation 11–3, if the input voltage is held constant, then the output voltage will be proportional to time. In other words, it will continue to increase or decrease linearly as time progresses. Of course, the output of a practical op amp integrator can never exceed the limits of the output swing, but a linear ramp can be generated within its limits.

Although not discussed extensively in this book, which focuses on computer electronics, it is important to know that op amps can also perform many other mathematical functions besides integration. Examples of other functions include addition, subtraction, multiplication, division, logarithmic amplification, antilog amplification, differentiation, squaring, and many others.

Active Filter

In an earlier chapter we discussed passive filter circuits. You will recall there are four basic types of filter configurations: low-pass, high-pass, bandpass, and band-reject. Op amps can be used in circuits that provide these same basic filter functions, but the performance of the filter is greatly enhanced. These are called active filters, and they have advantages over passive filter networks. First, if desired, the signal being filtered can also be amplified at the same time. Passive *RC*, *RL*, or *RLC* filters, by contrast, always introduce loss or signal amplitude reduction. Second, the sharpness of the filter (equivalent to Q of a resonant circuit) can be much greater with an op amp active filter. All of the following circuits have practical limitations imposed by the op amp. Thus, the expected performance is usually satisfied for frequencies where the op amp still has a high open-loop voltage gain but becomes less predictable after that point.

Low-Pass Active Filter

Figure 11–28(a) shows a basic low-pass filter circuit. The frequency response of the low-pass filter in Figure 11–28(a) is shown in Figure 11–28(b). As you can see, the voltage gain is approximately unity until the cutoff frequency (f_c) is approached. The voltage gain then drops off as the frequency increases (within practical limits of the op amp).

11.3 • OP AMP APPLICATIONS

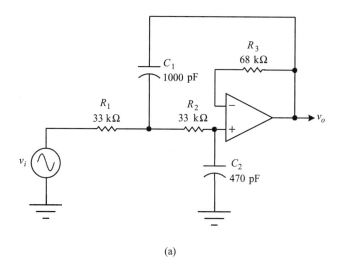

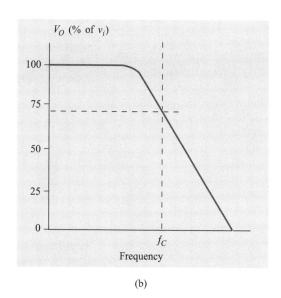

FIGURE 11–28 (a) A low-pass active filter circuit and (b) its frequency response curve.

To better understand the operation of the circuit, let us examine the circuit under two relatively extreme conditions: a frequency well below cutoff and a frequency well above cutoff. If the circuit is operating at a low frequency, then the capacitive reactance of C_1 and C_2 will be quite high. For our purposes, we will assume the reactances are so high as to appear open. In this case, the circuit will appear as shown in Figure 11–29, where the two capacitors have been removed from the circuit, and R_1 and R_2 (now in series) have been combined.

FIGURE 11–29 An equivalent circuit for low-frequency operation of the circuit in Figure 11–28(a).

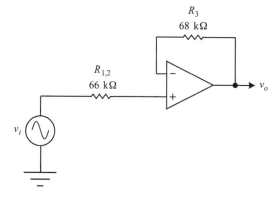

Inspection of the circuit in Figure 11–29 will reveal that it is operating as a simple voltage-follower circuit (i.e., $A_V = 1$). The two resistors are approximately equal to reduce the output offset voltage due to input bias currents. As long as the circuit is operated at a frequency much lower than the cutoff frequency for the circuit, it will act as a simple voltage follower, so the input voltage passes through to the output without attenuation.

Now, let's examine the operation at frequencies well above the cutoff frequency. In this case, the reactances of the two capacitors are relatively low. For simplification, let us assume the reactances are zero; the two capacitors are acting as short circuits. The equivalent circuit shown in Figure 11–30 on page 412, where C_1 and C_2 have been replaced by short circuits, represents this condition.

The low reactance of C_2 connects the input of the op amp to ground. Therefore, no significant signal can pass through the amplifier itself. Additionally, you will

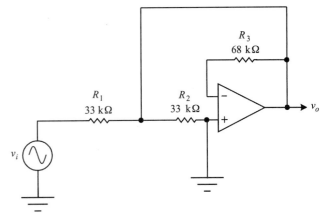

FIGURE 11-30 An equivalent circuit for high-frequency operation of the circuit in Figure 11-28(a).

recall that the output of an op amp has very low impedance, so the junction of R_1 and R_2 is essentially at ground potential for high-frequency signals. Here again, none of the incoming signal voltage is allowed to pass through the circuit to the output.

Now, the overall frequency response of the circuit ranges from the low-frequency case (represented in Figure 11-29) to the high-frequency case (illustrated in Figure 11-30). The cutoff frequency divides these two regions. For the active filter circuit shown in Figure 11-28(a), the cutoff frequency can be estimated with Equation 11-4.

$$f_c = \frac{1}{2\pi R\sqrt{C_1 C_2}} \qquad (11\text{-}4)$$

where R is the value of either R_1 or R_2.

•—EXAMPLE 11.5

Calculate the cutoff frequency for the circuit shown in Figure 11-28(a).

SOLUTION We apply Equation 11-4 as follows:

$$f_c = \frac{1}{2\pi R\sqrt{C_1 C_2}} = \frac{1}{2 \times 3.14 \times 33 \text{ k}\Omega \times \sqrt{1000 \text{ pF} \times 470 \text{ pF}}} = 7.04 \text{ kHz}$$

High-Pass Active Filter

Figure 11-31(a) shows the schematic diagram of a high-pass active filter circuit. The general frequency response is shown in Figure 11-31(b).

The high-pass circuit shown in Figure 11-31(a) results from interchanging the positions of the resistive and capacitive filter components from those shown for the low-pass filter in Figure 11-28(a). We can better understand its operation by considering how it responds to the extremes of its input frequency range.

If the input frequency is well below the cutoff frequency of the circuit, the reactances of C_1 and C_2 are quite high. For our initial analysis purposes, we will consider them to be open. If you mentally open C_1 and C_2 in Figure 11-31(a), you can readily see that the input signal is effectively disconnected from the amplifier circuit. Therefore, there will be no output voltage generated as a result of low-frequency inputs. The low-frequency equivalent circuit is shown in Figure 11-32(a).

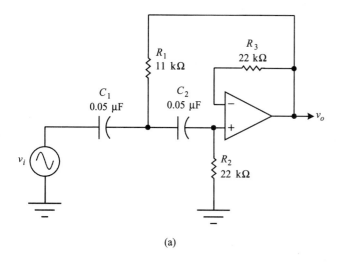
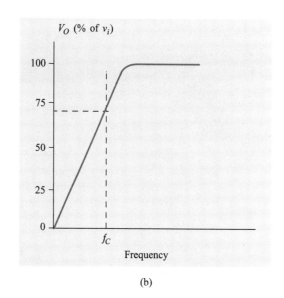

FIGURE 11–31 (a) A high-pass active filter circuit and (b) its frequency response curve.

Fig11_31(a).msm

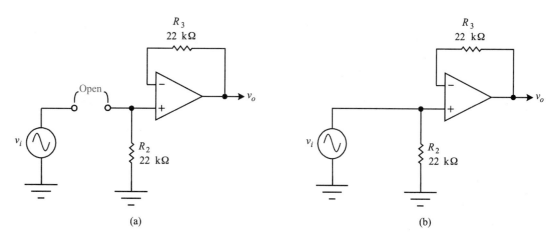

FIGURE 11–32 Equivalent circuits for the circuit in Figure 11–31 when operated at (a) low frequencies and (b) high frequencies.

At frequencies well above the cutoff frequency for the circuit, the reactances of C_1 and C_2 will be relatively low. For our immediate purposes, let's consider them to be effective short circuits. If C_1 and C_2 are shorted, the input will connect directly to the (+) input, and the overall circuit will act as a simple voltage follower. Additionally, with C_2 shorted, resistor R_1 will have essentially the same voltage on each end since it is connected between the input and output of a voltage follower. You can mentally remove R_1 because there will be no current through it during times when the input frequency is well above cutoff. The high-frequency equivalent circuit is shown in Figure 11–32(b).

The operation described in the preceding paragraphs is consistent with the frequency response curve expected of a high-pass filter circuit and represented in Figure 11–31(b). Between the two extremes outlined above, there is a region of transition near the cutoff frequency. For a circuit such as that shown in Figure 11–31(a), we can estimate the cutoff frequency with Equation 11–5.

$$f_c = \frac{1}{2\pi C \sqrt{R_1 R_2}} \qquad (11\text{–}5)$$

where C is the value of either C_1 or C_2.

•—EXAMPLE 11.6

Calculate the cutoff frequency for the high-pass filter circuit shown in Figure 11–31(a).

SOLUTION We apply Equation 11–5 as follows:

$$f_c = \frac{1}{2\pi C \sqrt{R_1 R_2}} = \frac{1}{2 \times 3.14 \times 0.05\ \mu F \times \sqrt{11\ k\Omega \times 22\ k\Omega}} \approx 205\ Hz$$

Bandpass Active Filter

You will recall that a bandpass filter allows a certain range of frequencies to pass with maximum gain (or minimum attenuation). Frequencies either lower or higher than this range (called the passband of the filter) are attenuated. Figure 11–33(a) shows the schematic of an active filter circuit that will provide a bandpass response. The generalized frequency response is shown in Figure 11–33(b).

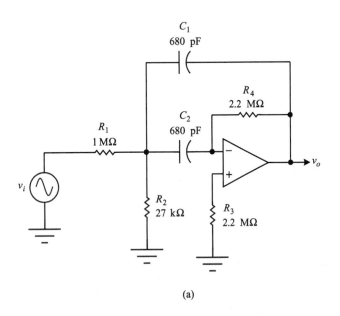
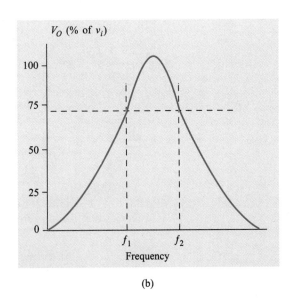

(a) (b)

FIGURE 11–33 (a) A bandpass active filter circuit and (b) its frequency response curve.

At frequencies much lower or much higher than the passband, we expect attenuation of the input signal. We can understand how the circuit achieves this reduction in signal amplitude by examining equivalent circuits for these two extremes. In the low-frequency case, we will assume the capacitors are open circuits due to their high capacitive reactance values. In the high-frequency case, we will assume the capacitors act like short circuits. Figures 11–34(a) and 11–34(b) show the low-frequency and high-frequency equivalent circuits, respectively, for the actual filter schematic shown in Figure 11–33(a).

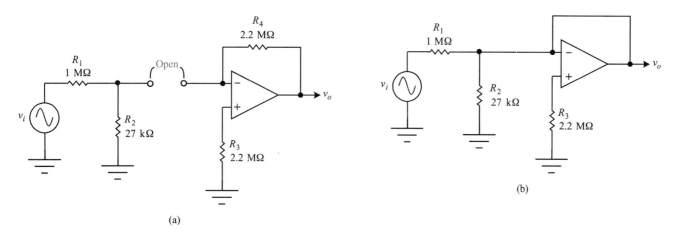

FIGURE 11–34 (a) The low-frequency and (b) high-frequency equivalent circuits for the bandpass filter shown in Figure 11–33.

For the low-frequency case illustrated in Figure 11–34(a), the input signal is isolated from the input of the amplifier, so no signal can be produced in the output. In the high-frequency case shown in Figure 11–34(b), the input signal is greatly attenuated by the voltage divider action of R_1 and R_2. Additionally, the amplifier has zero voltage gain, since its effective feedback resistance is 0 Ω. In either of these cases, there will be no significant signal voltage in the output of the filter circuit. This is what we would expect for a bandpass filter.

When the circuit is operated at frequencies that fall within the passband of the circuit, the losses introduced by the voltage divider action of R_1 and R_2 are offset by the closed-loop voltage gain of the op amp. Therefore, there is an output for frequencies that fall within the passband. The specific filter circuit shown in Figure 11–33(a) has a maximum gain of near unity; however, this circuit can be designed to provide additional voltage gain to the passband frequencies if desired.

Band-Reject Active Filter
A band-reject (also called bandstop) filter allows all frequencies above or below a certain band to pass through with gain or at least minimal attenuation. Frequencies that fall within the stopband of the filter are greatly attenuated. Figure 11–35(a) on page 416 shows the schematic diagram of an op amp circuit that provides the band-reject filter function. The general frequency response curve for a band-reject filter is shown in Figure 11–35(b).

The RC networks in Figure 11–35(a) form a twin-tee configuration. Let's first examine the behavior of the individual tee networks. The upper tee network acts as a simple low-pass filter. That is, low frequencies see a high reactance in C_3, so they can pass through R_1 and R_2 to reach the input of the op amp. High frequencies, though, see a low reactance in C_3 and are effectively shunted away from the op amp input. A similar but opposite action occurs in the lower tee network. Here C_1, C_2, and R_3 form a high-pass filter. Higher frequencies can easily pass through C_1 and C_2 to reach the input of the op amp, but low frequencies see these capacitors as high impedances. The two tee networks are essentially in parallel; we can see how very low or very high frequencies will be able to pass through the twin-tee network to reach the (+) input of the op amp.

Now, signals passing through the twin-tee network receive opposite polarity phase shifts. That is, in the upper network, the voltage is taken across the capacitor (C_3) in an RC circuit. This means the voltage across C_3 will lag the input voltage. But, in the lower network, the voltage is taken across R_3, and this voltage will lead the input voltage. At some specific frequency determined by component values, the phase shift of the two tee sections will be equal and opposite. Therefore, the

FIGURE 11–35 (a) A band-reject active filter circuit and (b) its frequency response curve.

Fig11_35(a).msm

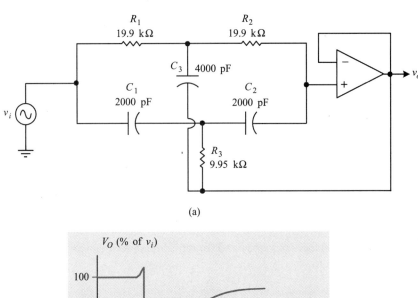

(a)

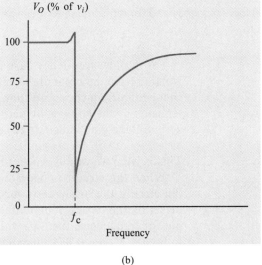

(b)

effective voltage applied to the (+) input of the op amp is zero at this frequency since the voltages from the two tee sections effectively cancel. This gives us the attenuation we need at the center of the stopband of the band-reject filter response.

The op amp itself is operating as a simple voltage-follower circuit. Its high input impedance reduces loading on the filter networks and allows a sharper filter response. Additionally, C_3 and R_3 are returned to the output of the op amp in a configuration known as bootstrapping. This has no significant effect at the center of the stopband, but it serves to narrow the bandwidth of the overall filter response. In fact, this particular filter design is characterized by a very narrow bandwidth, and for this reason, it is often called a notch filter. Care must be used when replacing parts in a twin-tee circuit because exact component values are required to maintain expected performance.

The center frequency of the notch in a twin-tee filter circuit such as the one shown in Figure 11–35(a) can be calculated with Equation 11–6.

$$f_R = \frac{1}{2\pi R_1 C_1} \qquad (11\text{–}6)$$

•EXAMPLE 11.7

What is the frequency of the notch for the filter circuit shown in Figure 11–35(a)?

SOLUTION We apply Equation 11–6 as follows:

$$f_R = \frac{1}{2\pi R_1 C_1} = \frac{1}{2 \times 3.14 \times 19.9 \text{ k}\Omega \times 2000 \text{ pF}} = 4 \text{ kHz}$$

Voltage Comparator

Many computer applications use voltage comparator circuits. Briefly, a voltage comparator has two analog inputs and one digital output. The voltage levels of the two inputs are continuously compared. The output of the voltage comparator is at one of two voltage levels, depending on the results of the comparison. Figure 11–36 shows a typical application that many notebook computers use to monitor battery voltage.

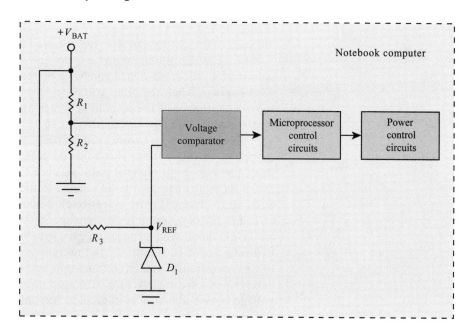

FIGURE 11–36 A voltage comparator can be used to sense a low battery in a notebook computer.

The circuit in Figure 11–36 uses R_1 and R_2 as a voltage divider. This provides a sample of the battery voltage (V_{BAT}) as one input of a voltage comparator circuit. The full battery voltage also goes to a zener voltage regulator circuit (R_3 and D_1). As long as the battery voltage is sufficiently high, the zener will maintain a constant voltage drop. The voltage across the zener (V_{REF}) serves as the second input to the voltage comparator circuit. The microprocessor control circuits monitor the output of the voltage comparator. If the battery voltage is above a predetermined level (set by the value of V_{REF} and the voltage divider R_1 and R_2), then the output of the voltage comparator is at one of two possible voltage levels. If the battery voltage falls below the specified level, then the output of the voltage comparator quickly changes to the second allowable voltage level or logic state. The microprocessor control circuits sense this new logic state and take appropriate action. This action varies with the system but might include deactivation of some devices (e.g., disk drives), dimming of the display to conserve power, activation of a low-voltage indicator, and so forth. More elaborate systems may automatically write critical information to the disk drive and then turn the system off, or place it in a power-saving sleep mode.

Figure 11–37 on page 418 shows the schematic diagram of a simple voltage comparator based on an op amp. Voltage comparators are also available as integrated devices. These self-contained circuits are simpler to utilize and generally have faster switching times than their op amp equivalents. Both implementation methods are common in industry.

FIGURE 11-37 An op amp configured as a voltage comparator circuit.

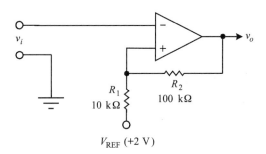

Let's begin by assuming that the input voltage (v_i) is well below the voltage on the (+) input of the op amp. In that case, v_o will be at its maximum positive extreme (positive saturation voltage). For illustrative purposes, let's consider the maximum output levels to be ±10 V. Under these conditions, the voltage at the (+) input can be shown to be +2.73 V by applying basic circuit analysis strategies to the voltage divider R_1 and R_2. This is called the upper threshold voltage (V_{UT}).

As long as the input remains less positive than +2.73 V, the output of the voltage comparator will remain at its positive extreme (+10 V in this example). Let's see what happens when the input voltage rises above +2.73 V. When this happens, the (−) input terminal of the op amp becomes more positive than the (+) input. Therefore, the output will switch to its other extreme (−10 V or negative saturation voltage). Even if the input barely exceeds the +2.73-V level, the output will still switch quickly to its other state. This behavior results from two things. First, the op amp itself has very high internal gain, so even the slightest difference in voltage between its inputs causes a large change in the output. Second, the op amp is using positive feedback. **Positive feedback** occurs when a portion of the output is returned to the (+) input pin. Any change in the output causes a regenerative change on the (+) input, which causes an even greater change in the output. This regenerative action continues until the output reaches one of its extreme limits. The length of time required to make the transition between output voltage states is primarily determined by the slew rate of the op amp. The use of positive feedback on a voltage comparator results in greater immunity to false triggering caused by noise.

With the output at −10 V, we could apply our voltage divider theory again to find that the voltage at the (+) input is now at +0.91 V. This is called the lower threshold voltage (V_{LT}). As long as the input is more positive than +0.91 V, the output will stay at its −10-V extreme. If the input falls below +0.91 V, the circuit will revert to its original state.

Figure 11-38 shows the relationship between input and output waveforms for the circuit in Figure 11-37 when a triangle waveform is applied.

FIGURE 11-38 The response of the voltage comparator in Figure 11-37 to a triangle input voltage.

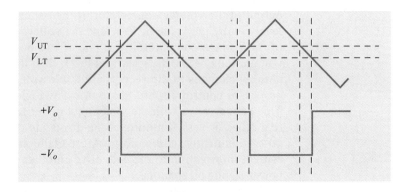

The difference between the upper and lower threshold voltages is called the hysteresis voltage. Hysteresis helps make the comparator less subject to small variations in input voltage such as might occur when noise is mixed with the desired signal.

✓ Exercise Problems 11.3

1. What is the typical voltage gain of a voltage-follower circuit?
2. A voltage follower is characterized as having (*low/high*) input impedance and (*low/high*) output impedance.
3. If the (+) input of the op amp in Figure 11–21(a) is +3 V with respect to ground, what is the voltage on the (–) input with respect to ground?
4. Refer to Figure 11–22. When the switch is in the upper position, the circuit acts as a voltage-follower circuit. (True/False)
5. Suppose the pressure transducer illustrated in Figure 11–24 is generating a 10-mA current. How much of that current will flow through the feedback resistor for the op amp?
6. What is the approximate voltage you would expect to measure across the pressure transducer in Figure 11–24?
7. During times when the input voltage is negative in Figure 11–26, the differential input voltage for the op amp may be substantially greater than zero. Explain why.
8. The circuit in Figure 11–27 is called an integrator. It can also operate as a low-pass filter. (True/False)
9. The output of the circuit in Figure 11–27 is proportional to both the amplitude and duration of the input waveform. (True/False)
10. Draw a basic frequency-response curve for a low-pass filter.
11. Draw a basic frequency-response curve for a high-pass filter.
12. Draw a basic frequency-response curve for a band-pass filter.
13. Draw a basic frequency-response curve for a band-reject filter.
14. Refer to the circuit in Figure 11–39. Calculate the cutoff frequency for this circuit.

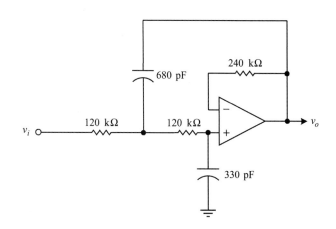

FIGURE 11–39 Calculate the cutoff frequency for this circuit.

Fig11_39.msm

15. Refer to Figure 11–39. If the input frequency is 100 kHz, explain how the reactance of the 330-pF capacitor affects circuit operation.
16. Refer to Figure 11–40. Assume that the maximum output voltage swing is ±10 V, and further assume that the output is at +10 V initially. Draw the output waveform (v_o) if the given waveform (v_i) is applied to the input of the circuit.

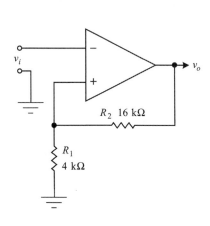

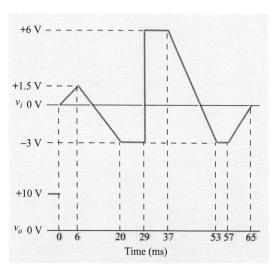

FIGURE 11–40 A voltage comparator circuit.

Fig11_40.msm

11.4 TROUBLESHOOTING OP AMP CIRCUITS

The difficulty in locating defects in op amp circuits ranges from nearly trivial to exceedingly complex depending on such things as circuit complexity, nature of the defect, frequency of operation, available test equipment, and your experience with troubleshooting op amp circuits. This section presents troubleshooting techniques and strategies that are applicable to op amp circuits.

Classification of Failure Mode

When localizing a defect in an op amp circuit, one preliminary goal should be to classify the defect as a dc problem or an ac problem. Successful classification of the defect into one of these two classes can dramatically reduce the list of possible defects, which allows you to more quickly identify the cause of the malfunction.

DC Failures

Measure the dc output voltage of the op amp and determine whether it is normal. If the dc output voltage is normal, then the defect must be an ac defect (discussed in the next section). By contrast, if the dc output voltage is abnormal, then the circuit has a dc defect.

A dc defect is one that upsets the normal dc operating voltage in the circuit. There are several things that can cause a dc defect including the following:

- Defective op amp
- Resistors that provide a dc path
- Shorted capacitors
- Other semiconductor components (e.g., diodes, transistors, and so forth)

AC Failures

If a check of the dc output voltage on the op amp reveals a normal dc level, then the circuit must have an ac defect. An ac defect is one that does not affect the dc levels in the circuit but upsets the ac behavior of the circuit. Some possible causes of ac defects include the following:

- Defective op amp (although this would be a rare situation)
- Resistors that have no dc path (e.g., in series with a capacitor)
- Open capacitors
- Shorted capacitors that have no dc path (e.g., two capacitors in series)

Examples of Failure Modes

To illustrate and contrast the two major classes of failure modes, let's consider a practical circuit. Figure 11–41 shows the schematic diagram of an active filter circuit. Table 11–1 lists a number of defects with a brief discussion of how they affect circuit operation and, therefore, how they would be classified.

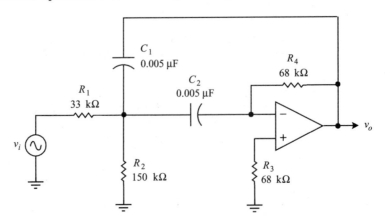

FIGURE 11–41 An active filter circuit.

11.4 • TROUBLESHOOTING OP AMP CIRCUITS

TABLE 11–1 Classification of failure mode for defects in the circuit shown in Figure 11–41.

Defect	Discussion	Failure Mode
R_1 open	This has no effect on the dc voltages in the op amp circuit because it is effectively isolated from the op amp by C_1 and C_2. Opens and shorts in R_2 would produce similar effects.	ac
R_1 shorted		ac
C_2 open	Capacitors are always open to dc, so this has no effect on the dc voltages in the circuit.	ac
C_2 shorted	This will likely cause a dc shift in the output of the op amp because there is now a dramatic difference between the resistances in the two op amp inputs. Therefore, the input bias current will produce an offset voltage in the output. Recall that we normally minimize this offset by making the resistance in the two branches equal.	dc
R_4 open	Since this defect removes the negative feedback path, we will expect a profound effect on dc output voltage. The op amp will be operating as an open-loop amplifier, so even the slightest input voltage or offset voltage will slam the output to one of its extremes.	dc
R_4 shorted	The op amp itself will still be configured as a voltage follower, but the short in R_4 will likely cause a dc shift in the output of the op amp because there is now a big difference between the resistances in the two op amp inputs. Therefore, the input bias current will produce a dc offset voltage in the output.	dc
C_1 shorted	More often than not, a shorted capacitor will alter the dc operating conditions in a circuit. In this particular case, however, the short may or may not affect the dc voltages. In essence, R_2 will now be connected between the output and ground. Since the normal dc output voltage is near zero, this will not likely upset the dc levels. The generator is also connected to the output via R_1 and the shorted C_1. If the generator has a dc level (e.g., the signal is coming directly from a common-emitter amplifier), then the dc voltage at the output of the op amp may be disturbed.	ac/dc
C_1 open	Capacitors are always open to dc, so this has no effect on the dc voltages in the circuit.	ac
R_3 shorted	This op amp itself will still be configured as a voltage follower, but the short in R_3 will likely cause a dc shift in the output of the op amp because there is now a big difference between the resistances in the two op amp inputs. Therefore, the input bias current will produce a dc offset voltage in the output.	dc
R_3 open	The (+) input has no path for input bias current, so the dc balance of the op amp will be upset. This will cause a dramatic shift in the dc output voltage causing the output to go to one of its extremes.	dc
Bad power supply	If one of the dc supply voltages is missing or extremely low, then it will have a significant effect on the dc output voltage (e.g., if the positive supply voltage is missing then the output will have a negative offset). A low or missing supply voltage will also affect the ac performance of the op amp, since the maximum excursions will be restricted.	ac/dc
Bad op amp	Some op amps can develop defects that only affect their ac performance, but this is unusual. In most cases, a defective op amp will result in the output voltage going to one of its extreme levels. This is so common that we will classify a bad op amp as a dc failure mode.	dc

General Troubleshooting Strategies

Once you have localized a system defect to an op amp circuit, you should apply a logical troubleshooting procedure to quickly identify the specific defect. The first general goal should be to classify the problem as an ac or dc failure as outlined in the preceding section. To accomplish this goal, you monitor the dc output of the op amp and contrast it with the normal dc value. In conjunction with this step, you should do a quick check of the dc power supply voltages.

If you have identified a dc defect, then check those components that have a complete dc path. By contrast, if you have identified an ac defect, then check any components that do not have a complete dc path. Remember that capacitors, transformers, relay contacts, switches, and some solid-state devices can break a dc path.

If you suspect a defective op amp, then perform the following procedure. Measure the output voltage and note the polarity. Measure the differential input voltage and note the polarity. Now, contrast your measured voltages with the expected behavior of an op amp. If the polarity of the differential input voltage is telling the output voltage to be one polarity but it actually measures the opposite polarity (regardless of magnitude), then there is a very good chance the op amp is defective. This assumes you have already verified the power supply voltage as previously recommended.

If, on the other hand, the output polarity agrees with the polarity dictated by the differential input voltage, then the defect probably lies elsewhere. That is, the op amp is probably doing what it is being told to do.

One of the most common dc defects is that the output voltage will be driven to one of its extremes. If you encounter this problem, quickly check the behavior of the op amp as described in the preceding paragraphs. If the op amp is thought to be good, verify that the dc input is correct. If the input voltage is correct and the op amp is good, then the problem most likely lies in the feedback path. If the feedback path is open, then the op amp operates as an open-loop (and very high gain) amplifier.

✓ Exercise Problems 11.4

1. Briefly explain the procedure for classifying an op amp circuit defect as an ac or dc failure.

Refer to Figure 11–42 for questions 2 through 5.

2. What is the approximate dc voltage that should be measured at the (−) input pin with respect to ground?
3. What is the approximate dc voltage that should be measured at the (+) output pin with respect to ground?
4. If resistor R_1 develops an open, will this cause an ac or a dc failure mode?
5. While troubleshooting, you make the following determinations: power supply voltages are good, v_o measures −13 V, the (−) input measures −13 V, and the (+) input measures 0 V. What is the most likely defect?

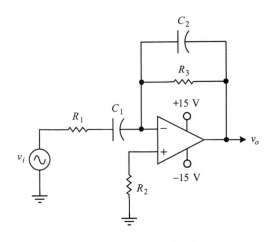

FIGURE 11–42 An op amp circuit.

•—SUMMARY

Operational amplifiers (op amps) are integrated amplifiers with extremely high open-loop voltage gain. Op amps have both inverting (−) and noninverting (+) inputs. The differential voltage between its inputs is amplified and determines the output voltage. Table 11–2 contrasts several parameters for ideal and practical op amps.

Negative feedback is required for practical op amps utilized in linear amplifier applications. The closed-loop voltage gain of the overall circuit is determined by feedback components and is much lower than the open-loop op amp voltage gain. With negative feedback, the differential input voltage of an op amp is very near zero as long as the output is not driven to one of its extremes. If the (+) input is grounded (even through a resistor), the (−) input is called a virtual ground as long as the amplifier has negative feedback and is not driven to its limits.

TABLE 11–2 Comparison of characteristics for ideal and practical op amps.

Characteristic	Ideal Value	Practical Value
Output voltage swing	$\pm V_{CC}$	$+V_{CC} - 2$ V and $-V_{CC} + 2$ V
Input bias current	Zero	From a few fA to 10s of μA
Input resistance	Infinite	1 MΩ to several TΩ
Output resistance	Zero	Fractional ohms (closed loop)
Open-loop voltage gain	Infinite	100,000 to > 1,000,000
Frequency response	dc to infinity (constant gain)	dc to 100s of MHz with decreasing gain
Slew rate	Infinite	Fractional V/μs to several kV/μs

There are endless op amp applications that utilize the high voltage gain, high input resistance, and low output resistance characteristics of an op amp. Several were discussed in this chapter. Most applications require negative feedback, which may be frequency selective. Voltage comparators, by contrast, utilize positive feedback, which adds hysteresis to the circuit. Hysteresis causes the circuit to have two different threshold voltages depending on the state of the output, and it makes the circuit less susceptible to noise impulses.

Defects in an op amp circuit can generally be classified as dc or ac failure modes based on whether or not the dc output voltage remains normal in the presence of the defect. Components with complete paths for dc (normally) and components that create a complete dc path when defective are the most probable causes for dc failure modes. Components with no complete dc path (e.g., capacitors) are often the cause of ac failure modes. If the op amp itself is defective, it usually causes a dc failure mode. It can often be identified as the defect by contrasting the polarity of the differential input voltage with the polarity of output voltage.

•—REVIEW QUESTIONS

Section 11.1: Op Amp Characteristics

1. For troubleshooting purposes, most op amp characteristics can be considered to be ideal. (True/False)
2. A typical op amp has _____ input terminal(s) and _____ output terminal(s).
3. All operational amplifier circuits require two power supplies (e.g., ±15 V). (True/False)
4. What is the maximum output voltage swing of an ideal op amp with ±15-V power supplies?
5. The output voltage of a practical op amp with MOSFET output drivers can swing to within a few millivolts of the full supply voltage. (True/False)
6. What is the value of input bias current for an ideal op amp?
7. What type of op amp input circuit draws the least input bias current?
8. An ideal op amp has ____ ohms input resistance and ____ ohms output resistance.
9. What type of feedback is required in order to make a stable and useful amplifier circuit based on an op amp?
10. Negative feedback arrives at the input as an (in-phase/out-of-phase) signal relative to the original input waveform.
11. An op amp can be operated in an open-loop configuration to stabilize its voltage gain. (True/False)
12. What is the name given to the distortion that occurs when the output of an op amp is driven to its extreme voltage excursions?
13. Most general-purpose op amps can respond to dc or very low-frequency inputs. (True/False)
14. What op amp characteristic or rating describes the maximum rate of change of output voltage?
15. What is the maximum rate of change of output voltage for an ideal op amp?

Section 11.2: Basic Amplifier Configurations

16. What is the input/output phase relationship in an inverting op amp circuit?
17. The input signal is applied to the [(−)/(+)] terminal of the op amp in a noninverting amplifier circuit.

18. If a portion of the output of an op amp circuit is returned to the (−) input through a resistive network, the circuit is using _____ feedback.

Refer to Figure 11–43 for questions 19 through 25.

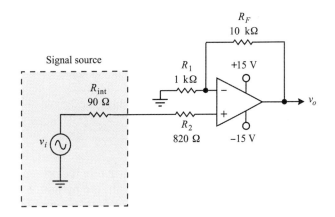

FIGURE 11–43 An op amp circuit.

19. This is (*an inverting/a noninverting*) amplifier circuit.
20. What is the voltage gain of this circuit?
21. What is the effect on voltage gain if resistor R_2 is increased to 2.7 kΩ?
22. The (+) input is a virtual-ground point. (True/False)
23. The (−) input is a virtual-ground point. (True/False)
24. There will be very little voltage dropped across the internal resistance of the signal source (R_{int}). Explain why this is true.
25. If the amplitude of the input signal (v_i) was increased sufficiently, then the output (v_o) would no longer be sinusoidal. Explain why this is true.

Refer to Figure 11–44 for questions 26 through 30.

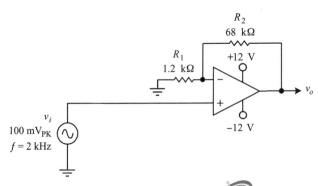

FIGURE 11–44 An op amp circuit. Fig11_44.msm

26. This is (*an inverting/a noninverting*) amplifier circuit.
27. What is the voltage gain of the circuit?
28. What is the value of peak output voltage?
29. Describe the approximate voltage and waveshape that you would expect to see between the (−) input and ground.
30. If the op amp is a bipolar device, then we might expect the output saturation voltages to be about ±10 V. (True/False)

Refer to Figure 11–45 for questions 31 through 35.

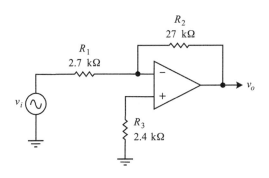

FIGURE 11–45 An op amp circuit.

31. This is (*an inverting/a noninverting*) amplifier circuit.
32. What is the voltage gain of the circuit?
33. What is the primary purpose of resistor R_3?
34. What happens to the voltage gain of the circuit if resistor R_2 increases in value?
35. The (−) input terminal of the op amp is a virtual-ground point. (True/False)

Section 11.3: Op Amp Applications

Refer to Figure 11–46 for questions 36 through 40.

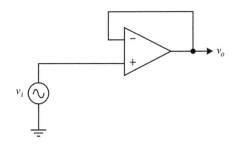

FIGURE 11–46 An op amp circuit.

36. What is the name of this circuit?
37. What is the voltage gain of this circuit?
38. Classify the input resistance of this circuit as low, medium, or high.
39. The (−) input terminal is a virtual-ground point. (True/False)
40. The output voltage is (*in phase/180° out of phase*) with the input voltage.

Refer to Figure 11–47 for questions 41 through 44.

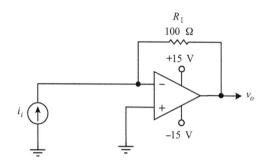

FIGURE 11–47 An op amp circuit.

41. What is the purpose of this circuit?
42. What is the approximate percentage of total input current (i_i) that flows through R_1?
43. What is the approximate output voltage if the input current is 50 mA?
44. If the input current (electron current) flows toward the op amp, what is the polarity of the output voltage?

Refer to Figure 11–48 for questions 45 through 47.

45. What is the overall purpose of this circuit?
46. What is the approximate voltage gain of the circuit at frequencies where the reactances of the three capacitors is quite low?
47. What is the approximate voltage gain of the circuit for extremely low-frequency inputs (e.g., dc)?

Section 11.4: Troubleshooting Op Amp Circuits

48. If the (+) input of an op amp is more positive than the (−) input, and the output is driven all the way to its positive saturation voltage, the op amp is probably defective. (True/False)
49. While troubleshooting an op amp circuit you determine that the (−) input measures +5 V and the (+) input measures +10 V (both with respect to ground). The output terminal measures −12 V with respect to ground. The ±15-V power supplies have the correct voltages. Is the op amp likely to be defective?

Refer to Figure 11–48 for questions 50 through 52.

50. If C_1 develops an open, it will cause a(n) (*dc/ac*) failure mode.
51. If resistor R_1 develops a short circuit, it will cause a(n) (*dc/ac*) failure mode.
52. If resistor R_3 develops a short circuit, it will cause a(n) (*dc/ac*) failure mode.

Refer to Figure 11–49 for questions 53 through 55.

53. If a component associated with amplifier U_2 causes a dc failure mode, what is the effect on the dc output voltage for U_3?

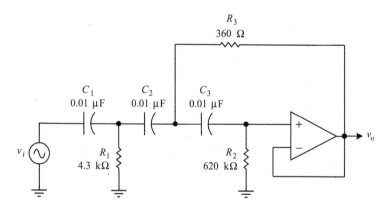

FIGURE 11–48 An op amp circuit.

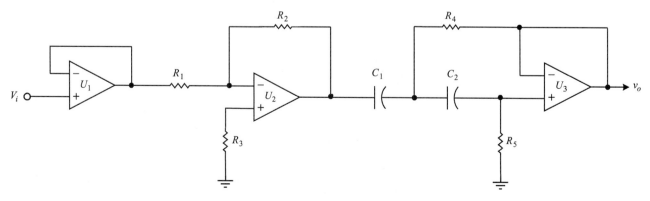

FIGURE 11–49 An op amp circuit.

54. If resistor R_3 opens, it will cause a(n) (*dc/ac*) failure mode in U_2 and a(n) (*dc/ac*) failure mode in U_3.
55. While troubleshooting this circuit, you discover a dc failure mode in U_3. Circle all of the following that could be a cause for these symptoms:
    a. C_2 open
    b. C_1 open
    c. R_1 shorted
    d. R_5 open
    e. U_3 defective
    f. U_2 defective
    g. R_4 shorted
    h. R_4 open
    i. R_1 increased

●—CIRCUIT EXPLORATION

One of the most important skills you can develop is the ability to understand and describe the operation of an unfamiliar circuit simply by inspecting the schematic diagram. This Circuit Exploration will provide you with an opportunity to do just that.

Figure 11–50 shows a circuit built around two operational amplifiers. Although you have not studied this specific circuit, you should be able to determine how it operates by applying the material presented in this chapter.

After you have studied the circuit and determined its operation, accomplish the following tasks:

- Draw a functional block diagram to represent the circuit.
- Draw a timing diagram to show the relationship between the waveforms at the outputs of U_1 and U_2.

- Build the circuit in the lab or in a circuit simulation environment and verify that you have correctly determined its operation. If you made mistakes, use the laboratory circuit or the simulation environment to help you fully understand the operation of the circuit and the purpose of every component.

Additional Circuit Exploration Opportunities

Use a circuit simulation package to build the various op amp circuits presented in this chapter. Verify that you fully understand circuit operation and then work on your troubleshooting skills by inserting defects and observing circuit behavior. Do your best to predict the circuit response to a given defect before you measure circuit values.

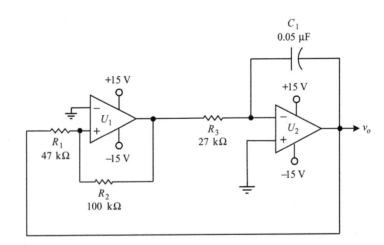

FIGURE 11–50 An unfamiliar op amp circuit.

CHAPTER 12
Power Supply and Voltage-Regulator Circuits

•—KEY TERMS
boost regulator
buck regulator
common-mode choke
crowbar circuit
error amplifier
feedback voltage
foldback current limiting
heat sink
line regulation
load regulation
pass transistor
pulse frequency modulation
pulse width modulation
shoot-through logic

•—OBJECTIVES
After studying this chapter, you should be able to:
- Describe and explain the operation of three basic types of voltage-regulation circuits: series, shunt, and switching.
- Understand the operation of integrated voltage regulators.
- Explain the need for voltage regulation.
- Distinguish, define, and calculate line regulation and load regulation.
- Discuss the operation of overcurrent and overvoltage protection circuits.
- Describe techniques for localizing defects in voltage-regulator circuits.

•—INTRODUCTION
Every electronic system requires a power supply to provide voltage to the various circuits. Similarly, nearly every electronic system requires a constant voltage source. In chapter 9, we discussed rectifier circuits that are used to convert alternating power line voltages into dc. We also discussed zener diodes, which can be used to provide a constant voltage. In this chapter, we will extend the discussion to include electronic voltage-regulation circuits. While its purpose is similar to that of a zener diode voltage regulator, an electronic voltage regulator can provide a much more constant voltage over a wider variation of conditions. We will also examine several circuits that are used to protect expensive or sensitive electronic systems from otherwise catastrophic power supply failures.

 SYSTEM PERSPECTIVE: **Power Supplies**

A notebook computer has a large number of functional blocks, as illustrated in the diagram. Some of these may be familiar to you (e.g., video display, microprocessor, and disk drives). Others, such as the AGP bus and PCI-to-ISA bridge, may not be as familiar. In any case, all of these functional blocks work together to provide the behavior required of a notebook computer. Each of these functional blocks consists of one or more integrated circuits. Each integrated circuit requires one or more dc voltages for its operation. The power supply circuits accept external raw power, typically low-voltage dc (e.g., 24 V, but sometimes 120 Vac), and produce several regulated output voltages for the various computer circuits. These voltages are distributed throughout the computer as required. The power supply circuits are essential parts of the computer; literally every circuit in the computer is dependent on proper operation of the power supply.

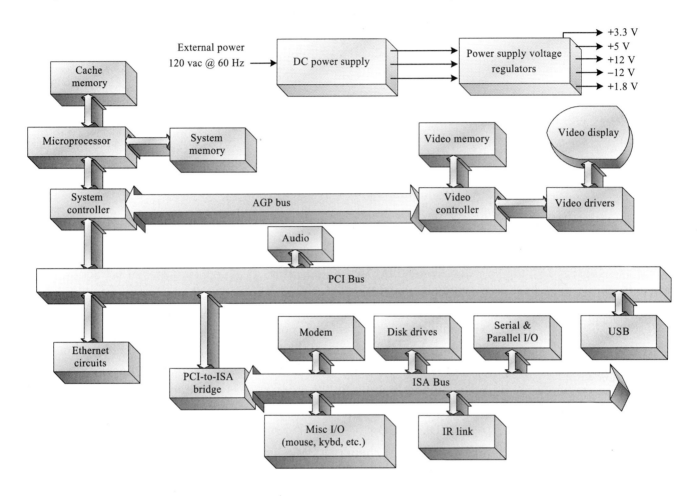

12.1 VOLTAGE REGULATION FUNDAMENTALS

Figure 12–1 illustrates the basic need for voltage-regulator circuits. The dashed box represents a source of dc voltage. In practice, this may be literally a battery, or it may be the result of passing the ac line voltage through a transformer and rectifier circuit. But in either case, it is a source of relatively smooth dc voltage.

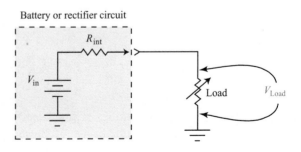

FIGURE 12–1 Why voltage regulation is needed.

Most integrated circuits and other electronic devices have fairly tight requirements with regard to dc supply voltage. Many integrated circuits used in computers, for example, require 3.3 V. While they may still work properly if the voltage drops to 3.25 V or increases to 3.35 V, they cannot be expected to work with large voltage variations.

In Figure 12–1, the dc supply (V_{in}) fluctuates with changes in power line voltage if it is developed from a rectifier circuit. If the source is a battery, then voltage will vary as the charge on the battery diminishes. In either case, the dc source cannot provide a constant voltage to a load.

Figure 12–1 shows the load as a variable resistance. This represents variations in current requirements as a system performs its intended function. For example, the current required by a notebook computer increases dramatically as the brightness of the display panel is increased. In Figure 12–1, it is clear that even if the dc supply voltage is constant, the actual load voltage will still vary as the load current changes. This is because all voltage sources have some value of internal resistance that appears in series with its output. R_{int} in Figure 12–1 represents this internal resistance. When load current increases, more of the source voltage is dropped across R_{int}. This causes a drop in the voltage across the load (V_{LOAD}).

A voltage-regulator circuit, then, will be a circuit that is positioned between the dc supply voltage and the variable load. It senses the actual load voltage and automatically compensates for changing dc supply voltage and/or changing load current. That is, it maintains a constant load voltage even if the line voltage and/or load current change. Now, let us examine these two variations more closely.

Line Regulation

Line regulation is used to describe a voltage regulator's ability to compensate for decreases and increases in line voltage. It is expressed as a percentage and is computed by using Equation 12–1.

$$\% \text{ line regulation} = \frac{V_{REG}(\max) - V_{REG}(\min)}{V_{LINE}(\max) - V_{LINE}(\min)} \times 100 \qquad (12\text{–}1)$$

Depending on the nature of the circuit being discussed, the term *line voltage* may mean the incoming ac power line voltage or the unregulated dc voltage. If you are describing the behavior of the voltage-regulator circuit itself, then the unregulated dc voltage is used as line voltage. On the other hand, if you are referencing a complete power supply circuit that includes rectifiers, filters, and voltage regulation components, then it is common to compute line regulation using the ac line voltage variations. For our purposes, the correct choice will be obvious from the context.

• EXAMPLE 12.1

The output of a voltage-regulator circuit changes from 11.95 V to 12.05 V when the line voltage changes from 105 V to 135 V. What is the line regulation specification for the voltage-regulator circuit?

SOLUTION   We compute line regulation with Equation 12–1 as follows:

$$\% \text{ line regulation} = \frac{V_{REG}(\max) - V_{REG}(\min)}{V_{LINE}(\max) - V_{LINE}(\min)} \times 100$$

$$= \frac{12.05 \text{ V} - 11.95 \text{ V}}{135 \text{ V} - 105 \text{ V}} \times 100 = 0.33\%$$

The smaller the percent of line regulation, the better the voltage regulator is at compensating for variations in line voltage.

Load Regulation

Load regulation describes a voltage regulator's ability to maintain a constant voltage when the load current changes. It is expressed as a percentage and is computed with Equation 12–2.

$$\% \text{ load regulation} = \frac{V_{\text{REG}}(\text{min load}) - V_{\text{REG}}(\text{full load})}{V_{\text{REG}}(\text{full load})} \times 100 \quad (12\text{–}2)$$

• EXAMPLE 12.2

The output of a voltage-regulator circuit changes from 23.95 V to 24.1 V when the load current changes from 510 mA to 225 mA. What is the load regulation specification for the voltage-regulator circuit?

SOLUTION   We compute load regulation with Equation 12–2 as follows:

$$\% \text{ load regulation} = \frac{V_{\text{REG}}(\text{min load}) - V_{\text{REG}}(\text{full load})}{V_{\text{REG}}(\text{full load})} \times 100$$

$$= \frac{24.1 \text{ V} - 23.95 \text{ V}}{23.95 \text{ V}} \times 100 = 0.63\%$$

As with line regulation, the lower the value of load regulation, the more constant the load voltage.

Voltage References and Feedback

An electronic voltage regulator continuously compares the actual load voltage against the desired load voltage and makes the necessary changes to maintain a constant load voltage. Figure 12–2 shows a functional block diagram of a voltage-regulator circuit. There are many types of electronic voltage-regulator circuits, but the concepts presented in Figure 12–2 are fairly universal.

As shown in Figure 12–2, the unregulated dc supply voltage is applied to a voltage-regulating element. This may be a transistor, an integrated circuit, or a complex network of electronic components. The regulating element decreases or increases the output voltage in response to a control voltage.

The control voltage comes from a voltage-comparator circuit. This is similar in some ways to the voltage comparators studied in chapter 11, but it is also different in some ways. It is similar in that its output voltage is determined by the

FIGURE 12–2 A functional block diagram of a voltage-regulator circuit.

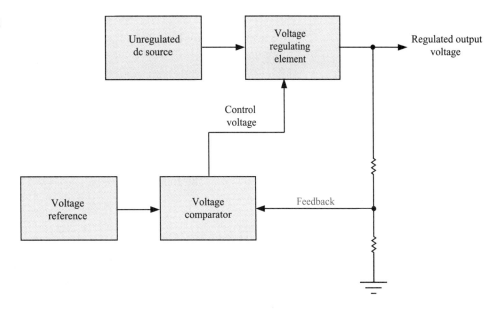

relative magnitudes of its two inputs. However, unlike the digital-output voltage comparators discussed in chapter 11, the voltage comparator represented in Figure 12–2 has an analog output. So, the control voltage shown in Figure 12–2 varies as an indication of how much voltage correction is needed at any given time.

The voltage comparator continuously monitors a sample of the load voltage (labeled feedback in Figure 12–2). It compares this sample with a constant reference voltage. Based on the relative values of these two voltages, a control voltage is provided to the voltage-regulating element that causes the load voltage to return to its correct value in the event of a slight load voltage change.

Now, it is imperative that the voltage reference be very stable. Any changes in the voltage reference will produce corresponding changes in the control voltage, which in turn will change the load voltage. So, although it sounds a bit peculiar, we need a stable voltage source (voltage reference) in order to produce a stable regulated voltage out of the overall circuit. The voltage reference can be a very low current device or circuit. Additionally, it can have a completely different voltage value than the overall regulated-output voltage.

There are many integrated voltage references available that are extremely stable. These are often used for the reference voltage in voltage-regulator circuits. These integrated circuits are often called micropower voltage-reference diodes. Although they are composed of transistors and resistors, they have only two external terminals, and they can generally be thought of as a very stable zener diode.

✓ Exercise Problems 12.1

1. If a system such as a computer is powered from the 120-Vac power line, why is a voltage regulator necessary?
2. The ability of a voltage regulator to eliminate load voltage changes in response to load current changes is called _____ _____.
3. The ability of a voltage regulator to eliminate load voltage changes in response to line voltage changes is called _____ _____.
4. Compute the line regulation percentage for a voltage regulator whose output changes from 4.98 V to 5.01 V when the unregulated dc input voltage changes from 11 V to 15 V.
5. Compute the load regulation for a voltage regulator in a computer whose output changes from 3.32 V to 3.3 V when the load current changes from 750 mA to 1.2 A.

12.2 SERIES VOLTAGE REGULATION

We are now ready to examine actual voltage-regulator circuits to understand how they operate. We will begin our exploration with series voltage regulation.

Basic Principles

Figure 12–3 shows a simplified way to view the operation of a series voltage-regulator circuit. Here, it is represented as a variable resistor. The voltage-regulator circuit is connected in series with the load and the unregulated dc supply voltage. We know by Kirchhoff's Voltage Law that part of the source voltage will be dropped across the load and the rest across the voltage-regulator circuit.

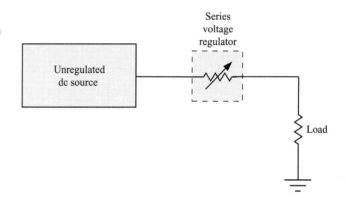

FIGURE 12–3 A simplified representation of a series voltage-regulator circuit.

For our immediate discussion, we will assume that the voltage regulator represented in Figure 12–3 is somehow able to sense the load voltage and to adjust its resistance as needed. We will examine the details of a series voltage-regulator circuit in the next section. First, let's assume that the unregulated dc supply and the load current in Figure 12–3 are momentarily stable and that the load voltage is correct. Now, suppose the unregulated supply voltage dropped slightly (perhaps in response to a dip in the 120-Vac power line voltage that supplies the rectifier and filter circuits used to produce the unregulated dc voltage). In any case, if the effective resistances of the load and the voltage regulator remain constant, then a drop in supply voltage will cause a corresponding drop in load voltage. But, suppose our voltage regulator circuit can sense the load voltage and detect the declining voltage as soon as it begins to drop. Further, suppose that the voltage regulator is able to decrease its own effective resistance. Can you see by inspection of Figure 12–3 that a decreasing regulator resistance would tend to increase load voltage? Therefore, if the regulator decreases its resistance the right amount, the net change in load voltage is barely detectable.

A similar action can compensate for changes in load current. Suppose, for example, that the load current suddenly decreases. If left uncompensated, this decrease in load current will cause less voltage drop across the voltage-regulator resistance, which will result in an increase in load voltage (Kirchhoff's Voltage Law). But, if the regulator senses the increasing load voltage and quickly responds by increasing its own series resistance, the load voltage can be held at a nearly constant value.

Representative Circuit

Figure 12–4 shows the schematic diagram of a series voltage-regulator circuit. In practice, the level of integration used may be substantially higher or lower than used in this circuit. That is, some regulator designs may use discrete transistors instead of an op amp. Other designs may employ a single integrated circuit that

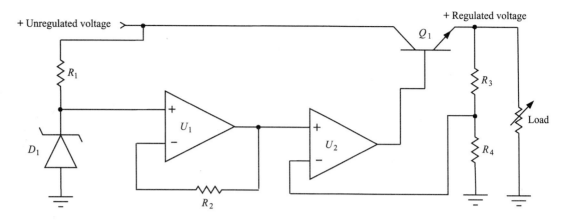

FIGURE 12-4 A series voltage-regulator circuit.

includes most of the circuitry in Figure 12–4. In either case, an understanding of the operation of the circuit in Figure 12–4 will be valuable to you.

Unregulated dc voltage serves as the input to this circuit. A transformer, bridge rectifier, and low-pass filter circuit operated from the 120-Vac power line might, for example, provide this source voltage. A zener voltage-regulator circuit consisting of R_1 and D_1 is placed across the unregulated source voltage. In most cases, we would not literally use a zener diode. We would more likely choose a voltage-reference diode mentioned in a previous section. This is conceptually similar to a zener, but provides a much more stable reference voltage. Additionally, we can get reference-voltage values that are well below available zener voltages. In any case, the voltage across D_1 provides a stable reference voltage for our circuit. The voltage across D_1 will remain constant over a very wide range of input voltages.

Amplifier U_1 is configured as a voltage-follower circuit. This provides isolation for the regulator diode and maintains a relatively constant load. In many circuits, the regulator diode would connect directly to U_2. Resistor R_2 in Figure 12–4 provides input bias current compensation as discussed in chapter 11. So, the purpose of the entire U_1 circuit is to provide an exceptionally stable reference voltage to the (+) input of amplifier U_2.

Transistor Q_1 is in series with the load. This transistor configuration is often called a series **pass transistor**. The unregulated supply voltage is dropped across the pass transistor and the load according to their relative resistances. Technically, the resistance of the voltage divider consisting of R_3 and R_4 is in parallel with the load, but in practice, the resistance of this branch is constant and much higher than the effective load resistance. Therefore, we can ignore its contribution to total load current. Q_1 is configured as an emitter-follower circuit. If we maintain a constant voltage on the base of Q_1, then we will have a constant voltage on the emitter that is about 0.7 V lower than the base voltage. This is the purpose of U_2.

U_2 is configured as a noninverting amplifier with the voltage reference serving as the signal input to the (+) terminal. The base-emitter junction of Q_1 and the voltage divider (R_3 and R_4) provide the feedback for the op amp. Figure 12–5 on page 434 shows this portion of the circuit redrawn so the noninverting amplifier circuit function is more apparent.

As revealed in Figure 12–5, the stable reference voltage is amplified by U_2. The voltage gain of this circuit is set by R_3 and R_4.

$$\text{i.e., } A_V = \frac{R_3}{R_4} + 1$$

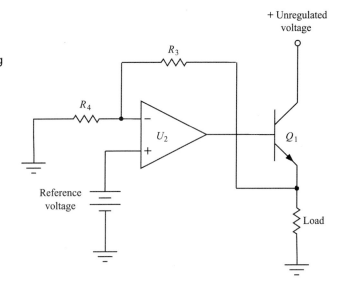

FIGURE 12–5 The U_2 amplifier circuit shown in Figure 12–4 redrawn to reveal its noninverting amplifier function.

Recall that the output voltage of the op amp will increase to whatever level is necessary in order to keep the differential-input voltage near zero. This means that the inclusion of the base-emitter junction of Q_1 within the feedback loop automatically compensates for changes in V_{BE} that may occur with temperature. As a specific example, suppose the reference voltage is +1.25 V and the voltage gain set by R_3 and R_4 is 8. This means the voltage at the emitter of Q_1 will be +10 V, and it will remain constant even if V_{BE} changes with temperature or as a result of a transistor replacement.

Once you recognize that the simplified circuit in Figure 12–5 represents the heart of Figure 12–4, circuit operation becomes clear. Nonetheless, let's examine the operation from yet another viewpoint as we refer to the original circuit shown in Figure 12–4. Suppose the load voltage tries to increase. This may be a result of a decrease in load current or perhaps an increase in the unregulated supply voltage. In either case, a higher load voltage will cause a higher voltage drop across R_4. This voltage is called the **feedback voltage**, and it is essentially a sample of the load voltage.

When the feedback voltage increases in the positive direction, this causes the (−) input of U_2 to become more positive. Since the (+) input remains stable at the reference level, we know that the output of the op amp (U_2) will become less positive. This reduces the voltage on the base of Q_1, and therefore, on the emitter of Q_1. Thus, the load voltage is decreased to compensate for the initial increase. If the circuit is well designed, this corrective action happens so quickly that for most practical purposes the load voltage doesn't make any significant changes. The feedback voltage is compared against the reference voltage in U_2, and a corrective action is sent to Q_1. For this reason, U_2 is sometimes called an **error amplifier**.

Integrated Series Voltage Regulators

Figure 12–6 shows a voltage-regulator circuit that uses a three-terminal integrated circuit as the heart of the circuit. In essence, the function of all of the circuitry shown in Figure 12–4 is included in a single package with only the unregulated input, the regulated output, and ground terminals provided for external connection.

Although three-terminal voltage regulators are available for several standard voltages, some designers add two external resistors to obtain voltage regulation at some nonstandard level. This is illustrated in Figure 12–7.

The easiest way to understand the technique illustrated in Figure 12–7 is by applying Ohm's Law and Kirchhoff's Voltage Law. The basic regulator package will always maintain a constant voltage between its ground and output terminals. Let's

FIGURE 12–6 A three-terminal voltage regulator.

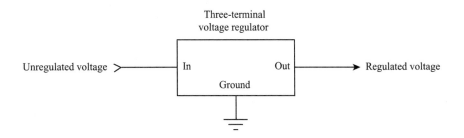

FIGURE 12–7 A three-terminal regulator with increased output voltage.

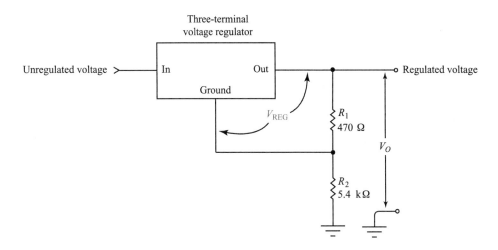

label this V_{REG} as shown in Figure 12–7. The current through R_1 can be expressed as V_{REG}/R_1 by Ohm's Law. Now, by design, this current will be substantially greater than the current flowing into the ground pin of the voltage regulator package. Therefore, for practical purposes, we can consider the current through R_1 and R_2 to be equal. The voltage drop across R_2 can now be expressed with Ohm's Law as $I_1 \times R_2$, since $I_1 = I_2$. We can also substitute the previous Ohm's Law calculation for I_1 to produce the expression $(V_{REG}/R_1)R_2$. Now, we can use Kirchhoff's Law to express the output (V_O) as the sum of the two resistive voltage drops. That is, $V_O = V_{R1} + V_{R2}$. By substitution, we can express this as Equation 12–3.

$$V_O = V_{REG}\left(\frac{R_2}{R_1} + 1\right) \qquad (12\text{–}3)$$

You might also note that we can make an adjustable, but regulated, voltage by replacing R_2 with a variable resistor.

• **EXAMPLE 12.3**

Suppose you are troubleshooting a communication circuit in a desktop computer that uses a voltage regulator like the one shown in Figure 12–7. The manufacturer's datasheet for the voltage regulator itself indicates that it is a 1.2-V device. What is the value of V_O with the values shown in Figure 12–7?

SOLUTION We apply Equation 12–3 as follows:

$$V_O = V_{REG}\left(\frac{R_2}{R_1} + 1\right) = 1.2\text{ V}\left(\frac{5.4\text{ k}\Omega}{470\text{ }\Omega} + 1\right) \approx 15\text{ V}$$

Calculator Application

Many three-terminal integrated-voltage regulators are available with current values ranging from 100 mA to more than 10 A. However, some designers prefer to add an external transistor when more current is needed. Figure 12–8 illustrates one way this can be done.

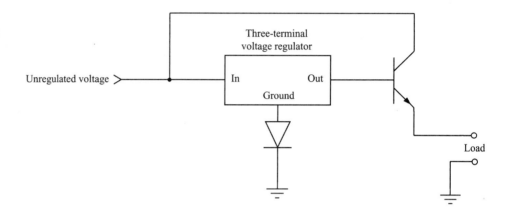

FIGURE 12–8 A three-terminal voltage regulator with a current boost transistor.

In Figure 12–8, a forward-biased diode is inserted in the ground terminal of the voltage regulator. This increases the output voltage by about 0.7 V with respect to ground. The output voltage is then connected to the base of an npn transistor configured as an emitter follower. The emitter voltage will be about 0.7 V lower than the base voltage, so the load voltage is equal to the normal output voltage of the voltage regulator. However, the current through the load can now be much larger than the regulator itself can handle, since the load current is actually the emitter current of the transistor. The three-terminal voltage regulator only has to supply base current, which is of course, substantially smaller than emitter current.

✓ Exercise Problems 12.2

1. Refer to Figure 12–3. If the current drawn by the load suddenly drops to a small percentage of its original value, what must happen to the effective resistance of the series voltage regulator in order to maintain a constant load voltage?

Refer to Figure 12–4 for questions 2 through 5.

2. What happens to the output voltage if resistor R_4 is increased?
3. If the unregulated supply voltage increases, what happens to the voltage on the (+) input of U_2?
4. If the unregulated input voltage increases, what happens to the voltage drop across the emitter-collector of Q_1?
5. What happens to the voltage on the base of Q_1 if resistor R_3 increases?
6. If resistor R_2 in Figure 12–7 is replaced with a variable resistor that could be adjusted from 0 Ω to 2500 Ω, what is the range of adjustment in output voltage (V_{REG} is 1.2 V)?

12.3 SHUNT VOLTAGE REGULATION

In chapter 9, we discussed zener diodes as voltage regulators. This was an example of a shunt voltage regulator. We will now extend this concept to include electronic circuits, which function as shunt voltage regulators.

Basic Principles

Figure 12–9 shows a simplified way to view the operation of a shunt voltage-regulator circuit. It is represented in this drawing as a variable resistor. The voltage-regulator circuit is connected in parallel (shunt) with the load. We know from basic circuit theory that the voltage across the load will be the same as the voltage across the voltage regulator since they are in parallel. Additionally, we know that the total unregulated source voltage will be distributed between R_S and the parallel combination of the load and voltage regulator.

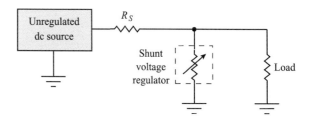

FIGURE 12–9 A simplified representation of a shunt voltage regulator.

Now, suppose the load voltage starts to decrease in response to a decrease in the unregulated supply voltage. What must the voltage regulator do in order to bring the load voltage back to its desired value? When the regulator circuit senses the drop in load voltage, it responds by increasing its resistance. This causes a greater percentage of the applied voltage to be dropped across the load and the regulator circuit, thus returning the load voltage to its proper level. This is very similar to the way a zener voltage regulator (discussed in chapter 9) works.

So, in short, the shunt voltage-regulator circuit will monitor the load voltage continuously. If an increase or decrease is sensed, then the voltage regulator adjusts its resistance such that the parallel combination of the load and the voltage regulator drop the correct amount of voltage. The sensing and correction activity is fast enough that for most practical purposes, the load voltage hardly changes at all.

Representative Circuit

Figure 12–10 on page 438 shows the schematic diagram of a shunt voltage-regulator circuit. Let's work to understand how the overall circuit is able to maintain a constant load voltage in spite of changes in the supply voltage and/or changes in load current.

Many of the components included in the schematic in Figure 12–10 serve functions identical to their counterparts in the series voltage-regulator circuit discussed in the preceding section. This includes D_1, R_1, R_2, U_1, R_3, R_4, and U_2. Details of the operation of these components will not be repeated here.

Resistor R_S is a series voltage-dropping resistor. It serves the same purpose as the series resistor that was required in a zener voltage-regulator circuit. That is, since the load voltage is necessarily lower than the unregulated dc input voltage, there has to be some series resistance to drop the rest of the input voltage.

Transistor Q_1 is the regulator transistor. By controlling the conduction (emitter-collector current) of Q_1, we can control the effective resistance of Q_1, and therefore, the relative distribution of voltage between R_S and the load. Transistor Q_1 is configured as an emitter follower.

To better understand the circuit operation, let's assume the load voltage tries to increase slightly and watch how the circuit responds. As soon as the load voltage

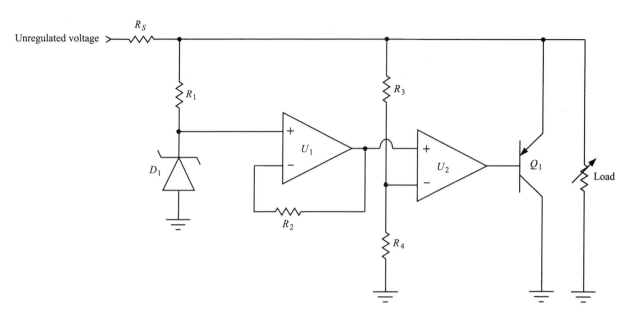

FIGURE 12–10 A shunt voltage-regulator circuit.

begins to increase, there will be a corresponding increase on the (−) input of U_2 as a result of the voltage divider action of R_3 and R_4. A positively increasing voltage on the (−) input of U_2 causes the output of U_2 to become less positive. This causes the base voltage of Q_1 to become less positive. And through emitter-follower action, the emitter (and load) voltage also decreases, returning the load voltage to its initial value. This balancing activity will occur regardless of why the load voltage increased. In other words, if the unregulated supply voltage increases or if the load current decreases, the circuit will respond as just described. Of course, if the load voltage decreases, then the compensatory action is similar, but the change is in the opposite direction.

✓ Exercise Problems 12.3

Refer to Figure 12–9 for questions 1 and 2.

1. In order to compensate for an increase in load voltage, the effective resistance of the voltage regulator must (*decrease/increase*).
2. In order to compensate for an increase in load current, the effective resistance of the voltage regulator must (*decrease/increase*).

Refer to Figure 12–10 for questions 3 through 5.

3. If the load current decreases, then the current through Q_1 will also decrease. (True/False)
4. If resistor R_4 increased in value, what effect would this have on the output voltage?
5. If the input voltage increases, what happens to the voltage drop across D_1?

12.4 SWITCHING VOLTAGE REGULATION

The series and shunt voltage-regulator circuits discussed in the earlier portions of this chapter fall into a category called linear voltage regulators. They are analog in nature. That is, all of the active devices (i.e., transistors and op amps) vary their conduction over a continuous range of operation. By contrast, the circuits presented in this section represent another category called switching voltage regulators. The actual regulating element in these circuits is characterized as being digital. That is, it is either fully on (e.g., saturated transistor) or fully off (e.g., cutoff transistor). As you will see, both categories of regulator circuits have relative advantages and disadvantages.

Basic Principles

The basic concept of switching voltage regulators can be illustrated with the simplified diagrams shown in Figure 12–11. Figures 12–11(a) and 12–11(b) show the same circuit at two different times. As we discuss this circuit, don't forget that the primary purpose of a voltage-regulator circuit is to provide a constant output voltage, regardless of changing load current or variations in supply voltage.

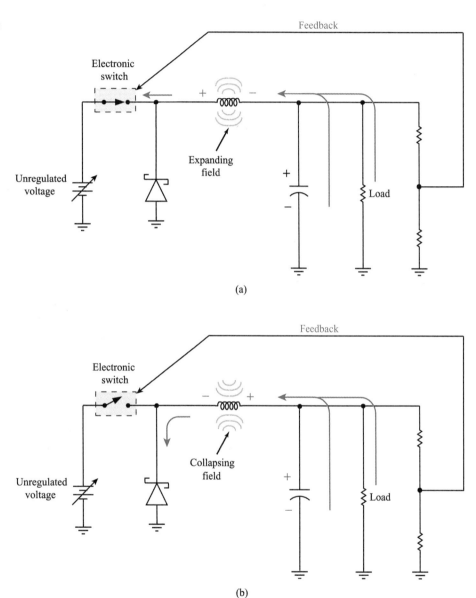

FIGURE 12–11 The operation of a simplified switching voltage regulator.

In Figure 12–11(a), the electronic switch is closed. For now, simply view this as a SPST switch. In practice, it will normally be a high-current MOSFET that is either saturated or cutoff. In any case, when the switch is closed as in Figure 12–11(a), electron current flows up through the load and the parallel-filter capacitor to accumulate a charge on the capacitor with the polarity shown. Note that current also flows up through the voltage-divider network used to obtain feedback, but we will ignore that current for now since it is a relatively small value compared to load current. The electron current continues through the series inductor. As it passes through the inductor, a magnetic field builds up around the coil.

Recall that an inductor opposes any change in current. It will have the polarity shown in Figure 12–11(a), which tends to retard the current buildup.

Now, suppose the electronic switch is opened, which interrupts the previous current path. As soon as the current through the coil tries to decrease, it is met with opposition. More specifically, the magnetic field surrounding the coil begins to collapse. As it collapses, it acts as a source of stored energy. The coil, then, is acting as a voltage source with the polarity as indicated in Figure 12–11(b). The current continues through the coil, through the Schottky diode, and up through the parallel combination of the load and filter capacitor. The polarity of voltage drop across the load and filter capacitor is the same as it was initially.

If we left the switch open long enough (i.e., for five time constants), we know that the magnetic field would totally collapse, and the current would decay to zero. But, what if we closed the switch again before the current had time to decay appreciably? And what if we continued to operate the switch at this rapid rate? There would be variations in current, but because of the smoothing action of the inductor (i.e., it won't allow sudden current changes), there would be an average value of current through the inductor and through the load-filter combination. The filter capacitor opposes any change in voltage. It tends to absorb (i.e., smooth) the voltage changes that would occur in the load voltage if the fluctuating coil current went strictly through the load. Therefore, with the filter capacitor in place, the voltage across the load is smooth dc.

Now, consider what would happen in this circuit if we alter the ratio of open time to closed time for the switch. If we leave the switch open for a longer period of time, then the current through the coil will decay to a lower value between cycles. This means the average current will be less. Additionally, the filter capacitor will charge to a lower value, which means that load voltage will be less than it was previously. By contrast, if we shorten the period of time that the switch is open, then the current will not have time to decay as much. In this case, the filtered load voltage will increase. The switching frequency of a real switching regulator is generally between 10 kHz and 1 MHz, with 50 kHz to 150 kHz being very widely used.

As indicated in Figure 12–11, a sample of the actual output voltage (feedback) is used to control the operation of the switch. If the sensed voltage is too low, then the on/off ratio of the switch is increased. If the load voltage goes too high, the on/off ratio of the switch is reduced. Thus, even though the regulating element (the electronic switch) is operated as a digital device that is fully on or fully off, the actual load voltage is smooth dc that is regulated to the desired value.

Notice that the diode used in this circuit is a Schottky diode. To achieve effective and efficient operation of this circuit, it is important for the diode to switch from on to off quickly. You will recall from an earlier chapter that this is the primary advantage of Schottky diodes. They are commonly used in switching regulator applications.

Switching versus Linear Regulation

Before we examine the actual schematic of a switching voltage regulator, let's compare the advantages and disadvantages of switching versus linear voltage-regulator circuits.

Power Dissipation

The primary advantage of switching regulators over linear regulators is lower power dissipation in the regulating element; a switching voltage regulator is more efficient than a linear regulator. Figure 12–12 shows why linear regulator circuits dissipate substantial power.

Figure 12–12(a) shows a simplified series voltage-regulator circuit. According to Kirchhoff's Voltage Law, we know that the voltage across the series transistor

FIGURE 12–12 A linear voltage regulator dissipates substantial power.

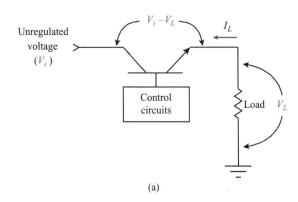

(a)

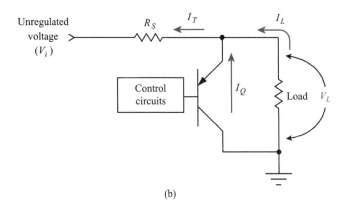

(b)

must be equal to the input voltage minus the load voltage. As a minimum, this generally has to be at least 2 to 3 V in order for the circuit to function properly, and in many cases, this voltage is considerably higher than 2 to 3 V. Additionally, the transistor must pass the full-load current. The power dissipated in the transistor is simply current times voltage, but since the voltage and current are both relatively high, the power dissipation in the transistor can be very high. A similar calculation would show high power dissipation for the transistor in the shunt voltage regulator illustrated in Figure 12–12(b). Whenever the load draws minimal current and/or the line voltage is at its highest level, the transistor will have substantial current flow. At the same time, it will have the full-load voltage across it. This produces high power dissipation in the transistor.

Because of the high power dissipations in linear voltage regulators, we typically need physically large transistors and provisions for cooling them. Cooling may be as simple as a large piece of metal thermally bonded to the transistor. This metal is called a **heat sink**. It helps remove the heat from the transistor and radiate it into the surrounding air. Other cooling means include fans, liquid-filled pouches (another form of a heat sink), and in extreme cases, cooling fluids. Clearly, we have to pay a high price for high levels of transistor power dissipation in terms of expensive transistors, heat sinks, weight, and physical space requirements. In terms of efficiency, a linear regulator is typically 30 to 40% efficient. This means that a regulator circuit with a +24-V output and a load current at 5 A would have to draw about 300 W from the power line in order to deliver 120 W (24 V @ 5 A) to the load. That is,

$$\text{load power} = V_L \times I_L = 24 \text{ V} \times 5 \text{ A} = 120 \text{ W}$$

And

$$\text{input power} = \frac{\text{load power}}{\text{efficiency}} = \frac{120 \text{ W}}{0.4} = 300 \text{ W}$$

By contrast, switching voltage regulators consume substantially less power for a given load; these circuits are more efficient. It is reasonable to expect efficiencies on the order of 75% or higher for switching regulators. So, to contrast with the previous example, a typical switching regulator could deliver 120 W (24 V @ 5 A) to a load with only 160 W drawn from the power line. Another way to view this is that 40 W would be dissipated in the transistor. The transistor in the linear regulator discussed previously would dissipate 180 W in order to deliver this same power to the load!

The regulating element in a switching regulator is normally a high-current MOSFET. It replaces the switch in Figure 12–11. As with any transistor application, the power dissipated in the transistor is equal to the current (source-to-drain current) times voltage (source-to-drain voltage). Remember, however, that the MOSFET is being used as a switch. It is either fully on or fully off. When it is off, the current is essentially zero, so the power dissipation ($P = VI$) is also zero. When it is fully on, the current may be quite high, but the voltage across the saturated MOSFET is extremely low (essentially zero). Again, the power dissipation is near zero. Unfortunately, it takes a finite time (typically measured in nanoseconds) to switch between the two operational states. During the short time the device is transitioning between cutoff and saturation (or vice versa), the transistor has current through it and voltage across it at the same time. This is the only time substantial power is dissipated. The average power, however, is considerably less than that of an equivalent linear regulator. Therefore, smaller transistors, smaller and lighter heat sinks, and reduced auxiliary cooling requirements are possible.

Electrical Noise

In order to maximize the efficiency of a switching regulator circuit, we want the switching transistor to change states as quickly as possible. That way the average power dissipation will be less. But, you will recall from our earlier studies, high edge rates on a waveform in the time domain correspond to high-frequency and high-energy harmonics in the frequency domain. For example, a square wave with faster rise and fall times has more energy in the higher-frequency harmonics than a square wave with slower rise and fall times. So it is with the transistors used in switching voltage regulators. The faster we make the transistors change states (i.e., shorter rise and fall times) the more efficient they are, but the more high-frequency noise they generate. This high-frequency energy can readily travel down the power cord and interfere with other electronic equipment. Additionally, as the high-frequency current flows through the power cord, some of the energy will escape the cord as radiated emissions. In other words, the power cord acts like a radiating antenna structure. This radiated energy can interfere with other nearby electronic equipment. For this reason, the Federal Communications Commission (FCC) in the United States and comparable agencies in other countries restrict the levels of conducted emissions (flowing in the power cord) and radiated emissions (transmitted through the air) that may be generated by an electronic system.

This is a disadvantage for switching regulators, since we often have to add additional circuitry (mostly low-pass filters) to suppress the high-frequency energy before it can reach the power cord. Alternatively, we can increase the rise and fall times, but then we begin to lose efficiency, which is the primary advantage of a switching regulator. A linear regulator does not generate high-frequency, high-energy noise.

Circuit Complexity

Switching regulators often require greater circuit complexity than comparable linear regulator circuits. This difference is not as critical as it once was, since much of the switching regulator circuitry can now be purchased as an integrated circuit. But, in addition to the potentially more complex circuitry needed for functionality, the switching regulator also has added circuitry that is required for suppression of high-frequency noise.

With reference to power supplies in computers and related equipment, however, nearly all utilize one or more switching voltage regulators. The smaller size, lower weight, less heat, and improved efficiency characteristics of switching regulators generally outweigh any potential advantages offered by linear regulators.

Step-Up versus Step-Down Conversion

The linear voltage regulators discussed in this chapter always require an input voltage that is higher than the output voltage. Switching regulators, by contrast, can be designed with output voltages both lower and higher than the input voltage. This can be a distinct advantage for switching voltage regulators.

Output Voltage Polarity

The linear voltage regulators discussed in this chapter always require an input voltage that has the same polarity as the output voltage. Although opposite polarity voltages can be generated by redefining the ground reference, this is not generally a straightforward technique. By contrast, switching voltage regulators can produce either positive or negative output voltages (or both simultaneously) with a single polarity input voltage. Again, this can be a distinct advantage for switching voltage regulators in many applications.

Representative Circuit

Let's now examine a representative switching voltage-regulator circuit. The circuit we will examine is a highly integrated design. This general type of switching regulator circuit is widely used in computers, personal data assistants (PDAs), digital cameras, and many other electronic products. Many digital products require multiple voltages for their operation with +3.3 V, +5 V, and ±12 V or ±15 V being common. The regulator circuit shown in Figure 12–13 uses +5 V as its input and generates a regulated +3.3 V as its output. A regulator application such as this is called step-down converter or a **buck regulator**. By contrast, a regulator whose output voltage is higher than the input voltage is called a step-up converter or a **boost regulator**. As you can see from inspection of Figure 12–13, the overall circuit requires few external components. Most of the circuitry is integrated into a single package.

The schematic diagram in Figure 12–13 is drawn as you would normally see it on the schematic diagram of a real product. However, to increase our understanding of circuit operation, we need to explore the internal working of the integrated

FIGURE 12–13 A switching voltage-regulator circuit. (*Courtesy of Texas Instruments*)

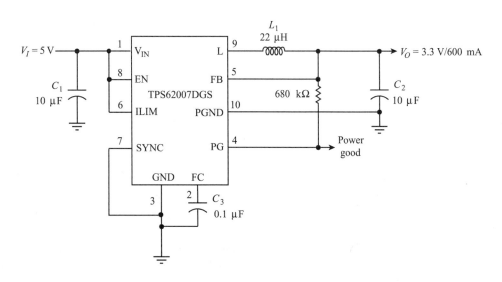

circuit in more detail. Figure 12–14 is the circuit from Figure 12–13, but the functional block diagram of the circuitry *internal* to the integrated circuit is shown. We will rely on this as we discuss circuit operation, but remember—if you were troubleshooting a circuit such as this, you would likely be using a schematic similar to Figure 12–13.

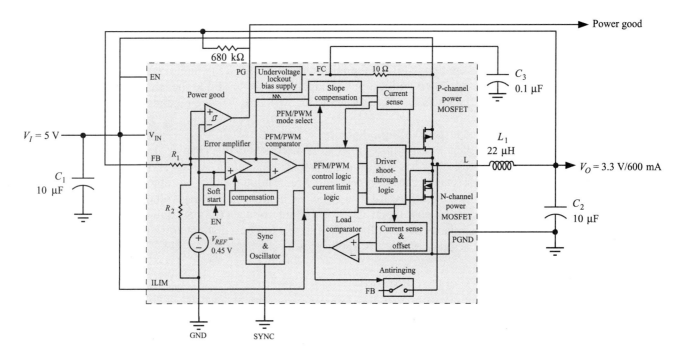

FIGURE 12–14 The voltage-regulator circuit (dc-to-dc converter) shown in Figure 12–13 with the functional block diagram of the controller IC. (*Courtesy of Texas Instruments*)

Overview

Inductor L_1 and capacitor C_2 in Figures 12–13 and 12–14 correspond to the inductor and capacitor in Figure 12–11. Instead of a voltage divider to provide feedback as shown in Figure 12–11, the circuit shown in Figures 12–13 and 12–14 uses the full output voltage for feedback via the FB input. Capacitor C_1 is a bulk-filter capacitor that is likely part of the rectifier-filter circuit used to produce the dc input voltage.

The regulator circuit has three basic modes of operation. For very low input voltages, the p-channel MOSFET is turned on continuously, which in effect, converts it to a series voltage regulator. The two normal modes of operation are **pulse width modulation** (PWM) and **pulse frequency modulation** (PFM). Recall from our previous discussion of switching regulators based on Figure 12–11 that the level of output voltage was controlled or regulated by altering the ratio of "on" time to "off" time for the switch. This is pulse width modulation. In essence, we vary the pulse width of the waveform that is being used to supply current to the external inductor (via pin L).

The regulator can also operate in a PFM mode. Here, the duty cycle of the current waveform is constant and greater than necessary to produce the desired output voltage. The frequency is then varied to control the actual output voltage. As the frequency is reduced, the average current used to charge the external inductor is also reduced. Thus, the output voltage decreases. Increasing the switching frequency causes the average inductor current and the output voltage to increase.

The basic operating cycle consists of turning on the p-channel MOSFET and allowing the current to build up through L_1. This is like closing the switch in the simplified circuit of Figure 12–11. Once a specified level of current is sensed by the internal circuitry, the p-channel MOSFET is turned off, and the n-channel MOSFET is turned on. This is equivalent to opening the switch in Figure 12–11. In this case, the inductor field collapses and forces current through the n-channel MOSFET and the load. This cycle repeats at approximately 750 kHz if the internal oscillator is used. Alternatively, the switching of the regulator can be synchronized to an external clock (via the SYNC pin). The circuit shown in Figures 12–13 and 12–14 does not utilize an external clock, so the switching frequency is nominally 750 kHz (when in the PWM mode).

Anti-Ringing
When the current through the n-channel MOSFET is interrupted near the end of a cycle, the inductor may still have some residual energy. If so, the abrupt interruption of current will cause the inductor and any stray circuit capacitance to exhibit a damped ringing waveform. That is, the inductor and circuit capacitance form a resonant circuit, which gets energized when the current is interrupted. If this damped ringing waveform is allowed to exist, then high-frequency emissions (both conducted and radiated) will be produced. To reduce this effect, the circuit includes an anti-ringing circuit that is essentially connected across the inductor when the n-channel MOSFET is turned off. This quickly dampens the ringing waveform.

Soft Start
When power is first applied to the circuit, the filter capacitances and any other capacitance associated with the load are discharged. If the circuit sensed this initially low voltage, it would try to provide maximum compensation. This would produce a high surge current that might damage components, blow fuses, and so forth. To minimize this effect, the circuit includes a soft start circuit that causes the load voltage and current to build up in a controlled manner. The ramp-up time for this device is about 1 ms.

Enable
The enable input can be used to effectively turn on and off the entire device. When it is on or enabled, it operates as described. When the enable pin is connected to ground, the device is disabled, and the total current requirement is about 1 µA. This feature is not used in the circuit shown in Figures 12–13 and 12–14.

Undervoltage Lockout
If the input voltage is below a specified level, the regulator will not function properly. To prevent unpredictable results, the device senses the input voltage with an undervoltage lockout circuit. If the input voltage is too low for proper operation, the device is not allowed to turn on.

Power Good
The circuit has a built-in voltage comparator that compares the output voltage (as sensed via the feedback input) with an internal reference voltage. If the output voltage is more than 94.5% of its nominal value, then the power good (PG) output goes to a high logic state. This can be used to activate other circuits, or more commonly, it can be monitored by a microprocessor as part of its built-in power management and control functions.

Driver Shoot-Through Logic
Since the same basic waveform is used to turn off one of the power MOSFETs and turn on the other one, there exists the possibility that both transistors will be on at the same time during transition. In effect, this would place a short circuit between

ground and the input voltage. To minimize this effect, the circuit has internal driver **shoot-through logic**, which ensures that one transistor is fully off before the other is allowed to turn on.

PFM/PWM Control Logic
The heart of the regulator lies within the PFM/PWM control logic functional block. PFM and PWM are abbreviations for pulse frequency modulation and pulse width modulation, respectively. This block accepts inputs such as a 750-kHz switching waveform, the output of an error amplifier indicating the relative level of the output voltage, and an indication of output current. From these inputs, it continuously adjusts the duty cycle (PWM mode) or frequency (PFM mode) to maintain the output voltage at the desired value.

The device can also be configured to automatically switch between operating modes as a function of load current in order to maintain the highest possible efficiency. More specifically, for relatively low-current loads, the circuit will operate in the PFM mode for greatest efficiency. Moderate to high-current loads, by contrast, can be controlled more efficiently in the PWM mode.

Performance Specifications
Line regulation, load regulation, and efficiency are three important specifications that provide insights into the effectiveness of this circuit. The integrated switching voltage regulator illustrated in Figures 12–13 and 12–14 achieves line regulation on the order of 0.165%, and load regulation of about 0.6%. Efficiencies of up to 95% are possible.

Alternative Configurations

The switching regulator configuration discussed in this section has been a step-down design, which is also called a buck regulator. Figure 12–15 shows two other configurations. Their basic concepts are similar to the boost regulators previously discussed, so we will only briefly examine their operation.

When the MOSFET is on in Figure 12–15(a), L_1 stores energy in its magnetic field. When the MOSFET quickly opens the circuit, the magnetic field around L_1 collapses. During this time, L_1 acts as a voltage source. This induction source is in series with the applied voltage. Diode D_1 and capacitor C_1 act as a half-wave rectifier, so C_1 charges to nearly the peak voltage (sum of L_1 induced voltage and the dc input voltage). This action continues on each switching cycle. As with previously discussed switching regulator circuits, feedback is used to alter the pulse width (PWM) or frequency (PFM) of the switching waveform. In any case, the dc output voltage is higher than the dc input voltage. This type of circuit is often called a boost regulator circuit.

When the MOSFET is on in Figure 12–15(b), L_1 stores energy in its magnetic field. When the MOSFET opens the circuit, the magnetic field around L_1 collapses and causes L_1 to act as a voltage source. The polarity of the inductor voltage as the field collapses makes the top of L_1 negative with respect to ground. Again, D_1 and C_1 act as a half-wave rectifier circuit, but since the effective input to the rectifier circuit is the inductive voltage across L_1, C_1 charges to the negative voltage developed by L_1. A circuit such as this is called an inverting regulator or a flyback regulator. As with other switching regulator circuits, feedback is used to control the switching waveform in order to maintain a constant output voltage.

Discrete versus Integrated Switching Regulators

While it is certainly possible to design switching regulator circuits based entirely on discrete components, it should be clear from Figures 12–13 and 12–14 why this is generally not done. As a minimum, all of the control circuitry is normally packaged as a single integrated circuit in modern designs. The major external

FIGURE 12-15 (a) A boost regulator and (b) an inverting regulator.

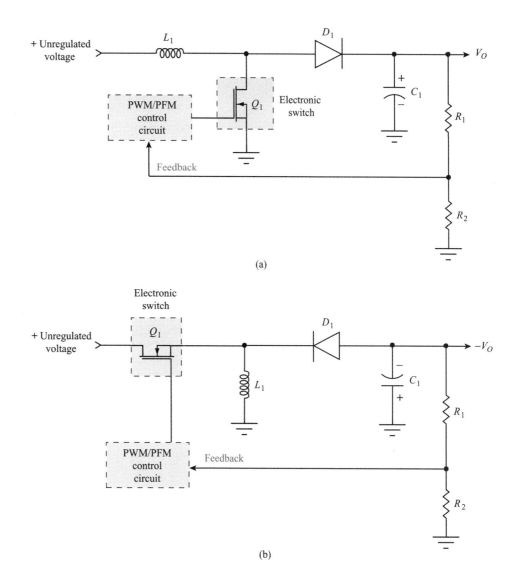

components then consist of the switching MOSFETS, filter capacitors, the storage inductor, and oftentimes a Schottky diode. As evidenced in Figures 12–13 and 12–14, even greater integration is possible in some designs when the power MOSFETs are also part of the integrated circuit.

✓ Exercise Problems 12.4

1. A switching voltage regulator is one example of a linear voltage regulator. (True/False)
2. The regulating element in a switching voltage regulator is most accurately classified as a (*digital/analog*) circuit.

Refer to Figure 12–11 for questions 3 through 5.

3. When the switch is closed, current flows through the diode. (True/False)
4. The field of the inductor is expanding during the time the switch is (*open/closed*).
5. When does the load have current flow?
   a. when the switch is open
   b. when the switch is closed
   c. at all times
6. Which class of voltage regulator is the least efficient?
7. Which class of voltage regulator is most prone to the generation of radiated emissions?
8. When the duty cycle of the switching waveform in a switching voltage regulator is altered as a means to control the output voltage, what mode of operation is being used? (PWM, PCM, or PFM)
9. A switching regulator whose output voltage exceeds its input voltage is called a _____ regulator.

12.5 POWER SUPPLY PROTECTION CIRCUITS

For our purposes, the term *power supply* shall include all of the circuitry required to convert the incoming line voltage (e.g., 120 Vac) into a regulated dc voltage at the levels required by an electronic system. For a typical computer application, this may include transformers, power switches, fuses, rectifier circuits, switching regulator circuits, and filtration circuits. This section discusses several protection circuits that are used in power supply circuits. Some protection circuits are intended to protect the power supply components, but many are designed to protect the electronic system in the event the power supply itself develops a defect. Consider, for example, the devastation that would result if the +3.3-V regulated voltage in your desktop computer suddenly went to +12 V as a result of a power supply defect. Every integrated circuit connected to the supply voltage would potentially be damaged.

Overcurrent Protection

If a short circuit occurs somewhere within an electronic system, then (according to Ohm's Law) a very high current will be drawn from the power supply. If left unchecked, this high current will likely damage one or more components in the power supply. Generally, the most susceptible components to this type of damaging event are the rectifier diodes and any series regulator transistors. A short can occur as a result of a defective component somewhere in the circuit. More commonly, however, short circuits can occur during troubleshooting or other servicing activities. For example, you might be measuring a waveform with an oscilloscope on a pin of a 100-pin microprocessor. Since the pins are only a few thousandths of an inch apart, it is fairly easy for a probe to slip and cause a momentary short. To prevent power supply damage from this sort of event, designers often include current-limit circuits. We will examine two types of overcurrent protection schemes.

Fuses

Figure 12–16 shows one of the simplest overcurrent protection strategies. Here, a fuse is included in the primary side of the power transformer. Its current rating is selected to be slightly higher than the normally expected current value.

Capacitors C_1, C_2, C_3, and inductor L_1 are filtration components that are generally required in all digital products that connect to the power line. Note that L_1 consists of two separate windings on a common core. It is called a **common-mode choke** and is indicated schematically by the circular arrow. It offers impedance to high-frequency noise that appears on both ac lines. It does not, however, offer any opposition to the ac power line current, which flows in opposite directions in the two ac lines. C_1 through C_3 also provide attenuation of high-frequency noise either from a switching voltage regulator or from high-speed digital circuits such as a microprocessor.

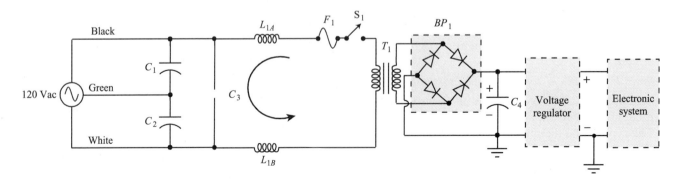

FIGURE 12–16 A fuse in the primary of a power transformer provides overcurrent protection.

The fuse (F_1) is in series with the power switch (S_1), so we know intuitively that an open fuse will have the same effect as turning off the power switch. If the fuse is good and the power switch is on, then 120 Vac is applied to the primary of transformer T_1. The secondary voltage provides a lower voltage to the bridge-rectifier circuit (BP_1) whose output ripple is filtered with C_4. This unregulated dc voltage is then sent to a voltage regulator, which provides a constant (i.e., regulated) voltage to the rest of the system.

Now, if a short circuit occurs at any point in the circuit beyond F_1 such that a destructive current begins to flow, then the current through F_1 will increase correspondingly. As soon as the current exceeds the rating of the fuse, the element burns open and interrupts the current path. This circuit will provide protection against many types of defects within the circuit, and it will prevent damage to many of the components. However, fuses cannot always respond fast enough to protect sensitive semiconductor devices. Therefore, if you find a blown fuse, it is always a good idea to be alert to the possibility of damaged rectifier diodes and/or regulator transistors.

Current-Limiting Circuits

Figure 12–17 shows a series voltage regulator with electronic current limiting. If the load current tries to increase beyond a preset limit, the circuit automatically restricts the maximum allowable current to a lower and safe level. This protects the power supply components from damage due to a short circuit in the load, and in many cases, it limits the degree of destruction to components in the system itself.

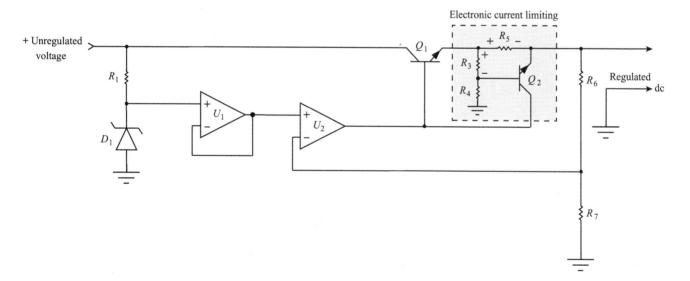

FIGURE 12–17 A series voltage regulator with electronic current limiting.

With the exception of the electronic current-limiting circuit identified by a dashed outline in Figure 12–17, the circuit shown is a basic series voltage regulator as previously shown in Figure 12–4. The operation of this portion of the circuit will not be repeated here. Under normal operating conditions, Q_2 is cutoff, and the circuit operates as a simple series regulator circuit.

The bias for the base emitter junction of Q_2 consists of the combined voltage drops across R_3 and R_5. The voltage drop across R_3 is determined by the output of the pass transistor and the voltage divider action of R_3 and R_4. The voltage across R_5, on the other hand, is determined by the value of load current, since the total load current must flow through R_5. Note that the polarities of the voltage drops across R_3 and R_5 oppose one another. The net voltage of these two resistor voltage drops is the base-emitter voltage of Q_2. As long as the load current is below the maximum allowed by design, the voltage drop across R_3 will dominate, which is why Q_2 remains cutoff.

Now, suppose the load current begins to increase beyond a safe level. As the current increases, so does the voltage drop across R_5. Eventually, the voltage across R_5 will be large enough to overcome the opposing voltage drop across R_3. When this happens, transistor Q_2 begins to conduct.

We know it takes about 0.7 V across the base-emitter junction of a transistor in order for it to have significant current in the emitter-collector circuit. So it is with the pass transistor (Q_1). Notice that the emitter-collector path for Q_2 is in parallel with the combination of the base-emitter of Q_1 and R_5. As Q_2 starts to conduct, we know its emitter-collector voltage will decrease. Since this is in parallel with the combination of the Q_1 base-emitter junction and R_5, we know these voltages must also decrease.

Now, when the base-emitter voltage of Q_1 decreases even slightly as a result of Q_2's increased conduction, this will lower the emitter voltage on Q_1. As this voltage drops, so does the voltage drop across R_3. Recall that the voltage drop across R_3 was opposing the voltage drop across R_5. With the R_3 voltage decreasing, Q_2 conducts even harder. So, as the resistance of the load decreases (perhaps as a result of a short), Q_2 will keep conducting harder, and Q_1 will conduct less until the circuit eventually stabilizes at some fairly low current value. This action is called **foldback current limiting**. Once the trip point is reached, the load current and output voltage decrease as the load resistance is made smaller. Of course, in the case of a short circuit, this action would happen very quickly. The circuit can be restored to normal operation by reducing load current (i.e., increasing load resistance). The graph in Figure 12–18 shows how the circuit responds to increasing load current.

As you can see from the graph in Figure 12–18, the output voltage remains at its regulated value (V_{REG}) as the current increases (i.e., load resistance decreases) from zero to the maximum allowable value or trip-point current (I_{TRIP}). Once the trip-point current is reached, both the output voltage and the load current drop if the load resistance decreases further. This action continues until the short-circuit current (I_{SC}) value is reached. This is the most current that can flow through the load even if the load resistance is zero. It is important to note that the short-circuit current is substantially less than the maximum normal current. This protects the power supply components (especially the pass transistor) from damage. It also makes the overall power supply circuit immune to accidental shorts.

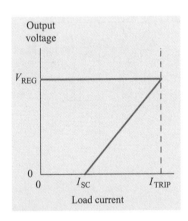

FIGURE 12–18 A graph of output voltage versus load current for a voltage regulator with foldback current limiting.

Current Limiting in Switching Regulators

As discussed in a previous section, the control circuitry for most modern switching power supplies is integrated into a single package as illustrated in Figure 12–13. One of the features that most integrated switching regulator controllers provide is the ability to sense and limit load current. As a specific example, consider the switching regulator circuit shown in Figure 12–13. It has a current limit pin (ILIM). When this pin is shorted to ground, the maximum switching current is 600 mA. When the pin is connected to V_{IN}, as shown in Figure 12–13, the current is internally limited to 1.2 A.

Overvoltage Protection

Try to imagine the devastation that would occur if a component in a 3.3-V power supply in a computer suddenly developed a defect that caused the output voltage to go to 12 V or more. Tens or even hundreds of integrated circuits could be destroyed nearly instantaneously. To reduce the risk of such a catastrophic event, some power supplies include an overvoltage protection circuit. Figure 12–19 is an example of such a circuit.

FIGURE 12-19 An overvoltage protection circuit for an electronic product.

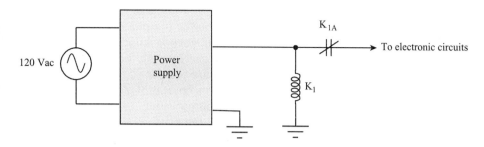

Assume the normal output voltage of the power supply is 3.3 V. In that case, the pull-in voltage for the relay might be chosen as 5 V. Under normal conditions, the relay is deenergized and the normally closed contacts (K_{1A}) are closed. This connects power to the load as with any power supply.

Now, if a defect in the power supply causes the output voltage to rise to the pickup voltage of the relay (5 V in this example), the relay will energize. When the relay energizes, its normally closed contacts (K_{1A}) open. This removes all voltage from the load and protects the electronic circuitry from the overvoltage condition.

Figure 12–20 shows another common technique that is used to provide overvoltage protection of electronic circuits. This particular strategy is called a **crowbar circuit**. It is so named because it protects the load in a manner similar to someone shorting across the load with a steel crowbar. The assumption here is that most of the current would flow through the low resistance of the parallel crowbar, and the load would be protected.

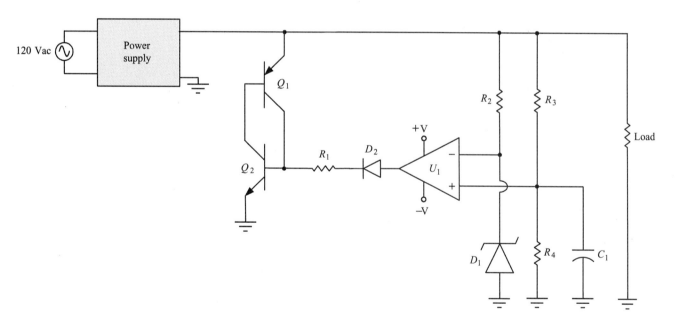

FIGURE 12-20 An electronic crowbar circuit to protect against overvoltage conditions.

Op amp U_1 is configured as a voltage comparator. Its inverting input is connected to a stable reference voltage established by R_2 and D_1. The (+) input of the op amp is a sample of the output voltage provided by the voltage divider (R_3 and R_4). Capacitor C_1 serves as a low-pass filter to prevent the circuit from responding to very-short duration noise pulses. Under normal conditions, the voltage on the (+) input is less than the reference voltage, so the output of the op amp is at its negative saturation level. This reverse biases D_2, which essentially disconnects the base current path for Q_2. Both Q_1 and Q_2 are off under normal circuit conditions.

If the output voltage of the power supply increases above the design threshold, the voltage on the (+) input exceeds the reference voltage, and the voltage comparator output goes to its positive saturation level. This forward biases D_2 and provides a base current path for Q_2, which goes into saturation. When Q_2 turns on, its collector circuit provides a base current path for Q_1. This drives Q_1 into saturation. When Q_1 turns on, it provides an alternate base current path for Q_2. So, even if the comparator output voltage were to change polarity or even go to zero, the two transistors would remain fully saturated since they provide bias paths for each other.

You will recall that a saturated bipolar transistor has on the order of 0.2 V across the emitter-collector circuit when it is saturated. The base-emitter voltage is of course near 0.7 V. Therefore, when Q_1 and Q_2 are both saturated, the voltage measured from the emitter of Q_1 to the emitter of Q_2 is about 0.9 V. Since this is in parallel with the load, the load voltage cannot be greater than 0.9 V. Thus the load is protected from a damaging overvoltage condition.

Now, the two saturated transistors are effectively acting as a short circuit across the output of the power supply. As you might expect, there will be a high current through the two transistors. First, this means that the transistors must be designed to carry this high value of current. Second, since the transistors place a short circuit across the power supply, the supply itself will have to respond to this shorted condition. In most cases, the result is a blown fuse in the power supply. In other words, when the output voltage of the supply goes too high, the crowbar circuit effectively shorts the power supply output and causes the power supply fuse to open.

In chapter 13, we will discuss another semiconductor device called a silicon-controlled rectifier (SCR). As you will then see, this three-terminal device behaves in the same manner as the cross-coupled transistors Q_1 and Q_2 in Figure 12–20. Additionally, SCRs can carry very high currents. Most crowbar circuits utilize SCRs, but the general operation is identical to that described for the circuit in Figure 12–20.

✓ Exercise Problems 12.5

1. Inclusion of a fuse is a common method for protecting an electronic circuit against overvoltage conditions in a power supply. (True/False)
2. If a fuse is blown in a power supply circuit, what should you do before you replace the open fuse?

Refer to Figure 12–17 for questions 3 through 6.

3. Under normal operating conditions, Q_2 is _____.
4. As the load current increases beyond the maximum design value, the conduction of Q_1 (*decreases/increases*) and the conduction of Q_2 (*decreases/increases*).
5. If resistor R_4 increased in value, what would be the effect on the value of trip-point current?
6. Suppose the circuit is designed such that a 500-Ω load resistor causes Q_2 to just begin to conduct. If a 400-Ω load is connected, will the load current be more or less than the current with a 500-Ω load?
7. Switching voltage regulators do not require any form of current limiting, since they are continuously switching from full on to full off. (True/False)

Refer to Figure 12–20 for questions 8 through 10.

8. If the circuit is operating normally with no overvoltage condition, the output of U_1 is (*negative/positive*).
9. If the circuit has responded to an overvoltage condition and you then measure the voltage across the load, what value of voltage would you expect to measure?
10. If this circuit operates as designed, a fuse may blow in the power supply. (True/False) Explain your answer.

12.6 TROUBLESHOOTING POWER SUPPLY CIRCUITS

Troubleshooting power supply circuits can range from simple to exceptionally challenging depending on the nature of the defect, the complexity of the design, and the level of repair that is applicable.

12.6 • TROUBLESHOOTING POWER SUPPLY CIRCUITS

Remove-and-Replace

Many electronic products utilize modular power supplies. These are self-contained systems. When a defect occurs in a modular power supply, it is unusual to troubleshoot the problem to a component level. Rather, the problem is resolved by replacing the entire power supply module. Most desktop and notebook computer systems use modular power supplies and fall into this category. Once you have localized a defect to the power supply module, you remove it from the product and replace it with a new power supply module.

Localizing a defect to a power supply module is generally a fairly simple troubleshooting task. Figure 12–21 illustrates the basic process.

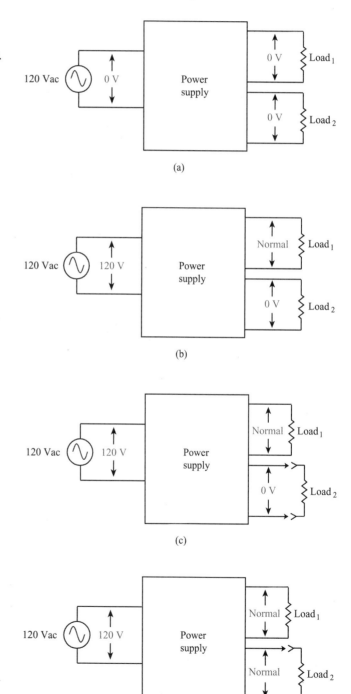

FIGURE 12–21
Troubleshooting a modular power supply.

In most cases, one or more of the output voltages will be zero or very low. In Figure 12–21(a), both outputs of a dual-output supply are zero. One cause for these symptoms is that the input voltage is missing. This case is illustrated in Figure 12–21(a). If the input voltage measures normal, then the defect is likely in the power supply module.

Figure 12–21(b) shows a case when only one output on a dual-output supply is zero. Here we know that the input must be normal or none of the outputs could be normal. The defect could lie within the power supply itself, or it could be within the load circuitry. Since modular supplies are often connected to the powered circuit with a connector, you can easily unplug the supply from the circuit (i.e., operate it in an unloaded condition). If the output voltage is still abnormal but the input line voltage is normal, the defect must lie with the power supply module. This case is shown in Figure 12–21(c). By contrast, if the output voltage of the power supply module returns to normal when the load is disconnected, then there is a fair chance that the defect is in the load. This situation is illustrated in Figure 12–21(d).

There are two precautions that must be observed while localizing power supply defects as illustrated in Figure 12–21. First, when a good power supply is disconnected from the load, we expect its output to be normal. For most modular supplies this is the case. However, some modular supplies have additional wires in the power connector that are used to control the power supply output. For example, if a computer remains idle for a certain period of time, it may be designed to enter a power-save mode. One way to achieve this is for the computer circuits to selectively disable one or more of the power supply outputs, while maintaining critical voltages. If you are troubleshooting a power supply such as this, you may get erroneous symptoms when you disconnect the load because you will be disconnecting one or more control lines at the same time. You can simulate normal signals to the control lines in some cases, but usually it is faster to simply substitute a known good supply.

The second precaution applies to the interpretation of the case illustrated in Figure 12–21(d). It is possible for a defective power supply to deliver a normal (or close to normal) output voltage under no-load conditions but fail to maintain that voltage under loaded conditions. If this type of defect exists, then you might mistakenly classify the defect as being in the load circuit. To guard against this type of error, you can simply replace the load with a resistor. The resistor should have a value that draws approximately the same current as the load normally draws. (Use Ohm's Law to choose a value.) Also, the power rating on the resistor should be high enough to withstand the power that will be dissipated. You can sometimes relax this last requirement a bit, since you can make your determination in a very brief test. Even if the resistor has a somewhat smaller power rating than needed, it can survive a brief test. In any case, if the output of the supply returns to normal with the resistor connected in place of the load, the supply is good and the defect lies in the load. Otherwise, the defect is within the power supply module.

Embedded Power Supplies

Many types of electronic equipment (e.g., some industrial computers and most consumer products) have the power supply circuits constructed as an integral part of the overall electronic system. In many cases, the supply is on the same physical circuit board as the load circuitry. In these cases, you may need to isolate a power supply defect to a specific component. Remove-and-replace may be inappropriate here, since it would require replacement of a larger circuit board with additional (and possibly expensive) components. The decision to repair or replace is generally an easy one and is largely determined by replacement cost. If a decision is made to repair the system, then the following procedure will help you isolate a defect within the power supply portion of the circuit. Of course, you can always apply your knowledge of general electronics troubleshooting to the solution of power supply malfunctions. For example, if you find a transistor with 5 V of forward bias on the base-emitter junction, the transistor is defective. We will not repeat this type of troubleshooting description here. Rather, we will focus on techniques that are somewhat peculiar to power supply circuits.

Step 1: Observation

Observation is the first step in any troubleshooting problem. It includes such considerations as the following:

- Interviewing the operator (e.g., Was there audible arcing? Was there any smoke?)
- Looking for evidence of damage (e.g., burned circuit board, pungent odors)
- Front-panel symptoms (e.g., state of all indicators, response to any controls)

Power supply defects are frequently catastrophic. One or more components and often one or more circuit board traces are vaporized when a defect occurs. Initial arcing and smoke generally accompany these events, and they leave visible evidence of damage. Additionally, you can generally smell a distinct odor long after such an event. If you are observant and detect some of these symptoms, your troubleshooting time can be dramatically reduced.

It is important, however, not to simply replace the visibly defective components and reapply input power. The visibly damaged components are often the effect rather than the cause. You should check for possible shorts before you reapply power. Otherwise, if a short exists, it will immediately destroy the newly-replaced components.

Step 2: Check Unregulated Voltage

If possible, isolate the regulator circuit from the unregulated supply circuit. This is easy if there is a series component between the two. Examples include fuses, connector pins, current-sensing resistors, and series regulator transistors. With the regulator disconnected, you can verify proper operation of the unregulated source. If it is found to be defective, then further isolate the defect as discussed in previous chapters.

Step 3: Check Regulator Circuit

Once the unregulated supply is known to be good, you are ready to reconnect the regulator and verify its operation. Before you do so, however, it is a good practice to be sure the regulator circuit itself is not shorted. Just a simple ohmmeter measurement between the two regulator inputs (usually the unregulated input voltage and ground) will reveal a low resistance if such a defect exists. If a short circuit is detected, then it must be located and corrected before power is reapplied to the regulator circuit. Unfortunately, your only practical tools for this exercise are an ohmmeter and your knowledge of electronics.

Once the regulated supply can be safely connected to the unregulated supply without causing damage (e.g., blown fuses and so on), then you can isolate any defects within the regulator circuit itself. Because most voltage regulator circuits are closed-loop (i.e., their operation relies on feedback), it is often difficult to distinguish between cause and effect. That is, a defect anywhere in the circuit may upset the normal voltage measurements everywhere in the circuit.

To overcome this problem you can employ either of two strategies:

- Force the circuit into open-loop operation
- Force extreme changes at one point and monitor results at another point

Forcing Open-Loop Operation Figure 12–22 on page 456 illustrates how a voltage regulator can be converted from a closed-loop to an open-loop operation. The best way to modify the circuit depends on the details of the circuit being considered, but the general strategy is illustrated in Figure 12–22.

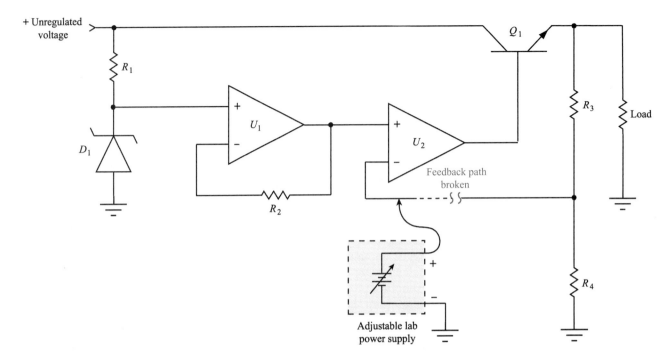

FIGURE 12-22 Interrupting the closed-loop nature of a voltage-regulator circuit.

In the case shown in Figure 12–22, the feedback path (a single wire) is open, and the output from a variable lab power supply is connected in its place. The output of the lab supply should be set to the nominal value of feedback voltage. Now, if the entire regulator circuit were functioning normally, then a small change in the lab supply voltage should cause a corresponding change at all other points in the regulator loop. For example, if the lab supply were increased slightly, we would expect to see a decrease at the output of U_2, a decrease on the emitter of Q_1, and a decrease across R_4. If a component is defective somewhere in the loop, then the change will not be passed on, and the defect will be revealed. Specifically, suppose you made a change to the lab supply voltage as previously described. Further, suppose you measure a corresponding decrease on the base of Q_1 but see no changes on the emitter of Q_1 or across R_4. In this case, you would suspect Q_1 as being defective.

Forcing Extreme Changes in a Closed-Loop Circuit If it is impractical to open the regulator loop, then a similar effect can be produced by forcing extreme changes at one point in the circuit while monitoring another point. In many cases, such a change can be induced by temporarily causing a short circuit. Clearly, you must understand circuit operation well enough to avoid shorting something that will cause further damage; but if care is used, this can be a valuable troubleshooting technique. Consider the regulator circuit shown in Figure 12–23.

Now, suppose the symptom is an output voltage that is lower than normal. We might connect a voltmeter across the load as a monitoring point. Then, we could briefly short R_4. If the circuit were operating normally, this would cause the monitored voltage to increase. If it does not, we could monitor some other intermediate point such as the output of U_2 and repeat the test.

In the circuit shown, we could introduce momentary shorts across D_1, R_3, R_4, and—provided we are using a resistor for a load and not the actual electronic circuits—across the emitter-collector of either Q_1 or Q_2. But, we must certainly avoid shorting R_1, for example, as this would surely destroy D_1.

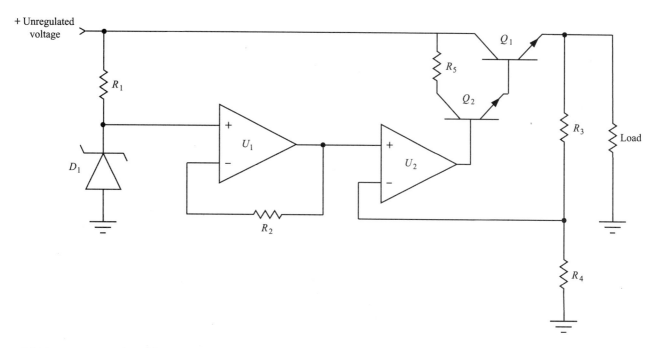

FIGURE 12–23 A closed-loop voltage-regulator circuit.

You must be exceptionally careful if you try to apply this technique to a switching power supply circuit. Recall that the switching transistor dissipates minimal power because it is continuously switching from full on to full off. If you are forcing conditions in the circuit and inadvertently cause the switching transistor to stop switching and remain in the ON state (especially a partially-ON state), then the transistor will surely be destroyed by the excessive power dissipation.

✓ Exercise Problems 12.6

1. What is the name of the general troubleshooting approach normally applied to modular power supply circuits in computers?
2. While troubleshooting a computer that uses a modular power supply, you discover that the output of the supply is zero volts. Name three major locations where the defect may be located.
3. If one of the outputs of a triple-output power supply is near zero but the other two outputs are normal, what can be said about the value of input voltage to the power supply?
4. If a modular power supply is disconnected from the load, is it possible for its output to be zero even if the input voltage is normal and the power supply has no defects? Explain.
5. The most efficient and cost-effective way to troubleshoot and repair an embedded power supply circuit is to utilize a remove-and-replace strategy. (True/False)

Refer to Figure 12–22 for questions 6 and 7.

6. As the lab supply voltage is adjusted, there is no corresponding change in the load voltage. Which of the following are possible defects? Circle all that apply.
   a. R_4 open
   b. R_2 shorted
   c. U_2 defective
   d. Q_1 defective
7. Could the troubleshooting strategy being illustrated in Figure 12–22 also be applied if the output of U_2 were disconnected and the lab supply connected directly to the base of Q_1?
8. Why must care be exercised when forcing extreme changes in a switching power supply?

SUMMARY

This chapter described the operation of three classes of voltage regulators: series, shunt, and switching. Series and shunt voltage regulators are classified as linear voltage regulators. The purpose of all voltage regulators is to maintain a constant output voltage even though the load current and/or the input voltage may change. The degree of immunity that output voltage has to changes in line voltage is called line regulation. The degree of immunity output voltage has to changes in load current is called load regulation. Both linear and switching regulators utilize feedback; they operate as closed-loop systems in order to maintain a constant output voltage.

Series voltage-regulator circuits utilize a pass transistor that is in series with the load. The pass transistor acts as a voltage divider in conjunction with the load. The effective resistance of the pass transistor is continuously adjusted to maintain the desired output voltage.

Shunt voltage-regulator circuits have the regulator transistor connected in parallel with the load. This parallel combination is then connected in series with a voltage-dropping resistor. The conduction of the regulator transistor is continuously adjusted to maintain the correct load voltage.

Switching voltage regulators operate the regulator transistors as digital devices in that they are fully on or fully off. The pulsating current of the switching transistor (generally a power MOSFET) is smoothed by a relatively large inductor and a low-pass filter network. The output voltage is smooth and regulated dc. Either the duty cycle or the frequency of the switching waveform is continuously altered to maintain the correct output voltage.

The major advantage of switching power supplies as compared to linear supplies is that they are far more efficient. That means less power is drawn from the power line to deliver a given amount of power to a load. Any power not delivered to the load is converted into heat. So, since switching regulators are more efficient, they require less cooling, they are physically smaller, and they are lighter in weight. A major disadvantage of switching regulators is that they inherently generate high levels of electromagnetic noise, which must be filtered in order to comply with government regulations. All classes of voltage regulators have all or major portions of the circuitry available as an integrated circuit.

This chapter also discussed the operation of several power supply protection circuits including overcurrent protection and overvoltage protection. Overcurrent protection may be as simple as a fuse that opens when excessive load current is drawn from a power supply. More elaborate strategies include electronic current limiting. In some cases, foldback current limiting is employed where short-circuit current is actually less than the full-rated current of the supply. The primary purpose of overcurrent protection circuitry is to protect the power supply circuitry.

Overvoltage protection is used to protect electronic circuitry from damage if the output of the power supply goes above its normal voltage value due to a defect. One common form of overvoltage protection is called a crowbar circuit. When an overvoltage condition is detected, the crowbar circuit essentially places a short circuit across the output of the power supply. This removes the overvoltage condition immediately and forces the power supply to respond to an overcurrent condition.

Troubleshooting power supply circuits requires application of basic electronic theory, but it can be somewhat complicated by the closed-loop nature of a voltage-regulator circuit. By opening the loop at some point and inserting a controllable voltage, troubleshooting can be greatly simplified. In the case of modular power supplies, the cost of the supply is often low relative to the labor costs required to locate a defective component. In these cases, a remove-and-replace strategy is generally employed. Defects in embedded power supplies, by contrast, must often be diagnosed to the component level.

REVIEW QUESTIONS

Section 12.1: Voltage Regulation Fundamentals

1. What is the main goal or purpose of a voltage-regulator circuit?
2. What specification for a voltage regulator describes how changes in line voltage affect the value of output voltage?
3. What specification for a voltage regulator describes how changes in load current affect the value of output voltage?
4. The output of a voltage-regulator circuit changes from 4.97 V to 5.05 V when the line voltage changes from 110 V to 130 V. What is the line regulation specification for the voltage-regulator circuit?
5. The output of a voltage-regulator circuit changes from 15.05 V to 15.059 V when the line voltage changes from 115 V to 125 V. What is the line regulation specification for the voltage-regulator circuit?

6. The output of a voltage-regulator circuit changes from 9.1 V to 8.9 V when the load current changes from 250 mA to 450 mA. What is the load regulation specification for the voltage-regulator circuit?
7. The output of a voltage-regulator circuit changes from 3.32 V to 3.29 V when the load current changes from 1.25 A to 1.5 A. What is the load regulation specification for the voltage-regulator circuit?
8. In order to provide a stable output voltage, a voltage regulator must utilize a stable voltage reference as part of its design. (True/False)

Section 12.2: Series Voltage Regulation

Refer to Figure 12–24 for questions 9 through 15.

9. What is the purpose of D_1?
10. If the load current changes from 750 mA to 375 mA, what effect does this change have on the voltage on the (+) input of U_1?
11. If the output voltage for U_1 is 5.7 V, what is the approximate voltage across the load?
12. If the regulated input voltage is +24 V and the load voltage is 5.0 V, what is the value of collector-to-emitter voltage of Q_1?
13. What is the purpose of R_2 and R_3?
14. By reversing the (−) and (+) input pins of the op amp, the circuit could be converted into a −5-V regulator circuit. (True/False)
15. If the unregulated input voltage varies from +20 V to +28 V, what is the ideal variation you would predict on the load voltage?

Refer to Figure 12–25 for questions 16 through 18.

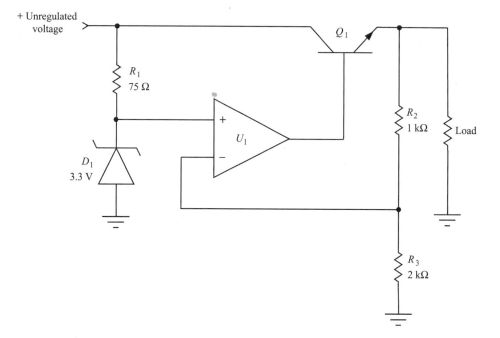

FIGURE 12–24 A series voltage-regulator circuit.

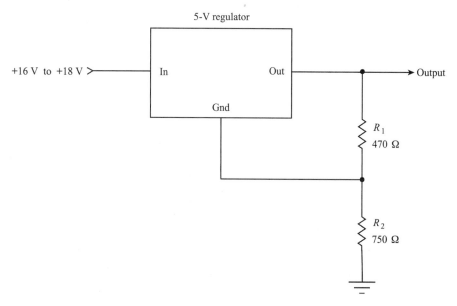

FIGURE 12–25 An integrated voltage-regulator circuit.

16. What is the normal voltage across R_1?
17. What is the value of output voltage with respect to ground?
18. If we wanted to convert this into a regulator with a 10-V output, what value would we make R_2 if R_1 remained at 470 Ω?

Section 12.3: Shunt Voltage Regulation

19. The voltage drop across the regulator transistor in a shunt voltage regulator is equal to the input voltage minus the output voltage. (True/False)

Refer to Figure 12–26 for questions 20 through 25.

20. If the input voltage changes from +22 V to +24 V, what happens to the voltage on the (+) input of U_1?
21. If the input voltage is 24 V and the load voltage is 11 V, what is the voltage across R_1?
22. If the load resistance increases, what happens to the current through Q_1?
23. If the input voltage increases, what happens to the current through the load?
24. If the input voltage increases, what happens to the current through Q_1?
25. What would happen to the value of load voltage if R_3 were increased in resistance?

Section 12.4: Switching Voltage Regulation

26. The regulator transistor in a switching voltage regulator is essentially (*digital/analog*) in nature.
27. A switching regulator is another name for a linear voltage regulator. (True/False)
28. For comparable loads, switching voltage regulators draw (*less/more*) power from the input source than do linear regulators.
29. Why do switching voltage regulators create high-frequency electrical interference (noise)?
30. The Federal Communications Commission dictates the switching frequencies that can be used in switching power supplies. (True/False)
31. A switching regulator whose output voltage is greater than its input voltage is called a step-_____ converter or a _____ regulator.
32. A switching regulator whose input voltage is greater than its output voltage is called a step-_____ converter or a _____ regulator.
33. A flyback switching voltage regulator can be used to generate an output voltage with a polarity that is opposite from the input voltage. (True/False)

Refer to Figure 12–27 for questions 34 through 37.

34. Diode D_1 conducts when Q_1 is (*on/off*).
35. What is the polarity of the output voltage?
36. What happens to the value of output voltage if the pulse width (on time for Q_1) is increased and the off time decreased?
37. What would happen to the value of output voltage if resistor R_2 were decreased?

Refer to Figure 12–28 for questions 38 through 41.

38. This is a (*boost/buck*) switching regulator circuit.
39. Diode D_1 conducts during the time Q_1 is cutoff. (True/False)
40. What is the purpose of resistors R_1 and R_2?
41. The polarity of the output is (*the same as/opposite*) that of the input voltage.

Refer to Figure 12–29 for questions 42 through 44.

42. The only time that current flows through L_1 is when Q_1 is on. (True/False)
43. If R_1 were increased, what would happen to the value of output voltage?
44. The polarity of the output is (*the same as/opposite*) that of the input voltage.

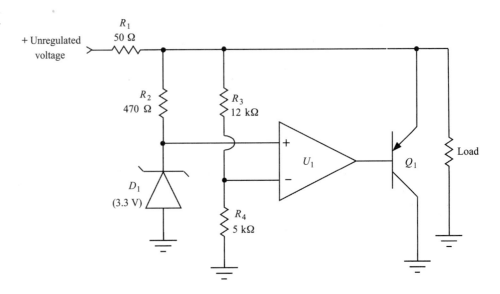

FIGURE 12–26 A shunt voltage-regulator circuit.

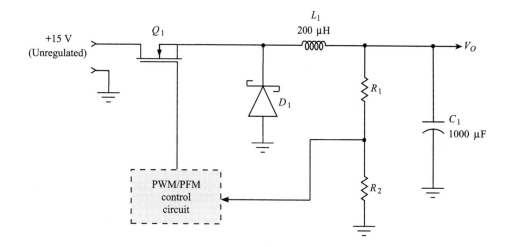

FIGURE 12–27 A switching regulator circuit.

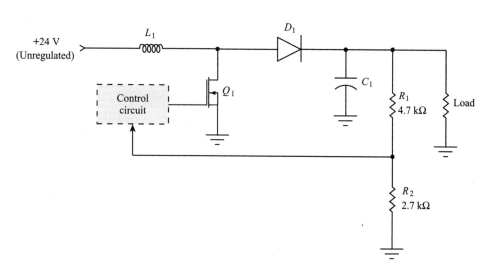

FIGURE 12–28 A switching regulator circuit.

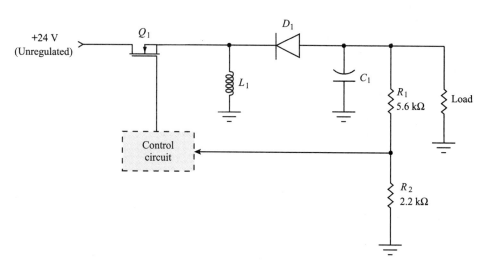

FIGURE 12–29 A switching regulator circuit

Section 12.5: Power Supply Protection Circuits

45. What general class of protective circuit is required to prevent damage when the output of a power supply is shorted?
46. A fuse is an example of what general type of protection scheme?
47. A crowbar circuit is an example of what general class of protection circuit?
48. What can be said about the relative values of full-load current and short-circuit current in a voltage-regulator circuit that employs foldback current limiting?

49. It is impractical to provide short-circuit protection for switching regulator circuits. (True/False)
50. The primary purpose of an overvoltage protection circuit is to prevent damage to components other than those in the voltage regulator itself. (True/False)

Section 12.6: Troubleshooting Power Supply Circuits

51. Power supply circuits are inherently simple and, therefore, easy to troubleshoot. (True/False)
52. When is the remove-and-replace strategy appropriate for power supply circuits?
53. If the output voltage of a power supply is zero, name at least two possible causes other than a defect in the power supply itself.
54. Name at least two important clues that could be uncovered during the observation phase of a troubleshooting exercise that might quickly lead you to the defective portion of the circuit.
55. One method for troubleshooting voltage-regulator circuits is to open the circuit at some point, inject an external voltage, and then troubleshoot the circuit as a normal open-loop system. (True/False)
56. Why is the "forcing of extreme changes" strategy risky when applied to switching voltage regulators?

•—CIRCUIT EXPLORATION

In this exercise, you will have the opportunity to contrast the behaviors of regulated and unregulated voltage sources. Additionally, you will gain valuable experience in troubleshooting closed-loop voltage-regulator circuits. First, build the circuit shown in Figure 12–30 either in the lab or with a circuit simulation package.

Perform the following tasks and respond to the related questions:
1. Substitute the values for Load 1 and Load 2 shown in Table 12–1, and complete the Load Voltage columns.
2. Calculate the load regulation percentage for each output.

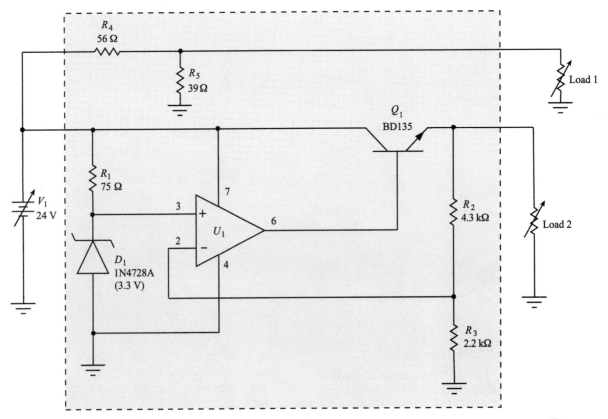

FIGURE 12–30 A power supply with both regulated and unregulated outputs.

Fig12_30.msm

TABLE 12-1 Comparison of output voltages with varying loads.

Load 1 and Load 2 Resistance (ohms)	Load Voltage	
	Load 1	Load 2
1000		
900		
800		
700		
600		
500		
400		
300		
200		
100		

TABLE 12-2 A comparison of output voltages with varying input voltages.

Input Voltage (volts)	Load Voltage	
	Load 1	Load 2
36		
32		
28		
24		
20		
16		

3. Replace both loads with 1000-Ω resistors.
4. Adjust the input voltage to the values indicated in Table 12–2 and complete the Load Voltage columns.
5. Calculate the line regulation percentage for each output.
6. Have your instructor or one of your friends insert several defects into the circuit (one at a time) without your knowledge of the nature of the defect. Then troubleshoot the circuit and identify the defect. Troubleshoot as many problems as you need to in order for you to feel confident in your ability to diagnose a defective regulator circuit such as the one shown in Figure 12–30.

CHAPTER 13
Thyristors and Optoelectronic Devices

•—KEY TERMS

absorption
breakover voltage
coherent light
collimation
critical angle
dark current
dynamic breakback voltage
firing angle
holding current
hot spots
index of refraction
latching current
photoconductive mode
photon
photovoltaic mode
population inversion
refraction
reverse breakdown voltage
spontaneous emission
stimulated emission

•—OBJECTIVES

After studying this chapter, you should be able to:

- Draw the schematic symbol and describe the general behavior of each of the following devices:

SCR	Triac	Diac
SCS	GTO	STS (SBS)
SUS	SIDAC	

- Sketch the characteristic curve for each of the following devices:

SCR	Triac	Diac
SBS		

  The sketch will include identification of key points.

- Describe a troubleshooting method that can be used to classify the condition of a given thyristor.
- Name three ways that a thyristor can be turned on and describe the normal turn-on method for a given thyristor.
- Describe what must be done in order to turn off a thyristor.
- Explain the operation of a thyristor phase-control circuit.
- Explain why a thyristor circuit generates electrical noise.
- Contrast the operation of light emitters and light sensors.
- Define and describe the following characteristics of light and, where applicable, interpret their units of measure:

Wavelength	Intensity	Reflection
Refraction		

- Explain the operation of each of the following optoelectronic devices:

Light-emitting diode	Laser diode	Photodiode
Phototransistor	Light-activated SCR	Optoisolator
Optointerrupter	Avalanche photodiode	

- Discuss the basic operating principles and typical applications for fiber optics.
- Describe a troubleshooting method that can be used to classify the condition of a light sensor, a light emitter, or a fiber-optic cable.

•—INTRODUCTION

This chapter introduces two new families of electronic components. Thyristors are widely used in many industries including computers. They are generally characterized as switching devices that are fully on or fully off. Some of them can switch very high power loads. Optoelectronics is another class of components that has widespread application. You are probably familiar with light-emitting diodes (LEDs) that are often used as indicators on computers. However, there are many other optoelectronic devices that will be important for you to understand.

13.1 THYRISTOR CHARACTERISTICS

Thyristors are semiconductor devices that are often designed for relatively high currents (up to several thousand amperes). They are multilayer (such as pnpn) bistable devices that are either fully on or fully off. There are several types of thyristors that vary in ways that include current and voltage capacity, method of turn on, and whether they conduct in one or two directions. Thyristors are used for many applications in electronics including light dimmers, motor speed controls, electronic ignition systems, industrial controls, computer power supplies, and high wattage adjustable power supplies.

Thyristor Family Characteristics

Our first step toward understanding thyristors is to identify some general characteristics of the entire class of devices. In this section, we will not refer to a specific device, but rather to behavior that characterizes thyristors as a class of semiconductor devices. In a subsequent section, we will narrow our discussion to the details of specific types of thyristors.

Figure 13–1 shows a simple circuit that can be used to illustrate some of the characteristics of a typical thyristor. This series circuit consists of a variable supply voltage, a thyristor, and a resistive load. Provisions have been made to monitor the current through the circuit and the voltage across the thyristor.

FIGURE 13–1 (a) A test setup for characterizing a thyristor and (b) the resulting characteristic curve for one polarity of supply voltage.

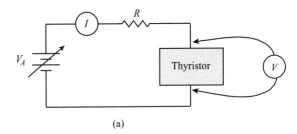

(a)

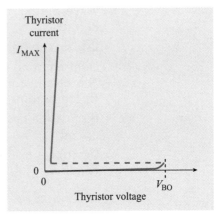

(b)

As the supply voltage is slowly increased, we note that the current remains at zero, but the voltage across the thyristor steadily increases. Clearly, the thyristor is acting like an open switch. As we continue to increase the supply voltage, the current remains at zero for practical purposes although there is a small leakage current through the device. As the applied voltage is increased still more, a point is reached where the thyristor loses its ability to block current. This value of voltage is called the **breakover voltage** and is labeled as V_{BO} in Figure 13–1(b). The value of breakover voltage varies between thyristors and can range from tens of volts to thousands of volts. When the thyristor experiences breakover, the current in the circuit increases rapidly (ideally instantly). Once the thyristor is on, it has no control over the amount of current that flows in the circuit. The current is strictly determined by the value of applied voltage and the resistance of the load (Ohm's Law). The on-state characteristics of an ideal thyristor include zero resistance and zero voltage drop. Practical thyristors have some small resistance and generally drop 1 to 2 V in the on state. As you can see from the nearly vertical slope of the characteristic curve, once the thyristor is on, the voltage across it remains fairly constant.

The sketch in Figure 13–1(b) only shows one quadrant of operation (i.e., one polarity of voltage and one direction of current). If we reversed the applied voltage and did a similar test, we could complete the sketch. Operation in the reverse direction, however, varies between different types of thyristors.

Bistable Operation

Thyristors are inherently bistable. That is, they have two stable conditions. One of the stable conditions is fully on. In this state, the thyristor acts like a closed switch. It drops very little voltage, and it can pass relatively large amounts of current with negligible opposition.

The second stable state shared by the various thyristors is the fully-off condition. Here, the thyristor acts like an open switch. The full applied voltage is felt across the thyristor, just as it would be across an open switch. When a thyristor is off, it has a very high resistance and prevents the flow of any significant current. Figure 13–2 illustrates the two states of a generic thyristor.

FIGURE 13–2 A thyristor has two stable states. (a) One is fully on and (b) the other is fully off.

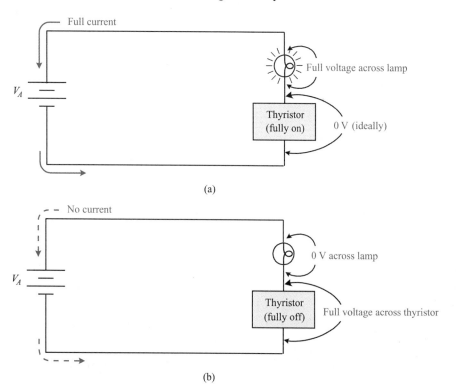

Negative Resistance

In a previous chapter we discussed negative resistance with reference to tunnel diodes. You will recall that a device exhibits negative resistance if the voltage across it is decreasing at the same time the current through it is increasing. Figure 13-3 shows the sketch of the simplified characteristic curve originally presented in Figure 13-1(b). Once the device experiences breakover, the voltage across it drops quickly at the same time the current through the device is increasing. This is, by definition, a region of negative resistance. One characteristic of this region is that it is inherently unstable; there is no way to force the thyristor to remain in this region. It simply passes through the region in the process of switching from its off state to its on state.

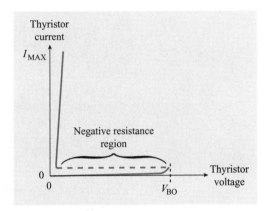

FIGURE 13-3 Thyristors have a region of operation that exhibits negative resistance.

Unidirectional versus Bidirectional

Thyristors that allow current to flow in only one direction are called unidirectional thyristors. Other thyristors permit current to flow in either direction and are called bidirectional thyristors. Clearly, bidirectional thyristors are well suited for operation with ac loads. For most practical purposes, a bidirectional thyristor can be considered to be symmetrical. That is, its operational characteristics are similar for both polarities of input voltage. Figure 13-4 shows the characteristic curve for a generic bidirectional thyristor.

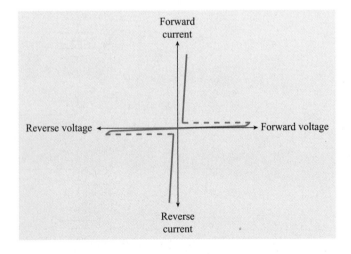

FIGURE 13-4 A bidirectional thyristor has a symmetrical characteristic curve that shows similar operation for both polarities of applied voltage.

Breakover versus Breakdown

It is easy to confuse the terms *voltage breakdown* and *voltage breakover*. At voltages lower than breakdown or breakover, the device acts like an open and blocks the current flow. In this regard, the two mechanisms are similar.

Once the breakdown or breakover voltage has been reached, the current in the circuit increases rapidly. Here again, the two phenomena have similar characteristics. What distinguishes breakover from breakdown (besides the physics involved) is the behavior of the voltage across the device. In the case of breakover, the voltage across the device drops quickly to a much lower value (typically 1 to 2 V). With breakdown, the voltage across the device remains relatively constant.

Although not a rigid rule, for most purposes we can consider breakover to be a desired or intentional event, whereas breakdown is undesired and often damaging. Figure 13–5 shows a generalized characteristic curve for a unidirectional thyristor. A unidirectional device normally passes current in only one direction. A rectifier diode is an example of a unidirectional device. However, as shown by the graph in Figure 13–5, if enough reverse voltage is applied, the thyristor will break down and allow reverse current to flow. A properly designed circuit will not allow this to happen. Contrast the relatively constant voltage drop associated with breakdown with the sudden decrease in voltage associated with breakover.

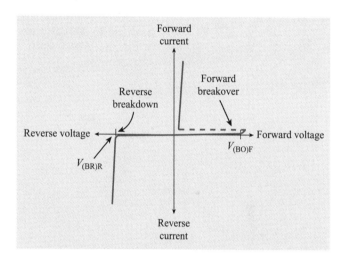

FIGURE 13–5 The voltage across a unidirectional thyristor drops rapidly when the forward breakover voltage is reached but remains fairly constant during reverse breakdown.

Basic Construction

Thyristors are basically four-layer devices—although there are many variations—made by alternating layers of p- and n-type material. Figure 13–6(a) shows how a single four-layer crystal can be viewed as equivalent pnp and npn transistors with shared regions. In Figure 13–6(b), the crystal has been split to highlight the regions associated with a particular transistor. The upper p-type region serves as the emitter of the equivalent pnp transistor. The n-type base and p-type collector of the pnp

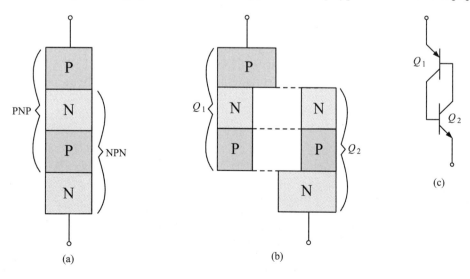

FIGURE 13–6 (a) A basic thyristor consists of four alternating layers of p- and n-type material. (b) The structure can be viewed in two parts to form (c) a transistor-equivalent circuit.

transistor are formed by the same crystal regions as the collector and base, respectively, of the npn transistor. Finally, the lower n-type region serves as the emitter of the equivalent npn transistor. Figure 13–6(c) shows the equivalent transistor circuit schematically.

Now, let's use this equivalent circuit to gain a better understanding of the forward characteristics of thyristors. Suppose we apply a voltage to the equivalent circuit shown in Figure 13–6(c) such that the emitter of the pnp transistor is positive and the emitter of the npn transistor is negative. This experiment is shown in Figure 13–7. As the applied voltage is increased from zero, the transistors act as an effective open. The base current of one is the collector current of the other. If there is no collector current in Q_1, there can be no base current in Q_2. If Q_2 has no base current, then it will have no collector current. Therefore, Q_1 will have no base current. Because Q_1 has no base current, we initially said Q_1 had no collector current.

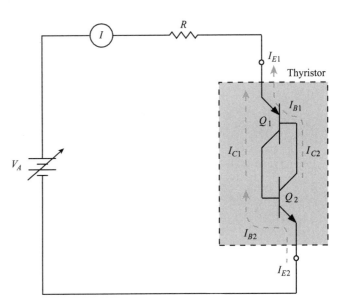

FIGURE 13–7 A transistor-equivalent circuit to demonstrate forward breakover in a thyristor.

As you know from our discussions of basic transistor operation, there are always leakage currents that can flow through the various transistor elements. A portion of the leakage current is voltage sensitive. As we increase the applied voltage in Figure 13–7, the leakage currents increase. Let's assume that at some point the leakage current that forms the collector current for Q_2 (and the base current for Q_1) increases to a point that Q_1 starts to conduct slightly. If Q_1 has a slight increase in collector current, then Q_2 will experience an increase in base current. With a higher base current, Q_2 will also show an increase in collector current. This causes even more base current for Q_1, which further increases the collector current in Q_1, which provides even more base current for Q_2 and so on. Clearly, the process is regenerative. Even though the applied voltage makes no further increase, the transistors will quickly go into saturation. This regenerative action occurs in a thyristor when the forward breakover voltage is reached.

Turn-On Methods

There are several ways that a thyristor can be made to switch from the off (blocking) state to the on (conducting) state, which is also called firing the thyristor. Not all methods apply to all thyristors, but at this point in our discussion we want to remain general.

Exceeding the Breakover Voltage

One way to turn on a thyristor is to exceed its forward breakover voltage. This mechanism was discussed in the preceding section. In the case of a bidirectional thyristor, we can turn it on by exceeding the breakover voltage in either polarity. Two-terminal thyristors always use this method of turn-on. Thyristors with three or more terminals rarely use this method. Rather, thyristors with three or more terminals are selected so that the forward breakover voltage rating is substantially higher than the maximum expected voltage under normal circuit conditions. In this way, the thyristor remains off until it is intentionally turned on by means of a control terminal called a gate.

Exceeding the *dV/dt* Rating

The expression dV/dt can be interpreted to mean "rate of change of voltage." The dV/dt rating of a thyristor indicates the maximum rate of change of voltage that can be applied across its main terminals without causing the device to switch on.

You will recall from our discussions of basic transistor theory that a capacitance exists across any reverse-biased pn junction. Consider the layer diagram for a thyristor shown in Figure 13–8(a). With the polarity of applied voltage shown, two of the junctions tend to be forward biased. Only the center junction is clearly reverse biased. The associated junction capacitance is represented in Figure 13–8(a) as a parallel capacitor.

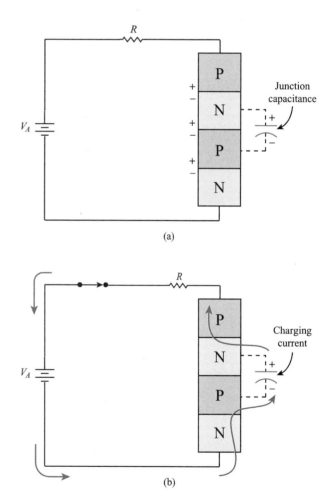

FIGURE 13–8 (a) The reverse-biased (blocking) junction has junction capacitance. (b) A charging current flows when the main voltage is applied.

Figure 13–8(b) shows that when power is first applied to the thyristor, a charging current flows through the thyristor as the junction capacitance is charged. Since the current is highly capacitive, it is largely determined by the rate of change of voltage across the thyristor. In the case of the thyristor, if the rate of change of voltage is too high, then the capacitive current that normally charges the junction capacitance as indicated in Figure 13–8(b) becomes high enough to provide forward bias for the two remaining junctions. If you briefly refer back to Figure 13–7, you will see that this charging current will forward bias the base-emitter junctions of both transistors in the equivalent circuit. This of course starts the regenerative process that quickly causes the thyristor to switch to its on state. As we progress, we will see circuit examples that illustrate methods for reducing the rate of rise of thyristor voltage to prevent premature turn-on.

Gate Triggering

Some thyristors have three or four electrical connections. Two of the connections provide a path for the primary thyristor current as described in the preceding sections. The remaining terminal(s) is called a gate. The gate (or gates for some thyristors) provides another mechanism for initiating the turn-on process. Refer to the transistor equivalent circuit for a basic thyristor that is repeated in Figure 13–9.

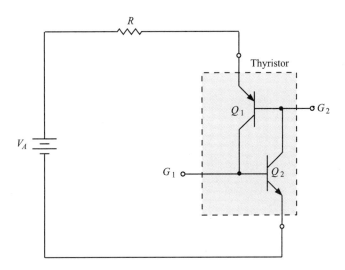

FIGURE 13–9 Gate terminals can be added to the bases of the transistors in an equivalent thyristor circuit.

Clearly, if we apply the correct voltage to either of the gate terminals (G_1 or G_2), then we can initiate turn-on. In the case of G_1, we must apply a voltage that is positive with respect to the negative main terminal. G_2 can initiate turn-on if we make it negative with respect to the positive main terminal. These are the polarities needed to forward bias the base-emitter junction of the associated transistor. Some single-gate thyristors (such as a silicon controlled rectifier or SCR) only have the G_1 connection. Other single-gate thyristors (such as the programmable unijunction transistor or PUT) use only the G_2 gate. Still other thyristors (such as the silicon controlled switch or SCS) have both gate connections brought out to external pins. Gate triggering is almost always the normal way to fire the device in practical three- and four-terminal thyristor circuits.

One extremely important characteristic of thyristors can be demonstrated with the equivalent circuit shown in Figure 13–9. Once a particular gate has caused turn-on of the thyristor, the gate voltage can be removed, and the thyristor will remain in the on state. This latching action is due to the regenerative feedback inherent in the four-layer pnpn construction and was described previously. To repeat, a current (or voltage) of the correct polarity can be applied to either gate connection to initiate turn-on. Once the thyristor is on, however, the gate current (or voltage) may be removed without affecting the conducting state of the thyristor. There is one condition to this last statement. During the time the gate current (or voltage) is applied, the primary current must rise above a minimum level called the **latching current**. Provided we meet this condition, the gate current can be removed without causing the thyristor to turn off.

Turn-Off Methods

We have seen several ways that can be used to cause a thyristor to switch from the blocking or off state to the on or conducting state. But how can we return the thyristor to its high-resistance off state? There are three common turn-off mechanisms or techniques:

- *Reduce primary current.* If the primary current ever falls below a value of current called the **holding current**, the thyristor quickly reverts to its off state.
- *Remove primary voltage.* By momentarily opening the circuit that supplies the primary voltage, we can cause the thyristor to enter its off state.
- *Reverse primary voltage.* If we momentarily reverse the polarity of the primary voltage to a thyristor, we can cause it to enter its off or blocking state.

In the prior section we saw that the gate of a thyristor can be used to trigger the thyristor into conduction. We also discussed how the gate current or voltage could be removed once the thyristor was fired without causing the thyristor to turn off. There is one other very important characteristic of the gate on a thyristor. Once the thyristor has entered the conducting state, the gate loses control of the thyristor. That is, once the primary current begins to flow through the thyristor, it cannot be stopped or altered by controlling the gate. So, the gate can be used to turn on a thyristor, but in most cases, it cannot subsequently turn it off. There are special thyristors whose primary current can be turned off by gate control, but these are exceptions to the behavior of most thyristors and will be discussed in a later section.

✓ Exercise Problems 13.1

1. Some thyristors can carry very high currents (e.g., thousands of amperes). (True/False)
2. How many stable states does a thyristor have?
3. When a thyristor experiences breakover, the voltage across it (*decreases/increases/remains fairly constant*) while the current (*decreases/increases/remains fairly constant*).
4. A thyristor exhibits negative resistance immediately (*before/after*) it is exposed to a voltage greater than its breakover voltage.
5. When a thyristor experiences breakdown, the voltage across it (*decreases/increases/remains fairly constant*) while the current (*decreases/increases/remains fairly constant*).
6. A thyristor can be turned on by applying the correct voltage to its gate terminal. (True/False)
7. A thyristor can be turned off by applying the correct voltage to its gate terminal. (True/False)
8. Once a thyristor has been fired, what minimum value of current must be reached initially in order for it to stay on?
9. If a thyristor is fully on, what is the minimum value of current required to keep it in the conducting state?

13.2 THYRISTOR TYPES AND APPLICATIONS

This section will refine the generalized characteristics previously discussed and introduce several types of thyristor devices. For each thyristor discussed, we will present its schematic symbol, operating characteristics, and a practical device part number, and describe a representative application.

Diac

The diac is a two-terminal thyristor that is used to trigger larger three- and four-terminal thyristors. Figure 13–10(a) shows the schematic symbol for a diac. Figure 13–10(b) shows a representative characteristic curve to describe the behavior of the diac. Based on the symmetrical appearance of the schematic symbol, you

FIGURE 13–10 (a) The schematic symbol and (b) characteristic curve for a representative diac.

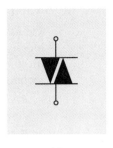

(a)

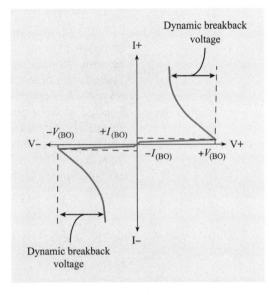

(b)

might infer that the diac is a bidirectional device. This conclusion is confirmed by the characteristic curve in Figure 13–10(b).

As shown in Figure 13–10(b), the diac acts like an open circuit (ideally) until the voltage across the device reaches the breakover voltage ($\pm V_{(BO)}$.) Once the breakover voltage has been reached, the current through the diac increases abruptly while the voltage across the thyristor decreases. This, of course, means that the diac is demonstrating negative resistance at turn-on. Unlike the generic thyristor that we studied in an earlier section, the voltage across the diac does not drop to between 1 to 2 V. Rather, it merely decreases by an amount called the **dynamic breakback voltage**. The breakover voltage for a typical diac is in the range of 30 to 35 V, and the breakback voltage is generally on the order of 5 to 10 V. Diacs are low-current devices with typical peak on-state currents of less than 250 mA. But, since the primary application is to provide the gate current for three- and four-terminal thyristors, only small currents are required.

Figure 13–11 shows a simplified circuit that uses a diac to generate a burst of gate current for the purpose of triggering a three-terminal bidirectional thyristor. For present purposes, our main interest should remain focused on the behavior of the diac. We will briefly discuss the application shown in Figure 13–11 to illustrate a practical use for the diac. In a later section, we will discuss a similar schematic diagram in more detail.

The circuit works as a light dimmer. As a first step, let's assume that neither the diac nor the main thyristor is ever triggered. That is, they continue to act as opens. Let's also assume that the resistance of L_1 is negligible compared to the variable resistance R. Under these conditions, the circuit reduces to a simple series RC circuit. Depending upon the relative values of the resistance of R and the reactance of

FIGURE 13–11 A light-dimming circuit that uses a diac to provide pulsed-gate current to trigger the main thyristor.

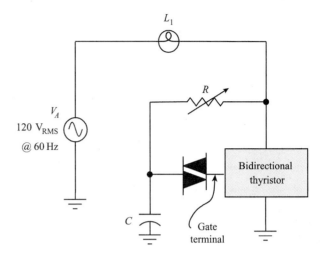

C, the sine wave of voltage across C will lag the sine wave of voltage across the main thyristor. Additionally, the amplitude of the sine wave of voltage across C will be less than the sine wave of voltage across the main thyristor.

Now, as long as the peak voltage across C is less than the breakover voltage of the diac, the diac will remain open. If the diac is open, then the main thyristor will have no gate current so it will also remain open. This is the condition that we initially assumed for our discussion. If we reduce the value of R, then the amplitude of the voltage across C will increase. When the peak voltage across C increases to the breakover voltage of the diac, it will fire. For the purpose of discussion, let's assume we are using a diac with a 30-V breakover and a 5-V breakback rating. Note that we have a closed loop consisting of capacitor C, the diac, and the gate junction of the main thyristor. Right up to the moment the diac fires, the instantaneous voltage on the capacitor is 30 V. This entire voltage is dropped across the diac since it is open. According to Kirchhoff's Voltage Law, this leaves 0 V for the gate junction. At the instant the diac fires, the voltage across the diac drops by an amount equal to the dynamic breakback voltage. In our example, the capacitor will have 30 V, and the diac will have 25 V. The remaining 5 V must be dropped across the gate junction of the thyristor. With the diac on, capacitor C can discharge through the gate junction and the diac. This discharge current is substantially more than what is needed for triggering of the main thyristor, so it is turned on. When the discharge current falls below the holding current of the diac, it will revert to its high-resistance state. As the main supply voltage passes through 0 V (at the 0° and 180° points), the main thyristor turns off since its main current falls below holding current. Because the circuit is symmetrical, the same action occurs on each subsequent alternation.

We cannot delay the firing of the diac beyond the 90° point of the capacitor wave, since this is the point of maximum voltage. However, due to the phase shift introduced by the RC circuit, the capacitor voltage lags the applied voltage. Therefore, the 90° point on the capacitor waveform can actually occur during the 90° to 180° portion of the supply voltage. This allows us to obtain a wide range of control; we can cause the main thyristor to fire throughout a wide range of angles. Once the main thyristor is on, it conducts for the rest of that alternation. The earlier in the cycle we fire the thyristor, the longer it will conduct. The longer it conducts, the higher the average current. And, for this specific example, a higher average current means a brighter lamp (L_1).

We will revisit this circuit in a later section. For now, be certain that you understand the behavior of the diac. Make sure you understand how it keeps the gate current of the main thyristor at zero until breakover occurs and why there is a surge of gate current when the diac turns on. The 1N5760 made by Motorola is a representative diac.

Silicon Trigger Switch

A silicon trigger switch (STS)—also called a silicon bilateral switch (SBS) or silicon bidirectional switch—is another bidirectional trigger device. It is used in the same way previously described for a diac but has superior performance for many applications. The schematic symbol for an STS is shown in Figure 13–12(a). The schematic symbol reveals that the STS is a three-terminal device, and the shape of the symbol implies that it is bidirectional. As we will soon see, the gate lead is often left open, effectively forming a two-lead device. From an external point of view, the STS is a simple two- or three-lead component. Actually, the STS package contains a small integrated circuit composed of zener diodes, resistors, and both npn and pnp transistors. Figure 13–12(b) shows the equivalent circuit for the STS. You may recognize the configuration of the Q_1/Q_2 and Q_3/Q_4 pairs as being similar to the regenerative connection we discussed in an earlier section. In essence, the two transistor pairs behave like two back-to-back thyristors: They conduct on opposite polarities of applied voltage.

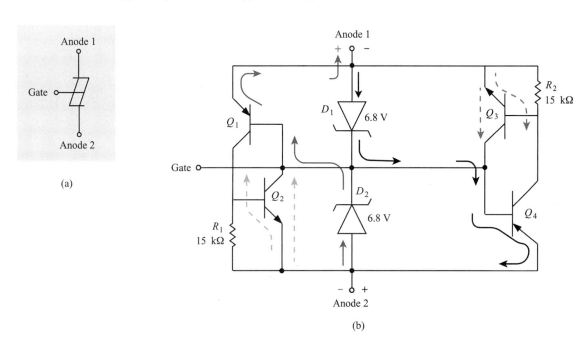

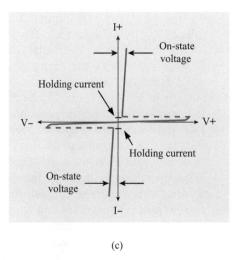

FIGURE 13–12 (a) The schematic symbol, (b) an equivalent circuit, and (c) the characteristic curve for a silicon trigger switch (STS).

For our present discussion, let us assume that the gate lead is left open (a common practice). Let's further assume that anode 1 in Figure 13–12(b) is becoming increasingly positive with respect to anode 2. Initially, there is no path for current flow other than leakage current. If the differential anode voltage exceeds the zener voltage (≈ 6.8 V) plus the V_{BE} of Q_1 (≈ 0.7 V), there will be a path for current from anode 2 through D_2 and through the base-emitter circuit of Q_1. This will forward bias Q_1, which provides base current for Q_2. When Q_2 turns on, it further increases the base current of Q_1 and begins the regenerative cycle that causes the device to quickly switch to the low-resistance on state. The on-state voltage is generally in the range of 1 to 2 V. The STS will remain in the on state until the current falls below holding current. When the anode current falls below holding current, the STS quickly reverts to its off state.

If the polarity of applied voltage is reversed (i.e., anode 1 is negative with respect to anode 2), then a similar action occurs in the other half of the symmetrical circuit. Because the STS is actually an integrated circuit (i.e., all of the components are formed simultaneously within a single crystal), the device exhibits excellent symmetry. Figure 13–12(c) shows the characteristic VI curve for a representative STS.

Because the breakover voltage of an STS is substantially lower than for a diac (≈ 8 V versus 30 to 35 V), it provides a greater range of control for thyristor phase-control circuits. For example, a light dimmer circuit operated from a 120-Vac power line like the one shown in Figure 13–11 might be able to achieve a firing range on the order of 15° to 165°. An STS in the same circuit might achieve a range of 3° to 177°. In effect, this means that the lamp could be adjusted to both dimmer and brighter settings when the STS is used as the trigger device.

In many thyristor trigger applications, the gate terminal is left open. However, it can be used for two purposes. First, the gate terminal can be used to fire the STS at an anode voltage lower than the normal breakover voltage. Second, external components can be added to alter the characteristics of the STS. You will note from the schematic in Figure 13–12(b) that any external component connected between the gate terminal and either of the anode terminals is electrically in parallel with the internal zeners. Recall that it is primarily the zener diodes that determine the value of breakover voltage. If, for example, we add lower voltage zeners (or even resistors) between the gate and anode terminals, we can lower the value of breakover voltage.

An STS is a relatively low-current device. A maximum on-state current rating of 200 mA or less is typical. But, since the primary application is to provide gate current for three- and four-terminal thyristors, only small currents are required. The BS08D device manufactured by Powerex (**www.pwrx.com**) is a representative SBS component.

Silicon Controlled Rectifier

We are now ready to discuss the operation of a silicon controlled rectifier (SCR), which is probably the most widely used thyristor. It is extremely probable that you will encounter SCRs, regardless of the branch of electronics you choose. Therefore, it is important for you to thoroughly understand its operation.

Overview

Figure 13–13(a) on page 478 shows the schematic symbol for an SCR. By inspection of the symbol, you could infer that the SCR is a unidirectional device. The symbol is not symmetrical, and it resembles the schematic symbol for a standard rectifier diode. The only difference (schematically) is the addition of the gate terminal. As with simple rectifiers, the primary electron current in an SCR flows against the arrow on the schematic symbol. That is, electron current (when it flows) goes from the cathode to the anode.

FIGURE 13–13 (a) The schematic symbol and (b) the layer diagram for an SCR. Both suggest that an SCR is a unidirectional device.

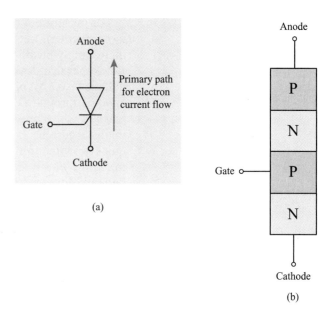

The most outstanding difference between a standard rectifier diode and a silicon controlled rectifier, is that the forward current may or may not be allowed, depending upon the gate voltage (or current). In general, the SCR remains in the off state until a proper amount and polarity of gate current is applied. In order to trigger an SCR, the gate must be made positive with reference to the cathode. Figure 13–13(b) shows the layer diagram of an SCR. Clearly, if the gate is positive with reference to the cathode, then the lower pn junction is forward biased. An SCR never conducts current in the reverse direction unless the **reverse breakdown voltage** is exceeded. Breakdown, you will recall, frequently results in damage to the device due to excessive power dissipation.

SCRs have the highest power switching capability of any thyristor. Some manufacturers make SCRs that can withstand off-state voltages of more than 1000 volts. Devices are also available that can switch currents of 2000 amperes or more. Smaller SCRs have applications in many types of circuits with switching currents of one ampere or less, but as a rule, SCRs are regarded as high-current, unidirectional switching devices.

Characteristics

Figure 13–14 shows the generalized characteristic curve for a representative SCR. The SCR blocks current in the forward direction until the forward breakover voltage ($V_{(BO)}$) is reached. The forward breakover voltage can be reduced by increasing values of gate current as indicated in Figure 13–14. When the SCR reaches breakover, it experiences negative resistance as it switches abruptly to the on state. In the on state, the SCR is characterized by having a low forward voltage drop (V_F) with 1 to 2 V being typical values. In order to remain in the on state, the forward cathode-to-anode current of the SCR must remain above its holding current (I_H) specification. When the primary current falls below holding current, the SCR reverts to its off state.

When the SCR is operated in its reverse mode, it blocks current regardless of the gate voltage. If the reverse voltage exceeds the reverse breakdown voltage rating ($V_{(BR)}$), the SCR may conduct heavily, but the voltage across the SCR remains relatively high. The simultaneous occurrence of high voltage and high current produces high power dissipation within the semiconductor crystal, which may damage the SCR.

SCRs are not generally considered to be high-speed components, although devices with faster switching times continue to appear in the industry. The switching

FIGURE 13–14 A family of characteristic curves for a representative SCR.

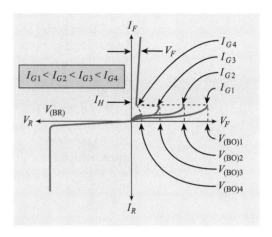

time of a typical SCR would frequently be in the range of 500 ns to 5 μs. The S1070W is a 70-ampere SCR made by Teccor (www.teccor.com).

Application

Silicon controlled rectifiers can be used in nearly every field of electronics. We can grossly classify their applications into two classes of usage: dc switching and phase control.

DC Switching Figure 13–15 illustrates an SCR used in a simple, but practical, dc switching application. This circuit could be used as a basic automobile theft alarm. Switch S_1 is a normally closed switch that is hidden from normal view. Additionally, it may require a key to operate. This switch is used to arm and disarm the alarm system.

FIGURE 13–15 An SCR can form the heart of an automobile theft alarm system.

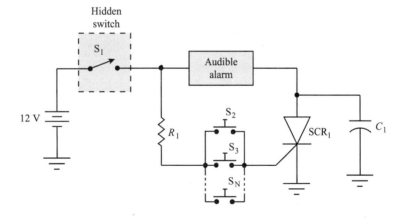

Switches S_2 through S_N represent normally open sensor switches that detect intrusion. There can be any number of switches in the parallel bank. A typical system may have a switch on each door, one to detect opening of the hood, and another to detect an open trunk lid. The system has an audible alarm or siren. This is intended to attract attention and thwart the auto theft.

Now, let's suppose the system has been armed by closing switch S_1. This applies power to the anode-cathode circuit of the SCR via the audible alarm. However, the SCR will remain in its off state until it is triggered. Initially, there is no gate current for the SCR because all of the sensor switches are open. This is the normal and stable condition for an armed system. Capacitor C_1 prevents accidental firing of the SCR due to the high rate-of-change-voltage that would otherwise occur when switch S_1 is first closed. In short, C_1 and the resistance of the alarm form an *RC* time constant, which slows the rise of voltage across the SCR.

If an intruder opens a door, one of the parallel switches (S_2–S_N) will close. This provides a path for gate current. We can trace the gate current path from the negative side of the power source, into the cathode of the SCR, out of the gate of the SCR, through the closed sensor switch, through R_1, and through S_1 to the positive side of the 12-volt source. This forward biases the gate-cathode junction and fires the SCR.

When the SCR fires, it acts like a short circuit and effectively connects the audible alarm across the 12-volt source. This sounds the alarm. Now, if the intruder closes the door in an effort to silence the alarm, the associated sensor switch will open, which interrupts the gate current path. However, the SCR has been triggered and is part of a dc series circuit. So, as long as the alarm current is greater than the holding current of the SCR, the SCR will remain on. It can be eventually turned off when the owner uses a key and momentarily interrupts the anode current by opening hidden switch S_1. The crowbar application discussed in chapter 12 is yet another dc application for SCRs.

Phase Control Phase-control circuits are one of the major applications for SCRs. As briefly described in a previous section, the SCR is held in the off state for a certain portion of the positive alternation of the input cycle. A pulse of gate current is timed to arrive at a specified point in each positive half-cycle. When the blast of gate current arrives, the SCR fires and continues to conduct for the remainder of the alternation. The earlier in the half-cycle the gate current arrives, the longer the SCR conducts and the higher the average current through the load circuit. By varying the average load current, we can change the brightness of a lamp, the speed of a motor, or perhaps the temperature of a heating element. A phase-control method similar to the one shown in Figure 13–11 is limited to a theoretical range of from 0° to 180° when using an SCR. This is because no conduction can occur on the negative alternation; the SCR is a unidirectional device.

Figure 13–16 shows one of several ways that can be used to allow an SCR to control the average current through an ac load on both alternations of the input cycle. Let's first consider the two extreme cases. First, if the SCR is never fired, then it always acts as an open circuit. Under this condition, there will be no complete path for current through the ac load. Due to the action of the bridge rectifier, the anode of the SCR will have a positive full-wave waveform with reference to the cathode. Therefore, it is capable of conducting on either half-cycle of the input if it receives a gate pulse.

The other extreme case occurs if the pulse generator circuit times the gate current pulses to arrive near the beginning of each alternation. In this case, the SCR would fire very early in each half-cycle and conduct for the remainder of the

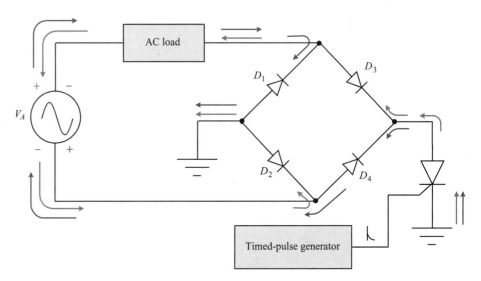

FIGURE 13–16 The current through an ac load can be varied by using an SCR phase-control circuit.

alternation. The net result is that the value of ac current through the load would be near maximum. For practical purposes, the ac load is effectively connected directly across the source under these conditions. Clearly, there will be small voltage drops (≈ 1.4 V) across the diodes (D_1 and D_4 on one alternation and D_2 and D_3 on the other alternation). There will also be a small voltage drop across the SCR (≈ 1 V).

By changing the timing of the gate pulses, we can adjust the current through the ac load from zero to maximum. The current paths for each alternation of the input are indicated on the schematic in Figure 13–16. Note that a bidirectional current flows through the ac load, but a unidirectional current flows through the SCR. The SCR operates as a fully-on or fully-off device. The average load current is varied by controlling the phase or timing of the gate current. This means that the waveform of current through the load (or voltage across it) will not be sinusoidal. But, for many ac loads such as lamps and motors, this nonsinusoidal waveform works well. Figure 13–17 shows the waveforms that would be seen across the SCR (anode-to-cathode) and across the load for a given **firing angle**, which is the angle on the sine wave where the SCR is triggered.

FIGURE 13–17
Waveforms in a phase-control circuit showing (a) applied voltage, (b) SCR (anode-to-cathode) voltage, and (c) the waveform across the load.

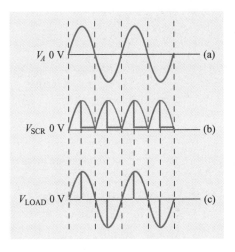

Triac

Next to the SCR, the triac is probably the most widely used thyristor. Applications for a triac are similar to those of an SCR, but the bidirectional characteristics of the triac makes it ideal for the control of ac loads.

Operation

Figure 13–18(a) on page 482 shows the schematic symbol for a triac. By inspection of the symbol, you could infer that a triac is a bidirectional device. The symbol is symmetrical, and it resembles the schematic symbols for two standard rectifier diodes connected in parallel (opposite directions). The only difference (schematically) is the addition of the gate terminal. The primary electron current in a triac flows between main terminals 1 and 2 (MT_1 and MT_2).

The triac is similar to the SCR in that the principal current may or may not be allowed depending upon the gate voltage (or current). In general, the triac remains in the off state until a proper amount and polarity of gate current is applied. There are four possible combinations of voltage polarity and all can be used to trigger the triac. Table 13–1 on page 482 lists the possible triggering polarities.

Triacs are relatively high-power devices. They are available with breakover voltages as high as 800 volts and rms current ratings of at least 40 amperes, although these are not absolute limits. The surge current ratings for some triacs are as high as 350 amperes. The gate current required to trigger a triac into conduction is generally in the 10- to 100-milliampere range, although neither of these values represents an absolute limit. A surge of gate current is the desired way to fire a

FIGURE 13–18 (a) The schematic symbol and (b) the characteristic VI curve for a triac both reveal that a triac is a bidirectional device.

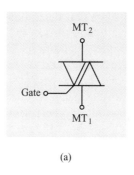

(a)

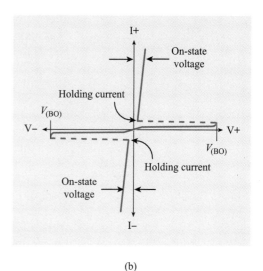

(b)

TABLE 13–1 Possible triggering combinations for a triac.

[1]MT$_2$	[1]Gate	Comment
Positive	Positive	Sensitive triggering
Negative	Negative	Sensitive triggering
Positive	Negative	Less sensitive triggering
Negative	Positive	Very insensitive triggering

[1] Voltages are measured with respect to MT$_1$.

triac. As with SCRs, a heavy gate current ensures reliable turn-on and helps reduce the risk of localized heating in the semiconductor material called **hot spots**.

Figure 13–18(b) shows the generalized characteristic curve for a representative triac. The triac blocks current in both directions until the breakover voltage ($V_{(BO)}$) is reached. The breakover voltage value can be reduced by increasing values of gate current as in a manner similar to an SCR. When the triac reaches breakover in either direction, it experiences negative resistance as it switches abruptly to the on state. In the on state, the triac is characterized by having a low on-state voltage drop, with 1 to 2 V being a typical value. In order to remain in the on state, the principal current (current between MT$_1$ and MT$_2$) of the triac must remain above its holding current (I_H) specification. If the principal current falls below holding current, then the triac reverts to its off state.

Triacs are very low-speed components. They are designed specifically for operation at the 60-Hz power line frequency. Since the triac can conduct in either direction, there is a very short time where the main terminal voltage drops to zero, and the device can recover its blocking condition. If the line frequency is too high, the triac doesn't have time to turn completely off before the voltage is reapplied in the opposite direction. The net result is that once the triac turns on, it can never be shut off until the power is turned off. That is, if we trigger the triac on one alternation, it will continue to fire on subsequent alternations without receiving a trigger on the gate. Clearly, this is not the intended behavior of the triac. The Q2010L5 is a 10-A triac made by Teccor (www.teccor.com).

Application

The primary application of triacs is in phase-control circuits for light dimmers, heating elements, and motor speed controls. Triacs have a distinct advantage over SCRs for many applications, since triacs are bidirectional. This means power can be applied to the load throughout the full ac cycle without the use of bridge circuits. The primary disadvantages of the triac as compared to the SCR are its lower power and frequency limitations.

Other Thyristors

We have discussed the two workhorses of the thyristor world: SCRs and triacs. There are several other thyristors that are used in industry. Table 13–2 lists several thyristors (including those previously discussed), their schematic symbols, triggering methods, brief summaries of their characteristics, and a typical application of each. This information, coupled with the detailed discussions previously presented, should give you a good understanding of the various thyristors and their operational characteristics.

TABLE 13–2 A summary of thyristors and their characteristics.

Name	Symbol	Triggering	Characteristics	Application
Diac		Exceed breakover voltage	Bidirectional, 30–35 V breakover, low current, 25–30 V on-state voltage	Triggering device for larger thyristors
Silicon trigger switch (STS). Also called silicon bilateral switch (SBS)			Bidirectional, ≈ 8 V breakover, low current, 1–2 V on-state voltage, gate often unused	Triggering device for larger thyristors
Silicon controlled rectifier (SCR)		Positive gate relative to cathode	Unidirectional, high voltage, high current, ≈ 1 V on-state voltage	Phase control for light dimmers, heaters, and motors
Schockley diode		Exceed breakover voltage	Unidirectional, ≈ 10 V breakover, low current, ≈ 1 V on-state voltage	Triggering device for SCRs in old designs
Silicon unilateral switch (SUS)		Exceed breakover voltage that is set by the gate	Essentially a Schockley diode with a gate	

TABLE 13–2 (continued)

Name	Symbol	Triggering	Characteristics	Application
Silicon controlled switch (SCS)		Positive on cathode gate relative to cathode, or negative on anode gate relative to anode	Unidirectional, low current (≈ 1 A max)	Triggering device for SCRs in old designs
Gate turn-off switch (GTO or GTS)		Positive gate relative to cathode. Negative gate relative to cathode to turn off	Unidirectional, low current	Used in place of a small SCR when gate control is needed
SIDAC		Exceed breakover voltage	Bidirectional, high current (≈ 20 A), high voltage (≈ 300 V), ≈ 1 V on-state voltage	Triggering device for large SCRs and triacs, lamp starters, and power oscillators
Triac		Positive and negative on gate relative to MT_1 when MT_2 is positive and negative, respectively	Bidirectional, high voltage, high current, ≈ 1 V on-state voltage	Phase control for light dimmers, heaters, and motors
Quadrac			Bidirectional, diac and triac packaged together	

✓ Exercise Problems 13.2

1. A diac is a _____-terminal device that is (*bidirectional/unidirectional*).
2. An SBS is primarily used as a triggering device for thyristors. (True/False)
3. An SBS is a (*bidirectional/unidirectional*) device.
4. An SCR is generally considered as a low-current component. (True/False)
5. In order to trigger an SCR, the _____ terminal must be made positive with respect to the _____ terminal.
6. When an SCR fires, the primary electron current path is from the _____ terminal to the _____ terminal.

7. Once an SCR has been fired, it can be turned off by applying a negative pulse to its gate terminal. (True/False)
8. Is it possible to make an SCR conduct when its anode is negative with respect to its cathode? Explain.
9. If the anode current of an SCR is reduced to a value that is less than _____ current, the device will revert to its off state.
10. When an SCR or triac is used in a phase-control application, the thyristor itself operates as a linear device and is no longer limited to on/off operation. (True/False)
11. Controlling the _____ angle of a series-connected thyristor can vary the average current through an ac load.
12. What is the name of the component that is essentially a diac and a triac in the same package?
13. An SCS would be a good choice to connect in series with a high-current heating element as a way to control the temperature. (True/False)
14. What is the triggering method for a diac?
15. A triac can be triggered with either polarity of gate voltage. (True/False)

13.3 TROUBLESHOOTING THYRISTORS

In this section, you will learn several new troubleshooting techniques that can help you locate defects in thyristor circuits. Because the general term *thyristor* includes devices with currents in the milliampere range, devices in the 1500-A range, and devices in both the 5-V and the 1000-V range, not all techniques are practical for every thyristor. However, your knowledge of thyristors and basic electronics should enable you to choose an appropriate technique for a particular troubleshooting situation. A thorough understanding of thyristor operation coupled with basic troubleshooting strategies provides an effective mix for diagnosing defects in thyristor circuits.

Classification of Defect

Since a thyristor has only two stable states, its defects also fall into two general classes. The thyristor can remain in the off state at all times. In the case of a lamp dimmer circuit, for example, this would cause the light to remain off regardless of the control setting.

A thyristor can also remain in the on state at all times. If we used the lamp dimmer circuit as an example, in this case the lamp would have full brightness regardless of the control setting.

If a thyristor is suspected of having a defect, your first task will be to classify it as being on all of the time or off all of the time. The actual defect may or may not be the thyristor, but we still need to classify the defect.

Cause versus Effect

The next step is to determine if the state of the thyristor (always on or always off) is due to a defective thyristor, a defective trigger circuit, or perhaps a problem with the power source. For example, suppose the trigger circuit had a defect that prevented the thyristor from receiving a gate pulse. In this case, the thyristor would always be in the off state, but this would be an effect and not the actual cause. By contrast, suppose the thyristor had an internal short between the anode and cathode. In this case, the thyristor would always be in the on state, and it would be the actual defect. In the following paragraphs, we will present some methods that can be used to localize the defect to the thyristor or to the external circuitry.

Thyristor Always On

Figure 13–19 shows a generalized thyristor circuit that is continuously in the on state. It applies equally well to unidirectional and bidirectional thyristors operated with either ac or dc loads. We first detect the "always on" condition when we note that the load is fully on (i.e., maximum brightness, speed, heat, and so forth) and cannot be reduced with the trigger control (represented by a rheostat in Figure 13–19). When we measure the voltages across the thyristor and across the load, we will get nearly zero and nearly the full supply voltage, respectively.

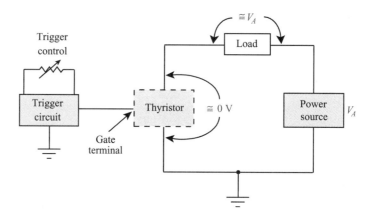

FIGURE 13–19 A defect in a thyristor circuit can result in continuous full-load power.

To isolate the problem, we can do one of two things depending on the specifics of the trigger circuit. If the trigger circuit is short-circuit tolerant (i.e., it can withstand a short to ground on the gate trigger line), then we can turn off power and temporarily connect a short circuit between the gate of the thyristor and ground. This will prevent the trigger pulse from reaching the gate. Now, reapply power and observe the symptoms. If the load voltage drops to zero (i.e., the lamp doesn't light, or the motor doesn't run), then the thyristor is good, but there is a defect in the trigger circuit, which is causing a continuous trigger. This defect can be localized using methods discussed in earlier chapters. If the load continues to receive full power with the gate of the thyristor shorted to ground, then the thyristor has an internal short and must be replaced. Before using this technique, examine the schematic carefully to be sure that a short circuit won't damage the trigger circuit. If you are uncertain, use the alternate method described in the next paragraph.

Another way to remove the trigger from the thyristor is to open the gate circuit. This normally requires a soldering iron or a wrench, depending upon the size of the thyristor. Once you have opened the gate circuit, reapply main power to the thyristor. If the load is still on continuously, the thyristor is defective (it has an internal short). If the open gate causes the load to turn off, the thyristor is good, and the defect lies in the trigger circuit.

Thyristor Always Off

Figure 13–20 shows a generalized thyristor circuit that is continuously in the off state. We first detect this condition when we observe that the load is off (i.e., no

FIGURE 13–20 A defect in a thyristor circuit can result in the load being off at all times.

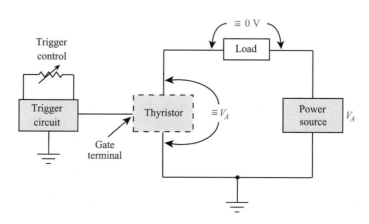

light, speed, heat, and so forth) and that the trigger control has no effect. First, verify that the main supply voltage is present. If it is not, you have located the defect. If the supply voltage is good, then circuit measurements will reveal one of the following conditions:

1. The full supply voltage is across the thyristor and none is across the load.
2. The full supply voltage is across the load and none is across the thyristor.

If you detect the second condition, the load itself is defective. It may be an open wire, a defective lamp, an open motor winding, and so forth, but the thyristor is not causing the problem.

The first condition cited is represented in Figure 13–20. These symptoms generally mean one of two things. First, the thyristor may be defective. Second, the thyristor is not receiving a proper trigger from the gate circuitry. If you monitor the gate voltage with an oscilloscope and see gate voltages that are adequate for proper triggering, the thyristor is probably defective. On the other hand, if no gate trigger can be seen, then we must make additional tests.

To localize the defect, open the gate of the thyristor and perform one of the following actions, depending on the details of the circuit being diagnosed:

1. Trigger the thyristor from an external source.
2. Monitor the trigger voltages at the output of the gate drive circuitry.

The first option is well suited to relatively low-power thyristors. In this case, the external trigger voltage can be provided from an adjustable dc bench supply. Just start with zero volts and increase it until the thyristor begins to fire. If the thyristor does begin to fire, the actual defect lies in the gate trigger circuitry. If the thyristor does not fire even with adequate gate voltage and current, the thyristor is defective.

If you are diagnosing a high-powered thyristor or one that uses dangerously high voltages and currents, you should employ alternative number two. Here we monitor the output of the gate circuit after the gate itself has been opened. If the gate voltages are adequate for normal thyristor firing, then the thyristor can be assumed to be defective. If no gate trigger is present at the output of the trigger circuit even with the thyristor gate disconnected, then the defect lies within the trigger generation circuitry.

A Quick Thyristor Test

Figure 13–21 shows another troubleshooting method that can be used to verify the condition of a thyristor. An SCR is illustrated in the drawing, but the method can be applied to other thyristors as well. To use the method, the thyristor must be removed from the circuit. An analog ohmmeter (e.g., VOM) is connected across the main terminals of the thyristor. If the device being tested is a unidirectional thyristor, then the polarity of the ohmmeter must be such that the thyristor is biased for forward conduction. For example, the ohmmeter in Figure 13–21 is connected such that the anode of the SCR is positive and the cathode is negative. The gate is left open for this portion of the test. The ohmmeter should read a very high ($\approx \infty$) resistance on the R × 1 scale. If it indicates a low resistance, the thyristor is defective.

FIGURE 13–21 An ohmmeter can be used to test the condition of many thyristors.

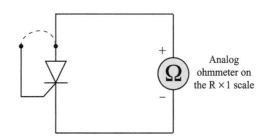

Assuming the thyristor indicates a high resistance between its main terminals, leave the ohmmeter connected and momentarily short the gate to the anode (unidirectional device) or to MT_2 (bidirectional device). If the measured resistance drops and stays low even after the gate connection is opened, then the thyristor can be assumed to be good. If the resistance drops, but returns to a high value when the gate is opened, then the thyristor may or may not be good. The only way to know is to compare the test results with another device of the same type that is known to be good. Finally, if no change in the resistance is noted, the thyristor is probably defective.

This test relies on the internal batteries of the ohmmeter to provide power to the thyristor. The resistance of the meter circuit acts as a load. The technique works best with an analog meter (VOM), since this device typically has higher voltages and currents available at the probes. Digital meters may not be able to supply adequate current to trigger or sustain the thyristor conduction. Although the test is simple, it can be an effective tool for verifying your suspicions when troubleshooting a thyristor circuit.

Thyristor Noise Problems

Since thyristors turn on quickly, the waveshapes of voltage and current in the circuit contain high rates of change. High edge rates in the time domain translate directly to high-frequency harmonics in the frequency domain. For example, when a thyristor is used in a phase-control circuit, the sharp turn-on transitions create high-frequency harmonics. These are often strong enough to create significant interference to nearby communication devices. Although the harmonic-related noise does not generally extend into FM radio and television bands, it does generate high levels of interference in the short-wave and AM radio bands. This radio-frequency interference (RFI) is normally suppressed by using a low-pass filter to isolate the ac power source from the thyristor circuit. The filter often takes the form of an inductor in series with the load and a capacitor directly across the incoming power lines.

✓ Exercise Problems 13.3

1. Defects in thyristor circuits can generally be classified into one of two groups. Name the two classes of defects.
2. If a thyristor load is fully on at all times and the thyristor is not defective, what else could cause the problem?
3. If a thyristor develops a main terminal short, the load will be full on at all times. (True/False)
4. If you remove (i.e., open) the gate connection to a thyristor but the load remains fully on, then the thyristor is probably defective. (True/False)
5. The condition of some SCRs can be checked with an ohmmeter. What general class of meter works best for this type of test?
6. If you are troubleshooting a thyristor circuit whose main problem is interference to nearby radio receivers, which of the following are possible causes for the interference?
   a. The gate is not being triggered (defective trigger circuit).
   b. Filter components on the incoming ac line are defective.
   c. The load is open.
   d. The thyristor is open.
   e. The thyristor is shorted.
7. Refer to Figure 13–22. If the thyristor is good, describe the indications on the ohmmeter for the polarity shown and for the gate (both connected and open as indicated).

FIGURE 13–22 What will the ohmmeter read?

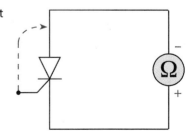

13.4 OPTOELECTRONIC DEVICES

Optoelectronic devices are electronic components that either produce light or respond to light. These devices are used extensively in nearly every field of electronics. Although some of the terminology and units of measurement used to describe optoelectronic devices are unique, many of the operational characteristics of the devices are similar to the operation of other solid-state devices such as diodes and transistors. In this section, we will focus on learning the operational characteristics of several solid-state optoelectronic components. Optoelectronic devices can be broadly grouped into three major categories: light-emitting, light-sensing, and combination devices.

Light-Emitting Devices

We will now examine the operational characteristics of two important semiconductor devices that are designed to emit light waves: standard light-emitting diodes and laser diodes.

Light-Emitting Diodes

A light-emitting diode or LED is a pn junction that releases light energy when it is forward biased. LEDs are available that emit infrared light and visible light. Within the visible light spectrum, devices are manufactured that emit red, blue, orange, green, and yellow light. An infrared (IR) LED is normally used as a light beam for communications (such as a TV remote control or the IR communications port on your computer) or for beam-interruption detection (such as a burglar alarm). Visible LEDs, by contrast, are generally used as indicators or displays. Figure 13–23(a) shows some representative light-emitting diodes. Figure 13–23(b) shows the schematic symbol.

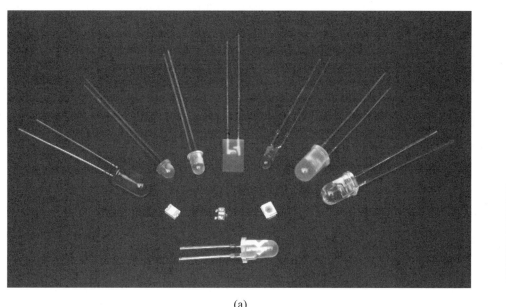

(a)   (b)

FIGURE 13–23 (a) Some representative light-emitting diodes (LEDs) and (b) the schematic symbol for an LED.

Since an LED is a diode, it must be properly connected into the circuit. For leaded devices such as those shown in Figure 13–23(a), the anode generally has a longer lead. Alternatively, some manufacturers put a flat side on the rim of the plastic dome to identify the cathode terminal. In any case, the diode must be connected into the circuit such that it is forward biased in order for it to emit light.

When it is forward biased, an LED generally drops 1.4 to 2.4 V across it. Current requirements vary between different LEDs, but 10 to 50 mA is a representative range. LEDs must always have a series current-limiting resistor to prevent damage. Never connect them directly across a voltage source. Although this caution is always true, it should be noted that some LED packages contain an integral series resistor. In these cases, no external resistor is required.

Laser Diodes

The term *laser* is an acronym of **L**ight **A**mplification by **S**timulated **E**mission of **R**adiation. The origin of the name will be more meaningful after we discuss laser fundamentals. We will restrict our discussion of lasers to the operation of a solid-state component called a laser diode. The laser diode is the most widely used laser source and has applications that include:

- Sensing the recorded information (music) on a compact audio disk
- Optical communications systems
- Laser printers
- Telemetry
- Remote alignment devices
- Bar code readers
- Surveying equipment
- Reading (and, in some cases, writing) data on optical storage disks for computers
- Transmission of cable television signals

Laser diodes are similar to standard LEDs in several ways:

- They are basically pn junctions.
- They emit light when forward biased.
- The wavelength of the light depends on the materials used in the diode.

How the light is produced in a laser diode, however, is an important difference between it and LEDS. We must understand the emission mechanism before we can appreciate some of the unique characteristics of a laser diode.

Laser Diode Fundamentals Previous chapters have revealed that when an electron moves between energy levels—such as from the conduction band to the valence band—energy must be absorbed or released. If an electron absorbs energy, it moves to a higher energy level. If it releases energy, it can move to a lower energy level within the crystal. We now need to examine this process of moving between energy levels more closely. There are actually three possible energy-level transitions that are important to the operation of optoelectronic devices. They are illustrated in Figure 13–24.

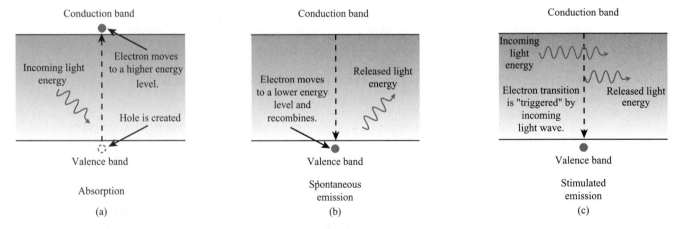

FIGURE 13–24 There are three possible types of transitions between energy levels that are important to the operation of optoelectronic devices.

In Figure 13–24(a), energy from an incoming light wave is absorbed by an electron in the valence band. This causes it to become excited and move to a higher energy level in the conduction band. This same basic process, called **absorption**, is what causes semiconductors to be sensitive to temperature. That is, heat energy can also be absorbed and cause electrons to move to the conduction band.

Figure 13–24(b) shows the inverse of absorption. Here, an electron in the conduction band gives up energy and recombines with a hole in the valence band. In the case of optoelectronic devices, we use materials that cause the frequency of the released energy to fall within the light spectrum. The process illustrated in Figure 13–24(b) is called **spontaneous emission**. This is the mechanism that is used in standard light-emitting diodes. The forward bias on an LED raises electrons to the excited state in the conduction band. As they cross over the pn junction, they recombine with valence-band holes and release light energy as illustrated in Figure 13–24(b). It is important to realize that the single electron transition represented in Figure 13–24(b) is being randomly replicated throughout a practical LED junction. That is, the transition of any given electron is unrelated to other transitions in the crystal.

Figure 13–24(c) illustrates yet another phenomenon that is important to the operation of optoelectronic devices and, in particular, laser diodes. Here, an electron in the conduction band makes its transition to the valence band and releases light energy in much the same way as described for spontaneous emission. The critical difference, however, is that the conduction-band electron is disturbed by an incoming light wave. This disturbance triggers the electron's transition to the valence band. Therefore, the released light energy is perfectly timed with the arrival of the incoming light energy. The net result is that one unit of light—called a **photon**—arrives in the region, but two photons of light energy leave the area. Additionally, and very importantly, the two exiting photons have identical wavelength (or frequency), amplitude, and phase. This process is called **stimulated emission**. It forms the basis for laser diode operation.

So, when a photon of light passes near an atom, either absorption or stimulated emission can occur depending upon the states of the electrons. In order for stimulated emission to produce practical sustained light emission, there has to be an abundance of electrons in the excited state. More specifically, there must be more atoms with electrons excited into the conduction band than atoms with complete bonds in the valence band. Since this condition is opposite the normal atomic condition, we call it **population inversion**. We can produce population inversion in a semiconductor diode if we use heavily-doped materials and operate the device with sufficient forward current values.

Figure 13–25 provides greater insight as to how a pn junction is constructed to form a laser diode. Initially, light energy that is emitted by spontaneous emission within the junction travels in random directions. Those light waves that travel

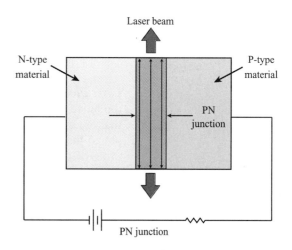

FIGURE 13–25 Light energy is reflected internally to the pn junction of a laser diode until stimulated emission contributes enough light energy to enable the beam to escape the crystal.

parallel to the junction are reflected at the surface of the crystal. As the light waves bounce between the surfaces of the crystal, they can trigger stimulated emission from excited atoms in the crystal. This amounts to amplification of the light energy, since each photon that causes stimulated emission ultimately produces two photons of light energy.

Any light energy that travels into the depths of the p- and n-materials is absorbed and converted to heat. Many of the light waves that are traveling neither parallel nor perpendicular to the junction are bent by the action of the materials used in the crystal. This bending tends to constrain the light energy to the junction area. The light energy continues to build within the junction as each reflected pass produces more photons by stimulated emission. Eventually, the light waves have sufficient energy, and a portion of the beam passes through the surface of the crystal. The illustration in Figure 13–25 shows the laser beam exiting from both ends of the crystal. Emissions from one end are allowed to escape the diode package, while emissions from the opposite are monitored by an internal photodiode used to control the light output from the laser diode. As the primary beam exits the crystal it begins to diverge (spread out). This diverging light detracts from the usefulness of the laser diode.

For most practical applications, we need to focus the laser light into a parallel beam. Figure 13–26 shows how an external lens system can bend the diverging light waves into a parallel beam. The process of converting a diverging light beam into a parallel beam is called **collimation**. Since the only light energy that escapes the diode was produced through stimulated emission, the resulting parallel beam consists only of light waves that have the same amplitude, frequency (or wavelength), and phase. This type of light energy is called **coherent light**.

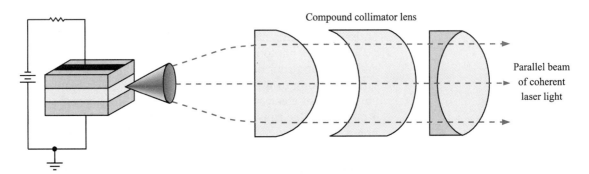

FIGURE 13–26 A collimator lens can convert a diverging light beam into a parallel light beam.

After passing through the collimator lens, the coherent light beam can travel great distances with minimal divergence. It can also be focused to a very fine point with an additional lens. For example, in order to read information from a disk in a standard compact disk (CD) player, the laser beam must be focused on a spot that is about one-millionth of a meter in diameter.

Laser Diode Characteristics The forward-voltage-versus-current curve for a laser diode is very similar to a light-emitting diode or any standard silicon diode except that the current values are typically much higher for a laser diode (can be several amps) than for a standard LED. The forward voltage drop, however, is still on the order of 1.4 to 2.4 V.

Laser diodes are available with power ratings that range from less than 1 mW to over 1 W. Stacked arrays of laser diodes can produce optical powers of several kilowatts. Devices are available that operate at wavelengths as short as 630 nm and as long as 1550 nm.

SYSTEM PERSPECTIVE: Laser Diodes

The adjacent illustration shows the functional block diagram of a typical laser diode control circuit. A photodiode (within the laser diode package) is used to sense the actual power output of the laser by monitoring a sample of the beam. The photodiode feedback is combined with two other inputs in an error amplifier. One of the inputs is a dc control voltage that establishes the normal operating power of the laser diode. The second input is a modulation or intelligence information that is added to the laser beam. Modulation frequencies well into the gigahertz range are possible with laser diodes. The output of the error amplifier in the figure passes through a current amplifier, which ultimately drives the laser diode. The figure also shows a control circuit to monitor and control the temperature of the laser diode. A thermistor mounted in thermal contact with the laser diode package provides feedback for the temperature control circuit. The feedback is combined with a dc voltage representing the desired operating temperature in an error amplifier. The output of the error amplifier controls the current through and temperature of a thermoelectric (TE) cooler, which is also mounted in thermal contact with the laser diode package.

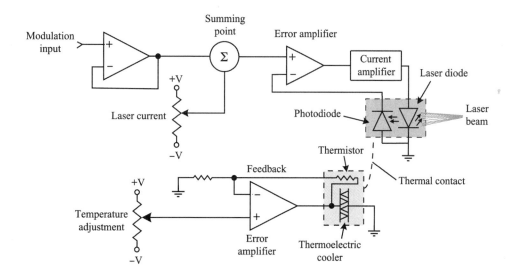

Laser Diode Application Although laser diodes are similar in many ways to LEDS and standard rectifier diodes, they do require special consideration for successful application. First, they are very susceptible to damage from excessive power dissipation. Improperly designed power supplies are a common cause of laser diode damage due to excessive power dissipation. When a standard power supply is turned on or off, it can produce current transients capable of destroying the laser diode. You should be certain to use only power supplies that are specifically designed for use with laser diodes. These supplies always have current limits to provide transient-free operation. They also act as a variable-current source, since the output power of the laser diode is proportional to the forward current. An associated cause of laser diode damage is loose or intermittent connections. If a connection opens momentarily, then the output of the current-source driver goes to maximum. When the connection is reestablished, the laser diode may be subjected to a transient current surge.

A laser diode can also be damaged from an electrostatic discharge (ESD) that lasts only a few nanoseconds. They are even more susceptible to ESD damage than many MOSFET devices because there is no internal protection in the laser diode package. Damage from ESD events is the primary cause of premature laser diode failure. This requires that you take care to avoid touching the laser diode with ungrounded objects (including the human hand).

Under normal operating conditions, the laser diode generates substantial heat. If the heat is not controlled, then the performance will degrade and the device may be damaged. Laser diodes generally require some type of cooling mechanism ranging from a simple heat sink to a very elaborate cooling system.

Another operational consideration you must always remember is that even low-power laser diodes can cause retinal or corneal damage to the human eye, even if the emissions are at a wavelength that is invisible. You must use extreme care to avoid injury either directly or indirectly. Following are a few common sense precautions regarding laser safety:

- Never look directly into a laser beam regardless of the power output.
- Never look into the end of a fiber-optic cable (discussed in a later section) that is powered by a laser.
- Never disable safety interlocks that are designed to prevent accidental exposure to a laser beam.
- Never allow a laser beam to travel into an uncontrolled area. It could be reflected back into your eyes.
- Be sure everyone in the immediate vicinity knows when a laser is operating in a potentially dangerous way, such as during a troubleshooting exercise.
- Use special safety goggles that pass visible light but block the laser light.

These warnings probably seem like common sense to you, but you must make a deliberate effort to avoid serious mistakes. For example, a fiber-optic communication cable is essentially a flexible glass fiber that is used to pass a laser beam. They are commonly used with computers and other systems in industry. Many fiber-optic cables have a protective outer sheath that makes them look the same as a standard electrical cable. The screw-on connector for many fiber-optic cables looks just like a standard electrical connector. If you unscrewed the connector—accidentally or intentionally—during a maintenance or repair operation, the laser beam would be free to shine out into the surrounding area.

A sign displayed in one laser laboratory expresses the seriousness of this danger in a humorous way. The sign reads, "Do not look into the laser with your remaining good eye."

Light-Sensing Devices

In this section, we will discuss the operation of several semiconductor devices that are designed to act as light sensors. For many applications, a light sensor (discussed in this section) is used in conjunction with a light emitter (discussed in the preceding section).

Photodiode

A pn junction can be used to sense the presence and relative intensity of light. Figure 13–27 illustrates the basic operation of a light-sensing diode or photodiode.

As shown in Figure 13–27, the diode is reverse biased. You will recall that one of the primary components of reverse current is due to the creation of electron-hole pairs in the depletion region. In ordinary diodes, the electron-hole pairs are thermally produced. You will remember that the reverse current increases with temperature. In the case of a photodiode, light energy is allowed to pass through the thin p-type portion of the crystal and is absorbed in the depletion layer. The absorption of light energy creates electron-hole pairs. Because of the strong electrostatic field in the depletion layer, the newly freed electrons are swept toward the positive source terminal and the holes are *effectively* moved toward the negative terminal. The greater the intensity of the incoming light energy, the greater the reverse current flow through the diode.

FIGURE 13–27 A pn junction will respond to incident light energy.

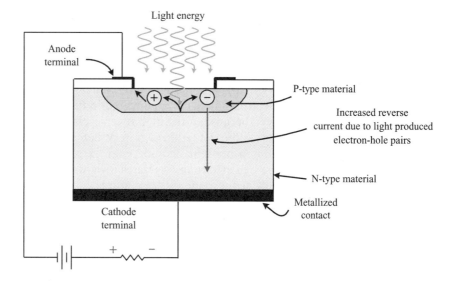

It is interesting to note that any pn junction will respond to light energy, just as it will also respond to thermal energy. However, standard diodes are generally packaged in a way to minimize the absorption of light energy (e.g., an opaque package), whereas photodiodes are specifically designed to allow light energy to reach the junction. Figure 13–28 shows three schematic symbols that can be used to represent a photodiode.

FIGURE 13–28 Three schematic symbols for a photodiode.

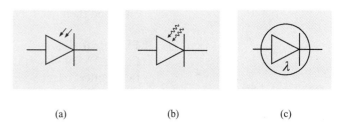

A pn junction photodiode like the one illustrated in Figure 13–27 can be operated in either of two modes. The mode shown in Figure 13–27 is called the **photoconductive mode**. That is, the conductivity of the diode changes with varying light energy. A photodiode can also be used with no bias. Even when the diode has no bias voltage, the light energy still creates electron-hole pairs in the depletion layer. They are still swept away from each other by the electrostatic field of the depletion layer. (Remember the barrier potential in an unbiased diode?) With the negative charges (electrons) going in one direction and the positive charges (holes) going in the opposite direction, a voltage source is created. That is, the photodiode is acting like a battery whose value is dependent upon the amount of light striking the pn junction. When operated in this manner, the diode is operating in the **photovoltaic mode**.

As a general rule, the photoconductive mode offers the following advantages relative to the photovoltaic mode:

- Improved stability
- Faster response times
- Less sensitivity to temperature with moderate light levels
- Response to a wider range of light intensities

The primary disadvantage of the photoconductive mode is that with no light striking the photodiode a current still flows. This current is called the **dark current** and is highly temperature dependent, since it exists due to thermally-produced electron-hole pairs.

Avalanche Photodiodes

An avalanche photodiode (APD) is a special type of silicon photodiode. It has an extremely high reverse-breakdown voltage with 2.5 kV being a typical value. The APD is operated under reverse-bias conditions. The normal reverse voltage is less than the reverse breakdown voltage, but is still very high (> 2 kV). When light energy strikes the junction of an avalanche photodiode, photons are absorbed and electron-hole pairs are created in the same way as in standard photodiodes. However, because of the high reverse-bias voltage, there is a very intense electric field across the depletion region. When an electron is freed as a result of absorbing a photon of light, it undergoes extreme acceleration from the electric field. As the free electron accelerates toward the positive terminal of the APD, it is likely to collide with one or more atoms. Each collision can cause the release of another electron. These secondary electrons are also accelerated by the intense electric field and can also have collisions that produce even more free electrons.

To put this action in perspective, let us look at the net results external to the APD. We see a single photon of light enter the junction area. As a direct result, we see as many as 500 to 1000 electrons exiting the diode and heading for the positive terminal of the power source. Clearly, the APD is substantially more sensitive than a standard photodiode. In essence, the APD has a built-in amplifier with a current gain of 500 to 1000.

Avalanche photodiodes are typically used in low-light applications. One specific example is laser ranging. In this application, a laser diode sends out a beam of laser light. It is reflected off of some distant object, and the reflected light is detected by an APD. Since light travels approximately one foot every nanosecond, we can measure the time it takes for the laser beam to travel to the distant object and back and then use this measured time to compute the actual distance to the object.

Phototransistor

Figure 13–29 shows the schematic symbol for a phototransistor. The functions of the emitter, base, and collector are identical to a standard npn transistor. The primary difference is that there are two ways of providing the base current needed for forward bias. First, the base lead can be biased using any of the biasing methods described for standard transistors. Allowing light energy to reach the pn junction can also provide forward bias. In a manner similar to a photodiode, the phototransistor package is designed to permit light to reach the pn junction. The light is absorbed, which creates electron-hole pairs. As the newly created carriers are swept out of the base region, they form current. The remaining portions of the transistor respond in the usual way. That is, regardless of how the current flow in the base region was created, the flow of electrons produces familiar results.

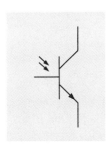

FIGURE 13–29 The schematic symbol for a phototransistor.

Figure 13–30 shows how a phototransistor might be used in conjunction with a light source to form an intruder-alarm circuit. As long as the phototransistor can "see" the light source, it has base current and remains in saturation ($V_{CE} \approx 0$ V). When an intruder walks between the light source and the phototransistor, the light beam is broken. This means that the amount of light striking the junction of the phototransistor decreases substantially. This causes the transistor to enter cutoff. When the transistor is cutoff, its collector voltage rises to the V_{CC} level (+5 V in this example). The output from the phototransistor can be used to activate an audible alarm, increment a counter, or perhaps serve as an input to a computer circuit. Phototransistors are also used in applications such as VCRs, floppy disk drives, photocopiers, smoke detectors, cassette players and recorders, and video cameras.

Light-Activated SCR

Figure 13–31 shows the schematic symbol for a light-activated SCR (LASCR). As you might expect, its operation is identical to that of a standard SCR with one exception. In addition to the usual ways of triggering an SCR, the LASCR is triggered when light energy strikes its gate-cathode junction. The LASCR package

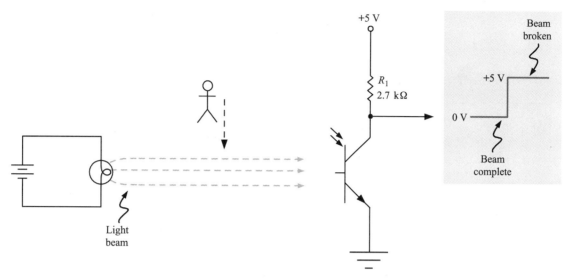

FIGURE 13–30 A phototransistor can be an integral part of an intruder alarm system.

FIGURE 13–31 The schematic symbol for a light-activated SCR (LASCR).

provides a means for light to reach the gate-cathode junction. With no light reaching the junction, the LASCR is off. When light strikes the junction, it is absorbed, causing generation of electron-hole pairs. As the electrons are swept away from the junction, they form gate current, which causes the LASCR to fire and enter its on state. As with standard SCRs, the LASCR will remain in the on state until the anode current falls below holding current.

Combination Optoelectronic Devices

The preceding sections discussed the two primary classes of optoelectronic devices: light emitters and light sensors. This section introduces two more devices that are actually emitter-sensor combinations called optoisolators and optointerrupters. Both of these devices are used in computer applications.

Optoisolator

An optoisolator—also called an optocoupler—is a device that houses a light emitter (generally an infrared LED) and a light sensor (generally a phototransistor) in the same package. Further, the light source and the light sensor are positioned such that the light from the emitter shines directly on the sensor. Also, the package for the optoisolator is opaque, which means that no outside light can reach the sensor. Figure 13–32 shows the schematic symbol for an optoisolator.

The operation of the optoisolator is precisely as you would expect based on the schematic symbol. The LED portion of the optoisolator is a standard LED and has all of the electrical characteristics of a discrete light-emitting diode. Similarly, the phototransistor performs like a discrete phototransistor. What makes the LED and phototransistor unique is that they are mounted within the same package and are optically linked to each other.

Figure 13–33 on page 498 shows how an optoisolator might be used in a practical application. Here, a manufacturing process is located a considerable distance away from an associated monitoring and/or control circuit such as a computer system. A switch closure (perhaps a limit switch) associated with the process must be sensed. As shown, the remote switch is in series with a dc power source and the LED portion of the optoisolator. When the switch is closed, current flows through the LED. Light from the LED strikes the phototransistor (internal to the optoisolator package) and causes it to go into saturation. The collector voltage on the phototransistor indicates the state of the remote switch (i.e., near 0 V = closed switch and +5 V = open switch).

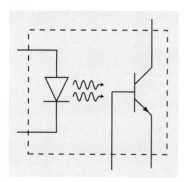

FIGURE 13–32 The schematic symbol for an optoisolator.

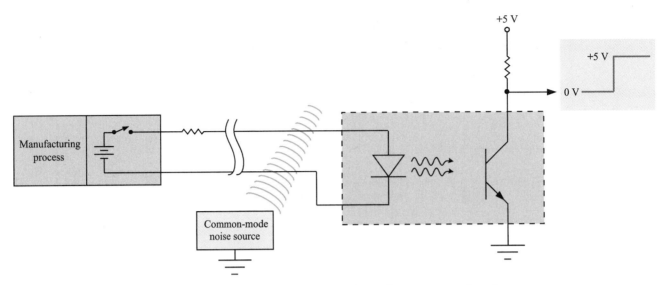

FIGURE 13–33 An optoisolator is often used to eliminate the effects of common-mode noise.

In a typical manufacturing environment, there are many devices that can cause high levels of common-mode noise. Common-mode noise consists of unwanted electrical signals that are coupled into both sides of a circuit. Some typical noise sources include nearby wires in the same conduit, electrical motors, welders, and other high-current devices. Even 60-Hz radiation from electrical wiring can cause problems if the cables are long and the input impedance is high. In any case, if common-mode noise is coupled to the wires connecting the optoisolator to the remote switch, there is no effect. That is, the LED responds only to the voltage difference between its cathode and anode terminals. If the voltage on both terminals is increased by 50 V, for example, due to a common-mode noise transient, the LED is unaffected. Now, if the LED is unaffected, then the phototransistor and all subsequent circuitry are unaffected.

Contrast this with the more obvious method of using the remote switch to provide voltages directly to a control circuit. If a 50-V transient were coupled directly into the input circuit, it would respond as a minimum, and it would very possibly be damaged. Many optoisolators have isolation between the LED circuit and the phototransistor circuit of more than a thousand volts. These devices are very often used to protect sensitive computer input circuitry against damage from common-mode voltage transients on the input lines. Optoisolator packages often look like a standard 6-pin IC package.

Some manufacturers also make optoisolators that have SCRs or triacs instead of the output transistor. The operation is similar to the standard optoisolator except that the output device is a thyristor. When the output device is a triac, the overall package is often called a solid-state relay (SSR). The LED leads are analogous to the coil leads of a standard relay, and the MT_1 and MT_2 connections to the triac are analogous to the contacts of a relay. The optoisolator provides isolation between the two circuits just as an electromechanical relay would do. Clearly, an SSR of this type will only operate with an ac load. If a dc load were used, then the triac would never turn off once it was triggered.

Optointerrupter

An optointerrupter is electrically identical to an optocoupler. The differences are physical and optical. More specifically, the light source and sensor in an optocoupler are optically coupled inside of a sealed package. The light source and sensor for an optointerrupter, by contrast, are optically coupled under some conditions, but the optical link is broken under other circumstances. When the optical link is complete, an optoisolator and an optocoupler have essentially identical performance. When the optical link is broken in the optoisolator, the output transistor is cutoff.

There are two general classes of optointerrupters: transmissive and reflective. Figure 13–34(a) shows a transmissive photointerrupter. The package between the LED and phototransistor in these devices is slotted. The LED shines light across the slot to the phototransistor via optical windows. If any opaque object passes through the slot, then the light beam is blocked. By connecting a dc source to the LED circuit and monitoring the output from the phototransistor, we can obtain a voltage level that indicates whether or not an object is in the slot. Many applications require an indication of the speed of a shaft or the relative timing of a mechanical system. By positioning the optoisolator such that the slot straddles the teeth on a sprocket, we can determine the speed of the sprocket. Similarly, if a rotating shaft has a protruding tab that can pass through the slot of an optointerrupter, we can determine a reference or home position for timing purposes. This same basic scheme can be used in a floppy disk drive to detect the position of the write-protect tab on a diskette.

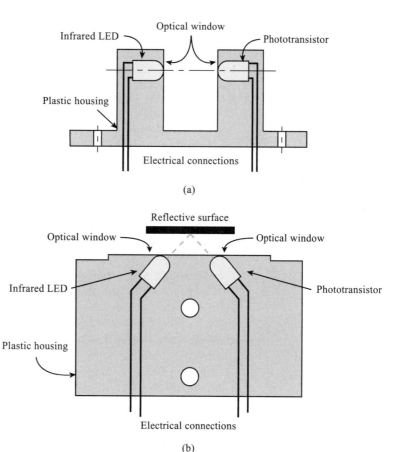

FIGURE 13–34 There are two classes of optointerrupters: (a) transmissive and (b) reflective.

Figure 13–34(b) illustrates a reflective optointerrupter. Here the LED and phototransistor cannot "see" each other directly, but they are mounted inside a common package. If a reflective object is brought within the range of the optointerrupter, then light from the LED strikes the reflective surface and bounces back to the phototransistor. The electrical operation is similar to the transmissive type of optointerrupter. These devices are often used to sense the presence of paper in a copy machine or printer. They are also used in some cassette players to detect a white strip on the end of the tape and subsequently activate the automatic-reversal mechanism.

Fiber Optics

This section offers an introduction to fiber optics. Although fiber optics is not a semiconductor device, it does play a key role in many applications that utilize semiconductor optoelectronic components.

Refraction of Light

For many purposes, we consider the speed of light to be a constant (300×10^6 meters per second). However, the actual speed that light travels through a material varies inversely with the optical density of the material. For example, light travels slower through water (more dense) than it does through air (less dense). When a light wave passes from a material with one optical density to a material with a higher or lower optical density at any angle other than perpendicular, the light wave is bent. That is, its angle of motion is altered. The bending of the light beam is called **refraction**. Figure 13–35 illustrates the refraction of a light wave. The ratio of the speed of light in a vacuum to the speed of light in a material is called the **index of refraction** for the material. Refraction is the key to the operation of fiber-optic cable.

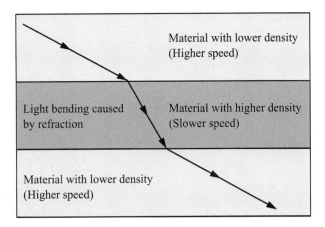

FIGURE 13–35 The bending of light as a beam travels between two materials with different optical densities is called refraction.

Construction and Operation

Fiber optics is the transmission of light through thin strands or fibers of glass. Although the glass fiber is not a semiconductor device, it is relative to our discussion of optoelectronic devices since it plays an integral role in many optical systems. Figure 13–36 illustrates the basic construction and operation of a fiber-optic cable.

Figure 13–36(a) shows that a fiber-optic cable consists of a glass core surrounded by a different type of glass or plastic called cladding. A protective outer covering encases both of the inner layers. Figure 13–36(b) shows how light is able to travel through the glass core. The inner glass core is made of ultrapure fused silica. The material must be exceedingly pure to minimize transmission losses. The index of refraction for the glass fiber is higher than the index for the cladding layer. When a light wave encounters the interface between the two materials, it is refracted as described in the previous section. As long as the angle of the incident light is greater (i.e., more in line with the axis of the fiber) than an angle called the **critical angle**, the light wave will be reflected back into the core. The light waves continue down the fiber core as they are reflected off of the core/cladding interface. The interface continues to refract and reflect the light even if the cable is bent (with a gradual radius). Essentially 100 percent of the light is constrained to the glass core. The amplitude of the light is reduced only slightly (fractional dBs per mile) due to imperfections and impurities as it bounces its way along the cable.

FIGURE 13–36 (a) The construction and (b) operation of a fiber-optic cable.

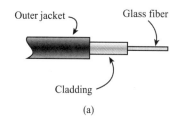

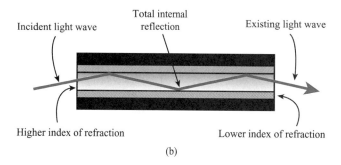

Application

Generally, a laser diode operating in the infrared range is used as the light source on the transmitter end of the fiber-optic cable. On the receiving end of the cable, the light is detected with an avalanche photodiode. The information is coded into pulses of light. The pulses can occur at rates as high as several billion per second. The amount of information that can be transmitted in this manner is astounding. For example, if the pulses on a particular fiber used to carry telephone signals occur at a 90-MHz rate, the fiber link is capable of carrying 1,340 *simultaneous* conversations. In the case of fiber-optic telephone links, many fiber links are bundled into a single physical cable. Amplifiers are placed every five or six miles along the cable to prevent deterioration of the signal.

Fiber optics are used for many other applications, but the uses can be divided into two general classes. One method of application was described previously; the cable is used as a link between a transmitter (laser diode) and a receiver (avalanche photodiode). The light beam carries encoded information. The other application class uses the fiber optics to transmit light, but no information is coded into the light beam. This includes a broad range of applications such as illumination of hard-to-reach places, imaging of relatively inaccessible places (such as inside the human body), and instrumentation illumination in automobiles.

Troubleshooting Optoelectronic Devices

Because optoelectronic devices are used in nearly every field of electronics, it is essential that you be able to effectively determine the condition of a particular component. Many optoelectronic components (including photodiodes, phototransistors, and LEDs) have basic characteristics that are similar to their nonoptical counterparts. In these cases, many defects can be found by using methods previously discussed. For example, the same basic troubleshooting procedures required to classify a rectifier diode as good or bad can often be used to classify the condition of an LED or photodiode. This section focuses on additional troubleshooting methods that are unique to optoelectronic devices.

Troubleshooting Light Emitters

Light emitters can generally develop three classes of defects: open, short, and reduced output. Clearly, the open and short conditions would result in no light output. Both open and short conditions in an LED—either standard or laser—can be detected with an ohmmeter. In the case of a laser diode, however, care must be exercised to avoid subjecting the diode to damaging transients. Additionally, the reverse breakdown voltage of most laser diodes is quite low, so be certain that the open terminal voltage of the ohmmeter is less than the breakdown voltage of the laser diode. In general, use the same basic precautions that are recommended for testing MOSFET devices.

The most definitive way to verify that a light emitter has insufficient light output is to replace it with a known good device. Test equipment manufacturers do make light monitors that can measure the actual light output from a light emitter. In most practical troubleshooting situations that you are likely to encounter, however, there will be no access to this type of test instrument. Therefore, substitution is generally the fastest procedure.

Another malfunction that can occur in an optical system is optical misalignment. Consider the case where the light output from an LED is supposed to shine on a phototransistor in a light-beam sensor application. If the mechanical alignment of the beam is incorrect, then the system will not work even though there are no electrical defects in the system. If the LED emits visible light, then alignment is rather straightforward and obvious. In most cases, however, the LED or laser diode for this type of application emits invisible infrared light. This makes alignment problems more difficult to detect. One good way to perform infrared beam alignment (or to judge beam characteristics such as shape and size) is to use infrared display cards. The card is placed so that it intercepts the infrared beam. When the infrared light strikes the display card, it activates an infrared-sensitive phosphor within the card. The energy of the absorbed infrared light energy is converted to visible light. This allows you to see where the beam is positioned and to make adjustments accordingly.

Troubleshooting Light Sensors

Light sensors such as photodiodes, phototransistors, and light-activated SCRs all have standard counterparts. Troubleshooting these devices can generally be done by using the same tests that would be applicable for a standard diode, transistor, or SCR. Exceptions to this rule include insensitivity and incorrect beam alignment. You could suspect an insensitive light sensor (or a light emitter with low output) if the circuit works when the distance between the source and sensor is reduced. As with low-output emitters, the most definitive way to confirm that a light sensor has become insensitive is to substitute a known good sensor.

Troubleshooting Fiber Optics

Since fiber-optic cable is made of glass, it can be broken if sharply bent, pinched, or subjected to other forms of mechanical stress. When the fiber is damaged, light can no longer travel from one end to the other. The system-level symptoms vary with the nature of the system, but in any case, the path provided by the optic cable will be interrupted. If you have easy access to both ends of the cable, then there is a simple and effective way to verify the integrity of the cable. Simply shine a flashlight or other light source on one end of the cable and look for it to exit at the other end. The end of the fiber will appear to be illuminated. Previously, you were advised *never* to look directly into a laser because of the potential for permanent eye damage. That advice should be extended to include fiber-optic cables. That is, *never* look directly into the end of a fiber-optic cable. Clearly, you would not expect to be injured by the light from a flashlight during the troubleshooting exercise previously described. However, if you inadvertently look at the cable end while a laser diode is driving the other end, you could receive serious eye damage. There is another way this situation could occur. Many systems have multiple fiber-optic cables terminating into one computer or equipment enclosure. You may shine a flashlight onto one cable, but when you check the other end, you may accidentally choose the wrong cable and expose your eyes to laser light. And finally, even if you are extremely careful, someone else may notice the cable is off on the transmit end and replace it while you are peering into the other end expecting to see light from a flashlight. So, the simple rule to avoid this type of injury is: Do not look directly into the end of the fiber-optic cable.

If the system you are troubleshooting uses infrared light within the cable, you can make use of infrared display cards to detect the presence of the light beam. Simply place the end of the fiber cable close to the card and look for indications that the light beam is striking the card.

✓ Exercise Problems 13.4

1. A light emitting diode must be (*forward/reverse*) biased to illuminate.
2. The cathode lead of an LED is sometimes (*shorter/longer*) than the anode lead for identification purposes.
3. Under normal operating conditions, an LED could have a forward voltage drop of about 1.5 V. (True/False)
4. Laser diodes must be reverse biased to emit coherent light energy. (True/False)
5. When the path of a photon intersects with an atom and causes a valence-band bond to be broken, the process is called _____.
6. Laser light is produced by a process known as spontaneous emission. (True/False)
7. What happens to the light energy that is produced in a laser diode but travels into the depths of the semiconductor material?
8. How many watts of light energy does a laser need to produce before you should avoid looking directly into the beam with your naked eye?
9. A photodiode is frequently operated with forward bias. (True/False)
10. An avalanche photodiode is generally operated with a reverse bias of over 2000 V. (True/False)
11. A phototransistor gets closer to saturation as the arriving light waves become more intense. (True/False)
12. A LASCR would be a good choice for an intruder alarm where breaking the light beam would sound an alarm. (True/False)
13. When the LED portion of an optocoupler has no current flow, the phototransistor portion is in cutoff. (True/False)
14. The LED and phototransistor in a typical optocoupler are separated by an opaque barrier to provide isolation. (True/False)
15. An optoisolator is often used to minimize the effects of common-mode noise on long cable runs. (True/False)
16. There are two basic types of optointerrupters called _____ and _____.
17. The core of a fiber-optic cable is made from a highly conductive silver or gold thread. (True/False)
18. The inner core and the cladding of a fiber-optic link must have different indexes of refraction in order to contain the light. (True/False)
19. The condition of an LED can be verified with an ohmmeter. (True/False)
20. What is the purpose of an infrared display card?
21. When is it advisable to look directly into the end of a fiber-optic cable?

•—SUMMARY

This chapter discussed the general operational characteristics of the thyristor family of devices. In most cases, thyristors are high-current and sometimes high-voltage devices. They are bistable components in that they are either fully off or fully on. They exhibit negative resistance during turn-on. Some thyristors are unidirectional, while others are bidirectional. Any thyristor can be switched from its off state to its on state by exceeding its breakover voltage rating or by excessive rates of change of voltage across the device. Some thyristors have one or more gate terminals in addition to the primary current elements. These devices can be made to switch from off to on at a voltage lower than the normal breakover voltage by applying a voltage to the gate. This is the normal triggering mode for most three- and four-terminal thyristors. Regardless of how it was initially turned on or fired, a thyristor will remain on until its primary current falls below a value called holding current. With only a few limited exceptions, the gate of three- and four-terminal thyristors loses control once the device has been fired.

We also presented the general characteristics of several members of the thyristor family including the following devices: diac, SBS, SCR, SCS, triac, Shockley diode, GTO, SIDAC, and quadrac. Thyristors like the diac, SBS, SIDAC, SUS, and SBS are primarily used to trigger larger thyristors such as SCRs and triacs. SCRs and triacs are often used in phase-control applications. Here, the triggering of the thyristor is synchronized or timed with the ac power line waveform. The average current through the thyristor (and its series-connected load) is determined by the relative timing of the gate pulse to the thyristor. Troubleshooting methods for locating defective thyristors were also presented.

A number of optoelectronic devices were discussed including light emitters, light sensors, and fiber optics. The operational characteristics of light-emitting diodes (LEDs) and laser diodes were shown. Light from LEDs and laser diodes may be visible, or it may be in the infrared range. Laser light is coherent and very intense. Exposure of the naked eye to laser light can cause permanent eye damage. Light from standard LEDs is produced by spontaneous emission, whereas laser light requires stimulated emission.

Photodiodes, avalanche photodiodes, phototransistors, and LASCRs were discussed as representative light sensors. Photodiodes can be operated with reverse bias (photoconductive mode) or with no bias (photovoltaic mode). Avalanche photodiodes require very high values of reverse bias, but they provide effective current gain, which makes them suitable for low light-level applications. LASCRs work just like standard SCRs, except that triggering can also be accomplished by exposing the optical window to light. Phototransistors work like standard bipolar transistors, except that they can be turned on by exposure to light instead of (or in addition to) normal base current. Optocouplers (optoisolators) and optointerrupters are devices that contain both a light emitter and a light sensor in the same package. The light path between source and sensor is sealed for an optocoupler but is exposed for an optointerrupter.

Fiber-optic cable consists of a core of ultrapure fused silica wrapped in a glass or plastic cladding and a protective outer covering. Due to refraction at the interface of the core and the cladding material, light is constrained to the core. That is, light that enters the core at one end will travel the length of the cable with very low losses. In most cases, laser diodes are used to drive the cable, and sensitive photosensors are used to detect the light at the receiver end of the cable.

•—REVIEW QUESTIONS

Section 13.1: Thyristor Characteristics

1. Many thyristors are classified as high-current devices. (True/False)
2. How many stable states does a thyristor have?
3. If the forward voltage on the main terminals of a thyristor is increased until the device switches to the conducting state, what is the name used to describe the value of voltage that was required?
4. The majority of thyristors drop 1 to 2 V across their main terminals when they are in the conducting state. (True/False)
5. Once a thyristor is switched on, the voltage across it is directly proportional to the current through it. (True/False)
6. By examining the characteristic current-versus-voltage curve for a device, how can you tell it exhibits negative resistance?
7. A thyristor that conducts current in only one direction is called _____, whereas a thyristor that passes current in both directions is called _____.
8. One common way to turn on a thyristor is to exceed its reverse breakdown voltage. (True/False)
9. When does a thyristor have the least voltage across it?
   a. immediately before it fires
   b. once it is fully on
   c. during reverse breakdown
   d. after it reverts to the off state
10. If either the forward breakover voltage or the reverse breakdown voltage rating of a thyristor is exceeded, the current through the device will increase. (True/False)
11. Which of the following phenomena is likely to cause damage to a thyristor?
    a. forward breakover
    b. reverse breakdown
12. In order for a thyristor to remain on after the gate voltage is removed, the primary current must have exceeded a value called the _____ current.
13. Once a thyristor is fully on, the primary current must remain above a value called the _____ current, or the thyristor will revert to its off state.
14. If a thyristor has a gate terminal, the primary current is nearly always turned on and off by controlling the gate voltage. (True/False)
15. If the primary terminals of a thyristor are connected through the load to a dc source, then the dc current through the load can be varied by adjusting the value of gate voltage. (True/False)

Section 13.2: Thyristor Types and Applications

16. What is the primary application for a diac?
17. A diac is a (*unidirectional/bidirectional*) device.
18. When a diac fires, the voltage across it drops to about 1 to 2 V. (True/False)
19. An STS is a (*unidirectional/bidirectional*) device.
20. An SBS is a (*unidirectional/bidirectional*) device.
21. For many applications, an SBS is used as a two-terminal component. (True/False)
22. When an STS fires, the voltage across it drops to about _____.
23. In a phase-control application, why does the use of an SBS for a triggering device instead of a diac allow a greater range of firing angles?
24. An SCR has _____ terminals.
25. An SCR is a (*unidirectional/bidirectional*) device.
26. SCRs are generally considered to be high-current devices. (True/False)
27. The primary current in an SCR can be switched (*on/off*) under gate control, but it cannot be switched (*on/off*) with the gate.
28. SCRs can be used in phase-control applications. (True/False)
29. Triacs can be used in phase-control applications. (True/False)
30. The current through the load in a phase-control application can be (*decreased/increased*) by firing the thyristor earlier on each alternation.
31. If the load in a triac phase-control circuit is a lamp, what happens to the brightness of the lamp if the firing angle of the triac is changed from 100° to 150°?
32. Is it possible to use an SCR as the phase-control element in a circuit that requires current flow on both alternations of the ac sine wave?
33. How many gates does a triac have?
34. Thyristors are generally considered to be (*low-speed/high-speed*) devices.
35. What is the primary advantage of triacs when compared to SCRs?
36. How many gates does an SCS have?
37. Name two ways to fire an SCS under gate control.
38. What is the name of the bidirectional thyristor device that is actually a triac and a diac in the same package?

Section 13.3: Troubleshooting Thyristors

39. Under normal conditions, thyristors are either fully on or fully off. When they develop defects, the device is generally in one of these two states permanently. (True/False)
40. If the current through the load on a thyristor circuit is maximum at all times, the defect is definitely a bad thyristor. (True/False)
41. If a triac developed a short between MT_1 and MT_2, what would be the symptoms in terms of load current?
42. If a triac develops a short between MT_1 and MT_2, what voltage would be measured across the triac during a troubleshooting exercise?
43. If you disconnect the gate of a thyristor but the load continues to receive full current at all times, the thyristor is most likely defective. (True/False)
44. If the load current in an SCR phase-control circuit is zero at all times, shorting the gate to the cathode is a good test. (True/False)
45. Why do thyristor phase-control circuits sometimes interfere with AM radios?

Section 13.4: Optoelectronic Devices

46. LEDs fall into the category of light-_____ devices.
47. Laser diodes fall into the general class of light-sensing devices. (True/False)
48. All LEDs emit visible light, but laser diodes may emit either visible or infrared light. (True/False)
49. When a standard LED is forward biased, what is the approximate voltage dropped across the pn junction?
50. In order to emit light, an LED must be (*forward/reverse*) biased.
51. Why does an LED require a series resistor when connected across a voltage source?
52. A process known as _____ emission generates the light from LEDs.
53. When an electron enters the region of an atom and causes a valence electron to move to the conduction band, we call the process _____.
54. Laser diodes rely on a process called _____ emission to produce the laser light.
55. What term is used to describe the condition of a material that has more atoms with electrons excited into the conduction band than atoms with complete bonds in the valence band?
56. The diverging light from a laser diode can be formed into a parallel beam by using a _____ lens.
57. The light from an LED is called coherent light. (True/False)
58. Why should you avoid looking directly into the beam from a laser diode?
59. Why should you avoid looking directly at the end of a fiber-optic cable?
60. A photodiode is operated with (*forward/reverse*) bias.
61. A photodiode can be operated with no bias at all. (True/False)
62. What is the main advantage of an avalanche photodiode over a standard photodiode?
63. When no light is allowed to reach the pn junction of a photodiode, the small current that still flows is called the _____ current.
64. The base of a phototransistor has no external lead. (True/False)

65. With no base connection and no incoming light, a phototransistor will be in (*cutoff/saturation*).
66. With no previous trigger, no gate connection, and no incoming light, a light-activated SCR is (*off/on*).
67. Once a LASCR has fired, it can be turned off by momentarily interrupting the light beam. (True/False)
68. What is the name of the optoelectronic component that houses a phototransistor and an LED in the same package and exposes the common light path?
69. An optocoupler could be used in an optointerrupter application. (True/False)
70. A transmissive optointerrupter could be used in an optocoupler application. (True/False)
71. The _____ __ _____ for a material indicates how fast light can travel through the material relative to its speed in a vacuum.
72. The material that immediately surrounds the glass core of a fiber-optic cable is called _____.
73. Why are light waves constrained to the core in a fiber-optic cable?
74. How can you verify the mechanical alignment of an optical link that uses infrared light?
75. If you are using a flashlight to verify the integrity of a fiber-optic cable, you should never look directly at the opposite end of the cable. Why not?

•—CIRCUIT EXPLORATION

This exercise will enable you to explore one of the most widely used thyristors—the SCR. Construct the circuit shown in Figure 13–37 in your circuit simulation environment or with actual parts in the laboratory.

Once you have constructed the circuit, accomplish the following tasks:

1. With the switch open, measure the waveform across the SCR (i.e., between anode and cathode). Explain why it looks the way it does. Also explain the behavior of the lamp.
2. As the circuit simulation progresses, press and release the push-button switch anytime *during the negative alternation of the ac waveform*. Explain the effects (or lack of effect) on the waveform between anode and cathode of the SCR. Also explain the behavior of the lamp.
3. Now, press and quickly release the switch anytime *during the positive alternation of the ac waveform*. Explain the effect on the waveform between the anode and cathode of the SCR. Also explain the behavior of the lamp.

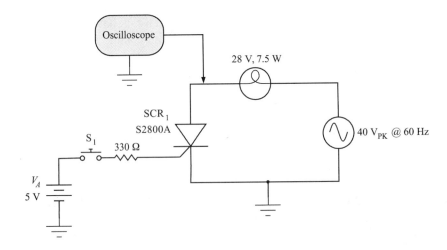

FIGURE 13–37 A circuit to explore the behavior of an SCR.

Fig13_37.msm

CHAPTER 14
Integrated Circuit Applications

•—KEY TERMS
crystal
decoupling capacitor
duty cycle
frequency stability
jitter
parasitic oscillation
piezoelectric effect
relaxation oscillator
transducer

•—OBJECTIVES
After studying this chapter, you should be able to:

- Describe the purpose of an oscillator circuit and explain the operation of each of the following oscillator circuits:

  Triangle generator   Wien-bridge oscillator
  555 timer   Oscillator module
  On-board oscillator with
     external crystal

- State the requirements for oscillation.
- Describe causes and cures for parasitic oscillations.
- Provide a functional description and give a specific example of the following types of integrated circuits:

  Voltage-to-frequency converters   Frequency-to-voltage converters
  Instrumentation amplifiers

- Explain the operation of a phase-locked loop and give an application example.
- Describe troubleshooting methods that can be used to localize defects in circuits that use integrated circuits.

•—INTRODUCTION
Oscillators and phase-locked loops are introduced in this chapter. These are two important types of circuits widely used in electronic products. We will examine the general operational characteristics of these circuits, and we will discuss representative circuits.

Integrated circuits (ICs) are used throughout industry to achieve complex circuit functions in relatively small and cost-effective packages. Some ICs perform digital functions, others perform analog functions, and some perform both digital and analog operations within the same physical package. This chapter will introduce you to voltage-to-frequency conversion, frequency-to-voltage conversion, and instrumentation amplifiers. These are all used for industrial computer applications. It is no more important for you to memorize the details for each type of IC than it is for you to memorize the electrical characteristics of every transistor that is manufactured. However, it is important for you to be aware of the wide range of ICs that are available.

14.1 OSCILLATOR CIRCUITS

Circuits and devices discussed so far in this text have accepted an input (such as a sine wave), performed some operation (such as amplification or voltage comparison), and generated an output. Additionally, all circuits with active components such as transistors and op amps require a dc power source for their operation. In this section, we will study oscillator circuits. They require a dc power source for their operation, and they generate an output waveform, but they require no input signal. That is, they create a signal waveform such as a triangle wave, a square wave, or a sine wave with no input other than the dc power supply. There are endless types of oscillator circuits, but they have many common requirements. We will first examine the basic requirements for oscillation and then explore several representative oscillator circuits.

Overview

Figure 14–1 shows a simplified functional block diagram of an oscillator circuit. The amplifier block can be a bipolar transistor, JFET, MOSFET, op amp, or any other device that can provide voltage gain at the required frequencies. The frequency selective filter is a circuit that allows only one frequency (ideally) or a narrow band of frequencies (practical circuit) to pass through with minimal attenuation. All frequencies that lie outside of this narrow band are severely attenuated. Additionally, the desired frequency receives the "correct" amount of phase shift, whereas all other frequencies receive more or less phase shift than is required. The correct amount of phase shift will soon be apparent to you.

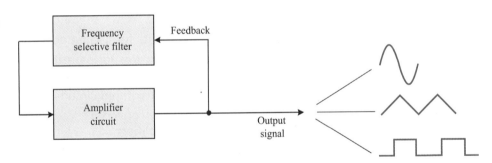

FIGURE 14–1 A functional block diagram for an oscillator circuit.

To begin, let us assume that electrical noise is present at the output of the amplifier when power is first applied to the circuit. Noise consists of energy at many random frequencies. Now, any noise energy that is of the correct frequency as determined by the frequency-selective filter will be allowed to pass through the filter. The output of the filter is applied to the input of the amplifier. It appears at the output of the amplifier and reinforces the original noise signal. Since the amplified signal is still the same frequency, it too can pass through the filter circuit and be amplified again. You will recognize the described activity as a regenerative process. In order for this to occur, we must have regenerative (i.e., positive) feedback, so that the returning signal reinforces the initial signal. To accomplish this, the frequency-selective filter must provide the correct phase shift to the desired frequency so that it will arrive in phase with the initial input to the amplifier. All other frequencies receive different phase shifts, so they are not regenerative, and they are not reinforced. For example, if the amplifier itself provides a 180° phase shift at the selected frequency, then the feedback path will need to provide another 180° phase shift. That is, the total required phase shift around the complete loop is 360°.

The operation of an oscillator circuit, then, is to essentially generate a waveform at a specific frequency with no input to the circuit other than a dc power supply. You probably know, for example, that your computer runs at 850 MHz or 1.5 GHz.

Within the computer, there is an oscillator circuit that requires dc and generates a continuous oscillation at the frequency your computer requires. In addition to the CPU clock frequency that is generally referenced in advertisements, there are many other oscillator circuits in a given computer.

If the activities just described were all that was required, the initial signal would continue to increase in amplitude each time it went around the loop. In a few iterations, the limits of the amplifier (i.e., cutoff and saturation) would be reached. From that point on, the oscillations would continue, but the waveform would be severely distorted or clipped. This may not be a problem if we want to generate rectangular waveforms such as would be required by a digital circuit. However, if we want to generate sine waves, then we cannot allow the amplitude to increase indefinitely. We must include provisions for amplitude control.

Figure 14–2 shows the functional block diagram for another class of waveform-generating circuit called a relaxation oscillator. A **relaxation oscillator** relies on the time delays provided by an RC circuit to determine its frequency of oscillation. Here, one of two current sources is used to charge a capacitor. Let us assume that I_{A2} is initially connected to the capacitor. When a constant current is provided to a capacitor, the resulting voltage across the capacitor is a linear ramp. A voltage comparator senses this ramp voltage. When the positive-going ramp exceeds the positive threshold of the comparator, the output changes state. The output of the comparator is used to change the position of an electronic switch. This switches in the I_{A1} current source. Now, the capacitor voltage begins to ramp in the other direction. When the negative-going ramp reaches the negative threshold of the voltage comparator, the output again changes state. This causes the original current source to be connected to the capacitor, so the cycle repeats.

FIGURE 14–2 The functional block diagram of a relaxation-oscillator circuit.

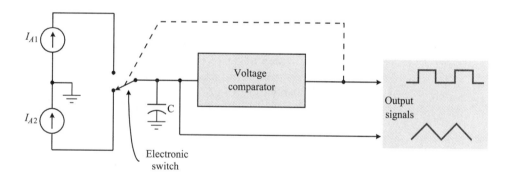

As indicated in Figure 14–2, a triangle wave output can be taken from the capacitor. The output of the voltage comparator provides a square wave signal. The frequency of either of these waveforms is determined by the charge and discharge times for capacitor C. If we use unequal current sources, then the positive and negative portions of the waveforms will be unequal. This produces a signal across the capacitor called a sawtooth waveform. The comparator output is still a digital waveform, but it will no longer be symmetrical.

Some relaxation oscillators use voltage sources to charge capacitor C through a series resistor. In this case, the charging (and discharging) is exponential according to the familiar RC charging curve. This is quite common in cases where the pulse waveform from the comparator is the desired signal.

We are now ready to examine some representative oscillator circuits. We will analyze the operation of relaxation oscillators and sine wave oscillators using op amps for the amplifier and voltage comparator functions. Additionally, we will study several other oscillators that are based on integrated circuits. We will not explore oscillators that use discrete transistors for the amplifying element, since these devices are rarely used in computer applications.

Triangle Generator

Figure 14–3 shows the schematic diagram of a relaxation oscillator based on operational amplifiers. U_1, R_1, and C_1 are configured as an integrator circuit. U_2, R_2, and R_3 form a noninverting voltage comparator. Details of integrator operation and voltage comparator operation were presented in chapter 11 and will not be repeated here.

FIGURE 14–3 A triangle wave and square wave generator.

Fig14_03.msm

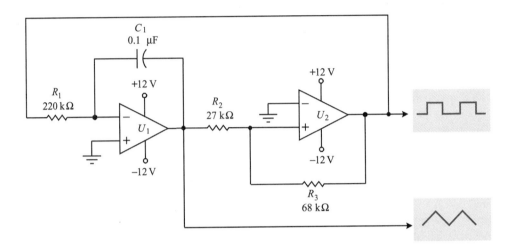

Suppose the output of U_2 is at its negative extreme (≈ -11 V). Since the (−) input of U_1 is a virtual-ground point, the full output of U_2 will be dropped across R_1. The current through R_1 will be constant under these conditions, since the voltage on both of its ends is constant. With the exception of a tiny bias current for the op amp, all of the current through R_1 is charging current for C_1. The capacitor charges with a negative on the left and a positive on the right. Further, since the charging current is constant, the voltage accumulation will be linear. So, at the output of U_1 we will see a positive-going linear ramp of voltage.

Eventually, the positive ramp will exceed the positive threshold (V_{UT}) of U_2. At this point, the output of U_2 switches quickly to its positive extreme (≈ 11 V). This changes the polarity of the voltage applied to the input of the integrator circuit. The current through C_1 now flows in the opposite direction, which will discharge C_1 and ultimately charge it in the opposite polarity. So, at this time, we will see a negative-going ramp on the output of U_1. This continues until the ramp voltage falls below the negative threshold (V_{LT}) of the voltage comparator, at which time the circuit reverts to its original state. This oscillatory action continues as long as power is applied to the circuit. Figure 14–4 shows the results of a circuit simulation for this circuit. The triangle wave is available from the output of U_1, and the square wave output is from the output of U_2.

The frequency of oscillation for the circuit shown in Figure 14–3 can be estimated with Equation 14–1.

$$f = \frac{R_3}{4R_1R_2C_1} \qquad (14\text{–}1)$$

•EXAMPLE 14.1

What is the frequency of oscillation for the circuit shown in Figure 14–3?

SOLUTION We apply Equation 14–1 as follows:

$$f = \frac{R_3}{4R_1R_2C_1} = \frac{68 \text{ k}\Omega}{4 \times 220 \text{ k}\Omega \times 27 \text{ k}\Omega \times 0.1\text{ μF}} = 28.6 \text{ Hz}$$

The results of this estimate can be easily verified by examining the timing of the waveforms in Figure 14–4. This type of circuit is generally well-suited to low-frequency applications, since its operation is restricted by the slew rate and bandwidth of the op amps.

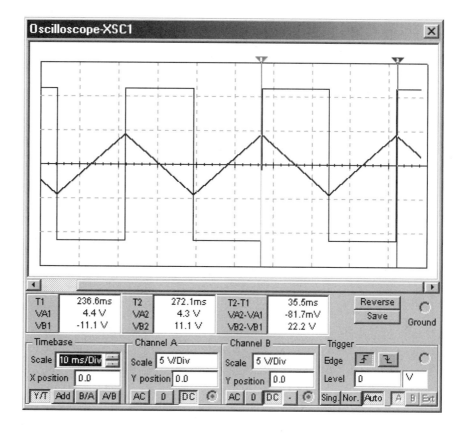

FIGURE 14–4 Results of circuit simulation for the circuit shown in Figure 14–3.

Wien-Bridge Oscillator

The Wien-bridge oscillator is a classic circuit that can generate good quality sine waves. It uses a series RC circuit and a parallel RC circuit in combination as the frequency-selective circuit. It also requires some sort of amplifier whose voltage gain can be changed dynamically in order to maintain an overall closed-loop circuit gain of exactly one. Recall that if the gain is higher than one, the amplifiers will eventually be driven to saturation. Also, if the voltage gain is less than one, the oscillations soon die out. Figure 14–5 on page 512 shows a representative Wien-bridge oscillator circuit based on an operational amplifier.

The dashed outlines in Figure 14–5 show that the Wien-bridge oscillator circuit actually consists of several smaller functional blocks. The parallel combination of Q_1 and R_1 form a variable resistor. If Q_1 is cutoff, the effective resistance is simply the value of R_1 (1.5 kΩ). If Q_1 is saturated, then the effective resistance is much lower (on-state resistance of Q_1). By controlling the voltage on the gate of Q_1, we can vary the effective resistance of the Q_1–R_1 network. In essence, they form a voltage-controlled resistance.

D_1, R_3, and C_2 are configured as a simple negative half-wave rectifier circuit with a filter. This network rectifies the output of U_1 and produces a negative dc

FIGURE 14–5 A Wien-bridge oscillator circuit.

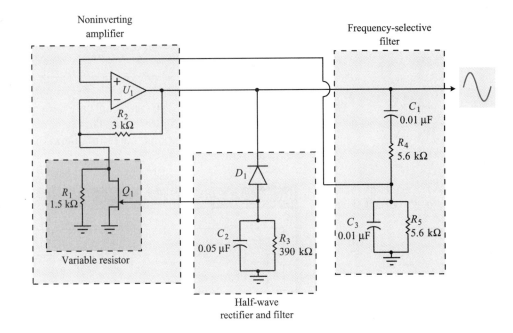

voltage on the gate of Q_1 that is proportional to the amplitude of the ac voltage on the output of U_1.

You can see that amplifier U_1 is configured as a noninverting amplifier. R_2 is the feedback resistor and, in conjunction with the effective resistance of R_1–Q_1, determines the voltage gain of the amplifier circuit. Although drawn differently, you should be able to recognize that the amplifier configuration is the same as the noninverting amplifier circuits discussed in chapter 11. Since R_2 is a fixed value, the voltage gain of the amplifier will change as the effective resistance of R_1–Q_1 changes. More specifically, if the ac output of U_1 increases, then the gate of Q_1 becomes more negative. This moves the operating point more toward cutoff, which increases its effective resistance and reduces the voltage gain of U_1.

The frequency-selective filter consists of C_1, R_4, C_3, and R_5. In some respects, this network functions as a bandpass filter. Very low frequencies will be blocked by the high reactance of C_1, while very high frequencies will be shunted to ground by the low reactance of C_3. Additionally, the phase of the waveform at the junction of the two RC circuits varies with respect to the waveform at the output of U_1 as frequency changes. Figure 14–6 reveals the behavior of the filter network through an ac analysis operation in a circuit simulation environment. Here, the frequency is swept from 100 Hz to 100 kHz. Both the amplitude and phase of the voltage across the C_3–R_5 network is plotted. As you can see, it behaves like a band-pass filter. At its center frequency, there is minimal attenuation. The voltage across the C_3–R_5 network at the center frequency is one-third of the voltage applied to the overall network. It is less at all other frequencies.

Figure 14–6 also reveals that the phase of the voltage across the C_3–R_5 network varies from −90° to +90°. But, importantly, it is 0° at only one frequency, which is the center frequency of the frequency-response curve where the voltage gain of the filter is one-third.

The output of U_1 is applied to the filter network. At the desired frequency, the voltage fed back to the input of U_1 will be one-third of the output amplitude and in phase with it. Now, if we cause U_1 to have a voltage gain of exactly 3, then the overall voltage gain of the complete loop is exactly 1. Therefore, oscillations will be sustained. If the circuit were to try to produce any other frequency, the voltage gain would be less than unity, and the feedback would not be in phase. Either of these conditions would prevent oscillation. However, at the center frequency of the filter circuit, conditions for sustained oscillations are present. If the waveform gets

FIGURE 14–6 The frequency and phase response of the frequency-selective filter network used in the Wien-bridge oscillator.

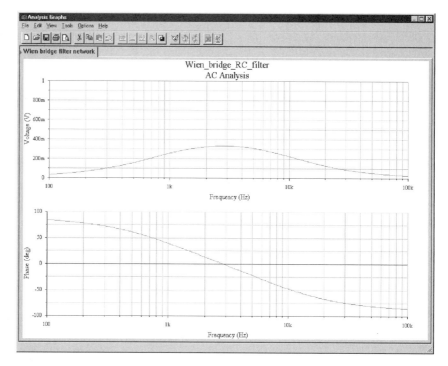

larger or smaller in amplitude over time, then the rectifier/filter circuit automatically adjusts the voltage on the gate of Q_1 to correct the gain of the amplifier such that a constant amplitude sine wave is produced. Figure 14–7 shows the results of a circuit simulation for the circuit shown in Figure 14–5.

FIGURE 14–7 Circuit simulation results for the circuit in Figure 14–5.

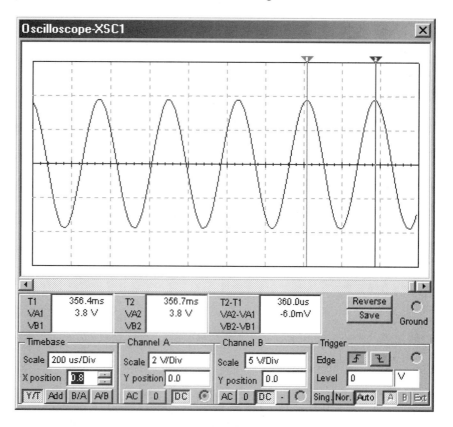

The frequency of oscillation for the Wien-bridge oscillator is determined with Equation 14–2.

$$f = \frac{1}{2\pi RC} \qquad (14\text{-}2)$$

where R and C are the values of resistance and capacitance used in the frequency-selective filter network.

EXAMPLE 14.2

What is the frequency of oscillation for the Wien-bridge oscillator shown in Figure 14–5?

SOLUTION We apply Equation 14–2 as follows:

$$f = \frac{1}{2\pi RC} = \frac{1}{2 \times 3.14 \times 5.6 \text{ k}\Omega \times 0.01 \text{ }\mu\text{F}} = 2.8 \text{ kHz}$$

The results of this calculation can be verified by examining the cycle time of the waveform in Figure 14–7.

555 Timer

The 555 integrated circuit (and several variations) has been around for more than two decades, but it continues to get widespread use in new designs. It is a versatile device that can perform several functions such as timer, time delay, oscillator, voltage-controlled oscillator, and more. Each of these applications requires just a few external components. For our immediate purposes, we will limit our discussion to the use of the 555 as a free-running oscillator. Figure 14–8 shows one way to configure the 555 timer as an oscillator circuit.

FIGURE 14–8 A 555 timer configured as a free-running oscillator circuit.

Fig14_08.msm

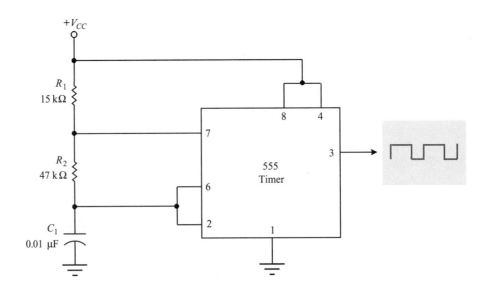

Initially, pins 2, 6, and 7 have a high internal impedance, so C_1 can charge through R_1 and R_2 toward the $+V_{CC}$ potential. During the charging time, the output (pin 3) is at its positive-most level (approximately equal to V_{CC}). C_1 continues to charge exponentially until it reaches a voltage equal to two-thirds V_{CC}. This level is sensed by an internal voltage comparator connected to pin 6.

When the two-thirds-V_{CC} point is reached, two important things happen internal to the 555. First, the output changes quickly to its lower extreme (near 0 V). Second, an internal switching transistor is turned on. The collector of this npn transistor is connected to pin 7. Its emitter connects directly to ground. So, while it was in cutoff (during the charge time for C_1), it had no effect. But, now that it is saturated, it provides a discharge path for the accumulated charge on C_1. More specifically, electrons leave the negative side (ground side) of C_1, flow through the internal npn transistor, come out of pin 7, and go through R_2 to return to the positive side of C_1. The voltage on C_1 decays exponentially with a time constant set by $R_2 C_1$.

When the voltage across C_1 has decayed to one-third V_{CC}, another internal comparator (connected to pin 2) senses this threshold. Again, two important things happen internal to the 555. First, the output switches back to its positive-most level. Second, the internal transistor connector to pin 7 is cutoff. This allows C_1 to begin another charge cycle. This process continues indefinitely. The output (pulses) is normally taken from pin 3 as shown in Figure 14–8. A nonlinear ramp waveform can be taken across C_1 if it is buffered with a high-impedance amplifier (e.g., a voltage follower). Figure 14–9 shows the results of a simulation exercise for the circuit shown in Figure 14–8. Since C_1 charges through both R_1 and R_2, but discharges through R_2 alone, it follows that the output pulse train will have more than a 50% duty cycle. **Duty cycle** describes the ratio of the positive pulse duration to the total period of the waveform. A square wave has a 50% duty cycle. The 555 waveform being discussed at this point has a longer positive pulse than the low-level portion of the waveform. Therefore, the duty cycle is more than 50%. This is evident from the waveforms shown in Figure 14–9.

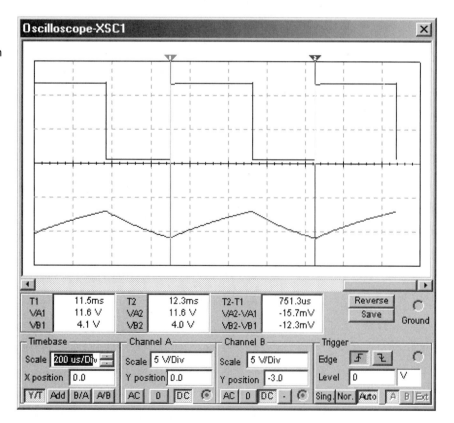

FIGURE 14–9 Circuit simulation results for the circuit shown in Figure 14–8. The upper waveform is from pin 3 of the 555, while the lower waveform shows the voltage across C_1.

The frequency of oscillation for the 555 timer when used as illustrated in Figure 14–8 can be found with Equation 14–3.

$$f = \frac{1.44}{(R_1 + 2R_2)C_1} \tag{14–3}$$

EXAMPLE 14.3

What is the frequency of oscillation for the 555 timer circuit shown in Figure 14–8?

SOLUTION We apply Equation 14–3 as follows:

$$f = \frac{1.44}{(R_1 + 2R_2)C_1} = \frac{1.44}{(15\ \text{k}\Omega + 2 \times 47\ \text{k}\Omega) \times 0.01\ \mu\text{F}} = 1.32\ \text{kHz}$$

Calculator Application

The results of this calculation can be verified by inspection of the timing for the waveforms in Figure 14–9.

Crystal Oscillators

Crystal oscillators are used in nearly all computer systems. They are characterized by a very stable oscillating frequency.

Crystal Construction

A **crystal** is a component that is constructed by placing a thin slice of quartz crystal between two metal plates. The basic construction of a crystal is shown in Figure 14–10(a). Crystals rely on the **piezoelectric effect**. The piezoelectric effect causes the crystal to produce a voltage when it is subjected to mechanical stress. A crystal microphone uses this principle. Conversely, the piezoelectric effect also causes a crystal to experience mechanical stress when it is subjected to an electric field. This latter behavior is used in crystal oscillators. The crystal—like any other physical object—has a tendency to vibrate at its own mechanical resonant frequency. This frequency is determined by physical characteristics of the crystal such as thickness and surface dimensions. In any case, if we apply an alternating voltage at the natural frequency of the crystal, then its mechanical resonance causes it to exhibit the characteristics of electrical resonance.

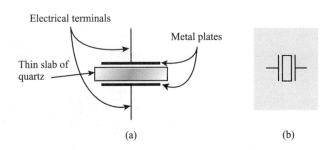

FIGURE 14–10 (a) The physical construction of a crystal and (b) the schematic representation.

Because crystals have exceedingly high values of Q (as high as one million), their impedance and phase characteristics vary abruptly on either side of resonance. The result is that the frequency of a crystal oscillator is extremely stable (called **frequency stability**). The frequency drift is primarily due to temperature variations, which cause changes in the physical size of the crystal, but values on the order of 0.0001% are typical. Frequency stability is the principal reason for using a crystal to control the frequency of an oscillator. Crystals are available for frequencies of less than 100 kHz and higher than 150 MHz. At low frequencies, they become impractical due to the large physical size of the crystal, and at frequencies above 100 MHz they become difficult to manufacture because the quartz slice has to be so thin.

A crystal can serve as the frequency-determining network for an oscillator based on a transistor, op amp, or other component capable of providing amplification. Through waveform shaping, we can get a variety of waveforms and still have the high stability provided by the crystal. Figure 14–10(b) shows the schematic representation for a crystal.

CMOS Inverter Crystal Oscillator

Figure 14–11 shows how a crystal can be used with a CMOS inverter (discussed in chapter 10) to produce a very stable square wave.

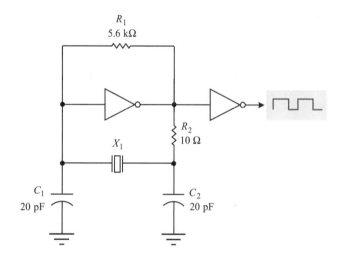

FIGURE 14–11 A crystal oscillator based on a CMOS inverter.

Resistor R_1 biases the first inverter to the middle or linear part of its operating curve. This causes the inverter to act as a high-gain inverting amplifier. The crystal X_1 provides a frequency-selective feedback path. It also provides the additional phase shift required to achieve positive feedback. Resistor R_2 limits the amount of feedback current applied to the crystal. C_1 and C_2 provide fine-tuning of the oscillator frequency and ensure reliable oscillation. The second inverter is merely a buffer to prevent load variations from affecting the oscillator frequency.

On-Board Oscillator with External Crystal

A very stable pulse waveform is required by many integrated circuits used in computers. Video controllers, Ethernet controllers, and the microprocessor itself are but a few of the devices in a computer system that require a pulse waveform at a stable frequency. To simplify the use of the device (and therefore increase sales), many integrated circuits include all of the circuitry except the crystal that is required for a crystal oscillator. That is, in addition to its intended function (e.g., an Ethernet control device), an integrated circuit may also provide the components necessary to build a crystal-controlled oscillator. The only external components that are required are the crystal and perhaps one or two small capacitors. Figure 14–12 on page 518 shows how an integrated circuit with an on-board oscillator circuit might appear on a schematic diagram. In essence, the two CMOS inverters shown in the circuit in Figure 14–11 are internal to the larger integrated circuit.

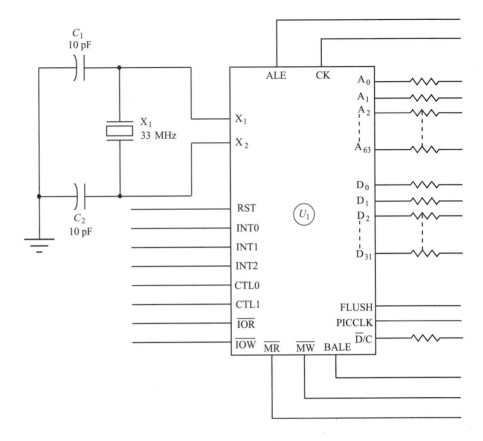

FIGURE 14–12 An integrated circuit with an on-board oscillator that uses an external crystal as the frequency-determining element.

This configuration provides considerable flexibility for the designer. First, the exact operating frequency can be customized by choice of external crystal. Second, if a pulse waveform is already available at the right frequency from some other part of the system, then it can be applied to one of the pins that normally connects to the external crystal. In that case, no external crystal is used. In essence, the externally-generated pulse train is being applied to the input of the first inverter shown in Figure 14–11.

Oscillator Modules

Many companies manufacture complete oscillator circuits that are packaged as a sealed module. All that is needed from the external circuit is a dc power source. These are available in a wide range of standard frequencies, as well as any custom frequency within the capabilities of the technology. Figure 14–13 shows representative oscillator modules. These are commonly used in computer applications.

Even though the oscillator modules are complete functioning circuits, they must be viewed as a single component with regard to troubleshooting. That is, if an oscillator module is found to be defective, it must be replaced. No repairs can be made to the module itself.

Parasitic Oscillations

To this point, an oscillator has been considered an intentional circuit. That is, by design we provided the requirements for oscillation, which include amplification, frequency selection, and positive feedback. However, many active components such as transistors, op amps, and some types of integrated circuits can satisfy the amplification requirement. It is possible that we can inadvertently provide the remaining two requirements needed for oscillation. When this occurs, a circuit will oscillate even though that is not its intended function. This undesired oscillation

FIGURE 14–13 Self-contained oscillator modules and crystals (2-leaded components). (*Courtesy of Connor-Winfield Corporation*)

is called **parasitic oscillation**. Parasitic components, such as stray capacitance between two adjacent circuit board traces or the inductance of a power supply trace, provide the feedback and frequency determination elements of the accidental oscillator circuit. Parasitic oscillations are normally more prevalent when developing new designs than when working with an existing product. In the case of computer systems, analog circuits are most prone to parasitic oscillations. For example, if the input and output wires for the amplifier that drives your computer's speakers are routed next to each other, then it is possible to couple enough energy between the wires to produce parasitic oscillation.

There are many things that can cause parasitic oscillations in a new design, but for an existing product the list of possible causes is more restricted. Poor routing of discrete input and output wires and cables is probably the most common cause. Beyond that, the most common causes might include poor ground connections or defective power supply **decoupling capacitors**. These are easily identified on a schematic diagram. They are small ceramic capacitors in the range of 1000 pF to 0.1 µF, and they connect between a power supply line and ground. Their purpose is to remove high-frequency energy from the power distribution line. If one of the capacitors becomes open, then high frequencies can sometimes use the power bus as a way to feed back energy from the output to the input of an amplifier. If the phase is correct, a parasitic oscillation will occur.

Phase-Locked Loops

A thorough discussion of phase-locked loops (PLLs) is beyond our current interests. In fact, entire books are dedicated to the subject. Rather, we will focus on the operation and use of a PLL as a frequency generator (i.e., signal source).

Functional Block Diagram

Figure 14–14 on page 520 shows the functional block diagram of a phase-locked loop. In short, a stable reference frequency is applied to the input of a PLL, and a multiple of the reference frequency is delivered at the output. Let's examine each internal functional block to understand how the PLL works.

The phase-locked loop utilizes a voltage-controlled oscillator (VCO). This is simply an oscillator whose output frequency can be varied or controlled by a dc

FIGURE 14–14 Functional block diagram of a phase-locked loop (PLL).

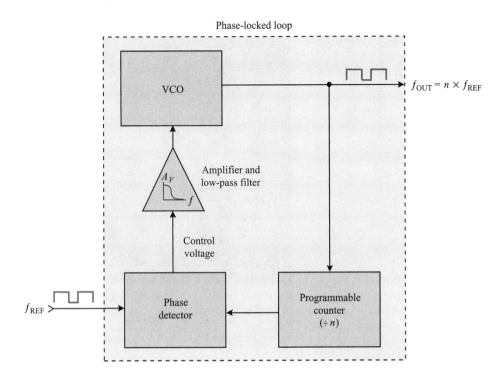

input voltage. In the example shown, the output of the VCO is a digital waveform, but the phase-locked loop concept can be applied to other waveforms as well. In any case, the output of the VCO serves as the output of the overall phase-locked loop. As a stand-alone circuit, a VCO is not a stable frequency source. However, as you will soon see, the overall operation of the PLL causes the VCO output to be just as stable as the reference frequency.

Figure 14–14 also shows a programmable counter, which is also called a divide-by-n circuit. We will study counters in more detail in chapter 15, but for now, we will examine its overall behavior. In short, we can program the circuit with some integer number, "n." Once programmed, the circuit will produce one pulse at its output for every "n" pulse at its input. As a specific example, let us assume that we program the counter to divide by 100. If its input is then 10 MHz, its output will be 10 MHz/100 or 100 kHz.

Finally, Figure 14–14 shows that a PLL also contains a phase-detector circuit. A phase-detector circuit accepts two ac inputs. The phase of the two inputs is compared by the circuit, and a voltage is produced at the output of the detector that is proportional to the phase difference between the two input signals. This will be a dc voltage if the two input signals are the same frequency. The value of the dc voltage is proportional to the phase difference. If the two signals are at different frequencies, then the output of the detector is a pulse whose frequency is equal to the difference between the two input frequencies. In either case, the output of the phase detector is routed through a low-pass filter and amplifier as shown in Figure 14–14. By the time it reaches the VCO, it is a dc voltage.

Now, let us suppose the reference frequency is 10 MHz and comes from a crystal-controlled oscillator. Let's further suppose that the programmable counter is programmed to divide by 10. The phase of the output of the counter is continuously compared to the phase of the reference input. The resulting dc control voltage is applied to the control input of the VCO. More specifically, the amplitude and polarity of the control voltage will be such that the VCO frequency is driven toward the desired frequency. The VCO output is continuously divided by the counter and returned to the phase detector. The circuit becomes stable when the output of the programmable counter is exactly in phase with the reference input.

14.1 • OSCILLATOR CIRCUITS

Being in phase necessarily implies being at the same frequency. Now, in order for the output of the programmable counter to be at the same frequency as the input, the input to the counter (VCO output) must be higher, as determined by the programmed count. In this example, the VCO output will have to be 100 MHz before the output of the counter will match the reference frequency. From that point on, the circuit is in a locked condition, and it will act as any closed-loop system. The VCO will stay locked to the reference frequency. If the VCO tries to drift off frequency in the slightest, the phase detector will sense the shift and produce a control voltage that brings the VCO back to the correct phase and frequency. The slight variation in output phase that results from continuous comparison and correction is called **jitter**.

Modern computer systems require many high-speed clocks (square wave signals used as timing inputs to the various integrated circuits). For proper operation of the overall computer system, many of these waveforms must have precise relationships. The relative frequencies are important, and in many instances, the phase between several different waveforms is critical. Multiple phase-locked loops (often integrated into a single package) can be used to produce the various clock signals. A single, stable reference signal serves as the input to all of the phase-locked loops. They each have their own programmed count, so each output is independent of the others with regard to frequency, but all outputs are a multiple of the reference frequency. Additionally, all are phase locked to the same reference frequency, so relative timing is exact.

Representative Circuit

Figure 14–15 shows the functional block diagram for an ICS9169C-46 Clock Synthesizer chip manufactured by Integrated Circuit Systems, Inc. This device generates a number of synchronized clocks for use with a microprocessor-based

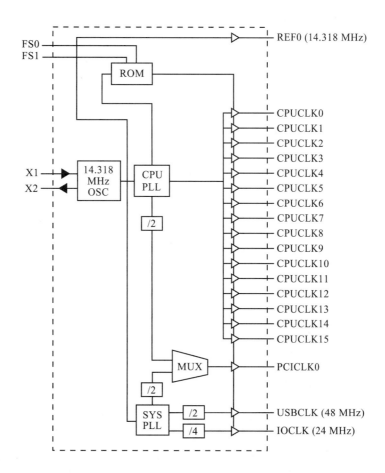

FIGURE 14–15 Functional block diagram of an ICS9169C-46 Clock Synthesizer chip. (*Courtesy of Integrated Circuit Systems, Inc.*)

computer system. More specifically, it uses a single 14.318-MHz external crystal, and it generates the following clock signals:

- *CPU clocks:* There are sixteen synchronized clocks at frequencies up to 66.66 MHz. These are generally used for the microprocessor and related circuitry that require in-phase clocks.
- *PCI clock:* A PCI clock that generally runs at 33.3 MHz. It is used to provide the timing signal for plug-in option boards used with computers.
- *USB clock:* The USB clock is a 48-MHz signal that is typically used to drive integrated circuits designed to interface to the Universal Serial Bus (USB) port found on many computers.
- *I/O clock:* This 24-MHz clock signal is generally used to drive the super I/O controller in a computer system. This controller provides the interface for such things as a keyboard, mouse, printer ports, standard serial ports, floppy disks, and so on.
- *Reference output:* The reference oscillator (14.318 MHz) is buffered and used to drive the video controller in a computer. Some plug-in option boards also require this signal.

As shown in the functional block diagram, the external crystal connects to the X1 and X2 pins of the IC. Alternatively, a 14.318-MHz clock signal can be connected to X1. The FS0 and FS1 inputs are programming inputs that specify the exact frequencies that will appear on the sixteen CPU clock outputs. With a 14.318-MHz reference clock, the choices for CPU clock frequency are 50 MHz, 60 MHz, 66.66 MHz, and disabled.

There are two PLLs shown in the functional block diagram in Figure 14–15. One is used to develop the CPU clock outputs, while the second generates the fixed-frequency USB clock and I/O clock. The PCI clock output is formed by dividing the CPU clock frequency by two.

As with any other integrated circuit, troubleshooting is necessarily restricted to verifying the correct inputs and outputs of the device. If the power supply voltage is correct (typically 3.3 V for this device), the FS0 and FS1 inputs are correct, and the crystal is good and connected properly, then the various clock signals should be present. If they are not, the device is likely to be defective.

✓ Exercise Problems 14.1

1. Oscillators create an alternating waveform with no input other than the dc power supply. (True/False)
2. What type of feedback is required for an oscillator to have sustained oscillations?
3. What general class of oscillator circuit relies on the charging and discharging of a capacitor through a resistor to determine the frequency of oscillation?
4. What waveshape is generally produced by a Wien-bridge oscillator?
5. A 555 timer can be configured as a relaxation oscillator. (True/False)
6. What is the most stable type of oscillator circuit?
7. What is the name of the phenomenon that makes crystals useful as a frequency-selective component?
8. Crystals are characterized as having extremely high Q values. (True/False)
9. Parasitic oscillations are normally considered to be a desirable circuit characteristic. (True/False)
10. If an oscillator module is defective, what options do you have to correct the defect?
11. If the input to a phase-locked loop were 10 MHz, then an output of 50 MHz would be reasonable. (True/False)
12. A phase-locked loop is a (*closed/open*) -loop electronic system.
13. The VCO used in phase-lock loops is an inherently stable functional block within the PLL circuit even if operated as an open-loop circuit. (True/False)

SYSTEM PERSPECTIVE: Integrated Circuits

Every practical computer system is composed of many interconnected integrated circuits. The majority of them are digital devices, since this is the fundamental nature of modern computers. However, analog integrated circuits are also used in computers. Portions of the sound circuits of some computers, and certain types of multimedia video circuits, utilize analog integrated circuits. Additionally, analog integrated circuits are used in industrial computers to monitor and control analog devices external to the computer (e.g., pressure sensors, proportional valves, and many others). Finally, some of the integrated circuits used in computers are called hybrid ICs, since they have both digital and analog functions. You will want to be very familiar with all types of integrated devices. (*Photograph courtesy of Agilent Technologies*)

14.2 INDUSTRIAL COMPUTER APPLICATIONS

This section will examine ICs often used for industrial computer applications. Again, be reminded that there are thousands of these devices, so it is impractical to learn the detailed behavior of each device. Rather, it is important to know that the general class of devices exists. Details on the operation of new or unfamiliar devices are always available from the manufacturers via the Internet.

Computers are used throughout industry for many purposes including the following:

- Monitoring manufacturing processes
- Controlling machines
- Inventory management
- Information processing (e.g., accounting, human resources, marketing, and so on)

Our immediate interest will be focused on computer applications that monitor manufacturing processes.

Transducers

A **transducer** is a device that allows the energy in one system to control the energy in a second system. Transducers are more commonly thought of as devices that convert energy from one form to another. In any case, transducers (also called sensors) are used to sense the values of process variables such as heat, pressure, light, weight, physical position, color, opacity, hardness, chemical content, moisture content, fluid level, and many other such quantities. Changes in these physical quantities cause the transducer to reflect proportional changes in an electrical characteristic of its output. The output electrical change may be in resistance, voltage, current, frequency, capacitance, or other such value. Examples of several common transducers are given in Table 14–1.

TABLE 14–1 Common transducers.

Transducer Class	Quantity Sensed (Input)	Proportional Electrical Output
Sound	Intensity (volume)	Voltage
Light	Intensity (brightness)	Current, voltage, or resistance
Pressure	Fluid or gas pressure	Voltage or current
Position	Linear movement	Voltage or current
Weight	Weight of an object	Voltage or resistance
Level	Fluid level	Voltage, resistance, or capacitance
Temperature	Temperature of a material	Current, voltage, or resistance

Instrumentation Amplifiers

The electrical changes in the output of a typical transducer are quite small. For example, an Entran EPI-411-1 pressure transducer has a 45-mV output with 15 psi of pressure. Similarly, the Entran EPI-411-7 pressure transducer produces a 100-mV output when exposed to a pressure of 100 psi. In either case, the transducers require a 5-Vdc input or excitation voltage. Transducer outputs require special consideration for several reasons including the following:

- The transducer output signals are so small.
- The distance between the sensor and computer may be great.
- The industrial environment is inherently harsh with regard to electrical noise and interference.

To overcome these problems and allow remote monitoring of manufacturing processes in the midst of high-level electrical noise, many systems utilize a special integrated amplifier called an instrumentation amplifier.

In some ways, an instrumentation amplifier can be thought of as a high-quality op amp. In particular, nonideal characteristics such as the following are optimized:

- Common-mode rejection ratio (i.e., the ratio of differential voltage gain to common-mode voltage gain). Recall that the ideal differential gain is infinite, and the ideal common-mode gain is zero. Therefore, a very high common-mode rejection ratio (CMRR) is desirable.
- Low dc offset voltages and voltage drifts with time and temperature
- Low-output impedance
- High-input impedance
- Stable gain characteristics

Figure 14–16(a) shows a representative symbol for a basic instrumentation amplifier. Note that it has a single external resistor. Also, observe that the input wires are shown as being shielded (e.g., coaxial cable). This is generally required to further reduce the susceptibility to external electrical noise that would otherwise corrupt the low-level transducer signals.

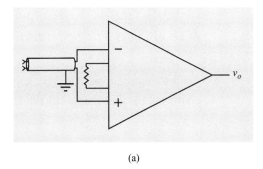

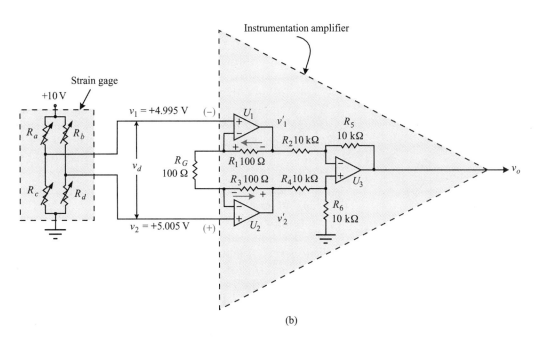

FIGURE 14–16 (a) An instrumentation amplifier symbol showing an external gain resistor and shielded inputs. (b) An equivalent circuit for an instrumentation amplifier using op amps with a strain gage as its input.

Figure 14–16(b) shows an exploded view of the instrumentation amplifier to further clarify its behavior. A strain gage is shown as a typical transducer. A strain gage consists of four resistive strips (shown schematically as resistors) connected as a bridge circuit. The entire assembly is glued to a surface of the object to be monitored. As the object is subjected to mechanical stress, the resistive strips (often made from p-type silicon) in the strain gage are stretched and compressed slightly, which causes a change in their individual resistances. With no distortion, the resistances in the strain gage are all equal (typically 350 to 1000 Ω). Under these conditions, there is no voltage difference across the output of the strain gage, since each resistor drops the same voltage. However, when mechanically distorted, the resistances become unbalanced, and a voltage appears at the output of the strain gage. Values from 2 mV to as high as 20 mV are representative maximum (i.e., full-scale) outputs. This low-level signal often serves as the input to an instrumentation amplifier.

Three standard op amps can provide the foundation for an instrumentation amplifier as shown in Figure 14–16(b). For a moment, ignore the external resistor R_G. Amplifiers U_1 and U_2 can now be recognized as simple voltage-follower circuits. Recall that voltage followers are characterized as having exceptionally high-input impedance. U_3 has an inverting voltage gain of -1 for voltages applied to R_2, and a voltage gain of +1 (noninverting) for voltages applied to R_4. The +1 gain is due to the combined effects of the voltage divider action of R_4 and R_6 (i.e., voltage gain of $\frac{1}{2}$) and the noninverting voltage gain of U_3 (+2). Thus, its output voltage is simply $v_2' - v_1'$.

The voltages on the inverting inputs of U_1 and U_2 are essentially the same as v_1 and v_2, respectively, because of the closed-loop behavior of the op amp circuits. This means that the voltage across R_G can be expressed as

$$|v_G| = |v_1 - v_2|$$

where the vertical bars indicate absolute value. For the values shown in Figure 14–16(b), the voltage across R_G will be |4.995 V − 5.005 V| = 10 mV. The actual polarity of the voltage drop is determined by which input is more positive. We can use Ohm's Law to compute the current through R_G as $I_G = v_G/R_G =$ 10 mV/100 Ω = 100 μA. Since none of this current can flow into the op amp input terminals (ideally), it must flow through R_1 and R_3 as indicated in Figure 14–16(b). Since these have the same resistance and the same current as R_G, it follows that each will have a 10-mV drop in the polarities indicated. We can apply Kirchhoff's Voltage Law beginning at the (−) input terminals of U_1 and U_2 and moving through R_1 and R_3 to determine the voltage levels at v_1' and v_2'. Those values are v_1' = +4.995 V − 10 mV = 4.985 V and v_2' = + 5.005 V + 10 mV = 5.015 V. These voltages are applied to U_3, which produces the overall output as $|v_2' - v_1'|$ = |+5.015 V − 4.985 V| = 30 mV. This may not seem especially remarkable, since the original input voltage (v_d) was 10 mV (i.e., a voltage gain of 3), but let's take a closer look.

First, the circuit offers extremely-high input impedance, so it will have negligible effects on transducer performance. Second, the output is totally dependent on the *differential* input voltage (v_d). It is unaffected by the absolute voltage of either input. That is to say, it is unaffected by common-mode voltages. Thus, we would have gotten the same output voltage if the input voltages had been 6.995 V and 7.005 V, for v_1 and v_2, respectively. We could have 2 V of noise (probably 60-Hz interference) present on both lines, and it would have no effect on the output even though the signal being amplified is only 10 mV. That is impressive!

This circuit has yet another interesting characteristic. Note that the voltage across R_G is determined by v_1 and v_2. It is not determined by the value of R_G. However, if we change the value of R_G, the current through it and through R_1 and R_3 will change. This in turn causes changes in v_1' and v_2', which of course is reflected in the output. By applying Ohm's and Kirchhoff's Laws and some algebraic manipulation, we can show the following relationship:

$$A_V = \frac{2R}{R_G} + 1 \qquad (14\text{--}4)$$

where A_V is the overall voltage gain of the instrumentation amplifier, R is the common value of the internal resistors R_1 and R_3 in Figure 14–16(b), and R_G is the external resistor.

As you can see, we can readily control the gain by changing the value of R_G. For this reason, it is often called a gain resistor. For the values listed in Figure 14–16(b), the circuit has a voltage gain of 3 (which was previously determined).

•–EXAMPLE 14.4

Compute the voltage gain of the circuit in Figure 14–16(b), if resistor R_G is changed to 10 Ω.

SOLUTION We apply Equation 14–4 as follows:

$$A_V = \frac{2R}{R_G} + 1$$

$$= \frac{2 \times 100\ \Omega}{10\ \Omega} + 1 = 21$$

The AMP02 instrumentation amplifier made by Analog Devices is a representative device that uses a single external gain resistor to set the voltage gain from 1 to 10,000. It also offers a CMRR of over 100 dB.

VFC and FVC

In this section, we will discuss two complementary circuit functions that are widely used for industrial electronics applications: voltage-to-frequency conversion (VFC) and frequency-to-voltage conversion (FVC). These two conversions have many industrial applications. One common use is to provide a convenient way to send transducer information (signals in the low millivolts range) over relatively long distances through a harsh electrical environment. Remember, transducers are devices that convert some physical quantity such as temperature, pressure, speed, acidity, light, and so forth to a proportional voltage or other electrical quantity.

Voltage-to-Frequency Conversion

Several techniques can be used to construct a circuit whose output frequency is proportional to its input voltage. For example, in a prior section we briefly discussed a voltage-controlled-oscillator (VCO) as an integral part of a phase-locked loop. In this section, we will discuss the operation of the VFC32, which is an integrated circuit manufactured by Burr-Brown Corporation. It uses a technique that gives greater linearity, faster operation, and better noise immunity than some voltage-to-frequency conversion methods. Let us now direct our attention to the functional block diagram and typical circuit configuration presented in Figure 14–17 on page 528.

As you can see in Figure 14–17, the VFC32 consists of three functional blocks and an output transistor. The amplifier and its associated components, R_1 and C_1, form a basic integrator circuit like we discussed in chapter 11. A positive input voltage will cause current to flow through R_1. Provided the switch is open, this same current will charge C_1. As C_1 accumulates a charge (negative on the right and positive on the left), the output of the amplifier will be ramping in the negative direction.

There is a second source of charging current for C_1. It is a 1.0-mA reference current source (I_R). This current source can be switched in or out of the circuit. If the switch is closed, the 1-mA current source overrides the input current. That is, according to Kirchhoff's Current Law, if the current source has a greater current than the current through R_1 (which it will have by design), then the current through capacitor C_1 will flow in the opposite direction. During this time, the output of the amplifier will be ramping in the positive direction. This action in the integrator circuit is a critical part of the overall operation, so let's examine it more closely. Figure 14–18 on pages 528 and 529 shows a numerical example for the two time intervals: switch open and switch closed.

The circuits in Figure 14–18 can be analyzed using the methods discussed in chapter 11. Be sure to note that the (–) input is a virtual-ground point. In any case, we find that $I_1 = I_2 = 0.125$ mA. This causes the output of the amplifier to ramp in the negative direction at the rate of 12.5 V/ms as shown in Figure 14–18(a). Now, when switch S closes, the current source forces a constant value of 1 mA into the node at the (–) input of the amplifier. According to Kirchhoff's Current Law, if

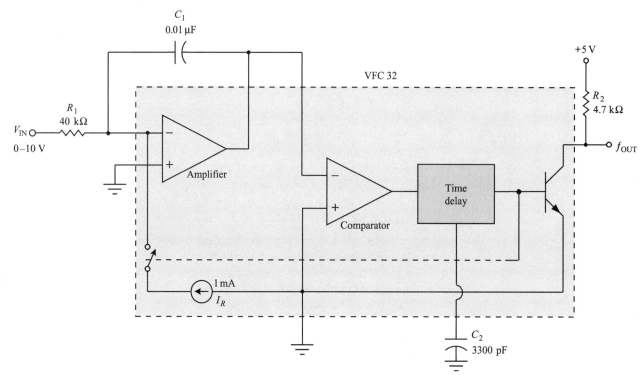

FIGURE 14–17 A functional block diagram of the VFC32 voltage-to-frequency converter and a typical circuit configuration. Redrawn with permission. (*Courtesy of Texas Instruments*)

1 mA is entering the node (I_R) and 0.125 mA is leaving the node (I_1), then I_2 must be 0.875 mA, and it must be flowing away from the (−) terminal as shown in Figure 14–18(b). We know from our earlier study of the integrator circuit that a charging current flowing in the direction of I_2 in Figure 14–18(b) will cause the output of the amplifier to ramp in a positive direction. The slope of the output ramp under these conditions can be shown to be 87.5 V/ms.

Let's shift our focus to the remaining functional blocks in Figure 14–17. The output of the integrator circuit is sent to one input of a voltage-comparator circuit.

FIGURE 14–18 A comparison of conditions in the integrator circuit with (a) the switched current source open and (b) the current source connected.

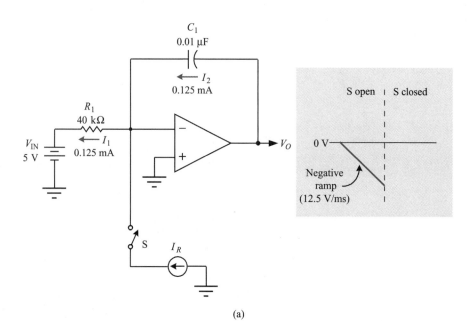

(a)

FIGURE 14–18 (continued)

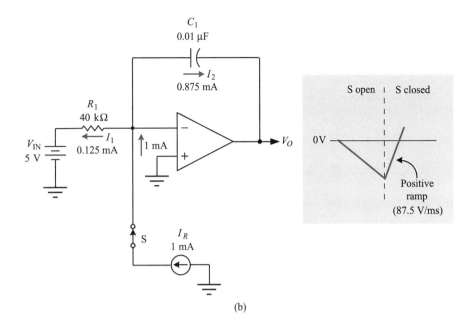

(b)

The other input of the comparator is connected to ground (0 V). The positive-going edge of the comparator output is used to initiate a time-delay circuit. The time-delay circuit has a positive output voltage until triggered by the comparator circuit. Once triggered, the output of the time-delay circuit goes to zero and remains there for a fixed time (i.e., the duration of the time delay). The length of the time delay is determined by capacitor C_2. The pulse waveform at the output of the time delay performs two functions. First, it controls the conduction of the output transistor, which operates in the switching mode. The collector waveform is the output of the voltage-to-frequency converter circuit (f_{OUT}). The output of the time-delay block also controls the position of the switch shown in Figure 14–17. During the positive duration of the output pulse (f_{OUT}), the switch connects the current source to the (−) terminal of the op amp. This is similar to the operation of a relay circuit, but clearly, the entire circuit is electronic and integrated within the IC package.

Figure 14–19 on page 530 shows some important circuit waveforms. The top waveform represents the output of the integrator. During the time the switch is closed, the output of the integrator ramps in a positive direction for a fixed time (t_d). When the time delay is over, the switch is opened, and the output of the integrator begins to ramp down at a slope determined by the input voltage. It will ramp down until it passes through zero, switches the comparator, and triggers the time delay for a second time. The length of time (t_x) required for the integrator to ramp to zero depends on the slope, which in turn depends on the input voltage. Since t_x varies with the input voltage, the total time for one cycle (T) also varies with the input voltage.

The output frequency, f_{OUT}, can be found with Equation 14–5.

$$f = \frac{V_{IN}}{R_1 t_d I_R} \qquad (14\text{–}5)$$

where t_d is the length of the time delay, V_{IN} is the input voltage, and I_R is the value of the current source.

The manufacturer's datasheet provides an equation to compute the length of the time delay. It is repeated here as Equation 14–6.

FIGURE 14–19 Circuit waveforms for the voltage-to-frequency converter shown in Figure 14–17.

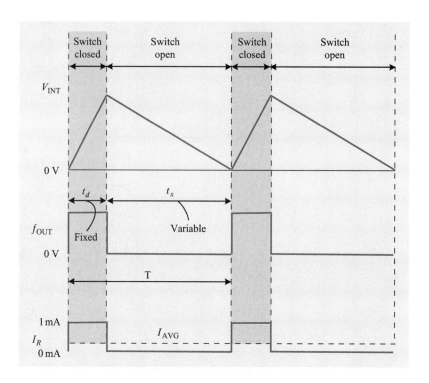

$$t_d = 7500 \times C_2 \qquad (14\text{–}6)$$

where t_d is measured in seconds and C_2 is given in farads. Using the values shown in Figure 14–17, we can compute t_d as follows:

$$t_d = 7500 \times C_2 = 7500 \times 3300 \text{ pF} = 24.75 \text{ μs}$$

●─EXAMPLE 14.5

Compute the output frequencies for the circuit shown in Figure 14–17 that correspond to input voltages of 2.5 V, 5 V, 7.5 V and 10 V.

SOLUTION   We apply Equation 14–5 for the various input voltages as follows:

$$f_{OUT} = \frac{V_{IN}}{R_1 t_d I_R} = \frac{2.5 \text{ V}}{40 \text{ k}\Omega \times 24.75 \text{ μs} \times 1 \text{ mA}} = 2.53 \text{ kHz}$$

$$f_{OUT} = \frac{V_{IN}}{R_1 t_d I_R} = \frac{5.0 \text{ V}}{40 \text{ k}\Omega \times 24.75 \text{ μs} \times 1 \text{ mA}} = 5.05 \text{ kHz}$$

$$f_{OUT} = \frac{V_{IN}}{R_1 t_d I_R} = \frac{7.5 \text{ V}}{40 \text{ k}\Omega \times 24.75 \text{ μs} \times 1 \text{ mA}} = 7.58 \text{ kHz}$$

$$f_{OUT} = \frac{V_{IN}}{R_1 t_d I_R} = \frac{10 \text{ V}}{40 \text{ k}\Omega \times 24.75 \text{ μs} \times 1 \text{ mA}} = 10.1 \text{ kHz}$$

If we use these four points to construct a graph as shown in Figure 14–20, the linear relationship between input voltage and output frequency becomes apparent.

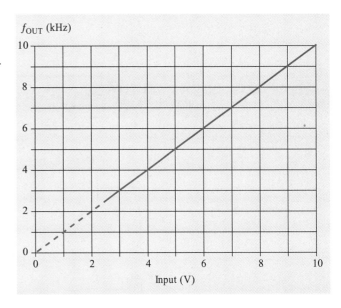

FIGURE 14–20 A graph showing the relationship between input voltage and output frequency for the voltage-to-frequency converter shown in Figure 14–17.

Frequency-to-Voltage Conversion

As you would expect from its name, a frequency-to-voltage converter performs the inverse operation of a voltage-to-frequency converter. That is, the input is a repetitive waveform, and the output is a dc voltage that is proportional to the input frequency. The VFC32 discussed in the previous section can also be configured to perform a frequency-to-voltage conversion. Figure 14–21 on page 532 shows how the integrated circuit would be connected to achieve this operation.

The operation of each of the internal functional blocks is the same as discussed with reference to the voltage-to-frequency converter. The dc input to the integrator circuit now comes from a potentiometer (P_1) that is used to calibrate the circuit. For now, we shall consider it a fixed dc voltage. The integrator operates in a manner similar to that previously described with some minor exceptions. The alternate charging and discharging of the integration capacitor (C_1) occurs when the internal switch opens and closes as previously discussed. The presence of resistor R_1 in parallel with the integration capacitor, however, causes the otherwise linear ramp voltage at the output of the integrator to become exponential. The average dc level of the exponential ramp corresponds to the frequency of the input. If necessary for a given application, the exponential waveform can be passed through a subsequent low-pass filter to obtain a smoother dc voltage.

The state of the internal switch is still controlled by the time-delay circuit. The time-delay circuit is still triggered by a positive-going transition at the output of the comparator. The comparator, however, is triggered by the incoming waveform (v_{in}). The $C_3/R_2/R_3$ network provides signal conditioning to ensure proper operation of the comparator. In order to trigger the comparator, the (−) input must go below about −0.7 V. The actual input voltage is a 0 to 5 V pulse. The voltage divider consisting of R_2 and R_3 sets the quiescent level of the (−) input to +2.5 V. Capacitor C_3 in conjunction with the resistances in the voltage divider, forms a differentiator circuit. Figure 14–22 on page 532 shows the waveform (v_{R_3}) that will be sent to the (−) input of the comparator. The basic circuit will work for nearly any input waveform, but the $C_3/R_2/R_3$ network might have to be altered for other input voltages and/or shapes.

So, in short, the negative-going transitions of the input waveform initiate the time delay. The time-delay circuit closes the switch and connects the current source to the integrator for a fixed time interval ($7500 \times C_2$). During this time, the

532 CHAPTER 14 • INTEGRATED CIRCUIT APPLICATIONS

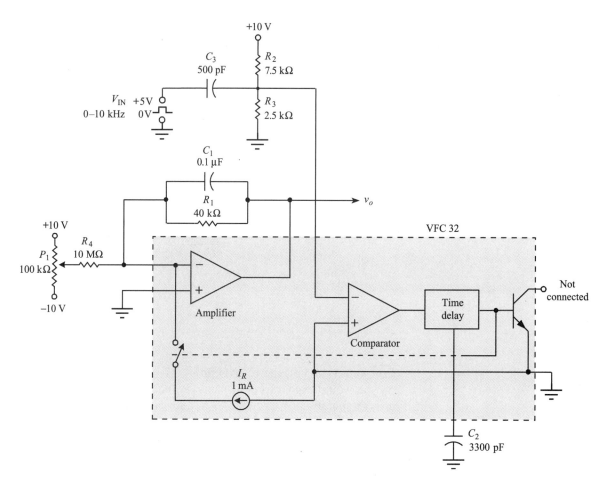

FIGURE 14–21 A frequency-to-voltage converter circuit using the Burr-Brown VFC32 integrated circuit.

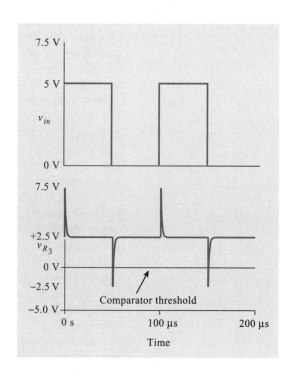

FIGURE 14–22 Capacitor C_3 and the R_2/R_3 voltage divider in Figure 14–21 form a biased-differentiator circuit.

integrator output is ramping exponentially in a positive direction. After the time delay is over, the switch opens, and the integrator output ramps exponentially in a negative direction. It will continue to ramp down until the next negative transition of the input signal. The sooner the next trigger arrives (i.e., the higher the input frequency), the less time the capacitor will have to discharge and the higher the average output voltage will be. The circuit maintains a linear relationship between input frequency and output voltage over its range of operation.

✓ Exercise Problems 14.2

1. An instrumentation amplifier is characterized by a very (*low/high*) CMRR.
2. Most transducers have outputs in the 1 to 5 V range. (True/False)
3. Details (e.g., pin outs, functional block diagrams, and applications information) for integrated circuits used in industrial applications can generally be obtained from the manufacturer via the Intranet. (True/False)
4. A VCO is an example of a frequency-to-voltage converter. (True/False)
5. When the reference current source (I_R) is connected to the (−) input of the op amp, electron current is flowing left-to-right through C_1. (True/False)
6. What is the purpose of capacitor C_2?
7. The frequency of the waveform on the f_{OUT} pin is linearly proportional to the dc voltage applied to V_{IN}. (True/False)
8. An FVC IC is bidirectional and can be operated as a VFC by simply connecting the IC in the reverse direction. (True/False)

Refer to Figure 14–17 for questions 5 through 7.

14.3 TROUBLESHOOTING CIRCUITS BASED ON ICs

With the limitless array of integrated circuits used in products today, combined with the rapid rate that new ICs are being developed, it is completely impractical to discuss a detailed troubleshooting strategy for each type of IC. However, we will discuss some techniques that are generally applicable to locating defects in oscillator circuits. We will also cite general procedures that are applicable to locating defects in circuits containing other types of integrated circuits.

Troubleshooting Oscillator Circuits

Problems with oscillator circuits generally fall into one of the following categories:

1. Completely inoperative (i.e., no output signal)
2. Distorted output wave form
3. Incorrect or unstable frequency of oscillation

In many cases, items 2 and 3 occur simultaneously. As always, your best troubleshooting tools are your basic electronics theory and your understanding of normal circuit operation.

The first thing to check if the oscillator is completely inoperative is the dc supply voltages. Be sure to measure the voltage directly at the oscillator circuit. Otherwise, an open PCB trace or wire might mislead you. If the supply voltages are normal, the next step is to classify the problem as an ac or a dc defect. This is a straightforward task and can be accomplished with a dc voltmeter. If the dc output voltage is abnormal, then it is a dc defect; otherwise, it is an ac defect. Once you have classified the problem, the list of possible defects will be extremely short. The specific defect can be located and verified by component testing or substitution.

Output distortion in an oscillator circuit is often caused by a shift in the dc operating point. This causes the op amp or transistor to move out of its normal range of operation and introduce distortion. Troubleshooting this type of problem requires verification of all dc circuit voltages. Once an incorrect level is located, isolation of the problem is similar to other circuits. It should be noted that some oscillator configurations can continue to oscillate with dramatic changes in power supply voltage. The symmetry and purity of the output waveform, however, may deteriorate.

For most oscillator circuits, very few components can cause a shift in the operating frequency without affecting the dc levels in the circuit. Identification is straightforward. In those cases where the frequency of operation is affected by numerous components, troubleshooting can often be streamlined by noting, but not focusing on, the frequency error. Rather, verify all other aspects of the oscillator's operation such as dc levels, wave shapes, duty cycle, and so on. If one of these other characteristics is found to be abnormal, concentrate your attention on this latter problem. The off-frequency problem is probably only a symptom, and it will be corrected when the other, more easily located, problem is corrected. If all other characteristics appear to be normal, then suspect the components whose sole purpose is for frequency determination. Specifically, look for frequency-selective components whose change in value would not alter the dc levels in the circuit. Very few components qualify for inclusion in this category.

Troubleshooting Other IC-Based Systems

A sound knowledge of how the system operates and a solid understanding of basic electronics are the most precious and powerful tools in your arsenal of troubleshooting strategies. Without these two essential prerequisites, your troubleshooting exercise will be reduced to a guessing game. On the other hand, if you confidently use these two tools together, you will be able to efficiently troubleshoot systems and identify specific defects.

We can break all integrated circuits into two rough categories: those with internal memory and those that have no internal memory. The term "internal memory," for our present purposes, means that the present behavior of the IC can be affected by events that occurred previously. You cannot, for example, monitor the inputs of a memory IC in a computer and determine what the correct outputs should be unless you know what has occurred previously. That is, you must know what information has been stored in the memory device. There are thousands of ICs with internal memory functions besides ICs whose primary function is memory. Many of these are called programmable devices: the electrical behavior of the device can be customized by a series of commands. Once the commands have been issued, there is no external record of those commands. Therefore, you cannot predict the correct behavior of the IC by measuring input voltages at a time after the device has been programmed. For the present discussion, we will exclude ICs with internal memory.

The outputs of all other ICs can be predicted by measuring the inputs and applying your knowledge of the system. By contrasting the predicted outputs with the actual outputs, you can quickly classify an IC as good or bad. That's about all you can or need to do, since you cannot make repairs inside the IC package. Most of the integrated circuits discussed to this point in the text are examples of ICs without internal memory. This includes op amps, linear-voltage regulators, oscillator modules, switching-regulator controllers, and many audio ICs. It is reasonably safe to say that all purely analog ICs fall into this category. Hybrid ICs that contain both digital and analog functions and ICs that are strictly digital devices generally (but not always) have internal memory functions.

Classifying a defective IC is a simple concept but a very important one, so let's look at a more familiar example. Suppose you had a television, and it was your job to determine whether the television was good or had a defect. You are not permitted to remove the enclosure of the TV. Further suppose there are other operational

TVs available to you. The only inputs to a television receiver are a 120-Vac power line and the television signal (from an antenna or cable connection). If you remove the cable connection from a working TV and connect it to the unit being diagnosed, then you know the input is good. Second, if you measure the 120 Vac with a voltmeter, you can verify that input as well. Now that you have established with certainty that the inputs are correct, you can use your knowledge of the system to determine what a correct output should be. In this case, you would expect sound and a picture as the normal outputs. If, after manipulating the controls, you were unable to verify a correct output, then you could confidently pronounce the receiver as defective. This is precisely the same logic that you can use to troubleshoot a system containing integrated circuits without internal memory functions.

✓ Exercise Problems 14.3

1. Name the three most common categories of defects for oscillator circuits.
2. If the power supply voltage is zero for an oscillator circuit, what symptoms will be present?
3. An open feedback capacitor in an oscillator circuit would create a(n) (*dc/ac*) defect.

Refer to Figure 14–23 for questions 4 through 6.

4. If diode D_1 became open, the circuit would continue to oscillate. Explain why and give the symptoms of the defect.
5. What is the effect on the dc voltage at the output of U_1 if C_1 develops an open?
6. If Q_1 develops an open on its drain terminal, will the circuit continue to oscillate?
7. Op amps are examples of ICs that have internal memory. (True/False)
8. Explain why the output of the PLL shown in Figure 14–14 not only depends on its inputs now, but on other events that occurred previously.

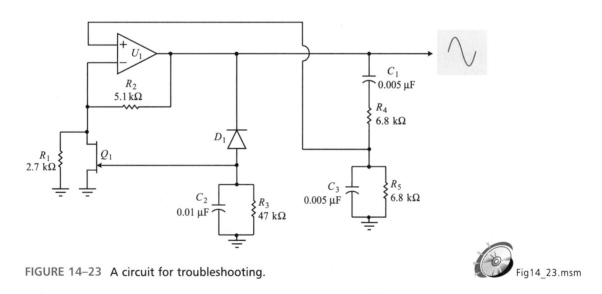

FIGURE 14–23 A circuit for troubleshooting.

Fig14_23.msm

•—SUMMARY

An oscillator is a circuit that requires only dc power as its input. It produces a continuous stream of alternating cycles at its output. The frequency and waveshape of the oscillator output are determined by the design of the specific circuit. Oscillator circuits require amplification, positive feedback, and a frequency-selective network.

A relaxation oscillator relies on the time delays provided by an *RC* circuit to determine its frequency of oscillation. Higher frequency oscillators often utilize frequency-selective networks composed of resonant circuits or crystals. Crystals use the piezoelectric effect to allow the mechanical resonance properties of the crystal to precisely control the frequency of an oscillator circuit.

Phase-locked loops were also introduced in this chapter. PLLs contain a VCO, a phase detector, and a programmable counter integrated into a single package. A PLL requires a stable reference frequency as its input. The output has the same degree of frequency stability as the input, but the output frequency can be substantially higher than the input frequency. The relationship between input and output frequency is determined by the programming of an internal counter. If, for example, the counter is programmed to divide by twenty, the output frequency will be twenty times higher than the input frequency.

We also discussed instrumentation amplifiers, and VFC and FVC integrated circuits as representative industrial applications. It is not necessary to memorize the details of these particular circuits, but rather work to gain an appreciation that integrated circuits have internal functional blocks just the same as circuits built from discrete components would have. So, armed with your general knowledge of electronics and the functional block diagram of an integrated circuit provided by the manufacturer, you can quickly understand the operation of any integrated circuit you are likely to encounter.

The first step in troubleshooting an oscillator circuit is to verify the dc supply voltages. If the dc supply is normal, then classify the defect as a dc or an ac defect. Once the defect is classified as ac or dc, the range of possible defects will be dramatically reduced. The specific defect can then be identified through component measurements.

ICs can be divided into two classes: one with internal memory and one without internal memory. Troubleshooting of ICs with internal memory will be discussed in a later chapter. Troubleshooting IC circuits with no internal memory consists of measuring the inputs and applying your knowledge of the system to predict the normal outputs. The actual outputs are then contrasted with the predicted outputs to classify the IC as good or defective.

•—REVIEW QUESTIONS

Section 14.1: Oscillator Circuits

1. Oscillators require _____ to overcome losses in the circuit.
2. What input(s) are required by an oscillator circuit?
3. Only square waves can be generated directly by an oscillator circuit, but other wave forms can be obtained through signal conditioning. (True/False)
4. What is the purpose of the frequency-selective filter network in the feedback path of an oscillator?
5. How much total phase shift must be applied to a signal as it makes the complete loop around an oscillator circuit?
6. The feedback in an oscillator circuit is (*degenerative/regenerative*).
7. The frequency of a relaxation oscillator is determined by an (*LC/RC/LR*) network.

Refer to Figure 14–24 for questions 8 through 12.

8. While the output of U_2 is at its positive-most level, the output of U_1 is ramping in a (*negative/positive*) direction.
9. While the output of U_2 is at its negative-most level, electron current is charging capacitor C_1 with a positive charge on its right side. (True/False)

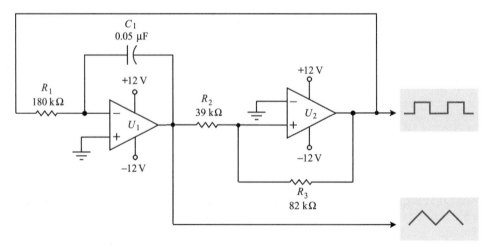

FIGURE 14–24 An oscillator circuit.

10. This circuit is an example of a relaxation oscillator. (True/False)
11. What is the approximate frequency of oscillation for this circuit?
12. If you temporarily connected a second 0.05-μF capacitor in parallel with C_1, what effect would this have on circuit operation?
13. What waveform is generated by a Wien-bridge oscillator?

Refer to Figure 14–25 for questions 14 through 20.

14. What is the approximate frequency of oscillation for this circuit?
15. What is the overall closed-loop voltage gain for the circuit (including loss in the feedback network)?
16. If the output amplitude increases slightly, what happens to the conduction of Q_1?
17. What is the basic waveform or signal type across R_3?
18. If you wanted to increase the frequency of oscillation, you could (*decrease/increase*) C_1 and C_3.
19. The time constants of C_2 and R_3 are the primary frequency-determining components in this circuit. (True/False)
20. What is the primary purpose of U_1?

Refer to Figure 14–26 for questions 21 through 25.

21. This is a relaxation-oscillator circuit. (True/False)
22. Describe the charge and discharge paths for C_1.
23. What is the approximate frequency of oscillation for this circuit?
24. The positive output for this particular circuit configuration will always be longer than the negative (0 V) excursion. Explain why this is true.
25. What would you expect to happen to the frequency of oscillation if resistor R_1 were increased in value?

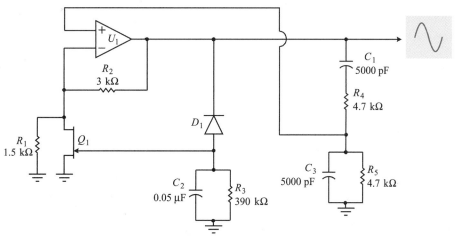

FIGURE 14–25 An oscillator circuit.

Fig14_25.msm

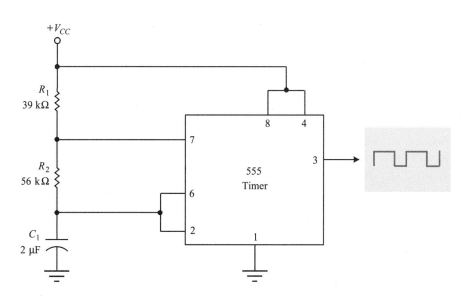

FIGURE 14–26 An oscillator circuit.

26. Which general class of oscillators provides the greatest degree of frequency stability?
27. The operation of crystal oscillators relies on the _____ effect of crystals.
28. Crystals have very (*low/high*) values of Q.
29. Crystals are generally only used for circuits requiring oscillations above 100 kHz or so. They are not practical for very low-frequency applications. (True/False)
30. Microprocessors and other high-speed integrated circuits often use external crystals to control the frequency of an internal oscillator circuit. (True/False)
31. If a modular oscillator circuit is shown to be defective, what must be done to repair it?
32. What is the name given to an oscillation that occurs in a circuit that was not specifically designed to be an oscillator?
33. What is the abbreviation (acronym) typically used to refer to a phase-locked loop?
34. The VCO in a phase-locked loop must be inherently stable with regard to frequency of oscillation. (True/False)
35. If the programmable counter in a phase-locked loop is set to count to a higher value, what effect does this have on the output frequency?
36. Once a phase-locked loop has achieved a locked condition, the output frequency is just as stable as the input reference frequency. (True/False)
37. A phase-locked loop achieves stable operation by employing (*open/closed*)-loop operation.
38. A certain microprocessor claims to have an operating speed of 1.5 GHz, but it only requires a 66.6-MHz crystal-controlled input waveform. Explain how this apparent discrepancy might be resolved if the microprocessor has an internal PLL.

Section 14.2: Industrial Computer Applications

39. What must be done to change the voltage gain of an instrumentation-amplifier circuit?
40. If the differential-voltage gain of a certain instrumentation amplifier is 100, you would expect the common-mode voltage gain to be (*much less/much greater*).
41. What is the input to a VFC circuit?
42. If the output of a VFC were connected to the input of an FVC, describe the resulting output from the FVC module.

Refer to Figure 14–17 for questions 43 through 46.

43. What is the purpose of capacitor C_2?
44. For the values shown in the circuit, what is the frequency of the output waveform if the input voltage is 6.25 V?
45. When the electronic switch is closed connecting the current source to the integrator, the output of the integrator ramps is in the (*negative/positive*) direction.
46. If the value of R_1 is increased, what effect does this have on the output frequency if the input voltage range remains the same?

Section 14.3: Troubleshooting Circuits Based on ICs

47. If an oscillator is completely inoperative, what is the first thing to check?
48. If the dc voltage on the output of an oscillator circuit is abnormal, then the defect can be classified as a(n) (*dc/ac*) defect.
49. A shift in the dc operating point of an oscillator circuit can cause a distorted waveform in the output. (True/False)
50. Why is it not possible to predict the output of an IC with internal memory by simply measuring its inputs?
51. An integrated audio amplifier based on bipolar transistors is an example of an IC with an internal memory function. (True/False)

•—CIRCUIT EXPLORATION

This section will give you an opportunity to become more familiar with the 555 timer IC that was briefly discussed earlier in the chapter. We will explore three different circuits that utilize the 555 timer.

First, construct the circuit shown in Figure 14–27 either in the lab or in a circuit simulation environment.

Accomplish the following tasks:
1. Measure the output frequency.
2. Determine the output duty cycle (high-level output/total cycle time).
3. Plot a graph that shows the relationship between the voltage across C_1 and the output voltage.
4. Reduce the supply voltage to 5 V and determine the effect on output frequency.

Next, construct the circuit shown in Figure 14–28 either in the lab or in a circuit simulation environment.

Accomplish the following tasks:

5. Measure the output frequency.
6. Determine the output duty cycle (high-level output/total cycle time).
7. Change R_1 to 27 kΩ.
8. Determine the new output duty cycle (high-level output/total cycle time).
9. Explain the advantage of this configuration over the one shown in Figure 14–27.

Now, construct the circuit shown in Figure 14–29 on page 540 either in the lab or in a circuit simulation environment.

Accomplish the following tasks:

10. What is the state of the LED before you close switch S_1?
11. Describe the action when switch S_1 is pressed and released (observe for at least 10 seconds).
12. Describe the action when switch S_1 is pressed and held (observe for at least 10 seconds).
13. Based on your knowledge of the 555 timer IC, write a brief theory of operation for this circuit configuration.

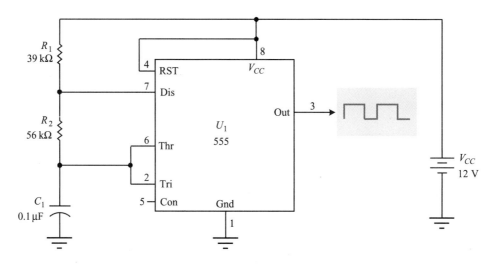

FIGURE 14–27 A 555 oscillator circuit.

Fig14_27.msm

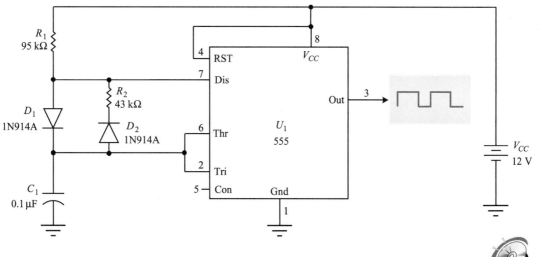

FIGURE 14–28 A variation of the basic 555 oscillator circuit.

Fig14_28.msm

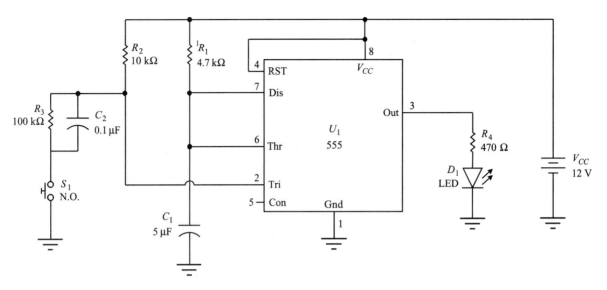

1. Change to 470 kΩ if built in lab (i.e., not simulated)

FIGURE 14–29 A time-delay circuit.

Fig14_29.msm

CHAPTER 15
Digital Electronics

•–KEY TERMS
asynchronous counter
asynchronous input
binary
bit
bubble
glitch
hexadecimal
logic analyzer
logic probe
logic pulser
maxterm
minterm
monotonic output
propagation delay
race condition
radix point
resolution
state symbol
synchronous counter
toggle
vinculum

•–OBJECTIVES
After studying this chapter, you should be able to:

- Make conversions between numbers expressed in binary, hexadecimal, and decimal formats.
- Determine the behavior of combinational and sequential logic circuits using truth tables, Boolean expressions, and logic diagrams. The logic circuits may contain any of the following logic devices:

Inverter	Buffer	AND gate
NAND gate	OR gate	NOR gate
Exclusive OR gate	Exclusive NOR gate	JK flip-flop
D flip-flop		

- Explain one D/A conversion technique and one A/D conversion technique.
- Troubleshoot circuits containing both combinational and sequential logic circuits. This includes selection of appropriate test equipment and application of an appropriate troubleshooting strategy.

•–INTRODUCTION
You no doubt realize that modern computers are composed mostly of digital circuits. This chapter will introduce you to the concepts and building blocks used in all digital electronics systems. Due to the nature of digital systems, you can easily forget that these impressive devices and circuits are composed of the same resistors, capacitors, transistors, and so forth that we have studied in earlier chapters of this book. Therefore, although the material may appear to be very new, related only slightly to the circuits and devices previously studied, rest assured that every digital device and circuit is built with familiar components.

15.1 DIGITAL CONCEPTS AND TERMINOLOGY

This section will acquaint you with some new terminology and tools that will provide the foundation needed to understand the operation of digital electronic devices and systems. While the nature of the material may seem like a departure from previous electronic studies, it is essential that you master the material in this section. This material provides the language that will allow us to discuss digital systems.

Binary Numbering System

Your experience with number systems is probably limited to the decimal number system (base 10). The operation of computers and all other digital systems, however, is based on the **binary** number system (base 2). You do not have to be an expert at doing manual arithmetic operations and manipulations with binary numbers. However, you must be able to interpret the binary system in order to understand the operation of digital circuits, which utilize this method of representing values.

Interpretation of Binary Numbers

To help you get a quick start, let's compare and contrast the binary numbering system with the familiar decimal system. Refer to Table 15–1.

TABLE 15–1 A comparison of decimal and binary number system characteristics.

Characteristic	Decimal System	Binary System
Indication of number system (base)	Subscript 10 (e.g., 259_{10})	Subscript 2 (e.g., 110011_2)
Number of digits	Ten: 0, 1, 2, 3, 4, 5, 6, 7, 8, and 9	Two: 0 and 1
Weights of digit positions	...10^3 10^2 10^1 10^0. 10^{-1} 10^{-2}..., or ...1000 100 10 1 . 1/10 1/100...	...2^4 2^3 2^2 2^1 2^0. 2^{-1} 2^{-2}..., or ...16 8 4 2 1 . ½ ¼
Determination of value	Sum of the products obtained by multiplying each digit by the weight of its position. *Example:* $327_{10} = 3 * 100 + 2 * 10 + 7 * 1$	Sum of the products obtained by multiplying each digit by the weight of its position. *Example:* $1101_2 = 1 * 8 + 1 * 4 + 0 * 2 + 1 * 1$ or 13_{10}

You can see from Table 15–1 that there is a lot of similarity between the two numbering systems. In essence, the basic rules are the same, it is just that we have fewer digits. The binary numbering system allows only two digits. A single binary digit is called a **bit** (**bi**nary digi**t**). So, for example, we might say that a certain logic device is an 8-bit circuit. In that case, it could manipulate numbers such as 11001010, 11100010, 11111011, 00010011, and so forth. Microprocessors are available in a variety of bit widths including 4, 8, 16, 32, 64, 128, and more. Your personal computer probably manipulates values that are at least 64 bits wide.

The **radix point** separates the integer portion of a number from the fractional portion. In the decimal system, the radix point is called the decimal point. The column to the immediate left of the radix point has a value of 1. Each higher column to the left has a weight that is equal to the previous column multiplied by the number base (2 for binary and 10 for decimal). Each column to the right is equal to the preceding column divided by the number base.

Numerical Sequences

Counting (i.e., incrementing to the next higher number) in binary is very simple and follows the same basic rule as counting in decimal numbers. In the decimal system, we count by incrementing the least-significant digit (the one to the immediate left of the decimal point) to the next higher digit value. If the initial digit value is already the largest possible value (9 in the case of the decimal system), we replace it with the smallest digit (0), and then increment the next higher digit position.

It works the same way with binary numbers, except the highest digit value is 1. Let's work some examples.

15.1 • DIGITAL CONCEPTS AND TERMINOLOGY

•—EXAMPLE 15.1

Increment each of the following binary numbers by one:

1. 1000   3. 1001   5. 1111
2. 1010   4. 1011

SOLUTION

Problem	Results	Explanation
1.	1000	Original number
	1001	Increment least-significant bit (LSB). Final result.
2.	1010	Original number
	1011	Increment least-significant bit (LSB). Final result.
3.	1001	Original number
	1000	Increment least-significant bit (LSB). Since it is already at the highest digit value (1), we replace it with the lowest digit value (0) and continue to the next higher column.
	1010	Increment the second column. Final result.
4.	1011	Original number
	1010	Increment least-significant bit (LSB). Since it is already at the highest digit value (1), we replace it with the lowest digit value (0) and continue to the next higher column.
	1000	Increment the second column. Since it is already at the highest digit value (1), we replace it with the lowest digit value (0) and continue to the next higher column.
	1100	Increment the third column. Final result.
5.	1111	Original number
	1110	Increment least-significant bit (LSB). Since it is already at the highest digit value (1), we replace it with the lowest digit value (0) and continue to the next higher column.
	1100	Increment the second column. Since it is already at the highest digit value (1), we replace it with the lowest digit value (0) and continue to the next higher column.
	1000	Increment the third column. Since it is already at the highest digit value (1), we replace it with the lowest digit value (0) and continue to the next higher column.
	0000	Increment the fourth column. Since it is already at the highest digit value (1), we replace it with the lowest digit value (0) and continue to the next higher column.
	10000	Increment the fifth column (leading zeros typically not shown). Final result.

•—Practice Problems

Increment each of the following binary numbers by one:

1. 1110   3. 1101   5. 0101   7. 1010011
2. 0110   4. 0011   6. 1001110   8. 1010100101010

Converting Binary to Decimal

It is important to be able to express an existing binary number as an equivalent decimal number. There are several ways this can be accomplished. We will examine one pencil and paper method, but most engineering calculators make these conversions almost trivial. The following procedure can be used to convert a binary number to an equivalent decimal value:

1. Multiply the weight of each column by the digit value in that column.
2. Sum the products obtained in step 1.

•—EXAMPLE 15.2

Convert the binary number 11011 to an equivalent decimal number.

SOLUTION Beginning at the radix point (assumed to be to the right of the number as with decimal numbers), we assign column weights of 1, 2, 4, 8, and 16. We then multiply each of these by the digit value in a given column. This process is accomplished as follows:

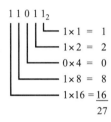

•—Practice Problems

Convert the following binary numbers to equivalent decimal numbers:

1. 110
2. 11100
3. 1010
4. 00110
5. 111111
6. 100001

Converting from binary format to decimal format with an engineering calculator is normally as straightforward as entering the number and then pressing a conversion button. You will need to refer to the operator's manual for your specific engineering calculator, but the conversion process should be very simple.

Converting Decimal to Binary

If we have a value expressed as a decimal number, we should be able to convert it to an equivalent binary value. Again, there are multiple ways to accomplish this (including the use of your engineering calculator's conversion functions). We will examine one pencil and paper method. The procedure is as follows:

1. Divide the decimal number by 2.
2. Write the remainder down as a bit in the converted number beginning with the least-significant bit (LSB).
3. Repeat the first two steps until a quotient of 0 is obtained.

•—EXAMPLE 15.3

Convert the decimal number 69 into an equivalent binary value.

SOLUTION We apply the conversion procedure as follows:

$\frac{69}{2}$ = 34 with a remainder of 1 →   1 (LSB)

$\frac{34}{2}$ = 17 with a remainder of 0 →   0

$\frac{17}{2}$ = 8 with a remainder of 1 →   1

$\frac{8}{2} = 4$ with a remainder of 0 → 0

$\frac{4}{2} = 2$ with a remainder of 0 → 0

$\frac{2}{2} = 1$ with a remainder of 0 → 0

$\frac{1}{2} = 0$ with a remainder of 1 → 1 (MSB)

So, the number 69_{10} can also be expressed as 1000101_2.

•—Practice Problems

Convert the following decimal numbers to equivalent binary numbers:

1. 15
2. 27
3. 7
4. 35
5. 75
6. 18

Converting Between Binary and Hexadecimal

Computers operate on binary numbers, but it is relatively burdensome for humans to directly interpret binary values. So, it is common for computer engineers and technicians to express binary values in hexadecimal format. **Hexadecimal** is simply the base 16 number system. The sixteen allowable digits in the hexadecimal system include 0 through 9 and A through F. Table 15–2 lists the decimal and binary equivalents for each of the hexadecimal digits.

TABLE 15–2 Decimal and binary equivalents for each hexadecimal digit.

Hexadecimal Digit	Binary Value	Decimal Value
0	0000	0
1	0001	1
2	0010	2
3	0011	3
4	0100	4
5	0101	5
6	0110	6
7	0111	7
8	1000	8
9	1001	9
A	1010	10
B	1011	11
C	1100	12
D	1101	13
E	1110	14
F	1111	15

Conversion between binary and hexadecimal values is very simple. Here is the procedure:

1. Begin at the radix point (or the far right for integer values) and mark off groups of four binary digits. Add leading zeroes as required to form the last group of four bits.

2. Replace each group of four bits with an equivalent hexadecimal digit (given in Table 15–2).

•—EXAMPLE 15.4

Convert the binary number 101010100101111101010_2 to an equivalent hexadecimal number.

SOLUTION First, we divide the original binary number into groups of four starting at the right-most bit (or the radix point for mixed numbers):

$$0001\ 0101\ 0100\ 1011\ 1110\ 1010$$

Note the addition of the three leading zeroes to form the left-most group of four. Second, we replace each of these 4-bit groups with an equivalent hexadecimal digit. Thus, the final converted value is $154BEA_{16}$. Many people represent hexadecimal numbers by appending the letter "h." In this case, we might write our result as 154BEAh.

Conversion from hexadecimal to binary is simply the reverse process. That is,

1. Replace each hexadecimal digit in the original number with an equivalent group of four binary digits.
2. Combine all bits to form the binary value.

•—EXAMPLE 15.5

Convert 5A72h to binary.

SOLUTION Straight substitution yields: 0101 1010 0111 0010. We merge these to form the converted binary value:

$$0101101001110010_2$$

Logic States

Digital devices operate on binary values, so it follows that there need be only two allowable digital states. One state will represent the binary digit 0. The other state will represent the binary digit 1. Each of these allowable states is called a logic state. Thus, digital devices and circuits have only two allowable logic states. Any number of distinguishing labels can be assigned to these states. Examples include 0/1, low/high, no/yes, false/true, and any number of other alternatives. The important thing to recognize is that there are only two states allowed.

Figure 15–1 shows an npn switching transistor. We know that a switching transistor is either cutoff or saturated. Since it has only two allowable states, it can be used as a logic device. When it is cutoff, the output will be +5 V. When it is saturated, the output will be near 0 V. We could assign other more meaningful labels to these voltage values. In one instance, we may choose to call 0 V false and +5 V true. We may also choose a more meaningful label for the output such as DISK BUSY. Once we have made these assignments, we no longer need to speak in terms of volts and other electrical terms. Rather, we can simply say something like, "The signal DISK BUSY is true."

Many present-day computers and other digital systems use logic devices that use 0 V/5 V or 0 V/3.3 V as their logic states. Some portions of the system (e.g., the serial communications port) may use −12 V/+12 V as the two logic levels. Many industrial systems use 0 V/+24 V as the allowable levels external to the computer proper. But, regardless of the electrical details of the circuit, all of these

15.1 • DIGITAL CONCEPTS AND TERMINOLOGY   547

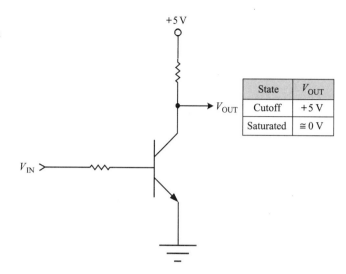

FIGURE 15–1 A switching transistor has two allowable states.

systems can refer to the various points in the circuit as being true/false, high/low, or some other nonelectrical designation. For the purposes of this text, we shall generally use the terms 0, low, and false to mean the same thing. Similarly, their counterparts of 1, high, and true will be considered as equivalent.

Truth Tables

Truth tables provide an orderly way to describe the relationship between inputs and outputs of many types of logic devices and circuits. Figure 15–2(a) shows a logic circuit with two inputs labeled A and B, and a single output labeled Y. Figure 15–2(b) shows a truth table to describe the behavior of the circuit. Throughout this chapter, we will consider A to be the least-significant variable (as in Figure 15–2), which is consistent with industry practice.

A truth table lists all possible combinations of the inputs along with the output that occurs under each of those conditions. The number of possible combinations of input conditions is given by Equation 15–1.

$$\text{Combinations} = 2^N \qquad (15\text{–}1)$$

where N is the number of input variables. In the example shown in Figure 15–2, there are two input variables, so there are 2^2 or 4 entries in the truth table. It is customary to list the input combinations in binary order. In other words, start with zero and increase by one with each successive entry. This lessens the chance of missing an entry, and it simplifies other more advanced design techniques.

The output column simply indicates the state of the output for the given input conditions. In the example shown in Figure 15–2, the output will be at a high (1) level, if and only if both inputs are low (0) at the same time. Truth tables are used extensively, so you must be able to interpret them. You will get additional practice as we progress through this chapter.

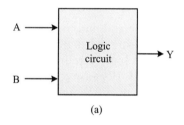

FIGURE 15–2 (a) A 2-input logic circuit and (b) its truth table.

Timing Diagrams

Timing diagrams are also used to show relationships between several digital signals. They are essential for understanding the detailed operation of a microprocessor or other complex digital IC. Figure 15–3(a) on page 548 shows a simple 3-input logic circuit with two outputs. A representative timing diagram is shown in Figure 15–3(b). The horizontal axis represents time. The vertical dashed lines identify discrete time intervals. All possible combinations of input variables are represented by one of the time intervals. In this particular case, the timing diagram

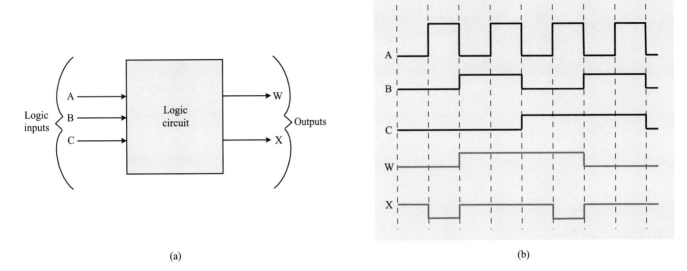

FIGURE 15–3 Timing diagram for a 3-input logic.

and an equivalent truth table would serve the same purpose. In more complex systems, however, timing diagrams are more widely used.

Boolean Algebra and Logic Gates

Boolean algebra provides us with another means to describe and understand the behavior of logic circuits. Boolean algebra is used to describe the logical relationship between input variables and output variables in a logic circuit. Logic diagrams present yet another way to describe a logic circuit. Because these two representations are so closely related, we will discuss them at the same time.

Logical Operators

There are only a few fundamental operators used for logical relationships. We will discuss the following operators and some combinations of these basic operators:

- NOT
- AND
- OR
- Exclusive OR

AND *Operator* Figure 15–4 illustrates the AND operator. Figure 15–4(a) shows the logic representation for an AND gate. Physically, the AND gate will be part of an integrated circuit that may contain additional AND gates or other types of logic

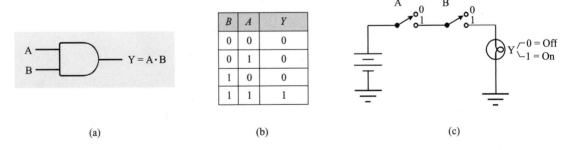

FIGURE 15–4 (a) The logic symbol for an AND gate along with its Boolean expression, (b) the truth table for a 2-input AND function, and (c) an electrical analogy.

circuits. The Boolean expression for a 2-input gate is shown as $Y = A \cdot B$ or simply $Y = AB$. Either of these expressions is verbalized as, "Y equals A and B."

Figure 15–4(b) shows the truth table for a 2-input AND function. As you can see, the output is high if and only if both inputs are high at the same time. That is, Y is high if A and B are high. Figure 15–4(c) shows that the AND function can be represented by series switches. If we assign 0 to an open switch, 1 to a closed switch, 0 to a dark lamp, and 1 to a lit lamp, then the lamp will be 1 only when both A and B are 1.

A 2-input AND function is shown in Figure 15–4 for simplicity, but in practice there can be any number of inputs. However, there can only be a single output. So, a 4-input AND gate, for example, might be represented as $Y = ABCD$. In this case, the output would be high only if all four inputs were high at the same time. In short, regardless of the number of inputs on an AND gate, they must all be high in order to get a high at the output. The 74HC11 by Philips (www.philipslogic.com) is a representative 3-input AND gate with three gates in the same IC package.

OR *Operator* Figure 15–5(a) shows the logic symbol and Boolean expression for a 2-input OR function. The expression is verbalized as, "Y equals A or B." Do not read it, say it, or even think of it as "A plus B," or you may get very confused on some circuits.

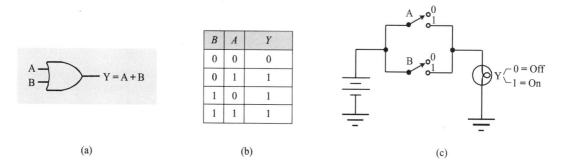

(a)   (b)   (c)

FIGURE 15–5 (a) The logic symbol for an OR gate along with its Boolean expression, (b) the truth table for a 2-input OR function, and (c) an electrical analogy.

The truth table for a 2-input OR function is shown in Figure 15–5(b). Here, we can see that the output of an OR function will be high if either or both of its inputs are high. Finally, Figure 15–5(c) illustrates an electrical analogy consisting of two parallel switches. Here, the lamp will illuminate (defined as 1) if either or both of the switches are closed (defined as 1).

As with AND gates, the OR gate can have any number of inputs, but it has only a single output. A 3-input OR gate might be represented with the expression $Y = A + B + C$. Here, the output will be high if any one or more of the inputs are high. The HEF4071B device from Philips (www.philipslogic.com) has four 2-input OR gates in a single package.

NOT *Operator* Figure 15–6(a) shows the logic symbol and Boolean expression for an inverter. This is a simple but important logic operation. The output of the inverter is always the opposite logic state from the input. The Boolean expression is read as, "Y equals NOT A." A line—called a **vinculum**—is drawn above a variable or an expression to indicate the NOT or complement function as shown in Figure 15–6(a) on page 550.

The inverter function is clearly revealed by the truth table in Figure 15–6(b). Whatever state is applied to the input causes the opposite state to appear on the output. An inverter can only have a single input and a single output.

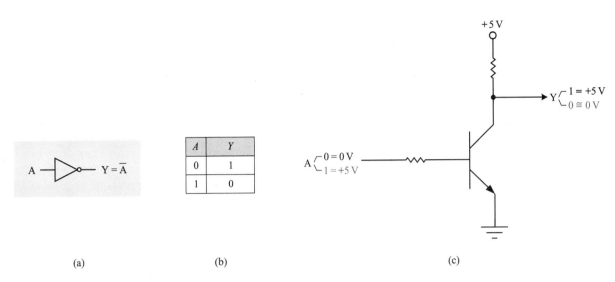

FIGURE 15–6 (a) The logic symbol for an inverter along with its Boolean expression, (b) the truth table for an inverter function, and (c) an electrical analogy.

Figure 15–6(c) illustrates one way to obtain a physical inverter. Here, we define 0 V and +5 V as logic levels 0 and 1, respectively. So, if a 0 is applied to the input of the inverter, the transistor is cutoff (no bias). This causes the collector (output) to rise to +5 V, which is a 1. By contrast, if a 1 (+5 V) is connected to the input, the transistor goes into saturation. This means the output will be near 0 V (logic level 0). Unlike analog circuits, we rarely care about the exact voltage level in a logic circuit. The logic gates are designed to correctly recognize logic states as long as they are within a defined range of allowable values. In a TTL gate, for example, any voltage between 0 V and +0.8 V will be interpreted by a gate input as a low, and any voltage in the range of 2.0 V to 5 V will be sensed as a high-level signal. The exact voltage levels that are allowed are determined by the technology used to fabricate the devices and the power supply voltage used. Appendix A contrasts several logic families. The SN7404 from Texas Instruments (www.ti.com) has six inverters in a single package.

NAND *Operator* A NAND gate is one of the most widely used logic functions. As its name implies, it is a combination of an AND gate and an inverter. You can initially think of a NAND function as a NOT AND operation, but because it is so commonly used, you should quickly work to recall the NAND operation directly without using the AND gate crutch. Figure 15–7(a) shows the logic symbol and Boolean expression for a 2-input NAND gate. The small circle (often called a **bubble** or **state symbol**) on the output indicates the inversion operation. A vinculum over the output expression reveals the presence of the NOT operation. That is, a vinculum is placed above the expression A AND B to form the overall NOT A AND B or NAND function. It is important to note that the input variables are ANDed first and then inverted. This is not the same result as inverting each variable and then ANDing them. That is, $\overline{AB} \neq \overline{A}\,\overline{B}$.

Figure 15–7(b) shows the truth table for a 2-input NAND gate. As you can readily see, the output is exactly opposite (i.e., inverted) from that of an AND gate. You can verbalize its operation by saying, "Any low in will give you a high out." The equivalent electrical circuit in Figure 15–7(c) agrees with this behavior. In order to get a high on the collector of the transistor (the Y output), the transistor must be cutoff. If either switch (or both) is open (logic 0), there will be no base current, and the transistor will be cutoff.

As with AND gates and OR gates, a NAND gate can have any number of inputs. Commercial devices, however, are generally limited to 2-, 3-, 4-, 8-, and 13-input

FIGURE 15-7 (a) The logic symbol for a 2-input NAND gate along with its Boolean expression, (b) the truth table for a 2-input NAND gate function, and (c) an electrical analogy.

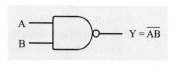

(a)

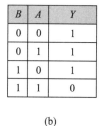

(b)

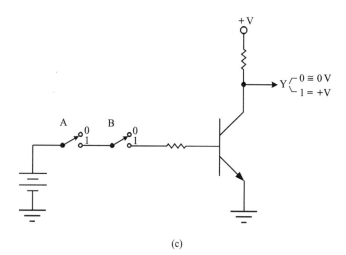

(c)

gates. These restrictions are a natural result of using standard IC packages with a fixed number of pins. Larger numbers of inputs are readily available in larger (generally custom-made) ICs. The SN7400 device from Texas Instruments (www.ti.com) provides four 2-input NAND gates in a single package.

NOR *Operator* The logic symbol and Boolean expression for a 2-input NOR gate are shown in Figure 15–8(a) on page 552. As indicated by the Boolean expression and the logic symbol, the input variables are ORed and the result is complemented. Note that the vinculum extends over both input variables and the OR operator. It is essential to realize that $\overline{A + B} \neq \overline{A} + \overline{B}$.

The truth table in Figure 15–8(b) reveals that a NOR operator produces the exact complement of the OR operator, which you should expect. You can verbalize this behavior by saying, "Any high in gives you a low out."

Finally, in Figure 15–8(c), we see an equivalent circuit for a 2-input NOR gate. If either switch is closed (logic 1), the transistor will be biased into saturation producing a low output. Only when both switches are open (logic 0) will the transistor be cutoff, allowing the output to rise to a high logic state.

As with AND, OR, and NAND operators, the NOR operator can have any number of inputs from a Boolean expression point of view. However, practical gates that are commercially available have limited inputs similar to the other basic logic gates. The CD74AC02 from Texas Instruments, for example, provides four 2-input NOR gates in a single integrated circuit package.

Exclusive OR *Operator* The logic symbol and Boolean expression for a 2-input exclusive OR function are shown in Figure 15–9(a) on page 552. The truth table for this logic function is shown in Figure 15–9(b). As evidenced in the truth table, the output will be high whenever either but not both of the inputs is high. If the inputs are the same (either low or high), the output will be low.

FIGURE 15-8 (a) The logic symbol for a 2-input NOR gate along with its Boolean expression, (b) the truth table for a 2-input NOR gate function, and (c) an electrical analogy.

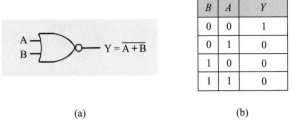

(a)

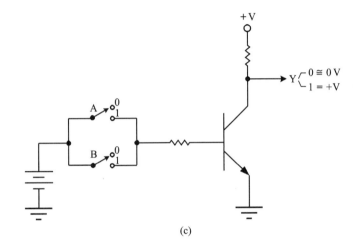

(b)

(c)

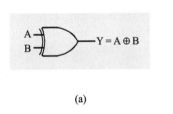

(a)

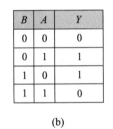

(b)

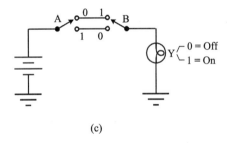

(c)

FIGURE 15-9 (a) The logic symbol for an exclusive OR gate along with its Boolean expression, (b) the truth table for an exclusive OR function, and (c) an electrical analogy.

Figure 15-9(c) shows one way to realize an equivalent circuit for a 2-input exclusive OR gate. With the logic levels as defined on the schematic in Figure 15-9(c), the output will be high (lamp on) anytime the two switches are in different logic states.

The exclusive OR expression is generally written $Y = A \oplus B$ as shown in Figure 15-9(a). However, the exclusive OR function can also be expressed as $Y = A\overline{B} + \overline{A}B$, or alternatively as $Y = (A + B)(\overline{A} + \overline{B})$.

The exclusive OR function can be extended to any number of inputs. The output will be high whenever there are an odd number of high inputs. Generally, commercially available gates are limited to two inputs, but they may be cascaded to accommodate any practical number of inputs. Additionally, exclusive OR functions with higher numbers of inputs are sometimes included as an integral part of a more complex integrated circuit. The SN74AC86 is an exclusive OR package made by Texas Instruments (www.ti.com) that has four 2-input gates.

Exclusive NOR *Operator* The logic symbol and Boolean expression for a 2-input exclusive NOR gate are shown in Figure 15-10(a). The symbol is the same as an exclusive OR gate with the addition of an inversion bubble on the output terminal.

Similarly, the Boolean expression is identical to that of an exclusive OR function except a vinculum is placed over the entire expression. Although the expression $Y = \overline{A \oplus B}$ is the most common way to express the exclusive NOR function, it can also be expressed as $Y = \overline{A}\,\overline{B} + AB$, or alternatively, as $Y = (\overline{A} + B)(A + \overline{B})$.

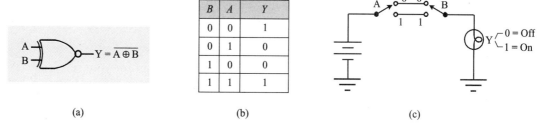

(a)   (b)   (c)

FIGURE 15–10 (a) The logic symbol for an exclusive NOR gate along with its Boolean expression, (b) the truth table for an exclusive NOR function, and (c) an electrical analogy.

The truth table for a 2-input exclusive NOR gate is shown in Figure 15–10(b). Note that the exclusive NOR truth table is the exact complement of an exclusive OR truth table. The output is high whenever the two inputs are the same (either both low or both high). If functions with higher numbers of inputs are implemented (often by interconnecting 2-input gates), the output will be high anytime there is an even number of high inputs.

Figure 15–10(c) shows a simple equivalent circuit for a 2-input exclusive NOR gate. This is identical to the equivalent circuit for an exclusive OR gate except for the logic states assigned to the various switch positions. With the assignments shown in Figure 15–10(c), the output will be high (lamp on) whenever the two inputs are both low or both high. The CD4077B from Texas Instruments (www.ti.com) provides four 2-input exclusive NOR gates in each package.

Switching Times If the inputs are instantaneously applied to a logic gate, there is a short delay before the output is correct. This short delay is called the **propagation delay** of the gate. Typical delays are from hundreds of picoseconds to a few nanoseconds. Unless specifically stated, we will assume that all gate propagation delays are zero. This same assumption is normally made during a troubleshooting exercise.

Boolean Algebra Postulates
Boolean algebra can be used to simplify complex expressions in order to reduce the complexity of a logic circuit. However, with modern designs, nearly all of the complex logic circuits are internal to ICs. That is, circuits made from individual logic gates are generally limited to two or three gates at most. Nonetheless, Boolean algebra can still be quite useful as an analytical tool. As with standard algebra, Boolean algebra has a number of basic rules called postulates or identities. Following are some of the more useful ones:

1. $A = 0$ if $A \ne 1$
2. $A = 1$ if $A \ne 0$
3. $0 \cdot 0 = 0$
4. $1 \cdot 1 = 1$
5. $0 + 0 = 0$
6. $1 + 1 = 1$
7. $1 \cdot 0 = 0$
8. $1 + 0 = 1$
9. $\overline{1} = 0$
10. $\overline{0} = 1$
11. $A + 0 = A$
12. $A + 1 = 1$
13. $A + A = A$
14. $A + \overline{A} = 1$
15. $A \cdot 0 = 0$
16. $A \cdot 1 = A$
17. $A \cdot A = A$
18. $A \cdot \overline{A} = 0$
19. $A = A$
20. $\overline{\overline{A}} = A$

21. $(A) = A$
22. $\overline{(\overline{A})} = \overline{A}$
23. $(AB)C = ABC$
24. $(A + B) + C = A + B + C$
25. $AB = BA$
26. $A + B = B + A$
27. $\overline{\overline{A}} = A$
28. $A(B + C) = AB + AC$
29. $A + BC = (A + B)(A + C)$
30. $A(A + B) = A$
31. $A + AB = A$
32. $A + \overline{A}B = A + B$
33. $A\,(\overline{A} + B) = AB$
34. $\overline{A + B + C \ldots + n} = \overline{A}\,\overline{B}\,\overline{C} \ldots \overline{n}$
35. $\overline{ABC \ldots n} = \overline{A} + \overline{B} + \overline{C} \ldots + \overline{n}$
36. $(A + B)(C + D) = AC + AD + BC + BD$

The rules are included at this point for your reference. We will refer to them as we work through the circuit problems in the next section. Rules 34 and 35 are very important and deserve some additional clarification. Here is a simple procedure that allows you to reduce the complexity of an expression having one or more vincula.

> ▶ **PROCEDURE:**
> 1. Count the number of vincula above each variable and above each operator. Variables with an odd number of vincula have a single vinculum in the final expression. Operators with an even number of vincula remain as they are, whereas operators with an odd number of vincula are changed (i.e., OR becomes AND, and AND becomes OR).
> 2. Use parentheses where necessary to ensure that all variables that were grouped in the initial expression remain grouped after the simplification process.

●—**EXAMPLE 15.6**

Simplify the following expression: $Y = \overline{A\overline{B} + \overline{CD}}$.

SOLUTION

$Y = \overline{A \cdot \overline{B} + \overline{C \cdot D}}$

$\overline{A} + B \cdot C \cdot D$ Combine/break vincula

$(\overline{A} + B)(CD)$ Maintain original groups

$(\overline{A} + B)CD$ Remove unnecessary parentheses

$\overline{A}CD + BCD$ Alternate form (Rule 28)

●—**Practice Problems**

Simplify the following expressions. For purposes of this exercise, an expression will be considered simplified when no two variables are joined by a single vinculum.

1. $Y = \overline{A + B + C + \overline{D}}$
2. $Y = \overline{(\overline{A} + B) \cdot (C + D)}$
3. $Y = \overline{\overline{A} + BC + D}$
4. $Y = \overline{AB + \overline{\overline{C}}}$
5. $Y = \overline{\overline{AB} + CD\overline{E}}$
6. $Y = \overline{(A + \overline{B}C) \cdot (D + \overline{\overline{E}})}$

Exercise Problems 15.1

1. A single binary digit is called a _____.
2. Write the next three sequential numbers after the binary number 10101.
3. Express each of the following binary numbers as an equivalent decimal number:
   a. 10010
   b. 10001
   c. 1001
   d. 1011
   e. 11001
   f. 10011
   g. 1101
   h. 1000101
   i. 11111
   j. 000011
4. Express each of the following decimal numbers as an equivalent binary number:
   a. 6
   b. 14
   c. 10
   d. 21
   e. 52
   f. 104
   g. 27
   h. 72
   i. 2
   j. 23
5. If one logic state in a digital circuit is defined as ON, then the other will likely be defined as _____.
6. Draw a truth table for a 3-input NAND gate.
7. How many entries would be required in a truth table that described a 5-input logic circuit?
8. What type of logic function produces a high output only if all of its inputs are high?
9. Write the Boolean expression for a 3-input OR gate whose inputs are A, B, and C. The output is labeled Y.
10. What type of logic function produces a low output anytime one or more of its inputs are high?
11. What type of logic function has only a single input?
12. Identify the type of logic gate that would be described by each of the following Boolean expressions:
    a. $Y = A + B + C$
    b. $Y = \overline{ABCD}$
    c. $Y = \overline{A \oplus B}$
    d. $Y = \overline{A + B}$
13. What logic function produces a high output anytime there is an odd number of high inputs?
14. Which Boolean rule will allow the expression $Y = \overline{AB}$ to be written as $Y = \overline{A} + \overline{B}$?
15. Which Boolean rule would tell you that two cascaded inverters is the same as no inverter at all?
16. Simplify the expression: $Y = \overline{AC} + \overline{AB} + DE$
17. Simplify the expression: $Y = \overline{ADE} + \overline{B + C} + \overline{F}$

15.2 COMBINATIONAL LOGIC

Figure 15–11 shows the general block diagram that represents a combinational logic circuit (also called combinatorial logic circuit). A combinational logic circuit can have any number of inputs and any number of outputs. However, the logic states of the inputs at any instant in time determine the state of the outputs. In the next section, we will study sequential logic circuits whose outputs at any given time are not determined solely by its inputs at that same time.

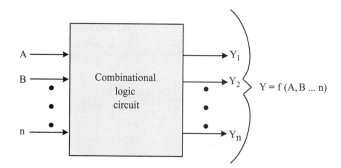

FIGURE 15–11 The generalized functional block diagram for a combinational logic circuit.

Logic Gates

All digital circuits are based on interconnections of the various logic gates discussed earlier in the chapter. In the case of complex circuits, the various gates are generally integrated into one or more integrated circuits. Nonetheless, the basic building blocks are the AND, OR, and NOT gates along with their derivatives (e.g., NAND, NOR, exclusive OR, and exclusive NOR). Figure 15–12 on page 556 shows several representative IC packages used for basic logic gates.

FIGURE 15–12 Physical package and logic symbols for (a) single 2-input OR gate, (b) hex inverter, and (c) quad 2-input NAND gate.

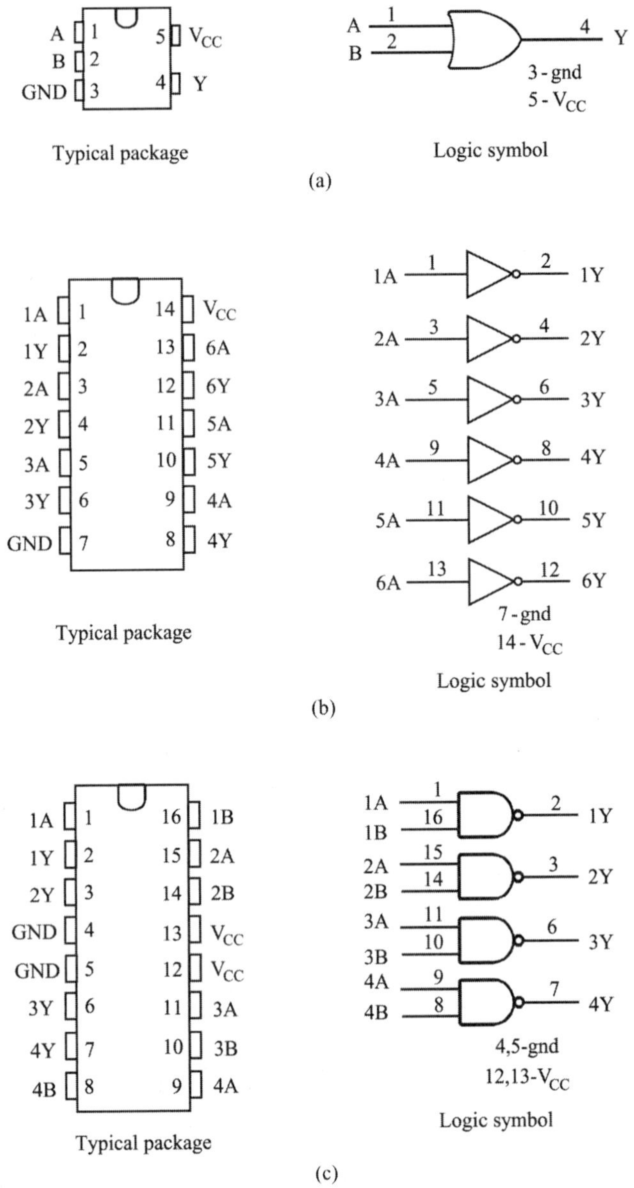

(a) Typical package and Logic symbol for single 2-input OR gate.
Pins: A-1, B-2, GND-3, Y-4, V_{CC}-5.
Logic symbol: A-1, B-2, Y-4; 3-gnd, 5-V_{CC}.

(b) Typical package and Logic symbol for hex inverter (14-pin DIP).
Pins: 1A-1, 1Y-2, 2A-3, 2Y-4, 3A-5, 3Y-6, GND-7, 4Y-8, 4A-9, 5Y-10, 5A-11, 6Y-12, 6A-13, V_{CC}-14.
Logic symbol: 1A(1)→1Y(2), 2A(3)→2Y(4), 3A(5)→3Y(6), 4A(9)→4Y(8), 5A(11)→5Y(10), 6A(13)→6Y(12); 7-gnd, 14-V_{CC}.

(c) Typical package and Logic symbol for quad 2-input NAND gate (16-pin DIP).
Pins: 1A-1, 1Y-2, 2Y-3, GND-4, GND-5, 3Y-6, 4Y-7, 4B-8, 4A-9, 3B-10, 3A-11, V_{CC}-12, V_{CC}-13, 2B-14, 2A-15, 1B-16.
Logic symbol: 1A(1),1B(16)→1Y(2); 2A(15),2B(14)→2Y(3); 3A(11),3B(10)→3Y(6); 4A(9),4B(8)→4Y(7); 4,5-gnd; 12,13-V_{CC}.

Alternative Symbols

The logic symbols presented earlier in the chapter have a distinctive shape for each logic function. These are the symbols used in most technical documentation, articles, and schematic diagrams for digital products.

The International Electrotechnical Commission (IEC) has been developing alternative symbols that describe the input-output relationships more completely. This work has been going on since the mid-1960s, and it is still in development today. Semiconductor manufacturers have embraced the new standards, and they utilize this new symbolic language in the datasheets for logic devices. However, for the most part, users of the logic devices (i.e., manufacturers of digital products) haven't been so willing to adopt the new symbols. Therefore, the schematics for computers and other digital products that you will work with in industry will most probably utilize the distinctive symbols previously presented. In fact, even semiconductor manufacturers include both types of symbols on their datasheets. Figure 15–13 shows the distinctive logic shapes and the new symbols for the basic logic gates. Details of the new standard are outlined in IEEE Std 91-1984. Information is also available from semiconductor manufacturers' Web pages such as Texas Instruments (www.ti.com).

15.2 • COMBINATIONAL LOGIC

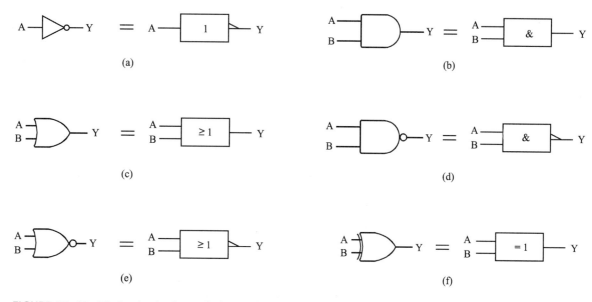

FIGURE 15-13 Distinctive logic symbols and associated new symbols for (a) inverter, (b) AND gate, (c) OR gate, (d) NAND gate, (e) NOR gate, and (f) exclusive OR gate.

Gate Conversions

Two or more logic gates can be interconnected to provide the same logic function as a different type of logic gate. This is often done to reduce the total number of IC packages in a product. Figure 15-14 provides a convenient memory tool to assist in remembering the various combinations.

FIGURE 15-14 Basic logic gate conversions.

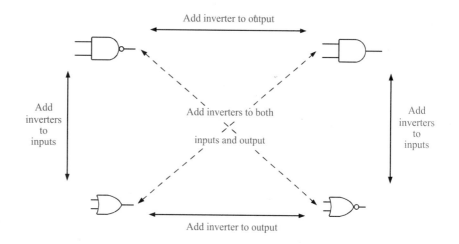

• EXAMPLE 15.7

How can you make a NOR gate perform the same logic function as an AND gate?

SOLUTION Figure 15-14 shows that to go from a NOR gate to an AND gate we must add inverters to the inputs. The equivalency is shown in Figure 15-15.

FIGURE 15-15 Conversion of a NOR gate into an AND function.

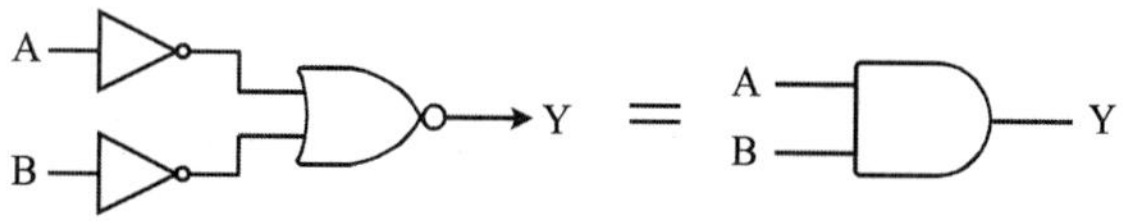

Practice Problems

1. Show how to use a NAND gate as an AND gate.
2. Show how to use an OR gate as a NAND gate.
3. Show how to convert a NOR gate into a NAND gate.
4. Show how to convert a 3-input AND gate into a 3-input OR gate.

Gate Array Logic

Many IC manufacturers produce devices called gate array logic (GAL), programmable array logic (PAL), programmable logic array (PLA), programmable logic device (PLD), complex programmable logic device (CPLD), and other similar names. These devices contain a large number of basic gates, but in many cases, the details of the interconnections are not defined at the time of manufacture. Therefore, from a manufacturing point of view, these are not custom devices. The design engineer for the final product can program the various internal interconnections to implement a specific logic circuit. In most cases, the programming is permanent and cannot be subsequently altered. These devices allow reasonably complex combinational logic circuits to be integrated into a single package without incurring the high costs associated with a truly custom IC. Figure 15–16 shows a representative device (PAL 16L8AM from Texas Instruments). This particular PAL comes in a 20-pin package as indicated in Figure 15–16(a). The internal logic diagram is shown in Figure 15–16(b).

This PAL provides ten dedicated logic inputs (I), two dedicated logic outputs (O), and six pins that can be programmed as either inputs or outputs as needed (I/O pins). Each dedicated input connects to a logic gate that provides both true and complemented outputs (e.g., A and $\overline{A}$). Each potential output (i.e., two dedicated and six programmable) has a 7-input OR gate followed by an inverter and preceded by seven AND gates. The AND gates appear to have only one input, but we will examine this in more detail.

Examination of the logic diagram in Figure 15–16(b) will show that the input to any given AND gate crosses the line connecting to every input variable and its complement. When the device is manufactured, all intersections within the device are connected by fusible links made of titanium-tungsten. To program a logic expression into the device, these fusible links are selectively burned open (permanently) by a controlled exposure to a higher-than-normal voltage. The remaining links define the logic expression. Figure 15–16(c) provides further clarification of the programming process. Here fuses have been blown to form the expression

$$Y = \overline{AB + A\overline{C} + \overline{B}\overline{C}}$$

As you can see, all variables connected to a particular AND gate are ANDed together. So, what is shown as a single input AND gate for simplicity can actually have many inputs depending on which fuses are left intact.

The output inverters shown in Figure 15–16(b) are called tri-state inverters. A tri-state output can be enabled or disabled by a control input (shown connected to the top side of the inverters in Figure 15–16(b)). When a tri-state device is enabled, it behaves as a normal logic gate (in this case an inverter). When it is disabled, the output is essentially disconnected or floating. Tri-state devices are extremely common in computer circuits. In the present case, illustrated in Figure 15–16(b), any one or more of the inputs can be used to activate or disable any of the outputs.

There are many types of programmable logic devices available, which have much more capability than the example discussed in this section. Many include sequential logic devices like those discussed in the next section. In any case, it is easy to see why computers and other products that require a lot of functionality in a relatively small space use programmable logic devices instead of individual logic gates whenever practical.

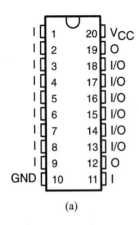

FIGURE 15–16 (a) The pin diagram for a representative programmable array logic (PAL), (b) its internal logic diagram, and (c) a programming illustration. Continues on pages 559 and 560. (Parts a and b Courtesy of Texas Instruments)

15.2 • COMBINATIONAL LOGIC

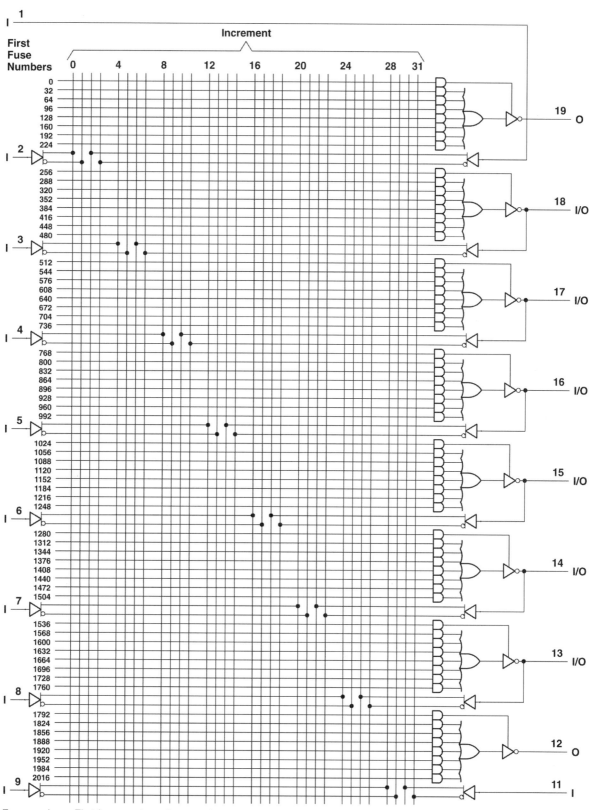

(b)

FIGURE 15–16 (*continued*)

FIGURE 15–16 (continued)

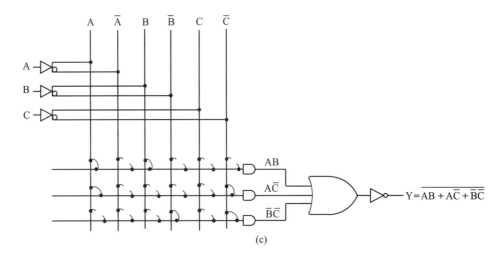

(c)

PALs and PLAs are the two most basic types of programmable logic devices (PLDs). A PAL is characterized by having programmable AND gates, but fixed OR gates, as discussed previously. PLAs, by contrast, are characterized as having both programmable AND gates and programmable OR gates in the same logic array. CPLDs are essentially two or more PLDs that are interconnected within the same IC package. They are capable of implementing both combinational logic and sequential logic (discussed in the next section). CPLDs can often be reprogrammed, which makes them ideal for development work where a design may have to be changed several times during initial development.

Boolean Expressions and Truth Tables for Logic Circuits

We will now focus on writing the Boolean expression and completing the truth table to describe a given logic circuit. This valuable skill will help you understand the operation of an unfamiliar logic circuit. It is also an important troubleshooting tool for digital circuits.

Writing Boolean Expressions

A Boolean expression to describe a logic circuit can be constructed by starting at the inputs and working toward the output. The expression to describe the output of each input gate is determined using the basic definitions of logic gate operation. This process is then repeated for any subsequent gates until the output is reached. The input variables to gates beyond the initial gates are simply the expressions produced by preceding gates. Let's work our way through some examples to illustrate the technique.

•—EXAMPLE 15.8

Determine the Boolean expression to describe the output for the logic circuit shown in Figure 15–17.

FIGURE 15–17 Write the Boolean expression for the output of this circuit.

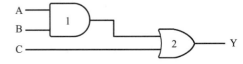

Fig15_17.msm

SOLUTION The output of the AND gate (1) is simply *AB*. It is a good practice to place parentheses around the intermediate output expressions. This is allowed by Boolean rule 21 presented in the previous section. The parentheses will help prevent confusion, and they can be removed later if they are found to be unnecessary. So, we will label the output of the AND gate as (AB).

The output of the AND gate is the ORed with input variable C by gate 2. The output of the OR gate (2) and the final output expression for the circuit is then $Y = (AB) + C$. In this case, the parentheses are not needed, so we can simply eliminate them (rule 21). Therefore, we can describe the behavior of the logic circuit as $Y = AB + C$.

•—EXAMPLE 15.9

Determine the Boolean expression to describe the output for the logic circuit shown in Figure 15–18.

FIGURE 15–18 Write the Boolean expression for the output of this circuit.

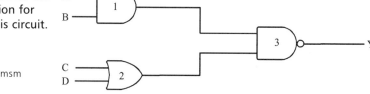

Fig15_18.msm

SOLUTION The output of gate 1 is AB, which we will write as (AB). The output of gate 2 is $C + D$, which we will write as $(C + D)$. The outputs of gates 1 and 2 serve as the inputs to gate 3. The output of gate 3 is simply the expression for a NAND gate whose inputs are (AB) and $(C + D)$. That is, $Y = \overline{(AB) \cdot (C + D)}$. Here is a case where we cannot simply remove the parentheses. With the parentheses in place, it is clear that the quantity AB is ANDed (ultimately NANDed) with the quantity $C + D$. However, if the parentheses are removed, it would appear (erroneously) that the quantity ABC was ORed (ultimately NORed) with D. We could remove the parentheses around AB according to rule 23, but when in doubt, leave them in for clarity. We could further alter the form of the equation by applying rules 34 and 35. The final expression becomes $Y = \overline{A} + \overline{B} + \overline{C}\,\overline{D}$.

•—Practice Problems

Determine the Boolean expression to describe the output for each of the circuits shown in Figure 15–19.

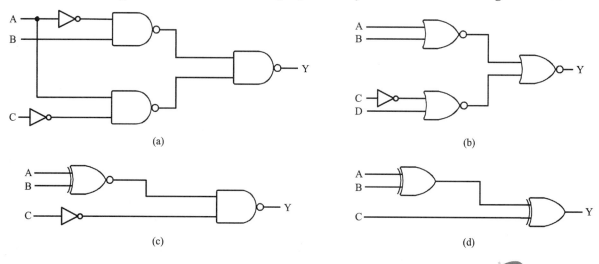

(a)   (b)   (c)   (d)

FIGURE 15–19 Write the Boolean expressions for these circuits.

Fig15_19.msm

Completing Truth Tables

It is often helpful to build a truth table for an unfamiliar combinational logic circuit. First, it helps you understand circuit operation. Second, a truth table can be a valuable troubleshooting tool. In the previous section, we practiced writing Boolean expressions to represent a given logic circuit. In this section, we will construct truth tables from a given Boolean expression. By combining the two techniques, you will be able to construct truth tables to describe logic circuits.

First, you must be able to express the Boolean equation in the correct form. If the expression is reduced until there are no two variables joined by a single vinculum, then there are only two basic types of expressions: minterm and maxterm. A **minterm** expression is also called the sum-of-the-products form. The following are examples of expressions written in minterm form:

1. $Y = AC + AB + B\overline{C} + AD$
2. $Y = \overline{B}\,\overline{D} + \overline{C}D + ABC$
3. $Y = AC + B\overline{C}\,\overline{D} + \overline{A}B$

Maxterm expressions are also called product-of-the-sums form. Following are some examples of expressions written in maxterm form:

1. $Y = (A + B)(B + \overline{C})$
2. $Y = (\overline{A} + B)(\overline{A} + \overline{C})(C + D)$
3. $Y = (B + D)(\overline{A} + D)$

One way to construct a truth table directly from a Boolean expression requires the equation to be written in minterm form. If it is in maxterm form, convert it to minterm by applying rule 36. You may be able to apply other rules to further simplify the expression, but this is not necessary to complete the truth table. Once the expression is in minterm form, you simply put 1's in the truth table on each entry that is satisfied by one or more of the terms in the minterm expression.

•EXAMPLE 15.10

Complete a truth table for the expression $Y = AB\overline{C} + \overline{A}\,\overline{B}\,\overline{C}$.

SOLUTION First, we sketch the blank truth table. Since there are three variables, we will need 2^3 or 8 entries in the table as follows:

C	B	A	Y
0	0	0	
0	0	1	
0	1	0	
0	1	1	
1	0	0	
1	0	1	
1	1	0	
1	1	1	

Next, we place a 1 in the Y column of every entry that corresponds to one of the terms in the Boolean expression. The $\overline{A}\,\overline{B}\,\overline{C}$ and $AB\overline{C}$ terms correspond to rows 1 and 4 in the table, respectively. We put 1's in these two rows and 0's in all other rows. The completed truth table is as follows:

C	B	A	Y
0	0	0	1
0	0	1	0
0	1	0	0
0	1	1	1
1	0	0	0
1	0	1	0
1	1	0	0
1	1	1	0

If one of the terms in the expression has one or more variables missing, then place a 1 in every row that contains the given term. For example, if our original expression had contained the term BC, then we would place a 1 on every line that has B and C high at the same time. In the preceding table, that would be lines 7 and 8, which contain the conditions 011 ($\overline{A}BC$) and 111 (ABC), respectively.

•—EXAMPLE 15.11

Complete a truth table for the Boolean expression $Y = \overline{A}\,\overline{B}\,\overline{C}\,\overline{D} + \overline{A}B\overline{C}D + AC$.

SOLUTION Since there are four variables (A, B, C, and D), we will need a truth table with 2^4 or 16 entries. The first two terms correspond to rows 1 and 11. In the case of the AC term, we will place a 1 in rows 6, 8, 14, and 16, since each of these has A and C high at the same time. The complete truth table is as follows:

D	C	B	A	Y
0	0	0	0	1
0	0	0	1	0
0	0	1	0	0
0	0	1	1	0
0	1	0	0	0
0	1	0	1	1
0	1	1	0	0
0	1	1	1	1
1	0	0	0	0
1	0	0	1	0
1	0	1	0	1
1	0	1	1	0
1	1	0	0	0
1	1	0	1	1
1	1	1	0	0
1	1	1	1	1

•—Practice Problems

Complete truth tables for each of the following Boolean expressions:

1. $Y = A\overline{B}C + \overline{A}B\overline{C}$
2. $Y = A\overline{B} + \overline{A}B\overline{C}$
3. $Y = (\overline{A} + B)(\overline{B} + C)$
4. $Y = \overline{AB + CD}$
5. $Y = \overline{(\overline{A} + \overline{B})(B + C)}$
6. $Y = \overline{AB} + A\overline{D}$

✓ Exercise Problems 15.2

1. The output of a combinational logic circuit at any instant in time is determined by its inputs at that same time. (True/False)
2. Draw the schematic symbol for each of the following logic devices:
   a. 2-input AND
   b. 2-input OR
   c. 2-input NAND
   d. 2-input NOR
   e. NOT
   f. Exclusive OR
3. Write the Boolean expression for each of the following logic devices:
   a. 2-input AND
   b. 2-input OR
   c. 4-input NAND
   d. 3-input NOR
   e. NOT
   f. Exclusive OR
4. Complete a truth table for each of the following logic devices:
   a. 2-input AND
   b. 2-input OR
   c. 3-input NAND
   d. 3-input NOR
   e. NOT
   f. Exclusive OR
5. Briefly explain the programming mechanism for a PAL.
6. Write the Boolean expression to describe each of the logic circuits shown below.
7. A truth table to describe an 8-input logic circuit would have _____ entries.
8. Complete truth tables for each of the following Boolean expressions:
   a. $Y = (\overline{A} + B)(A + \overline{B})$
   b. $Y = \overline{A}\,\overline{B} + A\overline{C} + ABC$
   c. $Y = \overline{B} + AC$
   d. $Y = \overline{A}\,\overline{D} + C + \overline{A}\,\overline{B}\,\overline{C}D$

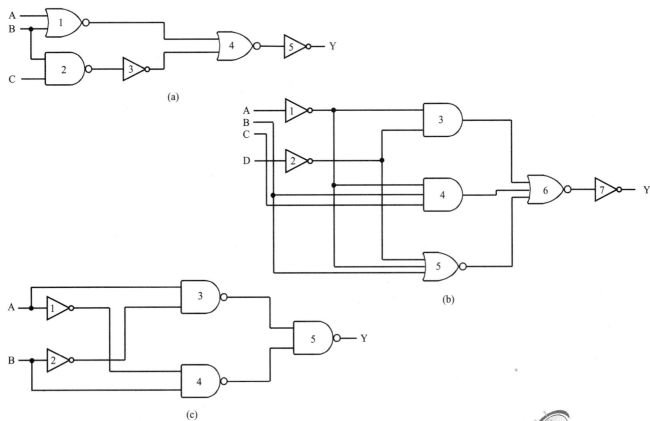

Write the Boolean expressions for circuits a, b, and c.

FigUNF15_6.msm

SYSTEM PERSPECTIVE: Glue Logic

The primary application of discrete combinational logic devices in a modern computer is to convert logic signals from one IC into a form compatible with a second IC. This general application for logic gates is called glue logic, since in effect, it is used to glue (i.e., connect) ICs together. Consider the adjacent illustration. Here, a low-level logic signal is required by an input on IC #2. But there is no such output on IC #1 that has the correct timing (i.e., that goes low at the proper time). However, at the time a low-level is needed at the input to IC #2, outputs A and B on IC #1 are high and low, respectively. Therefore, we can add some glue logic consisting in this case of an inverter and a 2-input NAND gate. Whenever output A is high at the same time output B is low, the NAND gate will have a low output. This satisfies the input requirements for the input of IC #2. Most glue logic applications require relatively few logic gates.

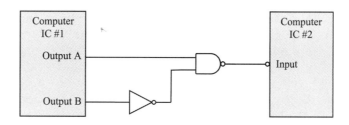

15.3 SEQUENTIAL LOGIC

A combinational logic circuit is characterized as having outputs that are solely determined by its inputs at any given time. A NAND gate is a good example. You can measure the logic levels on the inputs of a NAND gate at any time and immediately predict the output logic state. We are now ready to explore another class of digital circuitry called sequential logic. Figure 15–20 shows the general block diagram that represents a sequential logic circuit. A sequential logic circuit can have any number of inputs and any number of outputs.

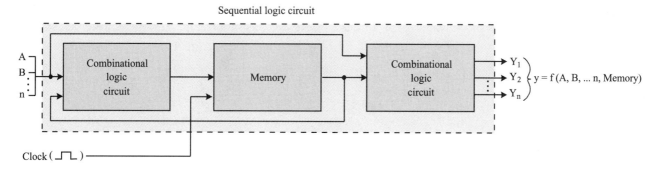

FIGURE 15–20 The generalized functional block diagram for a sequential logic circuit.

As indicated in Figure 15–20, a general sequential logic circuit consists of a memory element and two combinational logic circuits. The memory element is used to hold the state of the circuit. If the memory element can store one bit, the circuit can have two internal states. A 4-bit memory will allow the circuit to have sixteen internal states. In general, a memory element capable of storing k bits will provide a sequential logic circuit with 2^k internal states. The memory element in

Figure 15–20 also has a clock input. Technically, this is not required for a sequential state machine, but it is nearly always present for logic circuits used in computers. In any case, the clock input provides the overall timing that causes the memory element to change from one state to another.

When a clock pulse (simply a transition from one logic state to the other) arrives, the memory element has the opportunity to change states. The next state of the circuit is determined by two things: the present state and the state of the inputs. This is the purpose of the left-most combinational logic block in Figure 15–20.

Finally, the output(s) of the circuit are developed by the second combinational logic circuit. The outputs are determined by the inputs and by the present state of the internal memory. This means that you cannot examine the inputs to a sequential logic device and immediately predict the output states. You must know the internal memory state, which is determined by events that happened at some earlier time.

We will now explore some practical sequential logic circuits. First, we will learn the behavior of flip-flops, which are the basic building blocks of most sequential logic circuits. Next, we will examine counter circuits, which are composed of a number of flip-flops. Finally, we will examine more complex sequential logic systems that more completely demonstrate the functions illustrated in Figure 15–20.

Flip-Flops

Flip-flops are important devices used in the construction of computers and other digital circuits. Additionally, they serve as one of the basic building blocks for more complex integrated circuits.

D Flip-Flops

One of the two most common flip-flop types that are still widely used is called a D flip-flop. The SN74AC74 device from Texas Instruments (www.ti.com), for example, provides two D flip-flops in a single package. Figure 15–21 shows the logic symbol for a D flip-flop. It has four inputs (D, C_P, S_D, and R_D). It also has two outputs (Q and $\overline{Q}$). The outputs represent the state of the device. As you would expect, Q and $\overline{Q}$ must always be at opposite logic levels.

Asynchronous Operation The S_D and R_D inputs are **asynchronous inputs**. This means they are overriding inputs that perform their operation regardless of the timing input (C_P). The logic bubbles indicate that a low logic level is required to activate these inputs. These two asynchronous inputs behave as shown in Table 15–3.

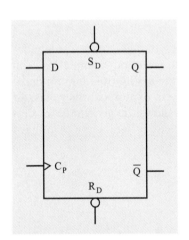

FIGURE 15–21 The logic symbol for a D flip-flop.

TABLE 15–3 A truth table for the asynchronous inputs S_D and R_D.

Inputs		Outputs	
S_D	R_D	Q	$\overline{Q}$
0	0	Disallowed	
0	1	1	0
1	0	0	1
1	1	No change	

When active (low) the S_D or Set Direct (also called Preset) input forces the Q output to a high state. Of course, $\overline{Q}$ must go to a low state. The R_D or Reset Direct (also called Clear) input forces Q to a low state (and $\overline{Q}$ to a high state). These inputs are generally used to force a flip-flop into a known state (for example, when power is first applied to a circuit or the reset button is pressed on a computer).

During normal synchronous operation, which is our primary interest now, the S_D and R_D inputs are both high. According to Table 15–3, when both asynchronous inputs are high, Q and $\overline{Q}$ are unaffected (i.e., no change from their present state). Unless otherwise shown or stated, the Set and Clear inputs are assumed to be inactive. This includes situations where these inputs are not even shown on a logic diagram.

Synchronous Operation   When the overriding Set and Clear inputs are disabled, the state of the flip-flop is controlled by the synchronous inputs (D and C_P). The clock pulse (C_P) input provides the timing for state transitions. The transition indicator (>) on the C_P input in Figure 15–21 means that a $0 \rightarrow 1$ transition is required to activate the input. Some flip-flops require a negative or falling edge (i.e., a $1 \rightarrow 0$ transition). An input requiring a negative transition for activation has a logic bubble in addition to the transition indicator.

Now, once the flip-flop receives the correct transition on the clock input, it will go to the next internal state. The next state is determined by the logic level on the D input at the time the correct transition is received on the C_P input. The relationship between these two inputs is shown in Table 15–4.

TABLE 15–4   A truth table for synchronous operation of a D flip-flop.

D	Q_{n+1}	$\overline{Q}_{n+1}$
0	0	1
1	1	0

First, note that the output columns are labeled Q_{n+1} and $\overline{Q}_{n+1}$. The subscript $n+1$ means the *next* internal state. That is, what the new state will be after the clock pulse has occurred. So, if the D input is high when a positive transition arrives on the C_P input, then the Q output will become high regardless of what it was previously. If it was previously a low, it will change to a high. When a flip-flop changes between two logic states, it is said to **toggle**. On the other hand, if Q was already high, then after the clock transition it would still be high. In other words, the flip-flop would not toggle.

•–EXAMPLE 15.12

Determine the state of the flip-flop in Figure 15–22 after three clock pulses. Assume it has been reset (i.e., $Q = 0$) initially.

FIGURE 15–22
Determine the state of this D flip-flop after three clocks (initially reset).

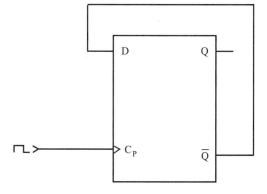

SOLUTION   During the initial reset condition, Q will be low (by definition), and $\overline{Q}$ will be high. D will be high, since it is connected to $\overline{Q}$, which is high. When the first clock pulse (positive transition) arrives, Q will go high (i.e., the flip-flop will toggle). This also causes $\overline{Q}$ to go low. When the second clock pulse arrives, it will find D low. Thus, Q will go low after the clock pulse. This returns the flip-flop to its initial state. Finally, the third clock pulse will once again toggle the flip-flop, leaving Q high and $\overline{Q}$ low. It is important to see that the timing of the state transitions is established by the clock input. If the clock is a 100-MHz pulse train, the state transitions will be 1/100 MHz or 10 ns apart.

JK Flip-Flops

The second type of commonly used flip-flop is called a *JK* flip-flop. The 74AC109 is a *JK* flip-flop made by Fairchild Semiconductor (www.fairchildsemi.com). Its logic symbol is shown in Figure 15–23. The asynchronous inputs S_D and R_D operate as previously discussed with *D* flip-flops and will not be repeated here. The *J* and *K* pins are synchronous inputs that determine the flip-flop's behavior when a clock transition arrives on C_P. Note the bubble on the clock input. This indicates that a negative (1 → 0) transition is required to activate this input.

The relationship between the *J* and *K* inputs and the clock is shown in Table 15–5.

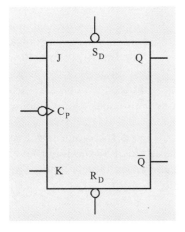

FIGURE 15–23 The logic symbol for a *JK* flip-flop.

TABLE 15–5 A truth table for synchronous operation of a *JK* flip-flop.

J	K	Q_{n+1}	$\overline{Q}_{n+1}$
0	0	Q_n	$\overline{Q}_n$
0	1	0	1
1	0	1	0
1	1	Toggles	

In this table, the subscript "*n*" refers to the present state, and the subscript *n* + 1 refers to the next state (i.e., the state after the clock transition occurs). As you can see, when *J* and *K* are both low, the next state will be the same as the present state. The flip-flop does not toggle. In essence, it is disabled with regard to the clock input. If *J* and *K* are in opposite logic states, then *Q* (after the clock transition) becomes whatever *J* is prior to the transition. This means the flip-flop may or may not actually toggle. Finally, if *J* and *K* are both high prior to the clock transition, the flip-flop will toggle regardless of its present state.

• EXAMPLE 15.13

Determine the state of the *JK* flip-flop in Figure 15–24 after three clock pulses. Assume it has been reset (i.e., $Q = 0$) initially.

FIGURE 15–24
Determine the state of this *JK* flip-flop after three clocks (initially reset).

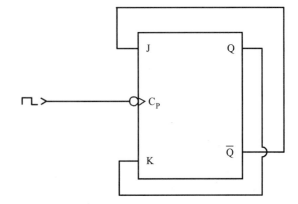

SOLUTION When the flip-flop is initially reset, Q will be low, and $\overline{Q}$ will be high by definition of the reset state. This means that $J = 1$ and $K = 0$ prior to the first clock. When the clock transition (1 → 0) arrives, Q will go high, and $\overline{Q}$ will go low according to Table 15–5. When the falling edge of the second clock arrives, $J = 0$ and $K = 1$. This will cause Q to go low and $\overline{Q}$ to go high. This is back to the original state. Finally, when the third clock pulse arrives with $J = 1$ and $K = 0$, the flip-flop will again toggle to make $Q = 1$ and $\overline{Q} = 0$. In short, always examine J and K *before* the clock in order to predict what Q and $\overline{Q}$ will be *after* the clock.

•—Practice Problem

Evaluate the behavior of the JK flip-flop circuit in Figure 15–25 and determine whether it could be used to replace a D-type flip-flop.

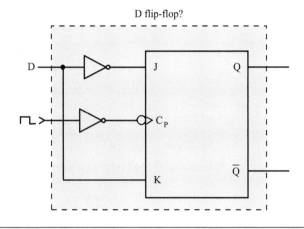

FIGURE 15–25 Could this *JK* flip-flop circuit replace a *D* flip-flop?

Fig15_25.msm

Counters

In essence, a flip-flop that is configured to toggle on every clock pulse is a 1-bit counter. It can store values (indicated by the Q output) of 0 and 1. If we connect multiple flip-flops together, we can build a counter of any length. For example, a 4-bit counter can count from 0000 (zero) to 1111 (fifteen). Even though counters are simple in principle, they play major roles in computer circuits and even the internal operation of microprocessors and support ICs.

Classes of Counters

There are two general classes of counters: asynchronous and synchronous. In an **asynchronous counter**, the clock pulses arrive at the various flip-flops at different times. Although the time differential may be short in some cases, the distinguishing characteristic is that the various flip-flops in the counter do not receive a clock at the same time, and therefore, cannot toggle at the same time.

All stages of a **synchronous counter**, by contrast, receive their clock pulses at the same time. This means that all flip-flops that can toggle on a particular clock transition will do so at the same time. Synchronous counters are the most widely used type of counter in high-speed computer applications, since they pose far fewer timing problems. For this reason, we will emphasize this class of counter.

Synchronous Flip-Flop Counters

Figure 15–26 shows a 3-bit synchronous counter made from *JK* flip-flops. It can be identified as a synchronous counter since the clock inputs all receive the same clock pulse. To determine the sequence of states, we apply the following procedure:

1. Determine the logic state for J and K (or D in the case of a D flip-flop) in the present state.

2. Based on the levels of *J* and *K* in the present state, determine which flip-flops will toggle. This will give you the next state.
3. Repeat these steps until you reach a previous state for the second time.

FIGURE 15–26 A 3-bit synchronous counter.

Fig15_26.msm

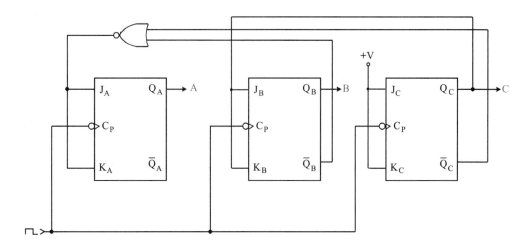

•–EXAMPLE 15.14

Determine the count sequence for the counter shown in Figure 15–26.

SOLUTION Unless otherwise stated, we will assume the counter is initially cleared or reset to state 000. Table 15–6 summarizes the process of determining the count sequence.

TABLE 15–6 Determination of the counting sequence for the counter in Figure 15–26.

Present State			Input Terminals						Next State		
C	*B*	*A*	J_C	K_C	J_B	K_B	J_A	K_A	*C*	*B*	*A*
0	0	0	1	1	0	0	0	0	1	0	0
1	0	0	1	1	1	1	0	0	0	1	0
0	1	0	1	1	0	0	0	0	1	1	0
1	1	0	1	1	1	1	1	1	0	0	1
0	0	1	1	1	0	0	0	0	1	0	1
1	0	1	1	1	1	1	0	0	0	1	1
0	1	1	1	1	0	0	0	0	1	1	1
1	1	1	1	1	1	1	1	1	0	0	0

Let's examine one of the states more closely. Suppose the present state of the counter is 101 ($C\overline{B}A$). This corresponds to row 6 (ignoring the column heading). J_A and K_A come from the output of a NOR gate whose inputs are the $\overline{Q}$ terminals for the *B* and *C* flip-flops. In the state 101, the $\overline{Q}$ terminals of the *B* and *C* flip-flops will be high and low, respectively, since the *Q* terminals are low and high, respectively. Any high into a NOR gate will produce a low output. Therefore, J_A and K_A are low. This will prevent the *A* flip-flop from toggling, so *A* will also be high in the next state. The *J* and *K* terminals of the *B* flip-flop are connected to the *Q* output of the *C* flip-flop, which is high in the present state. Therefore, J_B and K_B are high. This will allow the flip-flop to toggle, so in the next state we see that *B* has changed to a 1. Finally, J_C and K_C are high all of the time, so we know the C flip-flop will toggle on every clock. The next state for C must be 0. Thus, the overall counter transitioned from state 101 ($C\overline{B}A$) to state 011 ($\overline{C}BA$). This same basic procedure is repeated until we repeat a state. In the present example, we ultimately repeat state 000.

Although the reasons may not be apparent at this time, it is important to realize that not all counters count sequentially. You cannot assume that the next state will be one higher than the present state. You must go through the evaluation process just described.

•—Practice Problems

1. Determine the count sequence for the counter in Figure 15–27. Assume the counter begins in state 00.

2. Determine the count sequence for the counter in Figure 15–28. Assume the counter begins in state 000.

FIGURE 15–27 Determine the count sequence for this circuit.

Fig15_27.msm

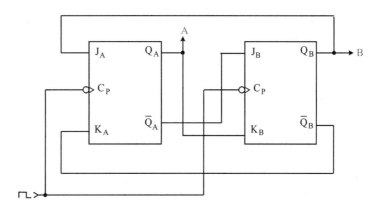

FIGURE 15–28 Determine the count sequence for this circuit.

Fig15_28.msm

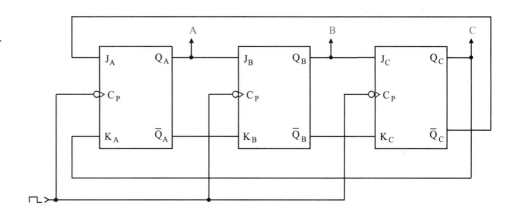

Asynchronous Flip-Flop Counters

Figure 15–29 on page 572 shows the logic diagram for a 3-bit asynchronous counter based on D flip-flops. You can recognize this as an asynchronous counter by noting that the clocks for the various flip-flops come from different places and necessarily arrive at different times. More specifically, flip-flop C would have to toggle in order to provide a clock transition to flip-flop B. Similarly, flip-flop B must toggle to provide a clock to device A. Therefore, the three flips-flops cannot change states at exactly the same time. As stated previously, most sequential logic

FIGURE 15–29 A 3-bit asynchronous counter.

Fig15_29.msm

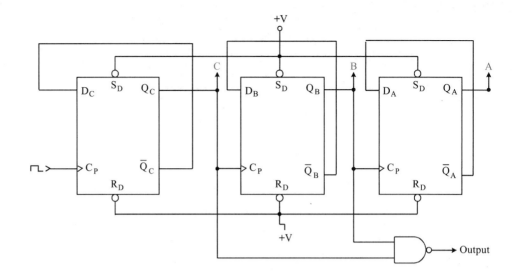

circuits used with computers are synchronous, but you may encounter some small asynchronous applications such as the one shown in Figure 15–29.

Asynchronous counters are analyzed in the same way as synchronous counters with one exception. In addition to determining the input logic levels in the present state, you must determine whether or not the proper clock transition is received. If the proper clock edge occurs, then the behavior of the flip-flop is determined by the input levels (i.e., D terminals in this case).

•EXAMPLE 15.15

Determine the count sequence for the circuit in Figure 15–29. Assume the counter begins in state 000.

SOLUTION Flip-flop C receives its clock directly so whether or not it toggles is strictly determined by the logic level on the D input. In state 000, the $\overline{Q}$ pin of the C flip-flop is high, so D_C is high. When the clock arrives, C will toggle. This causes the Q_C pin to make a $0 \rightarrow 1$ transition, which is the proper edge for a clock to B. Since the D_B pin is high, B will also toggle. A similar set of events causes A to also toggle. So, on the first clock pulse, the circuit advances from state 000 to state 111. If we continue with this same analytical procedure, we will find that the count sequence is as follows:

C	B	A
0	0	0
1	1	1
0	1	1
1	0	1
0	0	1
1	1	0
0	1	0
1	0	0
0	0	0

Now let's look at some of the shortcomings of asynchronous designs. First, they are inherently slower, since the flip-flops must toggle one after the other. This is not an issue in slow-speed systems, but it can render the technique useless in high-speed designs. In Figure 15–29 you can see that when the counter is in state 111 or state 110, the Q_B and Q_C terminals are both high. This causes the output of the NAND gate to go low. The output of the NAND serves as the input to some other circuit or device. But, for a moment, let's examine the circuit in slow motion. We will focus on the time when the circuit transitions from state 011 to 101. Both B and C flip-flops must toggle. However, the C device must toggle first. This means that the true state transition sequence is 011 → 111 → 101. State 111 is only there momentarily until the B flip-flop also completes its transition. Nonetheless, the NAND gate can recognize this as a valid state and produce a brief low output. This is an erroneous output that is only there for a few nanoseconds or even picoseconds, but it can cause errors in high-speed logic circuits. The erroneous transient on the output of the NAND gate is called a **glitch**. The condition whereby the inputs to the gate do not arrive at the same time is called a **race condition**.

Integrated Counters

Discrete flip-flop counters are normally used for very small counters (< 4 bits) or counters with very specialized count sequences. The majority of counter applications you are likely to encounter will be based in integrated counters. Standard integrated counters are available with many different count sequences, including standard binary as well as decade or BCD (counts 0 through 9); programmable counters that can start or stop at specified values; and bidirectional counters that can count up (increment) or down (decrement). The CD4060BC made by Fairchild Semiconductor (www.fairchildsemi.com) is a 14-stage binary counter. Of course, a custom IC (quite common for modern designs) can contain a counter with any arbitrary count sequence. Standard ICs are available with counters as small as four bits and as high as at least forty bits. The data sheet (normally available on the manufacturer's Web page) provides complete information of the count sequence for a particular IC counter. Figure 15–30 shows the pin diagram for one representative integrated counter.

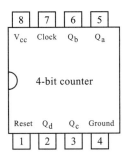

FIGURE 15–30 A representative integrated counter.

Counters as Frequency Dividers

Although the operation is identical to that previously discussed, counters are also viewed and used as frequency dividers. Each stage of a binary counter divides the input frequency by two. So, for example, if we use a 32-MHz clock as the input to a 4-bit binary counter, the outputs from the Q terminals of the four counter stages will be 16 MHz, 8 MHz, 4 MHz, and 2 MHz. The timing diagram in Figure 15–31 will clarify this relationship.

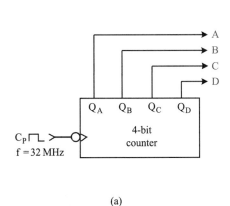

(a)

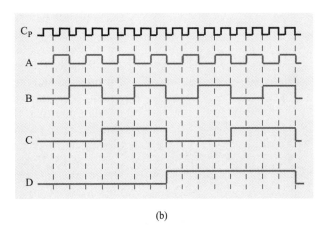

(b)

FIGURE 15–31 A counter is also a frequency divider.

SYSTEM PERSPECTIVE: Flip-Flops

Many of the operations within a computer must occur at the proper time. For example, all of the logic states on the inputs to a memory device are typically changing states at a high rate. There is a very small window of time when the logic states are correct and should be transferred into the memory device. If the timing is off as much as 2 to 3 ns, the system may not function properly. Computers use continuous, constant-frequency pulse trains called clocks to provide the timing signals for the various circuits. One way that computers achieve consistent timing throughout the system is to derive multiple clocks from a single clock source. As shown in the adjacent figure, a 48-MHz clock source can be passed through two flip-flops to obtain 24-MHz and 12-MHz clock signals. Since all of these signals are derived from the same clock source, they will all change states at nearly the same time. Thus, the critical timing requirements of the system can be satisfied. This is just one application of flip-flops in a practical computer circuit.

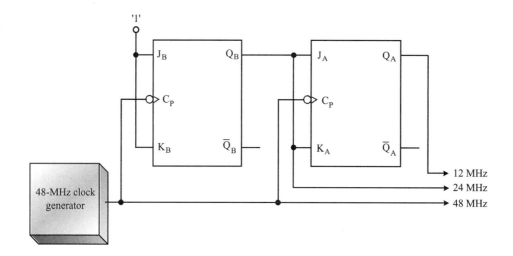

State Machines

State machines are being presented here as a separate topic because that is the way many people perceive them. However, it is important for you to understand that the flip-flops and counters previously discussed are examples of state machines. In short, these circuits have two or more internal states and some number of outputs. Additionally, they may have logic inputs beyond the required clock pulse. The present internal state and the states of any logic inputs determine both the next state and the level of any outputs. In the case of a simple counter, there were no external inputs other than the clock. Similarly, there were no outputs except for the state outputs. So, as we proceed, don't view this as a new topic. It is a simple extension of what has been previously discussed.

Figure 15–32 shows a state machine that might be used to control an electronic combination lock. It has three logic inputs that come from user-accessible push buttons. It also has two outputs. One is an LED intended to confuse anyone attempting to defeat the alarm by guessing the correct code. It serves no other functional purpose. The second output controls a solenoid, which is used to unlock a door if the correct code is entered.

This circuit demonstrates a full-fledged state machine with multiple internal states, multiple logic inputs, and multiple logic outputs. Analysis of the circuit is

FIGURE 15–32 An electronic combination lock system.

Fig15_32.msm

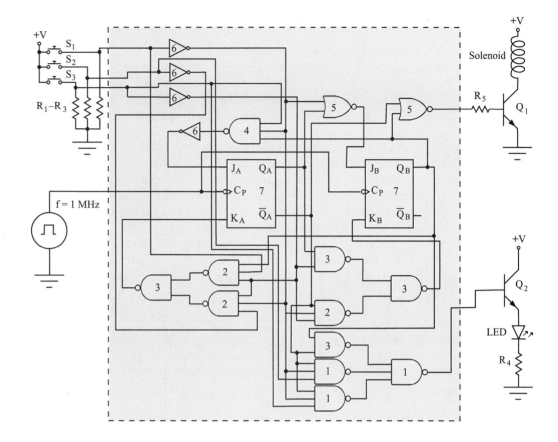

identical to the simple counter circuits previously analyzed, except that determination of the J and K inputs in a given state requires application of the rules for basic logic gates. One way to speed your analysis is to write the Boolean expression for each J and K input and for each logic output. If the expression is written in minterm form, then the expression will be high if any of the products occur in the present state. Let's consider the K_A input as an example.

The Boolean expression for the K_A input is $K_A = BS_1\overline{S}_3 + \overline{S}_1\overline{S}_2\overline{S}_3$. Therefore, if Q_B is ever high at the same time S_1 is high and S_3 is low (the first product), then K_A will be high. It will also be high anytime all three switch inputs are low (i.e., none of the switches are pressed). These are the only two times K_A will be high.

You will find it quite instructive to thoroughly analyze the circuit to determine how it operates. Any discrete sequential logic circuits that you are likely to encounter will be simpler than the one shown in Figure 15–32. More complex sequential logic circuits are quite common; however, they are normally part of an integrated circuit (either custom or standard). In this latter case, you only have access to the input and output terminals of the state machine.

•—Practice Problems

Refer to Figure 15–32 to answer the following questions:
1. How many states does the circuit have?
2. List all conditions that will cause the LED to illuminate.
3. List all states that will cause the solenoid to be energized.
4. Assume the machine is initially set to state 00. Describe the button sequence that is required in order to energize the solenoid.

Exercise Problems 15.3

1. The output of a sequential logic circuit at any instant in time is determined by its inputs at that same time. (True/False)
2. If the D input of a D flip-flop is permanently tied to a high logic level, it will toggle every time a clock pulse arrives. (True/False)
3. If the D input of a D flip-flop is low, the flip-flop is prevented from changing states. (True/False)
4. What does a logic bubble on the Clear direct input of a flip-flop mean?
5. Identify the logic states of the Q and $\overline{Q}$ outputs of a flip-flop for each of the following conditions: set, cleared, preset, and reset.
6. What does the transition indicator (>) on the C_P input of a flip-flop indicate? What if the input also has a logic bubble?
7. If the J and K inputs of a JK flip-flop are permanently tied to a high logic level, the flip-flop will toggle every time a clock pulse arrives. (True/False)
8. If the J and K inputs of a JK flip-flop are low, the flip-flop is prevented from changing states. (True/False)
9. How can you distinguish a synchronous counter from an asynchronous counter?
10. Determine the count sequence for the circuit shown in Figure 15–33. Assume an initial state of 000.
11. Asynchronous counters are often plagued by _____ conditions and associated _____.
12. If the clock input to an 8-bit binary counter is 128 MHz, what is the frequency of the signal on the Q terminal of the last stage of the counter?

FIGURE 15–33 Determine the count sequence for this circuit.

Fig15_33.msm

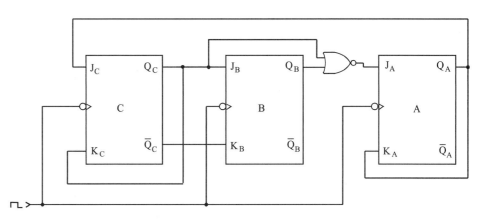

15.4 INTERFACING DIGITAL AND ANALOG SYSTEMS

Digital computers are used in nearly every aspect of our lives including business, industrial, recreational, and personal applications. But in many cases, particularly in control applications, the devices being monitored and/or controlled are analog in nature. Consider, for example, a computer used in an industrial manufacturing application. Inputs to the computer might be used to monitor pH, temperature, pressure, weight, viscosity, density, color, humidity, and other such values of a manufacturing process. These analog quantities can take on an infinite number of values within a specified range. Additionally, the computer may have outputs that control temperature, pressure, and other real world quantities that are analog in nature. Therefore, we must have a way to transform the analog inputs into digital equivalents so they can be interpreted and manipulated by the digital computer circuitry. Similarly, results from a digital computation must be transformed into equivalent output voltage and current values for controlling analog devices.

This section will discuss analog-to-digital (A/D) and digital-to-analog (D/A) conversion principles and circuits. There are many different techniques that can be used to accomplish these conversions. However, when analyzing or troubleshooting an existing design, the A/D or D/A converter circuit can generally be viewed on a functional block-diagram level. This assertion is further supported by the fact that most practical converters used with modern computer systems are based on integrated converter circuits. In the following sections, our primary goal will be to understand the functional operation of a D/A or an A/D converter without regard to details of the conversion process. To support this strategy, we will examine the circuit-level details of one representative conversion technique. This will serve to clarify the overall purpose and operation of a converter circuit.

Digital-to-Analog Conversion

Figure 15–34 shows the schematic diagram for one type of D/A converter (DAC). This particular technique utilizes a multiple-input op amp circuit configured as an inverting amplifier. It can be analyzed with techniques previously discussed.

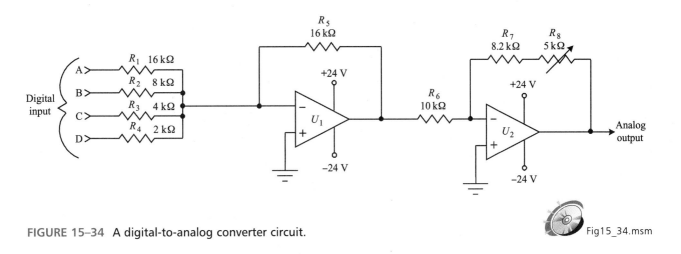

FIGURE 15–34 A digital-to-analog converter circuit.

For discussion purposes, let us assume that the defined logic states for the digital inputs are Low = 0 V and High = +1 V. The voltage gain for any given input is determined by the ratio of the feedback resistor to the input resistor. The various voltage gains are as follows:

$$\text{Voltage gain for } D \text{ input} = -\frac{R_5}{R_4} = \frac{16 \text{ k}\Omega}{2 \text{ k}\Omega} = 8$$

$$\text{Voltage gain for } C \text{ input} = -\frac{R_5}{R_3} = \frac{16 \text{ k}\Omega}{4 \text{ k}\Omega} = 4$$

$$\text{Voltage gain for } B \text{ input} = -\frac{R_5}{R_2} = \frac{16 \text{ k}\Omega}{8 \text{ k}\Omega} = 2$$

$$\text{Voltage gain for } A \text{ input} = -\frac{R_5}{R_1} = \frac{16 \text{ k}\Omega}{16 \text{ k}\Omega} = 1$$

U_2 is simply an inverting-buffer amplifier. Resistor R_8 is used to allow an exact voltage gain to be established. For our purposes, we will assume that U_2 has a voltage gain of unity. Now, by inspection, we can see that the (−) input of U_1 is a virtual-ground point. Therefore, the various inputs are effectively isolated and independent from each other. So, we can find the output voltage by simply summing the voltages caused by each individual input. For a binary count of 0000 ($\overline{D}\,\overline{C}\,\overline{B}\,\overline{A}$) to 1111 (DCBA), the output voltages are shown in Table 15–7 on page 578.

TABLE 15–7 Operation of the D/A converter shown in Figure 15–34.

Digital Input DCBA	Output of U_1	Output of U_2 (Analog Out)
0000	0 V	0 V
0001	−1 V	1 V
0010	−2 V	2 V
0011	−3 V	3 V
0100	−4 V	4 V
0101	−5 V	5 V
0110	−6 V	6 V
0111	−7 V	7 V
1000	−8 V	8 V
1001	−9 V	9 V
1010	−10 V	10 V
1011	−11 V	11 V
1100	−12 V	12 V
1101	−13 V	13 V
1110	−14 V	14 V
1111	−15 V	15 V

This same information is shown graphically in Figure 15–35. Here, the binary input value is shown as the *x*-axis of the graph. The vertical axis is the analog output voltage.

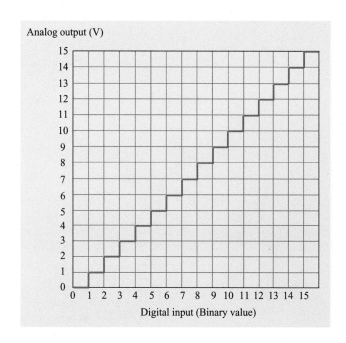

FIGURE 15–35 The performance of the DAC in Figure 15–34 displayed graphically.

The various steps in the staircase graph are exactly the same amplitude in an ideal DAC. In a practical DAC, they may vary slightly. Clearly, higher quality DACs have more uniform steps. The greater the number of steps for a given voltage span, the higher the **resolution** of the DAC. For a given number of bits (N) in the converter, there will be 2^N values and $2^N - 1$ steps. In the present case, the voltage span is 15 V, and it is divided into fifteen steps (sixteen values including zero). Therefore, each step equates to

$$\frac{15 \text{ V}}{15 \text{ steps}} = 1 \text{ V/step}$$

So, if the desired output voltage were 7.6 V, we would have at least a 0.4-V error. By contrast, a DAC with greater resolution would have finer steps. Suppose we had a 10-bit DAC with the same 15-V span. In this case, the improved resolution would yield

$$\frac{\text{Voltage span}}{\text{\#steps}} = \frac{15 \text{ V}}{2^{10} - 1} = \frac{15 \text{ V}}{1023 \text{ steps}} = 14.663 \text{ mV/step}$$

If the analog output is higher for each increasingly higher digital input, it is said to be a **monotonic output**. If, due to imperfections in the DAC, a higher digital input value at some point in the range of operation results in a lower analog output voltage, then the converter output is not monotonic.

Practical DACs are available with either voltage or current outputs that are proportional to the digital value at their inputs. DACs can be constructed from discrete components as shown in Figure 15–34, but because it is more difficult to get consistent precision with discrete components, greater circuit board space is required, and for other reasons, many designs utilize modular or integrated DACs. The AD566A device is a 12-bit DAC made by Analog Devices (**www.analogdevices.com**).

Analog-to-Digital Conversion

An analog-to-digital converter (ADC) is the functional opposite of a digital-to-analog converter (DAC). An analog voltage (or current) at the input to the ADC is transformed into an equivalent digital value at the outputs of the converter. As with DACs, there are many different techniques that can be used to accomplish A/D conversion. They vary in complexity, speed of conversion, and in other ways. Figure 15–36 on page 580 shows a representative ADC circuit. This particular technique is called a parallel A/D converter or more commonly a flash converter. It is the fastest method for converting analog to digital.

As you can see, the circuit uses a zener diode to obtain a steady 10-V reference. This reference voltage is applied to a voltage divider consisting of eight 10-kΩ resistors. The voltage drop across each resistor can be found with Ohm's Law to be 1.25 V. As you can see, the tapped voltages from 1.25 to 8.75 V are used as one input to several voltage comparators. The remaining input (noninverting input) of each comparator is connected to the analog input voltage. In this particular design, the analog input can span the range of 0 to 10 V.

Now, for any given analog input voltage, some number of comparators will have high-level outputs and some will have low-level outputs. For example, if the input voltage is +3 V, then the two lower comparators will have high outputs, since their (+) input is more positive than their (−) input. For similar reasons, the upper five comparators will have low outputs.

The logic gates connected to the output of the voltage comparators convert the results of the voltage comparison into standard binary values. Table 15–8 shows the relationships between input voltage, voltage-comparator outputs, and final converted outputs.

Since this is only a 3-bit converter, it has poor resolution. The 0- to 10-V span is divided into eight value ranges of 1.25 V each, so eight binary values represent this

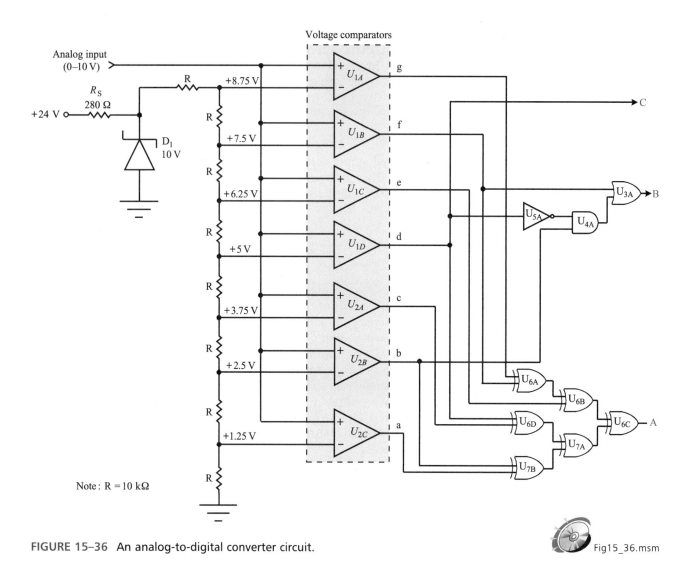

FIGURE 15–36 An analog-to-digital converter circuit.

entire span. This means that the input voltage must change by at least 1.25 V before a change will be seen in the converted output. This may be adequate for some non-critical applications, but in most cases, greater resolution is required. ADCs with 8-, 16-, and 64-bit resolutions are quite common. The MAX-104 ADC from Maxim (www.maxim-ic.com) is an 8-bit converter that provides 1 gigasample per second.

TABLE 15–8 Relationship between analog input and logic states in the flash ADC shown in Figure 15–36.

Analog Input Voltage	Comparator Outputs (abcdefg)	Binary Output (*CBA*)
$0 < 1.25$ V	0000000	000
1.25 V $< V_{IN} < 2.5$ V	1000000	001
2.5 V $< V_{IN} < 3.75$ V	1100000	010
3.75 V $< V_{IN} < 5$ V	1110000	011
5 V $< V_{IN} < 6.25$ V	1111000	100
6.25 V $< V_{IN} < 7.5$ V	1111100	101
7.5 V $< V_{IN} < 8.75$ V	1111110	110
8.75 V $< V_{IN} < 10$ V	1111111	111

✓ Exercise Problems 15.4

1. Many real-world quantities such as temperature, pressure, and light intensity are (*digital/analog*) values.

Refer to Figure 15–37 for questions 2 through 4 and assume the input voltages for low and high logic levels are 0 V and 500 mV, respectively. Note that D and A are the most-significant and least-significant bits, respectively.

2. If the binary input value is 0101 ($\overline{D}C\overline{B}A$), what is the voltage at the junction of R_5 and R_6?
3. If the binary input value is 1111 (*DCBA*), what is the voltage at the output of U_2?
4. If U_2, R_6, and R_7 were removed from the circuit, would the circuit still qualify as a D/A converter?
5. If a 6-bit D/A converter had an output range of 0 to 5 V, what is the smallest voltage change that would be seen on the output between two consecutive input values?

Refer to Figure 15–36 for questions 6 through 8.

6. If the analog input voltage is exactly 6.0 V, how many voltage comparators have high-level outputs?
7. If the analog input is +2.3 V, what is the binary output?
8. If the analog input voltage changes from 2 V to 6 V, what happens to the value of current through resistor R_S?
9. If we needed to distinguish between input voltage changes as small as 100 mV, what is the fewest number of bits that would be required?

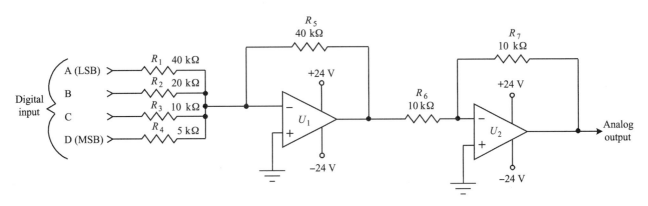

FIGURE 15–37   A digital-to-analog converter circuit.

Fig15_37.msm

15.5 TROUBLESHOOTING DIGITAL SYSTEMS

Troubleshooting digital systems is conceptually less difficult than troubleshooting analog circuits, since every point in the circuit has only two possible states. However, the complexity of digital systems makes the troubleshooting effort anything but simple in many cases. Regardless of how complex the system is, however, each device in the system conforms to some very basic rules.

In this section, we will examine some specific troubleshooting techniques that are applicable to digital circuits. These techniques, coupled with general troubleshooting strategies discussed in previous chapters, will allow you to successfully locate defects in digital systems.

Combinational Logic

In this section, we will examine ways to verify the condition of a basic logic gate. The ability to troubleshoot more complex combinational logic circuits is largely an extension of these methods.

Determining Relevant Inputs

One of the goals when troubleshooting is to locate the defect while making as few measurements as possible. Remember, some testpoints are not readily accessible, and each test takes time and increases the probability of accidental damage to the circuit board or other components.

There are many times when there is no need to test every pin on a gate while troubleshooting. Suppose, for example, we are troubleshooting a circuit with a NAND gate. We measure the output and find that it is high. We measure one of the inputs and find that it is low. At that point, there is absolutely no reason to measure the remaining inputs. Recall that any low input on a NAND gate will cause the output to go high. Therefore, knowing the state of the remaining inputs will provide no useful information. This is illustrated in Figure 15–38.

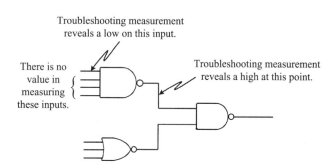

FIGURE 15–38 If the output of a NAND gate is high, then you need only locate one low input.

Now, this does not prove that the gate is good—although it most likely is—but testing additional inputs will not permit you to classify it with any more certainty. Table 15–9 summarizes this same concept for the other basic gates.

TABLE 15–9 A summary of troubleshooting actions applicable to logic gate testing.

Basic Gate	Output State	Troubleshooting Action
AND	Low	Don't measure additional inputs if one low input is found.
	High	Verify that all inputs are high.
NAND	Low	Verify that all inputs are high.
	High	Don't measure additional inputs if one low input is found.
OR	Low	Verify that all inputs are low.
	High	Don't measure additional inputs if one high input is found.
NOR	Low	Don't measure additional inputs if one high input is found.
	High	Verify that all inputs are low.

Probable Defects

As with any other solid-state device, there are many types of defects that can occur in basic logic gates. However, most defects produce symptoms that fall into one of the following categories:

- Output always low
- Output always high
- Output at a "bad" level
- Input shorted to ground or V_{CC}
- Input open

Each of these possibilities is summarized in Table 15–10.

TABLE 15–10 Possible defects that can be present when troubleshooting basic logic gates.

Symptom	Possible Defects
Output always low or high	If the output of a gate is always low or always high, this can be caused by one of three things. First, the gate may be responding to logic levels at the input in a normal way. For example, if even a single input of a NOR gate is high, that will cause the output to be low. In this case, you cannot isolate the defect until you change the input levels. If the PCB trace connected to the output pin is shorted to a low impedance point (such as V_{CC} or ground), then the output will remain low or high. Changing the input levels will have no effect. An internal defect in the gate can cause the output to remain low or high regardless of changes on the inputs.
Output at a "bad" level	If the output of a logic gate is at a voltage level that lies between the allowable levels for low and high logic signals, we label it as a "bad" logic state. This can be caused by any of three conditions. First, this can be normal operation for a special logic gate that has a tri-state output circuit. Second, the output of the gate can be shorted to another output or to some other voltage source. Third, the gate can have an internal defect.
Input shorted to ground or V_{CC}	A shorted input will remain low or high regardless of the state of the system.
Input open	An open input will appear to change states normally, but the changes will have no effect on the output state.

Identification of Specific Defects

A logic probe and a logic pulser are two useful tools for troubleshooting logic circuits. A **logic probe** is a small pencil-like device that contains LEDs, which provide one of the following indications when a point in a logic circuit is touched:

- High level
- Low level
- A pulsing logic signal (e.g., a clock pulse)
- A bad logic level

A **logic pulser** is also a pencil-like device, but unlike the probe, which senses a logic level, the pulser injects a logic level. More specifically, you place the pulser on a point in a logic circuit and press a button on the pulser. The pulser injects a logic signal that is opposite from the present logic state. In other words, it forces a change in the logic state of the point being pulsed.

Now, let's learn how to utilize the logic probe, logic pulser, and other equipment to identify specific defects in logic gates.

In general, the logic probe can be used to determine the state of the various input and output pins. If a pin (either input or output) is thought to be shorted, we can verify that possibility as follows:

1. Place the logic probe and the logic pulser on the suspected point at the same time.
2. Activate the logic pulser while monitoring the logic probe.
3. If the tested point is shorted to either ground or V_{CC}, then the probe will not respond to transients from the logic pulser.

This is illustrated in Figure 15–39.

FIGURE 15–39 Use of a logic probe and logic pulser in combination to detect shorts to ground and V_{CC}.

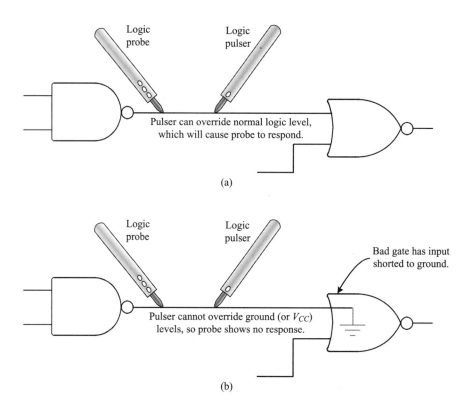

Shorts between two inputs or between two outputs can be detected by placing the probe on one of the inputs (or outputs) suspected of being shorted together and then pulsing the other one. If a pulse is detected, a short or low-resistance path exists between the two points. Classification of this type of defect is illustrated in Figure 15–40.

FIGURE 15–40 Detection of shorts between inputs (or outputs).

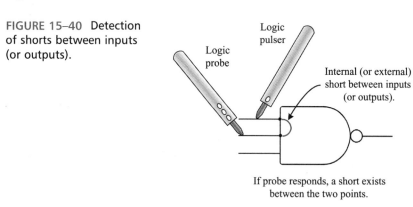

Another good troubleshooting tip is to feel the IC case with your finger. Certain types of internal defects can cause the IC package to become considerably warmer than normal. Use caution, however, because sometimes a defective device can become hot enough to burn you.

Sequential Logic

Sequential logic circuits can pose a significant troubleshooting challenge even to an experienced technician. In most cases, logic states throughout the system are continuously changing. In systems such as computers, the logic states are changing hundreds of millions of times per second. Not only must you measure the logic level at a given point, but you must also know the logic levels at several other points simultaneously. Suppose, for example, we were troubleshooting a circuit that contained a *JK* flip-flop. If the circuit were static (e.g., manually clocked), we could simply measure *J*, measure *K*, and then measure the set and clear pins. With this information, we would know what the flip-flop should do when it receives a clock. We could then monitor the output as we apply a single clock and quickly classify the results as good or bad.

But, consider a situation where all of the input terminals are changing millions of times per second. To evaluate the condition of the flip-flop, we need to essentially freeze time in order to examine the various points at our leisure. A **logic analyzer** is a digital test instrument that effectively freezes time for us.

In short, a logic analyzer has many inputs (as few as sixteen and as many as several hundred). You can connect the input wires to the various points of interest. You can also connect the system clock to the logic analyzer as a timing reference. Now, operate the circuit at normal speeds. The logic analyzer monitors and records the logic levels at the various points on every clock cycle. You can then scroll through the recorded data, examine the logic levels of all monitored points, and clearly see their relationships to each other. Logic analyzers can display the stored information as binary values or as a multi-trace timing diagram. You should be aware that logic analyzers are also powerful tools for troubleshooting high-speed combinational logic circuits. Here again, the analyzer allows you to monitor and record the logic states of several points simultaneously.

Troubleshooting with Software

You have utilized circuit simulation software in previous chapters to help you understand the operation of a circuit, to observe circuit waveforms, and to practice localizing defects. The use of software for localizing defects goes well beyond what we have discussed to this point. One of the more widely-used development tools is a software language called Very High-Speed Integrated Circuit (VHSIC) Hardware Description Language (HDL) or simply VHDL. This is a standardized way of describing a system such that it can be interpreted by many different computer applications. A circuit or system described by VHDL can be utilized in ways that include:

- Circuit simulation
- Printed circuit board layout
- Integrated circuit fabrication
- Troubleshooting and refinement of new designs *before they are fabricated*

VHDL was originally developed by the United States government in the early 1980s and formally adopted in 1987 by the Institute of Electrical and Electronic Engineers (IEEE) as a standardized language (Std. 1076). VHDL at that time was focused on digital circuitry. Since that time, the standard has been extended to include representation of analog systems and hybrid systems (digital and analog devices in the same system). The revised standard (Std. 1076.1) is referred to as VHDL-AMS. This extended version also allows for simulation of nonelectronic system components such as pressure, temperature, airflow, and so on.

✓ Exercise Problems 15.5

1. If a logic probe indicates a "bad" level on the output of a logic gate, then the gate is definitely defective. (True/False)
2. If the output of a NAND gate is low at all times, then there is no reason to measure any additional inputs if you know that at least one input is high. (True/False)
3. If the output and one input to an AND gate are found to be low, then there is no reason to measure any other inputs on the gate. (True/False)
4. While troubleshooting a digital circuit, you place a logic probe and a logic pulser on the same point and then activate the pulser. The probe indicates a steady

high-logic level and does not respond to the pulser. What condition does this probably indicate?
5. Briefly explain why troubleshooting a high-speed sequential logic system without a logic analyzer can be very difficult.
6. Under what conditions would a logic analyzer be a more appropriate troubleshooting tool for a combinational logic circuit than a logic probe and logic pulser combination?
7. At what point in the lifetime of a digital system does VHDL serve as a valuable troubleshooting tool?

SUMMARY

All digital systems operate on binary (2-state) values. This chapter contrasted the characteristics of binary and decimal number systems. Binary numbers can be incremented using the same basic rules as for decimal numbers.

We discussed the conversion of binary values into equivalent decimal numbers. A technique for converting a decimal value into an equivalent binary number was also presented.

There are many labels that are used to represent the two logic states in a digital system. Some typical examples are low/high, yes/no, true/false, and 0/1. Truth tables show the relationships between all possible input logic states and the resulting output logic state. There are 2^N entries in a truth table that represents a logic circuit with N inputs. Timing diagrams can also be used to show relationships between inputs and outputs of a logic circuit.

Boolean algebra allows us to represent a logic circuit mathematically. This is essential not only for the design of logic circuits, but also as a valuable aid to understanding and discussing circuit operation. The basic Boolean rules can all be demonstrated with the basic logic gates consisting of AND, OR, NOT, NAND, NOR, exclusive OR, and exclusive NOR gates.

Logic circuits can be classified into one of two types. Combinational logic circuits are characterized as having outputs whose logic states at any given time are solely determined by the inputs at the same time. Sequential logic circuits, by contrast, have outputs whose logic states are determined by its inputs at that time and by its present internal state. The internal state is essentially a memory function and is dependent on previous events.

Discrete combinational logic circuits are generally quite small in modern digital systems. More complex circuits are implemented with custom ICs or programmable devices such as PALs. The primary building blocks for sequential logic circuits are flip-flops. Both D and JK flip-flops are quite common. Many sequential logic circuits also require combinational logic devices to achieve the desired overall behavior. Counters are one common application of flip-flops and of sequential logic in general. Counters can be synchronous or asynchronous.

Digital-to-analog converters (DACs) convert digital values into equivalent analog voltages or currents. Analog-to-digital converters (ADCs) convert analog voltages and currents into equivalent digital values. DACs and ADCs provide the means to interface real-world analog devices with digital computers and other digital systems.

Troubleshooting digital systems (especially high-speed digital systems) can be quite challenging, but the use of proper tools simplifies the task. A logic probe, a logic pulser, and a logic analyzer are tools that are valuable for troubleshooting digital circuits.

REVIEW QUESTIONS

Section 15.1: Digital Concepts and Terminology

1. How many digits are allowed in the binary number system?
2. In the binary number 1001, what is the value of the left-most digit?
3. Express the number 1101_2 as an equivalent decimal value.
4. A single binary digit is called a _____.
5. Write the next two sequential binary numbers after 11010_2.
6. Arrange the following in numerical order (smallest to largest): 14_{10}, 1000_2, 1100_{10}, 100000000001_2, 1100_2.
7. Convert the number 26_{10} to an equivalent binary value.
8. Convert the number 49_{10} to an equivalent binary value.
9. Convert 1101010_2, 101001_2, and 111010110001_2 to equivalent hexadecimal numbers.

10. Convert 105h, A4h, and 5CD7h to equivalent binary numbers.
11. List at least three sets of names or labels that are commonly used to represent the two logic states in digital circuits.
12. How many entries are required in a truth table to describe a 3-input logic circuit?
13. How many entries are required in a truth table to describe a 6-input logic circuit?
14. What type of logic gate is represented by the timing diagram in Figure 15–41?
15. Using the *AB* input waveforms given in Figure 15–42, draw the output waveform that would represent an exclusive OR gate.
16. Complete Table 15–11.
17. What type of logic gate is described by the phrase, "Any high in gives a high out"?
18. What type of logic gate is described by the phrase, "Any low in gives a high out"?
19. Simplify the expression: $Y = \overline{\overline{ABC} + \overline{CD}}$.
20. Simplify the expression: $Y = (A + \overline{B})(C + D)$.

Section 15.2: Combinational Logic

21. The output of a combinational logic circuit at any given time is determined by its inputs at the same time. (True/False)
22. Draw a logic diagram to show how a NAND gate (plus any required inverters) can be used to implement the AND function.
23. Draw a logic diagram to show how a NOR gate (plus any required inverters) can be used to implement the NAND function.

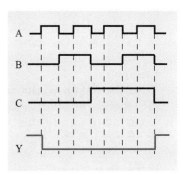

FIGURE 15–41 Timing diagram for a basic logic gate.

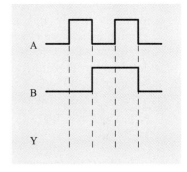

FIGURE 15–42 Draw the timing diagram for an exclusive OR gate.

24. Will the logic circuit in Figure 15–43 perform the same logic operation as a simple OR gate?

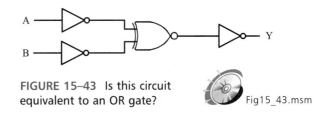

FIGURE 15–43 Is this circuit equivalent to an OR gate? Fig15_43.msm

TABLE 15–11 Complete this table.

Logic Gate	Logic Symbol	Logic Expression
	AND symbol	
		$Y = \overline{A + B}$
NAND		
		$Y = A \oplus B$
		$Y = \overline{A}$
	OR symbol	
Exclusive NOR		

25. What do the letters PAL and GAL represent when speaking of logic circuits?
26. A PAL that uses titanium-tungsten links can be reprogrammed as many times as necessary. (True/False)
27. Write the simplified Boolean expression for each of the logic circuits in Figure 15–44.
28. Complete a truth table to describe each of the circuits shown in Figure 15–44.
29. An expression such as $Y = A\overline{B}C + CD + \overline{E}$ is written in maxterm form. (True/False)

Section 15.3: Sequential Logic

30. How many inputs and how many outputs can a sequential logic circuit have?
31. A sequential logic circuit is characterized as having an internal memory function. (True/False)
32. A 7-bit sequential logic circuit can have a maximum of _____ internal states.
33. If you know the present level of all inputs on a sequential logic circuit, you can predict the current logic state of all outputs. (True/False)
34. If you know the present level of all inputs on a sequential logic circuit, and you know the present internal state, you can predict the next state of the circuit. (True/False)
35. If the D input of a D flip-flop is connected directly to the $\overline{Q}$ terminal of the same flip flop, it will toggle every time it receives a proper clock transition. (True/False)
36. A D flip-flop has _____ states and a JK flip-flop has _____ states.
37. If the reset input on a flip-flop has no logic bubble, what level is required to clear the flip-flop?
38. What is the state of the $\overline{Q}$ terminal of a flip-flop when it is in the set state?
39. What does the (>) symbol mean on the clock input of a flip-flop?
40. If the J and K inputs are both low, then a JK flip-flop is prevented from changing states in response to a clock input. (True/False)
41. If J and K are both tied high, a JK flip-flop will toggle when a clock arrives. (True/False)
42. If a binary counter has sixteen bits, how many states will it have?
43. If the clock inputs from all stages in a counter are connected directly together, the counter can be identified as a(n) (*synchronous/asynchronous*) counter.
44. Determine the count sequence for the two counters in Figure 15–45.
45. Race conditions are more likely to occur in (*synchronous/asynchronous*) counter circuits.
46. An erroneous transient on the output of a logic circuit that results from a timing difference between the inputs is called a _____.
47. Determine the output frequency for each of the indicated points in Figure 15–46. The circuit is configured such that the A flip-flop toggles the most often and the D flip-flop toggles the least often.

Section 15.4: Interfacing Digital and Analog Systems

48. What does the abbreviation ADC represent when used to label a type of integrated circuit?
49. What is the general nature of the output of a DAC?
50. Assuming they were properly scaled, could you connect the output of an ADC to the input of a DAC and expect to get the original ADC input signal at the output of the DAC?
51. Refer to the D/A converter circuit shown in Figure 15–47. This type of conversion technique would not be very practical for a 64-bit converter. Why not? (*Hint:* It has to do with the input resistors.)

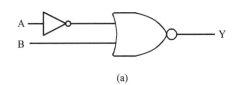

(a)

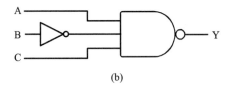

(b)

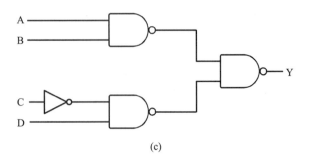

(c)

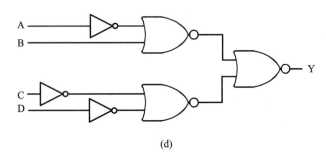

(d)

FIGURE 15–44 Write the Boolean expressions for these logic circuits.

REVIEW QUESTIONS 589

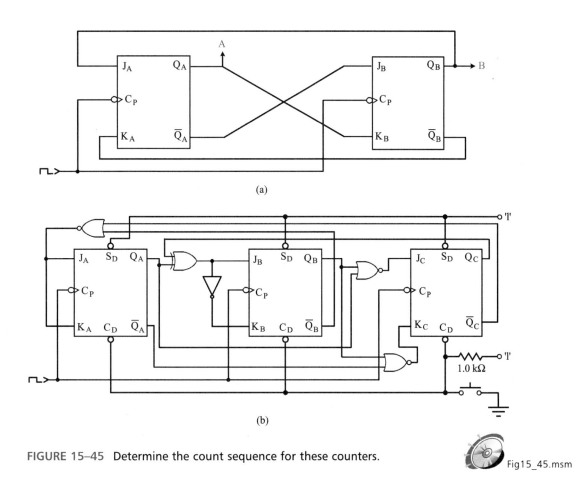

FIGURE 15–45 Determine the count sequence for these counters.

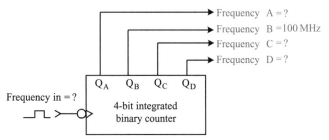

FIGURE 15–46 Determine the output frequencies.

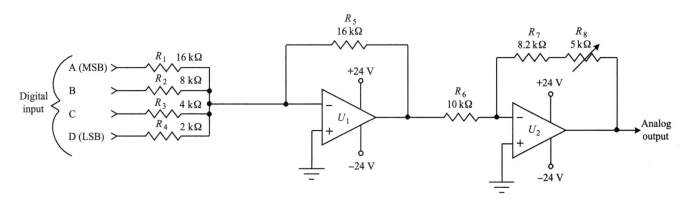

FIGURE 15–47 A D/A converter circuit.

52. A flash converter is one of the fastest forms of A/D conversion. (True/False)

Section 15.5: Troubleshooting Digital Systems

53. While troubleshooting, you find that one input of an 8-input NAND gate is high and the output is low. Can you gain any useful information by measuring the rest of the inputs?
54. While troubleshooting, you determine that one input of a 5-input NOR gate is high and that the output of the gate is low. Can you gain any useful information by measuring the rest of the inputs?
55. While troubleshooting, you measure a 1-kHz square wave on one input of a 2-input AND gate. The second input and the output are both low. What condition does this indicate?
56. While troubleshooting a digital circuit, you place a logic probe and a logic pulser on the same point and then activate the pulser. The probe blinks briefly. Is this an indication that the point is shorted to ground or V_{CC}?
57. During a troubleshooting exercise on a high-speed sequential logic circuit, you find it necessary to determine the logic levels on six different points at the same time. What test instrument would you choose?

●─CIRCUIT EXPLORATION

The exercise in this section will give you the opportunity to gain additional insights into the operation of sequential logic circuits. Figure 15–48 shows the logic diagram for a state machine.

It has two logic inputs in addition to the clock. It also has two logic outputs as well as the state outputs. The circuit is designed to function as a 2-bit binary counter. The DIR input determines the direction of count (i.e.,

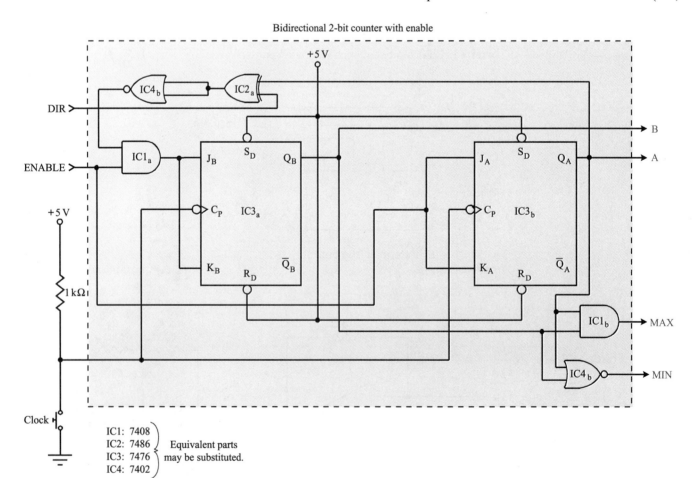

FIGURE 15–48 A state machine to operate as a bidirectional, 2-bit counter with enable.

Fig15_48.msm

increment or decrement to the next sequential binary value). The ENABLE input determines whether the counter is allowed to advance to the next state or remain in the present state.

The A and B outputs provide an indication of the present state. The MAX and MIN outputs go high to indicate that the counter has reached its highest (11) or lowest (00) state, respectively.

Study the circuit carefully and then complete the following tasks:

1. Build the circuit either in a lab or using circuit simulation software. The DIR and ENABLE inputs can come from a switch or other selectable source of 0-V and +5-V levels. The push button used for the clock is fine if you are in a simulation environment. If you physically build the circuit, then a bounceless switch should be used. Figure 15–49 shows one way to build a bounceless switch.
2. Exercise the circuit to verify and understand its operation. It will be helpful to attach logic indicators to the various outputs.
3. Enable the counter, and set it up to count down. Produce a timing diagram for the circuit using a logic analyzer. Monitor the clock input and the A and B outputs.

The following steps are optional:

4. Have your instructor or a friend insert a defect into the circuit and then troubleshoot it to identify the defect. Repeat until you are confident.
5. If you have physically constructed the circuit in a lab, then remove the push button used for the clock and replace it with a 1.0-kHz square wave (0-V to +5-V levels). Try your skill at troubleshooting the circuit with a fast clock.

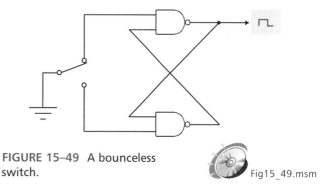

FIGURE 15–49 A bounceless switch.

Fig15_49.msm

ANSWERS TO PRACTICE PROBLEMS

Page 543
1. 1111
2. 0111
3. 1110
4. 0100
5. 0110
6. 1001111
7. 1010100
8. 1010100101011

Page 544
1. 6
2. 28
3. 10
4. 6
5. 63
6. 33

Pag 545
1. 1111
2. 11011
3. 111
4. 100011
5. 1001011
6. 10010

Page 554
1. $Y = \overline{A}\,\overline{B}\,\overline{C}D$
2. $Y = A + B + C + D$
3. $Y = A\overline{D}(\overline{B} + \overline{C})$ or $Y = A\overline{B}\,\overline{D} + A\overline{C}\,\overline{D}$
4. $Y = (\overline{A} + \overline{B})C$ or $Y = \overline{A}C + \overline{B}C$
5. $Y = (\overline{AB})(\overline{C} + \overline{D} + E)$ or $Y = \overline{A}\,\overline{B}\,\overline{C} + \overline{A}\,\overline{B}\,\overline{D} + \overline{A}\,\overline{B}E$
6. $Y = \overline{A}\,\overline{B}C + D + \overline{E}$

Page 558
1.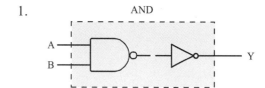

2.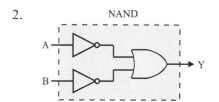

3.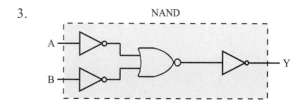

4.

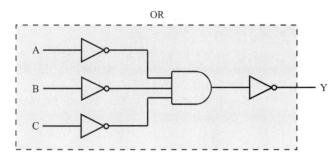

Page 561

1. $Y = \overline{A}B + A\overline{C}$
2. $Y = (A + B)(\overline{C} + D)$
3. $Y = A \oplus B + C$
4. $Y = A \oplus B \oplus C$

Page 564

1.

C	B	A	Y
0	0	0	0
0	0	1	0
0	1	0	1
0	1	1	0
1	0	0	0
1	0	1	1
1	1	0	0
1	1	1	0

2.

C	B	A	Y
0	0	0	0
0	0	1	1
0	1	0	1
0	1	1	0
1	0	0	0
1	0	1	1
1	1	0	0
1	1	1	0

3.

C	B	A	Y
0	0	0	1
0	0	1	0
0	1	0	0
0	1	1	0
1	0	0	1
1	0	1	0
1	1	0	1
1	1	1	1

4.

D	C	B	A	Y
0	0	0	0	0
0	0	0	1	0
0	0	1	0	0
0	0	1	1	1
0	1	0	0	0
0	1	0	1	0
0	1	1	0	0
0	1	1	1	1
1	0	0	0	0
1	0	0	1	0
1	0	1	0	0
1	0	1	1	1
1	1	0	0	1
1	1	0	1	1
1	1	1	0	1
1	1	1	1	1

5.

C	B	A	Y
0	0	0	1
0	0	1	1
0	1	0	0
0	1	1	1
1	0	0	0
1	0	1	0
1	1	0	0
1	1	1	1

6.

D	B	A	Y
0	0	0	1
0	0	1	0
0	1	0	1
0	1	1	0
1	0	0	1
1	0	1	1
1	1	0	1
1	1	1	0

Page 569

No. The inverter on the clock line does provide the inversion that is necessary to convert it from a falling edge $(1 \rightarrow 0)$ device to a rising edge $(0 \rightarrow 1)$ device, but the actual performance of the circuit is as follows:

D	Q_{n+1}	$\overline{Q}_{n+1}$
0	1	0
1	0	1

This is not consistent with the behavior of a D flip-flop. However, if connections to the J and K terminals were reversed, it could replace a D flip-flop.

Page 571

1. The count sequence is

    B A
    0 0
    1 0
    1 1
    0 1
    0 0

2. The count sequence is

    C B A
    0 0 0
    0 0 1
    0 1 1
    1 1 1
    1 1 0
    1 0 0
    0 0 0

Page 575

1. The machine has four states: 00 ($\overline{A}\,\overline{B}$), 01 ($\overline{A}B$), 11 ($AB$), and 10 ($A\overline{B}$).
2. The LED will illuminate under three conditions:
    a. Anytime the machine is in state 01 ($\overline{A}B$)
    b. When the machine is in state 00 ($\overline{A}\,\overline{B}$) and either of the following button conditions exist:
       i. S_1 not pressed and S_2 pressed
       ii. S_1 not pressed, S_2 not pressed, and S_3 pressed
3. The solenoid will be energized anytime the machine is in state 10 ($A\overline{B}$), regardless of which switches are pressed.
4. The following sequence of events must occur to energize the solenoid:
    a. Press and hold S_1. Machine advances to state 01 ($\overline{A}B$).
    b. Press and hold S_3 and then release S_1. Machine advances to state 11 (AB).
    c. Press and hold S_2 and then release S_3. Machine advances to state 10 ($A\overline{B}$), and the solenoid is energized. It will remain in this state until all three buttons are released. It will then return to state 00 ($\overline{A}\,\overline{B}$).

CHAPTER 16
Microprocessors and Computers

•—KEY TERMS
address bus
bidirectional bus driver
BIOS
bootstrap loader
byte
cache memory
clock skew
compiler
control bus
data bus
hardware
instruction register
instruction set
macro
nonvolatile
port
program
program branch
program counter
program label
program loop
RAM
read
register
ROM
shift register
software
volatile
word
write

•—OBJECTIVES
After studying this chapter, you should be able to:

- Explain the relationship between hardware and software in a computer system.
- Explain the primary purpose of each functional block in the functional block diagram of a computer.
- Name and describe the following levels of programming languages: machine language, assembly language, high-level language, and user-programming language.
- Highlight the general relationships between the following classes of computer software: BIOS, bootstrap loader, operating system, application software, and user-application programs.
- Name and state the general nature of the following standard computer buses: ISA, PCI, and AGP.
- List several types of I/O found on a typical computer.
- Contrast the general nature of serial and parallel data transmission.

•—INTRODUCTION

We are now prepared to study microprocessors and computers, which are exciting topics in digital electronics. Microprocessors serve as the "brains" of a computer. As you will see, a microprocessor is made up of nearly all the digital circuits discussed in chapter 15, yet can replace nearly all of them. That is, the microprocessor itself is an integrated circuit composed of counters, flip-flops, gates, and so forth, but the resulting chip is so powerful that it can replace many other integrated circuits. Our primary goal in this chapter is to talk about microprocessors in general and to begin to develop our vocabulary. As with all of our previous studies, we must expand our vocabulary and learn to interpret the manufacturer's data. If we can do this, then we are protected, to some degree, from being trampled in the runaway technology race.

16.1 BASIC CONCEPTS AND TERMINOLOGY

First, we will describe a microprocessor (µP) in a way that will provide perspective while we are learning the related vocabulary. The definition we will use, while grossly degrading to the µP, is surprisingly accurate after you see the whole picture. For now, we shall consider a μP to be "a high-speed instruction executer." That is all it is. You give it a list of things to do (instructions), and it simply executes or carries out those instructions.

Instruction Set

Each microprocessor has a certain set of instructions that it is capable of executing. This group of instructions is called the **instruction set**. Each type of μP will have a different instruction set, but there are usually many similarities. The instruction set, then, is a set of all the instructions or commands a particular μP can interpret. Typical instructions include the following:

- Add—one number to another.
- Move—a number from one storage location to another.
- Increment—increase the value of a counter by one.
- Decrement—decrease the value of a counter by one.
- Shift—all bits in a binary number one position to the left or right. This is equivalent to multiplying or dividing, respectively, the original number by two.

When we arrange the instructions in a particular sequence, we have written a **program**. As the μP executes the instructions sequentially, it executes our program. The result of executing a program can be anything from controlling your carburetor, to flying an aircraft, to calculating the price of three cheeseburgers, to monitoring and controlling a chemical process, to computing tax on your pay check, and so on. The list is endless and grows daily. The μP is merely a small computer. It can do nearly everything a large computer can do, only not as rapidly. Surprisingly, a microprocessor can actually outperform a full-size machine in some intensive control applications.

So far, we have taken certain instructions from the μP's instruction set and arranged them to form a program. The arrangement is such that when it is executed sequentially the program generates the desired results.

Memory

Now we need a place to store our program. This requires memory. The specific type of memory is completely dependent upon the particular application. For the present time we will assume we have a generic memory device capable of storing and retrieving information as needed.

How much memory will we need? First, we need to know how many bits (1s and 0s) are in each byte. A **byte** is a group of bits that move together as a unit. An instruction could qualify as a byte. In any case, there are eight bits per byte. Two bytes grouped together as a single unit are called a **word**. In general, the more bits that can be moved as a group, the faster and more powerful the microprocessor. Let us assume that we have a small 8-bit machine. This means that it deals with single bytes. It also means (with some exceptions) that the instructions are eight bits long. If the instructions are eight bits long, then you can see that we could have as many as 2^8 or 256 different instructions in the instruction set. For instance, 00110011 could mean ADD and 10111100 might mean SUBTRACT. We now have established that we need a memory that holds eight bits at each address location.

Now we must determine how many bytes of storage are required. The storage requirement is determined by two things: the amount of memory needed for the program and the amount of memory needed for data storage. Program storage is obviously dictated by the length of the program, which is determined by the application. Some microprocessor programs require less than 1000 bytes. Others require hundreds of millions of bytes (megabytes). Neither of these examples is a limit. The amount of data storage is also determined by the application. The memory requirement for data storage can vary from a few bytes to many terabytes.

Microprocessor Bus Networks

For our representative μP application, let us say that we need a 1k × 8 memory. This means we will have 1024 ($2^{10} = 1024$, but is routinely called 1k) storage locations with each location being able to store eight bits. Let us further assume that the program (sequential list of desired instructions) is already stored in memory. We must now develop a means whereby the μP can access memory. We will need address lines, so that the μP can identify what instruction it is ready for, and data lines for the actual instruction to move from memory into the μP where it can be interpreted. Each memory location in our system holds eight bits. Each time the μP stores or retrieves information it will do so eight bits at a time. This means that we will require eight data lines. These eight lines are collectively called the **data bus**.

How many address lines will we need? If our memory has 1024 (1k) locations, we will need 10 lines, since 2^{10} is 1024. These 10 lines make up the μP's **address bus**. A typical μP will have from 16 to 32 address lines and from 8 to 64 data lines. Generally, industrial control applications require smaller bus widths, while data manipulation applications (such as your home computer) require wider bus widths.

We still need one more set of lines to allow the μP and memory to communicate. We need control lines. The number and type of control lines are different with nearly every type of μP, but in each case they may be considered collectively as the **control bus**. Figure 16–1 shows a diagram of our μP system as we have developed it so far.

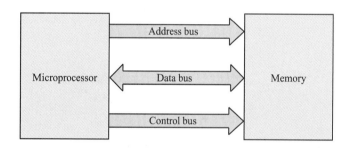

FIGURE 16–1 Microprocessor bus structure.

There is one additional and important point shown in Figure 16–1. Notice that the address bus and control bus are shown going from the μP to the memory, whereas the data bus is shown going both ways. This should seem sensible to you, since the μP must tell the memory where (address) it wants to **read** (retrieve data) or **write** (store data). It must also tell the memory when (control) to read or write. The actual data transfer, however, will be going into the μP during a read operation and from the μP to the memory during a write operation. To further define our buses, we say that the address and control buses are unidirectional, and the data bus is bidirectional. As you study various microprocessors, you will uncover some exceptions to these generalities. For instance, the control bus varies in number of lines and functions of lines and, depending on what you define as the control bus, may be bidirectional. At this point, it is important to understand the basic concepts as we progress. In later studies, you will have no trouble adapting these generalities to a specific μP.

Internal Microprocessor Structure

We shall now take our first peek inside the microprocessor chip itself. Let us begin by seeing what we need to drive the address bus.

Program Counter

The origin of the memory address for the stored instructions is a counter called the **program counter**. This is a presettable (can be force loaded to any value) counter that generally increments in a standard binary sequence, although some μPs use other sequences. The number of stages in the counter will be equal to the number of address lines.

The general process flow is to output an address, read the instruction stored there, execute the instruction, increment the program counter to the next address, and repeat indefinitely. As previously stated, the program counter is presettable. This is necessary for two major reasons. When power is first applied to the system, the various internal flip-flops and counters (including the program counter) come up in a random state. To begin executing our program at the proper starting point we preset or load the initial starting address into the program counter. This is usually done by activating an input pin on the microprocessor called reset. The actual starting address is dictated by the µP manufacturer. Consequently, you must organize your memory accordingly. For our representative system we will assume that reset causes the program counter to go to zero, and it increments sequentially from there.

The second reason for the program counter being presettable is to allow for program branches. Say, for example, that we are adding two numbers together. If the answer is positive, we may want to do one thing, but if the answer is negative, we may need to perform different actions. This means that our execution of instructions can no longer be sequential. The next instruction will be one of two different ones, in this case, depending on the results of the addition. This is called a **program branch**. It is accomplished by loading the program counter with the address of the next desired instruction. If the next instruction is in sequence, the counter is simply incremented. If the next instruction is out of sequence, its address is force-loaded into the program counter, and execution continues sequentially from that point.

Data Bus Drivers

Now let's examine the connections to the data bus inside the µP. First, we need a device capable of both read and write, transmit and receive, or in and out operation. This requires all of the lines in the data bus to be bidirectional. One method of accomplishing this is with a circuit equivalent to that shown in Figure 16–2. The circuit is aptly called a **bidirectional bus driver**. An even more descriptive name is a tri-state bidirectional bus driver. We will discuss the need for tri-state operation later on in our study.

FIGURE 16–2 Tri-state Bidirectional bus driver.

Fig16_02.msm

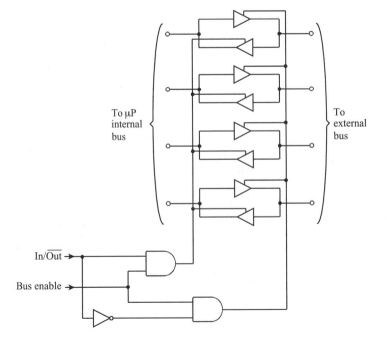

Figure 16–2 shows a 4-bit driver for simplicity, although we will need eight bits in our machine, since we have an 8-bit data bus. Figure 16–2 shows two control lines, IN/OUT and BUS ENABLE. The bus enable will place the buffers into the high-impedance tri-state condition if it is low. When the bus enable is high, activated, then one buffer in each pair will be activated, depending on the state of the IN/OUT signal. If the IN/OUT line is high, then the µP is in the read or receive mode and accepts data from the data bus. It passes information through to the internal µP bus structure. The output buffers are tri-stated during this operation.

When the µP wants to output data, the IN/OUT line is low. This causes the input buffers to go to the tri-state condition and the output buffers to turn on. Whatever data is on the µP's internal bus is now allowed to appear on the external data bus. Generally, control of the data bus drivers is done automatically inside the µP. It is not necessary for the user to be concerned with the moment-by-moment control of the data bus.

Internal Registers and ALU

So far in our discussion, we have managed to introduce one new concept at a time. Now we must look at where the data are going to or coming from on the internal µP data bus. Since the source/destination can be any one of several places, we will examine the internal data bus and its components as a whole. We will then continue to break it down even further and in more detail.

For our present discussion refer to Figure 16–3, which shows a simplified block diagram of the internal structure of a representative µP. We have previously discussed the program counter and its associated drivers, which provide the address for memory operations. We have discussed the control unit and drivers only to the extent of saying they drive the control bus, which controls, times, and sometimes monitors external activities.

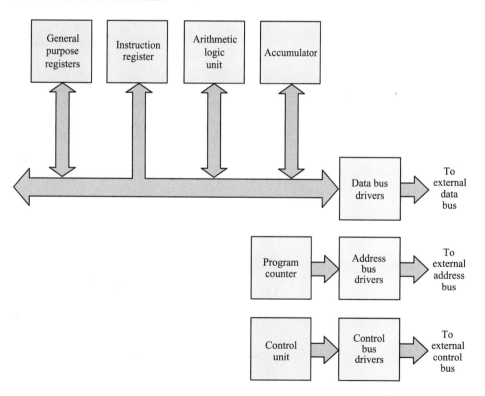

FIGURE 16–3 Simplified block diagram of the internal structure of a microprocessor.

There are four functional blocks shown in Figure 16–3 that are connected to the internal data bus. First is the general-purpose registers block. These registers are for temporary storage of data during program execution. A **register** is simply a series of flip-flops or a presettable counter into which a group of bits can be temporarily stored and subsequently retrieved. It is usually faster to store data in a register within the μP than it is to write it into external memory. These registers are usually the same length (number of bits) as the data bus, so we could rephrase this to say that each register can store one byte of data in a typical machine. In addition to storage, these registers can often be incremented or decremented like a counter. This, as you will see, greatly enhances their usefulness.

Instruction Register

Next, we have the **instruction register**. When the μP accesses memory to obtain the next instruction, the instruction goes from memory to the μP via the data bus. When it arrives on the internal μP data bus, it is loaded into the instruction register. Here the μP can decode and interpret what the instruction means and then carry out that command. The instruction register holds the instruction currently being executed.

Arithmetic Logic Unit

The arithmetic logic unit (ALU) is where all of the arithmetic functions such as ADD, SUB, A > B, and A = B are carried out. The ALU also provides the power necessary for logic operations such as A ⊕ B and so on. As you will see, the ALU is involved in the execution of nearly all instructions.

Accumulator

The accumulator is a register that has capabilities like the general-purpose registers. However, it is special in many ways. The accumulator usually holds one of the two operands (numbers, characters, and so on) during any ALU operation. It also receives the results of that operation. Suppose, for example, that we want to add two bytes together. To do so, we might put one of the bytes into the accumulator and the other byte either in memory or in one of the general-purpose registers. When the instruction is executed, the two numbers serve as inputs to the ALU. The output from the ALU, the sum in this case, is loaded back into the accumulator. The accumulator also has additional features, which make it unique, but the specific features vary considerably from μP to μP. In some μPs, the accumulator is the only register than can also function as a **shift register** (a register that can shift all bits either left or right one bit position). In some cases, all data being transmitted to external devices (such as printers, indicators, motor controls, and so forth) or received from external devices (such as limit switches, keyboards, and so forth) must enter or exit from or to the accumulator.

A Representative System

Now we are ready to see a μP in action. We will write a short program and follow the activity as the program is executed.

Example Instruction Set

Before we can write a program, we need an instruction set. Table 16–1 shows a portion of the instruction set for our μP. There are only three instructions shown, which will be enough for us to write a very simple program. You must realize that even the simplest μP will have many instructions in the instruction set.

The first instruction shown in Table 16–1 is the input instruction. When this instruction is executed, the following things occur. The second byte of the instruction, in this case 01h, is placed on the lower eight bits of the address bus. This identifies the device from which we are requesting data. Second, a signal on the control bus goes active to indicate that an external data input is taking place. Let us call that signal $\overline{IOR}$, which signifies that it is an active-low signal. The abbreviation I/O (or simply IO) is used for input/output. The letter R represents read or input as opposed to write or output. The third operation required to execute the IN instruction is to access the data bus and place its value into the accumulator. The

TABLE 16–1 A portion of the instruction set for a representative microprocessor.

Function	Mnemonic	Symbolic Description	Binary	Hexadecimal
Input	IN	A←data	1st 11011011 2nd 00000001	1st DBh 2nd 01h
Output	OUT	data←A	1st 11010011 2nd 00000010	1st D3h 2nd 02h
Jump (branch)	JMP	PC←byte 3 byte 2	1st 11000011 2nd Lower 3rd Upper	1st C3h 2nd … 3rd …

Note: PC = program counter; A = accumulator; data = external data

μP then increments the program counter and proceeds with the next sequential instruction stored in memory.

The second instruction in Table 16–1 is the output instruction. This is essentially the inverse of the input function. Here the accumulator contents are sent to the data bus, and the control bus signal called $\overline{\text{IOW}}$ (I/O write) goes active (low) to tell the external device that the data being output are now on the bus. Following this, the program counter is incremented, and the next instruction is executed.

The final instruction shown in Table 16–1 is the jump instruction. This causes the program counter to be forced to a particular state rather than the next sequential address. This is a three-byte instruction, and it is the second and third bytes that tell the program counter where to go.

System Hardware

We shall observe as a simple program is executed. For our first program, we will use the μP system shown in Figure 16–4(a) and the program shown in Figure 16–4(b) on page 602.

Figure 16–4 represents two major categories with reference to computers. The first (represented by the functional block diagram) is called **hardware**. Computer hardware consists of all of the physical boxes, circuit boards, integrated circuits, cables, and so forth required by the system. The second category (represented by the program listing) is called **software**. Computer software primarily refers to the computer program itself, although there are many subclasses of programs.

In Figure 16–4, we have added two external chips to serve as our I/O ports. A **port** is a path where data can exit (output) or enter (input) the computer. The input

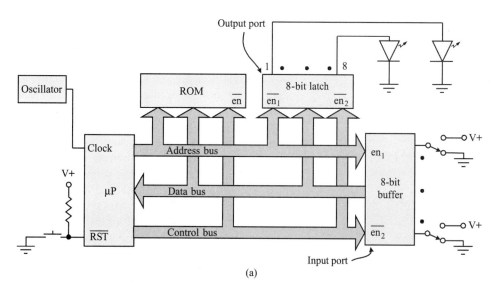

FIGURE 16–4 (a) A simplified microprocessor system and (b) a sample program.

FIGURE 16–4 (continued)

Label	Address (Hex)	Hexadecimal	Mnemonic	Comments
Start:	0000	DB	IN	Input data from device #1
	0001	01	Device #1	
	0002	D3	OUT	Send data to device #2
	0003	02	Device #2	
	0004	C3	JMP	Go to beginning
	0005	00	Start address	
	0006	00	"	
	0007	--	--	--

(b)

device is simply an 8-bit buffer. Whenever en_1 is high and en_2 is low, the inputs (state of the switches) are connected to its outputs (data bus). When the enable signals are not activated, the outputs will be in a tri-state condition, so they will not interfere with other devices connected to the same bus. The output port is an 8-bit latch or register. Whenever the two enables are active (low in this case), the input (data bus) is latched in the output. Our example system has the outputs connected to LEDs to provide a visual indication. The oscillator block provides the clock pulse necessary for the µP to operate. Depending on the particular µP, this may be single or polyphase and may vary from a few hertz to hundreds of megahertz. *Note:* Many µPs multiply the clock internally for higher speed operation. Finally, we have shown the reset ($\overline{RST}$) input. We shall assume that our representative µP goes to address 0000h when the reset pin is pulled low. That is to say, when we activate $\overline{RST}$, we will force the program counter to 0000h. This allows us to begin program execution at a known location—namely, address 0000h.

Figure 16–4 also shows a block labeled **ROM**. This is an abbreviation for read-only memory. This type of memory device retains its memory even with no power applied. For this example, we will assume our program is stored in ROM. The µP cannot store new data to a ROM (hence the name read-only).

Executing the Program
Now let's go inside the microprocessor and watch the action. Ready? Power on! As power is applied, we see all the registers, flip-flops, and even the program counter coming up in what appears to be a random state. The µP is outputting addresses and reading back memory and desperately trying to execute the program. However, since the PC (program counter) came up in a random state, the address that is being output is wrong, and the information being returned is of no value. Therefore, when the µP attempts to execute the retrieved instruction, no one knows what will happen. As you can see, the µP is running uncontrollably and serving no useful purpose, so let's get its attention.

We press the reset button and then release it. (This is automatic in most home computers.) The instant we press the button, the wild uncontrollable activity ceases, and the program counter goes to 0000h and stays there. When we release the $\overline{RST}$ button, the desired controlled program execution begins. It will progress as outlined in the following paragraphs.

First, the μP needs an instruction so it outputs the program counter, which is currently set to 0000h, to the address bus. It also activates a control bus signal called MEMR (or equivalent mnemonic indicating a memory-read cycle). The address goes to the memory and identifies the desired byte. The MEMR signal goes to the enable input of the ROM and causes it to enter the active state and place its selected data (our first instruction) on the data bus. After a short delay of a few nanoseconds, the μP latches the information currently on the data bus into its instruction register. It then deactivates the control bus and address bus lines and begins to decode the instruction to find out what it is expected to do next. Since our first instruction was IN, the instruction register now contains 11011011 or DBh. Upon decoding the instruction, the μP learns that it is an input instruction. However, to execute an input instruction requires an address of the input device. This information is part of the program and is stored in the next sequential memory location.

The μP then increments the program counter and fetches the next byte by sending the program counter to the address bus and activating MEMR on the control bus. Down the data bus from ROM comes 00000001 or 01h, since that is what we put there when we wrote the program. The μP takes this byte and stores it in a temporary latch internal to the μP.

It now knows what to do (input data) and where to do it (device number 1). It accomplishes this task by outputting 01h, the port number being held in temporary storage, to the address bus. It also causes IOR (or equivalent signal indicating an input operation) on the control bus to go low. This combination causes the 8-bit buffer to come out of the tri-state mode. It places the logic levels of the switches onto the data bus. Again, after a few nanoseconds of delay, the μP latches the data bus information into the accumulator.

Now that execution of the first instruction is complete, the buses are deactivated, the program counter is incremented, and we begin the fetch of the next instruction. Notice what has happened so far. The eight bits of data representing the current state of the switches are now being held in the accumulator. In addition, the program counter has been incremented twice, so we are now ready to address location 0002h, which contains our next instruction.

The μP now begins to access memory by outputting the program counter to the address bus and activating the control bus signal MEMR. The ROM places the information (D3h) on the data bus, and the μP latches it into the instruction register. Upon decoding the instruction, the μP must obtain another byte to determine which device should receive the data. Therefore, we increment the program counter and again access memory. This time 02h comes down the data bus, and we hold it in a temporary latch. The μP now has everything required to execute the instruction and proceeds to do so.

The device number (02h) is sent from the temporary latch to the address bus. The signal IOW is activated on the control bus, and the contents of the accumulator are sent down the data bus. The coincidence of IOW on the control bus and 02h on the address bus causes the 8-bit output latch to be enabled. This means that the information now on the data bus will be latched (stored) and presented to the LEDs. With the second instruction complete, the buses are deactivated, the program counter is incremented, and we are ready for another instruction fetch. Notice that the process so far has resulted in the state of the switches being displayed on the LEDs—there were, of course, a few nanoseconds or microseconds of delay in the accumulator.

The μP now outputs the program counter to the address bus and MEMR to the control bus. Down the data bus from ROM comes C3h, which is our next instruction. This byte is loaded into the instruction register and decoded. Upon decoding, the μP determines that this is a three-byte instruction, so it must access two more memory addresses. The program counter is incremented, the memory is accessed, and down the data bus comes the byte 00h. The μP holds this in a temporary storage register until it can get the last byte of the instruction. The program counter is again incremented and sent to the address bus. With 0006h on the address bus and $\overline{\text{MEMR}}$ on the control bus, the memory kicks out 00h to the data bus. The µP accepts this byte and is now ready to complete the execution phase of the instruction.

Instruction execution in this case is a matter of transferring the last two bytes received (and being held in temporary storage) into the program counter. Remember that the program counter always points to the address of the next instruction. Since we are now forcing it to 0000h, that will be the address of our next instruction, which happens to be where we began. The process will continue from here in an exact replay of our preceding discussion. A structure of this type is called a **program loop**—more specifically an infinite loop. Each time through the loop, we will input the status of the switches and transfer those data to the indicators. Although this is a trivial application of a microprocessor, it should serve as an adequate example of the sequential instruction execution of any µP. All communications are carried out via the three buses, which identify what, where, and when all data are transferred.

All of our programs, from a simple problem like transferring the states from switches to indicators, to the most complex industrial control application, will be composed of simple one-, two-, three-, or four-byte instructions. Furthermore, the µP will fetch and execute each of these instructions one at a time, as just described. This is perhaps one of the most serious limitations on microprocessors.

Since µPs are inherently limited to a one-step-at-a-time operation, there are some very real limits to the speed at which tasks can be accomplished. Even though a particular instruction may execute in a few nanoseconds, it may take hundreds or thousands of instructions to accomplish a practical task. Even if the µP is fast enough to keep up with a particular device like a video display, you can still run into problems if there are several devices being controlled by the µP.

Suppose, for example, that the µP must monitor 27 limit switches and one 20-key operator panel, and must control 14 solenoids and a 5-digit numeric display. While doing all of this, it must keep an accurate record of the time of day. As you can see, any one of these devices is exceptionally slow when compared to the µP. However, as the number of devices increases, the ability of the µP to adequately control each one becomes limited. As a final summarizing thought to this discussion, you may consider the combinational logic circuits studied earlier as parallel logic. This allows many events to occur at the same time. The µP is of a serial nature and must perform its actions on a one-at-a-time basis, which is inherently slower. Clearly, one way to minimize this limitation is to increase the speed of the µP. Ten years ago, a typical µP might have had operating speeds on the order of a few tens of megahertz. Today, even your home computer probably operates at 1.5 gigahertz or faster.

✓ Exercise Problems 16.1

1. List the following items in order of the relative sizes starting with the smallest:
    a. byte
    b. instruction
    c. bit
    d. program
2. Which of the following buses are bidirectional?
    a. control bus
    b. address bus
    c. data bus
3. How many bits must the program counter have in order to address 32k (32,768 bytes) of memory?
4. Describe how and why the contents of the program counter might be changed in a nonsequential manner.
5. What purpose does the address bus serve during execution of an input instruction?
6. The process of storing a value into a memory location is called a _____ operation.
7. The process of retrieving a value from a memory location is called a _____ operation.

8. Describe the characteristics of a tri-state bidirectional bus driver.
9. What is the name of the functional block inside a microprocessor where all of the logical and arithmetic operations are performed?
10. What is the name of the internal μP register that holds one of the operands during an arithmetic operation and receives the result of the operation?

16.2 HARDWARE

We will now take a more detailed look at the functional block diagram of a practical computer system. The primary learning goal in this section is to expand your vocabulary and broaden your understanding of overall computer system architecture.

System-Level Block Diagram

Figure 16–5 shows the block diagram for a representative computer system based on a microprocessor. The system described in this figure is probably very similar

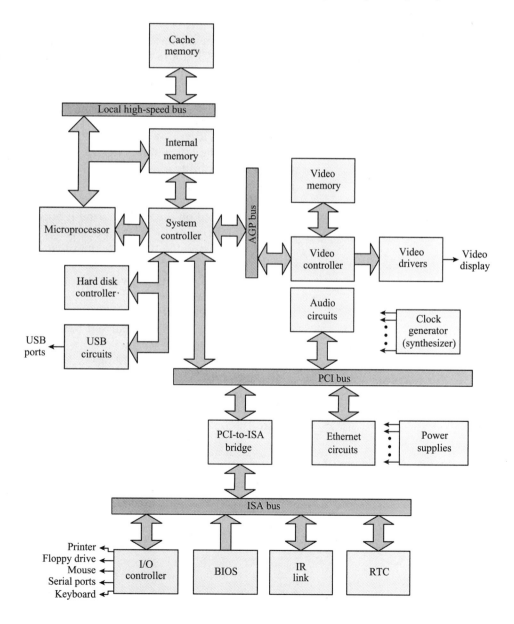

FIGURE 16–5 Functional block diagram of a representative microprocessor-based computer system.

to your home or school computer. We will discuss each of the functional blocks in Figure 16–5 in the following sections.

Standard Buses

As illustrated in Figure 16–5, the major functional blocks in a typical computer system communicate with each other via a number of buses. Recall that a bus is merely a group of lines that share the same general purpose and usually have similar electrical characteristics. Many bus configurations have been standardized in the computer industry. This enables different manufacturers to develop systems and expansion products that are compatible with each other.

As illustrated in Figure 16–5, the microprocessor communicates with its cache memory via a private high-speed bus. A cache memory is usually quite small, but very fast, and the μP can access it at the same time other activity is occurring on the other buses in the system.

There are three standard buses represented in Figure 16–5. The peripheral component interconnect (PCI) bus is a moderate-speed bus that provides communication between the system controller and several other major functional blocks. Data is transferred over this bus at a 33-MHz or 66-MHz rate. The width of the data portion of the PCI bus is generally 32 bits, but 64 bits are allowed in the standard. If you purchase an expansion board for your computer, it may be designed to be inserted into a PCI slot in the computer. When you insert the board, the connector pins on the board connect directly to the PCI bus in the system.

The industry standard architecture (ISA) bus is much older than the PCI bus, but it is still found in most computer systems. The purpose of the ISA bus is identical to that of the PCI bus in that it provides a standardized communication path for various devices. However, the data transfer rate is slower (8 MHz) on an ISA bus. Further, the bus width is less (8 or 16 data lines). Again, cards purchased to expand the capability of the computer are often designed to be inserted into an ISA slot in the computer.

Figure 16–5 also shows an accelerated graphics port (AGP) bus. This specialized bus is based on the PCI bus, and it provides for high-speed transfer of video information between the system controller and the video controller. By using a separate bus for video transfers, the other buses are able to provide effectively faster communication throughout the rest of the system. Several years ago, all data transfers were accomplished via a common bus. Since the μP can only do one instruction at a time, all activity within the system is inherently similar. Modern designs such as the one illustrated in Figure 16–5 use multiple buses and distributed controllers. This allows concurrent bus transfers, which increases the effective speed of the system.

The data portion of the AGP bus is 32 bits wide and runs at 66 MHz. This provides an effective data transfer rate of 266 MBps (megabytes per second). Further, the AGP bus allows optional rates up to 1.07 GBps.

Table 16–2 contrasts the basic capabilities of the three buses illustrated in Figure 16–5.

TABLE 16–2 A comparison of three common bus structures.

Bus Type	Width of Bus	Clock Speed (MHz)	Effective Transfer Rate (MBps)
ISA	8/16	8	1–2
PCI	32/64	33/66	133 (32-bit bus @ 33 MHz)
AGP	32	66	266 (optionally up to 1.07 GBps)

Clock Generator

Every computer system has a clock generator circuit. In most cases, this is a clock synthesizer circuit based on a phase-locked loop. We discussed such a device in chapter 14. Typically, a clock synthesizer accepts a relatively low frequency but stable input clock such as from a 14.318-MHz crystal oscillator. It then generates an array of clock signals that are utilized throughout the system. Most major functional blocks require a clock. Typical applications include PCI bus clock (33.3 MHz), ISA bus clock (8 MHz), USB clock (48 MHz), and many others depending on the specific design. When an IC requires a clock that is substantially higher than 100 MHz (such as a 1.5-GHz microprocessor), it usually accepts a lower frequency clock generated external to the IC. It then multiplies the external frequency inside the IC to obtain the required clock speed.

The clocks routed to the various ICs provide overall system timing. That is, activities within the various chips can be synchronized. This ensures, for example, that when one device sends data to another, the receiving device will be monitoring the data bus at the same time the information is presented. The timing for this type of transfer can be very critical, since the information is only present on the bus for a few nanoseconds. In many computers, a slight (measured in picoseconds) delay between the arrival of one clock and another can prevent proper system operation. The difference between the timing of two synchronous clocks is called **clock skew**.

Microprocessor

The microprocessor itself is a stand-alone functional block. This is the real "brains" of the system, although there is a lot of distributed control in a modern computer system. In any case, it is the microprocessor that reads memory, accesses the instructions that comprise the program, interprets the instructions, and ultimately directs the activity throughout the system.

In some cases, the details of a particular operation may be delegated to another IC, but the microprocessor still defines the activity that is to occur. For example, the video controller can handle many of the details associated with controlling a video display without microprocessor intervention. However, the microprocessor has to determine what information is to be displayed, where to position it, what color it should be, and so on.

Some microprocessors draw a lot of current and get very hot because of the high switching speeds of the millions of transistors inside of the IC. It is quite common for a microprocessor to require a metal heat sink (a piece of heavy metal that helps dissipate the heat and keep the device sufficiently cool). In many cases, μPs are also cooled with forced air from a fan. In fact, some of the faster μPs require a fan to be mounted on top of the μP package.

System Controller

The system controller is a very complex integrated circuit that takes orders from the μP and then carries out those orders with little or no further supervision by the μP. Only results of an action, requests for assistance, or status updates are communicated to the μP. Again, the strategy in using distributed control is that to some degree multiple activities can be occurring simultaneously in different parts of the system. A typical system controller IC will have hundreds of pins.

Memory

Many types of memory are used in a typical computer system. Figure 16–5 illustrates several of them.

Cache Memory
Cache memory is a small private memory that the μP can use as a scratch pad. It can read and write values to cache memory faster than the same values can be transferred to other larger memory devices. Additionally, when some μPs go to main memory to get the next instruction, they quickly read several sequential instructions into the cache memory. It can then fetch the next instruction from cache memory while the other system buses are transferring data to satisfy the requirements of a prior instruction.

Internal Memory
Internal memory is where the bulk of the information, programs, and partial results are stored during program execution. It is relatively fast (but not as fast as cache memory). The capacity varies greatly from a few kilobytes to several gigabytes. Your home or school computer probably has several hundred megabytes of internal memory.

Internal memory is often referred to as **RAM**. This is an acronym for random access memory. This means that any memory location can be accessed just as fast as any other. Contrast this to information stored on a cassette tape or a spinning CD. Here, the information (or at least part of it) must be read sequentially until you reach the data you actually want. RAM-type memory generally refers to memory that is capable of being both read and written. In most cases, RAM memory is **volatile**, which means it loses all stored information when the memory loses power.

Video Memory
Video memory consists of RAM devices that are dedicated to the storage of video information. Every point of light (called a pixel) that you see on your computer screen has one or more bytes of corresponding information stored in video memory. More specifically, if you have a color monitor, the apparently solid screen is actually composed of many tiny pixels. Any given pixel can be varying intensities of red, green, or blue. A good experiment is to carefully place a tiny droplet of water on the screen of a color display terminal. The droplet acts as a magnifier. If you look closely, you can see the individual red, green, and blue pixels. In any case, several bits are required for every pixel on the screen in order to specify its brightness. The relative brightness of one color pixel compared to adjacent pixels of different colors is how the computer appears to display thousands of colors.

BIOS
The BIOS functional block in Figure 16–5 is another type of computer memory. It is called the **BIOS**, which is an acronym for **B**asic **I**nput **O**utput **S**ystem. This is a ROM and a **nonvolatile** memory, which means that it retains its contents even when power is removed from the system. The BIOS holds programs that comprise the most basic level of computer operation. The programs stored in BIOS memory typically accomplish things that include loading more complex programs from external storage devices (e.g., hard or floppy disks) and managing the most basic forms of inputting and outputting bytes of data. When you first apply power to your computer, the program counter is force loaded to an address in the BIOS. The BIOS controls the startup of your computer until a more complex program (e.g., Windows or Unix) can be loaded. The BIOS programs are generally stored in a single IC package.

Mass Storage
Mass storage devices are generally slower than internal memory, but they are much less expensive on a per byte basis. Typical mass storage devices include floppy disks, hard disks, CD-ROMs, and tape storage units. Typical storage capacities for mass storage devices range from about one megabyte to many terabytes.

Video Controller/Driver

The video controller is an intelligent chip that is similar in many ways to the system controller. Just as the microprocessor delegates the details of many activities to the system controller, the system controller delegates the details of the video portions of the computer to the video controller. The video controller receives commands and data from the system controller via the AGP bus (in computers that use this type of high-speed bus). The video controller stores and retrieves display information from the video memory as required and sends the video signals to the video drivers. In some systems, the video controller is allowed to access main memory directly.

The video drivers are ICs or transistors that translate the electrical characteristics of the video signals into the voltage and current levels required by the particular type of video display being used with the system. In some cases, the video drivers are integrated within the video controller.

Audio

The audio circuits of a computer accept inputs such as microphones, audio CDs, sound tracks from video CDs, and the audio signals from many Web pages. These signals are processed and distributed by the audio circuits. Outputs include headphones and internal or external speakers.

The simplest of audio circuits for computers consists of nothing more than a simple driver circuit for a speaker. More up-to-date computers have more sophisticated audio circuits that generally comprise a separate plug-in printed circuit board. As shown in Figure 16–5, these cards usually connect to the PCI bus. In any case, a typical audio circuit board (or sound card as it is often called) manages all of the audio requirements and signal distribution in the computer. Additionally, many have internal programs that provide special effects to the sound. For example, it might add a slight delay to part of the audio signals, which makes it sound like you are in a large auditorium.

Input and Output

Although the terms input and output technically include the video and audio portions of the computer, it is common to exclude these from the general discussion of input/output or I/O. Figure 16–5 shows several other types of I/O. Each represents a path (often called a port) for data to enter and/or leave the computer proper.

The I/O controller is another smart integrated circuit that receives its direction from the system controller but then manages the details of the I/O devices assigned to it. A single I/O controller chip can manage a host of devices such as a printer, a floppy disk drive, a mouse, a keyboard, and one or more serial ports. A serial port is a communication path where data is sent over two or more wires one bit at a time. This technique is inherently slower than a parallel transmission method (such as the printer port) where all the bits of a given byte are transmitted at the same time over multiple wires. It is important to realize that not only can the I/O controller manage all of the devices previously listed, but the devices can operate at the normal speeds simultaneously. In short, the controller is so much faster than the data transfer rate of any of the devices being controlled that it can adequately service multiple devices.

Another common I/O path is a USB port. USB is an acronym for universal serial bus, which is common on home computers. A USB port, as its name implies, is serial in nature (i.e., one bit at a time). However, the rate at which the bits are transmitted is many times faster than from the standard serial ports on a computer. Its speed and its ability to connect multiple devices (as many as 127 from a single port) are two reasons for its extreme popularity.

Many computers—including home computers—have Ethernet ports. These are high-speed communication ports that allow data transfers at 10 MHz, 100 MHz, or higher rates. Ethernet ports are commonly used to connect multiple computers together to form a network. In most modern offices, for example, data can be transferred between any two computers just as easily as it can be moved from one location to another within the same computer. This characteristic exists because of the Ethernet links between the various computers.

The last type of I/O represented in Figure 16–5 is an infrared or IR link. Here, data is encoded and used to control an infrared light source. In short, the information is translated into flashes of infrared light. The light is formed into a beam by a lens. If another computer or other device with an IR receiver port is brought within range of the transmitting port, the information contained in the light beam can be received, decoded, and utilized by the second computer or device. The IR link is just another way to transfer data from one digital system to another (in this case without wires). You might note that the IR technology used for the IR computer link is essentially the same as that used in a remote-controlled television.

Power Supplies

The tens, hundreds, or thousands of integrated circuits and transistors in a computer require dc voltage for their operation. This is the purpose of the power supplies functional block in Figure 16–5. The input to the power supply circuits is frequently 120 Vac, but it may be other values including unregulated dc voltages such as 24 V. The supplies convert the incoming ac to dc, lower the voltage level to the required values, and then regulate the outputs to maintain constant voltages for the ICs. A typical computer power supply circuit will generate several output voltages such as +3.3 V, +5 V, and ±12 V.

Miscellaneous

Most computers (including the system represented in Figure 16–5) have a real-time clock or RTC. This circuit keeps track of the time of day and the date for use by the various programs in the computer. The RTC circuits normally have battery backup, which allows them to continue to function even when the primary power is removed from the computer. In essence, an RTC circuit consists of a stable oscillator circuit (generally operating at 32.768-kHz) followed by a counter circuit. A 15-bit binary counter, for example, would divide the 32.768-kHz oscillator frequency by 2^{15} or 32,768. The final stage of this counter would then have an output of a very stable pulse at the rate of one pulse per second. This can be divided by 60 to get minutes. Subsequent divisions produce hours, days, and so on.

Although the microprocessor itself is certainly capable of implementing the RTC function in software, most computers utilize a separate RTC chip. Or, in some cases, one of the controller ICs in the system may have a built-in real-time clock, which eliminates the need for a separate IC.

✓ Exercise Problems 16.2

1. Which is faster, a PCI bus or an ISA bus?
2. The majority of the data and programs used in a computer are stored in the cache memory until they are accessed by the system controller. (True/False)
3. Data transfers can occur simultaneously on two or more buses within a computer. (True/False)
4. RAM is a type of memory that supports both read and write operations. (True/False)
5. What word is used to describe a memory device that loses its information when power is removed from the system?
6. From a technical point of view, how many colors are actually used in a color video display for a computer?
7. When you very first apply power to a computer, what functional memory block contains the program that initializes the computer?

8. A CD-ROM would be classified as a mass storage device. (True/False)
9. Each path for data to enter or exit a computer is called an I/O _____.
10. Serial data transmission is (*slower/faster*) than parallel data transmission with all other factors being the same.

16.3 SOFTWARE

We will now focus on the software aspects of a microprocessor-based computer system. It is the software in a computer system that channels the power of the computer to accomplish useful tasks. As previously defined, a computer program consists of a number of instructions arranged in a specific sequence such that, when the instructions are individually executed in a quasi-sequential fashion, the desired goals of the program are accomplished. Useful program lengths vary from just a few instructions to several million instructions. In this section, we will discuss levels of programming and classes of software.

Levels of Programming

All of the software in a computer has to be developed one instruction at a time. Generically, this is the primary responsibility of a computer programmer. However, there are many different levels of programming. The various levels differ in ways that include ability to access system resources, speed of execution, amount of memory required for storage, portability to other computers, and skill level required by the programmer.

Machine Language

Machine language is the lowest level of programming. It requires the programmer to utilize the microprocessor's instruction set directly. That is, all instructions must be entered into the computer as a series of binary (or hexadecimal for convenience) numbers. Figure 16-6 shows a portion of a program written in machine language. The text entries to the right of the semicolons (;) are called comments. These do not affect the program itself but are included to help the programmer recall what is being accomplished by a given instruction.

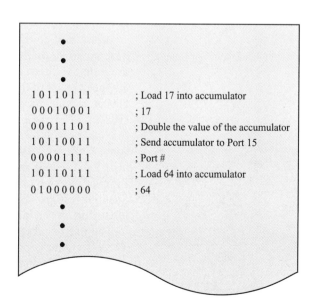

FIGURE 16-6 A program listing for a machine language program.

Programming in pure machine language is incredibly laborious and susceptible to error. Additionally, every microprocessor has a different instruction set. For these reasons, machine language is never used for any significant programming exercise. It is sometimes used on small training systems that allow students or designers to become acquainted with a new microprocessor.

Assembly Language

The next step upward from machine language is assembly language. Here, the programmer enters the instructions by using the mnemonic representation for each computer instruction. While still laborious, this represents a very practical level of computer programming. An accomplished assembly-language programmer can develop programs that are smaller and run faster than equivalent programs written in other higher-level languages. However, when programming at this level, you have full access to the computer's resources, so errors can produce devastating results. Figure 16–7 shows how a short segment from a program written in assembly language would appear. There is no need for you to attempt to decipher or memorize the individual instructions listed in Figure 16–7. The purpose here is to acquaint you with the general nature of assembly language programming.

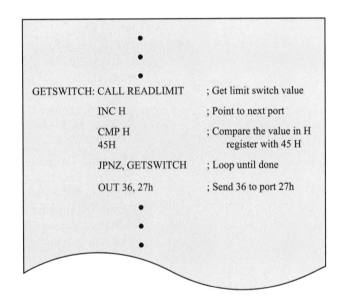

FIGURE 16–7 A program listing for an assembly language program.

Again, notice the use of comments to the right of the semicolon marks. These have no effect on the program itself; they merely serve to improve the clarity of the program. Additionally, note the use of a **program label**. A program label is generally positioned in the left-most column and terminated with a colon (:). This identifies a target for program branches. In the example shown in Figure 16–7, GETSWITCH is a label for the instruction CALL READLIMIT. Program execution can branch to here from other points in the program. In particular, the instruction JPNZ, GETSWITCH is a conditional instruction. This means a condition is tested and one of two alternate actions is taken depending on the results of the test. In this case, one result will cause program execution to continue in the normal sequential fashion. The other result will cause execution to branch to the instruction identified by the label GETSWITCH.

The use of assembly language requires another program called an assembler. The assembler program reads the text created by the programmer (such as shown in Figure 16–7) and creates equivalent machine language code that can be executed by the microprocessor. Assembly language mnemonics generally have a 1:1 correspondence with machine language instructions. This necessarily means that assembly language is unique to each microprocessor (or at least to each family of microprocessors).

High-Level Programming Languages

There is a long list of programming languages that are more easily interpreted by humans, provide the means for more efficient programming efforts, and are more portable between different types of computers. This general class of languages is called high-level programming language.

One instruction written in a high-level computer language may translate into many machine language instructions. The program required to make this translation is called a **compiler**. It is conceptually equivalent to an assembler program in that it accepts the text written by the programmer and converts it into machine language. Some common high-level languages are C++, Visual Basic, Fortran, Pascal, Lisp, and many others. Figure 16–8 shows a short program segment written in Visual Basic.

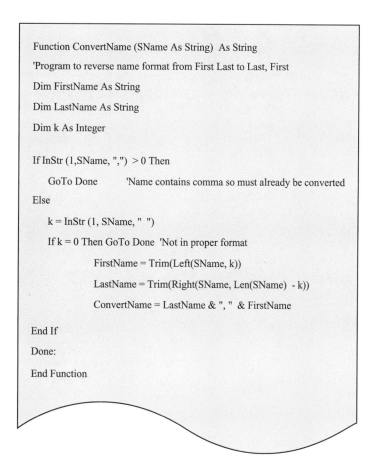

FIGURE 16–8 A listing for a program written in Visual Basic.

The listing in Figure 16–8 illustrates program labels (terminated by a colon), free form programmer comments (preceded by an apostrophe), and several Visual Basic instruction lines. Here again, it is not necessary for you to interpret the instructions at this point in your education; however, you should note that as we progress to higher languages, the instructions are becoming more readable and intuitive to humans.

User-Level Programming Languages

User-level languages are the highest language on the scale we are using here. They provide a means whereby a user with no significant programming skill can actually create practical computer programs to solve application problems. The software for most applications such as word processors, spreadsheets, computer-aided drafting and design, graphics creation, and many others include provisions for the creation of user programs.

Some programs are extremely easy to use and consist of nothing more than placing the computer into an effective record mode. The user then performs the desired action manually, while the computer records each user action. The resulting "program" that consists of stored keystrokes and mouse clicks is often called a **macro**. More complex user programs can also be written for many application software packages. In fact, the line between simple macro recording and conventional computer programming is almost indistinguishable in many cases. Often, the same actions and even the same computer code can be created either way. Figure 16–9 shows a short macro created by simply recording mouse clicks and keystrokes in the popular Microsoft® word-processing program, Word. You may see a resemblance to the Visual Basic listing in Figure 16–8, since Visual Basic is the underlying programming engine used by Word.

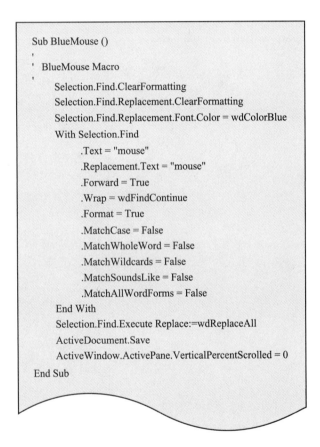

FIGURE 16–9 A macro created by recording keystrokes and mouse clicks in Microsoft Word.

The program listing shown in Figure 16–9 could be created directly by typing the various commands. While this gives the user-programmer access to a greater degree of program control, it also greatly increases the knowledge required to create a reliable program. A user who creates such a program by recording a macro never has to be aware of the meaning of the various instructions listed in Figure 16–9.

Classes of Software

In this section, we will examine an effective hierarchy of software. To some extent, this discussion overlaps the preceding section on programming languages. Here, we are more focused on the purpose or nature of the resulting program rather than on details of how the program was initially developed (e.g., assembly language versus C++).

BIOS

The BIOS memory was discussed in a previous section with a focus on hardware. The contents of the BIOS memory are the various program segments required to access and control the most basic elements of the computer. This generally includes such things as very simple input and output programs that allow data transfers to an I/O port, simple video display routines, and fundamental disk read and write programs. The BIOS software is a collection of very small utility programs that are normally stored in a single IC memory device. Creation of the BIOS code requires extreme familiarity with the details of the computer hardware.

Bootstrap Loader

When you turn on your computer, control is given to a program in the BIOS memory. One of the first actions accomplished by this initializing program is the accessing of a specific location on a floppy diskette or hard disk storage device. Another relatively short program called a **bootstrap loader** is stored on the disk at this specified location. As part of its initialization process, the BIOS reads the bootstrap loader program into memory and relinquishes control to it.

Once the bootstrap loader program takes control of the computer, it has the responsibility for loading a much larger and more complex program into memory from the disk storage device. The program that is loaded is called the operating system. The operating system software is discussed in the following section. Once the operating system has been read into internal memory, the bootstrap loader program passes computer control to the operating system.

Operating System

The operating system or OS is a very complex program consisting of millions of bytes of instructions. The operating system creates the overall environment in which all other software executes. You are probably familiar with the Windows operating system developed by Microsoft®. When you operate your computer, you may use it for computations, word processing, Internet access, or game playing. But, in all of these cases, it is the operating system that provides the overall umbrella or backdrop that defines the general nature of the computer behavior.

Another important purpose of the operating system is to provide a large library of general-purpose routines that other programs (e.g., word processing) can invoke. This dramatically reduces the required size of the other programs, since they can all utilize the same code that is provided by the operating system. Consider, for example, the instructions required to play a sound through your computer's audio system. This can require an amazingly long list of instructions when viewed from the perspective of actual microprocessor instructions. Further, nearly every program that runs on a computer needs to play sounds. This means that all of the code required to play the sound would be duplicated within each software package used on the computer. This would consume valuable and limited memory space. But, by providing a large pool of common routines within the operating system, all of the user's programs can share the same code. All that is required is for the program execution to temporarily branch into the operating system routines. Only a single instruction in the user's program is required to play the sound. The instruction identifies the OS routine that is to be executed, and it identifies the sound to be played.

Application Software

Application software is the name given to programs that run within the operating system environment and provide a specialized environment for a certain type of user activity. Word-processing programs, spreadsheet programs, drawing programs, and game programs are examples of application programs or application software.

The application software can tailor the user environment, but it is the OS that provides the underlying support. In a word-processing program, for example, the display may be configured to resemble a lined sheet of paper. But, the routines necessary to actually draw the paper and text on the screen are provided by the operating system. Deep within the operating system, BIOS routines are also called upon to perform some of the more basic tasks.

User-Application Programs
Finally, within the environment maintained by an application program, a user may be able to create user-application programs. The macros discussed in a previous section are a good example of this class of software. Instructions in the user-application program will be interpreted by the application program. Where appropriate, the operating system is called upon to perform common tasks. Finally, as the operating system is executing a particular instruction sequence, it may invoke routines that are part of the BIOS program collection. All of this transfer of control is transparent to the end user, but it has a profound effect on the reduction of program size and storage requirements.

✓ Exercise Problems 16.3

1. A computer program is composed of a number of _____, which are executed.
2. What is the job title of a person who arranges computer instructions in a specific sequence to accomplish a useful task?
3. What level of programming requires direct use of the microprocessor's instruction set in binary or hexadecimal format?
4. What level of programming uses mnemonic representations for the microprocessor's instructions?
5. How does a comment adjacent to an instruction in a computer program listing affect the operation of the microprocessor?
6. Visual Basic is an example of a high-level programming language. (True/False)
7. When is a bootstrap-loader program generally used?
8. What is the purpose of a program label in a computer program listing?

●─SUMMARY

We began this chapter by discussing a hypothetical microprocessor-based computer system. The μP executes instructions contained in the program. The allowable instructions are called the instruction set. Both instructions and data reside in memory. A program counter internal to the μP identifies the address in memory of where the next instruction is stored. There are many kinds of memory devices. Those that retain their information when power is removed are called nonvolatile. Data can be stored (written) and retrieved (read) from internal memory (generally called RAM). A memory that does not allow the μP to write new data is called a ROM (read-only memory).

The various functional blocks within a computer communicate via standard bus networks. PCI and ISA are two popular buses for home and school computers, with PCI being newer and faster. The AGP bus is sometimes used to achieve faster video transfers.

All computer systems and many of the ICs within a computer system require high-speed pulses called clocks. The clock signals provide overall timing of the various activities throughout the computer. Some clock timing is very critical, and differences between the arrival time of two clocks can prevent normal operation.

Two general classes of I/O operations are serial and parallel. Serial data transmission is inherently slower since data is transmitted on a bit-by-bit basis. Multiple bits are transmitted simultaneously in parallel-transmission schemes.

A programmer controls the way a computer behaves by arranging instructions in a particular sequence called a program. There are many different languages used to develop programs. With the exception of machine language, a program called a compiler or assembler translates a more human-readable language into the machine language required by the computer.

In a typical computer system, there are several layers of software. For example, the operator generally interacts with an application program. Various instructions within the application software result in program control being transferred to programs that are part of the operating system. Similarly, some instructions within the operating system's programs utilize other short utility programs that are part of the BIOS. This sharing of computer instructions for common tasks results in smaller application programs and smaller memory requirements.

REVIEW QUESTIONS

Section 16.1: Basic Concepts and Terminology

1. The list of allowable instructions for a given μP is called its _____ _____.
2. All μPs can interpret the same list of instructions. (True/False)
3. When computer instructions are arranged in a specific sequence to accomplish a desired task, the resulting sequence is called a computer _____.
4. What is the general name given to a group of wires (or printed circuit board traces) used to carry data?
5. What is the general name given to a group of wires (or printed circuit board traces) used to carry memory and I/O addresses?
6. What is the name of the functional element inside a μP that holds the address of the next instruction to be executed?
7. The instruction register is where instructions are stored after they have been executed. (True/False)
8. All logical and mathematical operations within the microprocessor are accomplished by a functional block called the _____ _____ _____.
9. All of the physical parts of a computer such as circuit boards, cables, and integrated circuits are collectively called _____.
10. In the process of executing a program, a microprocessor may store intermediate values in a ROM storage device. (True/False)
11. What is the name used to describe a program structure where the same sequence of instructions is repeated multiple times?

Section 16.2: Hardware

12. Which is the faster bus: PCI or ISA?
13. What internal bus is dedicated to high-speed transfers of video information?
14. Which standard computer bus operates at a 33-MHz (or 66-MHz) rate?
15. Explain why the use of multiple buses and distributed control in a computer system results in faster overall operation.
16. What functional block within a computer serves as the origin of all timing signals?
17. When two synchronous clocks arrive at their destinations at slightly different times, the time difference is called clock _____.
18. A typical system controller IC will have from 18 to 24 pins. (True/False)
19. RAM memory that is used for internal computer memory supports both read and write operations. (True/False)
20. BIOS memory normally supports both read and write operations. (True/False)
21. The smallest point of light on a color video display is called a _____.
22. What are the three basic colors used in a color video display?
23. BIOS programs are normally stored on a floppy diskette or a hard disk. (True/False)
24. A path where data can enter or leave a computer is called a _____.
25. If all other factors are the same, parallel data transfers are (*slower/faster*) than serial data transfers.
26. Most modern computers require more than one voltage for their operation. (True/False)
27. If you are troubleshooting a computer and discover a 32,768-Hz oscillator, there is a good chance it is part of a _____-_____ clock circuit.

Section 16.3: Software

28. Which is larger, a program or an instruction?
29. Machine-language programming is commonly used to develop user-application programs. (True/False)
30. If a programmer uses mnemonics that represent individual microprocessor instructions, what level of programming language is being used?
31. Would a macro normally be created by a computer user?
32. Would you consider it a normal activity if the operating system called upon the BIOS routines to perform certain operations?
33. What class of software creates the overall operating environment for a computer?
34. The OS generally provides a library of general-purpose utility programs. (True/False)
35. If you have a software game on your computer, what class of software would this represent?

●—CIRCUIT EXPLORATION

This exercise will allow you to examine how multiple devices can communicate on a common bus. Figure 16–10 shows the circuit to be explored.

U_1 and U_2 in Figure 16–10 represent the output drivers of two ICs that would be connected to the same bus in a microprocessor system. Switch Bank A and Switch Bank B select the data to be placed on the bus by U_1 and U_2, respectively. Switch S_1 selects the output that is to be placed on the bus at any given time. For this example, either Switch Bank A or Switch Bank B is on the bus at all times. In a real μP system, there may be more than two devices connected to a common bus, but there is never a time when more than one output is active (i.e., connected to the bus). There may be times when no outputs are connected to the bus.

Study the circuit to understand its operation, then accomplish the following tasks:

1. Construct the circuit in a laboratory or circuit simulation environment.
2. Verify that it operates according to your understanding. If not, investigate the discrepancy until you fully understand its behavior.
3. Figure 16–11 shows the logic symbol for a 3-to-8 decoder IC. This also called a 1-of-8 decoder. In short, a

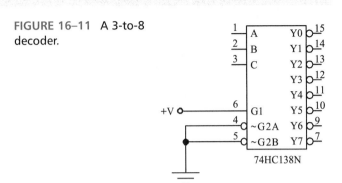

FIGURE 16–11 A 3-to-8 decoder.

3-bit binary number is applied to the A, B, and C inputs. One and only one of the outputs (Y_0 through Y_7) will go low according to Table 16–3 on page 619.

Draw a logic diagram showing how to connect the 3-to-8 decoder and some additional 74HC125 tri-state buffers to allow six switch banks to be connected to a common bus similar to the circuit shown in Figure 16–10. You will also need to have three logic switches to select the desired switch bank.

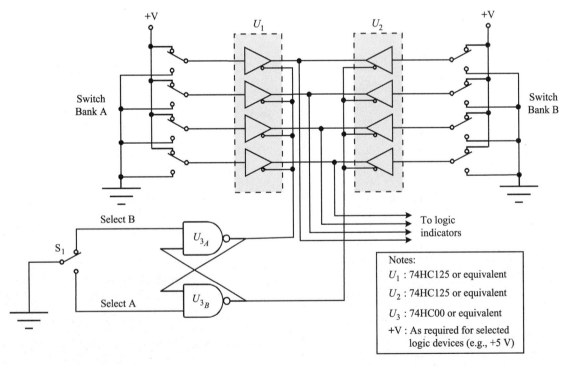

FIGURE 16–10 A demonstration of tri-state bus drivers connected to a common bus.

Fig16_10.msm

TABLE 16–3 Truth table for a 1-of-8 (or 3-to-8) decoder.

C	B	A	Y_0	Y_1	Y_2	Y_3	Y_4	Y_5	Y_6	Y_7
0	0	0	0	1	1	1	1	1	1	1
0	0	1	1	0	1	1	1	1	1	1
0	1	0	1	1	0	1	1	1	1	1
0	1	1	1	1	1	0	1	1	1	1
1	0	0	1	1	1	1	0	1	1	1
1	0	1	1	1	1	1	1	0	1	1
1	1	0	1	1	1	1	1	1	0	1
1	1	1	1	1	1	1	1	1	1	0

CHAPTER 17
Telecommunications

•—KEY TERMS
amplitude modulation
backbone
baseband
broadband
CAN
carrier
codec
demodulation
discriminator
duplex
frequency-hopping spread
    spectrum
full-duplex
half-duplex
hub
indirect FM
LAN
MAN
marking state
modulating signal
modulation
network operating system
node
overhead
parity
protocol
simplex
space state
start bit
TAN
telemetry
throughput
transponder
WAN

•—OBJECTIVES
After studying this chapter, you should be able to:

- Name and describe the four general classes of information that is transferred via telecommunication links.
- Describe the following network topologies: bus, ring, and star.
- Contrast peer-to-peer and client/server network architectures.
- Define and describe the following classifications of networks: TAN, CAN, LAN, MAN, WAN, and WLAN.
- Contrast communication systems by comparing the following parameters: bandwidth, broadband/baseband, serial/parallel, synchronous/asynchronous, digital/analog, and simplex/half-duplex/full-duplex.
- Describe each of the following modulation methods: AM, FM, PM, PAM, PWM, PPM, PCM, and delta.
- Discuss each of the following multiplexing techniques: FDM, TDM, and WDM.
- Describe the operational characteristics of each of the following wireless telecommunication applications: WLANs, satellite communications, and Bluetooth™ devices.
- Discuss the general operation of frequency-hopping spread-spectrum communication techniques.

•—INTRODUCTION
Telecommunications is a very broad and rapidly expanding field that has significant overlaps with computer technology. This chapter will serve as an introduction to telecommunications with emphasis on vocabulary building. In later courses, you will delve further into the technical details of this exciting field.

We will discuss a variety of telecommunication networks. We will emphasize those technologies that are directly related to computer technology with emphasis on data communications. We will also introduce an array of technical characteristics that distinguish the various communication methods. We will conclude with a discussion of wireless telecommunications, which is transforming many aspects of our lives.

17.1 TELECOMMUNICATION: INFORMATION AND NETWORKS

This section will introduce you to telecommunication networks. We will discuss network types and see how they can be categorized based on several different characteristics.

What Is Telecommunication

Communication is simply a system for transmitting and receiving information. Telecommunication is communication over distance. The information may be transported between two locations by standard wire, special high-speed cables, fiber-optic cables, or radio waves. Further, the distance between the two communicating points may be as small as a few feet or greater than 249,000,000 miles (the maximum distance to Mars).

Types of Telecommunications

Four classes of information are transferred via telecommunication systems. These include voice transmissions (or audio in general), data transfers, video transmissions, and control. Each of these is clarified in the following sections.

Voice and Audio Communications

There are many ways to achieve voice communications over great distances. You are well acquainted with many of these. Some of the more common techniques include the following:

- Standard telephone
- Wireless telephones and cell phones
- Radio transmission such as broadcast radio, two-way radio systems like those used by emergency teams, and satellite communications
- Voice communications over the Internet

Voice and other audio transmissions such as music may be transmitted in either of two very general forms: analog or digital. In many cases, however, both forms are used somewhere within the overall communications link. Figure 17–1 illustrates voice-based telecommunications.

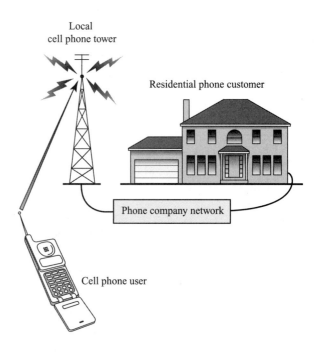

FIGURE 17–1 Voice-based telecommunications.

Data Transfers

Just as people can use telecommunications to effect long-distance voice communications, computers can transfer data over long distances using telecommunication techniques. For example, when you access the Internet to view a Web page in some remote city, many data are being transferred between computers via the telecommunications link.

Oftentimes, it is difficult to uniquely classify a particular communication into one of the four basic classes. For example, when analog voice information is converted to digital form prior to transmission, the transmitted information is essentially indistinguishable from a data transmission as it is routed to its destination. This type of overlap in classifications is common. It is important to be aware that category boundaries are often blurred, but nonetheless, they are useful for contrasting technologies. Figure 17–2 illustrates data transfers using telecommunications.

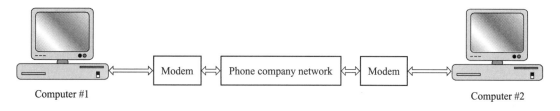

FIGURE 17–2 Data transfers via telecommunications.

Video Transfers

Video transmission simply means the transfer of video (i.e., live or recorded dynamic picture or image) information. This category includes communication services such as television, cable TV, and video conferencing. The category boundaries are once again blurred.

For example, if you set up a two-way video link over the Internet between your computer and a friend's computer, clearly you are transmitting video information. However, in this case, the video is converted to digital form before transmission, so during its travel on the Internet, it cannot be distinguished from any other data transmission. Figure 17–3 on page 624 illustrates video transmission via a telecommunications link.

Transfer of Control Information

The fourth category of information that is routinely transferred via telecommunication links is control information. Transfer of monitoring and control information over distance is often called **telemetry**, but for our purposes, we will use the broader classification of telecommunications. Here again, the line between control information and any other data transfer is unclear, but inclusion of this category is nonetheless valuable to us. Transferring control information over a telecommunications link allows us to have remote control of an enormous range of electrical, electromechanical, and electronic systems over huge distances. Figure 17–4 on page 624 illustrates transmission of control signals via telecommunications.

NASA, for example, has sent a number of robotic vehicles to remote destinations (e.g., the moon and Mars). Although these vehicles generally have on-board control systems that allow them to explore the terrain and report their findings (via telecommunication transfers of data and video), many can also be controlled by people on the ground. In this case, the ground station transmits control information (e.g., turn right 10°), and the remote robotic device responds.

A similar type of telecommunications control allows entire radio stations to be operated from remote cities. Thus, your "local announcer" who plays commercial jingles for local businesses, takes phone calls from local people, and so forth, may

FIGURE 17-3 Using telecommunications with video information.

FIGURE 17-4 Remote control via telecommunications.

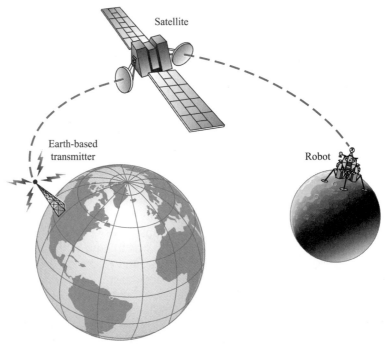

actually be sitting at a console in a city 2,000 miles away. The actual radio transmitting equipment and studio support equipment are indeed local, but they are controlled remotely. This arrangement allows one operator to control multiple radio stations simultaneously, while still maintaining the illusion of being a local announcer.

More examples of transmission of remote control information over telecommunication links include power companies, oil companies, and microwave communication companies. These and other companies use remote control techniques to adjust the performance of distant equipment. The power company may activate a large circuit breaker in a remote substation, an oil company may operate valves in a pipeline hundreds of miles away, and a microwave company might turn on an auxiliary power generation system at a microwave tower on a remote mountain.

Networks

A network is simply a number of transmitters and receivers interconnected by a common communication path. Each connection to the overall network is called a **node**. Thus, if all the computers in a business are interconnected via Ethernet cable, for example, then each computer is a node on the network. Any given network can connect to other networks to form a more expansive and complex connection of nodes. Let's look at ways to classify networks so that they can be distinguished from each other.

Topology

The topology of a network describes how the nodes are connected from a geometric point of view. Although there are variations of the basic topologies, networks can generally be classified into one of three topologies: bus, ring, and star.

Bus Network Topology Figure 17–5 shows how several computers can be connected into a network using a bus topology.

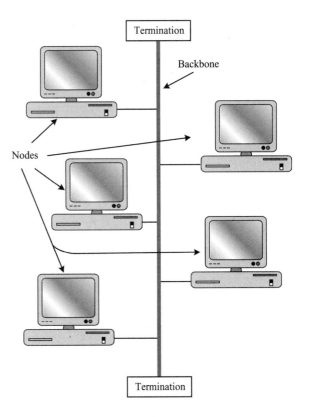

FIGURE 17–5 A group of computers forming a bus network.

As you might expect, the bus network configuration is bidirectional. That is, information can travel in either direction. In effect, a single communication line (electrically two or more wires) called the network **backbone** is used to link all the computers. This necessarily means that every data transmission on the bus is received by every node on the bus. Each device that is part of the network has a unique address or identification number. Each message that is sent contains the address of the device that is expected to accept the message. Although the entire message physically goes to all devices on the network, only the one that recognizes the identifier address will actually accept, interpret, and react to a particular message.

This configuration is generally only suited to small installations, since its performance becomes impractically slow as the number of users increases. Also, failure of the backbone (e.g., a cut or disconnected cable) will prevent communications on the entire network. Finally, adding or removing computers from the network can degrade network performance. For example, if a user disconnected the backbone from her computer to perform a maintenance task, it might disrupt other users even though the backbone is still connected. This unexpected behavior is caused by impedance mismatches on the connecting cable. In an ideal setup, a signal transmitted from one end of the cable will reach the far end of the cable and then be absorbed by a resistive termination device (e.g., a 50-Ω resistor connected across the transmission line at each end). In order for this to occur, the impedance must be consistent throughout the cable length. However, when loads are added (or removed) along the cable length, the impedance characteristics at the connection point can be affected. This can cause (or at least alter) signal reflections. The reflected signal combines algebraically with the original signal all along the length of the transmission line. Depending upon how the two signals interact near the connection point for a particular node, communication with that node (either transmit or receive or both) may be lost.

Ring Network Topology Figure 17–6 shows several computers connected in a ring network. A ring network is relatively robust. For example, if one of the computers on the network becomes inoperative, the rest of the computers can still communicate.

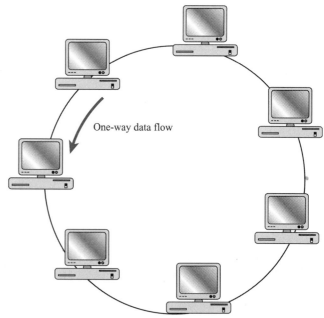

FIGURE 17–6 Computers configured in a ring network.

Perhaps the most common ring configuration is the token ring network originally developed by IBM. Here the messages are transmitted in a unidirectional manner around the ring. The bundled packet is called a token. Each node on the ring receives and inspects every message, but a given node only accepts those messages that are specifically addressed to it. In addition to receiving the message, the receiving node also marks the message as "read" and allows it to continue around the ring. Ultimately, the message returns to the sender where it is discarded.

Each node receives messages from the preceding node on a pair of twisted wires. Similarly, the messages are transmitted on to the next node via a pair of twisted wires.

Star Network Topology Figure 17–7 shows how computers can be configured in a star network. A star configuration provides advantages over the bus or ring configurations.

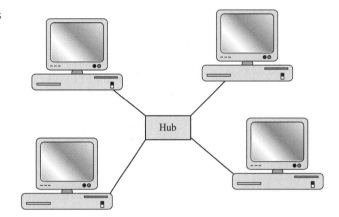

FIGURE 17–7 Computers connected in a star network configuration.

At the center of every star network is a central node called a **hub**. All computers on the network connect directly to the central hub. No other connections between computers are allowed. This necessarily means that communication between any two computers must involve the hub. In short, the transmitted information goes to the hub where it is retransmitted to all nodes on the network. As with other networking schemes, each message transfer has an integral identifier so that only the intended receiver actually accepts, interprets, and responds to the message.

One of the major advantages of the star configuration is that it is very robust. For example, even if all but two computers on a star configuration failed for some reason, the remaining two computers could continue communicating with no decrease in performance. This same characteristic makes adding or removing computers (either permanently or for maintenance) a simple matter, since the rest of the system is unaffected.

The star configuration also provides increased security opportunities for businesses. Since all communications must pass through the hub, all messages can be monitored. Further, communications to or from a particular node or nodes can be easily restricted.

The downside to the star topology is that a failure in the hub disrupts communication throughout the entire network. Thus, for example, if a hub needed to be replaced with a newer and faster device, the entire system would have to be shut down while the switch occurred.

This configuration (or at least a derivative of this configuration) can be extended beyond the capabilities of a single hub. One or more of the connections on a particular hub can actually be a connection to another hub. This expansion can continue to any practical depth. In essence, the overall network consists of many star networks with single links between adjacent hubs. When a particular hub becomes disabled in an extended star configuration, communication between some of the remaining nodes is possible. However, none of the computers connected to the defective hub can communicate with any other nodes. Additionally, no communication can occur between other star networks that pass through the defective hub.

Network Protocol

Network **protocol** is simply a list of rules that define allowable communications and data formats over a given network. In short, in order to establish and maintain meaningful communication, all transmitters and all receivers must be using and interpreting the same format. Protocol definitions include many things such as descriptions of how the information must be configured for transmission, how nodes are identified, and any responses that are required by receiving nodes.

As a specific protocol example, Figure 17–8 shows how a message could be constructed to allow devices to communicate over a popular network used to connect electronic test equipment (general-purpose interface bus or GPIB). This example showing the makeup of a representative message is only one aspect of the overall communication specifications. In the case of GPIB communications, the hardware interface is defined in a series of formal standards documents (IEEE488). The composition or content of the actual messages is defined by the test equipment manufacturers and is therefore device dependent.

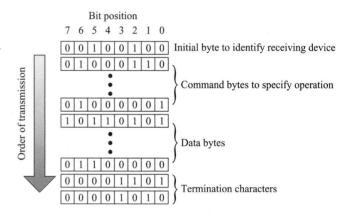

FIGURE 17–8 Message format for a representative network protocol.

Network Architecture

Network architecture can generally be classified into one of two types:

1. Peer-to-peer
2. Client/server

In a peer-to-peer architecture, all computers on the network are essentially equal. Any one of them can initiate communications, and they can all access files and programs physically stored on the other computers. Use of this architecture is usually limited to small networks.

A client/server architecture, by contrast, requires that one computer be defined as the server. All other computers are then clients. The software, which supervises all network activity and is called the **network operating system** or NOS, is stored on the server. Additionally, files and programs are stored in the server. A client station makes requests to the server for programs and files. Since the physical storage of the files and programs is on the server, each client station can be a less powerful machine with less storage capacity. Other system resources such as printers, Internet connections, and so forth are also managed by the server.

The client/server architecture is ideal for maintaining control over software, since there is essentially only one master copy of the software. Any changes made to the server software inherently flow to the clients the next time a link is established. This architecture is generally used by large corporations because of its inherent ability to promote network security (both from internal and external threats). The downside, of course, is that since all file and program storage is within the server, a malfunction in the server makes the entire network inoperative.

Classification by Physical Size

There are several general classes to describe the relative size of the network. We will discuss five common classifications.

Tiny Area Networks A tiny area network or **TAN** is intended for a small network consisting of no more than five or six nodes. Further, these nodes are in close physical proximity to each other (e.g., in the same room). This type of network is suitable for home use, but it is also used as a means to expand (i.e., add remote modules) to certain types of industrial control equipment such as programmable logic controllers.

Controller Area Network A controller area network or **CAN** is frequently used to connect computers and controllers (e.g., programmable logic controllers) in an industrial environment. It is a very robust network that allows high-integrity communications over relatively confined areas (e.g., it was originally developed for use in the automotive industry). Its robustness stems from extensive built-in error checking, which is of particular interest in an electrically-noisy industrial environment.

One interesting characteristic of a CAN configuration is that transmissions do not contain identifiers to specify the intended receiver. Rather, each message has an identifier that specifies the content of the message (e.g., conveyor belt stopped, pressure above limit, and so forth). When a message is transmitted, all nodes receive the message and examine the identifier. If the message content is relevant, then a particular node will take the appropriate action. Otherwise, the message is ignored.

Most CAN configurations are limited to less than 64 nodes by the current capacity of the cable drivers. However, this does not represent a theoretical limit. CAN buses have a maximum length restriction of roughly 125 feet when operated at full speed (1 Mbit per second). However, longer bus lengths can be used by lowering the data rates. For example, if the data rate is lowered to 125 kbit per second, then the bus may be extended to over 1500 feet.

Local Area Network A local area network or **LAN** is one intended to serve a small, well-defined community of nodes. For example, a single LAN might be used to connect all computers within a given building or even nearby buildings. The interconnection of computers in a single-location business would normally qualify as a LAN. In addition to connecting computers, the LAN is also used to connect limited resources such as color printers, fax machines, Internet access, and so forth. LANs typically have data transmission rates as high as 1000 Mbits per second.

Multiple LANs can be connected together over long distances through the use of satellite links, microwave transmissions, fiber-optic cable, or telephone lines. When LANs are connected together over long distances, the overall configuration is called a wide area network.

Metropolitan Area Network A metropolitan area network or **MAN** is a network intended to connect computers within a city or metropolitan area. Physically, this category falls between a local area network and a wide area network.

It is useful for sharing resources within a community. It also allows multiple users (generally business users) to share the expense of a high-speed connection to a wide area network. In other words, multiple MANs can each have a single link to wide area network. All of the users within the service area of a particular MAN share in the expense of the link to the wide area network. This brings the cost of a wide area network connection, which would otherwise be prohibitive to relatively small businesses, to a more practical level.

Wide Area Network A computer network that serves a large geographic area is called a wide area network or **WAN**. The most familiar WAN is the Internet. A

WAN is generally composed of multiple MANs and/or LANs connected through high-speed phone lines, satellite links, or microwave links.

Figure 17–9 illustrates the relationships between LAN, MAN, and WAN connectivity.

FIGURE 17–9 The relationship between LAN, MAN, and WAN connectivity.

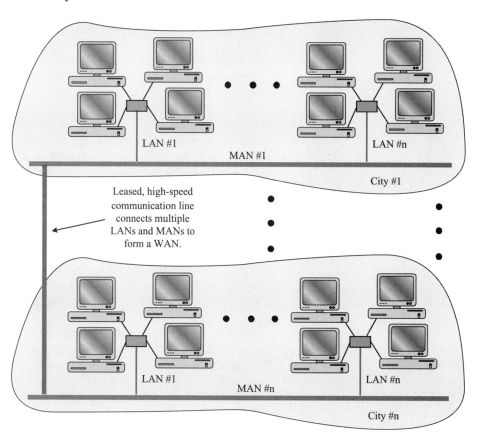

✓ Exercise Problems 17.1

1. What term is used to express the meaning of "communicating over distance?"
2. What four classes of information are transferred via telecommunication systems?
3. What name is used to refer to several transmitters and receivers interconnected by communication paths?
4. When several computers are connected together on the same network, each computer is referred to as a _____ on the network.
5. Name the three common network topologies.
6. What name is given to the central node in a star network configuration?
7. Which network is physically larger: CAN or MAN?
8. When several LANs and/or MANs are connected together via high-speed phone lines, microwave links, and so forth, the overall network is called a _____.

17.2 TECHNICAL CHARACTERISTICS

This section will introduce some of the more technical aspects of communication systems. This will not only allow you to expand your vocabulary, but it will remove some of the mystery of how telecommunication systems of all types actually operate.

Descriptive Characteristics of Communication Systems

There are many technical details that distinguish one form of communication from another. We will contrast several of these in the following section.

Information

As we discussed in the previous section, there are generally four classes of information that are transmitted over communication networks. These classes include voice (or audio in general), data, video, and control signals. The boundaries between these classes are indistinct. Therefore, much of the discussion in the following sections applies to all forms of information.

Communications System Bandwidth

The bandwidth of a communications system (or more typically, a given part of an overall communications system) refers to the range of frequencies that can be efficiently transmitted through the system. In effect, it describes the amount of information that can be transmitted through the system in a given period of time. The wider the bandwidth of the communications channel, the greater the amount of information that can be transmitted in a given amount of time (generally one second for comparison purposes). The amount of information transmitted per unit time is generically termed **throughput**.

Bandwidths of practical communication systems range from a few kilohertz in an audio system to 25 or more terahertz in an optical communications link. As you will see later in this section, multiple, unrelated messages are often transmitted over the same communications link. For example, two independent phone conversations might share the same physical wire at some point in the overall system. Since the bandwidth of a given system is determined by its physical and electrical characteristics, it is evident that the more users who try to utilize the same communications link, the less efficient (i.e., slower) the communications. Therefore, since the number of people needing to communicate remotely is increasing at a rapid rate, and each user has increasingly greater amounts of information that he or she wants to transmit, it follows that there is a continuing demand for communication systems with wider bandwidths.

There are many ways we can increase the bandwidth of a communication system. Figure 17–10 shows what is perhaps the simplest method. Here, multiple channels are used to achieve higher throughput of information (i.e., higher bandwidth). While this is a somewhat primitive way to increase bandwidth, it readily communicates the important concepts.

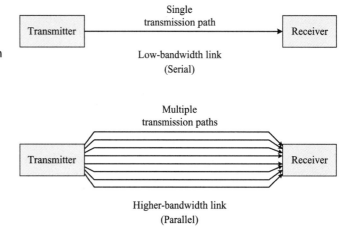

FIGURE 17–10 Higher bandwidths in a communications link allow more information to be transmitted in a given time period.

Broadband or Baseband

Baseband transmission utilizes the base or initial frequency range of a signal. It also describes a link that is dedicated to a single communication channel. **Broadband**, by contrast, refers to the simultaneous transmission of multiple communication channels on a common transmission path. Additionally, the original signals are converted to higher frequencies for increased efficiency.

One simple way to distinguish broadband from baseband communication links is to use the differences listed in Table 17–1.

TABLE 17–1   Differences between broadband and baseband communications.

Broadband	Baseband
Analog signals	Digital signals
Total bandwidth divided into multiple channels	Full bandwidth consumed by a single channel
One-way transmission	Two-way transmission
Good for long distances	Must be amplified and retransmitted after short distances

Cable television is a familiar communications system that uses broadband links. Here, many TV channels are brought to your house on a single cable. The modem in your computer is probably a baseband device, since the full bandwidth of the link is utilized for a single communication channel.

Serial versus Parallel

A serial-communication link uses a single channel. All parts of all messages travel over the same physical link. The transmission medium itself may be a pair of copper wires (a single channel consisting of forward and return currents), a coaxial cable, a fiber-optic cable, or other alternatives. The important characteristics are that there is only one channel, and all information is transmitted serially (i.e., one-after-the-other fashion). For example, assume that we wanted to transmit a digital information packet that consisted of 1024 bits. In the case of a serial transmission, only one of these 1024 bits could be on the communication link at any given time. The total throughput or bandwidth of the system would be limited by how fast (i.e., at what frequency) the bits could be transmitted. This is the same concept that was discussed in chapter 16 with reference to serial ports on computer systems.

A parallel-communication link (as with a parallel port on a computer), by contrast, provides multiple, parallel paths. Various portions of a given message can travel simultaneously over the various paths. Suppose, for example, our parallel communication system has sixteen parallel paths. Assuming the bandwidth of each path is similar to the serial links previously discussed, the parallel link would provide an effective throughput or bandwidth that is sixteen times higher than an equivalent serial channel. In the previous example of a 1024 bit transmission, the overall message could be transmitted sixteen bits at a time. With all other factors being equal, it is clear that parallel-communication links are faster than equivalent serial-communication links. However, they are inherently more expensive, since they must provide multiple physical paths. Figure 17–10 also illustrates serial- and parallel-communications links.

Synchronous or Asynchronous

Yet another way to distinguish between digital communication links is to classify them as synchronous or asynchronous.

Synchronous Communication   A synchronous communication requires that the transmitter and receiver be synchronized or timed. This is accomplished by using a separate timing signal or reference clock. The clock may be transmitted as a separate channel in the communication link, but in many cases it is embedded within the data. More specifically, the clock and data are encoded in such a way that the original clock signal can be recovered at the receiver. Figure 17–11 shows one method of combining the clock and data into an encoded signal suitable for a synchronous transmission link.

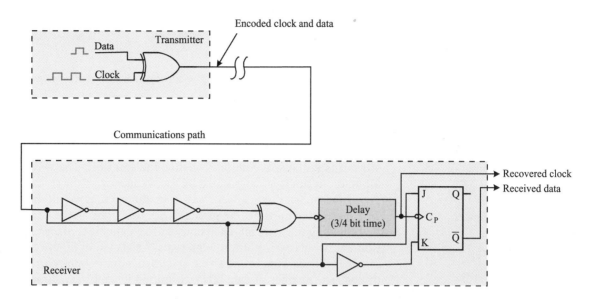

FIGURE 17–11 A simple encoder and decoder used for a synchronous data transmission.

The circuit in Figure 17–11 implements an encoding scheme called binary phase modulation. The data bits are sequentially applied to the exclusive OR gate in the transmitter on each rising edge of the clock waveform. If the data bit is high, the clock signal is inverted. If the data is low, the clock signal is transmitted without inversion.

In the receiver, a transition detector produces a short pulse on each edge of the incoming data stream. The particular transition detector shown relies on the propagation delay of three inverters to delay one of the two inputs to the exclusive OR gate in the receiver. In essence, the output of the exclusive OR gate is an intentional glitch. The output from the transition detector and the incoming data stream are routed to the remainder of the circuit consisting of a three-quarter bit-time delay and a *JK* flip-flop. The outputs of the last two functional blocks are the recovered clock and the received data. The received data, of course, exactly coincides with the original data applied to the transmitter.

Synchronous communication is most often used when the transfers involve large amounts of data. Special bit sequences are normally transmitted at the beginning and the end of a long transmission. These signify the beginning and end of a valid transmission as well as provide the opportunity for error checking. But, with the exception of these few added bits, all other transmitted bits correspond to message bits. For this reason, a synchronous transmission is efficient since it has a relatively low overhead. Any additional bits that are transmitted in order to maintain functionality, detect and correct errors, and so forth—but are not actually part of the transmitted message—are called **overhead**.

Asynchronous Communication Asynchronous communication links do not utilize the same clock at both transmit and receive ends of the link. Rather, the transmitter and receiver use clocks that are extremely close to the same frequency, but they are not synchronized. Asynchronous transmissions are character-oriented. That is, the message is transmitted one character (essentially one byte) at a time. Figure 17–12 on page 634 shows the construction of a typical asynchronous character transmission.

634 CHAPTER 17 • TELECOMMUNICATIONS

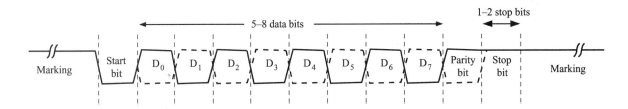

FIGURE 17-12 Character format for a representative asynchronous transmission scheme.

As shown in Figure 17–12, the communication line is held at a constant level called the **marking state** until a character is ready to be transmitted. The line is switched to the opposite level (**space state**) to indicate the beginning of a transmission. This initial space period is called the **start bit**. The width of the start bit and all other bit times is determined by the frequency of the transmit clock. For example, if the transmit clock were 128 kHz, the bit time would be $\frac{1}{128 \text{ kHz}}$ or 7.8125 μs. The shorter the bit time, the greater the throughput on the communication link.

Following the start bit, each consecutive data bit is placed onto the communication line at successive bit time intervals. As shown in Figure 17–12, the line can be in either the marking or the space state depending on the particular bit being transmitted. The number of data bits transmitted in each asynchronous character is usually in the range of five to eight bits, but it cannot change dynamically. In other words, once a communication link is set up, the format of the transmitted information must remain constant.

A parity bit follows the data bits. **Parity** is a method of error detection. A parity bit is used to implement a simple form of error checking. In short, the transmitter counts the number of binary 1s in the transmitted data and then sets the parity bit to either a 0 or a 1 as required for the particular parity scheme being used. Parity can be either even or odd. If odd parity is being used, then the parity bit will be set to a state that results in an odd number of 1s for the data bits and parity bit combined. The sum is an even number for even parity transmissions.

•—EXAMPLE 17.1

If the character to be transmitted in an asynchronous communication link is 10010110 and odd parity is being used, what will be the state of the parity bit?

SOLUTION Since there are four 1s in the initial character, the parity bit will be set to a 1. This results in a total of five 1s, which is an odd number of 1s.

As shown in Figure 17–12, the parity bit is followed by one or two stop bits. The stop bit(s) indicate the end of the transmitted character. They also establish the minimum time interval between two successive transmissions.

The receiver has an internal clock that is set to run at very nearly the same frequency as the transmitter clock. The receiver continuously monitors the communication line. When the line switches from a marking state to a space state, the receiver waits a one-half bit-time interval and then checks the line again. If it is still in the space state, it is interpreted as a legitimate start bit. Otherwise, it will be ignored as noise. Once a start bit is identified, the receiver then samples the line at full bit-time intervals. This means the sampling will occur near the middle of each bit time. Since the transmitter and receiver clocks are close to the same frequency, the sampling may drift slightly away from the bit-time center during a given character, but not enough to cause errors. In essence, the two systems are resynchronized by each start bit.

Analog versus Digital

Communication systems or links can be distinguished on the basis of whether the transmitted signals are digital or analog in nature. Here, it is important to clarify whether we are discussing the nature of the actual information or the nature of the transmitted signal. The actual information can be either digital or analog. Examples include computer data (digital) and voice signals (analog). In a similar manner, the transmitted signal can be either digital or analog. Transmission of digital information over a digital communications link is straightforward. Similarly, transmission of analog information over an analog communications link is also self-evident.

However, modern telecommunication systems almost always use digital transmission for the majority of the system regardless of the nature of the information itself. So, if analog information is to be transmitted over a digital communications path, it must first be converted to digital format. This is accomplished by using an A/D converter as discussed in chapter 15. Upon receiving the signal, the recovered digital information must then be applied to a D/A converter to recover the original analog information. The conversion devices are often called **codecs** (coder-decoder) in the telecommunications industry.

There are also situations that require the transmission of digital signals over an analog communications link. If you connect your home computer to the Internet via a modem and telephone lines, you are using this type of link. More specifically, the information being transmitted is digital in nature. It consists of the binary values used to represent all computer information. However, the portion of the phone line that actually enters your house is most likely an analog line. If you tried to send digital information over this line, it would be severely attenuated and would never reach its destination. Rather, your modem converts the binary zeroes and ones into corresponding tones that fall within the audio range. Because they are audio, they travel through the analog phone link. At the receiver, another modem distinguishes between the two tones and converts the received information back into standard logic levels.

Figure 17–13 illustrates these four transmission possibilities. Here, audio signals are used to represent analog information, but it is important to realize that analog information includes other frequencies such as video signals.

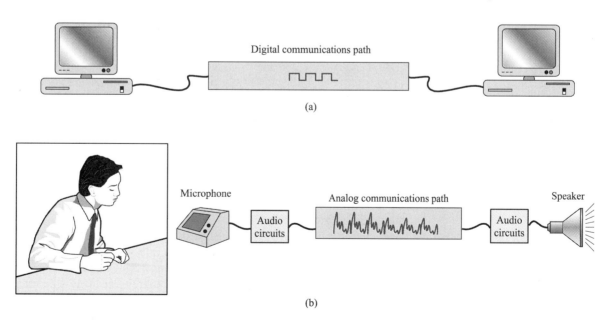

FIGURE 17–13 (a) Digital information on a digital communications path. (b) Analog information on an analog communications path. (c) Digital information on an analog communications path. (d) Analog information on a digital communications path.

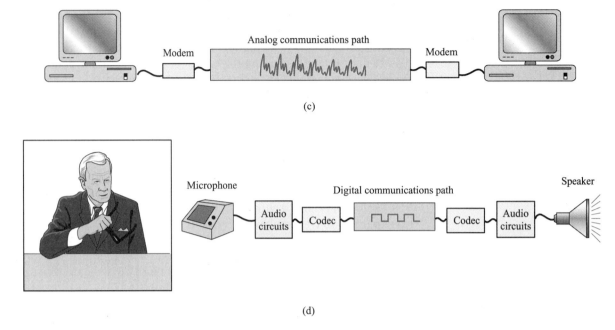

(c)

(d)

FIGURE 17–13 (*continued*)

Simplex, Half-Duplex, or Full-Duplex

The communications between two points can be either unidirectional (**simplex**) or bidirectional (**duplex**). Further, bidirectional communications may or may not be allowed in both directions simultaneously (**full-duplex** and **half-duplex**, respectively). The basic definitions for these three telecommunications terms are presented in Table 17–2 along with a familiar communication example that illustrates each transmission alternative.

TABLE 17–2 Comparison of simplex and duplex transmissions.

Transmission Method	Direction of Communication	Example
Simplex	One way only	Commercial broadcast stations
Half-duplex	Both directions, but only one direction at any given time	Walkie-talkies and other two-way radio systems where only one person can talk at a time
Full-duplex	Both directions simultaneously	Ethernet links typically used in offices and in many homes; also videoconferencing links

Modulation

Direct transmission of low frequency signals (such as audio) over great distances is very difficult because high powers are required to overcome losses, or multiple, distributed amplification must be used. In either case, the quality of the signal is degraded over long distances. Transmission of low frequencies using electromagnetic waves is also impractical due to the extremely large size that would be required for the antennas. For these and other reasons (e.g., greater bandwidth), it is preferable to use high frequencies for long distance communications.

Modulation is used to accomplish the transmission of relatively low-frequency signals that represent information, while still getting the advantages of high-frequency transmission. **Modulation** is the process whereby a lower-frequency signal (called the **modulating signal**) is used to alter the characteristics (e.g., amplitude, frequency, phase) of a higher-frequency signal called the **carrier**. The modulated high-frequency signal is then transmitted through the communications link (e.g., fiber-optic cable, radio link, and so on) to the receiver. The receiver then detects the changes in the carrier and extracts the original information signal. The information recovery process is called **demodulation**. The following sections will introduce you to several forms of modulation that are used for various telecommunication applications.

Amplitude Modulation

When the amplitude of a high-frequency carrier is varied by a modulating signal, we call the process **amplitude modulation** (AM). There are many variations of the basic amplitude modulation process, but the important concepts are illustrated in Figure 17–14.

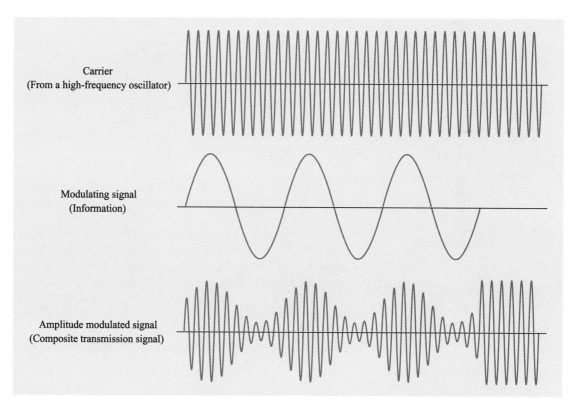

FIGURE 17–14 Amplitude modulation varies the amplitude of the carrier in response to a modulating information signal.

The difference between the frequencies of the carrier and the modulating signal are much greater in practice, but a lower ratio is shown in Figure 17–14 for clarity. As you can see from Figure 17–14, the amplitude of an otherwise constant carrier wave is varied in response to the instantaneous amplitude of the modulating signal (e.g., music, voice, video, and so on). The modulated signal can then be transmitted via radio link, fiber-optic cable, or some other medium. The amplitude of the modulating signal determines the amount of change in the amplitude of the carrier. The frequency of the modulating signal determines the rate at which the amplitude of the carrier is varied.

When the AM signal arrives at the receiver, the original information (i.e., modulating signal) must be recovered from the composite AM waveform. Figure 17–15 on page 638 shows a simplified, although functional, way that an AM signal can be demodulated.

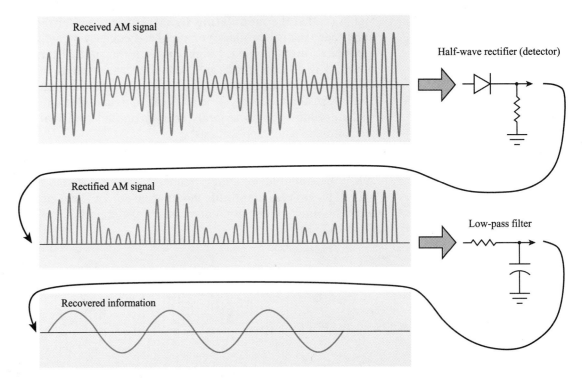

FIGURE 17–15 Demodulation of an AM waveform.

First, the received AM signal is sent to a half-wave rectifier circuit. This produces the center waveform. Next, the rectified signal is routed through a low-pass filter circuit. The cutoff frequency of the filter is such that the high-frequency carrier components of the signal cannot pass through, but the lower-frequency information (i.e., the variations in overall carrier amplitude) will pass through. Thus, the output of the low-pass filter is the original modulating signal—the information that was sent from the transmitter end of the communications link. In a more practical AM receiver, there are several other stages of signal conditioning and amplification, but the important principles relating to demodulation are represented in Figure 17–15.

AM broadcast transmissions rely on amplitude modulation. Similarly, the video portion of a standard television signal is amplitude modulated, as are some inexpensive two-way radios.

Frequency Modulation

Frequency modulation, as its name implies, causes the frequency of the high-frequency carrier waveform to be altered in response to a lower-frequency modulating signal (information signal). Figure 17–16 illustrates frequency modulation.

The top waveform in Figure 17–16 is the high-frequency carrier, whose instantaneous frequency is increased and decreased in response to the modulating information signal (the center waveform in Figure 17–16). The lower waveform in Figure 17–16 shows how the instantaneous frequency of the carrier has been modified by the information signal. The amplitude of the modulating signal determines the amount of frequency deviation, while the frequency of the modulation signal determines the rate at which the carrier frequency is varied.

At the receiver, a circuit called a **discriminator** senses the instantaneous frequency of the carrier and produces a voltage that is proportional to the amount of frequency deviation. Therefore, the output of the discriminator circuit is the recovered information signal.

FM broadcast stations utilize frequency modulation. In this particular case, the information (announcer's voice, music, and so on) causes a maximum frequency

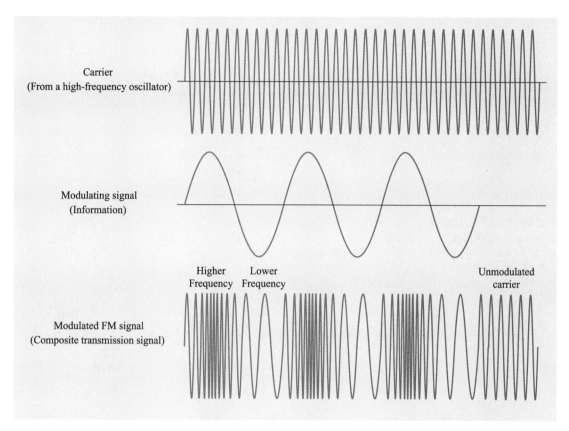

FIGURE 17-16 The modulating signal in a frequency-modulation system varies the frequency of a high-frequency carrier.

deviation of ±75 kHz. This combined with a 25-kHz guard band on either side is why FM stations are 200 kHz apart. That is, you may have a station at 106.7 MHz and another at 106.9 MHz (200 kHz apart). This separation ensures that they will not interfere with one another. The sound portion of a standard television signal is also frequency modulated, as are many two-way radio systems used by emergency crews and businesses.

Phase Modulation

Changing the instantaneous phase of a carrier wave in response to a modulating signal is analogous to changing the frequency of a carrier. That is, phase modulation (PM) produces a frequency-modulated signal. In fact, PM is often called **indirect FM**.

When the modulating signal is a digital waveform, phase modulation is called phase-shift keying or PSK. In this case, a specific phase shift in the carrier corresponds to a defined binary pattern. For example, we could make the assignments shown in Table 17–3 on page 640 for a PSK communications link.

By using an encoding scheme as shown in Table 17–3, multiple bits of the transmitted message can be represented by a single-phase shift change in the carrier. This translates into greater throughput of data for a given bandwidth communications link.

Pulse Modulation

The general category of pulse modulation includes several specific modulation methods. In general, some characteristic of a pulse (i.e., digital) waveform is altered in response to the modulating signal. Figure 17–17 on page 640 shows three possible pulse-modulation schemes.

TABLE 17–3 One possible encoding scheme for PSK modulation.

Information Bit Pattern		Carrier Phase Shift
Unmodulated Carrier		0°
0	0	45°
0	1	135°
1	0	225°
1	1	315°

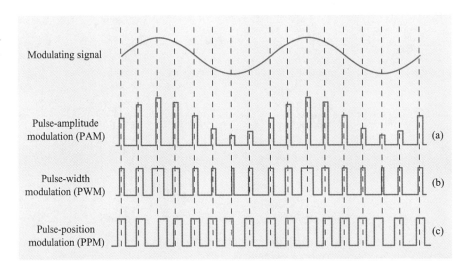

FIGURE 17–17 (a) Pulse-amplitude modulation, (b) pulse-width modulation, and (c) pulse-position modulation.

The amplitude of each fixed-position, fixed-width pulse is varied in accordance with the amplitude of the modulating waveform in pulse-amplitude modulation (PAM) as illustrated in Figure 17–17(a). The width of each fixed-position, fixed-amplitude pulse is varied in response to the modulating waveform to achieve pulse-width modulation (PWM) as shown in Figure 17–17(b). Finally, Figure 17–17(c) shows that the position (i.e., timing) of each fixed-amplitude, fixed-width pulse can be varied by the modulating signal to produce pulse-position modulation (PPM). In all cases, the receiver must be designed to respond to these variations in order to recover the original modulating waveform or information signal.

Pulse-Code Modulation

Pulse-code modulation (PCM) also relies on digital techniques, but the actual characteristics of the pulses (i.e., amplitude, position, and width) remain constant. Figure 17–18 shows a functional block diagram of a PCM transmitter.

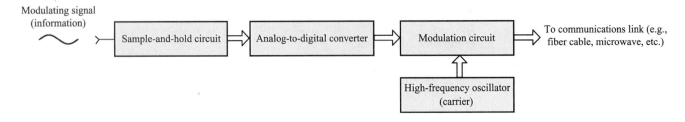

FIGURE 17–18 A simplified pulse-code modulator system.

Conversion from analog to digital cannot happen instantaneously. A certain minimum time is required even for fast circuits such as the flash converter presented in chapter 15. Now, since the analog signal may be continuously varying, errors or uncertainties can be introduced into the conversion process. That is, if the analog signal changes values during the conversion process, errors may be introduced. A sample-and-hold circuit can be used between the analog input and the A/D converter to eliminate this problem. A sample-and-hold circuit simply tracks the analog voltage (that is, the output is the same as the input) while it is in the sample mode. When the sample-and-hold circuit is switched to the hold mode, the output remains constant at the last sampled value. This constant value makes an ideal input for the A/D converter. In theory, we must sample at least twice as fast as the highest-frequency component in the analog signal in order to be able to reproduce the waveform. In practice, higher sampling rates are used.

The output of the A/D converter in Figure 17–18, then, is a series of binary values representing the level of the analog signal at each sampled point in time. We might, for example, use an 8-bit A/D converter and sample the analog signal at an 8-kHz rate. This means we will be generating one 8-bit value every 125 μs ($\frac{1}{8\,kHz}$). These parallel values are converted to serial format and used to modulate a high-frequency carrier as illustrated in Figure 17–18. In the example given, the actual data rate of the modulated carrier will be 64 kilobits per second.

The actual modulation process depends on the medium used. A microwave link, for example, might be frequency modulated by the binary data. Regardless of the method used to modulate the carrier, the receiver must be designed to respond to those changes in order to recover the signal. Once the binary information is recovered in the receiver, it is passed through a D/A converter to reproduce the original information (modulating signal).

Delta Modulation

Delta modulation is also a digital modulation technique. The transmitted bits (0s and 1s) in a delta-modulated system, however, do not represent the actual shape of the modulating signal. Rather, the individual bits indicate a decrease (binary 0) or increase (binary 1) relative to the previous sample. For example, if the modulating waveform were triangular, the transmitted bit stream would be a series of 1s (during the rising slope) followed by a series of 0s (during the falling slope).

One outstanding advantage of delta modulation is that it has a high degree of noise immunity. First, it is digital. Thus, in the presence of noise or weak reception, the receiver need only distinguish between the presence and absence of a pulse to achieve perfect recovery of the information. But, in extreme cases, one or more of the transmitted bits can be lost (due to noise or faded signal). If a single bit in a PCM system is lost, it can produce a dramatic effect at the output of the receiver. For instance, if the system is using 8-bit words to represent each transmitted value, and the most-significant bit (binary weight of 128) was interpreted incorrectly, the amplitude of the recovered analog would have a 50% error when that bit is decoded.

Contrast this with delta modulation. Here, each bit merely indicates the relative direction (decrease or increase) from the previous value. So, if a single bit is corrupted, the resulting output will make a small excursion in the wrong direction, but this is not likely to interfere with the fidelity of the reception. In fact, up to 10% of the transmitted bits can be corrupted in a voice transmission without making the received message unintelligible.

Multiplexing

Multiplexing refers to techniques that allow multiple, independent information streams or communication channels to exist simultaneously on a single transmission link. There are three multiplexing methods that we will discuss:

1. Frequency division
2. Time division
3. Wavelength division

All of these schemes have the net effect of increasing the effective data throughput of a communications link. It also has the effect of reducing the number of physical transmission paths in a large network of communicating devices (like the telephone system).

Frequency-Division Multiplexing

Frequency-division multiplexing (FDM) is largely used for the transmission of analog signals such as voice communications over the telephone system. Figure 17–19 illustrates the basic principles of FDM.

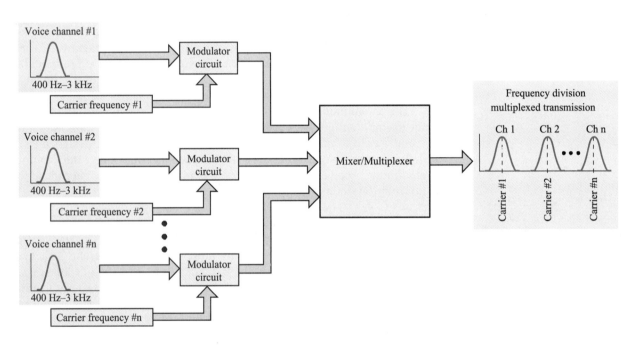

FIGURE 17–19 A frequency-division multiplexer.

As shown in Figure 17–19, multiple communication channels (voice channels in the illustration) are present, and they all occupy the same frequency range (400 Hz to 3 kHz in the illustration). If these were to be sent directly over a common communication path, the result would be unintelligible. Imagine hearing hundreds of separate phone conversations simultaneously in your telephone receiver!

As illustrated in Figure 17–19, each channel is used as the modulating signal to a higher-frequency carrier signal. Further, the carrier frequency for each channel is different. So, the output of each modulator contains a modulated carrier with the information for its respective channel. All of the carriers are now applied to a mixer (sometimes called a frequency multiplexer) circuit. The output of the mixer circuit is a composite signal made up of all of the original input channels. But, since each channel is impressed on a carrier of a different frequency, the channel signals do not interfere with each other during transmission.

When the FDM signal arrives at the receiver, it must be routed through a network of bandpass filters with each one designed to pass the range of frequencies allocated to a particular channel. The output of each bandpass filter is a single, modulated carrier. Now, the modulated carrier can be demodulated, and the original information (voice signals in this case) can be recovered.

FDM techniques can be extended to multiplex over 10,000 separate communication channels onto a single communication path. Its primary disadvantages stem from the fact that it is an analog system, so it suffers from the same weaknesses as all analog circuits (e.g., susceptibility to noise, signal degradation, and so on).

Time-Division Multiplexing

Time-division multiplexing (TDM) also enables multiple communication channels to be transmitted over a single communications link. Figure 17–20 shows a simple analogy that will help clarify the principles before we examine the technique more closely.

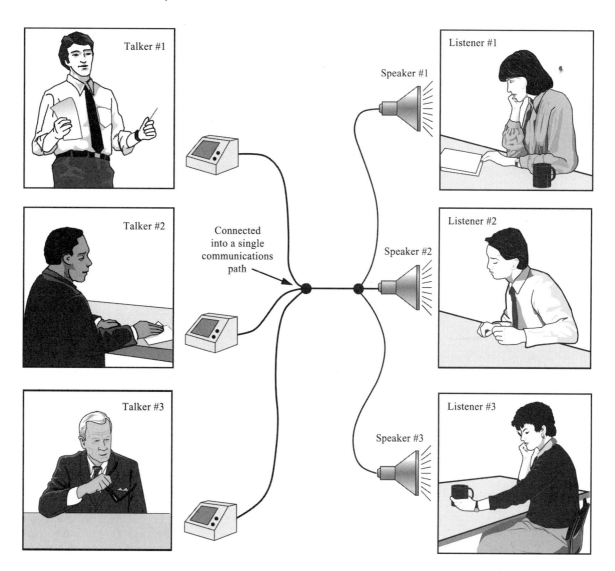

FIGURE 17–20 An intercom analogy to illustrate time-division multiplexing.

In Figure 17–20, there are three people wishing to communicate to three other people some distance away via an intercom system. The goal would be for listener #1 to be able to hear only those messages sent by talker #1. Similarly, listener #2 should only be able to hear talker #2. Finally, when talker #3 is speaking into the intercom system, only listener #3 should be able to hear the message.

Unfortunately, if we wired the system as illustrated in Figure 17–20, two very bad things would occur. First, if only one talker was communicating at any given time, the communication would be received clearly, but it would go to all three listeners. Second, if two or more talkers tried to use the intercom at the same time, the messages would be intermixed resulting in unintelligible noise for all three listeners. Let's work out a simple solution (although not especially practical) that will illustrate how time-division multiplexing could solve this problem.

We will break every hour into three 20-minute time segments. We will put a rule in place that restricts intercom usage according to the schedule in Table 17–4.

TABLE 17–4 Schedule of intercom usage.

Allowed User	Assigned Time Period
Talker #1	First 20 minutes of every hour
Talker #2	Middle 20 minutes of every hour
Talker #3	Last 20 minutes of every hour

For example, if it were 2:29 p.m., then only talker #2 would be authorized to use the intercom. With these rules in place, there would be no interference between messages. Unfortunately, all three listeners would still be able to hear all messages. We could solve this latter problem with a slight modification to the system design. We could install three-position rotary switches on both ends of the common communication line. A person on each end of the communication line will be assigned to rotate the switches to the next position every 20 minutes throughout the day. The switch on the transmit end will insure that only one talker's message is coupled onto the communication line. The switch at the receiver end will ensure that the received message is routed to the correct listener. The other listeners will be disconnected during that time.

Now, you can quickly see that this strategy does indeed allow multiple conversations to be routed over a single communication line and ultimately distributed to the intended receiver. You can also quickly see that it does not represent a practical system. However, we can make it practical very easily. First, we will use electronic switches to route the communication. Second, we will increase the rate of switching from one full-cycle per hour to one full-cycle every 125 μs. Now, talkers can use the intercom any time they wish. Technically, some of the information will be lost, but in practice, as long as the switching rate is much faster than the changes in the message signal, the received message will still be completely intelligible. This technique, which assigns each channel to a separate time slot on a common communication line, is called time-division multiplexing (TDM). TDM is very widely used in modern communication systems.

Wavelength-Division Multiplexing
Wavelength-division multiplexing (WDM) is used with fiber-optic cables and is analogous to FDM only at light frequencies. In short, rather than using a single light source (e.g., a laser diode) to transmit light (modulated with the intelligence information) over a fiber-optic link, we can use multiple laser beams. Each laser beam is configured for a different frequency or wavelength (i.e., color). Further, each is modulated with its own modulating signal (generally the composite output from a multiplexer). The various wavelengths of laser light pass through the common fiber without significant interaction. When they emerge at the receiver end of the cable, they are separated (based on wavelength), demultiplexed to recover the various channels within a single laser beam, and demodulated (to recover the original information on each channel).

✓ Exercise Problems 17.2

1. If the bandwidth of a system were reduced, what would happen to the throughput?
2. Which communication method—broadband or baseband—would most likely be used for cable television?
3. If all other factors were identical, which would be the fastest communication technique: serial or parallel?
4. What is the primary value of binary phase modulation?
5. If the binary value of 0100110 is being transmitted via an asynchronous communication link that employs odd parity, what will be the value of the parity bit?
6. One of the functions of a codec is to perform A/D and D/A conversions. (True/False)
7. If two-way communications are supported on a particular communications link, but the link only allows messages to travel in one direction at a time, we label the communications _____.
8. What name is used to describe a two-way communications system where information can travel in both directions simultaneously?
9. What characteristic of the carrier wave is modified by the modulating signal in an AM transmission?
10. _____ and _____ modulation both alter the frequency of a carrier in response to the modulating waveform.
11. PCM is well suited to TDM applications. (True/False)
12. Which modulation technique has the greater immunity to noise: PCM or delta?
13. Multiplexing (*decreases/increases*) the effective throughput of a communications system.

17.3 WIRELESS TELECOMMUNICATION

Many of the topics discussed in the preceding sections of this chapter also apply to wireless telecommunication. For example, a microwave link carrying multiplexed phone conversations is, by definition, wireless communication. There are many common applications of wireless telecommunication including the following:

- Intelligent or Smart phones
- Handheld computers
- Personal two-way pagers
- Personal data assistants (PDAs)
- Wireless Internet connections
- Wireless networks for home and business

In this section, we explore three specific applications of wireless telecommunication to better illustrate this important classification of devices. We will discuss the following applications:

1. Wireless LANs
2. Satellite communications
3. Bluetooth™ communication networks

Wireless LANs

A wireless LAN (WLAN) provides all the functionality of a standard wired LAN, but no interconnecting wires or cables are required. The data transfer rates on WLANs range from 1 Mbps to over 1 Gbps depending on the technology used.

WLANs cover a wide range of network sizes ranging from a small office environment to multi-location businesses. In this latter case, the various facilities would normally be within thirty miles of one another. Within a given facility, radio waves or infrared light transmitters and receivers would link the various users. Directed microwave transmissions or point-to-point lasers would likely link the remote facilities. Figure 17–21 on page 646 illustrates a representative WLAN application.

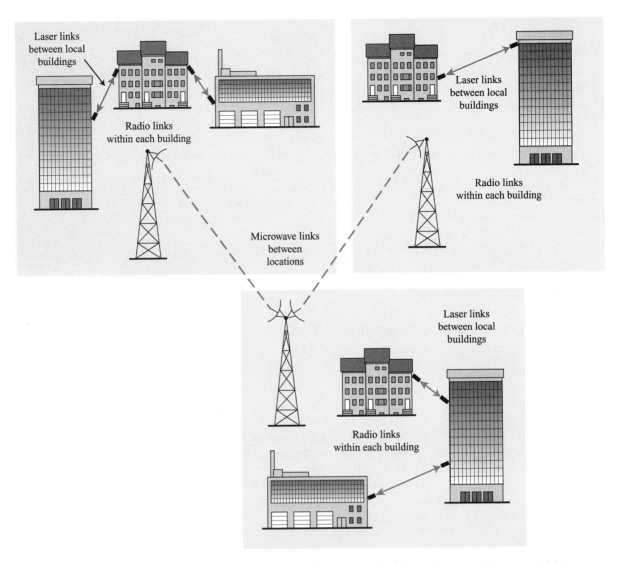

FIGURE 17-21 A WLAN application used to link users within local facilities and to provide connectivity between remote facilities.

Satellite Communications

Satellite communications refers to communication links between transmitting and receiving stations on earth that include an orbiting satellite as an integral part of the communications path. Figure 17-22 illustrates the basic concept of satellite communications.

Figure 17-22 shows two ground stations that are far removed from each other on the surface of the earth. For this reason, line-of-sight transmissions (such as lasers or microwave links) cannot be used, since the curvature of the earth would interrupt the communications path. However, a satellite orbiting the earth and physically located between the two ground stations can "see" both ground stations simultaneously. The satellite is equipped with antennas (both transmitting and receiving) and transponders. A **transponder** is a combination receiver, frequency converter, and transmitter. A receiving antenna on the satellite picks up a transmission from a ground station and routes it to the transponder circuits. The transponder shifts the carrier frequency (to prevent interference with the incoming signal) and retransmits the signal via a transmitting antenna. The second ground station can then receive this retransmitted signal and thus establish a communications link with the initial ground station. A satellite intended for communication

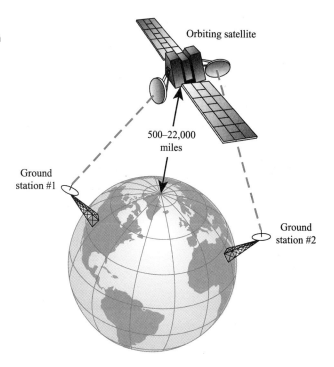

FIGURE 17–22 Satellite communications between two ground stations.

purposes will generally have between twelve and fifty transponders aboard. Having multiple transponders allows a greater number of simultaneous and unrelated communication links to be maintained.

The radiated power of a satellite transmitter is fairly low (on the order of 5 W). As previously stated, the transmitter and receiver frequencies are different, but both fall within the microwave range of 12 to 30 GHz, although neither of these frequencies should be viewed as a limit.

Bluetooth™ Communications

Bluetooth™ and other similar technologies are bringing a wide array of new telecommunication products to the business and personal lives of many people. These technologies are employed in a wide range of personal electronic products such as personal data assistants (PDAs), notebook computers, and cell phones.

Bluetooth Overview

Bluetooth-enabled devices are able to form small local networks with all other similarly enabled devices within the immediate area. Once a network has been established, communication and data transfer between devices in the shared network can occur without any intervention on the part of the user. Additionally, the same link eliminates wires and cables in more familiar communication links such as between your computer and a printer, or between computer and phone line, or between two or more computers. In short, Bluetooth technology is a complete two-way radio system consisting of receivers, transmitters, and an extensive underlying software structure.

Bluetooth Applications

Let's examine a few scenarios that will serve to clarify the types of products likely to be affected by Bluetooth, and to better illustrate the power of the technology on our daily lives. The handheld or notebook computer used by a traveling businessman could automatically update his computer upon arrival in the office. Information such as phone contacts, notes, e-mails, and other information would be synchronized automatically and invisibly. At the same time, the businessman can print out some new contracts on the office printer without having to physically connect to it.

A supervisor in an industrial plant can walk through the facility with her handheld computer. As she comes within close proximity of each control system used in the plant, a communication link is automatically established and her computer is updated to reflect the status of the control system and the machines it controls. If appropriate, she can choose to alter system parameters on the control system with a few keystrokes on her handheld computer.

Restaurants and lounges with Bluetooth-enabled systems will automatically link to the PDAs of arriving patrons. The specials for the day as well as the full menu can be immediately available. Orders can be routed directly to the kitchen from the patron's PDA. Additionally, patrons within the restaurant can communicate with each other, if so desired, or participate with others in group entertainment activities. Further, if a discussion or disagreement occurred during the restaurant visit, patrons would have immediate access to the Internet via their PDAs.

It is important to remember that at the time of the writing of this book, Bluetooth technology is still in its infancy. However, it promises to revolutionize the lives of most of us. The above examples, while somewhat contrived, are well within the capabilities of the technology.

Bluetooth Technical Characteristics
The transmissions in a Bluetooth system occur within the 2.4- to 2.5-GHz bands, although efforts are being made to secure other bands that will also permit short-range communications without requiring individual user licenses. The transmitter power ranges from 1 to 100 mW. These levels translate to a useful range of 30 to 300 feet.

Bluetooth devices are designed to automatically sense and establish communications with other Bluetooth-enabled devices within range. As many as eight devices can join to form what is called a piconet. One device in each piconet assumes the responsibility of master, while all others act as slaves within the piconet. Any given device can function as either a master or a slave. Further, any given device can be a member of more than one piconet. For example, a master in one piconet may be a slave in another piconet. Figure 17–23 illustrates the concept of piconets.

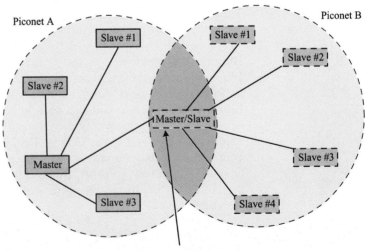

FIGURE 17–23 Bluetooth™ devices automatically link into piconets.

Bluetooth devices utilize a technique called **frequency-hopping spread spectrum** (FHSS). In a system that uses frequency hopping, the transmitter and all receivers periodically change to a different frequency according to a previously agreed pattern. All devices within a piconet change together in a pattern established by the master device. Since all devices within a piconet change frequencies at the same time, communication between the devices is continuous. The rate of frequency hopping for Bluetooth devices is on the order of 1600 hops per second. Any other devices that may intercept the transmission will not be able to interpret the communications. Rather, the reception will appear as a brief and random burst of noise. Frequency hopping allows multiple devices to establish multiple communication links simultaneously in the same frequency band without significant interference. Frequency hopping also adds a degree of security, since it is impossible to receive and interpret a complete message without prior knowledge of the hopping sequence.

Rather than transmitting on a single frequency like a broadcast station, a device using FHSS transmits on many different frequencies within a given band. Therefore, the total transmitted energy is spread out over a wider band of frequencies. This leads to the classification of spread spectrum. Bluetooth devices using FHSS have an effective data transfer rate of 1 Mbps using time-division multiplexing in a full-duplex mode.

Figure 17–24 shows a representative device that is Bluetooth-enabled.

FIGURE 17–24 A Bluetooth™ device that can be inserted into a notebook computer. (*Courtesy of Palm, Inc.*)

✓ Exercise Problems 17.3

1. WLAN is an acronym for what terminology?
2. How can two or more LANs be connected together without wires if the two LANs are 10 miles apart?
3. Explain how a satellite can allow locations on earth to communicate when they would otherwise be unable to communicate directly via a wireless communications link.
4. With reference to satellite communications, what is a transponder?
5. Any given satellite is limited to a maximum of one transponder. (True/False)
6. Bluetooth technology is designed to enable communications over (*short/long*) distances.
7. What is the maximum number of Bluetooth devices that can be in a single piconet?
8. What do the letters FHSS stand for with reference to Bluetooth devices?
9. Bluetooth devices operate in a simplex transmission mode. (True/False)

SUMMARY

Telecommunication is simply defined as communicating over distance. The transmission path may consist of wires, laser beams, fiber optics, microwave beams, or radio waves. The distance may range from a few feet to millions of miles. Information transferred over telecommunication links can be classified into four general classes: voice or audio, data, video, and control. Transfer of monitoring and control information is often called telemetry.

A network is a number of transmitters and receivers connected to a common communication path. Each connection to the network is called a node. Networks can be classified into three general topologies: bus, ring, and star. Computer network architecture can be classified as peer-to-peer or client/server. In a peer-to-peer network, all nodes are equivalent, and any computer can initiate communications and access programs and information stored on other computers in the network. A client/server configuration requires that one computer be designated as the server. All other computers in the network are clients. Client machines access programs and files that are physically stored on the server.

Networks can also be classified based on physical size. Tiny area networks (TAN) have no more than five or six nodes. Controller area networks (CAN) may have up to 64 nodes. A local area network (LAN) can connect hundreds of nodes within a single building or adjacent buildings. A metropolitan area network (MAN) is used to link computers throughout a city. A wide area network (WAN) consists of interconnected LANs. A WAN can literally span the earth.

The wider the bandwidth of a communications link, the more data throughput it can provide. Practical bandwidths for telecommunication systems range from a few kilohertz (e.g., standard phone system) to tens of terahertz (e.g., some optical communication links).

Serial communication paths are inherently slower than equivalent parallel communication paths. Any given digital communication can be classified as synchronous or asynchronous. Synchronous communication is generally used for large blocks of data. It requires that the transmitter clock be sent to the receiver for proper decoding of the received signal. The clock can be encoded with the data into a composite data stream. An asynchronous communication is byte-oriented. The system uses special bits that precede and follow the information bits to achieve synchronization for the duration of a single character. Parity is an error checking method where a parity bit is added to a transmitted string of information bits such that the total number of high-level bits is either odd (odd parity) or even (even parity). The receiver verifies the odd or even number of received high-level bits. If any one bit has been corrupted during its travel to the receiver, the error can be detected.

Transmissions can be either analog or digital—both in information content and transmission path. Communication links can be one-way (simplex), two-way but one direction at a time (half-duplex), or two-way simultaneously (full-duplex).

Modulation is the process whereby an information signal is used to alter some characteristic of a higher-frequency carrier. Modulation methods include amplitude modulation (AM), frequency modulation (FM), phase modulation (PM), and pulse modulation. There are several types of pulse modulation including pulse-amplitude modulation (PAM), pulse-width modulation (PWM), and pulse-position modulation (PPM). Pulse-code modulation (PCM) and delta modulation are widely used digital modulation techniques.

Multiplexing is used to allow multiple information streams to coexist on a common communications link without interference. Frequency-division multiplexing (FDM) uses separate carrier frequencies for each channel, but all channels are transmitted via a common communications path. Time-division multiplexing (TDM) breaks each communication signal into discrete packets or bursts. The packets from all transmitted channels are interleaved on a single communications path. Wavelength-division multiplexing (WDM) uses multiple wavelength lasers to carry multiple channels over a single optical path such as a fiber-optic cable.

A wireless LAN (WLAN) is functionally similar to a normal wired LAN, but the various nodes are linked without wires. The links are accomplished via radio waves or infrared light. Satellites are used to facilitate communications over greater distances than are possible between ground-based stations. Each satellite has a number of transponders that receive signals, shift the frequency, and then retransmit the signal to earth. Bluetooth is a system that utilizes radio communications and software that allow devices to automatically establish small wireless networks called piconets. Each piconet has a master and from one to seven slaves. Any Bluetooth device can be a master or a slave. Additionally, a given device can be a member of multiple piconets simultaneously. A piconet can include devices over a range of 30 to 300 feet depending on the power of the transmitter. Bluetooth devices use frequency-hopping spread-spectrum (FHSS) techniques to allow multiple communications to exist in the same physical space and on the same frequency without interference.

REVIEW QUESTIONS

Section 17.1: Telecommunication: Information and Networks

1. Telecommunication describes the process of communicating over a distance. (True/False)
2. Name the four general classes of information that are transferred via telecommunication links.
3. Transfer of monitoring and control information over distance is often called _____.
4. Each connection to a network is called a _____.
5. A network configured using a bus topology is (*unidirectional/bidirectional*).
6. In a bus network topology, every transmission is received by every node on the network. (True/False)
7. If one of the nodes on a ring network becomes inoperative, the rest of the nodes can still communicate. (True/False)
8. A network configured using a ring topology is (*unidirectional/bidirectional*).
9. The central node of a star network topology is called a _____.
10. Direct communication between two outer nodes on a star network is disallowed. (True/False)
11. Network _____ is essentially a list of rules that defines allowable communications and data formats.
12. Which of the following network architectures considers all nodes to be equal: peer-to-peer or client/server?
13. On a client/server network architecture, where do the majority of programs and data files reside?
14. If the server in a client/server network configuration becomes inoperative, the various clients can continue to function independently. (True/False)
15. What do the letters TAN represent when describing a network?
16. Which is generally larger, a TAN or a CAN?
17. List the following network classifications in the order of their physical size beginning with the smallest: CAN, WAN, MAN, TAN, and LAN.
18. Which would normally have the higher data transfer rate: CAN or LAN?
19. What is the acronym used to describe a network that is designed to service an entire city?
20. If two or more LANs are connected together over a long distance, the resulting configuration is called a _____.

Section 17.2: Technical Characteristics

21. Communications link A has greater bandwidth than communications link B. Which will likely have the greater throughput?
22. A transmission that utilizes the initial frequency range of a signal and limits communication to one channel per communications path is called a _____ transmission.
23. A transmission that converts the original frequencies to higher frequencies and allows simultaneous transmission of multiple channels on a single communication path is called a _____ transmission.
24. With all other factors being equal, which is the faster transmission method: serial or parallel?
25. A synchronous transmission requires that the same clock be used for both transmitting and receiving the data. (True/False)
26. Binary phase modulation is sometimes used with (*synchronous/asynchronous*) transmissions.
27. Synchronous communication is most often used for transferring (*small/large*) amounts of data.
28. Any bits that are transmitted in order to maintain functionality, detect and correct errors, and so forth, but are not actually part of the transmitted message, are called _____.
29. Asynchronous communication often transmits one character at a time. (True/False)
30. What is the name given to the first bit in an asynchronous data transmission?
31. Parity is a method of _____ _____.
32. If the character to be transmitted in an asynchronous communication link is 10000011 and odd parity is being used, what will be the state of the parity bit?
33. If the character to be transmitted in an asynchronous communication link is 11110100 and even parity is being used, what will be the state of the parity bit?
34. What is the name given to the bits at the end of an asynchronous transmission that establish the minimum time interval between two successive characters?
35. Which type of transmission is more immune to electrical noise and signal degradation: analog or digital?
36. What is the name given to a telecommunications functional block that converts analog to digital in a transmitter and digital to analog in a receiver?
37. When information can only be transferred in one direction over a communications link, the link is said to be a (*simplex/duplex*) link.
38. What name is used to describe a bidirectional communications system if data can only travel in one direction at a time?
39. What is the name of the process when a low-frequency information signal is used to alter the characteristics of a high-frequency carrier signal?

40. Refer to Figure 17–25. What type of modulation is represented by this waveform?
41. Refer to Figure 17–26. What type of modulation is represented by this waveform?
42. What is the purpose of a discriminator circuit?
43. What does the acronym PCM stand for?
44. If single bit errors occur in transmissions using PCM and delta modulation, which recovered information signal will likely have the most corruption?
45. Name three multiplexing techniques.
46. Which two multiplexing techniques discussed in this chapter are conceptually similar, but operate in very different frequency ranges?

Section 17.3: Wireless Telecommunication

47. What is the primary difference between a WLAN and a LAN?
48. What is the purpose of a transponder on a satellite used for communications?
49. What is the maximum number of Bluetooth devices that can share the same piconet?
50. What type of transmitted signals do Bluetooth devices use that enable multiple, independent, simultaneous transmissions in the same frequency band without interference to each other?

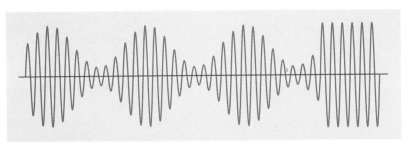

FIGURE 17–25 What type of modulation is represented by this waveform?

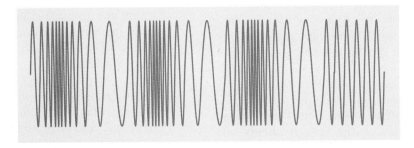

FIGURE 17–26 What type of modulation is represented by this waveform?

•—CIRCUIT EXPLORATION

The Circuit Exploration exercise will give you the opportunity to observe the operation of a simple time-division multiplexing scheme. This is a two-channel system that uses manually operated switches as the information to be transmitted. At the receiver, LEDs or any logic indicator can be used to indicate the received data for each channel. Complete the following tasks:

1. Construct the circuit shown in Figure 17–27. You can choose to simulate the circuit or construct it in a laboratory setting.
2. Exercise the circuit and verify that the Ch 1 switch operates the Ch 1 LED. Similarly, the Ch 2 switch should operate the Ch 2 LED. S_1 is shown as an alternative to the clock. By using S_1, you can slow the action down for easier analysis and observation. Once you understand circuit operation, replace S_1 with a free-running clock. The frequency is not critical, but 100 Hz should do nicely.
3. Once you have verified circuit operation, write a brief theory of operation to explain how it works.

You will note that in this simplified system the transmitter clock is actually sent to the receiver on a separate line. You might want to construct a circuit similar to the one shown in Figure 17–11 to combine the clock and data into a single line using binary phase modulation. This should eliminate the need to send the clock on a separate wire, but the overall circuit behavior should remain the same.

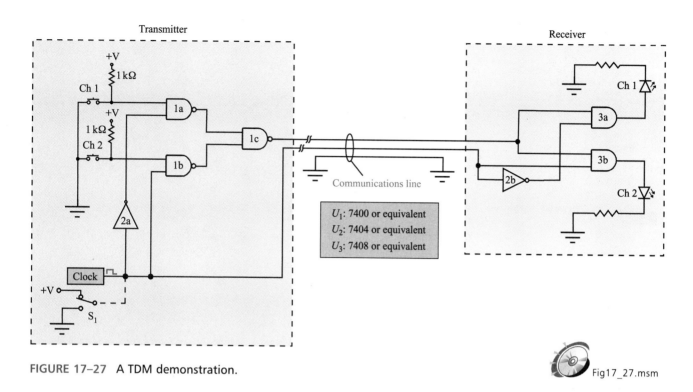

FIGURE 17–27 A TDM demonstration.

APPENDIX A
Logic Families

Table A–1 contrasts the basic electrical characteristics of several logic families. A more extensive list of logic families and greater detail on a particular characteristic are readily available on the various IC manufacturers' Web sites. Two representative Web sites with this type of information are Texas Instruments (www.ti.com) and Philips (www.philipslogic.com/datasheets/).

TABLE A–1 Comparison of logic family characteristics.

Parameter	Logic Family Technology								
	TTL					CMOS			ECL
	Standard	S	LS	ALS	ABT	Standard	HC	ACT	
Propagation delay (ns)	10	3	9.5	7	2–3	150	7	5	0.5
Toggle rate (MHz)	35	125	45	35	200	5	45	160	>1500
Low-level input voltage threshold (V)	0.8					30% of supply voltage	0.8		−1.48
High-level input voltage threshold (V)	2.0					70% of supply voltage	2.0		−1.17
Static supply current per gate	3 mA	3.8 mA	400 µA	240 µA	0.25 mA	10 nA	20 µA	20 µA	6 mA
Low-level drive current (mA)	16	20	8	8	64	1	25	24	40
High-level drive current	400 µA	1 mA	400 µA	400 µA	32 mA	2	25 mA	24 mA	40 mA
Supply voltage (V)	5					3–18	2–6	4.5–5.5	−4.22 to −5.7

APPENDIX B

Measurements with Electronic Test Equipment

Troubleshooting computers and other electronic systems generally requires the use of electronic test equipment. This appendix provides a brief description of how to make a number of measurements on an electronic circuit.

MEASURING CURRENT

Current can be measured with a current probe without breaking the circuit. However, to measure current with a standard current meter, you must physically open the circuit and insert the meter in series with the broken current path. Figure B–1 illustrates the measurement of current in a simple circuit, but the same process is required regardless of circuit complexity. An encircled A or I is used to represent a current meter on a schematic diagram.

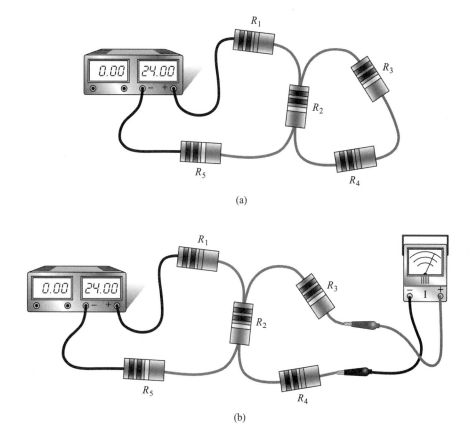

FIGURE B–1 Measuring current with a standard current meter requires breaking the circuit.

Current can be measured in a circuit by following these steps:

1. Remove power to the circuit.
2. Open the circuit at the point where current is to be measured.

3. Connect the current meter across the newly-formed open with the negative meter lead toward the negative source terminal and the positive meter lead toward the positive source terminal.
4. Set the meter for current measurements.
5. Select a current range higher than the expected value of the measured current, unless the meter is auto-ranging.
6. Reapply power to the circuit.

MEASURING VOLTAGE SOURCES AND VOLTAGE DROPS

Voltage measurements are probably the easiest type of circuit measurement to make. The following procedure can be used to measure voltage between any two points in a circuit:

- Select the correct voltage mode (ac or dc) on the meter.
- Select a range that is higher than the expected value of circuit voltage unless the meter is auto-ranging.
- Connect the leads of the meter between the two points being measured.

Figure B–2 shows the proper way to measure voltage in a circuit. An encircled V is used to represent a voltmeter on a schematic diagram. In most cases, the black (–) lead of the meter connects to the most negative of the two points being measured; the red (+) lead of the meter connects to the most positive potential. If an analog meter is connected in the reverse direction, it may be damaged since the pointer will try to deflect off the left side of the scale. A digital meter will normally operate correctly with either polarity; however, the polarity indications on the meter may be confusing unless the meter is used in a consistent manner (e.g., black lead to the most negative potential).

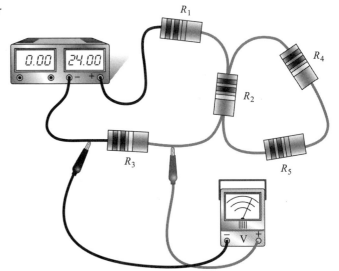

FIGURE B–2 A voltmeter is connected in parallel with the voltage being measured.

MEASURING RESISTANCE

The following procedure can be used to measure the resistance of a component in a circuit:

1. Remove power from the circuit.
2. Disconnect *at least* one end of the component to be measured.
3. Connect the ohmmeter across the component whose resistance is to be measured.

Failure to perform step 1 can cause damage to the meter. Failure to perform step 2 can result in erroneous measurements.

The range switch on digital ohmmeters simply sets the highest resistance that can be measured in a particular position. The value of resistance is displayed directly. You should choose the lowest resistance range that still allows the measurement to be taken. Figure B–3 shows the correct way to measure the resistance of a component.

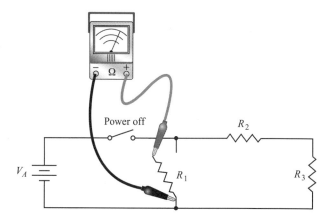

FIGURE B–3 After power is removed from the circuit, disconnecting one end of a two-lead component effectively isolates it from the rest of the circuit.

MEASURING WAVEFORMS

The available controls on an oscilloscope, and therefore the operation of the device, vary greatly between models and manufacturers. The following procedure describes the generic actions taken to display a waveform. The user manual for your oscilloscope will provide details for your specific oscilloscope.

1. Connect the ground lead of the oscilloscope probe to a ground point in the circuit. In the case of high-frequency waveforms (e.g., computer clocks in excess of a few megahertz), it is essential to use an extremely short (i.e., 0.5″) ground lead.
2. Connect the oscilloscope probe to the point being monitored.
3. Adjust the vertical (Y) gain and vertical (Y) position to obtain a display that is as large as possible but still fits completely on the screen.
4. Adjust the horizontal (X) sweep rate and horizontal (X) position to view at least one complete cycle of the waveform. If the waveform is not visible or if it changes continuously, adjust the trigger level to obtain a stable display.

When the above procedure fails to provide a clear display of the measured waveform, it is often due to one of the following problems:

- **Display is not visible.** May be due to one of the following causes:
  1. Brightness or intensity control misadjusted.
  2. Vertical and/or horizontal position controls set incorrectly.
  3. Triggering level set wrong.
  4. Trigger source set wrong. (Set the trigger source to be the same as the channel being utilized for the measurement.)

- **Display is unstable.** An unstable display is often caused by one of the following:
  1. Triggering level set wrong.
  2. Trigger source set wrong. (Set the trigger source to be the same as the channel being utilized for the measurement.)
  3. The waveform is not a periodic signal. That is, its characteristics such as pulse width, frequency, duty cycle, and so forth vary between consecutive cycles.

- **Waveform has unexpected distortion.** Waveform distortion can be caused by any of the following:
  1. Ungrounded oscilloscope probe.
  2. Ground lead of probe is too long.
  3. Selected probe is not appropriate for the frequency of waveform being viewed.

Nearly all practical oscilloscopes today provide an on-screen indication of the scope settings and the characteristics of the measured waveform. Displays that show frequency, period, rise time, pulse width, and so on are typical.

MEASURING STATIC LOGIC LEVELS

Logic levels in a static or very-low speed logic circuit can be monitored with a logic probe, a voltmeter, or an oscilloscope. This type of measurement is self-evident except to caution you to be alert to "bad" logic levels. That is, each logic family (see appendix A) has a certain range of voltage levels that it recognizes as a logic 0 and a certain range that it recognizes as a logic 1. The exact voltage levels depend on the logic family being considered. For example, a standard TTL logic device input responds to any voltage in the range of 0 to 0.8 V as a low logic level. Similarly, a standard TTL logic device input recognizes any voltage between 2.0 and 5 V as a high-level logic signal.

If you measure a voltage on an input that falls between the two defined levels, then the behavior of the gate is unpredictable. So, when measuring logic levels, be sure you know the voltage ranges for the logic family you are working with and then be alert for bad levels. A logic probe normally provides indications of low, high, and bad logic levels.

MEASURING DYNAMIC LOGIC LEVELS

An oscilloscope and a logic probe provide two ways to measure dynamic or rapidly changing logic signals. In the case of an oscilloscope, you simply measure the waveform as you would any other waveform, but always be certain that the two levels of the pulse waveform fall within the defined ranges of low and high logic levels for the logic family being monitored.

In the case of a logic probe, you cannot determine the frequency of the pulses, but most probes will flash to indicate the presence of a rapidly changing signal. Depending on the capabilities of the probe, you may also get an indication of bad logic levels.

CORRELATING TIME-RELATED LOGIC SIGNALS

In computers and other high-speed digital systems, proper operation relies on the relative timing of several (sometimes hundreds) of digital signals. You can monitor any one of these signals with an oscilloscope or logic probe, but you cannot readily determine the timing of one signal relative to the others.

If the waveforms are repetitive, then it is possible (although very cumbersome) to determine relative timing with an oscilloscope. This can be done by utilizing two or more channels of a multi-channel scope. Consider a four-channel oscilloscope for example. Four separate signals can be monitored simultaneously. Each has its own trace on the oscilloscope display.

A logic analyzer is the best tool to use for viewing the relative timing between multiple, high-speed digital waveforms. A logic analyzer provides many channels (the number of channels varies between models and manufacturers). Each channel can capture the logic activities of a different point in the circuit. Furthermore, the results are stored in a memory device, so you can view the various signals and their relative timing long after the actual events occurred.

APPENDIX C
Karnaugh Maps

Any practical logic circuit described by a Boolean expression can be simplified with Boolean algebra. However, this can be a very tedious and error-prone task for complex expressions. We can also use a Karnaugh map to describe the behavior of a logic circuit. We then apply some rather mechanical procedures and read the simplified Boolean expression directly from the map. A Karnaugh map is essentially a tool used to ease the task of simplifying Boolean expressions. Utilization of a Karnaugh map requires four basic sequential steps:

1. Drawing the map
2. Plotting the Boolean expression
3. Looping the map
4. Reading the simplified expression

DRAWING A KARNAUGH MAP

A Karnaugh map is simply an array of squares. There must be 2^N squares in the map, where N is the number of variables in the Boolean expression to be simplified. Figure C–1 shows Karnaugh maps for two-, three-, and four-variable expressions.

FIGURE C–1 Drawn and labeled Karnaugh maps for (a) two-, (b) three-, and (c) four-variable expressions.

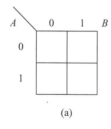

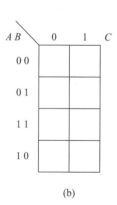

The labeling of the various rows and columns is *not* arbitrary. For example, the labeling of the rows on a three-variable map such as the one in Figure C–1(b) is *not* in binary order. Rather, the labeling is arranged such that the state of only one variable changes when going from one row to an adjacent row. A similar labeling scheme is used for both rows and columns on a four-variable map. Correct labeling is essential, or the map will not produce the correct simplified expression. Our present discussion will be limited to two-, three-, and four-variable maps, but maps for expressions as large as six variables are quite practical. Beyond that, simplification by a computer program is the most effective way to simplify a complex expression.

Plotting the Boolean Expression

The simplest way to transfer a Boolean expression to a Karnaugh map is to write the expression in its minterm form (i.e., sum-of-the-products form). Then, for each term in the expression, simply mark a 1 in each square on the map that contains the given term. This is called priming a cell. Figure C–2(a) shows how each of the cells is identified on a four-variable map. Figure C–2(b) shows how a Boolean expression is plotted.

FIGURE C–2
(a) Identification of cells and (b) plotting of an expression on a four-variable Karnaugh map.

AB \\ CD	00	01	11	10
00	$\overline{A}\overline{B}\overline{C}\overline{D}$	$\overline{A}\overline{B}\overline{C}D$	$\overline{A}\overline{B}CD$	$\overline{A}\overline{B}C\overline{D}$
01	$\overline{A}B\overline{C}\overline{D}$	$\overline{A}B\overline{C}D$	$\overline{A}BCD$	$\overline{A}BC\overline{D}$
11	$AB\overline{C}\overline{D}$	$AB\overline{C}D$	$ABCD$	$ABC\overline{D}$
10	$A\overline{B}\overline{C}\overline{D}$	$A\overline{B}\overline{C}D$	$A\overline{B}CD$	$A\overline{B}C\overline{D}$

(a)

AB \\ CD	00	01	11	10
00	1	1	1	1
01				
11				
10	1	1		1

$Y = \overline{A}\,\overline{B}\,C\overline{D} + A\,\overline{B}\,C + \overline{A}\,\overline{B}\,CD + A\,\overline{B}\,C\overline{D} + \overline{A}\,\overline{B}$

(b)

The first, third, and fourth terms transfer directly to their corresponding cells. The second term is missing one variable, so it can be found in two cells. Similarly, the last term is missing two variables, so it plots in four cells (the entire top row). A given cell can only be primed once, even if multiple terms map to the same cell.

Looping the Map

Looping consists of encircling two or more primed squares according to definite rules. First, the total number of looped cells must be a power of two (e.g., 2, 4, 8). Second, only logically adjacent cells can be looped. Logically adjacent means that the two cells differ by only one variable. Due to the way the map is labeled, physically adjacent cells (vertically or horizontally) are also logically adjacent. Additionally, the top and bottom rows are logically adjacent as are the two extreme columns. These last two options are more apparent if you view the map as being rolled into a cylinder (either vertically or horizontally), which brings the extreme rows or columns physically adjacent to each other.

The third looping rule requires us to loop as many cells as possible into a given loop. Finally, make as few loops as possible; however, all primed squares must be in at least one loop. A given cell may be looped more than once, but only if you include additional cells that were not previously looped. These four looping rules can be summarized as follows:

1. The number of primed squares in a loop must be a power of two.
2. Only logically adjacent squares can be looped together.
3. Loop as many primed cells as possible in each loop.
4. Make as few loops as possible to include all primed squares in at least one loop.

Figure C–3 shows several examples of properly looped maps.

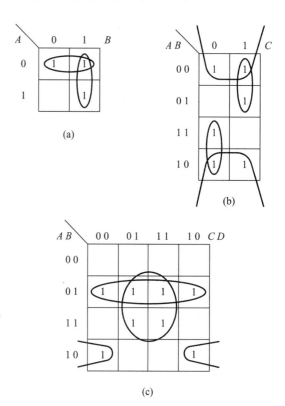

FIGURE C–3 Proper looping of Karnaugh maps.

Reading the Simplified Expression

Each loop on a properly looped Karnaugh map corresponds to a single term in the simplified minterm expression. The term to describe a given loop can be readily determined by inspection. If a variable and its complement (e.g., C and $\overline{C}$) are both in the same loop, then that variable does not appear in the simplified term that represents the loop. Applying this procedure to the maps shown in Figure C–3 yields the following simplified Boolean expressions:

a. $\overline{A} + B$
b. $\overline{B} + \overline{A}C + A\overline{C}$
c. $\overline{A}B + BD + A\overline{B}\,\overline{D}$

APPENDIX D

Solutions to Odd-Numbered Exercise Problems

CHAPTER 1

•—Exercise Problems

Section 1.1: Representative Systems

1. An electronic system is a complete working system whose basic operation is based on or controlled by a number of interconnected electronic components.

3. The instructor must verify this response.

Section 1.2: System Notations

1. False. A block diagram shows the general nature of a system along with the inputs and outputs of the various functional blocks. A schematic, on the other hand, shows the electrical interconnections for every component in the system.

3. 25%

5. Wiring diagram.

Section 1.3: Physical System Hierarchy

1. Integrated circuits contain the equivalent of thousands or millions of discrete components. Therefore, integrated circuits have far greater component density.

3. Nanotechnology is a general label that refers to the construction of machines and systems (both electrical and mechanical) using atomic-level building blocks.

Section 1.4: System Connectivity

1. The nonconductive outer covering of wire is called insulation, and its purpose is to prevent accidental connections from being made between the wire and some other point in an electrical system.

3. True. Electrical interference has no effect on the transmission of data and commands through a fiber-optic link.

Section 1.5: Elements of System-Level Troubleshooting

1. Block-diagram thinking provides the advantage of reducing apparent complexity. This makes it easier to follow the flow of the system without becoming entangled in details.

3. Front-panel milking involves intentional manipulation of all accessible controls while monitoring the system's behavior. Oftentimes, this provides clues as to the nature of the defect.

Section 1.6: Circuit Simulation

1. False. Circuit simulation software utilizes virtual instruments which simulate the operation of real test equipment.

3. True. A typical circuit simulation package contains many different types of analyses that can be performed.

CHAPTER 2

•—Exercise Problems

Section 2.1: Electrical Quantities

1. a. $V = 25.2$ V
   b. $f = 375$ Hz
   c. $P = 5$ W
   d. $Z = 3000$ Ω
   e. $t = 0.09$ s
   f. $I = 12.5$ A

3. Superconductor.

5. repel.

Section 2.2: Mechanical Quantities

1. True.

3. Revolutions per minute. This is a measure of speed of a rotating object such as a shaft, and it describes how many complete turns the device makes in one minute.

Section 2.3: Light and Other Waves

1. False. Intensity is a measure of the brightness of light.

3. $\lambda = \dfrac{v}{f}$

   $= \dfrac{300{,}000{,}000 \text{ m/s}}{100{,}000{,}000 \text{ Hz}} = 3$ meters, or

   $= \dfrac{186{,}411 \text{ mi/s}}{100{,}000{,}000 \text{ Hz}} = 0.00186411$ miles $= 9.84$ feet

5. $T = \dfrac{1}{f}$

   $= \dfrac{1}{50{,}000 \text{ Hz}} = 0.00002$ seconds

Section 2.4: Magnetism and Electromagnetism
1. Magnetic field.
3. The domains are randomly ordered in an unmagnetized material, but they are mostly oriented in the same direction when the material is magnetized.
5. High. The higher the relative permeability of a material, the easier it is to pass lines of flux.

CHAPTER 3
• Exercise Problems
Section 3.1: Technical Notation
1. a. 26.6×10^{-3}
   c. 0.07×10^{-6}
   b. 7.74×10^{9}
   d. 520×10^{3}
3. a. milli
   d. pico
   b. mega
   e. centi
   c. giga
5. a. 2200 µH
   c. 0.0008 µF
   b. 250 mW
   d. 59.9 kHz

Section 3.2: Wire and Cable
1. False.
3. directly
5. False.
7. $A_{\text{CMIL}} = d^2$

   $d = \sqrt{A_{\text{CMIL}}}$

   $= \sqrt{2025 \text{ cmil}} = 45$ mil
9. 0
11. 404 cmil from wire size table for annealed copper wire.
13. False. The breakdown voltage rating of a wire must exceed the actual voltage in the circuit to prevent breakdown from occurring.
15. Varnish.
17. A ribbon cable is a flat strap made by bonding a number of individually insulated wires together.

Section 3.3: Electronic Components
1. voltage sources.
3. A circuit breaker can be reset after the overload condition has been cleared. A fuse, by contrast, must be replaced once it has opened to protect a circuit.
5. False. Physical resistor size provides some indication of its power rating, but gives no clue about its resistance value.

7. No. The lowest it could be and still be within tolerance is determined as follows:

   maximum deviation = tolerance × marked value
   $= 0.05 \times 470{,}000\ \Omega = 23{,}500\ \Omega$

   lowest value = marked value − maximum deviation
   $= 470{,}000 - 23{,}500 = 446{,}500\ \Omega$

9. True.
11. Nominal value is 680 Ω with a 10% tolerance. Acceptable resistance range is determined as follows:

    maximum deviation = tolerance × marked value
    $= 0.1 \times 680\ \Omega = 68\ \Omega$

    resistance range = marked value ± maximum deviation

    lowest value $= 680 - 68 = 612\ \Omega$

    highest value $= 680 + 68 = 748\ \Omega$

13. 270 Ω
15. False. A rheostat is a two-terminal device. It cannot be used as a three-terminal potentiometer.
17. Two.
19. relay

CHAPTER 4
• Exercise Problems
Section 4.1: Basic Requirements for Current
1. Electromotive force and a complete path for current
3. Yes. Electron current always flows toward the more positive (or less negative) potential. In this case, current goes from the −2 V potential toward the +5 V potential.

Section 4.2: Series Circuits
1. A series circuit has only a single path for current.
3. True.
5. Total resistance in a series circuit is equal to the sum of the individual resistors.

   $R_T = R_1 + R_2 + R_3 + R_4 + R_5 + R_6$
   $= 5\ \Omega + 5\ \Omega + 5\ \Omega + 5\ \Omega + 5\ \Omega + 5\ \Omega = 30\ \Omega$

   This problem also points out a shortcut that can be used when all resistors are equal. Simply multiply the resistance value by the number of resistors. In this case, we would apply the shortcut method as follows:

   $R_T = nR_X$
   $= 6 \times 5\ \Omega = 30\ \Omega$

7. There are several ways to compute the voltage drop across R_2, but application of the voltage divider equation is perhaps the most straightforward.

$$R_2 = \left(\frac{R_2}{R_1 + R_2}\right) V_A$$

$$= \left(\frac{33 \text{ k}\Omega}{22 \text{ k}\Omega + 33 \text{ k}\Omega}\right) 5 \text{ V} = 3 \text{ V}$$

Section 4.3: Parallel Circuits

1. False. Noting that both ends of all components connect directly together identifies parallel circuits.
3. less than
5. 68 Ω. The simplest solution is to divide the common resistance by the number of resistors as follows:

$$R_T = \frac{R_N}{N}$$

$$= \frac{680 \text{ }\Omega}{10} = 68 \text{ }\Omega$$

7. We apply Ohm's Law to compute the branch currents as follows:

$$I_1 = \frac{V_1}{R_1} = \frac{25 \text{ V}}{5 \text{ k}\Omega} = 5 \text{ mA}$$

$$I_2 = \frac{V_2}{R_2} = \frac{25 \text{ V}}{10 \text{ k}\Omega} = 2.5 \text{ mA}$$

$$I_3 = \frac{V_3}{R_3} = \frac{25 \text{ V}}{50 \text{ k}\Omega} = 0.5 \text{ mA}$$

Next, we apply Equation 4–12 to calculate total current.

$$I_T = I_1 + I_2 + I_3$$

$$= 5 \text{ mA} + 2.5 \text{ mA} + 0.5 \text{ mA} = 8 \text{ mA}$$

9. False. The current entering and leaving a point must always be equal.

Section 4.4: Series-Parallel Circuits

1. False. Series-parallel is probably the most common circuit configuration used in practical electronic systems.
3. See Figure ANS4–3.

Section 4.5: Complex Circuits

1. The circuits in Figure 4–49 (b) and (d) are complex.
3. The original circuit is simplified by repeatedly replacing sets of series and parallel resistors with single equivalent resistors. If the simplification ultimately results in a single resistor, the original circuit was series-parallel. Otherwise, it was a complex circuit.

Section 4.6: Ground and Other Reference Points

1. $V_A = +17$ V, $V_B = +10$ V, $V_C = +8$ V, $V_D = -3$ V
   $V_E = -14$ V, $V_F = -18$ V

CHAPTER 5
•—Exercise Problems
Section 5.1: Troubleshooting Series Circuits

1. False. While it is true that purely series circuits are not found as frequently as other circuit configurations, it is essential to develop troubleshooting skills on series circuits. These same skills are transferable to troubleshooting more complex circuits (series-parallel in particular).
3. The current through all components goes to zero.
5. a. more positive. The voltage at TP2 goes more positive (+50 V).
7. The voltage at TP4 will go to zero if R_4 shorts since you will be measuring across a zero-ohm component.
9. If R_1 shorts, the current through all resistors including R_4 will increase because there is less total resistance in the circuit.

Section 5.2: Troubleshooting Parallel Circuits

1. False. A voltmeter is only marginally helpful when troubleshooting parallel circuits because the voltage is the same across all components regardless of the nature of the defect. It can be used to detect certain power supply defects (e.g., no-output voltage), but it is not the best tool for troubleshooting parallel circuits.
3. Both of these instruments provide a relative indication of current without the need to break the circuit. Many times a relative indication of current is all that is needed to locate a defect.

FIGURE ANS4–3

	R_1	R_2	R_3	R_4	Total
Resistance	6.8 kΩ	47 kΩ	33 kΩ	10 kΩ	36.188 kΩ
Voltage	4.7 V	13.4 V	13.4 V	6.91 V	25 V
Current	690.85 μA	285.11 μA	406 μA	690.85 μA	690.85 μA
Power	3.25 mW	3.82 mW	5.44 mW	4.77 mW	17.27 mW

Section 5.3: Troubleshooting Series-Parallel Circuits

1. Four. The series circuit would consist of the voltage source and R_1, $R_{2,3}$, R_4, and $R_{5,6}$.

3. All of the following defects would cause TP2 to measure higher than normal (but less than the full supply voltage):
   - Increased voltage source
   - R_1 decreased in value
   - Any of the resistors R_2 through R_6 increased in value

5. All of the following defects would cause TP5 to measure higher than normal (but less than the full supply voltage):
   - Defective voltage source (increased output)
   - Any of the resistors R_1 through R_6 shorted or reduced in resistance
   - R_7 increased

Section 5.4: Troubleshooting Strategies

1. Troubleshooting fences are used to separate components suspected of having defects from those that are thought to be good.

3. False. Equal-probability troubleshooting can only be utilized by someone who has experience on the specific type of equipment being diagnosed.

5. First, most components are very inexpensive. Second, the desoldering process can damage delicate parts without leaving any visible evidence of the damage.

CHAPTER 6

●─Exercise Problems

Section 6.1: Generation of Alternating Voltage

1. a. positive alternation: a
   b. negative alternation: d
   c. period: e
   d. time axis: c
   e. voltage or current axis: b

Section 6.2: Sine Wave Characteristics

1. period

3. Same as. The period of a sine wave is the time from any given point on the cycle to the same point on the following cycle.

5. 400 ps. We compute period from frequency as follows:
$$t = \frac{1}{f}$$
$$= \frac{1}{2.5 \text{ GHz}} = 400 \text{ ps}$$

7. 250 Hz. We compute frequency from period as follows:
$$f = \frac{1}{t}$$
$$= \frac{1}{4 \text{ ms}} = 250 \text{ Hz}$$

9. 3.185 mA. We find the average as follows:
$$I_{\text{avg}} = 0.637 \times I_P$$
$$= 0.637 \times 5 \text{ mA} = 3.185 \text{ mA}$$

Note that this is the average for one-half cycle, which is always assumed unless specifically stated otherwise, since the average for a full cycle is always zero.

11. 63.39 mV. We compute instantaneous voltage as follows:
$$v = V_P \sin \theta$$
$$= 150 \text{ mV} \sin 25°$$
$$= 150 \text{ mV} \times 0.4226 = 63.39 \text{ mV}$$

Section 6.3: Working with Phase Angles

1. False. Phase angles are measured from the right-most horizontal axis.

3. instantaneous

5. 75 V. Peak voltage is equal to the length of the phasor.

7. 46.66°. Following is one method for finding the angle:
$$\sin \theta = \frac{\text{opposite}}{\text{hypotenuse}} = \frac{200}{275} = 0.727$$
$$\arcsin 0.727 = 46.66°$$

Section 6.4: Circuit Analysis of AC Resistive Circuits

1. See Figure ANS6–1.

FIGURE ANS6–1

	Voltage (V)				Current (mA)			
	V_P	V_{PP}	V_{rms}	V_{avg}	I_P	I_{PP}	I_{rms}	I_{avg}
R_1	42.94	85.88	30.35	27.35	35.78	71.56	25.3	22.77
R_2	57.06	114.12	40.35	36.35	21.13	42.26	14.94	13.45
R_3	57.06	114.12	40.35	36.35	14.67	29.34	10.37	9.33
Total	100	200.0	70.7	63.7	35.78	71.56	25.3	22.77

Section 6.5: Alternating Voltage Applications

1. False. A battery is a source of dc.
3. 1,130 feet per second.
5. True. They both travel through air as pressure changes, moving at about 1,130 feet per second.
7. 100 meters. There are actually three power classes for Bluetooth devices. The most powerful devices (up to 100 mW) have an expected range of 100 meters.
9. piconet. Each piconet has one master (which can be any Bluetooth device) and up to seven actively communicating slaves.

CHAPTER 7

Exercise Problems

Section 7.1: Inductors

1. induction
3. False. The polarity of the magnetic field affects the polarity of the induced voltage.
5. opposes. An inductor has no effect on a steady current, but will oppose an increasing current and will aid a decaying current.
7. 188.4 kΩ. We compute inductive reactance as follows:

$$X_L = 2\pi f L = 2 \times 3.14 \times 300 \text{ MHz} \times 100 \text{ μH}$$
$$= 188.4 \text{ k}\Omega$$

9. lags; 90. In a pure inductor, sinusoidal current always lags the voltage by 90°.
11. 20 mH. We find total parallel inductance as follows:

$$L_T = \cfrac{1}{\cfrac{1}{L_1} + \cfrac{1}{L_2} + \cfrac{1}{L_3} + \cfrac{1}{L_4} + \cfrac{1}{L_5}}$$

$$= \cfrac{1}{\cfrac{1}{100 \text{ mH}} + \cfrac{1}{100 \text{ mH}} + \cfrac{1}{100 \text{ mH}} + \cfrac{1}{100 \text{ mH}} + \cfrac{1}{100 \text{ mH}}}$$

$$= 20 \text{ mH}$$

13. 142.8 mV(rms). There are many ways to solve this problem, but let us choose to first compute total inductance.

$$L_T = L_1 + L_2 = 250 \text{ mH} + 100 \text{ mH} = 350 \text{ mH}$$

We can now find total inductive reactance.

$$X_{LT} = 2\pi f L_T = 2 \times 3.14 \times 200 \text{ kHz} \times 350 \text{ mH}$$
$$= 439.6 \text{ k}\Omega$$

Next, we find total current in the circuit with Ohm's Law.

$$I_T = \frac{V_T}{X_{LT}} = \frac{500 \text{ mV}}{439.6 \text{ k}\Omega} = 1.137 \text{ μA}$$

Now, let us find the reactance of L_2.

$$X_{L2} = 2\pi f L_2 = 2 \times 3.14 \times 200 \text{ kHz} \times 100 \text{ mH}$$
$$= 125.6 \text{ k}\Omega$$

Finally, we use Ohm's Law to compute the voltage drop across L_2 (recognizing that the current is the same in all parts of a series circuit).

$$V_{L2} = I_{L2} X_{L2} = 1.137 \text{ μA} \times 125.6 \text{ k}\Omega$$
$$= 142.8 \text{ mV}$$

If desired, we could convert this rms voltage value to peak, average, or peak-to-peak by applying the conversion factors discussed for resistive circuits.

15. Infinite resistance.

Section 7.2: Transformers

1. False. Transformers are commonly used in computer circuits including LCD display panels, power supplies, modem circuits, and Ethernet communication ports.
3. higher. Greater coupling means more flux will be cutting the secondary winding, and greater voltages will be induced.
5. Phase relationships. Dot notation indicates the relative phase relationship between primary and secondary windings.
7. Air, iron, and ferrite.
9. Autotransformer.
11. 12 Vac. We compute secondary voltage as follows:

$$\frac{N_P}{N_S} = \frac{V_P}{V_S}$$

$$\frac{10 \text{ T}}{1 \text{ T}} = \frac{120 \text{ V}}{V_S} \text{ so}$$

$$V_S = \frac{120 \text{ V} \times 1 \text{ T}}{10 \text{ T}} = 12 \text{ V}$$

13. 24 mA. We compute primary current as follows:

$$\frac{N_P}{N_S} = \frac{I_S}{I_P} \text{ so}$$

$$I_P = \frac{N_S I_S}{N_P}$$

$$= \frac{1 \text{ T} \times 240 \text{ mA}}{10 \text{ T}} = 24 \text{ mA}$$

15. Any of the following are acceptable transformer defects:
    - Open primary
    - Open secondary(s)
    - Shorted turns or shorted winding

- Winding shorted to the core
- Short between primary and secondary

Section 7.3: Capacitors

1. False. A capacitor stores energy in an electrostatic field.

3. Zero. The capacitor will charge to a value equal to the applied voltage. Once it is fully charged, there will be no additional current flow (except leakage current).

5. decreases. Capacitive reactance is inversely proportional to frequency.

7. 3.98 MΩ. We compute the reactance as follows:

$$X_C = \frac{1}{2\pi f C} = \frac{1}{2 \times 3.14 \times 40 \text{ Hz} \times 1000 \text{ pF}}$$
$$= 3.98 \text{ M}\Omega$$

9. Surface mount. Surface-mount capacitors have no protruding leads and solder directly to pads on a printed circuit board. Their longest dimension can be as small as 20/1000 of an inch.

11. The circuit might operate incorrectly with changes in temperature. NPO capacitors have zero temperature coefficients, which means their values remain fairly constant over a wide range of temperatures. They are used in circuits where temperature stability is important. The N750 replacement would decrease in value as temperature increased.

13. First, realize that 0.005 µF is the same as 5,000 pF. So, we will need to connect two 0.005-µF capacitors in parallel to get the needed 10,000 pF, since the values of parallel-connected capacitors are additive.

15. If the ohmmeter indicates near zero resistance when placed across the capacitor, then the capacitor is shorted. At least one side of the capacitor must be removed from the circuit before connecting the ohmmeter to be absolutely certain the capacitor itself is defective. Otherwise, if a short elsewhere in the circuit is in parallel with the capacitor, the ohmmeter will still read zero even though the capacitor is good.

Section 7.4: *RC* and *RL* Circuits

1. False. The current and the resistance or reactance of the component determines the voltage drop across each component.

3. 35.36 V. We sum the component voltages using phasor addition.

$$V_A = \sqrt{V_R^2 + V_C^2} = \sqrt{(25 \text{ V})^2 + (25 \text{ V})^2}$$
$$= 35.36 \text{ V}$$

5. voltage drops. The voltages are the same across all components in any parallel circuit.

7. 44.72 Ω. First, we compute the branch currents using Ohm's Law.

$$I_R = \frac{V}{R} = \frac{100 \text{ V}}{50 \text{ }\Omega} = 2 \text{ A}$$

and

$$I_L = \frac{V}{X_L} = \frac{100 \text{ V}}{100 \text{ }\Omega} = 1 \text{ A}$$

Next, we compute total current in the circuit by summing (phasor addition) the branch currents.

$$I_T = \sqrt{I_R^2 + I_L^2} = \sqrt{(2 \text{ A})^2 + (1 \text{ A})^2}$$
$$= 2.236 \text{ A}$$

Finally, we apply Ohm's Law to compute the impedance of the circuit.

$$Z = \frac{V_A}{I_T} = \frac{100 \text{ V}}{2.236 \text{ A}} = 44.72 \text{ }\Omega$$

9. short

Section 7.5: *RLC* Circuits

1. False. The voltage drops across the various components are functions of the common current and the individual reactances and resistances.

3. $V_A = 5$ V. We sum the individual voltage drops with phasor addition.

$$V_A = \sqrt{V_R^2 + (V_L - V_C)^2} = \sqrt{(3 \text{ V})^2 + (21 \text{ V} - 25 \text{ V})^2}$$
$$= 5 \text{ V}$$

5. True. The voltage is the same across all components in any parallel circuit.

7. minimum. The impedance at resonance is equal to the resistance in the circuit. The impedance is higher on either side of resonance because the two reactances only partially cancel.

9. 12.43 MHz. We compute the resonant frequency as follows:

$$f_R = \frac{1}{2\pi\sqrt{LC}} = \frac{1}{2 \times 3.14 \times \sqrt{2 \text{ }\mu\text{H} \times 82 \text{ pF}}}$$
$$= 12.43 \text{ MHz}$$

11. Increase. The impedance of a series *RLC* circuit is minimum at resonance. Therefore, according to Ohm's Law, the current is maximum at resonance. Since the current is the same in all parts of a series circuit, this means the current through the resistor (and therefore the voltage across the resistor) is also maximum at resonance. So, as the operating frequency is moved closer to the resonant frequency of the circuit, the voltage across the resistor increases.

CHAPTER 8

•—Exercise Problems

Section 8.1: Basic Atomic Theory

1. insulators
3. True.
5. True. The higher the orbit of an electron, the higher the energy level.
7. conduction; valence

Section 8.2: Semiconductor Theory

1. False. A pure semiconductor material is called intrinsic silicon.
3. True. All electrons and all holes are EHPs.
5. hole
7. True.
9. True.
11. electrons; conduction

Section 8.3: Semiconductor Junctions

1. pn junction
3. False. The impurity atoms are held relatively immobile within the crystal structure.
5. 0.6 V–0.7 V
7. True. The majority carriers are pushed toward each other. The field established by the biasing source opposes the internal barrier potential.
9. True. Therefore, no significant current can flow.
11. True. The internal field is aided in the field setup by the biasing voltage.

Section 8.4: Troubleshooting Semiconductors

1. True.
3. True. If a lead is bent too close to the body of the component, either the lead or the component body may be damaged.
5. heat sink. This may take the form of pliers, hemostats, or other metal clamp with substantial mass.
7. Yes. A good junction should have at least a 1:10 front-to-back ratio.

CHAPTER 9

•—Exercise Problems

Section 9.1: Diode Characteristics

1. two. The two terminals or leads connect to the p- and n-type materials in the semiconductor material.
3. The colored band identifies the cathode terminal of the diode. Other methods of marking include printed diode symbols and specially-shaped diode packages.
5. open
7. forward biased. Since the anode is more positive than the cathode, we know the polarity is one of forward bias. Further, since the voltage is only 0.7 V, we know this is normal for a diode, so it is not defective.
9. The forward voltage drop is 0 V for an ideal diode, but it ranges from 0.1 V to as high as 1 V for a practical diode. Additionally, the voltage drop remains constant (at 0 V) for an ideal diode but increases with current in a practical diode. Finally, the forward current in an ideal diode can be any value, but a practical diode has maximum limits. The power dissipation in a practical diode will cause damage if the maximum forward-current rating is exceeded.

Section 9.2: Power Supply Applications

1. pulsating dc. One pulse occurs in the output for each cycle of the input. That is, the frequency of the pulsating dc is the same as the input frequency.
3. 50%. The diode conducts for one complete half-cycle (ideally). A practical diode will conduct slightly less than 50% of the full cycle since the input must be above the barrier voltage of the diode before any significant current can flow.
5. 12.5 V. Only one-half of the secondary voltage is supplied to the output of a center-tapped rectifier at any given time.
7. short; long. The charge time constant is short and consists primarily of the RC network formed by the filter capacitor and the forward resistance of the diode. The filter capacitor and the resistance of the load form the long discharge time constant.

Section 9.3: Miscellaneous Diode Applications

1. sinusoidal. As long as the input voltage does not go more negative than V_B, diode D_1 will be reverse biased, and the output will look the same as the input.
3. True. The polarity of the diode determines which polarity will be limited. With the diode reverse, the positive excursion will be limited to the value of V_B, which is negative in this case.
5. True. This circuit clips the negative peaks at a value of V_B. So, if V_B is 0 V, the entire negative alternation will be clipped. This is basically the same waveform as the output of a positive half-wave rectifier.

Section 9.4: Special Diodes

1. varactor
3. increases. Less reverse bias produces a narrower depletion layer. This, in effect, reduces the distance between the p- and n-type materials, which are acting as the plates of a capacitor. Capacitance increases as the plates are brought closer together.
5. Schottky or hot-carrier diode.
7. hot-carrier

9. Step-recovery diode. The sharp transition in current during turn-off is rich in harmonics. A resonant circuit tuned to a harmonic frequency can form the heart of a frequency-multiplier circuit.

11. False. In a device with negative resistance, the voltage decreases as current increases.

13. True.

Section 9.5: Troubleshooting Diode Circuits

1. An ohmmeter can be used to detect all of the listed defects.

3. The forward voltage drop of a good rectifier diode will generally be in the range of 0.3 V $< V_F <$ 1.0 V. If the forward voltage drop is within this range, the diode is probably good. If the forward voltage drop lies considerably outside of this range, then the diode is most certainly defective.

5. False. A diode tester can detect all of the failure mechanisms that an ohmmeter can detect plus others that an ohmmeter cannot readily measure.

7. These symptoms can be caused by a defective supply or a shorted component in the load circuit. By replacing the load with a resistor, the problem can be localized to the supply circuitry or to the load.

CHAPTER 10

Exercise Problems

Section 10.1: Bipolar Transistors

1. True. Both holes and electrons are used as current carriers in npn and pnp transistors.

3. The base is lightly doped relative to the emitter and collector regions.

5. False. The base is always lightly doped to limit recombination.

7. Cutoff. No substantial current will flow through the transistor during cutoff.

9. +9.3 V. The base-emitter junction must be forward biased. Therefore, the n-type emitter must be about 0.7 V less positive than the p-type base.

11. 500. We compute voltage gain as follows:

$$\text{Voltage gain} = \frac{\text{output voltage}}{\text{input voltage}}$$

$$\frac{5 \text{ V}}{10 \text{ mV}} = 500$$

13. 180

15. Input impedance. The lower the input impedance of an amplifier, the more prone it is to cause loading of a source. This loading effect could also be expressed in terms of input current. That is, higher input currents cause more loading of signal sources.

17. False. Common-emitter amplifiers have high current and voltage gains and very high power gains.

19. Common-emitter only.

Section 10.2: Junction Field-Effect Transistors

1. n-type. The channel material is p-type, but the gate is n-type.

3. negative. The gate material of an n-channel JFET is p-type, and the gate-source junctions must be reverse biased.

5. n-channel. Majority current carriers (electrons in the case of an n-channel device) always flow from source to drain.

7. increases. When reverse bias is increased, the depletion layer widens to consume a greater portion of the conductive channel. The depletion layer is effectively an insulator, since it is void of current carriers. With a wider depletion layer, the channel current is forced into a smaller region of the channel. This reduces channel current and increases the effective resistance of the channel.

9. saturation. With no bias on the gate-source junction, a JFET conducts maximum current.

Section 10.3: MOS Field-Effect Transistors

1. Oxide. A thin layer of silicon dioxide insulates the metalized gate from the semiconductor channel in a MOSFET.

3. Very high. For practical purposes, we consider the gate of a MOSFET to be an open circuit.

5. depletion. The channel of a depletion-mode MOSFET is complete, even with no bias, so drain current can flow.

7. Positive.

9. Negative.

11. False. Most digital integrated circuits utilize MOSFET transistors.

13. A MOSFET is easily damaged by ESD events. The static electricity accumulated on your body can puncture the gate insulation layer if a discharge occurs.

Section 10.4: Transistor Applications

1. False. Switching transistors operate fully on (saturated) or fully off (cutoff). They are not biased at midpoint.

3. 4.3 V. If the relay is energized, then the transistor must be conducting. If the transistor is conducting, then the base-emitter junction is forward biased and has a 0.7-V (approximately) voltage drop. Since the emitter is fixed at +5 V, the base will be 0.7 V less positive or 4.3 V.

5. Unity. Common-collector (emitter-follower) amplifiers always have a voltage gain that is very close to but slightly less than 1.0.

7. two; one; two. All difference amplifiers have two inputs. The circuit amplifies the difference between the two inputs. It is relatively immune to the absolute voltage on either input. There may be either one or two outputs available from a difference amplifier.

9. True. A cascode amplifier has a low input capacitance, which increases the effective input impedance at high frequencies. At high frequencies, part of the input impedance is determined by the shunting effects of the capacitive reactance of the input capacitance of the transistor. A smaller input capacitance results in higher capacitive reactance and, therefore, less shunting of the input signal (i.e., higher overall input impedance).

Section 10.5: Troubleshooting Transistor Circuits

1. Defective transistor. The transistor has 4.8 V of forward bias. It is definitely defective.

3. Defective transistor. The base-emitter junction is reverse biased, so the transistor should be cutoff. However, it is conducting as is evidenced by increased voltage drops across the emitter and collector resistances. Since the emitter and collector voltages are the same, we can infer that the transistor must be shorted between the emitter and collector. We would normally remove the transistor from the circuit and verify our suspicions with an ohmmeter.

5. TP3 would measure +24 V. With R_1 open, the base-emitter junction would have no forward bias. Therefore, there would be no collector current and no voltage drop across R_4. If there is no voltage drop across R_4, it will have the same voltage on both ends (+24 V).

CHAPTER 11

•—Exercise Problems

Section 11.1: Op Amp Characteristics

1. two; one

3. The inverting input is identified by a minus (−) sign.

5. False. Most bipolar op amps have an internal voltage drop of 1 to 3 V. So, for example, the output of a bipolar op amp operated from ±15-V supplies might be limited to a ±13-V swing.

7. negative

9. True.

11. Slew rate. The slew rate is measured in volts per microsecond (V/μs) and indicates the fastest rate of change that can occur at the output terminal.

Section 11.2: Basic Amplifier Configurations

1. negative. Both inverting and noninverting amplifiers rely on the effects of negative feedback for their operation.

3. noninverting

5. 0 V. Since the voltage gain of the op amp itself is so high, the effective input voltage (v_d) must be very close to 0 V, or the output will be driven to one of its extremes.

7. Noninverting. The input is applied to the (+) input, so it is a noninverting configuration.

9. The output voltage of any linear amplifier circuit is equal to the voltage gain times the input voltage. The voltage gain (4.7) was computed in the preceding problem. We compute the output voltage as follows:

$$v_o = v_i \times A_V = 1.0 \text{ V}_{PK} \times 4.7 = 4.7 \text{ V}_{PK}$$

11. less. The open-loop voltage gain of the op amp must be much higher than the closed-loop gain of the circuit in order for all of our assumptions and approximations to be valid.

13. True. The gain is reduced and stabilized, and the useful bandwidth is increased by the inclusion of negative feedback.

15. 0.301 V_{PK}. We compute the peak output voltage as follows:

$$v_o = v_i \times A_V = 25 \text{ mV}_{PK} \times 12.05 = 0.301 \text{ V}_{PK}$$

17. True.

Section 11.3: Op Amp Applications

1. A voltage follower has unity voltage gain.

3. Since the op amp utilizes negative feedback, we expect the differential input voltage to be very near zero. Therefore, the (−) input will be the same as the (+) input or +3 V with respect to ground.

5. 10 mA. The (−) input terminal of the op amp is essentially open (ignoring the small input bias current). Therefore, all of the transducer current must flow through the feedback resistor.

7. When the input voltage is negative, the output of the op amp is also negative. This reverse biases the diode, which eliminates the negative feedback path. With no feedback, the op amp is operating as an open-loop amplifier, so the voltages on the two input pins are totally independent.

9. True. Integration capacitor C_1 will charge higher in a given amount of time if the input voltage is increased. The capacitor will also charge higher if it is given more time to charge.

11. See Figure ANS11–11, page 674.

13. See Figure ANS11–13, page 674.

15. The reactance of the shunt 330-pF capacitor is low (4.8 kΩ) compared to the series resistances (240 kΩ). Therefore, the voltage divider action will severely attenuate the input signal before it reaches the (+) input.

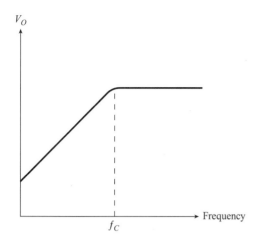

FIGURE ANS11-11

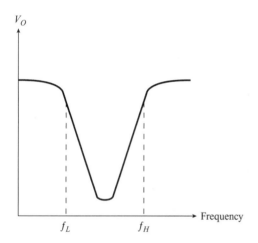

FIGURE ANS11-13

Section 11.4: Troubleshooting Op Amp Circuits

1. If the dc output voltage of the op amp is abnormal, we classify the defect as a dc failure mode. If the dc voltage is normal, then we have an ac failure mode.

3. 0 V. From a dc perspective, this is a voltage-follower circuit with no input voltage. To understand this, visualize the circuit with the capacitors open.

5. Bad op amp. The output polarity is negative, but the polarity of the differential input voltage is such that the output should be positive.

CHAPTER 12

•—Exercise Problems

Section 12.1: Voltage Regulation Fundamentals

1. Although we often speak of the 120-Vac power line voltage as being constant, it does in fact vary considerably. It varies in different parts of the country, and it varies throughout the day at a given location. Therefore, we need a voltage-regulator circuit to isolate the electronic circuitry from these voltage changes. Additionally, even if the power line did have a constant voltage, we would still need a voltage-regulator circuit to compensate for the effects of changing load currents.

3. line regulation

5. 0.61%. We compute load regulation as follows:

$$\% \text{ load regulation}$$
$$= \frac{V_{\text{REG}}(\text{min load}) - V_{\text{REG}}(\text{full load})}{V_{\text{REG}}(\text{full load})} \times 100$$
$$= \frac{3.32 \text{ V} - 3.3 \text{ V}}{3.3 \text{ V}} \times 100 = 0.61\%$$

Section 12.2: Series Voltage Regulation

1. Increase. A decrease in load current will cause the load voltage to increase. In order to bring the load voltage back to its initial value, the effective resistance of the series voltage regulator will have to increase. This increased resistance will now drop more voltage and return the load voltage to its proper value.

3. Remains the same. The (+) input of U_2 is provided by voltage follower U_1. The input to U_1 is the voltage across the reference diode. This is a constant voltage. The whole operation of the circuit relies on the (+) input of U_2 remaining constant regardless of line voltage or load current changes.

5. Goes more positive. If R_3 increases, then the feedback voltage fed to the (−) input of U_2 is less positive. This means the output of the inverting amplifier U_2 will become more positive. This supplies the base of Q_1.

Section 12.3: Shunt Voltage Regulation

1. decrease. By decreasing its effective resistance, the regulator causes more of the input voltage to be dropped across R_S, which means there will be less voltage available for the load.

3. False. When the load current decreases, the current through Q_1 must increase by a similar amount in order to maintain a constant voltage drop across R_S (assuming the input voltage has not changed).

5. Remains the same. The whole purpose of D_1 is to maintain a constant voltage for use as a stable reference.

Section 12.4: Switching Voltage Regulation

1. False. Switching regulators and linear regulators are two different classes of voltage-regulation techniques.

3. False. The diode is reverse biased whenever the switch is closed.

5. at all times. The whole purpose of the regulator is to maintain a constant voltage across the load, so we should expect current to flow through the load at all times.

7. Switching. A switching voltage regulator has high-current pulses with fast transition times. Short rise and fall

times in the time domain correspond to rich harmonics in the frequency domain.

9. boost or step-up

Section 12.5: Power Supply Protection Circuits

1. False. Fuses provide overcurrent protection. Although technically (according to Ohm's Law), an overvoltage condition would also produce an increase in current, in practice, the electronic circuits would likely be damaged by an overvoltage condition before the current increased enough to open a fuse.

3. cutoff. With normal values of load current, there is not enough voltage developed across R_5 to forward bias the base-emitter junction of Q_2.

5. Trip-point current would be lower. Resistors R_3 and R_4 form a voltage divider. If R_4 increases in resistance, it will have a greater percentage of the voltage. This means that the voltage across R_3 will decrease. Since this is the voltage that the R_5 voltage drop must overcome in order for current limiting to begin, the current limiting will occur at a lower value of load current.

7. False. An overcurrent condition will damage a switching regulator just as it would a linear regulator. Further, integrated-switching voltage regulators often have built-in current sensing and limiting.

9. ≈ 0.9 V. Once an overvoltage condition occurs, transistors Q_1 and Q_2 are latched on. They have a combined voltage drop of about 0.9 V. Since the transistor combination is in parallel with the load, the load voltage will also measure about 0.9 V.

Section 12.6: Troubleshooting Power Supply Circuits

1. Remove-and-replace. It is generally impractical and not cost-effective to troubleshoot modular power supplies down to the component level. For this reason, once a defect has been localized to the power supply module, it is removed from the system and replaced with a new one.

3. It is normal. If two of three outputs are normal, then you can generally assume that the input voltage is sufficient for operation. A possible exception to this general rule occurs when an output is just slightly low while the others are normal. In this case, it is possible that the input voltage is present, but it is lower than its expected value.

5. True or False. If the entire circuit board is relatively small and inexpensive, then remove-and-replace may be practical. Also, for purposes of repair activities at a customer's location, remove-and-replace can be a good strategy to minimize costly downtime. But, on expensive systems, embedded power supplies are often repaired to the component level. Even those that have been removed and replaced at a customer location may ultimately be repaired to the component level back at the shop or factory.

7. Yes. This would be an identical strategy and would work just as well. The important thing is to break the loop and to insert a controllable voltage at the open point.

CHAPTER 13

Exercise Problems

Section 13.1: Thyristor Characteristics

1. True.

3. decreases; increases. Once the breakover voltage for a thyristor has been exceeded, the voltage across the thyristor drops abruptly to 1 to 2 V, but the current through the device increases substantially. The current is limited only by the external circuitry.

5. remains fairly constant; increases. When a thyristor experiences breakdown, the voltage across it remains relatively constant, but the current through it can increase dramatically. The resulting power dissipation normally destroys the thyristor.

7. False. Once a thyristor enters the conducting state, the gate loses control. There are a few exceptions to this, but they have not been discussed up to this point.

9. holding current

Section 13.2: Thyristor Types and Applications

1. two; bidirectional

3. bidirectional

5. gate; cathode

7. False. The gate has no control over the conduction of the SCR once the device is on.

9. holding. This is the minimum value of primary current required to keep the SCR conducting once it has been fully switched on. Don't confuse this with latching current, which is the minimum value of primary current that must be achieved in order to fully turn it on in the first place. If the initial current is less than the latching current, the device turns off as soon as the gate voltage is removed.

11. firing. If the firing angle is reduced (i.e., fires earlier in the cycle), the average load current increases, since the thyristor is on for a greater percentage of the alternation. Increasing the firing angle so that turn-on occurs late in the alternation will reduce the average load current.

13. False. An SCS is a relatively low-current component.

15. True. However, when the gate and MT_2 have the same polarity relative to MT_1, triggering requires less gate current.

Section 13.3: Troubleshooting Thyristors

1. Most thyristor circuit defects result in the load being on all the time or being off all the time.

3. True. At least the load will have the full supply voltage. In the case of a dc load such as a motor operated on the ac line voltage but in series with a unidirectional thyristor such as an SCR, a shorted thyristor may not cause the load to act as though it had full voltage. Consider the case of a dc motor, for example. If a series SCR shorted and applied ac to the motor, it would not spin because it requires a single polarity for proper operation.

5. Analog meter (VOM). In order to test an SCR with an ohmmeter as described in the text, the meter must supply the voltage and current necessary to fire and maintain conduction in the SCR. A digital meter does not normally supply as much current and voltage at the probes as an analog meter does. Therefore, the test is generally more successful with a VOM.

7. The ohmmeter will read a high resistance regardless of the gate connection because the meter is connected with the wrong polarity. Thus, there can be no primary current through the SCR, and the gate-cathode junction is reverse biased when the gate is connected to the anode.

Section 13.4: Optoelectronic Devices

1. forward

3. True. A forward-biased LED generally has a voltage drop in the range of 1.4 to 2.4 V, depending largely on the amount of forward current.

5. absorption. The energy provided by the incoming photon is absorbed by the valence-band electron. An electron cannot absorb energy and stay in the valence band, so the covalent bond is broken, and the electron moves to the conduction band.

7. Any light traveling into the depths of the semiconductor material is absorbed and converted into heat within the crystal.

9. False. Photodiodes are operated with reverse bias. Light entering the pn junction and being absorbed breaks covalent bonds. The newly created carriers cause an increase in reverse current.

11. True. With no light striking the phototransistor, it is in cutoff. As the light intensity increases, more emitter-collector current is allowed to flow. If enough light is allowed to enter the junction area, the transistor will eventually enter its saturation state.

13. True. No current flow through the LED means no light will pass into the phototransistor. With no light on the phototransistor, there will be no emitter-collector current.

15. True. Noise that couples onto both leads of the LED portion of the optocoupler has no effect on the current through the device. Therefore, the light emitted from the LED is not affected, which means that the phototransistor within the package will be unaffected.

17. False. The inner glass core is made of ultrapure fused silica.

19. True. Although an ohmmeter test is not 100% foolproof, it can detect opens and shorts, which are two common failure mechanisms. An analog ohmmeter (VOM) should be used if possible. Alternatively, a digital ohmmeter can be used provided it has a special mode for testing pn junctions.

21. Never. You should always avoid looking directly into a fiber-optic cable because you cannot always be certain what is connected to the other end of the cable. If a laser diode is inadvertently driving the opposite end of the cable, you could receive serious and permanent eye damage.

CHAPTER 14

Exercise Problems

Section 14.1: Oscillator Circuits

1. True.

3. Relaxation oscillator. Relaxation oscillators use RC time constants in conjunction with positive feedback and amplification to achieve sustained oscillations.

5. True. The frequency of oscillation is determined by RC time constants.

7. Piezoelectric effect. The piezoelectric effect causes mechanical stress to the crystal when an electric field is applied across the crystal. Conversely, stressing the crystal mechanically will cause a voltage to be produced across the crystal.

9. False. Parasitic oscillations generally disrupt the intended operation of the circuit.

11. True. The output of a phase-locked loop is some multiple of the reference input frequency.

13. False. The VCO is not especially stable, but the closed-loop nature of a PLL forces stability by providing continuous feedback.

Section 14.2: Industrial Computer Applications

1. high. A high common-mode rejection ratio (CMRR) is one of the most dominant characteristics of an instrumentation amplifier.

3. True. Nearly every manufacturer of integrated circuits provides free datasheets for their ICs on the company's web site.

5. True. The current source is greater than the input current, so the remainder must flow through C_1 (left-to-right). This causes the output of the integrator to ramp in the positive direction.

7. True. That is the fundamental purpose of a VFC circuit.

Section 14.3: Troubleshooting Circuits Based on ICs

1. The three most common defect categories are: (1) no output, (2) distorted output, and (3) incorrect or unstable output frequency.
3. ac. A capacitor is normally an open to dc, so an open feedback capacitor would cause an ac defect.
5. No effect. Capacitor C_1 acts as an open to dc under normal conditions. No significant changes will occur to the dc levels in the circuit if the capacitor opens.
7. False. The output of an op amp at any given time is determined by its inputs at that same time.

CHAPTER 15

•—Exercise Problems

Section 15.1: Digital Concepts and Terminology

1. bit. This is an abbreviation for **bi**nary dig**it**.
3. a. 18   f. 19
   b. 17   g. 13
   c. 9   h. 69
   d. 11   i. 31
   e. 25   j. 3
5. OFF
7. Combinations = $2^N = 2^5 = 32$
9. $Y = A + B + C$
11. Inverter or NOT function or noninverting buffer.
13. Exclusive OR.
15. Rule 27. Figure ANS14–15 may help to clarify this relationship. Since the output is identical to the input, the two inverters may be eliminated.

   FIGURE ANS14–15

   A $\rightarrow\!\!\triangleright\!\circ\rightarrow$ $\overline{A}$ $\rightarrow\!\!\triangleright\!\circ\rightarrow$ $\overline{\overline{A}}$

   0/1   1/0   0/1

17. First, break the expression apart by counting the number of vincula above each variable and each operator. This gives us the following intermediate expression (which may not be correct if we stop at this point):

$$Y = \overline{A} + D + \overline{E} \cdot B + C + \overline{F}$$

Next, we add parentheses to maintain the original groupings. There were four groups in the given expression. This step gives us the following intermediate result:

$$Y = ((\overline{A} + D + \overline{E}) \cdot (B + C)) + \overline{F}$$

Finally, we remove any unnecessary parentheses to get the expression:

$$Y = (\overline{A} + D + \overline{E}) \cdot (B + C) + \overline{F}$$

This is an acceptable answer. If you prefer, you can further simplify the expression by applying rules 25 and 36 as follows:

$$Y = AB + \overline{A}C + BD + CD + B\overline{E} + C\overline{E} + \overline{F}$$

Section 15.2: Combinational Logic

1. True. Sequential circuits (studied in a later section) have outputs that are not solely determined by the inputs at any specific time.
3. a. $Y = AB$ d. $Y = \overline{A + B + C}$
   b. $Y = A + B$ e. $Y = \overline{A}$
   c. $Y = \overline{ABCD}$ f. $Y = A \oplus B$
5. The various programmable interconnections within the PAL are initially connected at the time of manufacture by fusible titanium-tungsten links. Selected links are then burned open by the user to program a specific logic expression into the device. Once a link has been burned open in a PAL, it cannot be subsequently changed. However, it is possible to open additional links.
7. 2^8 or 256 entries

Section 15.3: Sequential Logic

1. False. The inputs in conjunction with its internal state determine the outputs at any given time.
3. False. If D is low before the clock, then Q will be low after the clock. The flip-flop may or may not have to toggle to achieve this state.
5. Set and preset are terms to describe the condition where Q is high and $\overline{Q}$ is low. Clear and reset describe conditions where Q is low and $\overline{Q}$ is high.
7. True. The flip-flop will toggle every time a proper clock transition is received.
9. The clocks for all stages of a synchronous counter arrive at the same time, whereas the clocks of an asynchronous counter arrive at different times.
11. race; glitches. When decoding the states of an asynchronous counter, the decoding circuit can sometimes detect a false state as the counter transitions from one desired state to another. The race condition stems from the fact that the various stages in an asynchronous system do not change states at exactly the same time. They change in more of a domino fashion. In any case, the brief erroneous output of a decoding circuit that results from timing differences of the input signals is called a glitch.

Section 15.4: Interfacing Digital and Analog Systems

1. analog
3. +7.5 V. The inputs are all 0.5 V, and they receive voltage gains of 1, 2, 4, and 8. The net output voltage of U_1 will be −7.5 V. U_2 is an inverting amplifier with a voltage gain of −1, so the output of U_2 is +7.5 V.

5. 78.125 mV. The 0- to 5-V range would be divided into 2^6 or 64 steps. Each step would have a value of 5 V/64 or 78.125 mV.

7. 001 ($\overline{C}\overline{B}A$). Only the lower comparator will have a high-level output. Therefore, the C output will be low, since it connects directly to the output of U_{1D}. The B output will also be low, because both inputs to U_{3A} are low. One of these inputs comes directly from U_{1B}. The other comes from U_{4A}. U_{4A} has a low output because one of its inputs (the one connected to U_{2B}) is low. Finally, the A output will be high. Here is one way to determine the level at A. U_{6A} has two low-level inputs, so its output will be low. This, along with the low level from U_{1C}, means both inputs to U_{6B} are low, so its output will be low. Both inputs (and therefore the output) of U_{6D} are low. The output of U_{7B} is high since one of its inputs (U_{2B}) is low and one (U_{2C}) is high. This means that U_{7A} will have one low and one high input, which will produce a high output. This high, combined with the low-level output of U_{6B}, will produce a high-level output from U_{6C}, which is output A.

9. Seven bits. With seven bits, the 10-V input voltage span would be divided into 2^7 or 128 divisions. Each division would have a span of 10 V/128 or 78.125 mV.

Section 15.5: Troubleshooting Digital Systems

1. False. This can be caused by any of three conditions. First, this can be normal operation for gates with tri-state outputs that are not enabled. Second, the output of the gate can be shorted to another output or to some other voltage source. Finally, the gate can have an internal defect.

3. True. Any low input on an AND gate will cause a low output. Therefore, once you determine that at least one input is low, there is no value in measuring the remaining inputs.

5. You need to simultaneously examine the logic states of multiple points in the circuit in order to classify the behavior as good or bad. Without a logic analyzer, you are restricted to one point at a time (or possibly four points with a multichannel oscilloscope).

7. VHDL is a valuable tool for troubleshooting defects in systems before they are actually fabricated as physical components. This normally occurs in the very early stages in the lifetime of a product.

CHAPTER 16

Exercise Problems

Section 16.1: Basic Concepts and Terminology

1. The order based on size follows:
    c. bit
    a. byte
    b. instruction
    d. program

3. 15 bits. $2^{15} = 32,768$. Although not formally presented in the text material, this calculation can be expressed more formally as follows:

$$\text{Number of required bits} = \frac{\ln N}{\ln 2}$$

where N is the width of the bus. In the present example, we calculate as follows:

Number of required bits

$$= \frac{\ln N}{\ln 2} = \frac{\ln 32,768}{\ln 2} = \frac{10.3972}{0.693} = 15$$

5. During execution of the actual instruction, the address bus is used to identify the specific input device that is to be involved in the data transfer.

7. read

9. ALU or arithmetic and logic unit.

Section 16.2: Hardware

1. PCI bus. The PCI bus operates with a transfer rate of at least 33 MHz (66 MHz for some). The older ISA bus, by contrast, runs at 8 MHz.

3. True. This is the primary reason for having multiple buses and distributed control. This allows simultaneous activity to occur at different points in the system.

5. Volatile. A volatile memory device does not retain its stored information when power is removed.

7. Bios. These are very basic or primitive programs that allow the system to initialize. They also provide rudimentary communications and control of system devices (e.g., disk drives).

9. port

Section 16.3: Software

1. instructions

3. Machine language programming. This is not practical for anything other than a training environment because of the extreme tediousness and the likelihood of errors.

5. Comments are included in program listings to clarify and document the details of the program's operation. These are ignored by compiler or assembler programs and therefore by the microprocessor.

7. A bootstrap-loader program is only used during initial start-up of a computer system. Its purpose is to load the operating system into the computer and then relinquish control.

CHAPTER 17

Exercise Problems

Section 17.1: Telecommunication: Information and Networks

1. Telecommunication.

3. A group of interconnected transmitters and receivers is called a network.

5. Bus, ring and star topologies form the basis for most network connections.

7. A MAN covers a metropolitan area, which would generally be larger than the more confined area (< 1500 feet) served by a CAN link.

Section 17.2: Technical Characteristics

1. When the bandwidth of a communications link is reduced, the throughput is also reduced.

3. parallel. Several parts of a message (e.g., bits in a digital transmission) can be transmitted simultaneously with a parallel connection, whereas a serial link requires a slower one-after-the-other transmission.

5. Binary 0. Since the character 0100110 already has an odd number of high-level bits, the parity bit will be set to 0 so that the total number of high-level bits is an odd number.

7. half-duplex

9. Amplitude. In amplitude modulation (AM), the amplitude of the modulating signal determines the amplitude of the carrier. The rate at which the amplitude of the carrier is varied is determined by the frequency of the modulating waveform.

11. True. Since PCM is a digital representation of the message, it can easily be multiplexed into interleaved time slots along with other messages.

13. increases. More data can travel through the link in a given time period when multiplexing techniques are used.

Section 17.3: Wireless Telecommunication

1. Wireless LAN or wireless local area network.

3. The curvature of the earth interferes with direct radio links between widely separated ground-based locations. However, a satellite hundreds or thousands of miles above the earth can have an unobstructed communications path to both ground-based stations. Therefore, by simply receiving messages from the ground stations and retransmitting them, a satellite can enable wireless communications between distant locations on earth.

5. False. The more transponders a satellite has, the more simultaneous communication links it can maintain.

7. Eight. There is always one master, and there can be as many as seven slaves within the same piconet.

9. False. Bluetooth devices are capable of full-duplex operation.

Solutions to Odd-Numbered Review Questions and All Circuit Explorations

CHAPTER 1
Review Questions
1. Options a, b, and c are definitely electronic systems. A yardstick (d) is not an electronic system.
3. A block diagram provides the most general level of description.
5. Literally all forms of system notation can be useful as troubleshooting tools.
7. False. Integrated circuits can merge thousands of electronic components into a space that is smaller than even one discrete electronic component.
9. Lasers are the light source for optical communications links using fiber-optic cable. Light from the laser is coded (modulated) with information and instructions. The laser light then travels between the electronic systems by passing through a length of fiber-optic cable. At the receiving end, a light sensor and other electronic circuitry decode the original information.
11. Block-diagram thinking dramatically reduces system complexity while maintaining the relationships of the major functional blocks. Clearly, the value of block-diagram thinking in any specific situation is dependent upon the level of detail presented in the block diagram—the nature of the system—as well as the expertise and experience of the technician or engineer.
13. False. The method being described is signal tracing. Signal injection, by contrast, requires injection of a substitute signal for the normal signal.
15. False. Circuit simulation utilizes virtual test equipment connected to virtual circuits. There is no *physical* connection to real circuits or test equipment.

CHAPTER 2
Review Questions
1. Atom.
3. Repel. All electrons are negatively charged, and charged particles with like charges repel each other.
5. False. Once a valence electron leaves a stable atom, there will be more protons (positively-charged particles) in the nucleus than orbiting electrons (negatively-charged particles). Therefore, the net charge of the atom will be positive (i.e., it will be a positively-charged ion).
7. True. Unlike charges attract. If the charged particles on the two objects were provided with a conductive path between the two bodies, then the charges would eventually neutralize.
9. $R = \dfrac{V}{I} = \dfrac{3\text{ V}}{0.15\text{ A}} = 20\text{ }\Omega$
11. The first modem would have the higher resistance. We compare them as follows:

    $R = \dfrac{V}{I} = \dfrac{12\text{ V}}{0.1\text{ A}} = 120\text{ }\Omega$ for the first modem

    $R = \dfrac{V}{I} = \dfrac{5\text{ V}}{0.25\text{ A}} = 20\text{ }\Omega$ for the second modem
13. False. Conductance is a measure of how easily current can flow through a material. It is the opposite (reciprocal) of resistance. The greater the conductance, the easier the current can flow.
15. $R = 120\text{ }\Omega$
17. True. An example was presented in the chapter where three-dimensional positioning was used.
19. True. Rpm (revolutions per minute) is a measure of the speed of rotating objects.
21. True.
23. period. This is computed as the reciprocal of frequency (i.e., $T = 1/f$).
25. $f = \dfrac{1}{T} = \dfrac{1}{0.0005\text{ s}} = 2000\text{ Hz}$
27. flux.
29. flux density.

31. The material is unmagnetized or not magnetized. If it were magnetized, most of the domains would be aligned in the same direction.
33. Reluctance. Reluctance is a property that opposes the passage of magnetic flux, which is analogous to the way resistance opposes the flow of current in an electrical circuit.
35. Relative permeability is dimensionless (i.e., it has no units of measure).

- Circuit Exploration
    - The measured voltage across R_1 should be 12 V.
    - The calculated current through R_1 is
    $$I = \frac{V}{R} = \frac{12\ V}{10\ \Omega} = 1.2\ A$$
    - The measured current through R_1 should also be 1.2 A.
    - The calculated power dissipation for R_1 is
    $$P = VI = 12\ V \times 1.2\ A = 14.4\ W$$
    - The measured power in R_1 should also be 14.4 W.

CHAPTER 3
- Review Questions
1. False. It is the same as multiplying by 1/1000 or dividing by 1000.
3. 3
5. 3.4×10^3
7. 1.25×10^9
9. 36.5×10^{-3}
11. 25 pF
13. 390 nH
15. 50
17. 75 mΩ; 1,200 kΩ; 2.5 MΩ
19. 56,000,000 ohms can be expressed as either 56,000 kΩ or 56 MΩ. The 56 MΩ option is the best choice.
21. zero (or no)
23. True.
25. increases
27. 10 ga, 12 ga, 18 ga, and 30 ga. The higher the gauge number, the smaller the cross-sectional area of the wire, and the higher the resistance of the wire.
29. $A_{CMIL} = d^2$
    $= (35\ mil)^2 = 1,225\ cmils$
31. The 10-gauge wire has the largest diameter and, therefore, the highest ampacity rating.
33. cable.
35. Ribbon cable.
37. True.
39. True.
41. maximum deviation = tolerance × marked value
    $= 0.05 \times 560\ \Omega = 28\ \Omega$

    resistance range = marked value ± maximum deviation
    lowest value = 560 − 28 = 532 Ω
    highest value = 560 + 28 = 588 Ω

43. Surface-mount technology.
45. 3,900 Ω, ±5%
47. 22.5 kΩ, ±0.1%
49. three; two

- Circuit Exploration
1. Deenergized. The switch has NO contacts, so it is open until it is depressed. Because the switch is open, no current can flow through it or through the relay coil. Therefore, the relay is deenergized, and the contacts remain in their normal position.
2. L_1 is off, and L_2 is on. The relay is deenergized (see explanation for #1). Since the K_{1A} contacts are open, no current can flow through L_1, so it is off. The closed contacts, K_{1B} provide a current path for L_2, which causes it to illuminate.
3. Energized. When S_1 is depressed, its contacts close. This provides a current path for the relay coil, which causes it to energize.
4. L_1 is On, and L_2 is Off. The relay is energized (see explanation for #3). This means that K_{1A} contacts close, providing a current path for L_1. Similarly, the K_{1B} contacts open, which interrupts the current through L_2.
5. Deenergized. When S_2 is depressed, its contacts open. This interrupts the current path for the relay coil and the lamp. Since the relay is not energized, the K_{1A} contacts will be open. When S_2 is released, the current path for the relay coil and for the lamp is still broken, since neither the S_1 contacts nor the K_{1A} contacts are closed.
6. Off. (See explanation for #5.)
7. On. Pressing S_1 closes its contacts. This provides a current path for both the relay and the lamp.
8. Closed. Since the relay has a complete current path (see explanation for #7), the normally open contacts, K_{1A}, will be closed.
9. On. When S_1 is released, its contacts open. However, the current path for the relay and the lamp are maintained through the closed K_{1A} contacts.
10. Closed. Because the current path for the relay is maintained, the K_{1A} contacts remain closed.

CHAPTER 4

Review Questions

1. False. The requirements are an electromotive force and a complete path for current.

3. Zero. An open circuit has infinite resistance, so no current can flow.

5. Electron current always moves toward the more positive potential, so it will go toward the −10-V point in this case.

7. 250 kΩ. This is computed as follows:
$$R_T = R_1 + R_2 + R_3$$
$$= 100 \text{ k}\Omega + 75 \text{ k}\Omega + 75 \text{ k}\Omega = 250 \text{ k}\Omega$$

9. 7.5 V. We use Ohm's Law as follows:
$$V_3 = I_3 \times R_3$$
$$= 100 \text{ μA} \times 75 \text{ k}\Omega = 7.5 \text{ V}$$

11. No, the voltages are not correctly labeled. We write the loop equation (starting in the lower-left corner and moving clockwise) as follows:
$$+V_A - V_1 - V_2 - V_3 - V_4 - V_5 = 0$$
$$+24 \text{ V} - 3 \text{ V} - 2 \text{ V} - 16 \text{ V} - 2 \text{ V} - 2 \text{ V} = 0$$
$$-1 = 0 \quad [\text{error}]$$

13. The net voltage is 20 V with a polarity the same as V_{A_1}. We sum all of the series source algebraically to find the net voltage. In this case, we arbitrarily choose the polarity of V_{A_1} to be positive.
$$V_T = +V_{A_1} + V_{A_2} + V_{A_3} - V_{A_4}$$
$$= 10 \text{ V} + 20 \text{ V} + 5 \text{ V} - 15 \text{ V} = +20 \text{ V}$$

15. False. The applied voltage and the resistances of each branch (which aren't necessarily the same) determine the various branch currents.

17. Total current is greater than any branch current in a parallel circuit; it is the sum of all branch currents.

19. Yes. Total power has to be greater than the power in any one branch, since it is equal to the sum of all branch dissipations.

21. 0.5 Ω, computed as follows:
$$R_T = \frac{R_N}{N} = \frac{5 \text{ }\Omega}{10} = 0.5 \text{ }\Omega$$

23. Since total resistance is always less than the smallest branch resistance, the smallest overall resistance occurs when the 27-kΩ resistor is in the circuit. In order to accomplish this, SW_1 and SW_2 must be in positions B and A, respectively.

25. SW_1 in position A (SW_2 can be in either position). The least current in the circuit corresponds to the most resistance. This means fewest branches, so if we place SW_1 in position A, the circuit will only have the R_1 branch.

27. Minor change. R_3 is so much larger than the $R_1 \| R_2$ combination that the presence or absence of R_3 will have little effect on the total resistance.

29. 4.97 kΩ, computed as follows:
$$R_T = \frac{1}{\frac{1}{R_1} + \frac{1}{R_2} + \frac{1}{R_3}}$$
$$= \frac{1}{\frac{1}{10 \text{ k}\Omega} + \frac{1}{22 \text{ k}\Omega} + \frac{1}{18 \text{ k}\Omega}} = 4.97 \text{ k}\Omega$$

31. 16 kΩ. We apply the product-over-the-sum formula as follows:
$$R_T = \frac{R_1 \times R_2}{R_1 + R_2}$$
$$= \frac{27 \text{ k}\Omega \times 39 \text{ k}\Omega}{27 \text{ k}\Omega + 39 \text{ k}\Omega} = 16 \text{ k}\Omega$$

33. 2.7 kΩ. Compute the value as follows:
$$R_T = \frac{R_N}{N} = \frac{27 \text{ k}\Omega}{10} = 2.7 \text{ k}\Omega$$

35. 4 mA. We apply Ohm's Law as follows:
$$I_T = \frac{V_A}{R_T}$$
$$= \frac{100 \text{ V}}{25 \text{ k}\Omega} = 4 \text{ mA}$$

37. 70 mW. In any circuit configuration, total power is equal to the sum of the individual component power dissipations.
$$P_T = P_1 + P_2 + P_3$$
$$= 10 \text{ mW} + 50 \text{ mW} + 10 \text{ mW}$$
$$= 70 \text{ mW}$$

39. Series, parallel, series-parallel, and complex.

41. Two components are definitely in series if one end of each of the components ties together and no other connections are made to this common point.

43. one; total. A fully simplified series-parallel circuit will have only one resistor whose value is equal to the total resistance in the circuit.

45. 8.65 mA. We apply Ohm's Law as follows:
$$I_T = \frac{V_A}{R_T}$$
$$= \frac{50 \text{ V}}{5.78 \text{ k}\Omega} = 8.65 \text{ mA}$$

47. 887.54 Ω. First, find the equivalent resistance for the parallel combination of R_2 and R_3.

$$R_{2,3} = \frac{R_2 \times R_3}{R_2 + R_3}$$

$$= \frac{1.2 \text{ k}\Omega \times 910 \text{ }\Omega}{1.2 \text{ k}\Omega + 910 \text{ }\Omega} = 517.54 \text{ }\Omega$$

This gives us an equivalent series circuit consisting of R_1, $R_{2,3}$, R_4, and the voltage source. We sum the series resistances to find total resistance.

$$R_T = R_1 + R_{2,3} + R_4$$

$$= 100 \text{ }\Omega + 517.54 \text{ }\Omega + 270 \text{ }\Omega$$

$$= 887.54 \text{ }\Omega$$

49. 40.8 mW. Recall that the parallel resistance of the R_2 and R_3 network was 517.54 Ω in a previous problem. We can use the voltage divider equation (or alternatively an Ohm's Law calculation using $I_{2,3}$ and $R_{2,3}$) to compute the voltage across $R_{2,3}$. Since R_2 and R_3 are in parallel, they will each have the same voltage as $R_{2,3}$. Following is the voltage divider approach.

$$V_{2,3} = \frac{R_{2,3}}{R_T} \times V_A$$

$$= \frac{517.54 \text{ }\Omega}{887.54 \text{ }\Omega} \times 12 \text{ V}$$

$$= 7 \text{ V}$$

We can use one of the power equations to find the power dissipated in R_2.

$$P_2 = \frac{V_2^2}{R_2}$$

$$= \frac{(7 \text{ V})^2}{1.2 \text{ k}\Omega} = 40.8 \text{ mW}$$

51. R_3. Let's compute the powers and compare. First, we know the current through R_1 and R_4 from a previous problem, so we can compute power ($P = I^2R$). However, we need only compute P_4, since the larger resistor will have the most dissipation with equal currents.

$$P_4 = I_4^2 R_4$$

$$= (13.52 \text{ mA})^2 \times 270 \text{ }\Omega = 49.35 \text{ mW}$$

We computed P_2 in a previous problem, and it was smaller than 49.35 mW. So, we have only one other calculation, and that is P_3. The voltage across R_3 was previously computed as 7 V. We compute P_3 as follows:

$$P_3 = \frac{V_3^2}{R_3}$$

$$= \frac{(7 \text{ V})^2}{910 \text{ }\Omega} = 53.84 \text{ mW}$$

53. R_5. We could calculate all powers and compare, but there is an easier way in this case. In a previous problem, we found the equivalent resistance of R_1, R_2, and R_3 (809.17 Ω). Now, since $P = I^2R$ and the three series resistors ($R_{1,2,3}$, R_4, and R_5) have the same current, it follows that the larger resistance will dissipate the most power. In this case, that would be R_5.

55. 16.24 V. We computed $V_{1,2,3}$ in a previous problem. By inspection, this is the voltage V_1.

57. False. Complex problems can be readily solved on most simulation packages.

59. −2 V. This is the voltage drop across R_5.

61. +5 V. The voltage at point A with respect to point B is the voltage across R_1 with point A being the more positive point.

63. −35 V. This is simply the source voltage using point A as the reference.

65. −5 V. This is just the voltage drop across the parallel combination of R_6 and R_7 with point E being the more positive point.

●—Circuit Exploration

The corner-to-corner resistance of the cube network is 0.833 Ω. The simplest way to solve this is to construct the resistive cube and then connect a simulated ohmmeter to the network to measure its resistance. Alternatively, you might connect a voltage source across the corners of the cube and then measure total current. Ohm's Law could then be used to compute total resistance (0.833 Ω).

CHAPTER 5

●—Review Questions

1. False. Logical troubleshooting strategies should always be practiced, regardless of circuit complexity. Otherwise, the poor practices developed when troubleshooting simple circuits will hamper your success when diagnosing more complex circuits.

3. Each measurement takes time and that generally translates into money. Additionally, some components and testpoints in a practical system are relatively inaccessible. Therefore, you must learn to make fewer but smarter measurements.

5. Zero. Any open in a series circuit causes the current to drop to zero through all components.

7. a and e. Measuring at TP3 is essentially measuring the voltage drop across R_3 and R_4. If R_1 opens, there will be no current through R_3 and R_4, and therefore no voltage drop. If the voltage supply was defective and produced no output voltage, then TP3 would measure zero. All other defects either increase the voltage at TP3 or reduce it but not all the way to zero.

9. It increases. If R_3 increases in value, then a greater percentage of the circuit voltage will be dropped across the

part of the circuit consisting of R_2, R_3, and R_4. This is the voltage measured at TP2.

11. An increase. If R_4 shorts, there will be an increased current through R_2. This is the voltage that appears between TP2 and TP3.

13. It measures zero. If R_1 shorts, there can be no voltage across its terminals. This is the voltage measured between TP1 and TP2.

15. It decreases. If R_3 shorts, then the combined resistance of R_2 and R_3 decreases. This means the combined voltage of these two components will also decrease. This is the voltage measured between TP2 and TP4.

17. Since the branch currents in a parallel circuit are dependent on branch resistance, a current meter can readily detect defects. This is an advantage. The disadvantage of a current meter for troubleshooting parallel circuits, however, is that the various components must be disconnected (i.e., unsoldered) in order to insert the current meter into the circuit. This takes time, and it exposes the circuit board to possible damage.

19. If the branch has increased resistance, it will dissipate less power ($P = V^2/R$). This means that it will not generate as much heat as normal. This decrease can be sensed with a temperature probe if you know the relative temperatures of the various branches under normal conditions.

21. No effect. The current through any given branch (including R_4) is not affected by the resistance of the other branches.

23. No effect. Changes in resistance in one branch do not affect the voltages across any of the components, which is the same as the applied voltage.

25. True.

27. c and e. The voltage at TP3 will increase if R_3 shorts or if R_6 opens, because a greater percentage of the circuit voltage will be dropped across the resistance between TP3 and ground.

29. It goes to zero. If R_4 shorts, there will be no voltage across it. This is the voltage measured between TP3 and TP4.

31. Careful observation of the symptoms of a defective system can often lead you directly to the defect. As a minimum, observation will point the direction for your troubleshooting efforts. In most cases, observation will allow you to dramatically reduce the size of your troubleshooting fences at the very beginning of a troubleshooting exercise.

33. One. More and more components are excluded from the fences until only a single component remains. This is the component suspected as the cause of the system malfunction.

35. False. It will actually take a maximum of five measurements (four in some cases).

37. c. both

39. Yes. When switch S_1 is open, there are no parallel sneak paths to affect resistance measurements.

41. Yes. If R_3 opens, there will be a dramatic change in the combined resistance of R_3 and R_4. It could easily be detected with the meter, and there are no sneak paths to affect the measurement.

• Circuit Exploration

Figure ANS5–1 shows the normal voltage and resistance readings for the circuit presented in the Circuit Exploration.

Testpoint	Normal	
	Voltage	Resistance
TP1	20 V	8.53 kΩ
TP2	17.66 V	7.53 kΩ
TP3	13.39 V	5.71 kΩ
TP4	10.57 V	4.51 kΩ
TP5	4.22 V	1.8 kΩ
TP1–TP2	2.34 V	1.0 kΩ
TP2–TP3	4.27 V	1.82 kΩ
TP3–TP4	2.81 V	1.2 kΩ
TP4–TP5	6.35 V	2.71 kΩ

FIGURE ANS5–1

CHAPTER 6

• Review Questions

1. period
3. symmetrical
5. Alternator.
7. To maintain contact with the rotating slip rings.
9. peak
11. Seconds.
13. Hertz.
15. 13.33 MHz. We compute frequency as follows:
$$f = \frac{1}{t}$$
$$= \frac{1}{75 \text{ ns}} = 13.33 \text{ MHz}$$

17. 200 ns. We compute the period of a sine wave as follows:
$$t = \frac{1}{f}$$
$$= \frac{1}{5 \text{ MHz}} = 200 \text{ ns}$$

19. 16.67 ms. We compute the period as follows:

$$t = \frac{1}{f}$$
$$= \frac{1}{60 \text{ Hz}} = 16.67 \text{ ms}$$

21. Twice. A positive peak occurs at 90°, and a negative peak occurs at 270°.

23. True.

25. Zero. The positive and negative alternations of a sine wave are equal in amplitude but opposite in polarity, so the average for a full cycle is zero.

27. 15.925 V. Unless specifically stated otherwise, the average value of a sine wave refers to the average value for one half cycle. We compute the average voltage as follows:

$$V_{\text{avg}} = 0.637 \times V_P$$
$$= 0.637 \times 25 \text{ V} = 15.925 \text{ V}$$

29. False. An amount of dc equal to the rms value of a sine wave is required to produce the same amount of heat.

31. 166.5 mV. First, we compute the peak value.

$$V_P = \frac{V_{\text{avg}}}{0.637} = \frac{150 \text{ mV}}{0.637} = 235.48 \text{ mV}$$

Next we compute the rms value.

$$V_{\text{rms}} = 0.7071 \times V_P$$
$$= 0.7071 \times 235.48 \text{ mV} = 166.5 \text{ mV}$$

33. 20 V. Peak-to-peak voltage is computed as follows:
$$V_{PP} = 2 \times V_P = 2 \times 10 \text{ V} = 20 \text{ V}$$

35. degrees or radians

37. 70.71 V. We compute the instantaneous voltage as follows:
$$v = V_P \sin \theta = 100 \text{ V} \times \sin 45° = 70.71 \text{ V}$$

39. 136.6 V. First, we compute the peak voltage.

$$V_P = \frac{V_{\text{rms}}}{0.7071} = \frac{100 \text{ V}}{0.7071} = 141.42 \text{ V}$$

Now, we can find the instantaneous voltage.

$$v = V_P \sin \theta = 141.42 \text{ V} \times \sin 105°$$
$$= 141.42 \text{ V} \times 0.9659 = 136.6 \text{ V}$$

41. 243.7 V. First, we compute the peak voltage.

$$V_P = \frac{v}{\sin \theta} = \frac{120 \text{ V}}{\sin 100°} = \frac{120 \text{ V}}{0.9848} = 121.85 \text{ V}$$

Next, we compute the peak-to-peak voltage.

$$V_{PP} = 2 \times V_P = 2 \times 121.85 \text{ V} = 243.7 \text{ V}$$

43. third.

45. False. Phase angle is measured between the phasor and the right-most horizontal axis.

47. hypotenuse

49. opposite side; adjacent side

51. Arccos.

53. −0.0584

55. −0.0292

57. 1.34

59. 68.4. We utilize the sine function as follows:

$$\sin \theta = \frac{\text{opposite side}}{\text{hypotenuse}}$$

or

$$\text{opposite side} = \text{hypotenuse} \times \sin \theta$$
$$= 200 \times \sin 20° = 68.4$$

61. 15.62. We utilize the sine function as follows:

$$\sin \theta = \frac{\text{opposite side}}{\text{hypotenuse}}$$

and

$$\text{hypotenuse} = \frac{\text{opposite side}}{\sin \theta} = \frac{10}{\sin 39.8°}$$
$$= \frac{10}{0.6401} = 15.62$$

63. 154.87 V. We compute instantaneous voltage as follows:

$$v = V_P \sin \theta = 270 \text{ V} \times \sin 35°$$
$$= 270 \text{ V} \times 0.5736 = 154.87 \text{ V}$$

65. 64.16°. We utilize the sine function as follows:

$$\sin \theta = \frac{\text{opposite side}}{\text{hypotenuse}} = \frac{90 \text{ V}}{100 \text{ V}} = 0.9$$

Now, we can find the angle.

$$\theta = \arcsin 0.9 = 64.16°$$

67. 53.13°

69. 26.19 mA. First, we compute peak voltage.

$$V_P = 1.414 \, V_{\text{rms}} = 1.414 \times 50 \text{ V} = 70.7 \text{ V}$$

Next, we use Ohm's Law to compute current.

$$I_P = \frac{V_P}{R} = \frac{70.7 \text{ V}}{2.7 \text{ k}\Omega} = 26.19 \text{ mA}$$

71. 45.04 V. The peak voltage (70.7 V) was previously computed. We find average voltage as follows:

$$V_{avg} = V_P \times 0.637 = 70.7 \text{ V} \times 0.637$$
$$= 45.04 \text{ V}$$

73. 20.06 mA. Peak current was previously computed as 26.19 mA. We find the instantaneous current as follows:

$$i = I_P \sin \theta = 26.19 \text{ mA} \times \sin 50°$$
$$= 26.19 \text{ mA} \times 0.766 = 20.06 \text{ mA}$$

75. 3.55 mA. First, we compute the rms input voltage.

$$V_P = \frac{V_{PP}}{2} = \frac{500 \text{ mV}}{2} = 250 \text{ mV}$$

and

$$V_{rms} = 0.7071 \times V_P = 0.7071 \times 250 \text{ mV}$$
$$= 176.78 \text{ mV}$$

We now use Ohm's Law to compute total rms current.

$$I_{T(rms)} = \frac{V_{rms}}{R_T} = \frac{176.78 \text{ mV}}{49.75 \text{ }\Omega} = 3.55 \text{ mA}$$

77. 2.15 mA. First, we find the voltage across R_4 using Ohm's Law.

$$V_{4,5(rms)} = I_{4,5(rms)} R_{4,5}$$
$$= 3.55 \text{ mA} \times 22.22 \text{ }\Omega = 78.88 \text{ mV}$$

and

$$V_{4,5(peak)} = V_{4,5(rms)} \times 1.414$$
$$= 78.88 \text{ mV} \times 1.414 = 111.54 \text{ mV}$$

Now, we use Ohm's Law to find the current through R_4.

$$I_{4(peak)} = \frac{V_{4(peak)}}{R_4} = \frac{111.54 \text{ mV}}{33 \text{ }\Omega} = 3.38 \text{ mA}$$

and

$$I_{4(avg)} = 0.637 \, I_{4(peak)}$$
$$= 0.637 \times 3.38 \text{ mA} = 2.15 \text{ mA}$$

79. 3.55 mA. This is the same as total current, which was previously computed.

81. Zero. The average current for a full cycle of a sinusoidal waveform is always zero.

83. 2.39 mA. Peak current through R_4 (3.38 mA) was computed previously. We determine rms current as follows:

$$I_{4(rms)} = I_{4(peak)} \times 0.7071$$
$$= 3.38 \text{ mA} \times 0.7071 = 2.39 \text{ mA}$$

85. False.
87. True. They both travel as pressure changes in the air.
89. speaker
91. 300,000,000
93. True.

Circuit Exploration

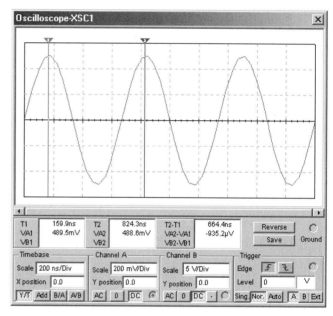

Sine wave with a frequency of 1.5 MHz and a peak amplitude of 500 mV.

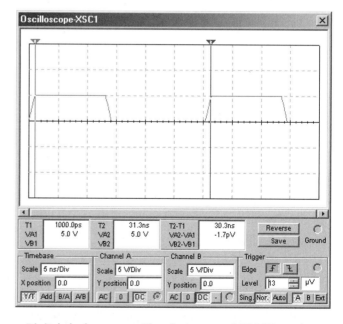

Digital clock source with a frequency of 33 MHz, a duty cycle of 40%, and voltage levels of 0 V and +5 V.

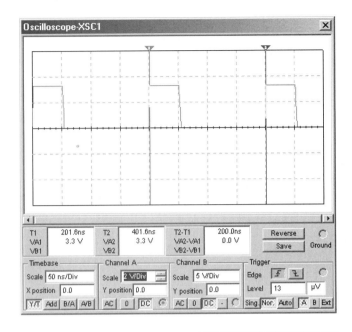

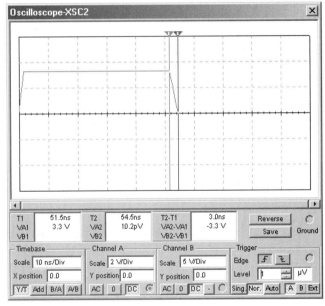

A pulsed voltage source with voltage levels of 0 V and 3.3 V, rise time of 1.5 ns, fall time of 3 ns, period of 200 ns, and a pulse width of 50 ns.

CHAPTER 7

Review Questions

1. coils or chokes
3. changes or variations
5. Ohms.
7. lags
9. True. It is interesting to note that both the resistive and inductive elements are functions of frequency. In the range of most practical interest, the inductance decreases and the resistance increases with increasing frequency.

11. 1.25 µH. We determine the total inductance of parallel-connected coils by using the reciprocal equation (as with parallel-connected resistors).

$$L_T = \frac{1}{\frac{1}{L_1} + \frac{1}{L_2} + \frac{1}{L_3} + \frac{1}{L_4}}$$

$$= \frac{1}{\frac{1}{5\ \mu H} + \frac{1}{5\ \mu H} + \frac{1}{5\ \mu H} + \frac{1}{5\ \mu H}} = 1.25\ \mu H$$

The shortcut calculations used with resistive circuits can also be applied to inductive circuits. In the present case, we could divide the common value by the number of coils to determine the total inductance.

$$L_T = \frac{L}{N} = \frac{5\ \mu H}{4} = 1.25\ \mu H$$

13. 15.79 kΩ. We compute total parallel inductive reactance as follows:

$$X_{L_T} = \frac{1}{\frac{1}{X_{L1}} + \frac{1}{X_{L2}} + \frac{1}{X_{L3}}}$$

$$= \frac{1}{\frac{1}{25\ k\Omega} + \frac{1}{100\ k\Omega} + \frac{1}{75\ k\Omega}} = 15.79\ k\Omega$$

15. Increases. Inductive reactance is directly proportional to frequency.
17. coefficient of coupling
19. step-down
21. Air-core. Air-core transformers are used for radio-frequency applications.
23. True. The turns ratios for transformers with multiple secondary windings are independent of each other. Therefore, one secondary can provide a higher voltage than the primary while a second winding provides a reduced voltage.
25. 41.67 mA. We solve the turns ratio problem as follows:

$$\frac{N_P}{N_S} = \frac{I_S}{I_P}$$

$$\frac{6}{1} = \frac{250\ mA}{I_P}$$

$$I_P = 41.67\ mA$$

27. 128 Ω. We compute the reflected impedance as follows:

$$\frac{N_P}{N_S} = \sqrt{\frac{Z_P}{Z_S}}$$

$$\frac{4}{1} = \sqrt{\frac{Z_P}{8\,\Omega}}$$

$$16 = \frac{Z_P}{8\,\Omega}$$

$$Z_P = 128\,\Omega$$

29. (c) possibly defective. A winding that measures 0.6 Ω may be good or it may have shorted turns. A standard ohmmeter does not generally provide enough resolution for low-resistance measurements to detect a few shorted turns.

31. electrostatic

33. decrease. Capacitance is directly proportional to the surface area of its plates.

35. Dielectric constant.

37. Zero. A capacitor acts like an open circuit to dc.

39. 7.49 kΩ. We compute capacitive reactance as follows:

$$X_C = \frac{1}{2\pi f C} = \frac{1}{2 \times 3.14 \times 850\text{ kHz} \times 25\text{ pF}}$$

$$= 7.49\text{ k}\Omega$$

41. The voltage rating of a capacitor indicates the maximum voltage that can be safely applied to the capacitor without risking damage to the dielectric. Greater voltage may cause the dielectric to break down, which means the capacitor will act like a short circuit.

43. 49.1 pF. We combine series-connected capacitors using the reciprocal formula.

$$C_T = \frac{1}{\frac{1}{C_1} + \frac{1}{C_2} + \frac{1}{C_3}}$$

$$= \frac{1}{\frac{1}{100\text{ pF}} + \frac{1}{270\text{ pF}} + \frac{1}{150\text{ pF}}} = 49.1\text{ pF}$$

45. 58.72 Ω. We compute total parallel reactance as follows:

$$X_{CT} = \frac{1}{\frac{1}{X_{C1}} + \frac{1}{X_{C2}} + \frac{1}{X_{C3}}}$$

$$= \frac{1}{\frac{1}{100\,\Omega} + \frac{1}{250\,\Omega} + \frac{1}{330\,\Omega}} = 58.72\,\Omega$$

47. Series. Regardless of how they are connected, each individual capacitor will have the same capacitive reactance. But, when reactances are connected in series, they add to produce a greater total.

49. A shorted capacitor will indicate near zero ohms on an ohmmeter.

51. In phase. Resistor current and resistor voltage are always in phase regardless of the connection or application.

53. Current lags voltage by 90°. Inductive current always lags inductive voltage by 90° regardless of the connection or application.

55. 901.39 mV. We combine the individual voltage drops using phasor addition.

$$V_A = \sqrt{V_R^2 + V_C^2} = \sqrt{(750\text{ mV})^2 + (500\text{ mV})^2}$$

$$= 901.39\text{ mV}$$

57. 5 V. All components in a parallel circuit have identical voltages.

59. 50.99 μA. We add the branch currents using phasor addition.

$$I_T = \sqrt{I_R^2 + I_C^2} = \sqrt{(10\text{ μA})^2 + (50\text{ μA})^2}$$

$$= 50.99\text{ μA}$$

61. 61 kΩ. We sum the resistance and reactance with phasor addition.

$$Z = \sqrt{R^2 + X_L^2} = \sqrt{(50\text{ k}\Omega)^2 + (35\text{ k}\Omega)^2}$$

$$= 61\text{ k}\Omega$$

63. 10.8 μs. First, we compute the time for one time constant.

$$\tau = RC = 1.2\text{ k}\Omega \times 1800\text{ pF} = 2.16\text{ μs}$$

Next, we multiply by five to determine the time required for five time constants, which is the time required to reach full charge.

$$\text{charge time} = 5\tau = 5 \times 2.16\text{ μs} = 10.8\text{ μs}$$

65. short; resistor

67. Low-pass filter.

69. Current lags by 90°. Current is common to all components, and the voltage across an inductor is always 90° ahead of the current through the inductor.

71. minimum. The reactive branch currents effectively cancel at resonance leaving only resistive branch current to determine total source current.

73. They remain in phase. The voltages across all components in a parallel circuit are identical. Therefore, they are continuously in phase with each other.

75. 19.38 MHz. We compute the resonant frequency as follows:

$$f_R = \frac{1}{2\pi\sqrt{LC}} = \frac{1}{2 \times 3.14 \times \sqrt{2.5\text{ μH} \times 27\text{ pF}}}$$

$$= 19.38\text{ MHz}$$

77. If the RLC circuit had poor selectivity, it would not be able to select one station and reject other stations on nearby frequencies. The net result is that you would receive multiple stations at the same time.

79. parallel

◆ Circuit Exploration

Solutions for Figure 7–73:

1. See Figure ANS7–73.
2. The cutoff frequency (measured) is near 16 kHz (15.9 kHz calculated). The cutoff frequency is calculated as the frequency where $R = X_C$.

$$f_{\text{CUTOFF}} = \frac{1}{2\pi RC} = \frac{1}{2 \times \pi \times 100\ \Omega \times 0.1\ \mu F}$$
$$= 15.9\ \text{kHz}$$

3. n/a
4. This is a high-pass filter, since frequencies higher than the cutoff frequency are allowed to pass with minimal attenuation, but lower frequencies are attenuated.

Solutions for Figure 7–74:

1. See Figure ANS7–74.
2. The cutoff frequency (measured) is near 165 Hz (159 Hz calculated). The cutoff frequency is calculated as the frequency where $R = X_L$.

$$f_{\text{CUTOFF}} = \frac{R}{2\pi L} = \frac{100\ \Omega}{2 \times 3.14 \times 100\ \text{mH}}$$
$$= 159.2\ \text{Hz}$$

3. n/a
4. This is a low-pass filter, since frequencies lower than the cutoff frequency are allowed to pass with minimal attenuation, but higher frequencies are attenuated.

Solutions for Figure 7–75:

1. See Figure ANS7–75.
2. The cutoff frequency (measured) is near 20.6 kHz (20.8 kHz calculated). The cutoff frequency is calculated as the frequency where $R = X_C$.

$$f_{\text{CUTOFF}} = \frac{1}{2\pi RC} = \frac{1}{2 \times \pi \times 150\ \Omega \times 0.051\ \mu F}$$
$$= 20.8\ \text{kHz}$$

3. n/a
4. This is a low-pass filter, since frequencies lower than the cutoff frequency are allowed to pass with minimal attenuation, but higher frequencies are attenuated.

Solutions for Figure 7–76:

1. See Figure ANS7–76A

FIGURE ANS7–73

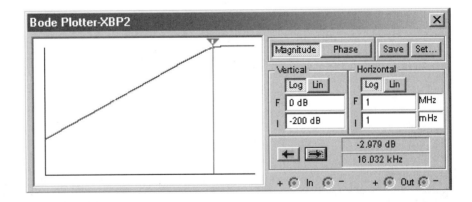

FIGURE ANS7–74

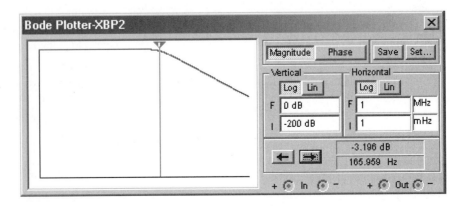

FIGURE ANS7–75

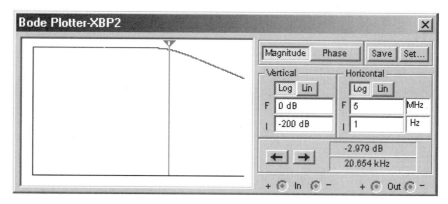

FIGURE ANS7–76A

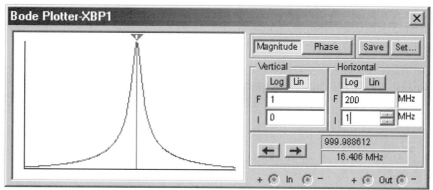

2. The resonant frequency is 16.4 MHz (measured). We can calculate the resonant frequency as follows:

$$f_R = \frac{1}{2\pi\sqrt{LC}} = \frac{1}{2 \times 3.14 \times \sqrt{2.0\ \mu H \times 47\ pF}}$$

$$= 16.4\ \text{MHz}$$

3. See Figures ANS7–76B and ANS7–76C.
4. This is a bandpass filter, since only those frequencies near resonance are passed without significant attenuation. Both higher and lower frequencies are attenuated.

FIGURE ANS7–76B Series resistance increased to 510 Ω.

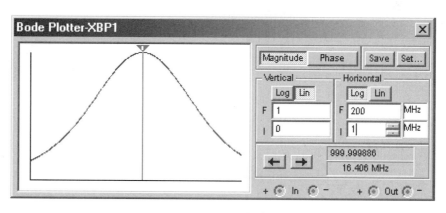

FIGURE ANS7–76C Series resistance reduced to 5.1 Ω.

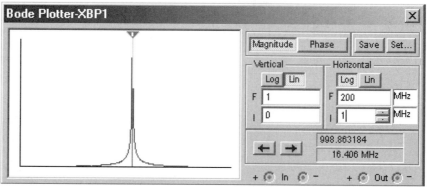

Solutions for Figure 7–77:

1. See Figure ANS7–77A
2. The resonant frequency is 173.8 kHz (measured). We can calculate the resonant frequency as follows:

$$f_R = \frac{1}{2\pi\sqrt{LC}} = \frac{1}{2 \times 3.14 \times \sqrt{1.5 \text{ mH} \times 560 \text{ pF}}}$$
$$= 173.7 \text{ kHz}$$

3. See Figures ANS7–77B and ANS7–77C.
4. This is a band-reject filter (also called a band-stop filter), since only those frequencies near resonance are attenuated significantly. Both higher and lower frequencies are passed with minimal attenuation.

CHAPTER 8

●━Review Questions

1. infinite
3. zero
5. True.
7. False. Neutrons are in the nucleus.
9. valence
11. False. Electrons must escape the valence band to become free electrons.
13. positive
15. False. Neutrons are neutral and have no charge at all.
17. True.
19. Covalent bonds

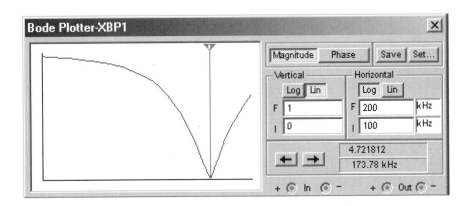

FIGURE ANS7–77A

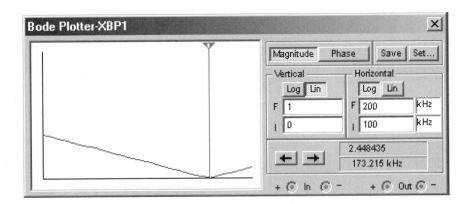

FIGURE ANS7–77B Series resistance increased to 5.1 kΩ.

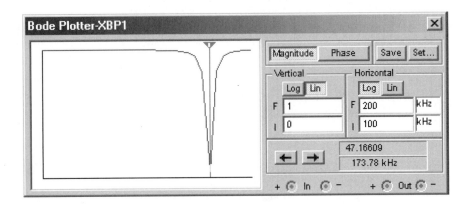

FIGURE ANS7–77C Series resistance decreased to 51 Ω.

21. hole
23. doping
25. trivalent
27. acceptor
29. p-type
31. False. P-type material always has more valence-band holes than conduction-band electrons.
33. False. There will always be more conduction-band electrons than valence-band holes in n-type material.
35. n-type
37. False.
39. True. It is called a depletion region because it is depleted of current carriers.
41. forward
43. True.
45. True.
47. False. The avalanche process does not inherently damage a pn junction. However, the resulting power dissipation does frequently destroy the junction. If the device is capable of dissipating the heat (e.g., a zener diode) or has a high resistance in series with it, then no damage will occur during avalanche.
49. True. A forward-biased junction cannot drop more than about 1 V without being damaged.
51. True. A good pn junction should have at least a 1:10 front-to-back ratio (higher in most cases).

• Circuit Exploration

See Crossword Puzzle below.

CHAPTER 9

• Review Questions

1. p-type
3. −10.7 V. The cathode must be more negative than the anode in order to be forward biased. Additionally, the forward voltage drop of a rectifier diode is on the order of 0.7 V.
5. An ideal forward-biased diode acts as a short circuit.
7. Yes. If the reverse voltage exceeds the reverse breakdown voltage, the diode can no longer stop current.

Crossword Puzzle:
- 1 Across: TRIVALENT
- 5 Across: DEPLETION
- 6 Across: DOPE
- 8 Across: PENTAVALENT
- 11 Across: HOLE
- 14 Across: PAIR
- 15 Across: COVALENT
- 16 Across: MAJORITY
- 20 Across: RECOMBINATION
- 2 Down: VALENCE
- 3 Down: LEE (LE...)
- 4 Down: GATE (GA...)
- 7 Down: EXTRINSIC
- 9 Down: ATOM
- 10 Down: ACCEPTOR
- 12 Down: LEA
- 13 Down: BINS
- 15 Down: COLLOID (COLLO...)
- 17 Down: JI...
- 19 Down: P

9. The ideal diode has no forward voltage drop, but a practical diode drops about 0.7 V when forward biased.

11. 18 V (ideal) or 17.3 V (with practical diode)

13. 50 V (ideal) and 49.3 (with practical diode)

15. False. The diodes conduct on opposite alternations of the input voltage.

17. Four.

19. 25 V (ideal) and 23.6 V (with practical diodes). Since two diodes are conducting at the same time, there are two forward voltage drops (0.7 V × 2 = 1.4 V). The peak output voltage will be 1.4 V less than the peak input voltage, if practical diodes are considered.

21. If the peak input voltage is 2 V, the diode never becomes forward biased. Therefore, the output will look identical to the input.

23. If the peak input voltage is 2 V, the diode never becomes forward biased. Therefore, the output will look identical to the input.

25. 0 V / −5 V   With a peak input voltage of 2 V, the diode will never be reverse biased. Therefore, the battery will be effectively connected across the output, so the output will be a constant −5-V level.

27. Diode D_1 affects the positive peaks. We want these to be clipped at +30 V, so we connect +30 V to the cathode of D_1. Similarly, D_2 affects the negative peaks, which should be clipped at −40 V. Therefore, V_{B_2} is oriented with the negative terminal on the anode of D_2. See Figure ANS9–27.

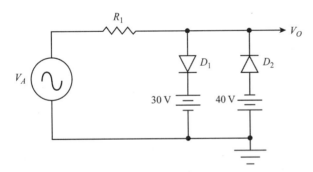

FIGURE ANS9–27

29. Zener diode.
31. reverse-biased
33. Varactor.
35. PIN diode
37. Step-recovery diode.
39. Voltage regulation.
41. True.

43. True. Good diodes normally have at least a 1:10 front-to-back ratio and generally have ratios in excess of 1:100.

45. It indicates that the diode is defective. The normal forward voltage drop is near 0.7 V.

47. True. There are other things that can cause these symptoms, but if a component is shorted, it will likely cause these effects.

Circuit Exploration

1. The waveform across the capacitor (and resistor) is essentially a straight line. That is, the capacitor charges to a 50-Vdc level and stays there with negligible discharge.

   The waveform across the diode is a sine wave whose positive peaks are at 0 V (≈ +0.7 V for practical diodes). The amplitude and frequency of the output sine wave is identical to the input waveform.

2. Figure ANS9–2 is an illustration of the circuit in the MultiSIM environment.

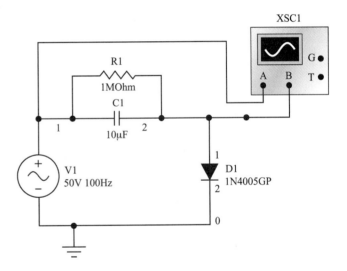

FIGURE ANS9–2

3. See Figure ANS9–3A. The upper waveform is the input voltage. The lower waveform is the output voltage (i.e., voltage across the diode). Note the positive output peaks go to +0.54 V (i.e., ≈ 0.7 V). Also note the input and output amplitudes are identical.

   Figure ANS9–3B shows the waveform (49.4-Vdc level) across the parallel RC network.

4. This is an important step if there were discrepancies.

5. During the first positive alternation, C_1 charges to the peak of the input voltage (less the forward voltage drop of the diode). From that point on, the capacitor remains charged and essentially acts as a dc source in series with the ac source. The diode is reverse biased at all times except on the positive-most peak of the input waveform. For just a few degrees on the positive peak, the diode

c. The positive peaks are clamped to a 0-V (or +0.7-V) level. The other characteristics of the waveform are unaltered. In effect, the dc level of the input waveform is shifted. If the diode were inverted, then the negative peaks would be clamped to 0 V. If the diode were returned to a dc voltage source instead of ground, then the peaks would be clamped to the level established by the dc voltage source.

CHAPTER 10

Review Questions

1. The operation of bipolar transistors relies on two types of current carriers: holes and electrons.
3. True.
5. True.
7. Cutoff.
9. Cutoff. Both base and emitter will be at zero volts, so the base-emitter junction will not be forward biased.
11. True. With −5 V at the input, the base-emitter junction will be forward biased. If the value of R_B is sufficiently low, the transistor will be saturated.
13. 0.7 V is a good approximation.
15. We convert the voltage gain ratio to dB as follows:

$$A_V(\text{dB}) = 20 \log \frac{\text{output voltage}}{\text{input voltage}} = 20 \log 80$$
$$= 20 \times 1.9 = 38 \text{ dB}$$

17. We convert the current gain ratio to dB as follows:

$$A_I(\text{dB}) = 20 \log \frac{\text{output current}}{\text{input current}} = 20 \log 90$$
$$= 20 \times 1.95 = 39 \text{ dB}$$

19. We convert the power gain ratio to dB as follows:

$$A_P(\text{dB}) = 10 \log \frac{\text{output power}}{\text{input power}} = 10 \log 250$$
$$= 10 \times 2.398 = 23.98 \text{ dB}$$

21. Zero. The input and output signals are in phase (for most practical purposes). It is interesting to note that at higher frequencies parasitic capacitances in the transistor and the circuit cause some degree of phase shift in the amplifier.
23. Zero. An emitter-follower circuit is the same as a common-collector circuit.
25. Common-collector (emitter-follower)
27. False. A JFET can also be used for digital applications.
29. True.
31. reverse

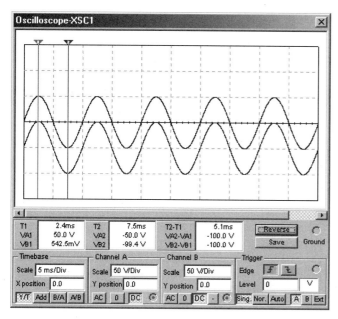

FIGURE ANS9–3A

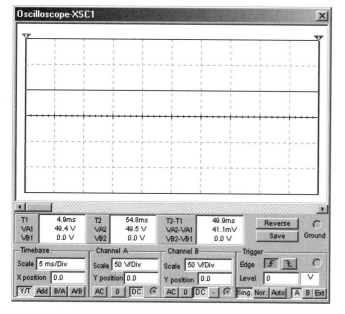

FIGURE ANS9–3B

conducts and replenishes the small amount of charge that C_1 has lost through R_1.

6. a. If the input voltage is doubled, the positive peaks will still be clamped to 0 V (ideally) or about +0.7 V (for a practical diode). The negative peaks of the output will be at −100 V (ideally), and the output waveform will still be sinusoidal.
   b. If the input were a square wave, then the output would also be a square wave with the same amplitude, shape, and frequency. The positive extremes of the output waveform would be clamped to 0 V (or about +0.7 V).

33. True. They have less recombination noise to interfere with amplification of the desired signal.

35. They are n-channel depletion-mode, p-channel depletion-mode, n-channel enhancement-mode, p-channel enhancement-mode.

37. Yes. Yes. Yes. A depletion-mode MOSFET is on with no bias. The source-to-drain current can be increased and decreased by making the gate positive and negative, respectively, with respect to the source. A linear amplifier application could, for example, be biased at 0 V for quiescent operation. The input signal would then cause the gate to go positive and negative with respect to the source as the gate voltage went above and below the 0-V reference level.

39. An enhancement-mode MOSFET (either p- or n-channel) has a broken line in the schematic symbol to represent the channel.

41. High-input resistance (same as saying low-input current) and low-noise operation.

43. To prevent damage from ESD. The insulation between the gate and channel in a MOSFET is a very thin oxide layer. The metal-oxide-semiconductor sandwich forms a capacitor, but since the oxide is so thin, the breakdown voltage of the effective capacitor is quite low. ESD, on the other hand, can be tens of thousands of volts. To prevent damage the leads are shipped with the leads electrically connected. That way, no difference of potential can exist between the gate and the channel, and the gate oxide will not be damaged.

45. two. In practical applications, these two states may represent such things as off/on, low/high, slow/fast, 0 V/5 V, −12 V/+12 V, and so on. In all cases, however, there are only two states used to represent information or control signals.

47. False. Most modern integrated circuits utilize MOSFETs either solely or in conjunction with another technology.

49. dc level. That is, there would be no ac signal in the output. An ideal difference amplifier only responds to the difference in voltage on its two inputs. If the inputs have identical sine waves, there will be no difference. Therefore, the output will remain at the quiescent dc level.

51. False. The forward-bias polarity is limited to about 0.7 V or so, but only the breakdown voltage of the base-emitter junction limits the reverse-bias polarity.

53. C_3 open. Capacitors act as an open circuit to dc under normal conditions, so an open coupling capacitor won't affect dc circuit values. Since the signal is normal at TP3 but absent from TP6, we can infer that the coupling capacitor is probably open. At first glance, you may think that a shorted R_6 would cause similar symptoms, but if R_6 were shorted, the ac signal at TP3 would be very reduced since the coupling action of C_3 would effectively place ground (for ac purposes) at TP3.

55. Defective transistor. The transistor has 2.55 V of forward bias! Anything much greater than 0.7 V to 1 V is a definite indication of a defective transistor.

●—Circuit Exploration
Defect 1: C_4 open

An open capacitor in a linear amplifier circuit has no effect on the dc voltages since a capacitor is normally open to dc. The ac values on TP1, TP2, and TP4 are relatively normal, but all points from the collector of Q_1 to the output are altered. Ordinarily the ac voltage on TP6 is fairly small (0.32 mV) due to the low reactance (15.9 Ω) of C_4. When it is open, however, the voltage at TP6 increases by nearly a factor of 4. As a general rule, an open capacitor that normally bypasses the emitter resistor of a common-emitter amplifier will cause the amplifier to have normal dc voltages, but the ac gain will be dramatically reduced. In the present case, the voltage gain was 117 (V_{TP7}/V_{TP5}) under normal circuit conditions, but less then 1 with C_4 open.

Defect 2: R_7 open

Since some of the dc voltages are dramatically altered, it is unlikely that the defect is an open capacitor. Further, since the dc voltages associated with the first amplifier stage Q_1 are normal, the problem most likely lies in the second stage. R_7 and R_8 form a voltage divider to set the base voltage for Q_2. Since the base voltage is 24 V with the defect, we can infer that the voltage divider action is not working properly. Either a shorted R_8 or an open R_7 could cause the base voltage on Q_2 to rise to 24 V. However, if R_8 were shorted, then the ac voltage at TP5 would be lower than normal (near zero). So, we infer that an open R_7 could cause these symptoms.

Defect 3: C_1 open

Since all dc voltages throughout the circuit are normal, we will initially suspect an open capacitor. Further, since the ac signal is present at TP1 but fails to couple to TP2, our focus immediately goes to an open in C_1.

Defect 4: Q_2 defective

Since the dc and ac voltages are relatively normal up to and including the base of Q_2 (although the ac voltages at TP3 and TP5 are slightly higher than normal), we can infer that the problem most likely lies beyond TP5. A closer examination of the dc voltage at TP5 and TP6 reveals that the transistor (Q_2) has 8 V of forward bias on the base-emitter junction. This is surely a defective transistor.

Defect 5: C_2 open

Since all dc voltages throughout the circuit are normal, we will initially suspect an open capacitor. Further, since the ac signal is present at TP3 but fails to couple to TP5, our focus immediately goes to an open in C_2.

Defect 6: R_2 open

The dc voltages are upset for Q_1 but normal for Q_2, so we will concentrate on the Q_1 portion of the circuit. Q_1 is either defective or it is operating in cutoff, since its collector-emitter voltage is the full supply voltage (24 V). The bias for the base-emitter of Q_1 is provided by R_2. If R_2 were open, the transistor would have no forward bias on the base-emitter junction, and the transistor would be in cutoff. These are the symptoms we have, so we identify R_2 as the defective component. You might also notice that the ac voltage at TP2 is higher than normal. This is because the transistor base-emitter junction provides less loading (i.e., acts as a higher resistance) when it is not forward biased.

Defect 7: C_3 open

Since all dc voltages throughout the circuit are normal, we will initially suspect an open capacitor. Further, since the ac signal is present (and relatively normal) at all points up to TP7 but fails to couple to TP8, our focus immediately goes to an open in C_3. You should also notice a slight rise in the ac voltage at TP7, since R_9 no longer loads the circuit.

CHAPTER 11

Review Questions

1. True. The errors caused by nonideal characteristics are not generally significant enough in a well-designed circuit to cause wrong decisions to be made during a troubleshooting exercise.

3. False. Any op amp can be operated from a single supply, and some op amps are specifically intended to be operated from a single supply.

5. True.

7. An op amp with a MOSFET input stage draws less current than either JFET or bipolar inputs. Bias currents of a few femtoamperes are possible with MOSFET inputs.

9. Negative. Without negative feedback, the voltage gain of an op amp is so high that it is unusable for practical purposes. Additionally, the voltage gain is unstable, varies with frequency, and varies between similar op amps. Further, the bandwidth of an op amp without negative feedback is often only a few hertz.

11. False. The voltage gain of an open-loop op amp is very high but neither constant nor consistent between similar devices.

13. True. The frequency response for most general-purpose op amps extends to include dc.

15. An ideal op amp has an infinite slew rate. In other words, there is no limit to the rate of change on the output of an ideal op amp.

17. (+). The noninverting input is identified by a plus symbol.

19. a noninverting. This can be determined by noting that the input is applied to the (+) input terminal.

21. No effect. Resistor R_2 is effectively in series with the input resistance of the op amp, which is extremely high. Therefore, the small resistance change will have no effect on overall voltage gain. This result can also be inferred by inspection of the equation for voltage gain in a noninverting amplifier circuit.

23. False. Since the amplifier uses negative feedback, we know that the (−) and (+) inputs will have essentially the same voltage. That is, the differential input voltage will be near zero. This means the (−) input will be following the input voltage waveform just as the (+) input does.

25. As the input signal voltage is increased, the output voltage tries to increase correspondingly ($v_o = v_i \times A_V$). However, the output of the op amp cannot increase beyond the output saturation voltages. Therefore, any input voltage that tries to make the output voltage go beyond one of the saturation voltages will cause clipping in the output waveform.

27. We compute voltage gain for a noninverting amplifier as follows:

$$A_V = \frac{R_F}{R_1} + 1 = \frac{68\ \text{k}\Omega}{1.2\ \text{k}\Omega} + 1 \approx 57.7$$

29. Since the differential input voltage must always be near zero, and since the (+) input is connected directly to the input signal, we will expect to see a 100-mV$_{PK}$ sine wave at the (−) input terminal. That is, it will look just like the input voltage.

31. an inverting.

33. Resistor R_3 compensates for the output voltage offset that would result from the voltage drop across R_1 and R_2 caused by the input bias current. The input bias current on the (+) input causes an opposite and approximately equal effect in the output as a result of the voltage drop across R_3.

35. True. Since the (+) input is essentially grounded (no significant voltage drop across R_3) and since the differential input voltage must be near zero, the (−) input must remain near ground potential.

37. Unity.

39. False. The (−) input will have the same voltage (v_i) as the (+) input, since the differential input voltage must always be near zero.

41. The purpose of the circuit is to produce an output voltage that is proportional to the input current. It is called a current-to-voltage converter.

43. 5 V. Since the left end of R_1 is grounded (virtual ground) and the right side is connected to the output, it follows

that the output voltage is equal to the voltage drop across R_1. We compute it as follows:

$$v_o = i_i \times R_1 = 50 \text{ mA} \times 100 \text{ }\Omega = 5 \text{ V}$$

45. High-pass filter.
47. Zero. The capacitors will be essentially open at very low frequencies (including dc), so the input signal cannot reach the (+) terminal of the op amp.
49. Yes. If the (+) input is more positive than the (−) input, the output should go positive. Since the output is actually negative and the power supplies are good, the op amp is probably defective.
51. ac. R_1 is isolated for dc purposes by C_1 and C_2, so shorting it will not affect the dc voltage on the output of the op amp.
53. No effect. The dc output voltage on U_2 has no effect on the dc output voltage of U_3, since the two stages are isolated for dc purposes by C_1 and C_2.
55. See Figure ANS11–1.

●—Circuit Exploration

The circuit consists of two major functional blocks that were discussed in this chapter: a voltage comparator and an integrator. The functional block diagram is shown in Figure ANS11–2.

The general timing diagram for this circuit is shown in Figure ANS11–3.

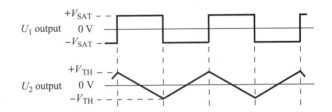

FIGURE ANS11–3

where $\pm V_{SAT}$ are the maximum allowable excursions for the output of U_1, and $\pm V_{TH}$ are the positive and negative threshold voltages for the comparator. The exact values for $\pm V_{SAT}$

FIGURE ANS11–1

Defect	Explanation
C_2 open	C_2 is always open to dc, so this defect does not change the normal dc voltages in U_3.
C_1 open	C_1 is always open to dc, so this defect does not change the normal dc voltages in U_3.
R_1 shorted	C_1 and C_2 isolate U_3 from any dc defects in stage U_1 or U_2, so this defect will not alter the dc voltages in U_3.
R_5 open	The (+) input of U_3 will have no bias current path, so the output will be driven to saturation, thus causing a dc failure mode.
U_3 defective	This will probably cause a dc failure mode in U_3.
U_2 defective	C_1 and C_2 isolate U_3 from any dc defects in stage U_1 or U_2, so this defect will not alter the dc voltages in U_3.
R_4 shorted	R_4 is isolated by C_1 and C_2, so its defects will not cause a dc failure mode in U_3.
R_4 open	R_4 is isolated by C_1 and C_2, so its defects will not cause a dc failure mode in U_3.
R_1 increased	C_1 and C_2 isolate U_3 from any dc defects in stage U_1 or U_2, so this defect will not alter the dc voltages in U_3.

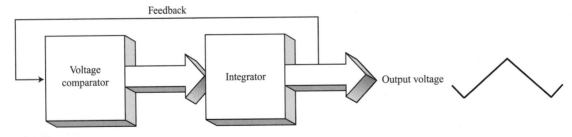

FIGURE ANS11–2

and $\pm V_{TH}$ depend on the specific op amp used to build the circuit, but values of ± 14 V for $\pm V_{SAT}$ and ± 7 V for $\pm V_{TH}$ are approximate values.

Figure ANS11–4 shows the result of a simulation exercise. The upper waveform (Channel B) is the output of U_1 and the lower waveform (Channel A) is the output of U_2. Notice that the finite slew rate of the op amp is evident in the upper waveform. In this particular simulation, the saturation and threshold voltages were ± 14.1 V and ± 7.2 V, respectively.

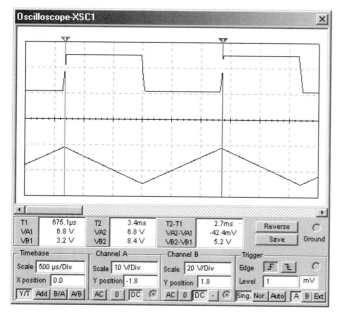

FIGURE ANS11–4

In short, the output of the voltage comparator provides the input to the integrator circuit. Similarly, the output of the integrator serves as the input to the voltage comparator. A circuit (such as this) that provides its own input is called an oscillator. We will discuss several types of oscillators in a later chapter. The voltage comparator is a noninverting circuit, which is unfamiliar to you, but you can recognize its general behavior by the presence of the positive feedback.

CHAPTER 12

Review Questions

1. The main purpose of a voltage-regulator circuit is to maintain a constant output voltage even though the input voltage and the load current may vary.

3. Load regulation.

5. 0.09%. We compute line regulation as follows:

% line regulation
$$= \frac{V_{REG}(\max) - V_{REG}(\min)}{V_{LINE}(\max) - V_{LINE}(\min)} \times 100$$
$$= \frac{15.059 \text{ V} - 15.05 \text{ V}}{125 \text{ V} - 115 \text{ V}} \times 100 = 0.09\%$$

7. 0.91%. We compute load regulation as follows:

% load regulation
$$= \frac{V_{REG}(\text{min load}) - V_{REG}(\text{full load})}{V_{REG}(\text{full load})} \times 100$$
$$= \frac{3.32 \text{ V} - 3.29 \text{ V}}{3.29 \text{ V}} \times 100 = 0.91\%$$

9. D_1 is a zener diode, and it serves as the reference voltage for the series voltage regulator. A sample of the output voltage (feedback) is continuously compared with the stable reference voltage in order to maintain the correct output voltage.

11. 5.0 V. The base-emitter junction of Q_1 drops approximately 0.7 V, so the emitter voltage (which is the voltage across the load) can be found with Kirchhoff's Voltage Law as 5.7 V − 0.7 V = 5.0 V.

13. Resistors R_2 and R_3 form a voltage divider across the load voltage. The portion of the voltage that is across R_3 serves as feedback for the regulator circuit. You might also view the purpose of the voltage-divider circuit as the gain-setting components for U_1, which therefore determine the value of the output voltage.

15. 0 V. Ideally, the regulator circuit should be able to compensate for variations in the input voltage and maintain a constant output voltage. Therefore, the ideal variation in output voltage would be zero.

17. 12.98 V. We compute the output voltage of a three-terminal regulator with external resistors as follows:

$$V_O = V_{REF}\left(\frac{R_2}{R_1} + 1\right) = 5 \text{ V}\left(\frac{750 \text{ }\Omega}{470 \text{ }\Omega} + 1\right)$$
$$= 12.98 \text{ V}$$

19. False. The voltage drop across the regulator transistor in a shunt regulator is identical to the load voltage, since the two are in parallel.

21. 13 V. According to Kirchhoff's Voltage Law, R_1 will drop whatever part of the input voltage that is not dropped across the load. We compute it as follows:

$$V_{R1} = V_{IN} - V_{LOAD}$$
$$= 24 \text{ V} - 11 \text{ V} = 13 \text{ V}$$

23. Remains the same. The purpose of the voltage-regulator circuit is to maintain a constant load voltage in spite of changes in input voltage. So, if the load voltage doesn't change and the load resistance doesn't change, then according to Ohm's Law, the load current must remain constant.

25. Increase. If R_3 were made larger, it would drop a greater percentage of the load voltage leaving less for R_4. This means the voltage at the (−) input of U_1 would decrease.

This would cause the output of U_1 to go more positive, which reduces the forward bias on Q_1 and also reduces its current flow. With less current through Q_1, there will be less voltage dropped across R_1 causing an increase in load voltage. You could also view Q_1 as an emitter follower; when the base voltage goes more positive, so does the emitter voltage (which is the same as the load voltage).

27. False. The series and shunt regulators discussed in this chapter are examples of linear regulators, but a switching regulator is not a linear regulator.

29. In order to achieve efficient operation, the switching transistor transitions between states as fast as possible. High transition rates on the rise and fall time in the time domain correspond directly to high-energy, high-frequency harmonics in the frequency domain.

31. up; boost

33. True.

35. Positive. When the magnetic field of L_1 collapses, electron current flows counterclockwise from L_1, through D_1 and C_1, and back to L_1. This charges C_1 with a positive on the upper terminal.

37. Increase. If resistor R_2 is made smaller, then there will be less feedback. This will have the same effect on the control circuit as if the actual output voltage decreased. The response of the control circuit will be to increase the output voltage until the feedback voltage returns to its initial value. Since R_2 is smaller, it drops a smaller percentage of the load voltage. Therefore, the load voltage will have to increase above its initial value in order to maintain the same value of feedback voltage.

39. True. When Q_1 turns off, the magnetic field around L_1 collapses. The resulting induced voltage in L_1 causes current to flow through diode D_1.

41. the same as

43. Increase. If R_1 is made larger, the feedback voltage will be smaller. The control circuit will sense the reduced feedback voltage and increase the output voltage to compensate.

45. Overcurrent protection. When the output of a power supply is shorted, the current can be very high. Components within the power supply can be damaged.

47. Overvoltage protection.

49. False. In many cases, current sensing and limiting is a built-in feature of integrated-switching regulator-control circuits.

51. False. Surprisingly, this is a common misconception perhaps because every system utilizes a power supply circuit, so they are perceived as "nothing special." In fact, some power supply circuits can be very challenging to troubleshoot to a component level. Much of the difficulty is a result of the closed-loop nature of regulator designs.

53. If the input voltage is zero (perhaps because of a blown fuse or defective switch), then the output of the power supply will also be zero. A defect (specifically a short) in the load circuitry can also cause the output voltage of a power supply to be zero. This latter case is true even if the supply is good and the input voltage is normal. As long as the power supply has electronic current-limiting protection, there may be no damage to the power circuits.

55. True.

Circuit Exploration

1. Your exact values may vary from those shown in Figure ANS12–1, but they should be fairly close. Most importantly, it should be apparent to you that the values for Load 1 vary considerably, since it is an unregulated output voltage. The output voltage for Load 2, by contrast, should remain fairly constant because it is regulated.

Load 1 and Load 2 Resistance (ohms)	Load Voltage (Volts)	
	Load 1	Load 2
1000	9.6	9.9
900	9.6	9.9
800	9.6	9.9
700	9.5	9.9
600	9.5	9.9
500	9.4	9.9
400	9.3	9.9
300	9.2	9.9
200	8.8	9.9
100	8.0	9.9

FIGURE ANS12–1

2. We compute load regulation for each output as follows:

$$\% \text{ load regulation (Load 1)}$$
$$= \frac{V_{REG}(\text{min load}) - V_{REG}(\text{full load})}{V_{REG}(\text{full load})} \times 100$$
$$= \frac{9.6 \text{ V} - 8.0 \text{ V}}{8.0 \text{ V}} \times 100 = 20\%$$

$$\% \text{ load regulation (Load 2)}$$
$$= \frac{V_{REG}(\text{min load}) - V_{REG}(\text{full load})}{V_{REG}(\text{full load})} \times 100$$
$$= \frac{9.9 \text{ V} - 9.9 \text{ V}}{9.9 \text{ V}} \times 100 = 0\%$$

The load regulation percentage for Load 2 is ideal in this particular example, since the output voltage remained constant. If you experience this situation in a practical application, you can do one of two things: Either measure the output voltage with greater precision or use a wider range of load values.

3. No response required.

4. Your exact values may vary from those shown in Figure ANS12–4, but they should be fairly close. Most importantly, it should be apparent to you that the values for Load 1 vary considerably, since it is an unregulated output voltage. The output voltage for Load 2, by contrast, should remain fairly constant because it is regulated. In order to compute more meaningful values, we have included three decimal places for the measurements on Load 2.

Input Voltage (Volts)	Load Voltage	
	Load 1	Load 2
36	14.4	9.887
32	12.8	9.877
28	11.2	9.865
24	9.6	9.851
20	8.0	9.835
16	6.4	9.814

FIGURE ANS12–4

5. We compute line regulation for the two outputs as follows:

$$\% \text{ line regulation}$$
$$= \frac{V_{REG}(\max) - V_{REG}(\min)}{V_{LINE}(\max) - V_{LINE}(\min)} \times 100$$
$$= \frac{14.4 \text{ V} - 6.4 \text{ V}}{36 \text{ V} - 16 \text{ V}} \times 100 = 40\%$$

$$\% \text{ line regulation}$$
$$= \frac{V_{REG}(\max) - V_{REG}(\min)}{V_{LINE}(\max) - V_{LINE}(\min)} \times 100$$
$$= \frac{9.887 \text{ V} - 9.814 \text{ V}}{36 \text{ V} - 16 \text{ V}} \times 100 = 0.365\%$$

CHAPTER 13

Review Questions

1. True. Some thyristors can switch thousands of amperes.
3. Forward breakover voltage. If you exceed the forward breakover voltage on a thyristor, it quickly switches to the on state.
5. False. When a thyristor is in the on state, the voltage across it remains fairly constant (1 to 2 V) regardless of the value of current through the device.
7. unidirectional; bidirectional
9. b. Once the thyristor is fully on, it drops 1 to 2 V in most cases.
11. b. Reverse breakdown is likely to damage a thyristor because the voltage across it remains high at the same time the current increases to a high level. The resulting power dissipation can destroy the thyristor.
13. holding
15. False. If the primary terminals connect through the load to a dc source, the thyristor will remain on once it has been fired. The gate voltage cannot control current; it merely determines when the thyristor will fire. Further, with a dc source, once the device has fired, it will remain in the conducting state until the primary terminal voltage is interrupted or the primary current is reduced to below the holding current.
17. bidirectional
19. bidirectional
21. True. The gate is often left unconnected.
23. An SBS has a lower breakover voltage. Therefore, the device can be fired earlier and later in the cycle than a diac.
25. unidirectional
27. on; off. The gate of an SCR loses control once the device has fired.
29. True. Phase control is a major class of applications for triacs.
31. It gets dimmer. When the thyristor is fired later in the cycle, the average current through the load will be less. Therefore, the brightness of the lamp will decrease.
33. One.
35. Triacs are bidirectional devices, so full load control can be obtained with a simpler circuit.
37. An SCS can be fired by a positive on the cathode gate relative to the cathode or by a negative on the anode gate relative to the anode.
39. True. A defective thyristor or thyristor circuit normally causes the thyristor to be either fully on or fully off at all times.
41. With a short between MT_1 and MT_2, the load would be connected directly across the power source. Thus, it would have maximum current flow at all times.
43. True. Assuming the circuit is properly designed and properly constructed, the load should receive no power with the thyristor gate disconnected. Since the load has maximum current, the thyristor is probably shorted.
45. The thyristor in a phase-control application switches quickly from off to on every alternation. The fast edges

associated with the current and voltage waveforms result in the generation of high-frequency harmonics. Some of these harmonics typically fall within the AM radio band, so they can easily cause interference. The actual interference may be radiated through the air, or it may travel to the radio via the power line.

47. False. Laser diodes are light emitters.

49. 1.4 to 2.4 V. The forward voltage drops of most LEDs fall within this range of voltages.

51. The voltage across the LED will remain relatively constant in the 1.4- to 2.4-V range. Any source voltage beyond this value must be dropped across a series resistor. Without the resistor, the LED would draw very high currents and would be quickly destroyed.

53. absorption

55. Population inversion.

57. False. Only a laser produces coherent light.

59. A laser may be driving the other end of the cable. When the laser light strikes your eyes, you can receive permanent eye damage.

61. True. The photovoltaic mode requires no bias. In essence, the photodiode acts as a light-dependent voltage source.

63. dark

65. cutoff

67. False. The gate of an LASCR loses control once the device has been triggered into the on state.

69. False. An optointerrupter application requires breaking of the optical path between the LED and the phototransistor. This path is sealed in an optocoupler, so it could not be used for an interrupter application.

71. index of refraction

73. Any light wave that tries to escape the core will be bent as it passes through the interface between the core and the cladding. The bending (refraction) causes the beam to return to the core of the fiber. This process works because the index of refraction is different for the core and the cladding materials.

75. Regardless of what you are using to illuminate a fiber-optic cable, it is a good practice never to look directly into the cable. You may have disconnected the wrong cable, and the one you are examining is still being driven by a laser diode that will injure your eyes.

Circuit Exploration

1. The oscilloscope display in Figure ANS13–1 shows that the waveform across the SCR is a full sine wave. That is because there is no path for gate current when the switch is open, so the SCR remains in its off state. When it is off, the SCR acts as an open circuit, so the full applied voltage is measured across the SCR.

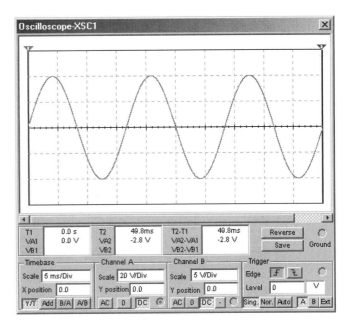

FIGURE ANS13–1

Because the SCR is open, there can be no current through the lamp, and it remains off.

2. When the switch is closed, the SCR has gate current and could trigger, except that the anode is negative with respect to the cathode. Therefore, when the switch is pressed anytime during the negative alternation, the SCR remains off, the waveform across the SCR remains at the full applied voltage (same as in step 1), and the lamp remains off.

3. When the switch is pressed during the positive alternation, the SCR immediately fires. See Figure ANS13–3. When the SCR is on, the voltage across it drops to a very low level (about 1 V). This is evident in the

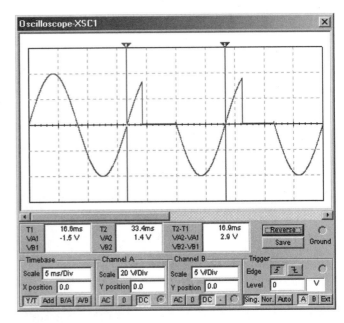

FIGURE ANS13–3

adjacent waveform. When the applied voltage enters the negative alternation, the SCR reverts to its off state, so we see the full applied voltage across the SCR. When the anode again goes positive with respect to the cathode, the SCR is capable of conducting, but it remains in its off state until it receives another trigger (i.e., the switch is pressed).

Once triggered, the SCR continues to conduct throughout the positive alternation. Therefore, if the trigger is applied earlier in the alternation, the average current (averaged over multiple cycles) through the SCR and the lamp is higher. This basic principle is what allows light dimmers, motor-speed control circuits, and other phase-controlled thyristor circuits to vary the current to their loads.

CHAPTER 14

Review Questions

1. amplification. The frequency-selective feedback path always has loss. Since the overall closed-loop gain of the circuit has to be at least unity to sustain oscillations, amplification is needed.

3. False. Triangle wave forms, sine waves, rectangular waves, and square waves as well as other shapes (e.g., parabolic ramps, sawtooth, reverse sawtooth, step followed by exponential ramp, etc.) can be directly generated by a properly designed oscillator circuit.

5. 0° or 360°. In some cases, the amplifier provides approximately 180° of phase shift, and the feedback circuit provides another 180°. In other cases, neither the amplifier nor the feedback network provides any significant phase shift to the oscillating frequency. However, in all cases, the oscillating frequency must arrive in phase after completing the loop through the oscillator circuit.

7. RC. It is the time delay associated with charging and discharging a capacitor through a resistor that determines the frequency of a relaxation oscillator.

9. True. As the current charges C_1 with a positive on the right side, the output voltage of U_1 continues to ramp in the positive direction. Eventually, it will be positive enough to cause U_2 to switch to a positive output level.

11. We estimate frequency for this relaxation-oscillator circuit as follows:

$$f = \frac{R_3}{4R_1R_2C_1}$$

$$= \frac{82\ k\Omega}{4 \times 180\ k\Omega \times 39\ k\Omega \times 0.05\ \mu F} = 58.4\ Hz$$

13. Sine wave. The ability to generate low-distortion sine waves is the primary advantage of the Wien-bridge circuit.

15. One. The overall closed-loop voltage gain must be exactly unity in order to maintain a stable sine-wave oscillation. Less voltage gain results in the oscillation decaying and dying out. More voltage gain results in a clipped output waveform.

17. dc. The sine wave output of U_1 is rectified by D_1 and then filtered into smooth dc voltage by C_2. Since R_3 is in parallel with C_2, it will have a dc voltage across it.

19. False. C_1, C_3, R_4, and R_5 determine the frequency of this circuit.

21. True. Its operation depends on the sequential charging and discharging of C_1 through resistors R_1 and R_2.

23. We can estimate the frequency of oscillation for a 555 timer circuit like the one in Figure 14–26 as follows:

$$f = \frac{1.44}{(R_1 + 2R_2)C_1}$$

$$= \frac{1.44}{(39\ k\Omega + 2 \times 56\ k\Omega) \times 2\ \mu F} = 4.77\ Hz$$

25. Decrease. If R_1 is larger, then it will take longer to charge C_1 to the required voltage. This delays switching of the internal circuitry, which lowers the frequency of oscillation.

27. piezoelectric. The piezoelectric effect produces a voltage across the crystal when it is subjected to mechanical shock. Conversely, an applied voltage mechanically distorts the crystal. If the voltage is applied at or near the natural mechanical resonant frequency of the crystal, the crystal can serve as the frequency-determining element in an oscillator circuit.

29. True. The required physical size of the crystal becomes impractical at low frequencies.

31. Replace it. Oscillator modules are not considered reparable. First, the sealed package would inherently be damaged in trying to gain access to the internal circuitry. Second, the internal circuitry is so small that ordinary tools are not useful. Finally, even if you could locate the actual defective component within the module, you could not likely buy a replacement part.

33. PLL

35. The output frequency will be higher (i.e., $f_{OUT} = n \times f_{in}$)

37. closed-loop. The output frequency is locked to the input frequency through closed-loop operation. If the output begins to drift, the feedback causes a corrective action to occur.

39. The voltage gain of an instrumentation-amplifier circuit can be changed by altering the value of a single external-gain resistor. It should also be noted that some instrumentation amplifiers have internal precision gain resistors that can be externally selected by connecting certain pins on the IC.

41. Variable dc voltage. A voltage-to-frequency converter (VFC) generates an alternating-output waveform whose frequency is proportional to the level of dc voltage at its input.

43. Capacitor C_2 determines the length of the time delay for the positive-going output pulse.

45. positive. Electrons flow from left to right through the integrating capacitor charging the capacitor such that the right side (output of the amplifier) is more positive. Since the left end is a virtual ground, this means the output voltage will be ramping in a positive direction.

47. Verifying that the supply voltage(s) are correct is a good first step. Without proper supply voltage, the oscillator cannot function correctly. Additionally, this is the only input to an oscillator circuit.

49. True. When the dc operating point shifts, the waveform is usually driven to either cutoff or saturation limits. Either of these will cause distortion unless the normal waveform is also driven to these limits.

51. False. The output of an audio amplifier at any given time is determined solely by its inputs at that same time.

•—Circuit Exploration

1. The measured frequency should be in the range of 95 to 100 Hz.

2. The high time is 6.4 ms, and the total time is 10.2 ms. We compute duty cycle as follows:

$$\text{duty cycle} = \frac{\text{high time}}{\text{cycle time}}$$
$$= \frac{6.4 \text{ ms}}{10.2 \text{ ms}} \approx 0.63 \text{ or } 63\%$$

3. The simplest way to accomplish this is to connect an oscilloscope to the two points of interest. In Figure ANS14–3, the upper waveform is the output voltage, and the lower waveform is the voltage across C_1.

4. When the supply voltage is changed, there should be no significant effect on the output frequency. This is one of the advantages of the 555 timer. Its timing is relatively independent of operating voltage.

5. The measured frequency should be in the range of 95 to 100 Hz.

6. The high time is about 7 ms, and the total time is approximately 10.4 ms. We compute duty cycle as follows:

$$\text{duty cycle} = \frac{\text{high time}}{\text{cycle time}}$$
$$= \frac{7 \text{ ms}}{10.4 \text{ ms}} \approx 0.67 \text{ or } 67\%$$

7. No response required.

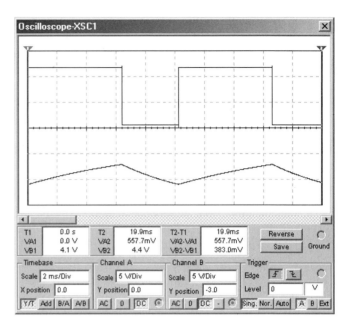

FIGURE ANS14–3

8. The high time is 2.1 ms, and the total time is 5.5 ms. We compute duty cycle as follows:

$$\text{duty cycle} = \frac{\text{high time}}{\text{cycle time}}$$
$$= \frac{2.1 \text{ ms}}{5.5 \text{ ms}} \approx 0.38 \text{ or } 38\%$$

9. The charge and discharge times for C_1 are essentially independent due to the steering action of D_1 and D_2. This allows duty cycles that are less then 50%. This is not possible with the configuration shown in Figure 14–27, since charge time is always longer than discharge time.

10. Off. The 555 output remains in the low (0 V) state indefinitely.

11. The LED turns on for about 25.5 ms (2.5 seconds if R_1 = 470 kΩ). It then goes out and remains out until the switch is pressed again.

12. The LED turns on for about 25.5 ms (2.5 seconds if R_1 = 470 kΩ). It then goes out and remains out until the switch is released and then pressed again.

13. When power is first applied, pin 2 is pulled to +12 V by R_2. Since this is more than the one-third-V_{CC} (or less) requirement to trigger the voltage comparator internal to pin 2, the internal npn transistor connected to pin 7 is on, and the output of the IC is low (0 V). With 0 V at the output, the LED is off. Because the internal transistor is connecting pin 7 to ground, C_1 is not able to charge. The circuit remains in this state until the switch is pressed.

When the switch is first closed, C_2 begins to charge through R_2 toward +12 V. However, at the first instant the capacitor has 0 V across it, which means that the voltage on pin 2 will be dropped to 0 V at the instant the switch is closed. Since this is less than one-third V_{CC}, the voltage comparator associated with pin 2 is triggered. This causes the output to go high (+12 V), which also provides power to the LED. Triggering of the internal voltage comparator also causes the internal npn transistor on pin 7 to be cutoff. This releases C_1, so it begins to charge through R_1 toward +12 V. The circuit remains in this condition while C_1 charges. In a much shorter period of time, C_2 becomes fully charged, which means that pin 2 is again higher than the one-third-V_{CC} level.

Eventually (after 25.5 ms), the voltage on C_1 reaches two-thirds V_{CC} and triggers the internal voltage comparator associated with pin 6. This causes the output to return to its low (0 V) level. It also causes the internal npn transistor connected to pin 7 to enter saturation. This quickly discharges C_1 and holds it down until pin 2 is triggered again. In order to trigger pin 2, the switch must be released momentarily. This allows C_2 to discharge (quickly) through R_3. When the switch is pressed again, the cycle repeats. Due to the action of C_2, R_2, and R_3, there is no difference in operation whether the switch is pressed momentarily or pressed and held.

CHAPTER 15

Review Questions

1. Two. The only allowable digits in the binary system are 0 and 1.
3. 13
5. 11010_2 (given), 11011_2, and 11100_2.
7. 11010_2
9. 6Ah, 29h, and 1D63h.
11. Acceptable responses include 0/1, yes/no, true/false, and on/off.
13. 2^6 or 64
15. An exclusive OR gate will have a high output anytime one and only one of its two inputs is high. See Figure ANS15–15.

    FIGURE ANS15–15

17. OR gate.
19. We break the vinculum and simplify as follows with rules 34 and 35, 23, and 17:

$$Y = \overline{\overline{\overline{ABC}} + \overline{\overline{CD}}}$$
$$= (\overline{\overline{ABC}})(\overline{\overline{CD}})$$
$$= \overline{ABCD}$$

21. True. Technically, there is a slight delay (propagation delay) between inputs and outputs on a combinational logic circuit, but for our purposes, we will ignore this.
23. A NOR gate with all of its inputs and outputs inverted can provide the same logic operation as a NAND gate, as shown in Figure ANS15–23.
25. Programmable array logic (PAL) and gate array logic (GAL).
27. The simplified output expressions follow:
    a. $Y = A\overline{B}$
    b. $Y = \overline{A} + B + \overline{C}$
    c. $Y = AB + \overline{C}D$
    d. $Y = (\overline{A} + B)(\overline{C} + \overline{D})$, or
    $= \overline{A}\,\overline{C} + \overline{A}\,\overline{D} + B\overline{C} + B\overline{D}$
29. False. The sum of the products is called the minterm form.
31. True.
33. False. The outputs are determined by the state of the inputs in conjunction with the present internal state.
35. True. According to the truth table for a D flip-flop, Q after the clock is the same as D before the clock. Therefore, Q after a given clock becomes what $\overline{Q}$ was before the clock, which is to say the flip-flop toggles every time.
37. A high level will reset or clear the flip-flop.
39. The transition indicator (>) means that the input is activated by a transition (i.e., a change from one logic level to the other). That is, the clock input is a dynamic input rather than a static input such as the set or clear inputs. The presence or absence of a logic bubble in conjunction with the transition indicator identifies a negative or positive edge of the clock, respectively.
41. True. With J and K high, the flip-flop will toggle every time a proper clock transition is received.
43. synchronous. With the clock inputs tied together, the various flip-flops will toggle at the same time when they change states. Therefore, this is a synchronous configuration.
45. asynchronous

FIGURE ANS15–23

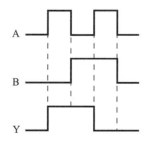

47. Clock = 400 MHz, A = 200 MHz, B = 100 MHz (given), C = 50 MHz, and D = 25 MHz. Each stage of the counter will divide the frequency of the previous stage by 2. So, if B is 100 MHz (given), then A must be 200 MHz. Similarly, the clock frequency must be 400 MHz. In a similar manner, the 100-MHz frequency at B is divided by two to produce 50-MHz at C. It is divided again to produce 25-MHz at D.

49. The output of a DAC is an analog signal (either voltage or current).

51. The resistor value for each additional input is half (or twice) the previous resistor value. Thus, the required range of resistors would be extreme. For example, if the smallest resistor used were 100 Ω, then the largest would have to be 6,553,600 Ω. Further, the tolerance and accuracy of the resistors would have to be impractically high.

53. Yes. The output could be low because all of the inputs are high. Alternatively, the gate could be defective or the output could be shorted to ground or to another output. In any case, you must determine the states of the remaining inputs before you can classify the condition of the gate.

55. Normal operation. Any low input on an AND gate will cause the output to be low, regardless of the level on the remaining inputs.

57. A logic analyzer is ideal for this type of problem. It can capture the logic levels at multiple points in the circuit simultaneously and store them in its memory. You can then examine them as static displays.

•—Circuit Exploration

The timing diagram for the circuit when it is enabled and in the Down mode and where traces 5 and 4 correspond to A and B, respectively, is shown in Figure ANS–15.

CHAPTER 16

•—Review Questions

1. instruction set
3. program
5. Address bus.
7. False. Instructions are held in the instruction register while they are being decoded and executed.
9. hardware
11. Loop. If the sequence is repeated without end, it is called an infinite loop.
13. AGP.
15. Multiple buses allow simultaneous data transfers in different parts of the system. A single bus would present a bottleneck, and the μP would have to play an active role in every data transfer.

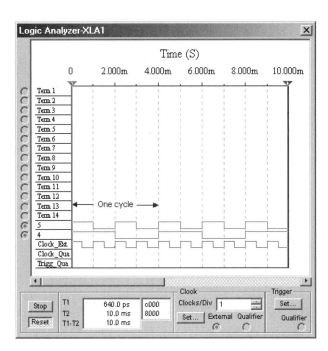

FIGURE ANS–15

17. skew. Clock skew results when two or more synchronous clock signals encounter differences in the paths to their destinations. Differences may include such things as different number of gates (each gate adds a propagation delay), different number of loads (each gate input adds capacitance which slows and delays the pulse), and different trace lengths.

19. True.
21. pixel
23. False. Bios must be stored in a nonvolatile IC. Programs within the bios set initialize the computer during power-up and control the loading of the operating system.
25. faster. Serial transfers can only move one bit at a time, whereas parallel transfers activate many bits at the same time on multiple wires.
27. real-time. The frequency of 32,768 Hz is selected, since it can easily be divided to produce seconds, minutes, hours, and so on.
29. False. Machine language is very laborious and error prone. It would never be used for serious programming and certainly not by a typical computer user.
31. Yes. Macros allow for the creation of user-application programs. Further, their development often requires little or no programming experience.
33. Operating system.
35. Application software.

•—Circuit Exploration
1. No response necessary.
2. No response necessary.
3. See Figure ANS16–3.

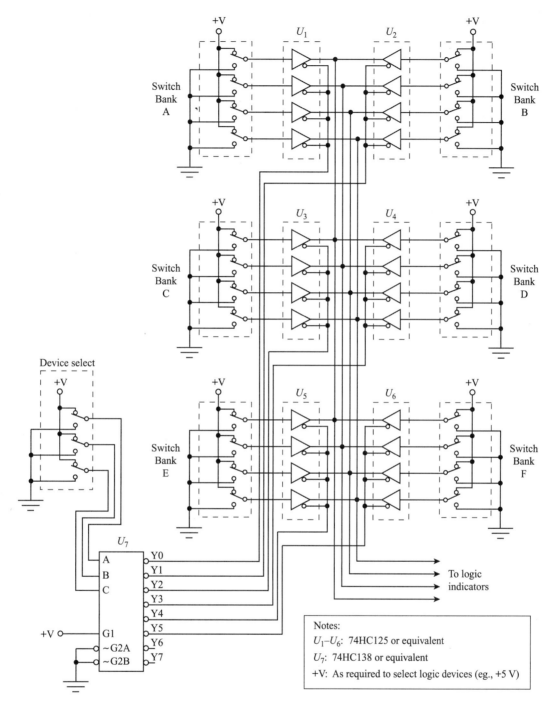

FIGURE ANS16-3

CHAPTER 17

Review Questions

1. True.
3. telemetry
5. bidirectional
7. True.
9. hub
11. protocol
13. Most programs and data files reside on the server. This centralized storage provides an opportunity for increased data security and simplifies software upgrades.
15. Tiny area network. This is a small network consisting of perhaps 5 or 6 nodes.
17. TAN, CAN, LAN, MAN, and WAN
19. MAN (metropolitan area network).
21. Throughput and bandwidth are directly related, so communications link A will have the greater throughput.

23. broadband
25. True. Oftentimes, the clock is encoded with the data into a single bitstream.
27. large
29. True.
31. error detection
33. High. Since the byte 11110100 has an odd number of high-level bits, the parity bit must be high in order to have an even number of high-level bits including the parity bit.
35. Digital transmissions are more immune to noise than analog transmissions. Unless the noise has enough amplitude to actually change the logic level of a bit, the data is unaffected. Analog, by contrast, can be corrupted with relatively small amounts of noise.
37. simplex
39. Modulation.
41. Frequency modulation (FM) or phase modulation (PM).
43. Pulse-code modulation.
45. Frequency-division multiplexing (FDM), time-division multiplexing (TDM), and wavelength-division multiplexing (WDM).
47. Both are functionally the same as a local area network (LAN), but the WLAN uses no wires. That is, WLAN is a wireless LAN, and the nodes are connected via radio or infrared links.
49. Eight. There can be only one master and as many as seven active slaves in any given piconet.

Circuit Exploration

1. No specific response needed.
2. No specific response needed.
3. When the clock is in the high state (e.g., S_1 connected to +V) then gate 1a is disabled, since it has one of its inputs low. With a low input, the output of 1a will be high. This enables gate 1c. Gate 1b is also enabled, because its clock input terminal is high. The output state of 1b (and subsequently 1c) is then determined by the state of the Ch2 information stream. As Ch2 changes states, so does the output of 1c.

    A similar selective-gating situation exists on the receiver end. During the high-level clock state, gate 3b is enabled and gate 3a is disabled (one input is low). When the received data arrives at gate 3b, it is allowed to pass through, and the Ch2 LED will indicate the received data.

    When the clock is in its low logic state, gates 1a and 3a are enabled while gates 1b and 3b are disabled. This allows changes from Ch1 to pass through the communications link and control the state of the Ch1 LED.

    If the clock rate is substantially faster than the Ch1 and Ch2 rates, then both channels of information can be effectively transmitted over a common path.

GLOSSARY

absorption A name given to the transfer of energy from an incoming high-energy electron to a valence-band electron, which causes the valence-band electron to move to a higher orbit (i.e., conduction band).

ac resistance The effective resistance of a component or device to alternating current.

acceleration A measure of the rate of change of speed (or velocity).

acceptor atom A trivalent impurity atom added to an intrinsic semiconductor material to create p-type material. Each acceptor atom contributes one hole to the crystal.

access time The time required for a memory device to produce a valid output after the address and control signals are stable.

accumulator A register in the ALU that holds one of the operands in an arithmetic or logical operation and stores the result of the operation. It is generally capable of other functions, such as shifting, rotating, and serving as the destination or source for input and output operations, respectively.

accuracy A meter specification that indicates the largest difference between actual and indicated values.

address An expression that identifies a specific location in a memory device. It may be either numeric or symbolic.

address bus A bus in a microprocessor system that transfers the memory address from the microprocessor (or other control device) to the memory address inputs.

address register A register in which an address is stored.

admittance (Y) The reciprocal of impedance. The ability of RC, RL, and RLC circuits to pass alternating current. Measured in ohms.

air gap Any portion of a magnetic circuit where the magnetic flux passes through air.

algorithm A term used to describe a list of steps or procedures by which a given result is obtained.

alpha The ratio of collector current to emitter current in a bipolar transistor.

alphanumeric Consisting of both digits (0–9) and letters (A–Z) and often including special characters (e.g., !, ~, &, %, etc.).

alternating current (ac) Current that periodically changes direction.

alternation Either the positive or negative half-cycle of a sinusoidal waveform.

alternator A device that converts mechanical energy into alternating voltage and current.

ALU Abbreviation for arithmetic and logic unit. A functional block within the microprocessor that performs the mathematical and logic operations.

amalgamation Used in the manufacture of carbon-zinc cells to reduce local action by coating the zinc electrode with mercury.

ammeter A device used to measure current.

ampacity The current-carrying capacity of a wire.

ampere (A) The unit of measure for current.

ampere-hour (A · h) The unit of measure for the capacity of a cell or battery.

ampere-turn (A · t or N · t) The unit of measure for magnetomotive force.

amplification The process of converting a small, changing signal into a larger, but corresponding signal.

amplitude The maximum (unless otherwise stated) value of a waveform.

amplitude modulation The process whereby the amplitude of a higher-frequency carrier signal is modified by a lower-frequency signal that represents information to be transmitted over a communications link.

analog A device, system, or quantity that is not limited to discrete values but is continuously variable.

analog-to-digital converter An interface device that converts an analog signal into its equivalent digital form.

anode The negative terminal of a cell. Also, the more positive terminal of a conducting diode.

apparent power (P_A) The phasor sum of true power and reactive power.

architecture The functional organization of a microprocessor system.

armature The moveable part of a motor or relay. Also refers to the winding of an alternator or generator where the voltage is induced.

ASCII American Standard Code for Information Interchange. The standard 7-bit code used to represent alphanumeric characters and control codes for transmitting information between two digital devices.

assembler A standard program that translates symbolic statements (assembly language) into a machine-language equivalent.

assembly language A programming language that uses the mnemonic representation for the microprocessor instructions, thereby relieving the programmer of the burden of remembering the bit patterns for each instruction.

asymmetrical Not symmetrical. Generally refers to a waveform with dissimilar half-cycles.

asynchronous A device or system in which the speed of operation is not directly related to any frequency in the system to which it is connected.

asynchronous counter A counter whose various stages receive their clocks from different sources.

asynchronous input An input that is activated by a logic level without the need for a clock.

atom The smallest part of an element that still has the properties of the element.

attenuation A reduction or loss of signal amplitude.

autotransformer A transformer with a single tapped winding that serves as both primary and secondary.

avalanche current The rapidly increasing current that flows through a reverse-biased pn junction after the reverse breakdown voltage is exceeded.

average value The sum of several instantaneous values of an electrical quantity divided by the number of values. In the case of a sinusoidal waveform, the average value is zero for a full cycle and $0.636\ V_{PEAK}$ for a half-cycle.

AWG An abbreviation for American Wire Gauge.

backbone The main trunk of a communications network that uses a bus topology.

backoff scale An ohmmeter scale that reads from right-to-left. The nonlinear scale has zero ohms on the far right and infinite ohms on the far left of the scale.

balanced bridge A condition in a bridge circuit where the two legs have equal resistance.

bandpass The band of frequencies between the half-power points of a circuit that pass through the circuit with at least 70.7% of the maximum amplitude.

bandpass filter A filter circuit that passes a band of frequencies with minimal attenuation but attenuates all frequencies lower or higher than the pass band.

band-reject filter A filter circuit that attenuates a band of frequencies but allows higher and lower frequencies to pass with minimal attenuation. Also called a bandstop filter.

bandstop filter See band-reject filter.

bandwidth The width (in hertz) of the range between the half-power points of a resonant circuit or a filter circuit. Also the range of frequencies that can be transmitted over a communications path. Often interpreted to mean the amount of data per unit time that can be transmitted over a communications link.

barrier potential The charges on the fixed impurity atoms in a pn-junction region that produce an effective difference of potential across the junction. Barrier potentials in pn junctions range from 0.1 to 1.5 volts depending on the materials used.

base One of the elements of a bipolar transistor. It is the center region of a three-layer semiconductor sandwich. The current through the remaining elements is proportional to the current through the base lead.

baseband A term used to describe a communications system that uses the original transmission frequencies and is limited to a single communications channel.

BASIC Beginner's All-purpose Symbolic Instruction Code. A popular high-level language for microprocessors that is easy to learn, yet provides sufficient power for fairly complex programming tasks.

battery A series and/or parallel connection of two or more cells.

BCD A numbering system that represents decimal numbers in groups of four bits. The groups of four bits range from 0000 to 1001 and represent the decimal digits of 0 through 9, respectively.

beta The ratio of collector current to base current in a bipolar transistor.

***BH* curve** A graph showing the relationship between flux density (B) and magnetizing force (H) for a given material.

bidirectional bus driver A digital device capable of amplifying (current amplification) logic signals from one computer bus and transferring the stronger signals to another bus. Transmission through the driver can be in either direction, but not simultaneously.

bimetallic A structure consisting of two dissimilar metals that bends in response to temperature changes due to the unequal coefficients of expansion.

binary The base 2 number system. Used extensively in all digital electronics devices.

BIOS **B**asic **I**nput **O**utput **S**ystem. A nonvolatile memory that contains low-level, general-purpose routines that can be used by other programs in a computer.

bipolar A device or process that relies on two types of current carriers such as holes and electrons.

bipolar transistor A transistor whose operation depends on two types of current carriers (holes and electrons). Two common types are npn and pnp.

bit **Bi**nary digi**t**. The smallest unit for information in a digital system.

bit manipulation The process of testing and/or altering individual bits within a byte.

bleeder current The current in a voltage divider that flows with or without a load connected.

block diagram A generalized sketch that represents the functional components of a system with graphic boxes. The boxes are interconnected with lines to show the flow (generally signal flow in block diagrams of electronic systems).

boost regulator A switching voltage regulator whose output voltage is higher than its input voltage.

bootstrap loader A short program, generally stored in ROM, that is used to load the larger operating system program from an external storage device, such as a floppy disk system.

branch 1. A current path in a circuit that is in parallel with one or more other paths. 2. The interruption of sequential program execution. The branch may be unconditional or conditional, which is based on the results of a test. Following the specified test, the program flow will resume at an address determined by the results of the test.

break frequency The frequency at which the output of a circuit falls to 70.7% of its maximum voltage or current response. This also corresponds to the half-power (−3 dB) point(s) on the frequency response curve.

breakdown voltage The amount of reverse voltage required to cause a pn junction to conduct in the reverse direction. The resulting power dissipation normally destroys the device.

breakover voltage The main terminal voltage required to cause a thyristor to enter the conduction state without being triggered by a gate element.

breakpoint A specific point in the program as set by the programmer at which normal execution is stopped and control is transferred to a debug routine. This is useful for debugging software problems.

bridge circuit A special series-parallel circuit consisting of two parallel branches with two series resistors in each branch. The output is taken between the midpoints of the two parallel branches. Bridge circuits are widely used in instrumentation circuits.

broadband A term used to describe a communications system that multiplexes several channels into a single communications link. Also, the original transmission frequencies are increased to higher and more efficient frequencies for transmission.

brush A graphite block that provides sliding contact to a slip ring (alternator or ac motor) or commutator (dc motor or generator).

bubble A small circle used on the input or output of a logic device symbol to indicate an active-low element.

buck regulator A switching voltage regulator whose output is lower than its input.

buffer 1. A hardware device or circuit used to provide isolation between two circuits. 2. A reserved portion of memory used to store information received from one device until a second device can access the information.

bug An error or defect in a microprocessor program.

bus A group of signal or control lines that all share a similar overall function.

byte A group of adjacent bits that are operated upon as a unit. The unit generally consists of eight bits.

cable Two or more individually-insulated wires bound in a common sheath.

cache memory A high-speed, local memory used by a microprocessor to increase overall system speed.

call A term used to describe the invocation of a subroutine. This operation will result in a program branch; however, normal sequential execution will resume after completion of the subroutine.

CAN Controller Area Network. A method of connecting computers or industrial controllers together in a small, local network. Generally limited to 64 nodes and 125 feet when operating at full speed.

candlepower A unit of measure of light intensity based on the light emitted by a standardized candle.

capacitance (C) The electrical property that allows a capacitor to store energy in an electrostatic field.

capacitive reactance (X_C) The opposition to alternating current flow offered by a capacitance. It is measured in ohms.

capacitor An electrical component consisting of two or more conductive plates separated by insulation and used to store energy in an electrostatic field.

carrier A high-frequency signal that is modulated by the information signal in a communications system. Characteristics of the carrier such as amplitude, frequency, and phase can be controlled by the information signal.

cathode 1. The positive terminal of a cell. 2. The more negative terminal of a conducting diode.

cell A single, stand-alone source of electrical energy (e.g., electrochemical cell, fuel cell, solar cell).

cemf Counter **emf**. The voltage induced in a conductor or inductor by a changing current. The induced voltage opposes the changing current.

center tap A coil or transformer winding that has a connection in the middle of the winding.

central processing unit (CPU) The functional part of a computer system that contains the ALU, general-purpose registers, and instruction decoding circuitry.

CGS An older system of measurement based on the centimeter, gram, and second.

channel The primary path for current in a field-effect transistor.

character A letter, digit, or control code that is used in the representation and transmission of data.

charge A term used to describe the condition that exists when a material or region has unequal numbers of electrons and protons.

chassis The metal frame or cabinet used to house an electronic system. The chassis is frequently, but not always, connected to ground.

chattering The audible noise that results when a normally closed contact of a relay is connected in series with its own coil.

checksum The logical sum of all data in a given record. This is used as part of an error-checking scheme to detect transmission errors.

chip The physical semiconductor material used to make integrated circuits and other solid-state devices.

choke A name given to an inductor when its primary purpose is to attenuate high frequencies.

circuit Any configuration of electrical and/or electronic devices interconnected with conductors.

circuit breaker A protective device that mechanically opens a circuit if the current through the circuit exceeds the trip current of the breaker.

circuit simulation A computer software package that is used to simulate the operation of an electronic circuit.

circular mil (Cmil) A unit of measure for wire area. One circular mil is the cross-sectional area of a round wire that has a diameter (d) of one mil (1/1000 of an inch). Equal to d^2.

cladding The glass surrounding the central core of a fiber-optic cable. Light waves are bent back into the central fiber when they enter the cladding.

clipping The amplitude distortion that occurs when the peaks of a waveform are limited to values less than the original peaks of the undistorted waveform.

clock A timing device or circuit that produces a periodic pulse train. The pulse train is used to synchronize all activity within a digital system.

clock skew The difference in time between corresponding points on two or more digital clocks that are ideally in phase.

closed circuit A circuit that has a complete path for current.

closed loop Circuit operation that results when a portion of the output is returned to the input of the system.

closed-loop differential input voltage The voltage between the input pins of an op amp when it is operated as a closed-loop amplifier.

closed-loop voltage gain The overall voltage gain of a device or circuit that has feedback (generally negative feedback).

coaxial cable A shielded, two-conductor cable in which the outer conductor completely surrounds the insulated inner conductor.

code A set of unambiguous rules that defines the way data are to be represented in a particular system.

codec **Co**der-**Dec**oder. The portion of a communications system that converts analog information signals to digital signals for transmission over a digital link. A second codec converts the received digital information back to its analog form.

coding The process of writing a program in a particular language.

coefficient of coupling (k) A dimensionless number that represents the percentage of total flux that is common between two coils or circuits.

coercive force The amount of magnetizing force needed to overcome or cancel the residual magnetism in a magnetic material.

coherent light Light composed of single frequency, in phase light energy.

coil An inductor or other device formed by winding multiple turns of wire on a core (even an air core).

coil resistance The dc resistance (measured with an ohmmeter) of a coil.

collector One of three elements of a bipolar transistor.

collimation The process of focusing a diverging laser beam into a parallel beam with minimal divergence.

color code A standardized scheme that uses colored markings to identify component values or part numbers. Also used to identify leads on transformers.

common-mode choke A coil (or coils) that is wound in such a way as to provide significant opposition to signals that are common to all conductors passing through the coil, but minimal opposition to differential currents—forward current in one wire and return current in another.

common-mode rejection ratio (CMRR) The ratio of differential voltage gain to common-mode voltage gain in a differential amplifier.

common-mode voltage gain The gain given to a voltage that appears on both inputs of a differential amplifier.

commutator A segmented conductor that is mounted on the rotor of a dc generator or motor and makes sliding contact with the brushes. Allows power to be applied (or removed) from a rotating coil.

compiler A computer program that translates high-level language commands into binary code that can be interpreted by the computer.

complex circuit A circuit configuration that cannot be simplified by replacing sets of series and parallel components with equivalent components.

complex number A number composed of both real (resistive) and imaginary (reactive) parts used to describe a phasor.

complex voltage source A circuit with both series-aiding and series-opposing cells or voltage sources.

compound A material that consists of two or more elements that are chemically bonded.

conditional jump A program jump that may or may not occur depending on the results of a specified test.

conductance (G) The ease with which a material permits current flow. The reciprocal of resistance. Measured in siemens or mhos and represented by the letter G.

conduction band The region which lies beyond the valence band in an atom. Electrons in this region are essentially free and have little or no association with a specific atom.

conductor A material with a very low resistance that readily permits current flow.

constant current A source of current that is constant at all times and is unaffected by changes in circuit resistance.

constant voltage A source of voltage that is constant at all times and is unaffected by changes in circuit resistance.

contact bounce The series of momentary opens and closures that occurs when mechanical switch or relay contacts are moved.

continuity A term used to indicate the presence of a complete path for current.

control bus A group of conductors in a microprocessor system that collectively provide control and timing functions for the system.

conventional current A convention that conceptualizes current as flowing in a direction opposite that of electron flow. Conventional current moves from positive to negative through a complete circuit.

copper loss The energy converted to heat by the resistance in the windings of a transformer or other electromagnetic device.

core The material used as the central part of an inductor or transformer. It supports and generally concentrates the magnetic flux, and it physically supports the coil windings.

core loss The energy converted to heat in the core material of an electromagnetic device. Core loss consists primarily of losses due to hysteresis and eddy currents.

cosine A trigonometric function of an angle in a right triangle that is equal to the length of the adjacent side divided by the length of the hypotenuse.

coulomb (c) The SI unit of measure for charge. One coulomb is the amount of charge on 6.25×10^{18} electrons.

Coulomb's law The law which relates the force between two charged bodies as a function of the strength of the two charges and the distance between them.

counter emf *See* cemf.

covalent bonding Covalent bonding occurs in a semiconductor crystal when a valence-band electron is shared by two atoms.

critical angle The maximum angle that light energy traveling down a fiber-optic cable can have and still result in the light wave being reflected back into the core when it enters the fiber/cladding interface.

crowbar circuit A protective circuit that acts as a short circuit across a power supply output if the power supply voltage goes above a specified value.

CRT An abbreviation for cathode-ray tube. It is the display tube used in most oscilloscopes, televisions, and radars.

cryogenics The study of the behavior of materials as they approach absolute zero (−273.2°C).

crystal A material whose atoms form a consistent lattice pattern throughout the bulk of the material.

current The directed movement of charged particles. Often used to describe the movement of electrons.

current divider A circuit that consists of two or more parallel branches. It is used to divide the total current into two or more components.

current flow The directed movement of charged particles.

current probe A test instrument used to sense, and sometimes measure, current in a conductor without having to open the circuit to insert the current probe.

current source *See* constant current.

cutoff One of two extreme conditions in a transistor. No primary current flows through the transistor when it is in cutoff mode.

cutoff frequency *See* break frequency.

cycle One complete repetition of a periodic waveform consisting of two alternations.

damping A technique used to prevent dramatic overswings in pointer movement on analog meters.

dark current The current that passes through a photodiode when it is not illuminated.

d'Arsonval movement A common meter movement consisting of a moving coil attached to a pointer and suspended in the field of a permanent magnet.

data Any group of bits or characters to which meaning is assigned.

data bus A group of lines that is used to transfer data throughout the microprocessor system.

dc resistance The ohmic resistance as measured by an ohmmeter.

debug To detect and correct defects in a microprocessor program. It is the equivalent of troubleshooting a hardware defect.

decade A tenfold (i.e., 10:1 or 1:10) change in the value of a quantity.

decibel (dB) The logarithmic unit of measure for a ratio.

decoder A device or circuit that accepts a small number of coded input lines and activates one of a larger number of output lines.

decoupling capacitor A capacitor connected between an ungrounded point and a grounded point in a circuit.

degauss Demagnetize.

degenerative feedback *See* negative feedback.

degree An angular measure equal to 1/360 of a full circle.

delta configuration A configuration where three components are connected in a loop with a connection made at each node. Also called a pi configuration. One of two common connections for three-phase transformer windings.

demodulation The process of recovering the information signal from a modulated carrier wave.

depletion mode A mode used in field-effect transistors that decreases the channel current below the no-bias value.

depletion region The region near a pn junction where recombination has eliminated most of the current carriers (holes and electrons).

diamagnetic A material with a relative permeability of less than one.

dielectric An insulator.

dielectric constant (k) A measure of the ability of a material (relative to air) to concentrate an electric field.

dielectric strength A measure of a dielectric material's ability to withstand high voltage.

difference of potential The value of voltage between two points.

differential voltage gain The gain given to a voltage that appears between the inputs of a differential amplifier.

differentiator A circuit whose output is proportional to the rate of change of input.

digital A device or system that utilizes discrete (noncontinuous) values.

digital-to-analog converter An interface device or circuit that converts digital numbers into the corresponding analog representation.

digitize To convert the analog representation of a physical quantity into an equivalent digital value.

diode A two-terminal electronic component that permits current flow in only one direction.

direct current (dc) Current that flows in only one direction and generally has a constant value.

directly proportional Two quantities are directly proportional when increases in one quantity cause corresponding increases in the second.

discrete components Individual electronic components as opposed to multiple components integrated into a single package.

discriminator A circuit used to recover the information from a frequency-modulated carrier wave.

DMM Digital multimeter. The digital equivalent of an analog VOM.

domain A region within a magnetic material that behaves like a small bar magnetic. Alignment of the domains in a material results in the material being magnetized.

donor atom A pentavalent impurity atom that contributes one conduction-band electron to a doped semiconductor crystal.

doping The process of adding impurity atoms to a semiconductor crystal.

Doppler effect The apparent shift in frequency that occurs when the source and the receiver are moving relative to each other.

dot notation A method used on schematics to identify in-phase points on a multiple-winding transformer.

drain One of three elements in a field-effect transistor. Analogous to the collector in a bipolar transistor.

drain wire An uninsulated wire running along the length of a coaxial cable that uses metal foil as the shield. The drain wire is positioned between the ribbon shield and the outer jacket, and it provides a means to connect to the fragile foil shield.

drift A long-term change in the dc characteristics of an op amp.

drop-out current The value of coil current which permits the contacts of a relay to return to their normal state after the relay has been energized.

drop-out voltage The value of coil voltage that permits the contacts of a relay to return to their normal state after the relay has been energized.

dropping resistor A name given to a resistor that is placed in series with another circuit and whose primary purpose is to drop a portion of the total voltage.

duplex A term used to describe a communications link where both ends are capable of transmitting and receiving information.

duty cycle The ratio of pulse width to period in a pulse waveform. Expressed either as a decimal (0 to 1) or a percentage (0 to 100%).

DVM Digital voltmeter. A digital meter that measures voltage.

dynamic breakback voltage The *change* in voltage across a diac when it switches from its nonconducting state to its conducting state.

dynamic resistance *See* ac resistance.

eddy current A circulating current in the core of an electromagnetic device that results from the conductive core material being cut by changing magnetic flux. Eddy currents produce a heat loss in the core.

effective value *See* root-mean-square value.

efficiency The ratio of output power to input power. Ideally 100%, but always less in practice.

electricity Energy produced by electron flow and capable of being converted into other forms of energy such as heat, light, and sound.

electrode An electrical terminal of a cell or other electrical device that permits current to enter/exit the device.

electrolyte A liquid or paste with mobile ions that reacts with the electrodes in an electrochemical cell and permits current flow between the anode and cathode of the cell.

electromagnet A magnet produced by the flow of current through a coil. The magnetic field is present only so long as the current is flowing.

electromagnetic field 1. A magnetic field produced by current flow through a conductor. 2. A composite field consisting of both electric and magnetic components in a high-frequency radiated wave.

electromagnetic induction The process that causes a voltage to be produced in a wire as it is intercepted by moving magnetic flux.

electromagnetic interference (EMI) The interference caused to an electronic system by electromagnetic fields emanating from a nearby system.

electromechanical A device whose operation is both electrical and mechanical.

electromotive force (emf) A voltage source or potential difference that is sustained as charges are moved through the circuit. The electrical pressure that causes sustained current to flow in an electrical circuit. Measured in volts.

electron A negatively-charged atomic particle that orbits the nucleus of all atoms.

electron current flow A convention that conceptualizes current as the movement of electrons. Electron current flows from negative to positive through a complete circuit.

electron-hole pair The valence-band hole and the conduction-band electron that are created when a covalent bond is broken in a semiconductor crystal.

electronic Components or devices that control or regulate the flow of electrons with active devices (e.g., transistors).

electronic system A complete functional unit whose operation relies on electronic components and electronic circuits.

electrostatic field The region around a charged body where another charged body would experience a force of attraction or repulsion.

element A material whose atoms are all identical. An element cannot be subdivided by chemical means.

emf *See* electromotive force.

EMI *See* electromagnetic interference.

emitter One of the elements of a bipolar transistor. It carries the total transistor current.

energy The ability to do work.

energy level Used to describe the energy content of an orbiting electron. Higher energy levels correspond to orbits that are farther from the nucleus.

engineering notation An application of powers of ten where all values are written in powers of ten format and the exponent is zero or evenly divisible by three.

enhancement mode A mode of operation in a MOSFET where the channel current is increased above the no-bias value.

equivalent circuit A simplified circuit that provides similar performance to a more complex circuit under a given set of conditions.

equivalent series inductance (ESL) The parasitic inductance of a capacitor caused primarily by the capacitor lead wires.

equivalent series resistance (ESR) The effective resistance (i.e., equivalent power loss) of a capacitor caused primarily by dielectric losses.

error amplifier An amplifier that amplifies the difference between a reference voltage and a feedback voltage in a closed-loop system.

ESL Equivalent series inductance. Generally used to describe the parasitic inductive characteristics of a capacitor.

ESR Equivalent series resistance. Generally used to describe the resistive characteristics in a capacitor that cause it to have an energy (heat) loss.

EVM Electronic voltmeter. An analog VOM that has electronic circuitry, which makes the meter appear more ideal.

extrinsic semiconductor A semiconductor that has been doped with impurities to make n- or p-type material.

farad (F) The unit of measure for capacitance.

Faraday shield A grounded conductive shield used to separate the primary and secondary windings of a transformer to reduce coupling via the interwinding capacitance.

fast-blo A type of fuse that is designed to open quickly.

feedback The portion of a signal in an electronic system that is returned to a prior stage of the same system.

ferrite A magnetic ceramic material used as a core material for transformers and coils. Also used in the form of a bead or clamp to attenuate high frequencies.

ferromagnetic A material with a relative permeability much greater than one.

fiber-optic link A communications link where the transmission medium is laser light passed through a fiber-optic cable.

field A group of bits within a record or within a byte that is used to represent a particular category of data.

field-effect transistor A type of transistor characterized by high input impedance and whose channel current is controlled by an internal electrostatic field.

field winding A coil in a motor, generator, or alternator that is used to create a steady magnetic field.

filter A circuit designed to pass certain frequencies but reject (i.e., attenuate) other frequencies.

firing angle The angle on a sine wave where a thyristor enters the conduction state.

firing voltage The minimum voltage required to ionize the gas in a neon lamp.

firmware Software that is stored in ROMs.

flag An indicator within a microprocessor that signals the occurrence of a specified condition.

flow chart A pictorial representation for describing the performance of a program or system.

flux Magnetic or electric lines of force.

flux density (B) The amount of magnetic or electric lines of force per unit area.

flux leakage Magnetic flux that is intended to, but does not, link two circuits (e.g., primary and secondary of a transformer).

foldback current limiting An overcurrent protection method for power supplies where the short-circuit current value is less than normal operating current.

forward bias The polarity of voltage on a pn junction that results in significant current flow. The p material is more positive than the n.

free electron An electron that has been disassociated from its parent atom and is no longer in orbit.

frequency (f) The number of complete cycles of a periodic waveform per unit time.

frequency-hopping spread spectrum (FHSS) A communication technique where the transmitter and receiver change frequencies (synchronously) on a regular basis. The transmitted energy is spread across a wider portion of the spectrum.

frequency response A description that reveals how the output of a circuit or system responds as a function of frequency.

frequency stability A measure of how well a signal (typically an oscillator) remains at a constant frequency.

fringing A phenomenon that causes magnetic flux lines to "bulge" when they pass between two high-permeability materials that are separated by a low-permeability material (e.g., an air gap in a magnetic core).

front-to-back ratio The ratio of the resistance of a forward-biased pn junction to its resistance when reverse biased.

FSK Frequency-Shift-Keying. A transmission technique that converts the two digital logic levels to corresponding audio tones.

fuel cell A special type of electrochemical cell that must have a continuous supply of external fuel (e.g., hydrogen and oxygen).

full-duplex A term used to describe a duplex communication system in which both devices can transmit and receive simultaneously.

full-scale current The amount of current required to fully deflect an analog meter movement.

full-scale voltage The amount of voltage required to fully deflect an analog meter movement.

fully-specified circuit A circuit with all component values given on the schematic diagram.

function generator An electronic instrument that generates periodic waveforms such as sine waves, rectangular waves, and triangle waves.

fundamental frequency The lowest frequency component in a periodic waveform. It is equal to the reciprocal of the period of the waveform.

fuse A protective device that burns open to protect a circuit if the current through the circuit exceeds the rating of the fuse.

ganged The mechanical connecting together of two or more adjustable components (e.g., switches or potentiometers) such that they operate simultaneously.

gate 1. A combinational logic device that implements a basic operation such as AND, OR, NAND, NOR, exclusive OR, exclusive NOR, and NOT. 2. The control element in a field-effect transistor.

gauss (G) The CGS unit of measure for magnetic flux density.

generator An electromechanical device that converts mechanical energy into electrical energy in the form of direct voltage and current.

giga (G) A prefix used to represent 10^9.

gilbert The CGS unit of measure for magnetomotive force.

glitch A transient or false output on a logic gate caused by a timing difference on its input signals.

graph A diagram showing the numerical relationship between two or more quantities.

graticle A screened grid on the face of an oscilloscope CRT used to increase reading accuracy.

ground The point in a circuit to which all voltage measurements are referenced. Sometimes used to refer to earth ground.

ground plane A metallic plane having zero potential with reference to ground.

half-cycle One alternation consisting of 180°. One half of a full sinusoidal cycle.

half-digit A digital indicator capable of displaying ±1.

half-duplex A term used to describe a duplex communication system that allows only one device to transmit at any given time.

half-power frequency *See* break frequency.

half-power point The point(s) on a frequency response curve where the output power drops to 50% of its maximum value. This corresponds to a 3-dB drop in power.

hardware Physical equipment in a computer system such as mechanical, electrical, or magnetic devices.

harmonics Any whole multiple of a base frequency called the fundamental frequency.

heat sink A thermal mass of metal or other material that is attached (thermally) to an electronic component to aid in the removal (i.e., dissipation) of heat generated within the component.

henry (H) The SI unit of measure for inductance.

hertz (Hz) The SI unit of measure for frequency.

hexadecimal The base 16 number system consisting of the digits 0–9 and the letters A–F.

high-level language Programming languages containing English-like procedural or function-oriented statements that produce several machine instructions when compiled or assembled.

high-pass filter A filter circuit that passes all frequencies above the cutoff frequency with minimal attenuation but attenuates all frequencies below the cutoff frequency.

holding current The minimum current through a thyristor that will cause it to remain in the on state.

hole A broken covalent bond in a semiconductor crystal.

horsepower A unit of mechanical power. One horsepower is equivalent to 746 watts.

hot Electrician's jargon for any wire that is 120 Vac or more with respect to earth ground.

hot-carrier diode A device such as a Schottky diode whose rectifying junction is formed by the interface between a semiconductor material (e.g., n-type silicon) and metal (e.g., gold or aluminum). They are characterized by fast switching times.

hot spots Regions within a semiconductor crystal where current is concentrated (i.e., not uniformly distributed), which cause localized (often damaging) heating.

hub The central node on a network using a star topology.

hydrometer A device used to measure the specific gravity of a liquid.

hypotenuse The longest side of a right triangle and opposite to the right angle.

hysteresis The process which causes the flux build up in a magnetic material to lag behind the magnetizing force.

hysteresis loss A heat loss in a magnetic material caused by the rapid switching of the magnetic domains.

impedance (Z) Total opposition to current flow in a circuit containing both resistance and reactance.

index of refraction The ratio of the speed of light in a vacuum to the speed of light in a material.

indirect FM Another name for phase modulation.

induced voltage A voltage that is created in a conductor as a result of being intercepted by a moving magnetic field.

inductance (L) The electrical property that allows a coil to store energy in a magnetic field. It is a characteristic or property which tends to oppose any change in current.

induction The process of producing an induced voltage. (*See* induced voltage.)

inductive reactance (X_L) The opposition to alternating current flow offered by an inductance. It is measured in ohms.

inductor A component (coil) that is specifically designed to have inductance.

input The voltage, current, or power applied to an electrical circuit.

input bias current The dc current that flows into or out of the input terminals of an op amp.

input/output device (I/O) The part of the microprocessor system hardware that allows data to be entered (input) and to be transmitted during output operations (output).

inrush current A transient current that is higher than normal but occurs only when power is initially applied to a device or circuit.

instantaneous value The value of a changing electrical quantity at a specific point in time.

instruction A statement that specifies a particular operation and the values or locations of its operands.

instruction register An internal microprocessor register that is used to hold the instructions while they are being executed.

instruction set The total set of instructions that is available for a particular microprocessor.

insulator A material that has a very high resistance and allows no practical current flow.

integrated circuit A semiconductor device that consists of thousands of transistors, resistors, and so on integrated into a single wafer of silicon.

integrator A circuit whose output is proportional to both amplitude and duration of the input.

intensity A measure of the brightness of a light generally expressed in candlepower (candela).

interface A device or circuit that is used to make two dissimilar devices compatible. The interface components generally translate the speed, voltage, and current requirements of one system to acceptable values for the other system.

internal resistance An effective resistance that appears to be internal to a device or component. Although it is not generally a physical resistance, it behaves like a series or parallel resistance.

interpolate A process used to estimate the indicated value when the pointer of an analog meter comes to rest between scale markings.

interpreter A program that fetches source code statements and executes them immediately. The interpreter does not produce object code like an assembler or compiler program.

intrinsic semiconductor A pure (not doped) semiconductor material.

inversely proportional Two quantities are inversely proportional when increases in one quantity produce corresponding decreases in the second.

inversion layer The region of the channel directly below the gate of a MOSFET where the type of semiconductor material is effectively changed by the effects of the charge on the gate-to-channel capacitance.

ion A charged particle. An atom that has unequal numbers of electrons and protons.

ionization The process of creating ions by adding or removing electrons from an otherwise neutral atom.

ionization voltage The voltage required to ionize the gas in a neon bulb and cause the indicator to emit light.

isolation transformer A 1:1 transformer used to isolate circuit grounds from earth grounds.

j operator A mathematical prefix or operation that indicates or produces a 90° phase shift.

jitter In a phase-locked loop, it is the slight variation in output phase that results from continuous comparison and correction.

joule The SI unit of measure for energy or work.

jump A nonrecoverable departure from the normal sequential instruction execution.

junction 1. A point in the circuit where two or more components are joined. 2. Internal to a semiconductor crystal where p and n material meet.

junction breakdown A condition that occurs when the voltage on a reverse-biased pn junction is so high that the junction can no longer block reverse current.

kilo (k) A prefix representing 10^3.

kilowatt-hour (kWh) The unit of energy commonly used by electrical power companies.

Kirchhoff's current law A fundamental law that states: the current entering any point in a circuit must be equal to the current leaving that same point.

Kirchhoff's voltage law A fundamental law that states: the sum of the voltage drops and voltage sources in any closed loop must be equal to zero.

knee current The value of reverse current that flows through a zener diode when the voltage across it is at a value called the knee voltage.

knee voltage The point where reverse current in a zener diode begins to increase sharply while the voltage remains constant.

L/R time constant The time required in an RL circuit for the current to increase or decrease by 63% of the total possible change. Measured in seconds.

label One or more characters used to provide a symbolic reference for a particular value or address in an assembly language or high-level language program.

lag To be behind. Generally refers to one sine wave that occurs later in time (out of phase) than a second sine wave.

laminated Built up from several layers. Material is usually different on adjacent layers.

LAN Local Area Network. A network of computers servicing, for example, all of the computers in a given facility.

latching current The minimum value of current that must flow through a thyristor at turn-on in order for it to remain on after the gate trigger is removed.

lead To be ahead of. Generally refers to one sine wave that occurs earlier in time (out of phase) than a second sine wave.

leakage current A small (ideally zero) current that flows through an insulator.

leakage flux *See* flux leakage.

leakage inductance An apparent inductance caused by leakage flux.

LED *See* light-emitting diode.

Lenz's law The law that states: the polarity of an induced voltage will oppose the current change that caused it.

light-emitting diode (LED) A semiconductor device that emits light. LEDs are used for indicators and digital displays.

line regulation A measure of the degree to which the output voltage of a power supply is immune to changes in input voltage.

linear Describes a relationship between two quantities that are directly proportional to each other.

linear scales Scales on an axis of a graph whose divisions are uniformly spaced and equally valued.

lines of flux Imaginary lines that indicate strength and direction of a magnetic or electric field.

Lissajous pattern A pattern formed on an oscilloscope by applying a sinusoidal waveform to both vertical and horizontal channels. Lissajous patterns can be used to measure frequency and phase.

load A device or component that draws current from a circuit.

load current The current that flows through a load connected to a circuit.

load regulation A measure of the degree to which the output voltage of a power supply is immune to changes in load current.

load resistor A resistor connected across the output of a circuit (e.g., a loaded voltage divider).

loaded voltage divider A voltage divider that supplies power to other circuits or devices.

loader A program that loads the operating system or an application program into memory from an external storage device.

loading The changing of a circuit quantity by the connection of another component, circuit, or device.

local action A localized chemical activity in an electrochemical cell that destroys the electrodes and eventually ruins the cell. Local action can lessen the shelf life of a cell.

logic analyzer A digital test instrument that can monitor the simultaneous activity on several lines and display this activity at a later time.

logic probe A test instrument used to sense the logic level at any given point in a digital circuit during troubleshooting.

logic pulser A test instrument used to inject a logic transition into a digital circuit during troubleshooting.

long time constant A time constant (RC or L/R) that is at least ten times greater than the period of the input pulse waveform.

loop 1. A sequence of instructions that is repeatedly executed two or more times. 2. Any closed path for current.

low-pass filter A filter circuit that passes all frequencies below the cutoff frequency with minimal attenuation but attenuates all frequencies above the cutoff frequency.

machine language A method of representing instructions that is directly executable by the microprocessor.

macro A small program often consisting of recorded keystrokes that serves to automate portions of a task. Commonly used with word processors, spreadsheets, drawing programs, and other software packages.

magnet A material that possesses magnetism.

magnet wire Small-gauge wire with a varnish insulation used to make coil windings in motors, generators, electromagnets, relays, inductors, and transformers.

magnetic domain *See* domain.

magnetic field The region around a magnet where another magnet would experience a force of attraction or repulsion.

magnetic field intensity *See* magnetizing force.

magnetic flux *See* lines of flux.

magnetic polarity The relative direction of a magnet's field with respect to the earth's magnetic field. The north (actually north-seeking) pole of a magnet is attracted by the earth's north pole.

magnetizing current The current that flows in the primary of a transformer with no load on the secondary. Power supplied by the magnetizing current is consumed as transformer core and winding losses.

magnetizing force (H) The magnetomotive force per unit length.

magnetomotive force ($\mathcal{F}$) The magnetic force produced by current flowing through a coil.

magnitude The value of a circuit quantity without consideration of phase angle.

majority carriers Conduction-band electrons in n-type semiconductor material and holes in p-type material.

make-before-break A type of switching contact where the moveable part of the contact connects to the new position before contact is broken with the original position.

MAN Metropolitan Area Network. A network of computers servicing a metropolitan area.

marking state The normal (no-signal) state of an asynchronous communications line.

matter Anything that occupies space and has weight.

maximum power transfer theorem A theorem that states that maximum power (50%) can be transferred between two circuits when their resistances are equal.

maxterm A form of a Boolean expression written as the product-of-the-sums.

maxwell (Mx) A CGS unit of measure for magnetic flux.

mega (M) A prefix representing 10^6.

memory The functional part of a microprocessor system that stores instructions and data. Each location in memory has a unique identifier called the address.

memory effect A characteristic of certain electrochemical cells (e.g., NiCad cells) that causes them to lose their ability to supply large amounts of power after an extended period of time with only modest load requirements.

mesh analysis A technique used to analyze complex circuit configurations.

meter shunt A resistance connected in parallel with a meter movement that bypasses part of the total current, thus extending the effective range of the movement to a value greater than full-scale current value of the meter movement itself.

micro (μ) A prefix representing 10^{-6}.

milli (m) A prefix representing 10^{-3}.

Millman's theorem A simplification technique that is particularly well-suited to circuits having several parallel (nonideal) voltage sources.

minority carriers Conduction-band electrons in p-type semiconductor material and holes in n-type material.

minterm A form of a Boolean expression written as the sum-of-the-products.

mixture A material composed of more than one element or compound but whose dissimilar atoms are not chemically bound together.

MKS A system of measurement based on the meter, kilogram, and second. Also called the Système International, or simply, SI.

mmf *See* magnetomotive force.

modem Modulator-demodulator. An interface device that is used to convert the digital representation of data within a microprocessor system to a corresponding form suitable for transmission over a communications link. The conversion is typically from microprocessor voltage levels to equivalent frequency-shift-keying (FSK) tones. The modem also converts the FSK at the receiver back into microprocessor compatible signals.

modulating signal The information or intelligence signal used to modulate a high-frequency carrier.

modulation The process of modifying characteristics (e.g., amplitude, frequency, or phase) of a high-frequency carrier wave with a lower-frequency information signal.

molecule The smallest part of a compound that still exhibits the properties of the compound.

momentary-contact switch A switch whose contacts change from their normal position when activated (e.g., a button is pressed) but return to their normal position as soon as the mechanical activation force is removed.

monotonic output A characteristic of a D/A converter when its output is continuously higher for increasingly higher digital input values.

motor A device that converts electrical energy into rotating mechanical energy.

multimeter A measuring instrument (e.g., VOM, EVM, VTVM, DVM) capable of measuring resistance, voltage, and current.

multiplier resistor A resistance connected in series with a meter movement to increase the effective value of voltage required for full-scale deflection.

mutual inductance Inductance that is common to two magnetically-linked coils.

nanotechnology The integration of electronics and mechanics into an atomic-level machine built from individual molecules or atoms.

negative (−) The polarity of charge represented by an excess of electrons.

negative feedback Feedback that is out of phase with the original signal and causes a decrease in amplitude. Also called degenerative feedback.

negative ion 1. An atom with fewer protons than electrons. 2. A negatively-charged particle.

negative resistance The property of certain devices that causes a decrease in voltage at the same time there is an increase in current.

network 1. A configuration of electrical components; a circuit. 2. Two or more interconnected computers.

network operating system (NOS) The software on the server in a client/server network that supervises all network activity.

neutral Electrician's terminology for the side of the 120-Vac power that is also connected to earth ground.

neutron An atomic particle located in the nucleus of an atom and having no charge. Its mass is equivalent to that of a proton.

nodal analysis A circuit analysis technique applicable to complex circuits.

node 1. Any point in a circuit where the current divides. 2. Any computer connected to a network.

nominal The ideal or expected value of a component or circuit quantity.

nonlinear Describes a relationship between two quantities that are not directly proportional to each other. Uniform changes in one quantity cause corresponding changes in the second, but the resulting changes are not uniform.

nonvolatile memory A memory device whose contents are not lost when power is removed.

normally closed The contacts of a switch or relay that are closed without activating the switch or relay.

normally open The contacts of a switch or relay that are open without activating the switch or relay.

Norton's theorem A simplification technique that converts a network into an equivalent circuit consisting of a current source and a parallel resistance.

n-type Semiconductor material that has been doped by adding pentavalent impurity atoms to create an abundance of conduction band electrons.

nybble A group of four bits treated as a unit. In an 8-bit machine, two nybbles make up a byte.

octave An eightfold (i.e., 8:1 or 1:8) change in the value of a quantity.

oersted (Oe) The CGS unit of measure for magnetizing force.

ohm (Ω) The unit of measure for resistance.

ohmic resistance *See* dc resistance.

ohmmeter An instrument used to measure resistance.

Ohm's law A fundamental law that describes the relationship between current, voltage, and resistance in an electrical circuit.

ohm's per volt *See* voltmeter sensitivity.

open circuit A circuit that has no complete path for current flow; no continuity.

open loop A circuit that operates without feedback.

open-loop voltage gain The overall voltage gain of a device or circuit that has no feedback.

operating system The system software that controls the overall operating environment of a computing system. The operating system is often transparent to the user but performs such tasks as memory allocation, disk controlling, task prioritizing, and so on.

optoelectronics Electronic components that either produce light or respond to light.

oscillator A circuit that produces alternating voltage waveforms when a dc voltage is applied.

oscilloscope An instrument used to display a graph of instantaneous circuit quantities as functions of time.

output The voltage, current, or power taken from a circuit.

output offset voltage The voltage at the output of an op amp when the differential input voltage is zero.

output saturation voltage The extreme limits of output voltage for an op amp.

overhead Noninformational bits or characters that are transmitted with a message.

pad A small area or region of copper on a printed circuit board where a component lead solders to the circuit board.

parallax error An interpretation error on an analog meter movement caused by viewing the pointer and scale marks at an angle.

parallel A method of connecting circuit components such that all components connect between the same two points.

parallel data transfer A method where multiple bits are transmitted simultaneously through individual lines.

paramagnetic Materials with a relative permeability slightly greater than one.

parameter Any electrical quantity or circuit characteristic.

parasitic oscillation An unintentional oscillation that occurs in a circuit, generally due to parasitic inductance and capacitance.

parity A method of error checking that appends a parity bit to a transmitted byte such that the total number of 1 bits is odd (odd parity) or even (even parity).

partially-specified circuit A circuit where one or more component values are unknown.

pass transistor The primary regulator transistor in a series-voltage regulator.

passband The band of frequencies that are passed with minimal attenuation by a filter circuit.

passive component An electrical component that cannot amplify or rectify. Resistors, inductors, and capacitors are passive components, whereas transistors and other solid-state devices are active components.

peak value The maximum instantaneous value of a waveform.

peak-to-peak value The measure of an alternating circuit quantity that describes the difference between the minimum and maximum levels.

pentavalent An atom that has five outer-orbit electrons.

period (t) The time required for one complete cycle of a periodic waveform.

periodic A waveform that repeats at regular intervals.

peripheral A machine or device that is external to a computer and used to input or output data.

permanent magnet A magnet that does not lose its magnetism when the magnetizing force is removed.

permeability (μ) The ability of a material to concentrate magnetic flux. Often expressed as a dimensionless ratio of the permeability of the material relative to air.

permeance ($\wp$) A measure of the ease with which a magnetic field may be established in a material. Analogous to conductance in an electrical circuit.

permittivity (ϵ) The ability of a dielectric material to concentrate an electric field.

phase The timing of a waveform relative to another waveform that has an identical frequency.

phase modulation The modulation that occurs when the instantaneous phase of a carrier wave is modulated by the information signal.

phasor A graphical representation of the magnitude and phase of an electrical quantity.

photoconductive mode A method of biasing photodiodes such that current through the device is proportional to incident light.

photon A unit or packet of light energy.

photovoltaic mode A method of operating photodiodes with no bias such that the voltage developed across the device is proportional to incident light.

pi configuration See delta configuration.

pick-up voltage The coil voltage required to energize a relay.

pico (p) A prefix representing 10^{-12}.

piconet A small wireless network formed by Bluetooth™-enabled products.

piezoelectric effect A phenomenon that causes certain materials to generate a voltage when mechanically stressed or to mechanically distort when a voltage is applied.

pn junction A region internal to a semiconductor crystal where p and n materials meet.

polar notation A method of expressing a complex circuit quantity that includes both magnitude and phase information (e.g., 50 ∠ 45°).

polarity A term used to describe the direction of the relative potential between two points in a circuit. Also used to describe the direction of a magnetic field.

polarization A buildup of hydrogen gas on the positive electrode of an electrochemical cell that degrades the cell's operation.

pole 1. The moveable contact of a relay or switch. 2. The area of maximum flux density in a magnet.

population inversion A condition within a material occurring when there are more atoms with electrons excited into the conduction band than atoms with complete bonds in the valence band.

port A point in a computer circuit where data enters or leaves the computer proper.

positive (+) The polarity of charge represented by a deficiency of electrons.

positive feedback Feedback that is in phase with the orginal signal and causes an increased amplitude. Also called regenerative feedback.

positive ion 1. An atom with fewer electrons than protons. 2. A positively-charged particle.

potential The ability of an electrical charge to do work by moving electrons.

potential difference Voltage.

potentiometer A three-terminal variable resistor. The end-to-end resistance remains constant, but the resistance from either end to the wiper varies as the wiper is moved.

power (P) The rate of energy consumption; the rate of doing work.

power factor (pf) The ratio of true power to apparent power in a circuit.

power of ten A method used to express large or small numbers as a modest number times a power of ten (i.e., ten raised to some exponent).

power supply A device or circuit used to supply electrical energy. In most cases, the power supply delivers direct current and voltage.

precision resistor A resistor whose value is designed to be within 2% of its nominal value.

pressure A measure of force distributed over an area.

primary The winding on a transformer where power is applied.

primary cell An electrochemical cell that is not designed to be recharged. The chemical processes are not generally reversible (practically).

printed circuit board A thin (e.g., 0.0625″) sheet of insulating material with an electrical circuit formed by applying thin copper strips to interconnect electronic components mounted on the sheet.

program A group of microprocessor instructions arranged in a sequence such that a desired result is produced when they are executed sequentially.

program branch See branch.

program counter A counter in a microprocessor that holds the address of the next instruction to be executed.

program label See label.

program loop See loop.

programmable logic controller (PLC) An industrial computer whose programming and status reporting are often based on relay logic, even though the computer itself is solid state. Commonly used to monitor and control industrial processes.

propagation delay The delay between a state change on the input of a logic gate and the corresponding state change on the output.

protocol The body of rules or standards defining the hardware and software requirements for a standardized communications scheme.

proton A positively-charged particle in the nucleus of an atom.

p-type Semiconductor material that has been doped by adding trivalent impurity atoms to create an abundance of holes in the valence band.

pull-in current The coil current required to energize a relay.

pulsating dc Unidirectional current with periodic fluctuations.

pulse-frequency modulation A form of pulse modulation in which the frequency of the pulses is varied in response to the modulating signal.

pulse-width modulation A form of pulse modulation in which the width of the pulses is varied in response to the modulating signal.

quality factor (Q) A dimensionless figure of merit for inductors, capacitors, and resonant circuits. Higher Q values correspond to less losses and more ideal performance.

quiescent operating point The operating point of a transistor or op amp with no signal applied.

race condition The condition whereby the inputs to a logic device do not arrive at the same time.

radian (rad) An angular measure equal to 57.3°.

radix point A mark which separates the integer portion of a mixed number from the fractional portion. A decimal point in the base 10 system.

RAM Random Access Memory. Generally refers to volatile semiconductor memory capable of both read and write operations.

RC time constant The time required for the capacitor voltage in an RC circuit to change by 63% of the total possible change.

reactive power (P_R) A measure of the power that is taken from and subsequently returned to a circuit by the inductance and capacitance in the circuit.

read The process of retrieving information stored in a memory device.

read-only memory A memory device that is programmed during the manufacturing process. It is nonvolatile and nonalterable.

real number Any rational or irrational number.

real power *See* true power.

recombination Occurs when a conduction-band electron loses energy and falls into a hole in the valence band. The electron-hole pair is eliminated.

rectangular notation A method of representing complex circuit quantities that includes both real (resistive) and imaginary (reactive) portions (e.g., $25 + j10$).

rectify To convert from alternating current or voltage into unidirectional current or unipolar voltage.

recurrent sweep A type of sweep circuit in an oscilloscope that causes the horizontal sweep to occur automatically (i.e., without a trigger).

reflected impedance The apparent impedance in the primary of a transformer that results from a load in the secondary.

refraction The bending of light as it passes through the interface of two materials having different optical densities.

regenerative feedback *See* positive feedback.

register An array of flip-flops used for temporary storage and manipulation of data within a digital system.

relative permeability (μ_R) A dimensionless measurement of the permeability of a material relative to the permeability of air (or vacuum).

relaxation oscillator An oscillator that relies on time delays provided by an RC circuit to determine its frequency of oscillation.

relay An electromechanical device that consists of an electromagnet and switching contacts. The switch contacts are activated by energizing the electromagnet.

reluctance ($\mathcal{R}$) A measure of a material's opposition (magnetic resistance) to magnetic flux.

residual magnetism The level of magnetism remaining in a material after the magnetizing force has been removed.

resistance (R) Opposition to current flow in a circuit; measured in ohms.

resistivity (ρ) The resistance of a given volume (e.g., one cubic meter) of a material.

resistor An electrical component designed to provide a given opposition to current flow.

resistor tolerance The maximum amount of deviation due to manufacturing tolerances between the nominal and actual values of a resistor.

resolution The smallest change in a measured value that can be resolved or displayed by a measuring instrument.

resonance A condition in an *LC* or *RLC* circuit where $X_L = X_C$ and the impedance of the network is maximum (parallel circuit) or minimum (series circuit).

resonant circuit An *LC* or *RLC* circuit that is operating at its resonant frequency.

resonant frequency The frequency in an *LC* or *RLC* circuit that causes the inductive reactance to be equal to the capacitive reactance.

response time The time required for a fuse to open in response to an overcurrent condition.

retentivity The property of a magnetic material that causes it to have residual magnetism.

reverse bias The bias on a pn junction when the anode is less positive than the cathode.

reverse breakdown voltage *See* breakdown voltage.

reverse current The current (ideally zero) that flows through a reverse-biased pn junction.

rheostat A two-terminal variable resistor.

ribbon cable A cable made by bonding several insulated wires into a flat strap. Generally terminated with crimp-on connectors that attach to all wires simultaneously.

right angle A 90° angle; an angle of $\pi/2$ radians.

ripple voltage The fluctuations that remain in the output of a power supply that rectifies and filters an ac input voltage.

roll-off The slope of the frequency response curve of a filter circuit beyond the cutoff frequency.

ROM Read-Only Memory. A type of computer memory that may be read (information retrieved), but it cannot be written (new information stored).

root-mean-square (rms) A unit of measure for alternating circuit quantities, which is an amount that produces the same heating effect in a resistance as a similar value of dc. It is numerically equal to 70.7% of the peak value in a sinusoidal circuit.

rotor The rotating coil of a motor, generator, or alternator.

sample-and-hold An interface circuit for an analog-to-digital converter. It tracks the analog signal for a specified time period and then holds the sampled value constant while the A/D converter accomplishes the actual conversion process.

saturation The condition of a transistor when it is fully on (i.e., maximum conduction).

schematic A diagram that depicts the various components and interconnections in an electrical circuit.

secondary A winding on a transformer from which power is removed.

secondary cell An electrochemical cell that is designed to be recharged. The chemical processes are reversible.

selectivity The ability of a circuit to pass certain frequencies and reject others.

self-holding contacts A relay circuit configuration where a normally open set of contacts is in parallel with the switch used to activate the relay. Once energized, the relay remains energized or latched.

self-inductance The property of a conductor or coil that opposes any change in current.

self-resonant frequency The frequency at which the parasitic components in a coil or capacitor resonate with the intended quantity.

semiconductors A class of materials whose conductivity lies between that of conductors and insulators.

sensitivity *See* voltmeter sensitivity.

serial data transmission The transmission of the bits of a data word one at a time via a single wire. The serial mode is slower than parallel methods, but the circuit is much less complex.

series A circuit configuration that results in a single path for current flow.

series-aiding Voltage sources connected in series with similar polarities are series-aiding. The net voltage is the sum of the individual sources.

series-opposing Voltage sources connected in series with opposite polarities are series-opposing. The net voltage is equal to the difference between the two source voltages.

series-parallel A circuit configuration that consists of groups of parallel and series components or networks.

seven-segment display A digital display consisting of seven illuminated segments. Selective illumination of the segments allows displaying of the numbers 0–9.

shelf life The length of time that an electrochemical cell can remain inactive and still be expected to deliver its rated characteristics.

shells Orbital levels of electrons in an atom. Each shell is a discrete region, but it may have electrons orbiting at multiple levels within the defined region.

shield A conductive or permeable sheet or enclosure used to provide electromagnetic isolation between two circuits.

shift register A register whose individual bits are shifted (left or right) into the adjacent bit position when a clock is received.

shoot-through logic Logic circuitry that reduces the effects of current transients (shoot-through) that occur when totem-pole logic outputs change states.

short circuit A low-resistance path that essentially bypasses the shorted component(s).

short time constant An *RC* or *L/R* time constant that is less than one-tenth the period of the input pulse.

shorting switch *See* make-before-break.

shunt 1. Parallel. 2. A resistor used to extend the current range of an ammeter.

SI Système International. A system of measurement based on the meter, kilogram, and second. Sometimes called the modern MKS system.

siemen (S) The SI unit of measure for conductance, admittance, and susceptance.

simplex A communication link that allows data transfers in only one direction.

sine A trigonometric function of an angle in a right triangle that is equal to the length of the opposite side divided by the length of the hypotenuse.

sine wave A periodic waveform whose amplitude-versus-time graph is the same shape as a graph of the trigonometric sine function versus degrees.

sinusoidal Any waveform (regardless of phase) that is generally shaped like a sine wave.

skin effect A phenomenon that causes high-frequency currents to flow near the surface of a conductor, which reduces the effective cross-sectional area of the wire and increases the effective (ac) resistance of the wire.

slew rate The maximum rate of change of voltage on the output of an op amp.

slip rings Metal rings on the rotor of an alternator or ac motor that make sliding contact with the brushes and provide a means of connecting or removing power from a rotating coil.

slo-blo A type of fuse that can withstand currents greater than its rating as long as the excessive current is only momentary.

SMT *See* Surface-Mount Technology.

sneak path A nonobvious current path that is in parallel with a component whose resistance is being measured.

software The programs, documentation, operating procedures, and the like, associated with a computer system.

solar cell A solid-state device that converts light energy into electricity.

solder A tin and lead alloy with a low melting point that is used to permanently bond electrical connections.

solenoid 1. Any coil wound on a long coil form. 2. An electromagnetic device that converts electrical energy into a pushing, pulling, or twisting motion.

solid state A device constructed of a crystalline semiconductor material such as silicon or germanium.

source One of three elements in a field-effect transistor. Analogous to the emitter in a bipolar transistor.

source code The statements comprising an assembly language or high-level language program. The source code is used as input to the assembler, interpreter, or compiler program.

space state The logic level on a digital, asynchronous communications line that is opposite from the marking or no-signal state.

specific gravity The ratio of the weights of equal quantities of some liquid and water.

speed A measure of how fast an object changes position without regard to specific direction.

spiraling A manufacturing process used to trim film resistors to the correct value.

spontaneous emission Energy (often in the form of visible light) that is released when a conduction-band electron recombines with a valence-band hole.

start bit A single bit used to identify the beginning of transmission in an asynchronous serial communications link.

state symbol *See* bubble.

stator The stationary or nonrotating part of a motor, alternator, or generator.

step-down transformer A transformer whose secondary voltage is lower than its primary voltage.

step-up transformer A transformer whose secondary voltage is higher than its primary voltage.

stimulated emission Light energy emission in a material occurring when recombination of electron-hole pairs is timed or triggered by incoming light energy. It has a multiplying effect, and the released light energy has the same wavelength and phase as the incoming light. Fundamental to laser operation.

stop bit One or more bits used to identify the end of a transmission in an asynchronous serial communications link.

subroutine A program segment that can be called from any point in a program and will then transfer control back to the original point in the program upon completion.

subroutine nesting The result of subroutines being called by other subroutines.

substrate 1. The insulating base material used as the core of film resistors. 2. The base semiconductor layer in a transistor or integrated circuit.

superconductivity A characteristic of certain materials that causes their effective resistance to drop to zero as their temperature approaches absolute zero.

superconductors Materials that exhibit superconductivity.

superposition A simplification procedure for multiple-source, linear circuits.

surface-mount technology A type of leadless packaging for electronic components that allows the fabrication of high-density circuit boards.

surge A short-duration transient or current burst.

susceptance (B) The ease with which current flows through a reactive component; the reciprocal of reactance. Measured in siemens.

switch An electrical component used to open and close a current path in a circuit.

symmetrical A waveform whose positive and negative alternations are equal in amplitude, duration, and shape.

synchronous counter A counter whose various stages all receive a clock from the same source. Any toggling stages toggle at the same time.

syntax The exact rules describing the allowable structures and statements in assembly language and high-level language programming.

TAN Tiny Area Network. A small computer network scheme often used for industrial controllers.

tangent A trigonometric function of an angle in a right triangle that is equal to the length of the opposite side divided by the length of the adjacent side.

tank circuit A parallel LC circuit.

taper Describes the relationship (linear or logarithmic) between the resistance and the angle of rotation of a variable resistor.

tee configuration A circuit configuration in which one end of three components connects to a common point. Also called a wye configuration. One of two common connections for three-phase transformer windings.

telecommunications Describes any of the various methods for transmitting data over distances, such as by radio, telephone, and the like.

telemetry Controlling or monitoring equipment remotely.

temperature coefficient Describes the relationship between a circuit parameter and its temperature. May be positive, negative, or zero.

temporary magnets A magnetic material with minimal residual magnetism.

tesla The SI unit of measure for flux density.

tetravalent An atom with four outer-orbit electrons.

thermistor An electrical device whose resistance varies with temperature in a specific way.

thermocouple A bimetallic junction that converts heat energy into electrical energy.

thermopile A series connection of several thermocouples.

Thevenin's theorem A simplification technique that converts a network into an equivalent circuit consisting of a voltage source and a series resistance.

throughput A relative term used to describe the amount of information that can be transmitted over a given communications path in a certain time interval (generally one second).

throw Identifies the number of circuits that are opened or closed *by each pole* of a switch.

time constant A fixed time interval determined by the RC or RL values in an RC or RL circuit that sets the time required for the voltage and current to change by 63% of the total possible change.

toggle The changing of states in a bistable digital device such as a flip-flop.

tolerance The maximum amount of deviation due to manufacturing tolerances between the nominal and actual values of a component.

toroid A donut-shaped object. Usually refers to a toroidal-shaped core used for a transformer or coil.

totem-pole output A configuration of two series transistors in the output stage of an integrated circuit. The output is taken from the common connection between the transistors. One transistor connects to the negative source, and the other connects to the positive source.

traces Strips of copper on a printed circuit board that provide the interconnections between the various electronic components.

transducer A device used to sense the energy in one system and proportionally control the energy in another. Generally considered to be a device that converts energy from one form to another.

transformer A device that couples electrical energy from one circuit (primary winding) to another (secondary winding) via magnetic flux linkage.

transient A momentary, short-duration voltage or current surge.

transponder A circuit that receives a transmission, alters its frequency, and then retransmits the signal.

triggered sweep A type of sweep circuit used in oscilloscopes that produces no horizontal sweep until a specific set of conditions exists (e.g., the input waveform is at a given voltage).

trimmer A variable component used to adjust a circuit parameter to a precise value. Generally provides a very limited range of adjustment (i.e., fine tuning).

tripped A term used to describe the condition of a circuit breaker after it has opened the circuit.

trivalent An atom with three electrons in the outer orbit.

troubleshooting The process of locating the defective components in an electrical circuit.

true power (P_T) The power, measured in watts, that is dissipated by the resistance in a circuit.

truth table A table that lists all possible combinations of inputs and the corresponding outputs for a combinational logic circuit.

turns ratio The ratio of primary-to-secondary turns in a transformer.

unipolar Single polarity. Also refers to semiconductor devices whose operation depends upon only one type of majority current carrier (either holes or electrons).

universal shunt A method of connecting and switching an ammeter shunt to provide multiple ranges.

unloaded voltage divider A voltage divider that does not provide power to other circuits or devices.

valence band The outermost orbit of an atom.

valence electrons Electrons in the outermost orbit of an atom.

VAR Abbreviation for volt-ampere-reactive. The unit of measure for reactive power.

vector A graphical representation of the magnitude and phase of an electrical quantity or the magnitude and direction of a mechanical force.

velocity Rate of change of position in a specific direction. Loosely interpreted as speed.

vinculum The bar drawn above variables in Boolean algebra to signify the NOT or complement operation.

virtual ground A point in a circuit that behaves as a ground connection but is not actually connected to ground.

volatility The property of certain memory devices that results in loss of data when power is removed.

volt (V) The SI unit of measure for electromotive force.

voltage (V) The amount of potential difference between two points that can be used to cause current flow.

voltage breakdown The condition that exists when an insulator material is subjected to a sufficiently high voltage to destroy the high-resistance characteristics of the insulator. Breakdown results in a high current that may or may not permanently damage the insulator.

voltage divider A series circuit used to reduce a supply voltage to one or more lower voltages.

voltage divider equation An equation that determines the amount of voltage (V_X) across a specific resistor (R_X) in a multi-resistor voltage divider. It is expressed as follows:

$$V_X = \left(\frac{R_X}{R_T}\right) V_A$$

voltage drop (V) The voltage developed across a component as a result of the current flowing through it.

voltage regulation The process of maintaining a relatively constant output from a voltage source even when the load current or input voltage varies.

voltage source A circuit or device that provides a relatively constant voltage to other circuits or devices.

voltmeter An instrument used to measure voltage.

voltmeter loading Occurs when the internal resistance of a voltmeter causes the measured voltage to be less than the actual (unloaded) voltage in the circuit. The higher the internal resistance of a voltmeter, the less it disrupts or loads the circuit.

voltmeter sensitivity   A rating that indicates how much series resistance is required to limit the current through the meter movement of a voltmeter to the full-scale value. Higher sensitivities produce less voltmeter loading.

VOM   Volt-ohm-milliameter. An analog meter that measures voltage, current, and resistance.

WAN   Wide Area Network. A computer network that covers a very large geographical area and is generally composed of multiple LANs and/or MANs.

watt (W)   The SI unit of measure for power.

wattmeter   An instrument used to measure true power.

Watt's law   A fundamental law that describes the relationships between power, current, voltage, and resistance.

waveform   A graph of voltage, current, or other circuit parameter versus time.

wavelength   The physical distance that an acoustical or electromagnetic wave travels in the time required for one complete cycle of the waveform.

weber (Wb)   The SI unit of measure for magnetic flux.

Wheatstone bridge   *See* bridge circuit.

winding   The turns of wire on an electromagnetic device such as a coil, transformer, relay, or motor.

wiper   A moveable, sliding contact; the center terminal of a potentiometer.

wireless link   A communications path that does not require physical wires. Generally refers to radio or infrared communication links.

wiring diagram   A diagram that shows the electrical connections between major components in the circuit. It is closely related to the physical nature of the system (e.g., wire colors are often shown).

word   A unit of data in a computer consisting of two bytes.

work   The expenditure of energy. Mechanical force × distance. Electrical power × time.

write   The process of storing information into a memory device.

wye configuration   *See* tee configuration.

z-axis   Refers to intensity control of the CRT display in an oscilloscope.

zener voltage   The rated voltage of a zener.

zeroed   The condition of an ohmmeter that has been calibrated such that when the leads are shorted together, the pointer indicates zero ohms.

INDEX

–A–

Abbreviations, for electrical quantities, 37
Absolute permeability, 44
Absolute permittivity, 219
Absorption, 491
ABT. *See* Advanced BiCMOS technology (ABT)
AC. *See* Alternating current (ac)
AC and DC waveforms, combined, 169
Accelerated graphics port (AGP), 606. *See also* AGP busses
Acceleration, 39
Acceptor atom, 268
AC circuits, current flow in, 220
Accumulator, 600
AC failures, 420
AC generators, 155
Acoustical waves, 41, 46
ACOS. *See* Arccos operation (ACOS)
AC resistive circuits, circuit analysis of, 176–180
Active filter, 410–417
Active rectifier, 407–409
AC voltmeters, 163
A/D. *See* Analog-to-digital (A/D) converter (ADC)
AD566A device, 579
ADC. *See* Analog-to-digital (A/D) converter (ADC)
Address busses, 3, 597
Advanced BiCMOS technology (ABT), 366
Advice, bad, 285
AGP. *See* Accelerated graphics port (AGP)
AGP busses, 609
Aircraft electronics (avionics), 2–3
Air-core transformers, 207–208
Alpha (α), of a transistor, 337
Alternating current (ac), 153–186. *See also* AC entries
Alternating voltage, 185
  applications of, 181–185, 186
  generation of, 153–157
Alternations, 154
Alternator, construction of, 155

ALU. *See* Arithmetic logic unit (ALU)
Aluminum electrolytic capacitors, 224
AM. *See* Amplitude modulation (AM)
AM detector, 310
AM detector circuit, 309–311
American Wire Gauge (AWG) numbers, 58
AMP02 instrumentation amplifier, 527
Ampacity, 58
Amp circuits, identification of, 393
Ampere-hour (Ah) ratings, 64
Amperes, 29
Ampere-turns per meter, 43
AM receiver, 310
Amplification, gain and, 339
Amplifier circuit components, purpose of, 398–399
Amplifier circuits, operation of, 394–396
Amplifiers, 377
  biasing, 358
  classes of, 342
  configurations of, 342–346, 367–372, 392–403
  general characteristics of, 339–342
  JFET, 350–352
  low-noise, 358–359
  MOSFET, 358–359
  types of, 372
Amplitude modulation (AM), 309, 637–638, 650
Analog and digital systems, interfacing, 576–581
Analog information, 635
Analog integrated circuits, 523
Analog meter, 373–374
Analog ohmmeter, 228–230, 487, 488
Analog signal, 307
Analog-to-digital conversion, 579–580, 641
Analog-to-digital (A/D) converter (ADC), 307, 586, 641
Analog transmissions, 650
Analog VOM, 286
Analysis software, 118
AND operator, 548–549
Anode lead, 292

Anode-cathode circuit, 479
Anode-to-cathode open, 321
Anode-to-cathode short, 321
Anti-ringing circuit, 445
APDs. *See* Avalanche photodiodes (APDs)
Application software, 615–616
Approximations, mental, 131
Arccos function, 173
Arccos operation (ACOS), 173
Arcsin function, 174
Arcsin operation (ASIN), 173
Arctan function, 175
Arctan operation (ATAN), 173
Arithmetic logic unit (ALU), 599–600
Armature, 79–80
ASIN. *See* Arcsin operation (ASIN)
Assembly language, 612–613
Asymmetrical time constants, 308–309
Asymmetrical waveforms, 154, 185
Asynchronous communication, 633–634, 650
Asynchronous counters, 569
Asynchronous flip-flop counters, 571–573
Asynchronous inputs, 566–567
ATAN. *See* Arctan operation (ATAN)
Atomic model, 262
Atomic structure, 23–25
Atomic theory, 261–264
Atoms, 46
Audio circuits, 609
Audio communications, 622
Audio transformers, 206
Automobile car analyzer, 17
Auto theft alarms, 479
Autotransformer, 209–210
Avalanche current, 280–281
Avalanche photodiodes (APDs), 496, 504
Average value, 160–162
"Average value of a sine wave," 161–162
Avionics, 2–3
AWG. *See* American Wire Gauge (AWG) numbers
Axial lead capacitors, 221

731

–B–

Backbone, network, 625
Bandpass, 250
Bandpass active filter, 414–415
Bandpass filters, 239
Band-reject active filter, 415–417
Band-reject filters, 239
B&S. *See* Brown and Sharpe (B&S) gauge
Bandstop filter, 415
Bandwidth, 251, 650
Barrier potential, 277, 279, 293
    forward bias and, 277–278
    reverse bias and, 279
Base, of a transistor, 332
Baseband transmission, 631–632
Base-collector junction, 335
Base current, 470
Base-emitter junction, 333–335, 342, 377
Base voltage, 369
Basic Input Output System (BIOS), 608. *See also* BIOS memory
Batteries, 64
Battery backup, in computers, 308
BE junctions, forward- and reverse-biased, 336–338
Beta (β), of a transistor, 338
Bias current effects, 398
Biasing
    of junction field-effect transistors, 349–350
    of pn junctions, 333
    requirements for, 338
    simultaneous, 335–338
Biasing amplifier, 358
Bias voltage, polarities of, 334
Bidirectional bus driver, 598
Bidirectional busses, 597
Bidirectional thyristors, 468
Binary digit (bit), 366, 542
Binary number system, 542–546, 586
Binary to decimal conversion, 544
Binary to hexadecimal conversion, 545–546
BIOS. *See* Basic Input Output System (BIOS)
BIOS memory, 615
Biphase half-wave rectifier, 299, 325
Bipolar device, 274
Bipolar logic device, 363–365
Bipolar transistor amplifiers, 338–347
Bipolar transistors, 332–348, 377
    characteristics of, 347
    identifying defects in, 373
    versus JFETs, 350
Bistable operations, 467
Bit. *See* Binary digit (bit)
Block diagrams, 3–4, 16
    of a phase-locked loop, 519–521
Block-diagram thinking, 15–16

Bluetooth communications, 647–649, 650
Bluetooth devices, 184, 251
Bluetooth global specification, 183–184
Bluetooth-enabled devices, 647
Boolean algebra, 548–554, 586
    postulates of, 553–554
Boolean expressions, for logic circuits, 560–561
Boost regulator, 443
Bootstrap loader, 615
Branch currents, 234
    calculating, 105–106
Branches, 91. *See also* Parallel branches
    current through, 103–104
    of parallel circuits, 141
Breakdown voltage, 60
Breakover voltage, 467, 474, 475, 477
    exceeding, 470–471
Bridge rectifier, 300–301, 325
Broadband transmission, 631–632
Brown and Sharpe (B&S) gauge, 58
BS08D device, 477
Bubble, 550
Buck regulator, 443
Bus configurations, 606
Bus networks microprocessor, 597
Bus network topology, 625–626
Busses, 3
    standard, 606
Bypass capacitor, 350
Byte, 596

–C–

Cable, 61–63, 82
    Ethernet, 207
    troubleshooting, 62–63
    types of, 58–62
    versus single-conductor wires, 61
Cable connectors, 62
Cable television, 632
Cache memory, 608
Calculations, sequence of, 114
Calculator operation, 172
Calculator sequences, 52–53
Candela, 40
Candlepower, 40
CANs. *See* Controller area networks (CANs)
Capacitance, 215
    defined, 215
    factors affecting, 218–219
    testers for, 228
    unit of measure for, 218
Capacitance value, 223
Capacitive circuits
    circuit analysis of, 228
    current flow in, 219–220
Capacitive reactance, 220–221, 254

Capacitor charges, 216–218
Capacitor connections, series and parallel, 225–228
Capacitor filters, 302–303
Capacitor plates, 216
Capacitor ratings, 223–224
Capacitors, 74–75, 82, 215–230, 254, 355
    construction of, 216
    discharging, 218
    fixed and variable, 222
    nonideal, 253–254
    polarized and nonpolarized, 223
    schematic symbols for, 75
    shorted, 230
    troubleshooting, 228–230
    types of, 221–223
Capacitor technologies, 224–225
Capacitor voltage, timing of, 236–237
Capacitor voltage phasor, 232
Car battery, 27
Carbon-composition resistors, 69
Carrier, 309
Carrier migration, 275–277
Carrier wave, 309, 637
Cascode amplifiers, 370–371
Cathode lead, 292
CB junctions
    forward-biased, 338
    reverse-biased, 336–338
CDS. *See* Compact disk (CD) players
Cells, 64
Center-tapped rectifier circuit, 299
Ceramic capacitors, 224, 230
Channels, 348
Charge (Q)
    electrical, 25
    unit of measure for, 26
Charges, attraction and repulsion by, 26
"Charging the capacitor," 216, 217
Chips, 10
Chokes, 192
    common-mode, 207
Circuit analysis, 20
    of AC resistive circuits, 176–180
    of capacitive circuits, 228
    of inductive circuits, 201
    software for, 118
Circuit analysis problems, importance of, 118
Circuit behavior, 191
    intuitive understanding of, 131, 142
Circuit board traces
    defects in, 137
    repairing, 323
Circuit breakers, 65, 66
Circuit complexity, 442–443
Circuit connectivity problems, 137–138
Circuit construction, 21

Circuit elements, parasitic, 252–254
Circuit ground, 122
Circuit operation
  of a bridge rectifier, 300–301
  of a half-wave rectifier, 296–297
Circuit quality, 249–250
Circuits, 87–123
  basic requirements for current, 87–90
  classes of, 108
  complex, 118–119
  electrical power in, 35
  ground and other reference points in, 119–122
  integrated logic, 363–366
  interconnecting components in, 87
  mathematical analyses of, 93
  parallel, 101–108
  RC and RL, 231–241
  series, 90–101
  series-parallel, 108–117
  troubleshooting, 149
Circuit simplification, 178
  steps in, 114
Circuit simulation, 20, 21, 22, 511, 513, 515
  environment for, 130–131
  software for, 20, 22, 126, 151–152
  user manual for, 48
Circuit theory, understanding, 323
Circuit troubleshooting, 129–148
  of series circuits, 130–138
Circuit waveforms, 238–239, 529, 530
Circular mils (cmil), 57
Cladding, 14, 500
Client/server architecture, 628
Clipper circuits, 305–307, 325
Clipping, 389
Clock generator circuit, 607
Clock pulse, 566
Clocks, computer, 574
Clock skew, 607
Closed-loop circuits, forcing extreme changes in, 456–457
Closed-loop differential input voltage, 394
Closed-loop equations, 95–96
Closed loops, 387
Closed-loop voltage gain, 389, 390–391, 396, 397, 401, 402
CMOS. See Complementary metal oxide semiconductor (CMOS)
CMOS devices, 366
CMOS inverter crystal oscillator, 517
CMRR. See Common-mode rejection ratio (CMRR)
Coaxial cable, 61–62, 82
Coefficient of coupling, 205
Coherent light, 492
Coils, 76, 192, 193, 194

overheated, 202
shorted, 201–202
Collector-base junction, 343
Collector-clamping diode, 364
Collector current, 470
Collector region, 332
Collimation, 492
Combinational logic, 555–565, 581–584
Combinational logic circuits, 586
Combination lock system, electronic, 575
Common-base amplifier, 345–346
Common-collector amplifier, 344–345, 366–367
Common-drain amplifier, 351
Common-emitter amplifier, 342–344
Common-gate amplifier, 351
Common-mode choke, 448
Common-mode noise, 207, 498
Common-mode rejection ratio (CMRR), 524
Common-mode voltage gain, 389
Common-source amplifiers, 350–352, 370
  MOSFET, 358
Communication, wireless, 251–252. See also Telecommunication
Communication circuits, 317
Communication links, 635
Communication services, 623
Communications satellites, 646–647
Communication systems
  bandwidth for, 631
  characteristics of, 630–636
Compact disk (CD) players, 492
Compensation pins, 385
Compilers, 613, 616
Complementary metal oxide semiconductor (CMOS), 365. See also CMOS entries
Complete path, 123
Complete temperature coefficient, 223
Complex circuits, 118–119, 123
  versus series-parallel circuits, 108–110
Complex programmable logic device (CPLD), 558
Complex voltage sources, 100
Component leads, bending, 283
Components
  desoldering, 142
  in-circuit testing of, 147–148
  incorrectly valued, 136
  nonideal, 253
  open versus shorted, 135
  optoelectronic, 40
  parallel, 101, 102
  power dissipation in, 140
  removing from circuit boards, 147
  repeated defects in, 144

replacing, 323
symbols used to represent, 8
test instruments for, 146
troubleshooting fences and, 144
Component substitutions, 18–19, 146–147, 375
Component tolerances, effects of, 135–136
Computer "brains," 595, 607
Computer circuits, 167
  high-speed, 220
Computer clocks, high-speed, 521
Computer-controlled machines, 361
Computer hardware, 605–611
Computer keyboard, troubleshooting, 17–18
Computer memory, 607–608, 616
  defects in, 17
Computer memory circuit, isolation diodes for, 307
Computer oscillator circuit, 509
Computer power supplies, 610
Computers, 2, 595–619. See also Industrial computers; Notebook computers
  clippers in, 305–307
  external cooling in, 344
  ferrite beads in, 197
  industrial, 241
  relays and, 80
  transistors as current amplifiers in, 366–367
Computer software, 611–616
Computer systems, 616
  block diagrams for, 4, 605–606
  integrated circuits in, 81
  transformers in, 204
  troubleshooting, 5
Conductance, 36–37
Conduction band, 263–264
Conduction-band electrons, 271, 272
  migration of, 275–276
Conductive wire, 13
Conductors, 25, 46, 262
Constant current source, 369
Control busses, 597
Control information, 623–625
Controller area networks (CANs), 629, 650
Control lines, 4
Conventional current, 28–29
Conversion, step-up versus step-down, 443
Copper atom, 27–28
Copper wire, sizes for, 59
Core material, of inductors, 196
COS button, 172
Cosine function, 172
Coulomb, 26
Counters, 569–573
  as frequency dividers, 573

Counting sequence, determination of, 570
Coupling capacitor, 343
Covalent bonds, 266, 287
 broken, 268
CPLD. See Complex programmable logic device (CPLD)
CPU clocks, 522
Crests, wave, 40
Critical angle, 500
Crowbar circuit, 451, 458
Crystal lattice, 287
Crystal oscillators, 516–518
Crystal pattern, 265
Crystals, 535–536
 construction of, 516–517
Current (I), 46. See also Kirchhoff's Current Law
 basic requirements for, 87–90
 calculating the value of, 33
 complete path for, 88–89
 conventional, 28–29
 direction of, 89–90
 electron, 28–29
 induced, 193–194
 mathematical expression for, 93
 relationship to resistance, 32
 relationship to voltage, 31–32
 in a series-parallel circuit, 113–116
 unit of measure for, 29–30
 visualizing, 28
Current amplifiers, 342
 transistors as, 366–367
Current capacity, 107
Current carriers, majority versus minority, 274
Current flow, 25, 27–28, 87, 89
 maximum, 278–279
 minimum, 280
 in a semiconductor, 272–274
 in a series circuit, 90
Current gain, 340
Current limiting, in switching regulators, 450
Current-limiting circuits, 449–450
Current limit pin, 450
Current meter, 138
Current peaks, 195–196
Current rating, of a fuse, 65
Current ratio, 211–212
Current-regulator diodes, 316–317
Currents, equality of, 92–93
Current-to-voltage converter, 406–407
Current-versus-voltage curve, 293, 295
Cutoff frequency, 250, 412, 413
Cutoff state, 336
Cycles, 154. See also Half-cycle waveform, 154
Cycles per second (cps), 159

—D—

D/A. See Digital-to-analog (D/A) converter (DAC)
DAC. See Digital-to-analog (D/A) converter (DAC)
Damped ringing waveform, 445
Dark current, 495
Darlington amplifiers, 377
Darlington pairs, 367–368
Data bus drivers, 598–599
Data busses, 597
Datasheet ratings, manufacturer's, 324
Data transfers, 623
dB gain, 339–340
DC. See Direct current (dc); Pulsating dc
DC capacitive circuit, current flow in, 219
DC drain voltage, 375
DC failures, 420
DC switching, silicon controlled rectifiers and, 479–480
DC voltage, unregulated, 433
Decibels (dB), 40
 gain and, 339
Decimal to binary conversion, 544–545
Decoupling capacitors, 220, 345, 519
Deenergized state, of a relay, 79
Defective components
 discarding, 147
 identifying, 144
 as substitutes, 228
Defective series circuits, relative voltage values in, 142
Defects, in solid-state devices, 582–584
Degrees, 165
Delta modulation, 641
Demodulation, 310, 637
Depletion layers, 337, 349
Depletion-mode devices, 356
Depletion-mode MOSFETs, 353–354, 358
 biasing and operation of, 355–356
Depletion region, 277, 287
 forward bias and, 278
 reverse bias and, 280
D flip-flops, 566–568
Diacs, 473–475
Diagnostic software, 17
Dielectric, 216
 capacitor, 230, 359
Dielectric constant, 219
Dielectric material, 219
 thickness of, 225
Difference of potential, 27
Differential amplifiers, 368–370
Differential-mode voltage gain, 389
Differentiated waveform, 239
Differentiator circuit, 239

Digital and analog systems, interfacing, 576–581
Digital circuits, transistor use in, 361–366
Digital devices, binary values and, 546
Digital electronics, 541–593
 concepts and terminology concerning, 541–555
Digital information, 635
Digital multi-meters (DMMs), 139
Digital ohmmeters (DMMs), 286, 374
Digital signals, 154, 307
Digital systems, troubleshooting, 581–585, 586
Digital-to-analog conversion, 577–579
Digital-to-analog (D/A) converter (DAC), 366–367, 406, 577–579, 586
Digital transmissions, 635, 650
Digital voltmeter, 131
Diode applications, 304–311
Diode circuits, troubleshooting, 320–325
Diode elements, 292
Diode orientation, 305
Diode resistance checks, in-circuit, 324
Diodes, 81, 82, 275. See also Isolation diodes; Special diodes
 characteristics of, 291–295
 damage to, 295
 defective, 294, 321, 324
 forward biasing, 305
 front-to-back resistance ratios of, 321
 ideal, 292–293, 325
 identifying, 292
 isolation, 307–308, 325
 maximum-to-minimum capacitance for, 315
 practical, 293–295, 325
 special-purpose, 325
 summary of, 325–326
 testing, 321, 374
Diode testers, 321
Direct current (dc), 153. See also DC entries
Discharge current, 237
Discrete combinational logic circuits, 586
Discrete combinational logic devices, 565
Discrete components, 9
Discrete flip-flop counters, 573
Discrete switching regulators, 446–447
Discriminator, 638
Divide-by-n circuit, 520
DMMs. See Digital multi-meters (DMMs); Digital ohmmeters (DMMs)
Donor atoms, 269

Doped crystal, energy levels in, 271
Doped silicon, 275
Doping
　level of, 318
　nonuniform, 317
　pentavalent, 269–270
　semiconductor, 267–268
　trivalent, 268–269
Doppler effect, 182
Dot notation, 205, 206
Double-throw switch, 78
Drain terminal, 348, 350
Drain wire, 62
DRG button, 173
Driver shoot-through logic, 445–446
"Driving the output to the rails" operation, 389
Dual power supplies, 384
Dual-gate MOSFET, 371
Duty cycles, 309, 515
dV/dt rating, exceeding, 471–472
Dynamic breakback voltage, 474

—E—

EHPs. See Electron-hole pairs (EHPs)
EIA. See Electronics Industries Association (EIA) Color Code
Electrical charge, 25
Electrical law, 26
Electrical noise, 442, 508
Electrical potentials, scale of, 90
Electrical power, 34
　unit of measurement for, 34
Electrical pressure, 104
Electrical quantities, 23–37
Electric fields, 216–218
Electromagnetic fields, 44–45, 46
Electromagnetic induction, 192
Electromagnetic interference (EMI), 15
Electromagnetic waves, 41, 46, 183
Electromagnetism, 42–46
Electromechanical systems, 21
Electromotive force (E, emf), 27, 43, 88, 123
Electromotive source, 89
Electron current, 28–29
Electron distribution, 266–267
Electron-hole pairs (EHPs), 267, 280
Electronic circuits, 9–10
Electronic components, 9, 63–82
　symbols used to represent, 8
Electronic devices, light and, 40
Electronic products, reducing the size of, 70
Electronics, basic, 23–46
Electronics Industries Association (EIA) Color Code, 70
Electronics Workbench, 20
Electronic systems, 1–21
　of aircraft, 2–3

components of, 82
defined, 1
diagnosing defects in, 97
troubleshooting, 131
Electronic tuning, 315
Electronic voltage source, 64
Electron orbits, 263
Electrons, 24
　valence-band, 263
Electrostatic discharge (ESD), 359, 493
Electrostatic field, 26, 216
Embedded power supplies, 454–457
Emf. See Electromotive force (E, emf)
EMI. See Electromagnetic interference (EMI)
Emitter-base junctions, 361
Emitter-collector circuit, 373
Emitter-follower amplifier, 344–345, 366–367
Emitter-follower circuit, 368
Emitter region, 332
Emitter terminal, 333
Enable input, 445
Energy levels, 263–264, 270–272
Energy-level transitions, 490
Engineering calculator, 12, 52–53, 165
Engineering notation, 49–50, 51–52, 82
　converting between, 52–53
Enhancement mode, 353
Enhancement mode MOSFETs, 354–355
　biasing and operation of, 356–358
Equal-probability troubleshooting, 144
Equal-valued resistors, computing the total resistance of, 105
Equivalent capacitance, for series-connected capacitors, 226
Equivalent circuits, 178, 413
Equivalent resistances, expanding, 113
Equivalent series inductance (ESL), 253
Equivalent series resistance (ESR), 253
Error amplifier, 434
ESD. See Electrostatic discharge (ESD)
ESL. See Equivalent series inductance (ESL)
ESR. See Equivalent series resistance (ESR)
Ethernet cables, 57
Ethernet communications link, 153
Ethernet connections, 207
Ethernet ports, 609
Ethernet transformer, 207
Exclusive NOR operator, 552–553
Exclusive OR operator, 551–552
External flux lines, 42
External load circuitry, defect in, 322
Extrinsic semiconductor, 355
　material for, 268

—F—

Failure modes
　classification of, 420
　examples of, 420–421
Farad (F), 218
Fast-attack slow-decay circuit, 309
Fast-Blo fuses, 66
FCC. See Federal Communications Commission (FCC)
FDM. See Frequency-division multiplexing (FDM)
Federal Communications Commission (FCC), 442
Feedback, 430–431
Feedback voltage, 434
Feet per second (fps), 39
Fences, troubleshooting, 143–144
Ferrite beads, 196–197
Ferrite-core transformers, 208
FET amplifiers, 359
FETs. See Field-effect transistors (FETs)
FHSS. See Frequency-hopping spread spectrum (FHSS)
Fiber-optic cables, 494, 500, 504
Fiber-optic links, 14–15
Fiber optics, 500–501
　troubleshooting, 502–503
Fiber-optic telephone links, 501
Field-effect transistors (FETs), 317, 348
Film resistors, 69, 70
Filter capacitors, shorted, 323
Filter configurations, 239
Filtered load voltage, 440
Filter networks, 302–304
Filters, 296–297
Fires, from incorrect wire selection, 56
Firing angle, 481
"Firing the thyristor," 470
Five-band color code resistor, 73
555 timer, 514–516
Fixed capacitors, 222
Fixed resistors, 66
　schematic symbols for, 73
Fixed-value resistors, 82
Flip-flop counters
　asynchronous, 571–573
　synchronous, 569–571
Flip-flops, 566–569, 574
Flowcharts, 4–5
Flux, 46
Flux density, 43
Flux lines, 42
　magnetic, 155
FM. See Frequency modulation (FM)
Foldback current limiting, 450, 458
Force, 38
Forward bias, 292–293, 293–294, 335
　barrier potential and, 277–278
　depletion region and, 278

Forward bias, *continued*
   maximum current flow with, 278–279
Forward biased base-emitter
   junction, 373
Forward biased diode, 325
Forward-biased junctions, 277, 336–338
Forward-biased pn junctions, 281–282
   voltage drop across, 285–286
Forward voltage drop, measurement of, 321
Fourier, Jean Baptiste Joseph, 167
Fourier analysis, 168
Free electrons, 24, 28
Freezing mist sprays, 375
Frequency, 40–42, 158–159
   shift in, 182
Frequency-division multiplexing (FDM), 642–643, 650
Frequency domain, 167–168
Frequency-hopping spread spectrum (FHSS), 648–649, 650
Frequency modulation (FM), 638–639, 650
Frequency-multiplier circuits, 317
Frequency response, 390–391
Frequency-selective filter, 512
Frequency stability, 516–517
Frequency-to-period conversions, 159
Frequency-to-voltage conversion (FVC), 531–533
Frequency-to-voltage converter, 532
Front-panel milking, 18
Front-to-back ratio, 286
Full sine wave, average value of, 160–161
Full-duplex communications, 636
Full-wave bridge rectifier circuit, 300, 301
Full-wave center-tapped rectifier, 325
Full-wave rectifier, 299–300
   with filter, 303
Function generators, 156
Functional block diagram, for an oscillator circuit, 508
Functional blocks, in notebook computers, 427–428
Functional modules, 11–12
Fundamental frequency, 168
Fuses, 64–66, 448–449
   categories of, 65–66
   current rating of, 65
   response time of, 65
   voltage rating on, 65
FVC. *See* Frequency-to-voltage conversion (FVC)

–G–

Gain, 339
GAL. *See* Gate array logic (GAL)
Ganged switches, 78
Gate array logic (GAL), 558–560
Gate capacitance, 356–357
Gate "capacitor," 355
Gate pulses, timing of, 481
Gate terminals, 348, 472
   in thyristor trigger applications, 477
   negative, 356
Gate-to-drain measurements, 376
Gate-to-source measurements, 376
Gate triggering, 472
General-purpose interface bus (GPIB) communications, 628
Germanium, 264
Germanium atoms, 265
GETSWITCH instruction, 612
Glitch, 573
Global positioning system (GPS) devices, 12
Glue logic, 565
GPIB. *See* General-purpose interface bus (GPIB) communications
GPS. *See* Global positioning system (GPS) devices
Graph, defined, 5
Graphical data, 5–7
Ground, 119–122, 123
Ground plane, 119
Ground references, 122
Ground symbols, 119
   alternative, 122

–H–

Half-cycle, 156
   average value for, 161–162
Half-duplex communications, 636
Half-power points, 250
Half-wave rectifier, 296–298, 325
   with filter, 302–303
Half-wave rectifier circuit, 298
Hand-held computers, 648
Hardware, 601–602, 605–611
Hardware Description Language (HDL), 585
Harmonic frequencies, 168, 317
Harmonics, 168–169
HDL. *See* Hardware Description Language (HDL)
Heat sinks, 284–285, 441, 607
HEF4071B device, 549
Hemostats, 284
Henry (H), 194
Hertz (Hz), 159
Hexadecimal system, 545
High-frequency circuits, 197
High-frequency waveforms, 240
High input impedance, 404
High-level programming languages, 613
High-pass active filter, 412–414
High-powered thyristors, 487
High-speed computer circuits, 220
High-speed computer clocks, 521
High-speed digital circuits, 119
Holding current, 473, 475
Holes, 267, 268, 273–274, 287
Hoover Dam, 181
Hot spots, 482
Hot-carrier diodes, 316
Hub, 627
100-stage system, split-half troubleshooting applied to, 145, 146
Hybrid integrated circuits, 366, 523, 534
Hypotenuse, 171
Hysteresis voltage, 418

–I–

IC-based systems, troubleshooting, 534–535
ICs. *See* Integrated circuits (ICs)
ICS9169C-46 Clock Synthesizer chip, 521–522
Ideal circuit components, 252–253
Ideal diodes, 292–293, 325
IEC. *See* International Electrotechnical Commission (IEC)
IEEE. *See* Institute of Electrical and Electronic Engineers (IEEE)
IEEE 802.11 standard, 183
Impedance (Z), 212, 233
   calculating, 212–213
   input and output, 341
   matching, 342
Impedance ratio, 212–214
Impurities, semiconductor, 268
Impurity atoms, 276
In-circuit component testing, 147–148
In-circuit diode resistance checks, 324
Increased voltage symbol, 134
Index of refraction, 500
Indirect FM, 639
Induced current, effects of, 193–194
Induced voltage, amount and polarity of, 192–193
Inductance (L), factors associated with, 194
Inductance meter, 202
Induction
   electromagnetic, 192
   factors affecting, 192–193
Inductive circuits, analysis of, 201
Inductive reactance, 195, 254
Inductor defects, summary of, 203
Inductors, 76, 82, 192–204
   current changes in, 194
   nonideal, 253
   series and parallel, 198–200
   troubleshooting, 201–203
   types and applications of, 196–198
Industrial computers. *See also* Computers

battery backup in, 308
clippers in, 305–307
integrators in, 410
transistors as current amplifiers in, 366–367
Industrial controllers, 80
Industrial robotics, 2
Industry application
of a bridge rectifier, 301
of a half-wave rectifier circuit, 298
Industry standard architecture (ISA), 606
Infinite resistance, 131
Information, classes of, 631
Infrared display cards, 502, 503
Infrared (IR) LED, 489
Infrared (IR) link, 14, 610
In-phase sine waves, 164
Input bias current, 386
Input capacitance, 220, 370
Input circuit, 363–364
Input impedance, 341
high, 526
Input/output (I/O), 609–610. *See also* I/O entries
Input resistance, 340–341
Input voltage, 238
Instantaneous value, 166–167
Instantaneous voltages, 160, 185
Institute of Electrical and Electronic Engineers (IEEE), 183, 585
Instruction register, 600
Instruction set, 596
Instrumentation amplifiers, 524–527, 536
Instruments, virtual, 20
Insulation wire, 13
Insulators, 25, 46, 262
Integrated capacitors, 222
Integrated circuit functions, 11
Integrated circuit packages, 11
Integrated circuits (ICs), 10–11, 81, 331–332, 363. *See also* Hybrid integrated circuits; IC-based systems
applications for, 507–540
classes of, 536
industrial computer applications for, 523–533
integration complexity of, 11
troubleshooting circuits based on, 533–535
Integrated counters, 573
Integrated linear amplifiers, 371–372
Integrated logic circuits, 363–366
Integrated series voltage regulators, 434–436
Integrated switching regulators, 446–447
Integrated voltage references, 431
Integrator circuit, 239

Integrators, 409–410
Intel Pentium microprocessor, 363
Intensity
of light, 40
of magnetic field, 43
Intentional circuit, 518
Interface functions, of transistor applications, 366
Interfacing, of digital and analog systems, 576–581
Interference, rejection of, 252
Intermediate frequencies, 206
Intermediate frequency (i.f.) transformers, 206
Internal electrical noise, 358–359
Internal insulator, 219
Internal memory, 534, 608
Internal registers, 599–600
Internal source voltage, 388
International Electrotechnical Commission (IEC), 556
Internet, 3
Internet links, wireless, 184–185
Interpolation, 6
Intrinsic semiconductors, 264, 267
Intrinsic silicon, 287
Intruder-alarm circuit, 496
Inverse trigonometric functions, 173. *See also* Trigonometric functions
Inversion layer, 357
Inverter, CMOS, 365
Inverting amplifier circuits, 399–402
identification of, 399
operation of, 399–400
purpose of components in, 402
voltage gain in, 400–402
Inverting-buffer amplifier, 577
I/O. *See* Input/output (I/O)
I/O clock, 522
I/O controller, 4, 609
Ionization, 25
Ions, 25, 46
I/O operations, 616
$\overline{\text{IOW}}$ (I/O write) control bus signal, 601, 603
Iron-core coils, 202
Iron-core transformers, 208, 214
ISA. *See* Industry standard architecture (ISA)
Isolation diodes, 307–308, 325
Isolation transformer, 207

–J–
JFET amplifiers, 350–352
JFET characteristics, summary of, 352
JFETs. *See* Junction field-effect transistors (JFETs)
Jitter, 521
JK flip-flops, 568–569
Junction breakdown, 280
reverse current and, 280

Junction capacitance, 316, 318–319, 471
Junction field-effect transistors (JFETs), 348–353, 377. *See also* JFET entries
identifying defects in, 375–376
versus bipolar transistors, 350
Junctions
reverse-biased, 336–338
semiconductor, 275–283
Junction voltage measurements, 285–286

–K–
Kirchhoff's Current Law, 104, 106, 123, 180, 234, 335, 400, 401
Kirchhoff's Laws, 110, 115–116, 176, 179
Kirchhoff's Voltage Law, 94–97, 123, 210, 232, 346, 432, 475, 526
Knee current, 312
Knee voltage, 312, 320

–L–
Lagging phase, 165
LANs. *See* Local area networks (LANs)
LASCRs. *See* Light-activated SCRs (LASCRs)
Laser diodes, 490–494, 504
Laser light, 504
eye exposure to, 502–503
Laser printer, 97
Lasers, 490
Laser safety, 494
Latching current, 472
LC circuits. *See* *RLC* circuits
LCR tester, 228, 229
Lead formation, 283–284
Leading phase, 165
Lead style capacitors, 221–222
Leakage currents, 470
Leakage flux, 208
Leaky capacitors, 230
LEDs. *See* Light-emitting diodes (LEDs)
Left-hand rule for conductors, 45
Lenz's Law, 193, 194
L filter, 303–304
Light Amplification by Stimulated Emission of Radiation (laser), 490. *See also* Laser entries
Light emitters, troubleshooting, 501–502
Light
intensity of, 40
refraction of, 500
Light sensors, troubleshooting, 502
Light waves, 39–40
Light-activated SCRs (LASCRs), 496–497, 504

738 INDEX

Light-emitting devices, 489–494
Light-emitting diodes (LEDs), 6, 9, 466, 489–490, 504, 603
Light-sensing devices, 494–497
Limiter circuit, 305, 325
Linear integrated circuits, 369–370
Linearly proportional relationship, 31
Linear regulation, versus switching, 440–443
Linear scales, 6
Linear transistor applications, 366–372
Line graphs, 6
  family of, 7
Line regulation, 429–430
Lines of flux, 42
Line voltage, 429
Load current changes, 432
Load regulation, 430
Load resistor, 296
Local area networks (LANs), 629, 650
  wireless, 645
Logarithmic amplifier, 342
Logarithmic scales, 6
Logbook, 19
  maintaining, 18
Logical operators, 548–553
Logical troubleshooting procedures, 141
Logic analyzer, 585
Logic bubbles, 566
Logic circuits, 586
  Boolean expressions and truth tables for, 560–563
Logic devices, 546
Logic gate conversions, 557
Logic gates, 11, 548–554, 555–558
  defects in, 583
  IC packages used for, 555–556
Logic gate testing, actions applicable to, 582
Logic probe, 583, 584
Logic pulser, 583–584
Logic signal, 362
Logic states, 546–547
Logic symbols, alternative, 556–557
Long time constant, 239
Low front-to-back ratio, 321
Low-noise amplifiers, 358–359
Low output impedance, 404–405
Low-pass active filter, 410–412
Low-pass filter circuit, 240
Low-power thyristors, 487
Lower threshold voltage, 418
L-type filters, 325

–M–

Machine language, 611–612
Machinery speed, monitoring, 39
Machines, acceleration of, 39
Macros, 614
Magnetic circuits, Ohm's Law for, 44
Magnetic coupling, 204–205

Magnetic domains, 43
Magnetic field intensity (H), 43
Magnetic field polarity, 45
Magnetic fields, 42, 44–45, 46, 193, 194, 204
Magnetic flux (ϕ), 42
Magnetic poles, 42–43
Magnetism, 42–46
Magnetizing force (H), 43
Magnetomotive force, 43, 46
Magnet wire, 60–61
Maintenance history, 18
Majority carriers, 274
Malfunctions, capacitor, 228
MANs. *See* Metropolitan area networks (MANs)
Manufacturers' datasheet ratings, 324
Manufacturers' wire data, 58
Manufacturing processes, industrial computers in, 307
Marking state, 634
Mass storage, 608
Materials, categories of, 25, 287, 262
Matter, building blocks of, 24
MAX-104 ADC, 580
Maximum current flow, 278–279
Maximum deviation, 68
Maxterm expression, 562
Measurements
  making, 130
  minimizing the number of, 130
Mechanical quantities, 38–39
Mechanical resonance, 245
Megabytes, 596
Memory devices, 366
Memory microprocessor, 596
MEMR control bus signal, 603
Metallic conductors, 30
Metal-oxide-semiconductor-field-effect transistors. *See* MOS entries; MOS field-effect transistors (MOSFETs)
Metal-silicon junctions, 316
Metric prefixes, 53, 82
Metropolitan area networks (MANs), 629, 650
Mica capacitors, 225
Microprocessor block diagram, 599
Microprocessors, 595–619
  basic concepts and terminology concerning, 595
  internal structure of, 597–600
Microwave devices, low-power, 324
Microwave oscillator, 324
Miles per hour (mph), 39
Minimum acceptable resistance value, 68
Minimum current flow, with reverse bias, 280
Minority carriers, 274
Minterm expression, 562

Modular power supplies, 453–454
Modulating signal, 637
Modulation, 309, 636–637, 650
Modulation frequencies, 493
Modules, functional, 12
Momentary-contact switches, 78
Monotonic output, 579
MOS devices, precautions when working with, 359–360
MOSFET amplifiers, 358–359
MOS field-effect transistors (MOSFETs), 353–361, 377
  biasing and operation of, 355–358
  characteristics of, 360
  enhancement-mode, 357
  handling precautions for, 359–360
  high-current, 442
  identifying defects in, 376
  in integrated circuits, 365
Multiconductor cables, 61
Multiple inductors, 198
Multiple LANs, 629
Multiple phase-locked loops, 521
Multiple voltage sources, connecting in parallel, 107
Multiplexing, 641–642, 650
MultiSIM package, 20, 151
  test circuit constructed with, 48

–N–

NAND gate, 565, 582
NAND operator, 550–551
Nanotechnology, 12
National Electrical Code, 58
N-channel depletion-mode MOSFET, 353
N-channel enhancement-mode MOSFET, 356–357
N-channel JFET, 348, 349
  biasing, 349–350
N-channel MOSFET, 445
NDI. *See* Nondestructive inspection (NDI) equipment
Negative battery terminal, 92
Negative charge, 28
Negative feedback, 387, 409, 422
  effects of, 397–398
Negative gain, 405
Negative gate voltage, 357
Negative half-wave rectifier, 297
Negative ions, 25
Negative resistance, 318, 468
Negative temperature coefficient, 267
Negative tolerance, 68
Networks, 625–630, 650
  architecture of, 628
  backbones of, 625
  protocols in, 628
  sizes of, 629–630
  topologies of, 625–627
Network operating system (NOS), 628

Neutrons, 24
No-bias drain current, 356
Node, 625
Noise
   common-mode, 498
   differential amplifiers and, 370
   electrical, 508
   high-frequency, 207
   internal electrical, 358–359
Nondestructive inspection (NDI) equipment, 182
Nonelectronic voltmeter, 131
Nonground references, 121, 123
Nonideal capacitors, 253–254
Nonideal inductors, 253
Noninverting amplifier circuits, 391, 393–399
Noninverting amplifiers, voltage gain of, 396–397
Nonpolarized capacitors, 223
Nonresonant *RLC* circuits, 241–245
Nonvolatile memory, 608
No-output voltage, 136
Normal-Blo fuses, 66
Normal state
   of a relay, 79
   of a switch, 78
NOR operator, 551
North-seeking (north) pole, 43
NOS. *See* Network operating system (NOS)
Notebook computers, 427–428, 453
NOT operator, 549–550
Npn switching transistor, 546, 547
N-type crystal, energy levels in, 272
N-type material, 355, 357
   current flow through, 272–273
N-type semiconductor material, 270
N-type semiconductors, 355
   energy distribution of, 271
Numerical sequences, 542–543

–O–

Observation, power of, 17–18
Observation skills, 149
Offset null inputs, 386
Ohm (Ω), 30, 221
Ohm, Georg Simon, 30
Ohmmeter checks, 286
   in-circuit, 148
Ohmmeters, 62–63, 138, 201, 202–203
   for thyristor testing, 487–488
Ohmmeter tests, 214–215, 228–230, 324
   of diodes, 321
Ohm's Law, 30–33, 88, 93, 98, 103, 105, 106, 110, 115, 123, 176, 178, 201, 234, 318
   alternative forms of, 32–33
   for magnetic circuits, 44, 46
   power calculations based on, 35–36

Ohm's Law equation, 30, 32
On-board oscillator, with external crystal, 517–518
1N5760 diac, 475
Op amp circuits, troubleshooting, 420–422
Op amp integrator circuit, 409
Op amp rules, 394
Op amps, 422
   analysis rules for, 393–394
   applications for, 423
   characteristics of, 383–392
   defective, 422, 423
   industrial applications for, 403–419
   input bias current and input resistance for, 386–387
   internal circuitry of, 385–386
   negative feedback for, 387
   output resistance for, 387–388
   output voltage swing for, 388–389
   schematic representation of, 384–385
OPA502 integrated amplifier, 372
Open branch, 149
Open circuit, 88, 89, 219
   effects of, 131–132
Open components, 135
   in a parallel circuit, 141
Open-loop operations, forcing, 455–456
Open loops, 387
Open-loop voltage gain, 389, 393
Operating system, 615
Operational amplifiers. *See* Op amps
Optical misalignment, 502
Optocouplers, 497, 504
Optoelectronic devices, 489–503, 504
   troubleshooting, 501–503
Optoelectronics, 40
Optointerrupters, 498–499, 504
Optoisolators, 497–498, 504
OR gate, 633
OR operator, 549
Oscillation, frequency of, 245
Oscillator circuits, 508–523, 535
   troubleshooting, 533–534
Oscillator modules, 518
Oscillators, 535–536
Oscilloscopes, 97, 164, 167
Out-of-phase sine waves, 164
Out-of-tolerance parameters, 321
Output current, 406–407
Output distortion, oscillator-circuit, 533–534
Output impedance, 341
Output offset voltage, 385
Output resistance, 340–341
Outputs
   computer, 2
   IC, 534
Output saturation voltage, 389

Output slew rate, 391–392
Output voltage swing, 388–389, 391
Output voltage polarity, 443
Overcurrent protection, 448, 458
Overhead, 633
Overvoltage protection, 450–452, 458
Overvoltage protection circuit, 451

–P–

PA. *See* Public address (PA) system
Pads, 9
PAL. *See* Programmable array logic (PAL)
PAL 16L8AM device, 558
PAM. *See* Pulse-amplitude modulation (PAM)
Parallel branches, monitoring currents in, 139–140
Parallel capacitances, 226–227
Parallel capacitive reactances, 227–228
Parallel capacitor connections, 226–228
Parallel circuits, 101–108
   current paths in, 103
   defects in, 149
   effects of a short in, 141
   intuitive observations about, 103–104
   mathematical relationships for, 104–107
   troubleshooting, 138–141
Parallel-communication links, 632
Parallel communication paths, 650
Parallel components, 102
Parallel data transmission, 616
Parallel inductive reactances, 200
Parallel inductors, 198
Parallel networks, 123
Parallel-plate capacitor, 216
Parallel *RC* circuits, 233–234
   phase relationships of, 233
Parallel resistance equation, 111
Parallel resistances, relative values of, 107
Parallel resonance, 246
Parallel *RLC* circuits, 243–245, 255
Parallel troubleshooting techniques, 149
Parallel voltage sources, 107
Parasitic circuit elements, 252–254, 372
Parasitic oscillations, 518–519
Parity, 634, 650
Parity bit, 634
Parts, discarding, 147
Pass band, 250
Passive filter circuits, 239–240
Pass transistors, 433
PCBs. *See* Printed circuit boards (PCBs)
P-channel depletion-mode MOSFET, 354, 355

P-channel enhancement-mode MOSFET, 357
P-channel JFET, 348, 349
P-channel MOSFET, 445
PCI. *See* Peripheral component interconnect (PCI)
PCI clock, 522
PCM. *See* Pulse-code modulation (PCM)
PDAs. *See* Personal data assistants (PDAs)
Peak, of a sine wave, 185
Peak input voltage, 302
Peak-to-peak value, 163–164
Peak-to-peak voltage, 185
Peak value, 156, 160
Peak voltage, on a phasor diagram, 171
Peer-to-peer architecture, 628
Pentavalent atoms, 268
Pentavalent doping, 269–270, 275
Period, 157–158
  of a wave, 41
  of a waveform, 154, 185
Period-to-frequency conversions, 159
Peripheral component interconnect (PCI), 606. *See also* PCI clock
Permanent magnet, 44
Permeability (μ), 44, 192
Permittivity, 219
Personal data assistants (PDAs), 14, 647, 648
PFM. *See* Pulse frequency modulation (PFM)
PFM/PWM control logic, 446
PG. *See* Power good (PG) output
Phase, 164–165
  of a sine wave, 185
Phase angles, 165, 170–176, 185
Phase-control circuits
  silicon controlled rectifiers and, 480–481
  triacs as, 482
Phase-detector circuit, 520
Phase inversion, 341
Phase-locked loops, 519–522, 536
Phase modulation (PM), 639, 650
Phase relationships, 195–196, 232, 340
Phase-shift keying (PSK), 639
Phasor addition, 232
Phasor calculations, 174–176, 185
Phasor diagrams, 170–171, 185
Phasors, simplification of, 242, 243
Photoconductive mode, 495
Photodiodes, 493, 494–495, 504
Photons, 491
Phototransistors, 406, 496, 504
Photovoltaic mode, 495
Physical system hierarchy, 9–12
Piconet, 184, 648, 649, 650
Piezoelectric effect, 516, 535–536

PIN. *See* P-type intrinsic and n-type (PIN) materials
PIN crystal switching, 319
PIN diodes, 318–320
Pi-type filters, 304, 325
Pixels, 608
PLA. *See* Programmable logic array (PLA)
Plastic film capacitors, 225
Plate area, 219
Plate separation, 219
PLC. *See* Programmable logic controller (PLC)
PLDs. *See* Programmable logic devices (PLDs)
PM. *See* Phase modulation (PM)
PN junction diodes, 332, 325
PN junctions, 275
  biasing, 333
  energy levels in, 281–282
  testing, 286
PNP transistors, 332–333, 338
Polar graphs, 6–7
Polarity
  of capacitors, 223
  in secondary windings, 205
  of temperature coefficients, 223
  of voltage drops, 95, 96, 120
Polarized capacitors, 223
Poles, 46
  of a magnet, 42–43
  of a switch, 77–78
Population inversion, 491
Ports, 601
Position, 38–39
Positive charge, 28
Positive feedback, 418
Positive gate voltage, 357
Positive half-wave rectifier, 297
Positive ions, 25, 28
Positive tolerance, 68
Potential difference, 27
Potentiometers, 74, 531
  schematic symbols for, 75
Pound (lb), 38
Pounds per square inch (psi), 38
Power (*P*), 46, 97–98
  electrical, 34
Power amplifier, 342
Power calculations, for series-parallel circuits, 116
Power dissipation, 177, 440–442
  summing, 107
Power equations, memory aid for, 46
Power formulas, 35
Power gain, 340
Power-generating plants, 181
Power good (PG) output, 445
Power-In jacks, 301
Power rating, of a resistor, 67
Power ratio, 211

Power relationships, 35
Powers of ten, 50–51
Power supplies, 64, 136, 427–463
  embedded, 454–457
Power supply applications, 296–304
Power supply circuits, 239
  troubleshooting, 452–457, 458
Power supply defects, 455
  effects of, 136–137
  localizing, 454
Power supply lines, reversed polarity, 308
Power supply protection circuits, 448–452, 458
Power supply symptoms, 137
Power transformers, 206, 208
PPM. *See* pulse-position modulation (PPM)
Practical diodes, 293–295, 325
Precision resistors, 68, 73
Prefixes, converting between, 54–55
Pressure, 38
Pressure transducer, 524
Primary-to-secondary short, 215
Primary winding, 205
Printed circuit boards (PCBs), 9–10, 11, 119
  damaging, 19
  troubleshooting, 139
Printed circuit board traces, 13–14
Printed wiring assembly, 9
Printers, troubleshooting, 18
Product-of-the-sums form, 562
Product-over-the-sum formula, 105, 111, 114
Program branch, 598
Program counter, 597–598
Program execution, 602–604
Program label, 612
Program loop, 604
Programmable array logic (PAL), 558
Programmable counter, 520
Programmable logic array (PLA), 558
Programmable logic controller (PLC), 80
Programmable logic devices (PLDs), 558–560
Programmable unijunction transistor (PUT), 472
Programmers, 616
Programming languages, 613–614
Programming levels, 611–614
Programs, 596. *See also* Software
Propagation delay, 553
Protection diodes, 359
Protocol, network, 628
Protons, 24
PSK. *See* Phase-shift keying (PSK)
P-type crystal, energy levels in, 271

P-type intrinsic and n-type (PIN) materials, 318–319. *See also* PIN entries
P-type material, 269, 355
   current flow through, 273–274
P-type semiconductors, energy distribution of, 271
Public address (PA) system, 16
Pulsating dc, 296–297
Pulse-amplitude modulation (PAM), 640
Pulse circuit response, 234
Pulse-code modulation (PCM), 640–641, 650
Pulse frequency modulation (PFM), 444
Pulse modulation, 639–640
Pulse-position modulation (PPM), 640
Pulse transformers, 207
Pulse waveform, 240
   stable, 517
Pulse width modulation (PWM), 444, 640, 650
Push-button switch, 78
PUT. *See* Programmable unijunction transistor (PUT)
PWM. *See* Pulse-width modulation (PWM)

–Q–

Q factors, 249–250
Q2010L5 triac, 482
Quality, of a circuit, 249–250
Quiescent operating point, 344

–R–

Race condition, 573
RAD button, 173
Radial lead capacitors, 222
Radian mode, 172, 173
Radians, 165
Radiation, 40
Radio frequency (rf) circuits, 315
Radio-frequency (rf) devices, 183
Radio frequency interference (RFI), 488
Radio frequency (r.f.) transformers, 206
Radio frequency (RF) wireless links, 15
Radio links, 15
Radio receiver, 252
Radio waves, 40, 183
Radix point, 542, 544
RAM (random access memory), 608
Ramp voltage, 509
Random access memory (RAM), 608
"Rate of change of voltage," 471
RC circuits, 231–234
   applications of, 239–241
RC coupling circuit, 239
RC networks, 415
RC time constants, 235–237
Read (retrieve data), 597

Read-only memory (ROM), 602
Receiver clocks, 634
Receivers, 650
   infrared, 14
   ultrasonic, 182
Reciprocal equation, 104
Reciprocal formula, 111
Recombination, 276, 280
Rectangular waveshapes, 153–154
Rectifier circuits, 296–304, 325
   defects in, 321–323
Rectifier filter circuit, 325
Rectifiers
   bridge, 300–301
   full-wave, 299–300
   full-wave with filter, 303
   half-wave, 296–298
   half-wave with filter, 302–303
   malfunctioning, 322
Rectifying, 296
Reference output signal, 552
Reference point, specifying, 121
Reflected impedance, 212
Reflective optointerrupter, 499
Refraction, 500
   of light, 500
Registers, 599
   general-purpose, 600
   internal, 599–600
Regulator circuits, checking, 455–457
Relative permeability (μ_r), 44
Relative permittivity, 219
Relaxation oscillator, 535
Relay circuits, analyzing and explaining, 84
Relays, 79, 84
   computers and, 80, 362–363
   construction of, 79–80
   schematic symbols for, 80
   transistors and, 362–363
Relevant inputs, troubleshooting and, 581–583
Reliability band, on a five-band resistor code, 73
Reluctance ($\mathfrak{R}$), 44
Remote control techniques, 625
Remove-and-replace power supply, 453–454
Replacement components, substituting, 18–19
Representative systems, 1–3
Reset input pin, 598
Resistance (*R*), 30
   computing, 33–34
   expanding, 113
   input and output, 340–341
   larger, 93
   mathematical expression for, 94
   measurements of, 374
   range of, 68
   relationship to current, 32

in a series-parallel circuit, 110–112
   unit of measure for, 30
   values of, 201
   of wire, 56–58
Resistance checks, for bipolar transistors, 373–374
Resistance line, 32
Resistive circuits, 186
   power in, 107
   troubleshooting, 146
Resistive feedback, 399
Resistor color codes, 70–73
   interpreting in bands, 71
Resistor current, 234
Resistor power rating, 66–67
Resistors, 9, 10, 66
   surface-mount, 70, 71
Resistor technology, 69–70
Resistor tolerance, 68
Resistor value, decoding, 72
Resolution, 579
Resonance, 241, 245
Resonant circuit measurements, 249–251
Resonant frequency, 245, 248–249
   computing, 248–249
Resonant rise in voltage, 248
Resonant *RLC* circuits, 245–251
Response time, 65
Reverse bias, 287, 293, 294, 335
   barrier potential and, 279
   depletion region and, 280
   minimum current flow with, 280
Reverse-biased diodes, 315
Reverse-biased junctions, 279, 336–338, 471
Reverse-biased pn junction, 282
Reverse breakdown, 295, 469
Reverse breakdown voltage, 280, 478
   of laser diodes, 502
Reverse current
   junction breakdown and, 280
   temperature dependence and, 280
Reverse diode currents, 294
Reverse Polish Notation (RPN) calculator, 53, 172
Revolutions per minute (rpm), 39
RF, rf. *See* Radio frequency (RF) entries
RF resistance, of a PIN diode, 319–320
RFI. *See* Radio frequency interference (RFI)
Rheostats, 74
   schematic symbols for, 75
Ribbon cables, 62, 82
Right-triangle relationships, 171–172
Ring network topology, 626–627
Ripple voltage, 302–303, 304, 322
RLC circuits, 241–254, 255
   applications for, 251–252

RLC Circuits, *continued*
  selectivity of, 250
RL circuits
  applications of, 239–241
  characteristics of, 234, 235
RL time constants, 238
Rms (root-mean-square) value, 162–163, 176, 177
Rms voltage, 181
Robots, 361
Rocker switch, 78
ROM (read-only memory), 602
Room-temperature superconductors, 25
Root-mean-square (rms), 162–163. *See also* Rms (root-mean-square) value
Rotary switches, 78
RPN. *See* Reverse Polish Notation (RPN) calculator
RTC circuits, 610

–S–

S1070W SCR, 479
Sample-and-hold circuit, 641
Satellite communications, 646–647
Saturated transistors, 452
Saturation, 338
SBS. *See* Silicon bilateral switch (SBS)
Schematic capture, 20
Schematic diagrams, 7–9
Schematic symbols, 73
Schottky diodes, 316, 364, 440
Scientific calculators, 166, 172
SCR phase-control circuit, 480–481
SCRs. *See* Silicon-controlled rectifiers (SCRs)
SCS. *See* Silicon controlled switch (SCS)
Secondary impedance, 212
Secondary windings, 205
  multiple, 209
Second harmonic frequency, 168
Selective inversion circuit, 405–406
Selectivity, of an *RLC* circuit, 250
Self-contained circuits, 417
Self-induced voltage, 194, 195, 198, 199
Self-resonant frequency, 253, 254
Semiconductor atoms, 265
  isolated, 264
Semiconductor crystals, 265–266
Semiconductor devices, 261
  soldering/desoldering, 284–285
  working with, 287
Semiconductor diodes, 291
Semiconductor doping, 267–268
Semiconductor junctions, 275–283
Semiconductors, 25, 46, 81, 262
  current flow in, 272–274
  intrinsic, 264
  packaging for, 283
  troubleshooting, 283–287

Semiconductor technology, 261–287
Semiconductor testers, 286
Semiconductor theory, 264–274
Sensitive devices, protecting from heat, 284
Sequential logic, 565–576, 584–585
Serial communication, 4
Serial-communication links, 632
Serial communication paths, 650
Serial data transmission, 616
Series-aiding voltage sources, 99
Series capacitances, 225–226
Series capacitive reactances, 226
Series capacitor connections, 225–226
Series circuits, 90–101
  basic, 92
  shorts in, 134
  troubleshooting, 130–138
  troubleshooting rules for, 136
  visualizing, 142
  voltage drops in, 131–132
Series components, 91
Series inductive reactances, 199
Series inductors, 198
Series networks, 123
Series-opposing voltage sources, 99–100
Series-parallel circuits, 108–117, 123, 149, 177
  numerical analysis of, 110–112
  power calculations for, 116
  troubleshooting, 141–143
  troubleshooting chart for, 142
  versus complex circuits, 108–110
Series-parallel resistive circuits, 147
Series *RC* circuits, 231–233
  phase relationships of, 231
Series resistance, relative values of, 98–99
Series resistance equation, 111, 112
Series resistive circuit, shorted, 133
Series resonance, 247–248
Series *RLC* circuits, 242–243
  characteristics of, 248
Series voltage regulation, 432–436
Series voltage-regulator circuits, 432–434, 440–441, 458
Series voltage regulators, 449
Series voltage sources, 99–100
Shells, 24
Shift register, 600
Shoot-through logic, 445–446
Short circuits, 64, 131, 448–449
  effects of, 133–135
Shorted branch, 149
Shorted components, 135
  checking for, 323
Shorts
  detecting, 63
  in parallel circuits, 141
  in series circuits, 149

Short time constant, 238
Shunt voltage regulation, 437–438
Shunt voltage-regulator circuits, 437–438, 458
Siemens (S), 36
Signal generator, 156
Signal injection, 17, 145–146
Signal response, in MOSFET amplifiers, 358
Signals
  forcing, 17
  monitoring, 16–17
Signal tracing, 16–17, 145–146
Silicon, 264, 268
  intrinsic, 277, 287
Silicon atoms, 265
Silicon bilateral switch (SBS), 476
Silicon chips, 10
  electromechanical systems on, 12
Silicon-controlled rectifiers (SCRs), 452, 472, 477–481. *See also* SCR phase-control circuit
  applications for, 481
Silicon controlled switch (SCS), 472
Silicon diode, 303
Silicon trigger switch (STS), 476–477
Simplex communications, 636
Simplification, steps in, 114
Simultaneous biasing, 335–338
SIN button, 172
Sine, of an angle, 166
Sine function, 172
Sine wave, 156. *See also* Sinusoidal waveforms
  characteristics of, 157–169
  frequency of, 158–159
  full, 160–161
  full cycle of, 161
  graph of, 160
  half-cycle of, 161–162
  peak-to-peak voltage of, 163–164
  peak value of, 160
  period of, 157–158
  rms value of, 162–163
Single-conductor wires, 61
Single-pole double-throw (SPDT) switches, 78
Single-pole single-throw (SPST) switches, 77
Sinusoidal voltage, 196
Sinusoidal waveforms, 155, 169, 170, 185. *See also* Sine wave
60-Hz power distribution, 181
Slew rate, 391–392
Slew-rate limiting effect, 392
Slide switch, 78
Sliding contacts, 74
Slip rings, 155
Slo-Blo fuses, 65
SMT. *See* Surface-mount technology (SMT)

SN7400 device, 551
SN74AC74 device, 566
SN74AC86 exclusive OR package, 552
Sneak paths, 148
Soft start circuit, 445
Software, 601
　circuit simulation, 20, 22, 126
　classes of, 614–616
　diagnostic, 17
　troubleshooting with, 585
Soldering/desoldering, precautions for, 284–285
Soldering iron, choosing, 284
Solid-conductor wire, 13
Solid-state components, 81
Solid-state devices, 262, 283, 358
Solid-state relay (SSR), 498
Solid wire, 59–60
Sound card, 609
Sound waves, 40, 182
Source current, 244–245
Source-swamping resistor, 352
Source terminal, 348, 350
Source-to-drain current, 349, 350
Source voltage, 242, 243. *See also* Power supply
South pole, 43
Space state, 634
SPDT. *See* Single-pole double-throw (SPDT) switches
Special-purpose diodes, 311–320, 325
　defects in, 324
Spectrum analyzer, 167–168
Speed, 39
Speed per unit time, 39
Spiraling, 69
Split-half troubleshooting, 140, 145, 149
Spontaneous emission, 491
SPST. *See* Single-pole single-throw (SPST) switches
Square mils, 57
Square wave, 168
Square wave generator, 510
Squelch circuits, 361
SSR. *See* Solid-state relay (SSR)
Stand-alone systems, 11
Star network topology, 627
Start bit, 634
State machines, 574–575
State symbol, 550
Step-down conversion, 443
Step-down transformer, 206, 363
Step-recovery diodes, 317
Step-up conversion, 443
Step-up transformer, 206
Stimulated emission, 491
Strain gage, 525
Stranded wire, 13, 59–60
STS. *See* Silicon trigger switch (STS)
Subatomic particles, 24

Subscripts, use of, 110
Substitution
　for testing transistors, 375
　in troubleshooting, 228
Sum-of-the-products form, 562
Superconductivity, 25
Superconductors, 25, 262
Surface-mount capacitors, 222, 228
Surface-mount resistor markings, 73
Surface-mount resistors, 71
Surface-mount technology (SMT), 70
Surge, 66
Surge current ratings, for triacs, 481
Sustained current flow, 87
Switches, 76–77, 82
　operation of, 77–78
　specifications for, 79
　transistor, 361–362
Switching, versus linear regulation, 443
Switching mechanisms, 78
Switching power supply circuit, 457
Switching regulators
　current limiting in, 450
　discrete versus integrated, 446–447
Switching times, 553
Switching voltage regulation, 438–447
　principles of, 439–440
Switching voltage-regulator circuit, 443
　design of, 443–446
Switching voltage regulators, 458
　operation of, 439
Symbols
　for electrical quantities, 37
　manufacturers' use of, 122
Symmetrical waveforms, 154, 185
Synchronous communication, 632–633, 650
Synchronous flip-flop counters, 569–571
Synchronous inputs, 567–568
System connectivity, 13–15
System controller, 607
System hardware, 601–602
System-level troubleshooting, 15–20
System maintenance history, 18
System notations, 3–9
System resources, shared, 207

–T–

TAN button, 172
Tangent function, 172
TANs. *See* Tiny area networks (TANs)
Tantalum capacitors, 224
Tapped transformers, 209
TDM. *See* Time-division multiplexing (TDM)
TE. *See* Thermoelectric (TE) cooler
Technical notation, 49–55, 82
Technologies. *See also* Capacitor technologies; Nanotechnology;

Resistor technology; Semiconductor technology
　capacitor, 224–225
　semiconductor, 261–287
Telecommunication networks, 622–630
Telecommunications, 621–653, 650. *See also* Wireless telecommunication
Telecommunications control, 623–624
Telecommunication systems, technical characteristics of, 630–645
Telemetry, 623
Television, 1
Television receiver, 535
Temperature, of a component, 140
Temperature coefficient, 57
　of a capacitor, 223
Temperature probes, troubleshooting with, 140–141
Temperature variations, transistor defects and, 375
Temporary magnets, 44
10:1 probe, 97
Teraohms, 356
Tesla (T), 43
Test instruments, 8
　specialized, 146
Testpoint monitoring, 146
Tetravalent atoms, 264
Texas Instruments, 556
Thermoelectric (TE) cooler, 493
3-bit asynchronous counter, 572
Three-phase alternator, 165
Three-terminal bidirectional thyristor, 474
Three-terminal voltage regulator, 435, 436
Throughput, 631
Throw, 78
Thyristor noise problems, 488
Thyristors, 466–488, 503–504
　characteristics of, 466–473, 483–484
　construction of, 469–470
　defects in, 485
　off state of, 486–487
　on state of, 485–486
　troubleshooting, 485–488
　turn-off methods for, 473
　turn-on methods for, 470–472
　types and applications of, 473–485
Thyristor test, 487–488
Time constants, 234, 255
　asymmetrical, 308–309
　RC, 235–237
　RL, 238
Time-delay circuits, 240–241, 529, 531
Time-division multiplexing (TDM), 643–644, 650
Time domain, 167
Timing diagrams, 547–548
Tiny area networks (TANs), 629, 650

744 INDEX

Toggle switch, 78
Toggling, 567
Tolerance band values, for a five-band resistor code, 73
Tools, selecting, 130–131
Total capacitance, in series-connected capacitors, 225
Total circuit inductance, 200
Total current, 103–104, 113, 114
Total inductive reactance, 199, 200
Total power, 93, 104
Total power dissipation, computing, 179
Total reactance, for parallel circuits, 227
Total resistance, 93, 94, 104, 111, 113, 114, 123
  computing, 104–105
Total voltage, 93
Totem-pole output, 364, 365
Traces, 9
Transducers, 307, 523–524
  common, 524
Transformer action, 204–205
Transformer circuits, analysis of, 210–214
Transformers, 76, 204–215, 254
  common defects in, 214
  overheating, 214
  step-up and step-down, 206
  terminology for, 205–206
  troubleshooting, 214–215
  types and applications of, 206–210
Transformer windings, altering the connection of, 209
Transients, 66
Transistor amplifiers, bipolar, 338–347
Transistor applications, biasing requirements for, 338
Transistor circuits
  troubleshooting, 373–377
  ratios of, 343
Transistor currents, relationships among, 338
Transistors, 81, 82, 275
  applications for, 361–373
  bipolar, 332–348
  as current amplifiers, 366–367
  discrete, 366, 373, 377
  junction field-effect, 348–353
  MOS field-effect, 353–361
  relays and, 362–363
Transistor switches, 361–362
Transistor testers, 375
Transmission, simplex versus duplex, 636
Transmissive photointerrupter, 499
Transmitters, infrared, 14
Transponders, 646
Tri-state bidirectional bus driver, 598
Tri-state inverters, 558
Triacs, 481–483, 504

Triangle generator, 510–511
Trigonometric functions, 172–173. *See also* Inverse trigonometric functions
Trivalent atoms, 268
Trivalent doping, 268–269
Trivalent impurity atoms, 269, 271, 276
Troubleshooting, 129. *See also* Circuit troubleshooting
  basic rules for, 131
  of capacitors, 228
  of digital systems, 581–585
  of diode circuits, 320–325
  of electronic systems, 1
  of fences, 149
  focus in, 148
  of IC-based systems, 534–535
  of inductors, 201–203
  intuitive, 131
  of op amp circuits, 420–422
  of optoelectronic devices, 501–503
  of oscillator circuits, 533–534
  of parallel circuits, 138–141
  of power supply circuits, 452–457, 458
  of semiconductors, 283–287
  of series-parallel circuits, 141–143
  strategies for, 143–148
  system-level, 15–20
  of thyristors, 485–488
  of transformers, 214–215
  of transistor circuits, 373–377
  of wire and cable, 62–63
Troubleshooting chart, 142
Troubleshooting methods
  logical, 130, 141
  unprofessional, 138
Troubleshooting tools
  diagnostic software as, 17
  schematics as, 8
Truth tables, 547
  for logic circuits, 562–563
TTL gate, 550
Tunnel diodes, 318
Turns ratio, 210, 211–212

–U–

Ultrasonic waves, 40, 182
Ultrasonic frequencies, 182
Ultrasound imaging systems, 182
Unbiased junction, 275
Unbiased pn junction, 281
Undervoltage lockout circuit, 445
Unequal voltage sources, connecting in parallel, 107
Unidirectional busses, 597
Unidirectional thyristors, 468
Unipolar devices, 274, 316, 359
Units of measure, 23–46
  for charge, 26

  for current, 29–30
  for electrical quantities, 37
  for electrical power, 34
  prefixes for, 53–54
  for resistance, 30
  for voltage, 27
Up-arrow symbol, 134
USB (Universal Serial Bus) clock, 522
USB port, 609
User-application programs, 616
User-level programming languages, 613–614
User/operator interviewing, 18

–V–

Valence band, 263
  electrons in, 265–266, 281
Valence electrons, 24, 28, 287
Valleys, wave, 40
Varactor diodes, 315
Variable autotransformer, 323
Variable capacitors, 222
Variable input voltage, 313–314
Variable resistors, 66, 74, 82
  schematic symbols for, 75
Varying voltage, 366
VCO. *See* Voltage-controlled oscillator (VCO)
Velocity, 39
Very High-Speed Integrated Circuit Hardware Description Language (VHDL), 585
Very high-speed integrated circuits (VHSICs), 585
VFC. *See* Voltage-to-frequency conversion (VFC)
VFC32 integrated circuit, 527–528, 531, 532
VHSICs. *See* Very high-speed integrated circuits (VHSICs)
Video controller, 609
Video display device, 167
Video drivers, 609
Video memory, 608
Video transfers, 623
Vinculum (vincula), 549, 554
Virtual ground, 399, 409
Virtual-ground point, 400
Virtual instruments, 20
Visible LEDs, 489
Visual Basic, 614
Voice-based telecommunications, 622
Voice communications, 622
Volatile memory, 608
Voltage (V), 27, 30, 46. *See also* Kirchhoff's Voltage Law; Voltages
  calculating, 34
  capacitor, 216
  determining at any point in a circuit, 120
  instantaneous, 166

mathematical relationship for, 94–97
peak-to-peak, 163
relationship to current, 31–32
relative changes in, 134–135
resonant rise in, 248
in a series-parallel circuit, 113–116
unit of measure for, 27
Voltage amplifier, 342
Voltage breakdown, 468–469
Voltage breakover, 468–469
Voltage changes, short circuits and, 134
Voltage checks, for bipolar transistors, 373
Voltage-comparator circuits, 430–431
Voltage comparators, 417–418, 423
Voltage-controlled oscillator (VCO), 519–521, 527
Voltage divider, 417
Voltage divider equation, 96–97
Voltage drops, 93, 94, 232, 233
  polarity of, 95
Voltage-follower circuit, 404–405, 411, 416
Voltage followers, 525
Voltage gain, 340, 350–352
  common-mode versus differential-mode, 389
  open-loop versus closed-loop, 389
Voltage measurements
  for diodes, 321
  for series-parallel circuits, 141–142
Voltage peaks, 195–196
Voltage rating
  of a capacitor, 223
  of a fuse, 65
Voltage ratio, 210–211
Voltage references, 430–431
  indicating, 121
Voltage regulation, 296
  fundamentals of, 428–431
  shunt, 437–438
Voltage-regulator circuits, 427–463
  output from, 430

Voltage regulators
  classes of, 458
  integrated series, 434–436
  zener diodes as, 312–314
Voltages. *See also* Voltage (*V*)
  damaging, 359
  ground and, 119–120
  labeling, 96
  mathematical relationships between, 106–107
  ratio between primary and secondary, 211
Voltage sensor, 241
Voltage sources, 63–64, 82, 88
  negative side of, 132
  series, 99–100
Voltage-to-frequency conversion (VFC), 527–531
Voltage-to-frequency converter circuit, 529
Voltage transducer, 307
Voltage values, computing, 178
Voltage-variable capacitance diodes, 315
Voltage waveform, 195
Voltmeter, 120, 138, 141–142
  troubleshooting circuits with, 132
Voltmeter loading, 130
Voltmeter tests, 149, 214
VOM. *See* Analog VOM

–W–
WANs. *See* Wide area networks (WANs)
Watchdog timer, 241
Watts (W), 34
Waveforms, 154
  combined AC and DC, 169
  differentiated, 239
  rectangular, 153–154
  sinusoidal, 155–156
Waveform distortion, 398
Wavelength, 40–42

Wavelength-division multiplexing (WDM), 644, 650
Waves, 39–40, 46
Waveshapes. *See* Waveforms
WDM. *See* Wavelength-division multiplexing (WDM)
Weber, 42
Wide area networks (WANs), 629–630, 650
Wien-bridge oscillator, 511–514
Winding-to-core short, 215
Windings, shorted, 202
Wire, 55–61, 82
  conductive, 13
  defects in, 137
  gauge numbers of, 58
  solid versus stranded, 59–60
  troubleshooting, 62–63
  types of, 58–62
Wire insulating materials
  comparison of, 61
  types of, 60–61
Wireless communication, 251–252
Wireless Internet links, 184–185
Wireless LANs (WLANs), 645, 650
Wireless links, 15
Wireless technology, 183–184
Wireless telecommunication, 645–649
Wire resistance, 56–58
Wire sizes, 58, 59
Wirewound resistors, 69
Wiring diagrams, 7, 9
WLANs. *See* Wireless LANs (WLANs)
Word, 596
Write (store data), 597
Wrong-output voltage, 136–137

–Z–
Zener diodes, 312–314, 579
Zener voltage, 313
Zener voltage regulator circuit, 312–314
Zero resistance, 133
Zero voltage, 141